Mathematical Analysis of Groundwater Flow Models

Mathematical Analysis of Groundwater Flow Models

Edited by
Abdon Atangana

CRC Press
Taylor & Francis Group
Boca Raton London New York

CRC Press is an imprint of the
Taylor & Francis Group, an **informa** business

First edition published 2022
by CRC Press
6000 Broken Sound Parkway NW, Suite 300, Boca Raton, FL 33487-2742

and by CRC Press
2 Park Square, Milton Park, Abingdon, Oxon, OX14 4RN

Library of Congress Cataloging-in-Publication Data
Names: Atangana, Abdon, editor.
Title: Mathematical analysis of groundwater flow models / edited by Abdon Atangana.
Description: First edition. | Boca Raton : CRC Press, [2022] | Includes
bibliographical references and index.
Identifiers: LCCN 2021045765 (print) | LCCN 2021045766 (ebook) | ISBN
9781032209944 (hbk) | ISBN 9781032209951 (pbk) | ISBN 9781003266266 (ebk)
Subjects: LCSH: Groundwater flow--Mathematical models.
Classification: LCC GB1197.7 .M388 2022 (print) | LCC GB1197.7 (ebook) |
DDC 551.4901/5118--dc23/eng/20211119
LC record available at https://lccn.loc.gov/2021045765
LC ebook record available at https://lccn.loc.gov/2021045766

ISBN: 978-1-032-20994-4 (hbk)
ISBN: 978-1-032-20995-1 (pbk)
ISBN: 978-1-003-26626-6 (ebk)

DOI: 10.1201/9781003266266

Typeset in Times
by SPi Technologies India Pvt Ltd (Straive)

Contents

Preface

Groundwater is a natural resource present below the Earth's surface; stored within rocks and soil pore spaces; and it makes up the largest portion of the existing body of freshwater on Earth and is highly useful to sustain life for both humans and other ecosystems. Groundwater is recharged from the surface; and is naturally discharged on to the surface through springs, seeps and/or rivers forming important sources of water known as oases. Because of its availability and accessibility, it is often withdrawn via boreholes for various uses including agricultural, industrial, mining, municipal and domestic. This limited resource has been for decades subjected to over-use, over-abstraction or overdraft leading to significant problems encountered by human users in different parts of the world. In addition, some groundwater sources are affected significantly by pollution, which reduces the availability of clean and healthy water. As a result several related environmental issues have been observed on a large scale around the world. The worst case scenario is that subsurface water pollution is hardly noticeable, and more difficult to purify, than pollution occurring in surface water. Hence the protection, regulation and monitoring of these sources of fresh water has recently become a focal point of human beings to ensure their sustainability. However, the realization of this process incorporates several steps needing to be performed. The first step includes data collection, and the second consists of analyzing collected data to identify which law the recorded data may follow. The last step is based on the conversion from observation to mathematical model, derivation of solutions, and finally the comparison of obtained solutions with experimental data. In the event, an agreement between collected data and the solution of mathematical models is obtained, and a good prediction can be performed.

The book is devoted to discussions underpinning modeling groundwater problems such as flow in different geological formations, and artificial and natural recharge as well as the flow of contamination plumes. Existing mathematical models are analyzed and modified using new concepts to include the models in mathematical equations related to complexities of geological formations. Classical differential and integral operators are considered in some cases to model local behaviors observed in groundwater flow, recharge and pollution problems. Additionally, different types of non-local operators, including fractal derivative and integral, fractional derivatives and integral based on power law kernel, fractional derivative and integral based on exponential decay, and the generalized Mittag–Leffler functions are used to include into mathematical equations heterogeneous properties of subsurface formation. Thus several analytical and numerical techniques are accordingly utilized to derive exact and approximated solutions, in which numerical solutions are depicted using different software such as maple, Mathematica and Matlab.

MATLAB® is a registered trademark of The MathWorks, Inc. For product information, please contact:
The MathWorks, Inc.
3 Apple Hill Drive
Natick, MA, 01760-2098 USA
Tel: 508-647-7000
Fax: 508-647-7001
E-mail: info@mathworks.com
Web: www.mathworks.com

Editor

Abdon Atangana works at the Institute for Groundwater Studies, University of the Free State, Bloemfontein, South Africa as a full Professor. His research interests are, but not limited to, fractional calculus and applications, numerical and analytical methods, and modeling. He is the author of more than 250 research papers and four books in top tier journals of applied mathematics and groundwater modeling. He was elected Highly Cited Mathematician in 2019 and Highly Cited Mathematician with Crossfield Impact in 2020. He is a recipient of the World Academia of Science Award for Mathematics 2020. He serves as editor in top tier journals in various fields of study.

Contributors

Abdon Atangana
Institute of Groundwater Studies
University of the Free State
Bloemfontein, South Africa
and
Department of Medical Research,
China Medical University Hospital
China Medical University
Taichung, Taiwan

Amanda Ramotsho
Institute of Groundwater Studies
University of the Free State
Bloemfontein, South Africa

Awodwa Magingi
Institute of Groundwater Studies
University of the Free State
Bloemfontein, South Africa

Dineo Ramakatsa
Institute of Groundwater Studies
University of the Free State
Bloemfontein, South Africa

Disebo Venoliah Chaka
Institute of Groundwater Studies
University of the Free State
Bloemfontein, South Africa

Hans Tah Mbah
Institute of Groundwater Studies
University of the Free State
Bloemfontein, South Africa

Mashudu Mathobo
Institute of Groundwater Studies
University of the Free State
Bloemfontein, South Africa

Mohau Mahantane
Institute of Groundwater Studies
University of the Free State
Bloemfontein, South Africa

Makosha Ishmaeline Charlotte Morakaladi
Institute of Groundwater Studies
University of the Free State
Bloemfontein, South Africa

Mpafane Deyi
Institute of Groundwater Studies
University of the Free State
Bloemfontein, South Africa

Palesa Myeko
Institute of Groundwater Studies
University of the Free State
Bloemfontein, South Africa

Rendani Vele Makahane
Institute of Groundwater Studies
University of the Free State
Bloemfontein, South Africa

Sarti Amakali
Institute of Groundwater Studies
University of the Free State
Bloemfontein, South Africa

Siphokazi Simnikiwe Manundu
Institute of Groundwater Studies
University of the Free State
Bloemfontein, South Africa

Tshanduko Mutandanyi
Institute of Groundwater Studies
University of the Free State
Bloemfontein, South Africa

1 Analysis of the Existing Model for the Vertical Flow of Groundwater in Saturated–Unsaturated Zones

Rendani Vele Makahane and Abdon Atangana
University of the Free State, Bloemfontein, South Africa

CONTENTS

1.1 INTRODUCTION

Since studying the interaction between geology and the movement of groundwater can be quite complex, a model expressing the nature of the system must be introduced. A model is viewed as an approximation and not an exact solution of the physical process; nonetheless, even as an approximation, it can be a useful investigation tool (Atangana & Botha, 2012, 2013). A groundwater model describing the movement of water in a porous media is defined in mathematical terms by combining the law of mass conservation and Darcy's law. The advantage of this equation is that it can be used for the whole flow region and can handle both unconfined and confined saturated aquifers (Freeze, 1971). The resulting equation is linear for a saturated flow and nonlinear for the unsaturated flow (Nishigaki & Kono, 1980; Cheng & Gulliksson, 2003). This chapter focuses on the mathematical modeling of problems related to groundwater flow in the saturated–unsaturated zone. The main objective is to accomplish both analytical and numerical solutions for the classical saturated and

DOI: 10.1201/9781003266266-1

unsaturated groundwater flow equation, where applicable. Different solutions will be compared to see which one can best describe saturated–unsaturated groundwater flow problems.

1.2   BACKGROUND REVIEW

The rate at which water flows through a porous medium is associated with the properties of a porous medium, the properties of the water and change in hydraulic head. This relationship is well described by Darcy's law (Konikow, 1996). By considering that the air pressure is always constant, i.e., zero, the movement of water through the saturated–unsaturated porous media can be described mathematically (List & Radu, 2015; Zimmerman & Bodvarsson, 1989). The saturated–unsaturated equation is obtained by merging the mass conservation equation with Darcy's law (Danesfáraz & Kaya, 2009). The equation includes the hydraulic properties of the soil, which are a function of the suction head of the soil and therefore are nonlinear (Allepalli & Govindaraju, 2009). The 1-d saturated–unsaturated groundwater flow equation is given in (1.1):

$$\left[S_S S_a \left(\psi \right) + C \left(\psi \right) \right] \frac{\partial \psi}{\partial t} = \frac{\partial}{\partial z} \left[K_z \left(\psi \right) \left(\frac{\partial \psi}{\partial z} - 1 \right) \right] \tag{1.1}$$

where:
  ψ = pressure head
  n = porosity
  S_S = specific storage of the soil
  S_a = saturation of the aqueous phase
  $C(\psi)$ = capillary capacity of soil = $d\theta/d\psi$
  K_z = hydraulic conductivity
  z = vertical co-ordinate.

Since the movement of water in the unsaturated zone must be distinguished from the movement of water in the saturated zone, we shall provide a brief literature on how flow in these two zones will differ, with the most complex being the unsaturated zone. Overcoming this complexity will require knowledge of the nonlinear relationship that exists between the soil hydraulic function (Cattaneo et al., 2016). Soil hydraulic functions refer to the hydraulic conductivity function, $K(\psi)$, and the soil water content function, $\theta(\psi)$, that are required to explain the movement of water. Numerous functions have been suggested to define the soil hydraulic properties empirically. Popular models are the equations of Brook and Corey (Allepalli & Gavindraraju, 1996):

$$K \left(\psi \right) = K_s \left(\frac{\psi_b}{\psi} \right)^{2+3\lambda} \qquad\qquad \psi \leq 0 \tag{1.2}$$

$$K \left(\psi \right) = K_s \qquad\qquad \psi > 0 \tag{1.3}$$

$$\theta \left(\psi \right) = \theta_0 + \left(\theta_s - \theta_0 \right) \left(\frac{\psi_b}{\psi} \right)^{\lambda} \qquad\qquad \psi \leq 0 \tag{1.4}$$

where:
  ψ = pressure head
  ψ_b = air entry suction pressure head

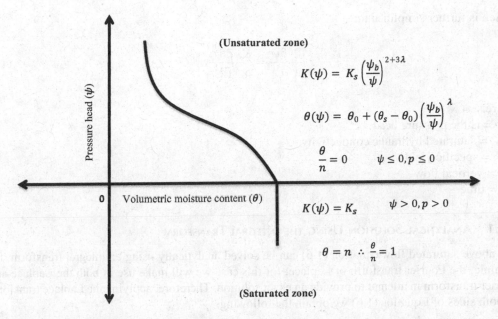

FIGURE 1.1 Relationship between the pressure head and volumetric water content for saturated–unsaturated flow (Modified after Nishigaki & Kono, 1980).

$\theta(\psi)$ = soil water content
θ_s = saturated water content
θ_0 = residual water content
λ = pore size distribution index
K_s = saturated hydraulic conductivity.

The functional relationship between ψ and θ is called the water retention curve and is shown in Figure 1.1. Incorporated in this figure are the soil hydraulic functions ($K(\psi)$ and $\theta(\psi)$) showing how they differ between unsaturated and saturated zones. This figure is modified after Nishigaki & Kono, 1980.

Now that the flow in the unsaturated and saturated zones has been distinguished, classical equations suitable for each zone are derived using powerful methods that result in explicit forms of a solution. In the case of the saturated zone, analytical and numerical solutions will be derived using integral transform, the methods of separation of variables and numerical schemes, respectively. The conditions under which the numerical method used converges will be derived and presented in detail. Because of the complexity of the unsaturated model, we relied only on the numerical method to derive an approximate solution. Detailed results of the analytical and numerical solutions are documented below for the classical equations where applicable.

1.3 GOVERNING SATURATED GROUNDWATER FLOW EQUATION

When considering flow in the saturated zone only, it is assumed that the volumetric water content is equal to the porosity, the hydraulic conductivity is constant, and the capillary capacity becomes zero. Therefore, Equation (1.1) is written such that (Maslouhi et al., 2009):

$$\left(S_S\right)\frac{\partial \psi}{\partial t} = \frac{\partial}{\partial z}\left[K_s\left(\frac{\partial \psi}{\partial z}-1\right)\right] \tag{1.5}$$

Which is further simplified to:

$$\frac{\partial \psi}{\partial t} = \frac{K_s}{S_s} \frac{\partial^2 \psi}{\partial z^2} \tag{1.6}$$

where:

ψ = is the pressure head
K_s = saturated hydraulic conductivity
S_s = specific yield
z = vertical flow
t = time.

1.3.1 ANALYTICAL SOLUTION USING THE INTEGRAL TRANSFORM

The above saturated flow equation (1.6) can be solved analytically using an integral transform, for example, the Fourier transform or Laplace; for this case we will make use of both the Laplace and Fourier transform in attempt to provide an exact solution. Therefore, applying the Laplace transform on both sides of Equation (1.6) we obtain the following:

$$\mathcal{L}\left(\frac{\partial \psi}{\partial t}\right) = \mathcal{L}\left(\frac{K_s}{S_s} \frac{\partial^2 \psi}{\partial z^2}\right) \tag{1.7}$$

Applying the Laplace transform of a derivative into Equation (1.7) we obtain the following with s being considered as the Laplace variable:

$$s\tilde{\psi}(z,s) - \psi(z,0) = \left(\frac{K_s}{S_s} \frac{\partial^2 \psi}{\partial z^2}\right)(z,s) \tag{1.8}$$

To eliminate the space component in order to obtain an algebraic, we can use the Fourier transform on the above equation:

$$\mathcal{F}\left(s\tilde{\psi}(z,s) - \psi(z,0)\right) = \frac{K_s}{S_s} \mathcal{F}\left(\frac{\partial^2 \tilde{\psi}(z,s)}{\partial z^2}\right) \tag{1.9}$$

Using the properties of the Fourier transform, the right-hand side of the above equation can be simplified further into:

$$s\tilde{\psi}_{\mathcal{F}}(\omega,s) - \psi_{\mathcal{F}}(\omega,0) = \frac{K_s}{S_s}(i\omega)^2 \tilde{\psi}_{\mathcal{F}}(\omega,s) \tag{1.10}$$

The above equation can be factorized to obtain the equation below:

$$\left(s - \frac{K_s}{S_s}(i\omega)^2\right)\tilde{\psi}_{\mathcal{F}}(\omega,s) = \psi_{\mathcal{F}}(\omega,0) \tag{1.11}$$

We further simplify the above equation into:

$$\tilde{\psi}_{\mathcal{F}}(\omega,s) = \frac{\psi_{\mathcal{F}}(\omega,0)}{\left(s - \dfrac{K_s}{S_s}(i\omega)^2\right)} \tag{1.12}$$

And then:

$$\tilde{\psi}_{\mathcal{F}}(\omega,s) = \frac{\psi_{\mathcal{F}}(\omega,0)}{\left(\dfrac{S_s}{K_s}\,s + (\omega)^2\right)\dfrac{K_s}{S_s}} \tag{1.13}$$

The above formula is the multiplication of two Fourier transforms of two functions; thus one can use the convolution theorem to obtain the inverse transform as follow:

$$\mathcal{F}^{-1}\big(\psi_{\mathcal{F}}(\omega,0)\big) = \psi(z,0) \tag{1.14}$$

$$\psi(z,s) = \mathcal{F}^{-1}\left(\frac{1}{\omega^2 + \left(\sqrt{\dfrac{S_s}{K_s}}\,s\right)^2}\right) \tag{1.15}$$

$$\psi(z,s) = \frac{2\sqrt{\dfrac{S_s}{K_s}}\,s}{2\sqrt{\dfrac{S_s}{K_s}}\,s\left(\omega^2 + \left(\sqrt{\dfrac{S_s}{K_s}}\,s\right)^2\right)} \tag{1.16}$$

$$\psi(z,s) = \frac{1}{2\sqrt{\dfrac{S_s}{K_s}}\,s\,\exp\left(-\sqrt{\dfrac{S_s}{K_s}}\,s(z)\right)} \tag{1.17}$$

By integrating the above equation, we obtain:

$$\psi(z,s) = \frac{1}{2\sqrt{\dfrac{S_s}{K_s}}\,s}\int_0^z \psi(\lambda,0)\exp\left[-\sqrt{\dfrac{S_s}{K_s}}\,s\,(z-\lambda)\right]d\lambda \tag{1.18}$$

To obtain the exact solution of the saturated zone, we apply the inverse Laplace transform on both sides of Equation (1.18):

$$\psi(z,t) = \mathcal{L}^{-1}\big(\psi(z,s)\big) \tag{1.19}$$

1.3.2 ANALYTICAL SOLUTION USING THE METHOD OF SEPARATION OF VARIABLES

We shall now solve the above equation (1.6) using the analytical method of separation of variables. This method is used to solve a wide variety of linear and homogeneous differential equations, such as the saturated groundwater flow equation. The boundary conditions are linear and homogenous, such that:

$$\psi\big|_{z=0} = \psi\big|_{z=L} = 0 \qquad (1.20)$$

The dependence of ψ on z and t can be written as a product of function z and function t, such that:

$$\psi(z,t) = F(z)G(t) \qquad (1.21)$$

Substitution ψ back into Equation (1.6) using the product rule, we obtain:

$$\frac{\partial\big(F(z)G(t)\big)}{\partial t} = \frac{K_s}{S_s}\frac{\partial^2\big(F(z)G(t)\big)}{\partial z^2} \qquad (1.22)$$

Now we can separate the variables as follows:

$$F(z)\frac{S_s}{K_s}\frac{dG(t)}{dt} = G(t)\frac{d^2F(z)}{dz^2} \qquad (1.23)$$

Variables can be further separated to obtain the following:

$$\frac{\dfrac{S_s}{K_s}\dfrac{dG(t)}{dt}}{G(t)} = \frac{\dfrac{d^2F(z)}{dz^2}}{F(z)} \qquad (1.24)$$

The RHS and the LHS of Equation (1.24) can be a function of z and t, respectively, if they both equate to a constant value α, such that:

$$\frac{S_s}{K_s}\frac{dG(t)}{dt} = \alpha G(t) \qquad (1.25)$$

and:

$$\frac{d^2F(z)}{dz^2} = \alpha F(z) \qquad (1.26)$$

To find a solution that satisfies boundary conditions and is not identically zero, let's assume that for the constant $\alpha < 0$ there exist real numbers B, C such that:

$$F(z) = Be^{\sqrt{-\alpha}z} + Ce^{-\sqrt{-\alpha}z} \qquad (1.27)$$

From the boundary conditions given in Equation (1.20), we get:

$$F(0) = 0 = F(L) \qquad (1.28)$$

$B = 0 = C$, implies that ψ is identically 0; therefore, we reject this case. Now for the second try, we assume that when $\lambda = 0$ there exist real numbers B, C such that:

$$F(z) = B(z) + C \qquad (1.29)$$

We draw the same conclusion, that ψ is identically 0, and again we reject this case. The last example is when $\alpha > 0$. Then there exists the real number A, B, C such that:

$$G(t) = Ae^{-\alpha \frac{S_s}{K_s} t} \qquad (1.30)$$

and

$$F(z) = B\sin\left(\sqrt{\alpha}\, z\right) + C\cos\left(\sqrt{\alpha}\, z\right) \qquad (1.31)$$

From the above information $C = 0$ and that for some positive integer n:

$$\sqrt{\alpha} = n\frac{\pi}{L} \qquad (1.32)$$

Therefore, a general solution can be given as:

$$\psi(z,t) = \sum_{n=1}^{\infty} D_n \sin\frac{n\pi z}{L} \exp\left(-\frac{n^2\pi^2 \frac{S_s}{K_s} t}{L^2}\right) \qquad (1.33)$$

where D_n can be evaluated using the Fourier series, given the following initial condition:

$$\psi\big|_{t=0} = f(z) \qquad (1.34)$$

So, we obtain:

$$v(z) = \sum_{n=1}^{\infty} D_n \sin\frac{n\pi z}{L} \qquad (1.35)$$

Multiplying both sides with $\sin\frac{n\pi z}{L}$ and integrating over $[0, L]$ results in:

$$D_n = \frac{2}{L}\int_0^L f(z)\sin\frac{n\pi z}{L}\, dz \qquad (1.36)$$

Hence the complete solution for Equation (1.33) is given by:

$$\psi(z,t) = \sum_{n=1}^{\infty}\left(\frac{2}{L}\int_0^L f(z)\sin\frac{n\pi z}{L}\, dz\right)\sin\frac{n\pi z}{L}\exp\left(-\frac{n^2\pi^2 \frac{S_s}{K_s} t}{L^2}\right) \qquad (1.37)$$

1.4   NUMERICAL SOLUTION

The analytical solution of our model is limited to less complexity such as the assumption of homogeneity, isotropy, simple initial condition, and simple geological formation. However, natural systems can have a more complex geological formation, a complex initial condition and they can be heterogeneous and anisotropic (Zhang, 2016). Such complexities require numerical solutions (Igboekwe & Amos-Uhegbu, 2011; Konikow, 1996). Depending on how ψ_t is approximated, we have three basic finite difference schemes: implicit Crank–Nicolson scheme, explicit, and implicit. Using these schemes, we attempt to find numerical solutions for the 1-dimension saturated groundwater flow equation (1.6).

1.4.1   NUMERICAL SOLUTION USING THE FORWARD EULER METHOD (FTCS)

Appling the explicit forward Euler method to Equation (1.6), we obtain the following numerical solution:

$$\frac{\psi_i^{n+1} - \psi_i^n}{\Delta t} = \frac{K_s}{S_s}\left[\frac{\psi_{i+1}^n - 2\psi_i^n + \psi_{i-1}^n}{(\Delta z)^2}\right]. \tag{1.38}$$

1.4.2   NUMERICAL SOLUTION USING THE BACKWARD EULER METHOD (BTCS)

We obtain another numerical solution by applying the implicit backward Euler method on Equation (1.6):

$$\frac{\psi_i^{n+1} - \psi_i^n}{\Delta t} = \frac{K_s}{S_s}\left[\frac{\psi_{i+1}^{n+1} - 2\psi_i^{n+1} + \psi_{i-1}^{n+1}}{(\Delta z)^2}\right]. \tag{1.39}$$

1.4.3   NUMERICAL SOLUTION USING THE CRANK–NICOLSON METHOD

Now, we apply the implicit Crank–Nicolson discretization on Equation (1.6) such that:

$$\frac{\psi_i^{n+1} - \psi_i^n}{\Delta t} = \frac{K_s}{S_s}\left[\frac{\psi_{i+1}^{n+1} - 2\psi_i^{n+1} + \psi_{i-1}^{n+1}}{2(\Delta z)^2} + \frac{\psi_{i+1}^n - 2\psi_i^n + \psi_{i-1}^n}{(2\Delta z)^2}\right]. \tag{1.40}$$

1.5   NUMERICAL STABILITY ANALYSIS

In physical problems such as groundwater flow, stability analysis of an equation is fundamental. A procedure called von Neumann stability analysis based on the Fourier series is used to analyze the stability of the finite difference schemes. The stability of finite difference schemes is linked to numerical errors. A scheme is said to be von Neumann stable if its amplification factor is less or equal to 1. Accuracy, however, requires that the amplification factor be as close to 1 as possible. In this section, the numerical solutions provided in Equations (1.38)–(1.40) for the 1-d saturated groundwater flow Equation (1.6) are subjected under von Neumann stability analysis. A detailed procedure for each solution is given below.

1.5.1 STABILITY ANALYSIS OF A FORWARD EULER METHOD (FTCS)

Let us recall the forward Euler method (Equation 1.38), which can also be written as:

$$\psi_i^{n+1} = \psi_i^n \left(1 - 2\alpha\right) + \alpha \left(\psi_{i+1}^n + \psi_{i-1}^n\right) \tag{1.41}$$

where $\alpha = \dfrac{K_s}{S_s}\left(\dfrac{\Delta t}{\left(\Delta z\right)^2}\right)$

If we consider an initial harmonic perturbation:

$$\psi_i^0 = e^{ik_i \Delta z}$$

which with time evolves as:

$$\psi_i^n = \sigma^n e^{ik_m z}$$

To analyze the stability of this scheme, let us find the amplification factor σ by inserting the above assumption into Equation (1.41) such that:

$$\sigma^{n+1} e^{ik_m \Delta z} = \sigma^n e^{ik_m z}\left(1 - 2\alpha\right) + \alpha \left(\sigma^n e^{ik_m\left(z + \Delta z\right)} + \sigma^n e^{ik_m\left(z - \Delta z\right)}\right) \tag{1.42}$$

We can simplify Equation (1.42) by pulling out the common factor, such that:

$$\sigma^{n+1} e^{ik_m \Delta z} = \sigma^n e^{ik_m z}\left[1 - 2\alpha + \alpha \left(e^{ik_m \Delta z} + e^{-ik_m \Delta z}\right)\right] \tag{1.43}$$

$$\sigma^{n+1} e^{ik_m \Delta z} = \sigma^n e^{ik_m z}\left[1 + 2\alpha \left(\frac{e^{ik_m \Delta z} + e^{ik_m \Delta z}}{2} - 1\right)\right] \tag{1.44}$$

Considering the definition for a hyperbolic cosine, we can rewrite the above equation (1.44) as:

$$\sigma^{n+1} e^{ik_m \Delta z} = \sigma^n e^{ik_m z}\left[1 + 2\alpha \left(\cos\left(k_m \Delta z\right) - 1\right)\right] \tag{1.45}$$

Using the double angle identity for cos, Equation (1.45) is written such that:

$$\sigma^{n+1} e^{ik_m \Delta z} = \sigma^n e^{ik_m \Delta z}\left[1 + 2\alpha \left(1 - 2\sin^2\left(\frac{k_m \Delta z}{2}\right) - 1\right)\right] \tag{1.46}$$

$$\sigma^{n+1} e^{ik_m \Delta z} = \sigma^n e^{ik_m \Delta z}\left[1 - 4\alpha \sin^2 \frac{k_m \Delta z}{2}\right] \tag{1.47}$$

Dividing both sides by $\sigma^n e^{ik_i \Delta z}$ we get:

$$\sigma = 1 - 4\alpha \sin^2 \frac{k_m \Delta z}{2}. \tag{1.48}$$

The above equation gives us the amplification factor σ for the forward Euler method, and the stability condition is written as:

$$|\sigma| \le 1 \tag{1.49}$$

However, if $k_m \Delta z = \pi \Rightarrow \sigma = 1 - 4\alpha$ the stability of the above scheme (1.41) for all k will only hold if $\alpha \le \frac{1}{2}$. Therefore, the forward Euler method (1.41) is conditionally stable or von Neumann unstable as applied to the 1-d saturated groundwater flow equation (1.6).

1.5.2 STABILITY ANALYSIS OF A BACKWARD EULER METHOD (BTCS)

For the above backward Euler method (1.39), we can obtain the following from solving a system of linear equations:

$$-\psi_i^n = \alpha \psi_{i+1}^{n+1} - (1 + 2\alpha)\psi_i^{n+1} + \alpha \psi_{i-1}^{n+1} \tag{1.50}$$

where $\alpha = \dfrac{K_s}{S_s} \left(\dfrac{\Delta t}{(\Delta z)^2} \right)$.

If we consider an initial harmonic perturbation from the previous section, we can find the amplification factor σ for this scheme by inserting it into Equation (1.50), such that:

$$-\sigma^n e^{ik_i \Delta z} = \alpha \sigma^{n+1} e^{ik_{i+1} \Delta z} - (1 + 2\alpha)\sigma^{n+1} e^{ik_i \Delta z} + \alpha \sigma^{n+1} e^{ik_{i-1} \Delta z} \tag{1.51}$$

We can simplify Equation (5.13) by pulling out the common factor, such that:

$$-\sigma^n e^{ik_i \Delta z} = \sigma^{n+1} e^{ik_i \Delta z} \left[-(1 + 2\alpha) + \alpha \left(e^{ik\Delta z} + e^{-ik\Delta z} \right) \right] \tag{1.52}$$

Further simplification of the above equation results in the following:

$$-\sigma^n e^{ik_i \Delta z} = \sigma^{n+1} e^{ik_i \Delta z} \left[-(1 + 2\alpha) + 2\alpha \left(\frac{e^{ik\Delta z} + e^{-ik\Delta z}}{2} \right) \right] \tag{1.53}$$

Considering the definition for a hyperbolic cosine, we can rewrite the above equation (1.53) as:

$$-\sigma^{n+1} e^{ik_i \Delta z} = \sigma^n e^{ik_i \Delta z} \left[-1 - 2\alpha + 2\alpha \cos(k\Delta z) \right] \tag{1.54}$$

Using the double angle identity for cos, Equation (1.55) is written such that:

$$\sigma^{n+1} e^{ik_i \Delta z} = \sigma^n e^{ik_i \Delta z} \left[1 + 2\alpha - 2\alpha - 4\alpha \sin^2 \left(\frac{k\Delta z}{2} \right) \right] \tag{1.55}$$

$$\sigma^{n+1}e^{ik_i\Delta z} = \sigma^n e^{ik_i\Delta z}\left[1 + 4\alpha\sin^2\frac{k\Delta z}{2}\right] \tag{1.56}$$

Dividing both sides by $\sigma^n e^{iki\Delta z}$:

$$\sigma = \left(1 + 4\alpha\sin^2\frac{k\Delta z}{2}\right)^{-1} \tag{1.57}$$

The above equation gives us the amplification factor σ for the backward Euler method, and the stability condition is written as:

$$|\sigma| \leq 1 \tag{1.58}$$

This equation is unconditionally stable, since the value of σ will always be less than or equal to 1.

1.5.3 Stability Analysis of the Crank–Nicolson Method

So far, we have considered the forward Euler scheme and the backward Euler scheme for the 1-dimension saturated groundwater flow equation. The forward Euler scheme is an explicit method, therefore easy to implement; however, the results obtained for the forward Euler scheme indicates that it is only stable under the condition that $\frac{\alpha\Delta t}{\Delta x^2} \leq \frac{1}{2}$. The backward Euler scheme, on the other hand, is an implicit method and is unconditionally stable; however, it requires more arithmetic operations to find values at a specific time step (Grigoryan, 2012). It is also essential to note that the two schemes use different sets of points in the computation of ψ_i^{n+1}. The accuracy and stability of the two schemes can be improved by developing a single implicit scheme which will be a combination of the two schemes with different weights. With this method, a broader set of points can be used to compute the same values (Narasimhan, 2011; Grigoryan, 2012). For us to investigate the possible limits of this single scheme, let's consider:

$$\psi_i^{n+1} - \psi_i^n = \alpha\left[\left(\psi_{i+1}^{n+1} - 2\psi_i^{n+1} + \psi_{i-1}^{n+1}\right) + \left(\psi_{i+1}^n - 2\psi_i^n + \psi_{i-1}^n\right)\right] \tag{1.59}$$

We substitute the initial harmonic perturbation from the previous section into Equation (1.59), which leads to:

$$\sigma^{n+1}e^{ik_iz} - \sigma^n e^{ik_iz} = \alpha\left[\left(\sigma^{n+1}e^{ik_m(z+\Delta z)} - 2\sigma^{n+1}e^{ik_m z} + \sigma^{n+1}e^{ik_m(z-\Delta z)}\right)\right.$$
$$\left. + \left(\sigma^n e^{ik_m(z+\Delta z)} - 2\sigma^n e^{ik_m z} + \sigma^n e^{ik_m(z-\Delta z)}\right)\right] \tag{1.60}$$

We can simplify the above equation to get:

$$\sigma^n e^{ik_iz}(\sigma - 1) = \alpha\left[\sigma^n e^{ik_m z}\sigma\left(e^{ik_m\Delta z} - 2 + e^{-ik_m\Delta z}\right) + \sigma^n e^{ik_m z}\left(e^{ik_m\Delta z} - 2 + e^{-ik_m\Delta z}\right)\right] \tag{1.61}$$

Solving for σ, we obtain the following growth factor:

$$\sigma = \frac{1 - 2\alpha(1 - \cos k\Delta z)}{1 + 2\alpha(1 - \cos k\Delta z)} \tag{1.62}$$

Then the stability condition for this scheme is given by:

$$\left| \frac{1 - 2\alpha \left(1 - \cos k\Delta z\right)}{1 + 2\alpha \left(1 - \cos k\Delta z\right)} \right| \leq 1 \tag{1.63}$$

This scheme was developed in 1947 by John Crank and Phyllis Nicolson, and it is shown as the average of the backward and forward Euler schemes. The denominator of Equation (1.63) will always be greater than the numerator, since α and $1 - \cos k\Delta z$ are positive. This also means that under every condition, the value of σ will always is less than 1; therefore, the Crank–Nicolson method for the 1-d saturated groundwater flow equation (1.6) is unconditionally stable. It is clear that the Crank–Nicolson scheme is the most stable and accurate of the three schemes.

1.6 GOVERNING UNSATURATED GROUNDWATER FLOW EQUATION

In heterogeneous soils, the water content is uneven across layer boundaries because of exceptional unsaturated capillary head relations in different soil layers (Assouline, 2013). Relatively, the capillary head (ψ) is continuous, and can be represented by an equation with ψ as the dependent variable and the moisture content in terms of ψ, $\theta = \theta(\psi)$ (Farthing & Ogden, 2017). Equation (1.1) can be arranged so as to illustrate the complexity of the unsaturated flow using the equation below:

$$\frac{\partial \theta(\psi)}{\partial t} = \frac{\partial}{\partial z} \left[K_z(\psi) \left(\frac{\partial \psi}{\partial z} - 1 \right) \right] \tag{1.64}$$

The above Equation (1.64) consists of the soil hydraulic functions ($K(\psi)$ and $\theta(\psi)$). We shall use the famous soil hydraulic property Equations (2 and 5) of Brooks and Corey (1966) to describe these functions. We can now write the above equation as:

$$\frac{\partial \left(\theta_0 + \left(\theta_s - \theta_0\right) \left(\dfrac{\psi_b}{\psi} \right)^{\lambda} \right)}{\partial t} = \frac{\partial}{\partial z} \left(K_s \left(\frac{\psi_b}{\psi} \right)^{2+3\lambda} - 1 \right) \tag{1.65}$$

Simplifying the equation results in Equation (1.66) below:

$$\frac{\partial \theta_0}{\partial t} + \left(\theta_s - \theta_0\right) \frac{\partial}{\partial t} \left(\frac{\psi_b}{\psi} \right)^{\lambda} = \frac{\partial}{\partial z} \left[K_s \left(\frac{\psi_b}{\psi} \right)^{2+3\lambda} \frac{\partial \psi}{\partial z} - K_s \left(\frac{\psi_b}{\psi} \right)^{2+3\lambda} \right] \tag{1.66}$$

$$\left(\theta_s - \theta_0\right) \psi_b^{\lambda} \frac{\partial}{\partial t} \left(\frac{1}{\psi} \right)^{\lambda} = K_s \frac{\partial}{\partial z} \left(\frac{\psi_b}{\psi} \right)^{2+3\lambda} \frac{\partial \psi}{\partial z} + K_s \left(\frac{\psi_b}{\psi} \right)^{2+3\lambda} \frac{\partial^2 \psi}{\partial z^2} - K_s \frac{\partial}{\partial z} \left(\frac{\psi_b}{\psi} \right)^{2+3\lambda} \tag{1.67}$$

$$\left(\theta_s - \theta_0\right) \psi_b^{\lambda} \left(-\lambda \psi^{\,n\psi^{-\lambda-1}} \right) = K_s \frac{\partial}{\partial z} \left(\frac{\psi_b}{\psi} \right)^{2+3\lambda} \frac{\partial \psi}{\partial z} + K_s \left(\frac{\psi_b}{\psi} \right)^{2+3\lambda} \frac{\partial^2 \psi}{\partial z^2} - K_s \frac{\partial}{\partial z} \left(\frac{\psi_b}{\psi} \right)^{2+3\lambda} \tag{1.68}$$

The final unsaturated groundwater flow equation is given by:

$$-(\theta_s - \theta_0)\psi_b^\lambda \frac{\partial \psi}{\partial t} \frac{\lambda}{\psi_i^{\lambda+1}} = K_s \frac{\partial}{\partial z}\left(\frac{\psi_b}{\psi}\right)^{2+3\lambda} \frac{\partial \psi}{\partial z} + K_s \left(\frac{\psi_b}{\psi}\right)^{2+3\lambda} \frac{\partial^2 \psi}{\partial z^2} - K_s \frac{\partial}{\partial z}\left(\frac{\psi_b}{\psi}\right)^{2+3\lambda} \quad (1.69)$$

The above equation is nonlinear and cannot be handled analytically; thus we rely only on numerical methods to provide the numerical solution. This will be done in the next section.

1.6.1 NUMERICAL SOLUTION FOR THE UNSATURATED GROUNDWATER FLOW MODEL

This section aims to provide a numerical solution to the nonlinear partial differential equation representing the dynamical system underlying the movement of sub-surface water in an unsaturated zone. To achieve this, we substitute the intervals [0, T] to $0 = t_0 < t_1 < t_2 < t_3 < t_4 < \cdots < t_n$ and the intervals [0, R] to $0 = z_0 < z_1 < z_2 < z_3 < z_4 < \cdots < z_m$. Also, we recall that:

$$\frac{\partial \psi}{\partial t}(z_i, t_{n+1}) = \frac{\psi_{i+1}^{n+1} - \psi_i^{n+1}}{\Delta t} \quad (1.70)$$

And the second derivative gives us the following:

$$\frac{\partial^2 \psi}{\partial z^2}(z_i, t_{n+1}) = \frac{\psi_{i+1}^{n+1} - 2\psi_i^{n+1} + \psi_{i-1}^{n+1}}{(\Delta z)^2} \quad (1.71)$$

By substituting the above Equations (1.68) and (1.69) into Equation (1.67), the groundwater flow equation representing the unsaturated flow can be written in the discrete form given below:

$$-(\theta_s - \theta_0)\psi_b^\lambda \frac{\lambda}{\psi_i^{\lambda+1}}\left(\frac{\psi_{i+1}^{n+1} - \psi_i^{n+1}}{\Delta t}\right) = K_s \left(\frac{\left(\frac{\psi_b}{\psi_{i+1}^{n+1}}\right)^{2+3\lambda} - \left(\frac{\psi_b}{\psi_i^{n+1}}\right)^{2+3\lambda}}{\Delta z}\right)\frac{\psi_{i+1}^{n+1} - \psi_i^{n+1}}{\Delta z} + K_s \left(\frac{\psi_b}{\psi_i^{n+1}}\right)^{2+3\lambda}$$

$$\times \frac{\psi_{i+1}^{n+1} - 2\psi_i^{n+1} + \psi_{i-1}^{n+1}}{(\Delta z)^2} - K_s \left(\frac{\left(\frac{\psi_b}{\psi_{i+1}^{n+1}}\right)^{2+3\lambda} - \left(\frac{\psi_b}{\psi_i^{n+1}}\right)^{2+3\lambda}}{\Delta z}\right)$$

$$(1.72)$$

The above equation (1.72) is the numerical solution for modeling the flow of groundwater through the unsaturated zone.

1.7 NUMERICAL SIMULATIONS

In this section, we introduce an example illustrating the performance of the suggested methods. We only demonstrate the classical saturated groundwater flow equation using the Crank–Nicolson scheme. The resulting figures are presented below to show their use and effectiveness in the application of groundwater flow through the saturated zone. Matlab software was used to plot the figures.

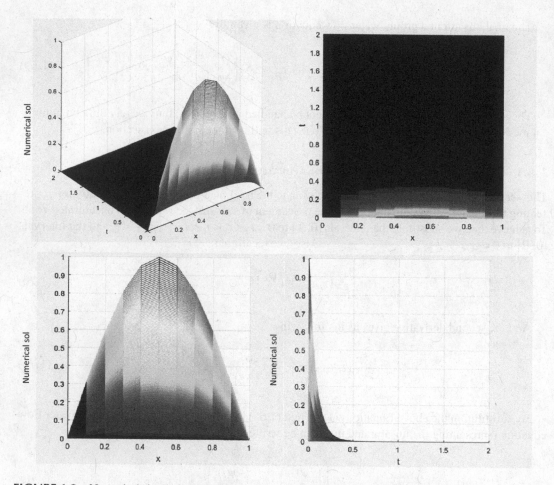

FIGURE 1.2   Numerical simulations for the classical saturated groundwater flow equation.

The numerical simulations presented below will enable us to follow the evolution of groundwater flow in the saturated zone. Let us consider the equation with the following boundary conditions (Figure 1.2):

$$\psi\left(0,t\right)=0, \quad \text{and} \quad \psi\left(1,t\right)=0, t>0$$

And the following initial condition:

$$\psi\left(z,0\right)=\sin\left(n\pi z\right), z \in \left[0,L\right]$$

For computation work, we use the following theoretical parameters $K_s = 2.5$ cm/hr, $S_s = 0.023$ cm, $\Delta t = 0.001$, h (*space step*) = 0.1.

1.8   CONCLUSION

The study of groundwater flow within the saturated–unsaturated zone has attracted the attention of few scholars in the recent decades, perhaps because of the complexity when modeling the passage from unsaturated to the saturated zone. The most attention was not devoted to modeling such a dynamical process using a mathematical equation, in particular, to provide a solution to the already existing mathematical models. While much work has not been done in this direction, it is worth noting that that this process is important for the recharge process. In this work, we considered some existing models. We derived an exact solution in the case of saturated flow using the method of separation of variables, together with Laplace and the Fourier transform operators. We provided a numerical solution and presented in detail the conditions under which the numerical scheme used was stable. Since the unsaturated equation is highly nonlinear, we relied on the numerical scheme to provide a numerical solution. Some numerical simulations are presented to access the efficiency of the mathematical model and the numerical schemes used.

REFERENCES

Allepalli, P.K., & Govindaraju, R.S., 2009. *Modeling fate and transport of atrazine in the saturated–unsaturated zone of soil.* Kansas State University, Department of Civil Engineering.

Assouline, S., 2013. Infiltration into soil: Conceptual approaches and solutions. *Water Resources Research*, 49: 1755–1772.

Atangana, A., & Botha, J.F., 2012. Analytical solution of the groundwater flow equation obtained via homotopy decomposition method. *Journal of Earth Science and Climate Change*, 3(2): 115.

Atangana, A., & Botha, J.F., 2013. A generalized groundwater flow equation using the concept of variable-order derivative. *Boundary Value Problems*, (53).

Brooks, R.H., & Corey, A.T., 1966. Properties of porous media affecting fluid flow. *Journal of the Irrigation and Drainage Division, ASCE*, 92(IR2): 61–88.

Cattaneo, L., Comunian, A., de Filippis, G., Giudici, M., & Vassena, C., 2016. Modeling groundwater flow in heterogeneous porous media with YAGMod. *Computation*, 4(1): 2.

Cheng, A., & Gulliksson, M., 2003. Finite difference methods for saturated–unsaturated flow in porous media. *FSCN/System Analysis and Mathematical Modelling*: 29.

Danesfaraz, R., & Kaya, B., 2009. A numerical solution of the one dimensional groundwater flow by transfer matrix method. *Journal of Science and Technology*, 3(2): 229–240.

Farthing, M.W., & Ogden, F.L., 2017. Numerical solution of Richards' equation: A review of advances and challenges. *Soil Science Society of America Journal*, 81: 1257–1269.

Grigoryan, V., 2012. Finite differences for the heat equation. *Math 124B, Lecturer 17*: 4.

Igboekwe, M.U., & Amos-Uhegbu, C., 2011. Fundamental approach in groundwater flow and solute transport modelling using the finite difference method. *Earth and Environmental Science*: 302–327.

List, F., & Radu, F.A., 2015. A study on iterative methods for solving Richards' equation, *Computers Geoscience*, 20. 341–353.

Konikow, L.F., 1996. *Use of numerical models to simulate groundwater flow and transport.* US Geological Survey, Reston, Virginia, USA. Ch. 4.

Maslouhi A., Lemacha H., & Razach M., 2009. Modelling of water flow and solute transport in saturated–unsaturated media using a self adapting mesh. *Hydroinformatics in Hydrology, Hydrogeology and Water Resources*, 331: 480–487.

Narasimhan, T.N., 2011. *A unified numerical model for saturated–unsaturated groundwater flow.*

Nishigaki, M., & Kono, I., 1980. A consideration on physics of saturated–unsaturated groundwater motion. *Memoirs of the School of Engineering, Okayama University.* 14(2).

Zimmerman, R.W., & Bodvarsson, G.S., 1989. *A simple approximate solution for absorption in a Brooks–Corey medium.* U.S. Department of Energy

Zhang, Y., 2016. *Groundwater flow and solute transport modelling.* Draft lecture note. Dept. of Geology & Geophysics, University of Wyoming.

2 New Model of the Saturated–Unsaturated Groundwater Flow with Power Law and Scale-Invariant Mean Square Displacement

Rendani Vele Makahane and Abdon Atangana
University of the Free State, Bloemfontein, South Africa

CONTENTS

2.1 INTRODUCTION

Modeling the movement of water within a geological formation appears to be a very difficult task as one is unable to see how the flow takes place within the aquifer. This is perhaps one among many other physical problems observed in nature. It is worth noting that, similar problems can be found in different fields of science, engineering and technology. While researchers are interested in understanding, analyzing and predicting such behavior, they make use of some important mathematical tools called differential and integral operators. Previously they relied on differential and integral operators based on the concept of rate of change. While these operators have been used very successfully in many fields, researchers observed that only classical mechanical problems with no memory could be depicted using these mathematical tools. Therefore, a need to formulate new mathematical tools that can be used to depict non-conventional behavior was urgent, as some physical problems could not be understood. For example, the flow of sub-surface water within a geological formation with a fracture network, such a situation follows a process known as a power law. Its behavior was observed

DOI: 10.1201/9781003266266-2

in many areas such as biology, computer science, physics, demography, information theory language, and astronomy, among others, and has been applied successfully (Ghoshdastidar and Dukkipati, 2013; Pinto et al., 2014). It has been considered in a lot of work through its ability to describe material moving in singular media and its long-range memory (Ghoshdastidar and Dukkipati, 2013). In diffusion, power law is very important because it reveals the underlying regularities in the properties of a system. Some derivatives associated with the power-law kernel include the Caputo and Riemann–Liouville (R–L) fractional derivative. The R–L fractional derivative is historically the first definition in literature and has been used by researchers for years; however, there are difficulties in applying it to certain real-life problems. To overcome these limitations, the Caputo fractional derivative was proposed (Demirci & Ozalp, 2012). Because of its impact in modeling complex problems, we apply the Caputo fractional derivative to the classical model of groundwater flow in the saturated–unsaturated media to develop new models. Alternative solutions using the Toufik and Atangana numerical scheme are also presented.

2.2 NUMERICAL SOLUTION FOR THE SATURATED–UNSATURATED ZONE USING THE CAPUTO FRACTIONAL DERIVATIVE

The Caputo fractional operator is suitable for the study of differential equations of fractional order. It was introduced by M. Caputo in 1967. Using this derivative has the advantage that it does not only consider simply interpreted initial conditions, but also the derivative of a constant is zero, meaning that this fractional derivative is bounded (Sontakke & Shaikh, 2015). Suppose that $\alpha > 0$, $t > \alpha$, α, $t \in R$ For: $n - 1 < \alpha < n$, then the Caputo fractional derivative is given as:

$$
{}_0^C D_0^\alpha f\left(t\right) = \frac{1}{\Gamma\left(n-\alpha\right)} \int_0^t \frac{f^{(n)}\left(\tau\right)}{\left(t-\tau\right)^{\alpha+1-n}} \, d\tau \tag{2.1}
$$

Looking at the convolution integral equation, we can define the Caputo fractional derivative for $0 < \alpha \leq 1$ as:

$$
{}_0^C D_t^\alpha f\left(t\right) = \frac{1}{\Gamma\left(1-\alpha\right)} \int_0^t \frac{d}{d\tau} f\left(\tau\right)\left(t-\tau\right)^{-\alpha} \, d\tau \tag{2.2}
$$

$$
{}_0^C D_t^\alpha f\left(t\right) = \frac{d}{dt} f\left(t\right) * \frac{t^{-\alpha}}{\Gamma\left(1-\alpha\right)}. \tag{2.3}
$$

2.2.1 NUMERICAL SOLUTION OF THE CAPUTO FRACTIONAL DERIVATIVE

In this section, the Caputo Equation (2.2) will be discretized to make it suitable for numerical evaluation. To achieve this we substitute the interval $[0, T]$ to $0 = t_0 < t_1 < t_2 < t_3 < t_4 < \cdots . < t_n$. The Caputo fractional differential operator can be discretized as follows:

$$
{}_0^C D_t^\alpha \psi\left(t_n\right) = \frac{1}{\Gamma\left(1-\alpha\right)} \int_0^{t_n} \psi'\left(\tau\right)\left(t_n-\tau\right)^{-\alpha} \, d\tau \tag{2.4}
$$

Using the numerical approximation of the first derivative, the above equation is converted to:

$$
{}_0^C D_t^\alpha \psi\left(t_n\right) = \frac{1}{\Gamma\left(1-\alpha\right)} \sum_{j=0}^n \int_{t_j}^{t_{j+1}} \frac{\psi^{j+1}-\psi^j}{\Delta t}\left(t_n-\tau\right)^{-\alpha} \, d\tau \tag{2.5}
$$

$$\,_0^C D_t^\alpha \psi\left(t_n\right) = \frac{1}{\Gamma\left(1-\alpha\right)} \sum_{j=0}^{n} \frac{\psi^{j+1} - \psi^j}{\Delta t} \int_{t_j}^{t_{j+1}} \left(t_n - \tau\right)^{-\alpha} d\tau \tag{2.6}$$

To integrate this equation easily, we substitute $t_n - \tau = y$, and then $d\tau = -dy$. We further substitute τ with the lower and upper boundary in this manner $t_n - (t_{j+1})$ and $t_n - (t_j)$ which gives us the following:

$$\,_0^C D_t^\alpha \psi\left(t_n\right) = \frac{1}{\Gamma\left(1-\alpha\right)} \sum_{j=0}^{n} \frac{\psi^{j+1} - \psi^j}{\Delta t} \int_{t_n - t_j}^{t_n - t_{j+1}} \left(y\right)^{-\alpha} \left(-dy\right) \tag{2.7}$$

Now that the equation has been integrated, we can substitute $(\Delta t_n - \Delta t_j)$ and $(\Delta t_n - \Delta t_{(j+1)})$ back into the equation to yield the following:

$$\,_0^C D_t^\alpha \psi\left(t_n\right) = \frac{1}{\Gamma\left(1-\alpha\right)} \sum_{j=0}^{n} \frac{\psi^{j+1} - \psi^j}{\Delta t} \Delta t^{1-\alpha} \left\{ \frac{\left(n-j\right)^{1-\alpha} - \left(n-j-1\right)^{1-\alpha}}{1-\alpha} \right\} \tag{2.8}$$

Finally, we obtain:

$$\,_0^C D_t^\alpha \psi\left(t_n\right) = \frac{\left(\Delta t\right)^{-\alpha}}{\Gamma\left(2-\alpha\right)} \sum_{j=0}^{n} \left(\psi^{j+1} - \psi^j\right)\left(\left(n-j\right)^\alpha - \left(n-j-1\right)^\alpha\right). \tag{2.9}$$

2.2.2 Numerical Solution of the 1-d Saturated–Unsaturated Groundwater Flow Equation Using the Caputo Fractional Derivative

The 1-d saturated–unsaturated groundwater flow model is based on the classical equation given as:

$$\left[S_S S_a\left(\psi\right) + C\left(\psi\right)\right] \frac{\partial \psi}{\partial t} = \frac{\partial}{\partial z}\left[K_z\left(\psi\right)\left(\frac{\partial \psi}{\partial z} - 1\right)\right] \tag{2.10}$$

where:
  ψ = pressure head
  n = porosity
  S_S = specific storage of the soil
  S_a = saturation of the aqueous phase
  $C(\psi)$ = capillary capacity of soil = $d\theta/d\psi$
  K_z = hydraulic conductivity
  z = vertical coordinate.

Equation (2.10) has the soil hydraulic conductivity function, $K(\psi)$, and the soil water content function, $\theta(\psi)$, that are required to explain the movement of water. One of the popular models is the equations of Brook and Corey (Allepalli & Govindaraju, 2009):

$$K\left(\psi\right) = K_s \left(\frac{\psi_b}{\psi}\right)^{2+3\lambda} \qquad \psi \leq 0 \tag{2.11}$$

$$K\left(\psi\right) = K_s \qquad \psi > 0 \tag{2.12}$$

$$\theta(\psi) = \theta_0 + (\theta_s - \theta_0)\left(\frac{\psi_b}{\psi}\right)^\lambda \qquad\qquad \psi \leq 0 \qquad\qquad (2.13)$$

where:

K_s = saturated hydraulic conductivity

ψ = pressure head

ψ_b = air entry suction pressure head

$\theta(\psi)$ = soil water content

θ_s = saturated water content

λ = pore size distribution index

θ_0 = residual water content.

By substituting the above equations into Equation (2.10), we obtain the following classical equations for the saturated and unsaturated zones, respectively:

$$\frac{\partial \psi}{\partial t} = \frac{K_s}{S_s}\frac{\partial^2 \psi}{\partial z^2} \qquad\qquad (2.14)$$

and,

$$\frac{\partial \theta(\psi)}{\partial t} = \frac{\partial}{\partial z}\left[K_z(\psi)\left(\frac{\partial \psi}{\partial z} - 1\right)\right] \qquad\qquad (2.15)$$

Using the Brooks and Corey equations, Equation (2.9), becomes:

$$-(\theta_s - \theta_0)\psi_b^\lambda \frac{\partial \psi}{\partial t}\frac{\lambda}{\psi_i^{\lambda+1}} = K_s \frac{\partial}{\partial z}\left(\frac{\psi_b}{\psi}\right)^{2+3\lambda}\frac{\partial \psi}{\partial z} + K_s\left(\frac{\psi_b}{\psi}\right)^{2+3\lambda}\frac{\partial^2 \psi}{\partial z^2} - K_s\frac{\partial}{\partial z}\left(\frac{\psi_b}{\psi}\right)^{2+3\lambda}$$

With the benefits and application of the power law, we now use the discretized Caputo fractional derivative to convert the above classical models to models with the waiting time distribution and scale invariant mean square displacement. The stability of the numerical solution is also presented.

2.2.2.1 Numerical Solution of the 1-d Saturated Groundwater Flow Equation Using the Caputo Fractional Derivative

To achieve this, the classical groundwater flow equation can be extended to fractional differential equation by replacing the time classical derivative with Caputo derivative. Then the saturated zone classical model with the Caputo fractional derivative becomes:

$$_0^C D_t^\alpha \psi(t_n) = \frac{K_s}{S_s}\frac{\partial^2 \psi}{\partial z^2} \qquad\qquad (2.16)$$

Now, applying the explicit forward Euler method on the equation, we obtain the following numerical solution:

$$\frac{(\Delta t)^{-\alpha}}{\Gamma(2-\alpha)}\sum_{j=0}^{n}\left(\psi_i^{j+1} - \psi_i^j\right)\left((n-j)^\alpha - (n-j-1)^\alpha\right) = \frac{K_s}{S_s}\left[\frac{\psi_{i+1}^n - 2\psi_i^n + \psi_{i-1}^n}{(\Delta z)^2}\right] \qquad (2.17)$$

Before we can perform stability analysis for the above explicit forward Euler method, lets simplify the equation using the following parameters:

$$a = \frac{(\Delta t)^{-\alpha}}{\Gamma(2-\alpha)},$$

$$b = \frac{K_s}{S_s \left(\Delta z\right)^2}$$

$$R_{n,j}^{\alpha} = \left(\left(n-j\right)^{\alpha} - \left(n-j-1\right)^{\alpha}\right)$$

Now we can rewrite the equation with parameter:

$$\sum_{j=0}^{n} \left(\psi_i^{j+1} - \psi_i^{j}\right) aR_{n,j}^{\alpha} = b\left(\psi_{i+1}^{n} - 2\psi_i^{n} + \psi_{i-1}^{n}\right) \tag{2.18}$$

The above equation is then simplified into:

$$\left(\psi_i^{n+1} - \psi_i^{n}\right) aR_{n,n}^{\alpha} + \sum_{j=0}^{n-1} \left(\psi_i^{j+1} - \psi_i^{j}\right) aR_{n,j}^{\alpha} = b\left(\psi_{i+1}^{n} - 2\psi_i^{n} + \psi_{i-1}^{n}\right) \tag{2.19}$$

Now we have this equation for a numerical solution:

$$\psi_i^{n+1} aR_{n,n}^{\alpha} = \psi_i^{n}\left(aR_{n,n}^{\alpha} - 2b\right) + b\left(\psi_{i+1}^{n} + \psi_{i-1}^{n}\right) - \sum_{j=0}^{n-1} \left(\psi_i^{j+1} - \psi_i^{j}\right) aR_{n,j}^{\alpha} \tag{2.20}$$

For stability analysis, we assume that:

$$\psi_i^{n} - \sigma_n e^{ik_m z}$$

Substituting the above into Equation (2.20) we obtain the following:

$$\sigma_{n+1} e^{ik_m z} aR_{n,n}^{\alpha} = \sigma_n e^{ik_m z}\left(aR_{n,n}^{\alpha} - 2b\right) + be^{ik_m z}\left(\sigma_i e^{ik_m \Delta z} + \sigma_n e^{ik_m \Delta z}\right) - \sum_{j=0}^{n-1} \left(\sigma_{j+1} - \sigma_j\right) e^{ik_m z} aR_{n,j}^{\alpha} \tag{2.21}$$

If we simplify the above equation, we obtain the following:

$$\sigma_{n+1} aR_{n,n}^{\alpha} = \sigma_n\left[aR_{n,n}^{\alpha} - 2b + b\left(e^{ik_m \Delta z} + e^{-ik_m \Delta z}\right)\right] - \sum_{j=0}^{n-1} \left(\sigma_{j+1} - \sigma_j\right) aR_{n,j}^{\alpha} \tag{2.22}$$

$$\sigma_{n+1} aR_{n,n}^{\alpha} = \sigma_n\left[aR_{n,n}^{\alpha} + 2b\left(\frac{e^{ik_m \Delta z} + e^{-ik_m \Delta z}}{2} - 1\right)\right] - \sum_{j=0}^{n-1} \left(\sigma_{j+1} - \sigma_j\right) aR_{n,j}^{\alpha} \tag{2.23}$$

$$\sigma_{n+1} aR_{n,n}^{\alpha} = \sigma_n\left[aR_{n,n}^{\alpha} + 2b\left(\cos\left(k_m \Delta z\right) - 1\right)\right] - \sum_{j=0}^{n-1} \left(\sigma_{n+1} - \sigma_n\right) aR_{n,j}^{\alpha} \tag{2.24}$$

Using the double angle for cos we can simplify the equation into:

$$\sigma_{n+1} aR_{n,n}^{\alpha} = \sigma_n\left[aR_{n,n}^{\alpha} + 2b\left(1 - 2\sin^2\left(\frac{k_m \Delta z}{2}\right) - 1\right)\right] - \sum_{j=0}^{n-1} \left(\sigma_{j+1} - \sigma_j\right) aR_{n,j}^{\alpha} \tag{2.25}$$

$$\sigma_{n+1} a R_{n,n}^{\alpha} = \sigma_n \left[a R_{n,n}^{\alpha} - 4b \sin^2 \left(\frac{k_m \Delta z}{2} \right) \right] - \sum_{j=0}^{n-1} \left(\sigma_{j+1} - \sigma_j \right) a R_{n,j}^{\alpha} \tag{2.26}$$

If $n = 0$ the above equation becomes:

$$a\sigma_1 R_{0,n}^{\alpha} = \sigma_0 \left[a R_{0,n}^{\alpha} - 4b \sin^2 \left(\frac{k_m \Delta z}{2} \right) \right] \tag{2.27}$$

$$\left| \frac{\sigma_1}{\sigma_0} \right| < 1 \rightarrow \left| \frac{\sigma_1}{\sigma_0} \right| = \left| \frac{a R_{0,n}^{\alpha} - 4b \sin^2 \left(\frac{k_m \Delta z}{2} \right)}{a R_{0,n}^{\alpha}} \right| < 1 \tag{2.28}$$

$$a R_{0,n}^{\alpha} - 4b \sin^2 \left(\frac{k_m \Delta z}{2} \right) < a R_{0,n}^{\alpha} \tag{2.29}$$

$$0 < 4b \sin^2 \left(\frac{k_m \Delta z}{2} \right) \tag{2.30}$$

$$0 < 4 \frac{K_s}{S_s \left(\Delta z \right)^2} \sin^2 \left(\frac{k_m \Delta z}{2} \right) \tag{2.31}$$

We assume that $\forall n > 0$:

$$\left| \frac{\sigma_n}{\sigma_0} \right| < 1$$

Using the above assumption, we will prove that:

$$\left| \frac{\sigma_{n+1}}{\sigma_0} \right| < 1$$

$$\sigma_{n+1} a R_{n,n}^{\alpha} = \sigma_n \left[a R_{n,n}^{\alpha} - 4b \sin^2 \left(\frac{k_m \Delta z}{2} \right) \right] - \sum_{j=0}^{n-1} \left(\sigma_{j+1} - \sigma_j \right) a R_{n,j}^{\alpha} \tag{2.32}$$

$$\sigma_{n+1} = \frac{\sigma_n \left[a R_{n,n}^{\alpha} - 4b \sin^2 \left(\frac{k_m \Delta z}{2} \right) \right]}{a R_{n,n}^{\alpha}} - \frac{\sum_{j=0}^{n-1} \sigma_{n-j} a R_{n,j}^{\alpha}}{a R_{n,n}^{\alpha}} \tag{2.33}$$

$$\left| \sigma_{n+1} \right| \le \left| \sigma_n \right| \left| \frac{\left[a R_{n,n}^{\alpha} - 4b \sin^2 \left(\frac{k_m \Delta z}{2} \right) \right]}{a R_{n,n}^{\alpha}} \right| - \frac{\sum_{j=0}^{n-1} \left| \sigma_{n-j} \right| a R_{n,j}^{\alpha}}{a R_{n,n}^{\alpha}} \tag{2.34}$$

Using the recursive hypothesis:

$$|\sigma_{n+1}| < |\sigma_0| \left[\frac{aR_{n,n}^\alpha - 4b\sin^2\left(\frac{k_m\Delta z}{2}\right)}{aR_{n,n}^\alpha} \right] + \frac{\sum_{j=0}^{n-1}|\sigma_0|aR_{n,j}^\alpha}{aR_{n,n}^\alpha} \qquad (2.35)$$

$$|\sigma_{n+1}| < |\sigma_0| \left[\frac{\left[aR_{n,n}^\alpha - 4b\sin^2\left(\frac{k_m\Delta z}{2}\right)\right]}{aR_{n,n}^\alpha} + \frac{\sum_{j=0}^{n-1}aR_{n,j}^\alpha}{aR_{n,n}^\alpha} \right] \qquad (2.36)$$

$$\frac{|\sigma_{n+1}|}{|\sigma_0|} < 1$$

$$\left| \frac{aR_{n,n}^\alpha - 4b\sin^2\left(\frac{k_m\Delta z}{2}\right) + \sum_{j=0}^{n-1}aR_{n,j}^\alpha}{aR_{n,n}^\alpha} \right| < 1 \qquad (2.37)$$

$$aR_{n,n}^\alpha - 4b\sin^2\left(\frac{k_m\Delta z}{2}\right) + \sum_{j=0}^{n-1}aR_{n,j}^\alpha < aR_{n,n}^\alpha \qquad (2.38)$$

The stability condition is given as:

$$\sum_{j=0}^{n-1}aR_{n,j}^\alpha < 4b\sin^2\left(\frac{k_m\Delta z}{2}\right) \qquad (2.39)$$

We obtain another numerical solution by applying the implicit backward Euler method on Equation (2.14), such that:

$$\frac{(\Delta t)^{-\alpha}}{\Gamma(2-\alpha)}\sum_{j=0}^{n}\left(\psi_i^{n+1}-\psi_i^n\right)\left((n-j)^\alpha-(n-j-1)^\alpha\right) = \frac{K_s}{S_s}\left[\frac{\psi_{i+1}^{n+1}-2\psi_i^{n+1}+\psi_{i-1}^{n+1}}{(\Delta z)^2}\right] \qquad (2.40)$$

Before we can perform stability analysis for the above explicit forward Euler method, lets simplify the equation using the following parameters:

$$a = \frac{(\Delta t)^{-\alpha}}{\Gamma(2-\alpha)},$$

$$b = \frac{K_s}{S_s(\Delta z)^2}$$

$$R_{n,j}^{\alpha} = \left((n-j)^{\alpha} - (n-j-1)^{\alpha} \right)$$

Now we can rewrite the equation with parameter:

$$\sum_{j=0}^{n} \left(\psi_i^{n+1} - \psi_i^{n} \right) a R_{n,j}^{\alpha} = b \left(\psi_{i+1}^{n+1} - 2\psi_i^{n+1} + \psi_{i-1}^{n+1} \right) \tag{2.41}$$

$$\left(\psi_i^{n+1} - \psi_i^{n} \right) a R_{n,n}^{\alpha} + \sum_{j=0}^{n-1} \left(\psi_i^{j+1} - \psi_i^{j} \right) a R_{n,j}^{\alpha} = b \left(\psi_{i+1}^{n+1} - 2\psi_i^{n+1} + \psi_{i-1}^{n+1} \right) \tag{2.42}$$

Now we have the above equation as a numerical scheme:

$$\left(\psi_i^{n+1} - \psi_i^{n} \right) a R_{n,n}^{\alpha} = b \left(\psi_{i+1}^{n+1} - 2\psi_i^{n+1} + \psi_{i-1}^{n+1} \right) - \sum_{j=0}^{n-1} \left(\psi_i^{j+1} - \psi_i^{j} \right) a R_{n,j}^{\alpha} \tag{2.43}$$

Which can be further simplified:

$$-\psi_i^{n} a R_{n,n}^{\alpha} = -\psi_i^{n+1} \left(a R_{n,j}^{\alpha} + 2b \right) + b \left(\psi_{i+1}^{n+1} + \psi_{i-1}^{n+1} \right) - \sum_{j=0}^{n-1} \left(\psi_i^{j+1} - \psi_i^{j} \right) a R_{n,j}^{\alpha} \tag{2.44}$$

For stability analysis, we assume that:

$$\psi_i^{n} = \sigma_n e^{ik_m z}$$

Substituting the above into Equation (2.44) we obtain the following:

$$-\sigma_n e^{ik_m z} a R_{n,n}^{\alpha} = -\sigma_{n+1} e^{ik_m z} \left(a R_{n,n}^{\alpha} + 2b \right) + b \left(\sigma_{n+1} e^{ik_m (z+\Delta z)} + \sigma_{n+1} e^{ik_m (z-\Delta z)} \right)$$
$$- \sum_{j=0}^{n-1} \left(\sigma_{j+1} - \sigma_j \right) e^{ik_m z} a R_{n,j}^{\alpha} \tag{2.45}$$

$$-\sigma_n e^{ik_m z} a R_{n,n}^{\alpha} = -\sigma_{n+1} e^{ik_m z} \left(a R_{n,n}^{\alpha} + 2b \right) + b \sigma_{n+1} e^{ik_m z} \left(e^{ik_m \Delta z} + e^{-ik_m \Delta z} \right) - \sum_{j=0}^{n-1} \left(\sigma_{j+1} - \sigma_j \right) e^{ik_m z} a R_{n,j}^{\alpha} \tag{2.46}$$

$$-\sigma_n e^{ik_m z} a R_{n,n}^{\alpha} = -\sigma_{n+1} e^{ik_m z} \left(a R_{n,n}^{\alpha} + 2b - b \left(e^{ik_m \Delta z} + e^{-ik_m \Delta z} \right) \right) - \sum_{j=0}^{n-1} \left(\sigma_{j+1} - \sigma_j \right) e^{ik_m z} a R_{n,j}^{\alpha} \tag{2.47}$$

$$-\sigma_n e^{ik m z} a R_{n,n}^{\alpha} = -\sigma_{n+1} e^{ik_m z} \left(a R_{n,n}^{\alpha} + 2b - 2b \left(\frac{e^{ik_m \Delta z} + e^{-ik_m \Delta z}}{2} \right) \right) - \sum_{j=0}^{n-1} \left(\sigma_{j+1} - \sigma_j \right) e^{ik_m z} a R_{n,j}^{\alpha} \tag{2.48}$$

$$-\sigma_n e^{ik_m z} a R_{n,n}^{\alpha} = -\sigma_{n+1} e^{ik_m z} \left(a R_{n,n}^{\alpha} + 2b - 2b \left(\cos(k_m \Delta z) \right) - \sum_{j=0}^{n-1} \left(\sigma_{j+1} - \sigma_j \right) e^{ik_m z} a R_{n,j}^{\alpha} \tag{2.49}$$

$$-\sigma_n e^{ik_m z} a R_{n,n}^\alpha = -\sigma_{n+1} e^{ik_m z}\left(a R_{n,n}^\alpha + 2b - 2b\left(1 - 2i\sin^2\left(\frac{k_m\Delta z}{2}\right)\right)\right) - \sum_{j=0}^{n-1}\left(\sigma_{j+1} - \sigma_j\right) e^{ik_m z} a R_{n,j}^\alpha \quad (2.50)$$

$$-\sigma_n a R_{n,n}^\alpha = -\sigma_{n+1}\left(a R_{n,n}^\alpha + 4bi\sin^2\left(\frac{k_m\Delta z}{2}\right)\right) - \sum_{j=0}^{n-1}\left(\sigma_{j+1} - \sigma_j\right) a R_{n,j}^\alpha \quad (2.51)$$

If $n = 0$ then the above equation becomes:

$$\sigma_1\left(a R_{0,n}^\alpha + 4bi\sin^2\left(\frac{k_m\Delta z}{2}\right)\right) = \sigma_0 a R_{0,n}^\alpha \quad (2.52)$$

$$\left|\frac{\sigma_1}{\sigma_0}\right| < 1 \rightarrow \left|\frac{\sigma_1}{\sigma_0}\right| = \left|\frac{a R_{0,n}^\alpha}{a R_{0,n}^\alpha + 4bi\sin^2\left(\frac{k_m\Delta z}{2}\right)}\right| < 1 \quad (2.53)$$

$$a R_{0,n}^\alpha < a R_{0,n}^\alpha + 4bi\sin^2\left(\frac{k_m\Delta z}{2}\right) \quad (2.54)$$

$$0 < 4bi\sin^2\left(\frac{k_m\Delta z}{2}\right) \quad (2.55)$$

We assume that $\forall n > 0$:

$$\left|\frac{\sigma_n}{\sigma_0}\right| < 1$$

Using the above assumption, we will prove that:

$$\left|\frac{\sigma_{n+1}}{\sigma_0}\right| < 1$$

$$\sigma_{n+1}\left(a R_{n,n}^\alpha + 4bi\sin^2\left(\frac{k_m\Delta z}{2}\right)\right) + \sum_{j=0}^{n-1}\left(\sigma_{j+1} - \sigma_j\right) a R_{n,j}^\alpha = \sigma_n a R_{n,n}^\alpha \quad (2.56)$$

$$\sigma_{n+1} = \frac{\sigma_n a R_{n,n}^\alpha}{\left[a R_{n,n}^\alpha + 4bi\sin^2\left(\frac{k_m\Delta z}{2}\right)\right]} - \frac{\sum_{j=0}^{n-1}\sigma_{n-j} a R_{n,j}^\alpha}{a R_{n,n}^\alpha + 4bi\sin^2\left(\frac{k_m\Delta z}{2}\right)} \quad (2.57)$$

$$\left|\sigma_{n+1}\right| \le \left|\sigma_n\right|\left|\frac{a R_{n,n}^\alpha}{\left[a R_{n,n}^\alpha + 4bi\sin^2\left(\frac{k_m\Delta z}{2}\right)\right]}\right| - \frac{\sum_{j=0}^{n-1}\left|\sigma_{n-j}\right| a R_{n,j}^\alpha}{a R_{n,n}^\alpha + 4bi\sin^2\left(\frac{k_m\Delta z}{2}\right)} \quad (2.58)$$

Using the recursive hypothesis:

$$|\sigma_{n+1}| < |\sigma_0| \left[\frac{aR_{n,n}^{\alpha}}{aR_{n,n}^{\alpha} + 4b\sin^2\left(\frac{k_m\Delta z}{2}\right)} \right] + \frac{\sum_{j=0}^{n-1}|\sigma_0| aR_{n,j}^{\alpha}}{aR_{n,n}^{\alpha} + 4bi\sin^2\left(\frac{k_m\Delta z}{2}\right)} \quad (2.59)$$

$$|\sigma_{n+1}| < |\sigma_0| \left[\frac{aR_{n,n}^{\alpha}}{aR_{n,n}^{\alpha} + 4b\,i\sin^2\left(\frac{k_m\Delta z}{2}\right)} + \frac{\sum_{j=0}^{n-1} aR_{n,j}^{\alpha}}{aR_{n,n}^{\alpha} + 4b\,i\sin^2\left(\frac{k_m\Delta z}{2}\right)} \right] \quad (2.60)$$

$$\frac{|\sigma_{n+1}|}{|\sigma_0|} < 1$$

$$\left| \frac{aR_{n,n}^{\alpha} + \sum_{j=0}^{n-1} aR_{n,j}^{\alpha}}{aR_{n,n}^{\alpha} + 4bi\sin^2\left(\frac{k_m\Delta z}{2}\right)} \right| < 1 \quad (2.61)$$

$$aR_{n,n}^{\alpha} + \sum_{j=0}^{n-1} aR_{n,j}^{\alpha} < aR_{n,n}^{\alpha} + 4b\,i\sin^2\left(\frac{k_m\Delta z}{2}\right) \quad (2.62)$$

The stability condition is given as:

$$\sum_{j=0}^{n-1} aR_{n,j}^{\alpha} < 4bi\sin^2\left(\frac{k_m\Delta z}{2}\right) \quad (2.63)$$

Now we solve the model in Equation (2.14) numerically using the implicit Crank–Nicolson method:

$$\frac{(\Delta t)^{-\alpha}}{\Gamma(2-\alpha)}\left(\psi_i^{j+1} - \psi_i^{j}\right)\left((n-j)^{\alpha} - (n-j-1)^{\alpha}\right) = \frac{K_s}{S_s}\left[\frac{\psi_{i+1}^{n+1} - 2\psi_i^{n+1} + \psi_{i-1}^{n+1}}{2(\Delta z)^2} \right.$$
$$\left. + \frac{\psi_{i+1}^{n} - 2\psi_i^{n} + \psi_{i-1}^{n}}{2(\Delta z)^2} \right] \quad (2.64)$$

For simplicity, let us use the following parameters:

$$a = \frac{(\Delta t)^{-\alpha}}{\Gamma(2-\alpha)},$$

$$c = \frac{K_s}{S_s 2 (\Delta z)^2}$$

$$R_{n,j}^{\alpha} = \left((n-j)^{\alpha} - (n-j-1)^{\alpha} \right)$$

Now we can rewrite the equation as:

$$a R_{n,j}^{\alpha} = c \left[\left(\psi_{i+1}^{n+1} - 2\psi_i^{n+1} + \psi_{i-1}^{n+1} \right) + \left(\psi_{i+1}^n - 2\psi_i^n + \psi_{i-1}^n \right) \right] \tag{2.65}$$

$$\sum_{j=0}^{n} \left(\psi_i^{j+1} - \psi_i^j \right) a R_{n,j}^{\alpha} = c \left[\left(\psi_{i+1}^{n+1} - 2\psi_i^{n+1} + \psi_{i-1}^{n+1} \right) + \left(\psi_{i+1}^n - 2\psi_i^n + \psi_{i-1}^n \right) \right] \tag{2.66}$$

$$\left(\psi_i^{n+1} - \psi_i^n \right) a R_{n,n}^{\alpha} + \sum_{j=0}^{n-1} \left(\psi_i^{j+1} - \psi_i^j \right) a R_{n,j}^{\alpha} = c \left[\left(\psi_{i+1}^{n+1} - 2\psi_i^{n+1} + \psi_{i-1}^{n+1} \right) \right.$$
$$\left. + \left(\psi_{i+1}^n - 2\psi_i^n + \psi_{i-1}^n \right) \right] \tag{2.67}$$

$$\psi_i^{n+1} a R_{n,n}^{\alpha} = c \left(\psi_{i+1}^{n+1} - 2\psi_i^{n+1} + \psi_{i-1}^{n+1} \right) + c \left(\psi_{i+1}^n - 2\psi_i^n + \psi_{i-1}^n \right)$$
$$+ \psi_i^n a R_{n,n}^{\alpha} - \sum_{j=0}^{n-1} \left(\psi_i^{j+1} - \psi_i^j \right) a R_{n,j}^{\alpha} \tag{2.68}$$

For stability analysis, we assume that:

$$\psi_i^n = \sigma_n e^{ik_m z}$$

$$\sigma_{n+1} e^{ik_m z} a R_{n,n}^{\alpha} = c \left(\sigma_{n+1} e^{ik_m (z+\Delta z)} - 2\sigma_{n+1} e^{ik_m z} + \sigma_{n+1} e^{ik_m (z-\Delta z)} \right)$$
$$+ c \left(\sigma_n e^{ik_m (z+\Delta z)} - 2\sigma_n e^{ik_m z} + \sigma_n e^{ik_m (z-\Delta z)} \right)$$
$$+ \sigma_n e^{ik_m z} a R_{n,n}^{\alpha} - \sum_{j=0}^{n-1} \left(\sigma_i^{j+1} - \sigma_i^j \right) e^{ik_m z} a R_{n,j}^{\alpha} \tag{2.69}$$

$$\sigma_{n+1} e^{ik_m z} a R_{n,n}^{\alpha} - c\sigma_{n+1} e^{ik_m z} \left(e^{ik_m \Delta z} - 2 + e^{-ik_m \Delta z} \right) = c\sigma_n e^{ik_m z} \left(e^{ik_m \Delta z} + \sigma_n e^{-ik_m \Delta z} \right) + \sigma_n e^{ik_m z} \left(a R_{n,n}^{\alpha} - 2c \right)$$
$$- \sum_{j=0}^{n-1} \left(\sigma_i^{j+1} - \sigma_i^j \right) e^{ik_m z} a R_{n,j}^{\alpha} \tag{2.70}$$

$$\sigma_{n+1} a R_{n,n}^{\alpha} - c\sigma_{n+1} \left(e^{ik_m \Delta z} - 2 + e^{-ik_m \Delta z} \right) = c\sigma_n \left(e^{ik_m \Delta z} + e^{-ik_m \Delta z} \right) + \sigma_n \left(a R_{n,n}^{\alpha} - 2c \right)$$
$$- \sum_{j=0}^{n-1} \left(\sigma_i^{j+1} - \sigma_i^j \right) a R_{n,j}^{\alpha} \tag{2.71}$$

$$\sigma_{n+1}\left(aR_{n,n}^{\alpha}-2c\left(\frac{e^{ik_m\Delta z}+e^{-ik_m\Delta z}}{2}-1\right)\right)=\sigma_n\left(aR_{n,n}^{\alpha}+2c\left(\frac{e^{ik_m\Delta z}+e^{-ik_m\Delta z}}{2}-1\right)\right)$$

$$-\sum_{j=0}^{n-1}\left(\sigma_i^{j+1}-\sigma_i^{j}\right)aR_{n,j}^{\alpha} \qquad (2.72)$$

$$\sigma_{n+1}\left(aR_{n,n}^{\alpha}-2c\left(\cos\left(k_m\Delta z\right)-1\right)\right)=\sigma_n\left(aR_{n,n}^{\alpha}+2c\left(\cos\left(k_m\Delta z\right)-1\right)\right)-\sum_{j=0}^{n-1}\left(\sigma_i^{j+1}-\sigma_i^{j}\right)aR_{n,j}^{\alpha} \qquad (2.73)$$

If $n = 0$ the above equation becomes:

$$\sigma_1\left(aR_{n,n}^{\alpha}-2c\left(\cos\left(k_m\Delta z\right)-1\right)\right)=\sigma_0\left(aR_{n,n}^{\alpha}+2c\left(\cos\left(k_m\Delta z\right)-1\right)\right) \qquad (2.74)$$

$$\left|\frac{\sigma_1}{\sigma_0}\right|<1\rightarrow\left|\frac{\sigma_1}{\sigma_0}\right|=\left|\frac{aR_{n,n}^{\alpha}+2c\left(\cos\left(k_m\Delta z\right)-1\right)}{aR_{n,n}^{\alpha}-2c\left(\cos\left(k_m\Delta z\right)-1\right)}\right|<1 \qquad (2.75)$$

$$aR_{n,n}^{\alpha}+2c\left(\cos\left(ik_m\Delta z\right)-1\right)<aR_{n,n}^{\alpha}-2c\left(\cos\left(k_m\Delta z\right)-1\right) \qquad (2.76)$$

$$0<-4c\left(\cos\left(k_m\Delta z\right)-1\right) \qquad (2.77)$$

$$4c\cos\left(k_m\Delta z\right)<4c \qquad (2.78)$$

$$\cos\left(k_m\Delta z\right)<1 \qquad (2.79)$$

We can write 1 as $\cos 2\pi$ such that:

$$\cos\left(k_m\Delta z\right)<\cos 2\pi \qquad (2.80)$$

$$ik_m\Delta z<2\pi \qquad (2.81)$$

Therefore:

$$k_m<\frac{2\pi}{\Delta z} \qquad (2.82)$$

We assume that $\forall n>0$:

$$\left|\frac{\sigma_n}{\sigma_0}\right|<1$$

Using the above assumption, we will prove that:

$$\left| \frac{\sigma_{n+1}}{\sigma_0} \right| < 1$$

$$\sigma_{n+1}\left(aR_{n,n}^{\alpha} - 2c\left(\cos\left(k_m\Delta z\right) - 1\right)\right) = \sigma_n\left(aR_{n,n}^{\alpha} + 2c\left(\cos\left(k_m\Delta z\right) - 1\right)\right) - \sum_{j=0}^{n-1}\sigma_{n-j}aR_{n,j}^{\alpha} \tag{2.83}$$

$$\sigma_{n+1} - \sigma_n\left[\frac{aR_{n,n}^{\alpha} + 2c\left(\cos\left(ik_m\Delta z\right) - 1\right)}{aR_{n,n}^{\alpha} - 2c\left(\cos\left(k_m\Delta z\right) - 1\right)} \right] - \frac{\sum_{j=0}^{n-1}\sigma_{n-j}aR_{n,j}^{\alpha}}{aR_{n,n}^{\alpha} - 2c\left(\cos\left(k_m\Delta z\right) - 1\right)} \tag{2.84}$$

$$\left|\sigma_{n+1}\right| \le \left|\sigma_n\right|\left| \frac{aR_{n,n}^{\alpha} + 2c\left(\cos\left(k_m\Delta z\right) - 1\right)}{aR_{n,n}^{\alpha} - 2c\left(\cos\left(k_m\Delta z\right) - 1\right)} \right| + \frac{\sum_{j=0}^{n-1}\left|\sigma_{n-j}\right|aR_{n,j}^{\alpha}}{aR_{n,n}^{\alpha} - 2c\left(\cos\left(k_m\Delta z\right) - 1\right)} \tag{2.85}$$

Using the recursive hypothesis:

$$\left|\sigma_{n+1}\right| < \left|\sigma_0\right|\left| \frac{aR_{n,n}^{\alpha} + 2c\left(\cos\left(k_m\Delta z\right) - 1\right)}{aR_{n,n}^{\alpha} - 2c\left(\cos\left(k_m\Delta z\right) - 1\right)} \right| + \frac{\sum_{j=0}^{n-1}\left|\sigma_0\right|aR_{n,j}^{\alpha}}{aR_{n,n}^{\alpha} - 2c\left(\cos\left(k_m\Delta z\right) - 1\right)} \tag{2.86}$$

$$\left|\sigma_{n+1}\right| < \left|\sigma_0\right|\left[\frac{aR_{n,n}^{\alpha} + 2c\left(\cos\left(k_m\Delta z\right) - 1\right)}{aR_{n,n}^{\alpha} - 2c\left(\cos\left(k_m\Delta z\right) - 1\right)} + \frac{\sum_{j=0}^{n-1}aR_{n,j}^{\alpha}}{aR_{n,n}^{\alpha} - 2c\left(\cos\left(k_m\Delta z\right) - 1\right)} \right] \tag{2.87}$$

$$\left| \frac{aR_{n,n}^{\alpha} + 2c\left(\cos\left(k_m\Delta z\right) - 1\right) + \sum_{j=0}^{n-1}aR_{n,j}^{\alpha}}{aR_{n,n}^{\alpha} - 2c\left(\cos\left(k_m\Delta z\right) - 1\right)} \right| < 1 \tag{2.88}$$

$$aR_{n,n}^{\alpha} + 2c\left(\cos\left(km\Delta z\right) - 1\right) + \sum_{j=0}^{n-1}aR_{n,j}^{\alpha} < aR_{n,n}^{\alpha} - 2c\left(\cos\left(k_m\Delta z\right) - 1\right) \tag{2.89}$$

$$\sum_{j=0}^{n-1}aR_{n,j}^{\alpha} < -4c\left(\cos\left(k_m\Delta z\right) - 1\right) \tag{2.90}$$

The stability condition is given as:

$$c4\cos\left(k_m\Delta z\right) + \sum_{j=0}^{n-1}aR_{n,j}^{\alpha} < 4c \tag{2.91}$$

The above stability condition can be simplified into:

$$c4\cos\left(k_m\Delta z\right) < 4c - \sum_{j=0}^{n-1} aR_{n,j}^{\alpha} \tag{2.92}$$

$$k_m\Delta z < \cos^{-1}\left(1 - \frac{\sum_{j=0}^{n-1} aR_{n,j}^{\alpha}}{4c}\right) \tag{2.93}$$

To illustrate the complexity of the unsaturated flow, let us rewrite Equation (2.15) using the first and second order derivatives; therefore, the discrete form of the groundwater flow equation representing the unsaturated zone is given by:

$$-\left(\theta_s - \theta_0\right)\psi_b^{\lambda}\frac{\lambda}{\psi_i^{\lambda+1}}\frac{\psi_{i+1}^{n+1} - \psi_i^{n+1}}{\Delta t} = K_s\left(\frac{\left(\frac{\psi_b}{\psi_{i+1}^{n+1}}\right)^{2+3\lambda} - \left(\frac{\psi_b}{\psi_i^{n+1}}\right)^{2+3\lambda}}{\Delta z}\right)\frac{\psi_{i+1}^{n+1} - \psi_i^{n+1}}{\Delta z} + K_s\left(\frac{\psi_b}{\psi_i^{n+1}}\right)^{2+3\lambda}$$

$$\times\frac{\psi_{i+1}^{n+1} - 2\psi_i^{n+1} + \psi_{i-1}^{n+1}}{\left(\Delta z\right)^2} - K_s\left(\frac{\left(\frac{\psi_b}{\psi_{i+1}^{n+1}}\right)^{2+3\lambda} - \left(\frac{\psi_b}{\psi_i^{n+1}}\right)^{2+3\lambda}}{\Delta z}\right) \tag{2.94}$$

Now, the classical groundwater flow equation can be extended to a fractional differential equation by replacing the time classical derivative with the Caputo derivative, giving the following new equation derived from modeling the unsaturated zone:

$$-\left(\theta_s - \theta_0\right)\psi_b^{\lambda}\sum_{j=1}^{n}\frac{\left(\Delta t\right)^{-\alpha}}{\Gamma\left(2-\alpha\right)}\left(\psi_i^{j+1} - \psi_i^{j}\right)\left(\left(n-j\right)^{\alpha} - \left(n-j-1\right)^{\alpha}\right)$$

$$= K_s\frac{\left(\frac{\psi_b}{\psi_{i+1}^{n+1}}\right)^{2+3\lambda} - \left(\frac{\psi_b}{\psi_i^{n+1}}\right)^{2+3\lambda}}{\Delta z}\frac{\psi_{i+1}^{n+1} - \psi_i^{n+1}}{\Delta z} + K_s\left(\frac{\psi_b}{\psi_i^{n+1}}\right)^{2+3\lambda}$$

$$\times\frac{\psi_{i+1}^{n+1} - 2\psi_i^{n+1} + \psi_{i-1}^{n+1}}{\left(\Delta z\right)^2} - K_s\frac{\left(\frac{\psi_b}{\psi_{i+1}^{n+1}}\right)^{2+3\lambda} - \left(\frac{\psi_b}{\psi_i^{n+1}}\right)^{2+3\lambda}}{\Delta z} \tag{2.95}$$

To make further simplification on the above equation, let us say:

$$\wp = -\left(\theta_s - \theta_0\right)\psi_b^{\lambda}\frac{\lambda}{\psi_i^{\lambda+1}}$$

$$a = \frac{\left(\Delta t\right)^{-\alpha}}{\Gamma\left(2-\alpha\right)},$$

$$b = \frac{K_s}{\Delta z},$$

$$c = \frac{1}{\Delta z},$$

$$d = \frac{1}{(\Delta z)^2}$$

$$R_{n,j}^{\alpha} = \left((n-j)^{\alpha} - (n-j-1)^{\alpha} \right)$$

Now we rewrite the equation as:

$$\wp \sum_{j=0}^{n} \left(\psi_i^{j+1} - \psi_i^{j} \right) a R_{n,j}^{\alpha} = b \left(\left(\frac{\psi_b}{\psi_{i+1}^{n+1}} \right)^{2+3\lambda} - \left(\frac{\psi_b}{\psi_i^{n+1}} \right)^{2+3\lambda} \right) c \left(\psi_{i+1}^{n+1} - \psi_i^{n+1} \right) + K_s \left(\frac{\psi_b}{\psi_i^{n+1}} \right)^{2+3\lambda}$$

$$\times d \left(\psi_{i+1}^{n+1} - 2\psi_i^{n+1} + \psi_{i-1}^{n+1} \right) - b \left(\left(\frac{\psi_b}{\psi_{i+1}^{n+1}} \right)^{2+3\lambda} - \left(\frac{\psi_b}{\psi_i^{n+1}} \right)^{2+3\lambda} \right) \tag{2.96}$$

The above numerical scheme is highly non-linear; therefore, it is not possible to perform a stability analysis.

2.3 NUMERICAL SOLUTION OF THE NEW SATURATED–UNSATURATED GROUNDWATER FLOW MODEL USING THE NEW NUMERICAL SCHEME

The Adams–Bashforth scheme has been documented in many publications as a powerful tool when solving non-linear equations; however, it has proven not to be effective for fractional differential equations with non-singular and non-local properties. To extend these limitations, Toufik and Atangana introduced a new numerical scheme for non-linear fractional differential equations with fractional derivatives with the non-singular and non-local kernel. This numerical scheme is the combination of the two-step Lagrange polynomial and the fundamental theorem of fractional calculus, it also extends the limitations of the Adams–Bashforth method (Toufik & Atangana, 2017). In this section, we are going to develop new numerical solutions for the 1-d saturated–unsaturated groundwater equation using the new Toufik and Atangana numerical scheme.

2.3.1 NUMERICAL SOLUTION OF THE SATURATED ZONE MODEL USING THE NEW NUMERICAL SCHEME

The classical groundwater flow equation can be extended to fractional differential equation by replacing the time classical derivative with the Caputo derivative, giving the following new equation for the saturated zone:

$$_0^C D^{\alpha} \psi(z,t) = \frac{K_s}{S_s} \frac{\partial^2 \psi}{\partial z^2} \tag{2.97}$$

where

$$_0^C D^{\alpha} \psi(z,t) = F\left(z,t,\psi(z,t)\right)$$

Then we have the numerical scheme as:

$$\psi\left(z_i, t\right) - \psi\left(z_i, 0\right) = \frac{1}{\Gamma(\alpha)} \int_0^t F\left(z_i, \tau, \psi\left(z_i, \tau\right)\right)\left(t - \tau\right)^{\alpha-1} d\tau \tag{2.98}$$

At point t_{n+1},

$$\psi\left(z_i, t_{n+1}\right) - \psi\left(z_i, 0\right) = \frac{1}{\Gamma(\alpha)} \int_0^{t_{n+1}} F\left(z_i, \tau, \psi\left(z_i, \tau\right)\right)\left(t_{n+1} - \tau\right)^{\alpha-1} d\tau \tag{2.99}$$

$$\begin{aligned}
\psi\left(z_i, t_{n+1}\right) - \psi\left(z_i, 0\right) = \frac{(\Delta t)^\alpha}{\Gamma(2+\alpha)} \sum_{j=0}^n \Big[& F\left(z_i, t_j, \psi_i^j\right)\Big(\left(n-j+1\right)^\alpha\left(n-j+2+\alpha\right) \\
& - \left(n-j\right)^\alpha\left(n-j+2+2\alpha\right)\Big) - F\left(z_i, t_{j-1}, \psi_i^{j-1}\right)\Big(\left(n-j+1\right)^{\alpha+1} \\
& - \left(n-j\right)^\alpha\left(n-j+1+\alpha\right)\Big)\Big]
\end{aligned} \tag{2.100}$$

where:

$$F\left(z_i, t_j, \psi_i^j\right) = \frac{K_s}{S_s} \frac{\psi_{i+1}^j - 2\psi_i^j + \psi_{i-1}^j}{(\Delta z)^2} \tag{2.101}$$

$$F\left(z_i, t_{j-1}, \psi_i^{j-1}\right) = \frac{K_s}{S_s} \frac{\psi_{i+1}^{j-1} - 2\psi_i^{j-1} + \psi_{i-1}^{j-1}}{(\Delta z)^2} \tag{2.102}$$

The final numerical solution for the saturated groundwater flow equation is given as:

$$\begin{aligned}
\psi\left(z_i, t_{n+1}\right) - \psi\left(z_i, 0\right) = \frac{(\Delta t)^\alpha}{\Gamma(2+\alpha)} \sum_{j=0}^n \Big[& \frac{K_s}{S_s} \frac{\psi_{i+1}^j - 2\psi_i^j + \psi_{i-1}^j}{(\Delta z)^2}\Big(\left(n-j+1\right)^\alpha\left(n-j+2+\alpha\right) \\
& - \left(n-j\right)^\alpha\left(n-j+2+2\alpha\right)\Big) - \frac{K_s}{S_s} \frac{\psi_{i+1}^{j-1} - 2\psi_i^{j-1} + \psi_{i-1}^{j-1}}{(\Delta z)^2}\Big(\left(n-j+1\right)^{\alpha+1} \\
& - \left(n-j\right)^\alpha\left(n-j+1+\alpha\right)\Big)\Big]
\end{aligned} \tag{2.103}$$

2.3.2 Numerical Solution of the Unsaturated Zone Using the New Numerical Scheme

Now, using the new numerical scheme, the new model for the unsaturated zone is given by the following equation:

$$
\begin{aligned}
{}_0^C D_t^\alpha \psi\left(z_i, t_{n+1}\right) = {} & \frac{1}{-\left(\theta_s-\theta_0\right)\psi_b^\lambda \dfrac{\lambda}{\psi_i^{\lambda+1}}} \left[K_s \left(\frac{\left(\dfrac{\psi_b}{\psi_{i+1}^{n+1}}\right)^{2+3\lambda} - \left(\dfrac{\psi_b}{\psi_i^{n+1}}\right)^{2+3\lambda}}{\Delta z} \right) \frac{\psi_{i+1}^{n+1}-\psi_i^{n+1}}{\Delta z} \right. \\[2ex]
& + K_s \left(\frac{\psi_b}{\psi_i^{n+1}}\right)^{2+3\lambda} \times \frac{\psi_{i+1}^{n+1}-2\psi_i^{n+1}+\psi_{i-1}^{n+1}}{(\Delta z)^2} - K_s \left(\frac{\left(\dfrac{\psi_b}{\psi_{i+1}^{n+1}}\right)^{2+3\lambda} - \left(\dfrac{\psi_b}{\psi_i^{n+1}}\right)^{2+3\lambda}}{\Delta z} \right) \right]
\end{aligned}
\tag{2.104}
$$

where

$$
{}_0^C D_t^\alpha \psi\left(z_i, t_{n+1}\right) = F\left(z, t, \psi\left(z, t\right)\right)
$$

Then we have the numerical scheme as:

$$
\psi\left(z_i, t\right) - \psi\left(z_i, 0\right) = \frac{1}{\Gamma(\alpha)} \int_0^t F\left(z_i, \tau, \psi\left(z_i, \tau\right)\right)\left(t-\tau\right)^{\alpha-1} d\tau
\tag{2.105}
$$

At point t_{n+1}

$$
\psi\left(z_i, t_{n+1}\right) - \psi\left(z_i, 0\right) = \frac{1}{\Gamma(\alpha)} \int_0^{t_{n+1}} F\left(z_i, \tau, \psi\left(z_i, \tau\right)\right)\left(t_{n+1}-\tau\right)^{\alpha-1} d\tau
\tag{2.106}
$$

$$
\begin{aligned}
\psi\left(z_i, t_{n+1}\right) - \psi\left(z_i, 0\right) = {} & \frac{(\Delta t)^\alpha}{\Gamma(2+\alpha)} \sum_{j=0}^n \left[F\left(z_i, t_j, \psi_i^j\right)\left((n-j+1)^\alpha (n-j+2+\alpha)\right) \right. \\
& \left. -(n-j)^\alpha (n-j+2+2\alpha)\right) - F\left(z_i, t_{j-1}, \psi_i^{j-1}\right)\left((n-j+1)^{\alpha+1}\right. \\
& \left. -(n-j)^\alpha (n-j+1+\alpha)\right) \Big]
\end{aligned}
\tag{2.107}
$$

where:

$$F\left(z_i,t_j,\psi_i^j\right)=\frac{1}{-\left(\theta_s-\theta_0\right)\psi_b^\lambda\dfrac{\lambda}{\psi_i^{\lambda+1}}}\left(K_s\left(\frac{\left(\dfrac{\psi_b}{\psi_{i+1}^j}\right)^{2+3\lambda}-\left(\dfrac{\psi_b}{\psi_i^j}\right)^{2+3\lambda}}{\Delta z}\right)\frac{\psi_{i+1}^j-\psi_i^j}{\Delta z}+K_s\left(\frac{\psi_b}{\psi_i^j}\right)^{2+3\lambda}\right.$$

$$\left.\times\frac{\psi_{i+1}^j-2\psi_i^j+\psi_{i-1}^j}{\left(\Delta z\right)^2}-K_s\left(\frac{\left(\dfrac{\psi_b}{\psi_{i+1}^j}\right)^{2+3\lambda}-\left(\dfrac{\psi_b}{\psi_i^j}\right)^{2+3\lambda}}{\Delta z}\right)\right)\qquad(2.108)$$

$$F\left(z_i,t_{j-1},\psi_i^{j-1}\right)=\frac{1}{-\left(\theta_s-\theta_0\right)\psi_b^\lambda\dfrac{\lambda}{\psi_i^{\lambda+1}}}\left(K_s\left(\frac{\left(\dfrac{\psi_b}{\psi_{i+1}^{j-1}}\right)^{2+3\lambda}-\left(\dfrac{\psi_b}{\psi_i^{j-1}}\right)^{2+3\lambda}}{\Delta z}\right)\frac{\psi_{i+1}^{j-1}-\psi_i^{j-1}}{\Delta z}+K_s\left(\frac{\psi_b}{\psi_i^{j-1}}\right)^{2+3\lambda}\right.$$

$$\left.\times\frac{\psi_{i+1}^{j-1}-2\psi_i^{j-1}+\psi_{i-1}^{j-1}}{\left(\Delta z\right)^2}-K_s\left(\frac{\left(\dfrac{\psi_b}{\psi_{i+1}^{j-1}}\right)^{2+3\lambda}-\left(\dfrac{\psi_b}{\psi_i^{j-1}}\right)^{2+3\lambda}}{\Delta z}\right)\right)\qquad(2.109)$$

The final numerical solution for the unsaturated groundwater flow equation is given as:

$$\left(\psi\left(z_i,t_{n+1}\right)-\psi\left(z_i,0\right)\right)$$

$$=\frac{\left(\Delta t\right)^\alpha}{\Gamma\left(2+\alpha\right)}\sum_{j=0}^n\left[\frac{1}{-\left(\theta_s-\theta_0\right)\psi_b^\lambda\dfrac{\lambda}{\psi_i^{\lambda+1}}}\left(K_s\left(\frac{\left(\dfrac{\psi_b}{\psi_{i+1}^j}\right)^{2+3\lambda}-\left(\dfrac{\psi_b}{\psi_i^j}\right)^{2+3\lambda}}{\Delta z}\right)\frac{\psi_{i+1}^j-\psi_i^j}{\Delta z}+K_s\left(\frac{\psi_b}{\psi_i^j}\right)^{2+3\lambda}\right.\right.$$

$$\left.\times\frac{\psi_{i+1}^j-2\psi_i^j+\psi_{i-1}^j}{\left(\Delta z\right)^2}-K_s\left(\frac{\left(\dfrac{\psi_b}{\psi_{i+1}^j}\right)^{2+3\lambda}-\left(\dfrac{\psi_b}{\psi_i^j}\right)^{2+3\lambda}}{\Delta z}\right)\right)\left(\left(n-j+1\right)^\alpha\left(n-j+2+\alpha\right)-\left(n-j\right)^\alpha\left(n-j+2+2\alpha\right)\right)$$

$$-\frac{1}{-\left(\theta_s-\theta_0\right)\psi_b^\lambda\dfrac{\lambda}{\psi_i^{\lambda+1}}}\left(K_s\left(\frac{\left(\dfrac{\psi_b}{\psi_{i+1}^{j-1}}\right)^{2+3\lambda}-\left(\dfrac{\psi_b}{\psi_i^{j-1}}\right)^{2+3\lambda}}{\Delta z}\right)\frac{\psi_{i+1}^{j-1}-\psi_i^{j-1}}{\Delta z}+K_s\left(\frac{\psi_b}{\psi_i^{j-1}}\right)^{2+3\lambda}\right.$$

$$\left.\left.\times\frac{\psi_{i+1}^{j-1}-2\psi_i^{j-1}+\psi_{i-1}^{j-1}}{\left(\Delta z\right)^2}-K_s\left(\frac{\left(\dfrac{\psi_b}{\psi_{i+1}^{j-1}}\right)^{2+3\lambda}-\left(\dfrac{\psi_b}{\psi_i^{j-1}}\right)^{2+3\lambda}}{\Delta z}\right)\right)\left(\left(n-j+1\right)^{\alpha+1}-\left(n-j\right)^\alpha\left(n-j+1+\alpha\right)\right)\right]\qquad(2.110)$$

For the simplicity of the above equation, let:

$$\wp = -\left(\theta_s - \theta_0\right)\psi_b^\lambda \frac{\lambda}{\psi_i^{\lambda+1}}$$

$$a = \frac{\left(\Delta t\right)^\alpha}{\Gamma\left(2-\alpha\right)}$$

$$b = \frac{K_s}{\Delta z}$$

$$c = K_s$$

$$d = \frac{1}{\left(\Delta z\right)^2}$$

$$e = \frac{1}{\Delta z}$$

$$\delta_{\alpha,j}^{1,n} = \left(n-j+1\right)^\alpha \left(n-j+2+\alpha\right) - \left(n-j\right)^\alpha \left(n-j+2+2\alpha\right)$$

$$\delta_{\alpha,j}^{2,n} = \left(n-j+1\right)^{\alpha+1} - \left(n-j\right)^\alpha \left(n-j+1+\alpha\right)$$

$$
\begin{aligned}
\left(\psi\left(z_i, t_{n+1}\right) - \psi\left(z_i, 0\right)\right) = a \sum_{j=0}^{n} &\left[\frac{1}{\wp}\left(b\left(\left(\frac{\psi_b}{\psi_{i+1}^j}\right)^{2+3\lambda} - \left(\frac{\psi_b}{\psi_i^j}\right)^{2+3\lambda}\right)e\left(\psi_{i+1}^j - \psi_i^j\right) + c\left(\frac{\psi_b}{\psi_i^j}\right)^{2+3\lambda}\right. \right. \\
&\left. \times d\left(\psi_{i+1}^j - 2\psi_i^j + \psi_{i-1}^j\right) - b\left(\left(\frac{\psi_b}{\psi_{i+1}^j}\right)^{2+3\lambda} - \left(\frac{\psi_b}{\psi_i^j}\right)^{2+3\lambda}\right)\right)\delta_{\alpha,j}^{1,n} \\
&- \frac{1}{\wp}\left(b\left(\left(\frac{\psi_b}{\psi_{i+1}^{j-1}}\right)^{2+3\lambda} - \left(\frac{\psi_b}{\psi_i^{j-1}}\right)^{2+3\lambda}\right)e\left(\psi_{i+1}^{j-1} - \psi_i^{j-1}\right) + c\left(\frac{\psi_b}{\psi_i^{j-1}}\right)^{2+3\lambda}\right. \\
&\left.\left. \times d\left(\psi_{i+1}^{j-1} - 2\psi_i^{j-1} + \psi_{i-1}^{j-1}\right) - b\left(\left(\frac{\psi_b}{\psi_{i+1}^{j-1}}\right)^{2+3\lambda} - \left(\frac{\psi_b}{\psi_i^{j-1}}\right)^{2+3\lambda}\right)\right)\delta_{\alpha,j}^{2,n}\right]
\end{aligned}
\tag{2.111}
$$

2.4 CONCLUSION

The commonly used groundwater flow equation is based on the classical differential operator using the concept of rate of change. This operator is suitable to model real-world problems exhibiting mechanical and classical laws. It is believed that most complex real-world problems follow the power law kernel. For this reason, we replaced the time derivative on the groundwater flow equation with a fractional derivative based on the power law kernel. Using the power law kernel enable us to capture more complex problems such as problems with the long-tailed, which can be assimilated with the flow within a fracture.

REFERENCES

Allepalli, P.K., & Govindaraju, R.S., 2009. *Modeling Fate and Transport of Atrazine in the Saturated–Unsaturated Zone of Soil*. Kansas State University, Department of Civil Engineering, Manhattan, Kansas.

Atangana, A., 2017. Fractal-fractional differentiation and integration: Connecting fractal calculus and fractional calculus to predict complex system. *Chaos, Solutions and Fractals*. 102: 396–406.

Brooks, R.H., & Corey, A.T., 1966. Properties of porous media affecting fluid flow. *Journal of the Irrigation and Drainage Division, ASCE*, 92(IR2): 61–88.

Demirci, E., & Ozalp, N., 2012. A method for solving differential equations of fractional order. *Journal of Computational and Applied Mathematics*, 236: 2754–2762.

Ghoshdastidar, D., & Dukkipati, A., 2013. *On Power law Kernels, Corresponding Reproducing Kernel Hilbert Space and Application. Department of Computer Science and Automation*. Indian Institute of Science, Bangalore. 7p.

List, F., & Radu, F.A., 2015. A study on iterative methods for solving Richards' equation. *Computer Geoscience*, 20: 341–353.

Pinto, C.M.A., Lopes, A.M., & Machado, J.A.T., 2014. Double power laws, fractals, and self-similarity. *Applied Mathematical Modelling*, 38: 4019–4026.

Sontakke, B.R., & Shaikh, A.S., 2015. Properties of Caputo operator and its applications to linear fractional differential equations. *International Journal of Engineering Research and Applications*, 5(5): 22–27.

Toufik, M., & Atangana, A., 2017. New numerical approximation of fractional derivative with non-local and non-singular kernel: Application to chaotic models. *The European Physical Journal Plus*, 132: 444.

3 New Model of the 1-d Unsaturated–Saturated Groundwater Flow with Crossover from Usual to Confined Flow Mean Square Displacement

Rendani Vele Makahane and Abdon Atangana
University of the Free State, Bloemfontein, South Africa

CONTENTS

3.1 INTRODUCTION

In recent years, many researchers have modeled natural problems in their particular fields using fractional order derivatives because of their efficiency in describing such problems. These derivatives include the well-known Riemann–Liouville and the recently proposed Caputo–Fabrizio derivatives. The definition presented by Riemann–Liouville and the first Caputo version has a limitation of its singularity. With this limitation, the two fractal operators cannot fully describe the memory effect (Atangana and Alkahtani, 2015). To improve the full explanation of memory, Fabrizio and Caputo proposed a new definition of a fractional operator built on the exponential decay law with no singular kernel (Sheikh et al., 2017). This fractional operator is based on the convolution of a first-order derivative and the exponential function. Some advantages of this derivative are that it can be used when exponential law is observed in nature (Alkahtani, 2016; Gomez-Aguilar et al., 2017; Atanackovic et al., 2018). A straightforward example of a real-world process that follows the exponential decay law is the cooling of hot water. The speed at which the hot water cools is proportional to the change in temperature between the cooling body and its environment, therefore, if the

DOI: 10.1201/9781003266266-3

temperature difference is higher, then the cooling follows an exponential decay. Another distribution of theoretical and practical importance is Poisson. The Poisson distribution allows us to find probabilities for random points in time for a process; for example, modeling a number of cars that pass a certain place in a given time period (Rice, 2006). The relation between these two distributions is that, if the Poisson provides a suitable description of events occurring per interval of time, then the exponential will describe the length of time between events (Cooper, 2005). Groundwater flow problems can sometimes be explained using both of these distributions. Other properties that are observed in groundwater flow problems include the fading memory and the fatigue effect. The fading memory states that the probability of something happening in the future has no relation to whether or not it has occurred in the past, for example, should anything cause a temporary pause in groundwater flow, whatever happens after the break has no relation to whether or not it has happened before the break. On the other hand, the fatigue effect would explain why groundwater cannot flow endlessly; at some point, it would stop, be it due to no pressure or force exerted on it. Both these properties, exponential decay law, and Poisson distribution can be observed in groundwater flow problems and are necessary to incorporate in a groundwater mathematical model. Therefore, in this chapter, we shall generate new 1-d saturated–unsaturated groundwater flow equations by replacing the ordinary time derivative on the original equation with the Caputo–Fabrizio fractional-order derivative. Having said that, useful information on the derivative will be presented to inform readers who are not familiar with it.

3.2 THE CAPUTO–FABRIZIO FRACTIONAL-ORDER DERIVATIVE

Singularity is the main problem in the first definition of fractional order derivative. To avoid this problem, Caputo and Fabrizio presented a fractional-order derivative without singularity. The definition of this derivative is given as follows (Caputo & Fabrizio, 2015; Atangana & Alkahtani, 2015; Atangana & Baleanu, 2016):

Definition

Let $f \in H^1(a,b)$, $b > a$, $\alpha \in [0,1]$ then, the new Caputo fractional derivative is:

$$D_t^\alpha \left(f(t) \right) = \frac{M(\alpha)}{1-\alpha} \int_b^t f'(\tau) \exp\left[-\alpha \frac{t-\tau}{1-\alpha} \right] d\tau \tag{3.1}$$

where $M(\alpha)$ denotes a normalization function obeying $M(0) = M(1) = 1$. However, if the function doesn't belong to $H^1(a,b)$ the derivative becomes:

$$D_t^\alpha \left(f(t) \right) = \frac{\alpha M(\alpha)}{1-\alpha} \int_b^t \left(f(t) - f(\tau) \right) \exp\left[-\alpha \frac{t-\tau}{1-\alpha} \right] d\tau \tag{3.2}$$

If $\sigma = \dfrac{1-\alpha}{\alpha} \in [0,\infty]$, $\alpha = \dfrac{1}{1+\sigma} \in [0,1]$, then Equation (4.4) above assumes the following form:

$$D_t^\sigma \left(f(t) \right) = \frac{N(\sigma)}{\sigma} \int_b^t f'(\tau) \exp\left[-\frac{t-\tau}{\sigma} \right] d\tau \, N(0) = N(\infty) = 1 \tag{3.3}$$

3.3 GOVERNING EQUATION

The equation under consideration describes the continuous flow between the saturated–unsaturated soil. The equation is the combination of Darcy's law and the continuity law (Pelka, 1983; List and Radu, 2015). It involves the hydraulic properties of the soil that are a function of the suction head of the soil and are therefore non-linear (Allepalli and Govindaraju, 2009). The 1-d saturated–unsaturated groundwater flow equation is given below:

$$\left[S_S S_a \left(\psi \right) + C \left(\psi \right) \right] \frac{\partial \psi}{\partial t} = \frac{\partial}{\partial z} \left[K_z \left(\psi \right) \left(\frac{\partial \psi}{\partial z} - 1 \right) \right] \tag{3.4}$$

where:

ψ = pressure head

n = porosity

S_S = specific storage of the soil

S_a = saturation of the aqueous phase $= \dfrac{\theta}{n}$

$C(\psi)$ = capillary capacity of the soil $= d\theta/d\psi$

$q(t)$ = groundwater flux $= -K_z \left(\psi \right) \left(\dfrac{\partial \psi}{\partial z} - 1 \right)$

K_z = hydraulic conductivity

z = vertical coordinate.

Research development in the area of unsaturated soils has resulted in the development of relationships between the pore-water pressure head and the coefficient of permeability (Papagianakis & Fredlund, 1984). For the unsaturated zone, $\psi \le 0$ and the saturated zone $\psi > 0$. Brooks and Corey (1966) derived a mathematical relationship between the coefficient of permeability and the capillary pressure

$$K \left(\psi \right) = K_s \left(\frac{\psi_b}{\psi} \right)^{2+3\lambda} \qquad \psi \le 0 \tag{3.5}$$

$$K \left(\psi \right) = K_s \qquad \psi > 0 \tag{3.6}$$

$$\theta \left(\psi \right) = \theta_0 + \left(\theta_s - \theta_0 \right) \left(\frac{\psi_b}{\psi} \right)^{\lambda} \qquad \psi \le 0 \tag{3.7}$$

Therefore, the corresponding equations for the saturated and unsaturated zone are as follows:

$$\frac{\partial \psi}{\partial t} = \frac{K_s}{S_s} \frac{\partial^2 \psi}{\partial z^2} \tag{3.8}$$

and,

$$\frac{\partial \theta \left(\psi \right)}{\partial t} = \frac{\partial}{\partial z} \left[K_z \left(\psi \right) \left(\frac{\partial \psi}{\partial z} - 1 \right) \right] \tag{3.9}$$

Using the Brooks and Corey equations, Equation (3.9) become:

$$-\left(\theta_s - \theta_0 \right) \psi_b^{\lambda} \frac{\partial \psi}{\partial t} \frac{\lambda}{\psi^{\lambda+1}} = K_s \frac{\partial}{\partial z} \left(\frac{\psi_b}{\psi} \right)^{2+3\lambda} \frac{\partial \psi}{\partial z} + K_s \left(\frac{\psi_b}{\psi} \right)^{2+3\lambda} \frac{\partial^2 \psi}{\partial z^2} - K_s \frac{\partial}{\partial z} \left(\frac{\psi_b}{\psi} \right)^{2+3\lambda} \tag{9.1}$$

3.4 NUMERICAL SOLUTIONS FOR THE SATURATED–UNSATURATED ZONE USING THE CAPUTO–FABRIZIO FRACTIONAL DERIVATIVE

In this section, we apply the Caputo–Fabrizio fractional derivative to our classical saturated–unsaturated equation to generate a new numerical model. We begin by creating a discretized version of the Caputo–Fabrizio fractional derivative. For time t_n the Caputo–Fabrizio fractional derivative is written as:

$$
{}^{CF}_{0}D^{\alpha}_{t}\big(\psi(t)\big) = \frac{M(\alpha)}{1-\alpha} \int_{0}^{t_n} \psi'(\tau) \exp\left[-\alpha \frac{t_n-\tau}{1-\alpha}\right] d\tau \tag{3.10}
$$

If $n \geq 0$, Equation (3.10) becomes:

$$
{}^{CF}_{0}D^{\alpha}_{t}\psi(t_n) = \frac{M(\alpha)}{1-\alpha} \sum_{j=0}^{n} \int_{t_j}^{t_{j+1}} \frac{\psi^{j+1}-\psi^{j}}{\Delta t} \exp\left[-\frac{\alpha}{1-\alpha}(t_n-\tau)\right] d\tau \tag{3.11}
$$

$$
{}^{CF}_{0}D^{\alpha}_{t}\psi(t_n) = \frac{M(\alpha)}{1-\alpha} \sum_{j=0}^{n} \frac{\psi^{j+1}-\psi^{j}}{\Delta t} \int_{t_j}^{t_{j+1}} \exp\left[-\frac{\alpha}{1-\alpha}(t_n-\tau)\right] d\tau \tag{3.12}
$$

To easily integrate this equation, we shall integrate by substituting $t_n - \tau = y$, $d\tau = -dy$ and $\mu = \dfrac{\alpha}{1-\alpha}$ we further substitute τ with the lower and upper boundary in this manner $t_n - (t_{j+1})$ and $t_n - (t_j)$ which gives us the following:

$$
{}^{CF}_{0}D^{\alpha}_{t}\psi(t_n) = \frac{M(\alpha)}{1-\alpha} \sum_{j=0}^{n} \frac{\psi^{j+1}-\psi^{j}}{\Delta t} \int_{t_n-t_{j+1}}^{t_n-t_j} \exp\left[-\mu y\right] dy \tag{3.13}
$$

$$
{}^{CF}_{0}D^{\alpha}_{t}\psi(t_n) = \frac{M(\alpha)}{1-\alpha} \sum_{j=0}^{n} \frac{\psi^{j+1}-\psi^{j}}{\Delta t} \left[-\frac{1}{\mu}\exp(-\mu y)\, {}^{t_n-t_j}_{t_n-t_{j+1}}\right] \tag{3.14}
$$

Now that the equation has been integrated, we can substitute $(\Delta t_n - \Delta t_j)$ and $(\Delta t_n - \Delta t_{(j+1)})$ back into the equation to yield the following:

$$
{}^{CF}_{0}D^{\alpha}_{t}\psi(t_n) = \frac{M(\alpha)}{1-\alpha} \sum_{j=0}^{n} \frac{\psi^{j+1}-\psi^{j}}{\Delta t} \left[-\frac{1}{\mu}\exp\big(-\mu(t_n-t_j)\big) + \frac{1}{\mu}\exp\big(-\mu(t_n-t_{j+1})\big)\right] \tag{3.15}
$$

Our final discretized version of the Caputo–Fabrizio fractional derivative is given by:

$$
{}^{CF}_{0}D^{\alpha}_{t}\psi(t_n) = \frac{M(\alpha)}{1-\alpha} \sum_{j=0}^{n} \frac{\psi^{j+1}-\psi^{j}}{\mu\Delta t} \left[\exp\big(-\mu\Delta t(n-j-1)\big) - \exp\big(-\mu\Delta t(n-j)\big)\right] \tag{3.16}
$$

The Caputo–Fabrizio fractional integral of order α of a function $\psi(x,t)$ is given as:

$$
{}^{CF}_{0}I^{\alpha}_{t}\psi(z,t) = \frac{1-\alpha}{M(\alpha)}\psi(z,t) + \frac{\alpha}{M(\alpha)} \int_{0}^{t} \psi(z,\tau) d\tau \tag{3.17}
$$

At point (z_i, t_n) we have,

$$ {}_0^{CF}I_t^\alpha \psi (z_i, t_n) = \frac{1-\alpha}{M(\alpha)} \psi (z_i, t_n) + \frac{\alpha}{M(\alpha)} \int_0^{t_n} \psi (z_i, \tau) d\tau \tag{3.18}$$

$$ {}_0^{CF}I_t^\alpha \psi (z_i, t_n) = \frac{1-\alpha}{M(\alpha)} \psi_i^n + \frac{\alpha}{M(\alpha)} \sum_{j=0}^{n-1} \int_{t_j}^{t_{j+1}} \psi_i^j d\tau \tag{3.19}$$

$$ {}_0^{CF}I_t^\alpha \psi (z_i, t_n) = \frac{1-\alpha}{M(\alpha)} \psi_i^n + \frac{\alpha}{M(\alpha)} \sum_{j=0}^{n-1} \psi_i^j \int_{t_j}^{t_{j+1}} d\tau \tag{3.20}$$

$$ {}_0^{CF}I_t^\alpha \psi (z_i, t_n) = \frac{1-\alpha}{M(\alpha)} \psi_i^n + \frac{\alpha}{M(\alpha)} \sum_{j=0}^{n-1} \psi_i^j (t_{j+1} - t_j) \tag{3.21}$$

$$ {}_0^{CF}I_t^\alpha \psi (z_i, t_n) = \frac{1-\alpha}{M(\alpha)} \psi_i^n + \frac{\alpha}{M(\alpha)} \sum_{j=0}^{n-1} \psi_i^j \Delta t \tag{3.22}$$

3.4.1 NUMERICAL SOLUTION FOR THE SATURATED ZONE USING THE CAPUTO–FABRIZIO FRACTIONAL DERIVATIVE

For our numerical solution, we replace the time derivative on our saturated equation (3.8) with the Caputo–Fabrizio fractional derivative, such that:

$$ {}_0^{CF}D^\alpha \psi (z, t) = \frac{K_s}{S_s} \frac{\partial^2 \psi}{\partial z^2} \tag{3.23}$$

where

$$ {}_0^{CF}D^\alpha \psi (z, t) = f \big(z, t, \psi (z, t) \big) \tag{3.24}$$

We apply the Caputo–Fabrizio integral to the above equation:

$$ \psi (z, t) = \psi (z, 0) + \frac{1-\alpha}{M(\alpha)} f \big(z, t, \psi (z, t) \big) + \frac{\alpha}{M(\alpha)} \int_0^t f \big(z, \tau, \psi (z, \tau) \big) d\tau \tag{3.25}$$

When $t = t_{n+1}$ we have the following equation:

$$ \psi (z_i, t_{n+1}) = \psi (z_i, 0) + \frac{1-\alpha}{M(\alpha)} f \big(z_i, t_n, \psi (z_i, t_n) \big) + \frac{\alpha}{M(\alpha)} \int_0^{t_{n+1}} f \big(z_i, \tau \, \psi (z_i, \tau) \big) d\tau \tag{3.26}$$

When $= t_n$:

$$ \psi (z_i, t_n) = \psi (z_i, 0) + \frac{1-\alpha}{M(\alpha)} f \big(z_i, t_{n-1}, \psi (z_i, t_{n-1}) \big) + \frac{\alpha}{M(\alpha)} \int_0^{t_n} f \big(z_i, \tau \, \psi (z_i, \tau) \big) d\tau \tag{3.27}$$

Subtracting the above solution for t_n from t_{n+1}, we obtain the following:

$$\psi_i^{n+1} - \psi_i^n = \frac{1-\alpha}{M(\alpha)}\left[f\left(z_i,t_n,\psi_i^n\right) - f\left(z_i,t_{n-1},\psi_i^{n-1}\right)\right] + \frac{\alpha}{M(\alpha)}\int_{t_n}^{t_{n+1}} f\left(z_i,\tau,\psi\left(x_i,\tau\right)\right)d\tau \quad (3.28)$$

$$\psi_i^{n+1} = \psi_i^n + \frac{1-\alpha}{M(\alpha)}\left[f\left(z_i,t_n,\psi_i^n\right) - f\left(z_i,t_{n-1},\psi_i^{n-1}\right)\right]$$

$$+ \frac{\alpha}{M(\alpha)}\left[\frac{3}{2}\Delta t f\left(z_i,t_n,\psi_i^n\right) - \frac{\Delta t}{2} f\left(z_i,t_{n-1},\psi_i^{n-1}\right)\right] \quad (3.29)$$

where:

$$F\left(z_i,t_j,\psi_i^n\right) = \frac{K_s}{S_s}\frac{\psi_{i+1}^n - 2\psi_i^n + \psi_{i-1}^n}{\left(\Delta z\right)^2} \quad (3.30)$$

$$F\left(z_i,t_{j-1},\psi_i^{n-1}\right) = \frac{K_s}{S_s}\frac{\psi_{i+1}^{n-1} - 2\psi_i^{n-1} + \psi_{i-1}^{n-1}}{\left(\Delta z\right)^2} \quad (3.31)$$

We substitute the above Equations (3.30) and (3.31) into Equation (3.29) to obtain the final numerical solution for the saturated groundwater flow equation:

$$\psi_i^{n+1} = \psi_i^n + \frac{1-\alpha}{M(\alpha)}\left[\frac{K_s}{S_s}\frac{\psi_{i+1}^n - 2\psi_i^n + \psi_{i-1}^n}{\left(\Delta z\right)^2} - \frac{K_s}{S_s}\frac{\psi_{i+1}^{n-1} - 2\psi_i^{n-1} + \psi_{i-1}^{n-1}}{\left(\Delta z\right)^2}\right]$$

$$+ \frac{\alpha}{M(\alpha)}\left[\frac{3\Delta t}{2}\left(\frac{K_s}{S_s}\frac{\psi_{i+1}^n - 2\psi_i^n + \psi_{i-1}^n}{\left(\Delta z\right)^2}\right) - \frac{\Delta t}{2}\left(\frac{K_s}{S_s}\frac{\psi_{i+1}^{n-1} - 2\psi_i^{n-1} + \psi_{i-1}^{n-1}}{\left(\Delta z\right)^2}\right)\right]. \quad (3.32)$$

3.4.2 Stability Analysis Using Von Neumann

To check the stability of the above numerical solution, we use Von Neumann by assuming a Fourier expansion in space of:

$$\psi_i^n = \sigma^n e^{ik_m z}$$

We also assume the following parameters:

$$a = \frac{1-\alpha}{M(\alpha)}$$

$$b = \frac{\alpha}{M(\alpha)}$$

$$c = \frac{K_s}{S_s (\Delta z)^2}$$

$$d = \frac{\Delta t}{2}$$

Then Equation (3.32) becomes:

$$\sigma^{n+1} e^{ik_m z} = \sigma^n e^{ik_m z} + a \left[c \left(\sigma^n e^{ik_m(z+\Delta z)} - 2\sigma^n e^{ik_m z} + \sigma^n e^{ik_m(z-\Delta z)} \right) \right.$$
$$-c \left(\sigma^{n-1} e^{ik_m(z+\Delta z)} - 2\sigma^{n-1} e^{ik_m z} + \sigma^{n-1} e^{ik_m(z-\Delta z)} \right) \right]$$
$$+b \left[3dc \left(\sigma^n e^{ik_m(z+\Delta z)} - 2\sigma^n e^{ik_m z} + \sigma^n e^{ik_m(z-\Delta z)} \right) \right.$$
$$\left. -dc \left(\sigma^{n-1} e^{ik_m(z+\Delta z)} - 2\sigma^{n-1} e^{ik_m z} + \sigma^{n-1} e^{ik_m(z-\Delta z)} \right) \right] \quad (3.33)$$

$$\sigma^{n+1} e^{ik_m z} = \sigma^n e^{ik_m z} + a \left[\sigma^n e^{ik_m z} c \left(e^{ik_m \Delta z} - 2 + e^{-ik_m \Delta z} \right) - \sigma^{n-1} e^{ik_m z} c \left(e^{ik_m \Delta z} - 2 + e^{-ik_m \Delta z} \right) \right]$$
$$+b \left[3\sigma^n e^{ik_m z} dc \left(e^{ik_m \Delta z} - 2 + e^{-ik_m \Delta z} \right) - \sigma^{n-1} e^{ik_m z} dc \left(e^{ik_m \Delta z} - 2 + e^{-ik_m \Delta z} \right) \right] \quad (3.34)$$

$$\sigma^{n+1} = \sigma^n + a \left[\sigma^n c \left(2\cos(k_m \Delta z) - 2 \right) - \sigma^{n-1} c \left(2\cos(k_m \Delta z) - 2 \right) \right]$$
$$+b \left[3\sigma^n dc \left(2\cos(k_m \Delta z) - 2 \right) - \sigma^{n-1} dc \left(2\cos(k_m \Delta z) - 2 \right) \right] \quad (3.35)$$

$$\frac{\sigma^{n+1}}{\sigma^n} = a \left[2c \left(\cos(k_m \Delta z) - 1 \right) - 2c \frac{\sigma^{n-1}}{\sigma^n} \left(\cos(k_m \Delta z) - 1 \right) \right]$$
$$+b \left[3dc \left(\cos(k_m \Delta z) - 2 \right) - 2dc \frac{\sigma^{n-1}}{\sigma^n} \left(\cos(k_m \Delta z) - 2 \right) \right] \quad (3.36)$$

$$\frac{\sigma^{n+1}}{\sigma^n} = 2ac \left(\cos(k_m \Delta z) - 1 \right) \left[1 - \frac{\sigma^{n-1}}{\sigma^n} \right] + bc \left(\cos(k_m \Delta z) - 1 \right) \left[3 - 2\frac{\sigma^{n-1}}{\sigma^n} \right] \quad (3.37)$$

For $n = 0$ the equation becomes:

$$\left| \frac{\sigma^1}{\sigma^0} \right| = \left| 2ac \left(\cos(k_m \Delta z) - 1 \right) + 3bdc \left(\cos(k_m \Delta z) - 2 \right) \right| \quad (3.38)$$

The stability is given by:

$$\left| 2ac \left(\cos(k_m \Delta z) - 1 \right) + 3bdc \left(\cos(k_m \Delta z) - 2 \right) \right| < 1 \quad (3.39)$$

Now we assume that $\forall n > 0$:

$$\left| \frac{\sigma^n}{\sigma^0} \right| < 1$$

Using the above assumption, we will prove that:

$$\left|\frac{\sigma^{n+1}}{\sigma^0}\right| < 1$$

The equation then becomes:

$$\sigma^{n+1} = \sigma^n + a\left[\sigma^n 2c\left(\cos\left(k_m\Delta z\right)-1\right) - \sigma^{n-1}2c\left(\cos\left(k_m\Delta z\right)-1\right)\right]$$
$$+b\left[\sigma^n 3dc\left(\cos\left(k_m\Delta z\right)-2\right) - \sigma^{n-1}2dc\left(\cos\left(k_m\Delta z\right)-2\right)\right] \qquad (3.40)$$

which implies that:

$$\left|\sigma^{n+1}\right| = \left|\sigma^n\right|\left(1+2ac\left(\cos\left(k_m\Delta z\right)-1\right)\right) - \left|\sigma^{n-1}\right|2ac\left(\cos\left(k_m\Delta z\right)-1\right)$$
$$+\left|\sigma^n\right|3bdc\left(\cos\left(k_m\Delta z\right)-2\right) - \left|\sigma^{n-1}\right|bdc\left(2\cos\left(k_m\Delta z\right)-2\right) \qquad (3.41)$$

Using recursive hypothesis, the equation becomes:

$$\left|\sigma^{n+1}\right| < \left|\sigma^0\right|\left[\left(1+2ac\left(\cos\left(k_m\Delta z\right)-1\right)\right) - 2ac\left(\cos\left(k_m\Delta z\right)-1\right)\right.$$
$$\left. +3bdc\left(\cos\left(k_m\Delta z\right)-2\right) - bdc\left(2\cos\left(k_m\Delta z\right)-2\right)\right] \qquad (3.42)$$

$$\left|\sigma^{n+1}\right| < \left|\sigma^0\right|\left[1+2bdc\left(2\cos\left(k_m\Delta z\right)-2\right)\right] \qquad (3.43)$$

$$1+4bdc\left(\cos\left(k_m\Delta z\right)-1\right) < 1 \qquad (3.44)$$

$$4bdc\cos\left(k_m\Delta z\right) < 4bdc \qquad (3.45)$$

The stability condition is given by:

$$\cos\left(k_m\Delta z\right) < 1 \qquad (3.46)$$

3.4.3 NUMERICAL SOLUTION FOR THE UNSATURATED ZONE USING CAPUTO–FABRIZIO FRACTIONAL DERIVATIVE

For our numerical solution, we replace the time derivative on the unsaturated equation (3.9) with the Caputo–Fabrizio fractional derivative, and used the first and second order derivative on the equation, such that:

$$-\left(\theta_s-\theta_0\right)\psi_b^\lambda\frac{\lambda}{\psi_l^{\lambda+1}}{}_0^{CF}D^\alpha\psi\left(z,t\right) = K_s\frac{\partial}{\partial z}\left(\frac{\psi_b}{\psi}\right)^{2+3\lambda}\frac{\partial\psi}{\partial z} + K_s\left(\frac{\psi_b}{\psi}\right)^{2+3\lambda}\frac{\partial^2\psi}{\partial z^2} - K_s\frac{\partial}{\partial z}\left(\frac{\psi_b}{\psi}\right)^{2+3\lambda} \qquad (3.47)$$

$$
{}_0^{CF}D^\alpha \psi(z,t) = \frac{1}{-(\theta_s - \theta_0)\psi_b^\lambda \dfrac{\lambda}{\psi_l^{\lambda+1}}} \left(K_s \frac{\partial}{\partial z}\left(\frac{\psi_b}{\psi}\right)^{2+3\lambda} \frac{\partial \psi}{\partial z} \right.
$$

$$
\left. + K_s\left(\frac{\psi_b}{\psi}\right)^{2+3\lambda} \frac{\partial^2 \psi}{\partial z^2} - K_s \frac{\partial}{\partial z}\left(\frac{\psi_b}{\psi}\right)^{2+3\lambda} \right) \tag{3.48}
$$

where

$$
{}_0^{CF}D^\alpha \psi(z,t) = f(z,t,\psi(z,t)) \tag{3.49}
$$

We apply the Caputo–Fabrizio integral to the above equation:

$$
\psi(z,t) = \psi(z,0) + \frac{1-\alpha}{M(\alpha)}f(z,t,\psi(z,t)) + \frac{\alpha}{M(\alpha)}\int_0^t f(z,\tau,\psi(z,\tau))d\tau \tag{3.50}
$$

When $t = t_{n+1}$ we have the following equation:

$$
\psi(z_i,t_{n+1}) = \psi(z_i,0) + \frac{1-\alpha}{M(\alpha)}f(z_i,t_j,\psi(z_i,t_j)) + \frac{\alpha}{M(\alpha)}\int_0^{t_{n+1}} f(z_i,\tau,\psi(z_i,\tau))d\tau \tag{3.51}
$$

When $= t_n$:

$$
\psi(z_i,t_n) = \psi(z_i,0) + \frac{1-\alpha}{M(\alpha)}f(z_i,t_{n-1},\psi(z_i,t_{n-1})) + \frac{\alpha}{M(\alpha)}\int_0^{t_n} f(z_i,\tau,\psi(z_i,\tau))d\tau \tag{3.52}
$$

Subtracting the above solution for t_n from t_{n+1}, we obtain the following:

$$
\psi_i^{n+1} - \psi_i^n = \frac{1-\alpha}{M(\alpha)}\left[f(z_i,t_n,\psi_i^n) - f(z_i,t_{n-1},\psi_i^{n-1}) \right] + \frac{\alpha}{M(\alpha)}\int_{t_n}^{t_{n+1}} f(z_i,\tau,\psi(x_i,\tau))d\tau \tag{3.53}
$$

$$
\psi_i^{n+1} = \psi_i^n + \frac{1-\alpha}{M(\alpha)}\left[f(z_i,t_n,\psi_i^n) - f(z_i,t_{n-1},\psi_i^{n-1}) \right]
$$

$$
+ \frac{\alpha}{M(\alpha)}\left[\frac{3}{2}\Delta t f(z_i,t_n,\psi_i^n) - \frac{\Delta t}{2}f(z_i,t_{n-1},\psi_i^{n-1}) \right] \tag{3.54}
$$

where:

$$
F(z_i,t_n,\psi_i^n) = \frac{1}{-(\theta_s - \theta_0)\psi_b^\lambda \dfrac{\lambda}{\psi_i^{\lambda+1}}} \left(K_s \left(\frac{\left(\dfrac{\psi_b}{\psi_{i+1}^n}\right)^{2+3\lambda} - \left(\dfrac{\psi_b}{\psi_i^n}\right)^{2+3\lambda}}{\Delta z} \right) \frac{\psi_{i+1}^n - \psi_i^n}{\Delta z} \right.
$$

$$
\left. + K_s\left(\frac{\psi_b}{\psi_i^n}\right)^{2+3\lambda} \times \frac{\psi_{i+1}^n - 2\psi_i^n + \psi_{i-1}^n}{(\Delta z)^2} - K_s\left(\frac{\left(\dfrac{\psi_b}{\psi_{i+1}^n}\right)^{2+3\lambda} - \left(\dfrac{\psi_b}{\psi_i^n}\right)^{2+3\lambda}}{\Delta z} \right) \right) \tag{3.55}
$$

$$F\left(z_i, t_{n-1}, \psi_i^{n-1}\right) = \frac{1}{-\left(\theta_s - \theta_0\right)\psi_b^\lambda \dfrac{\lambda}{\psi_i^{\lambda+1}}}\left(K_s\left(\frac{\left(\dfrac{\psi_b}{\psi_{i+1}^{n-1}}\right)^{2+3\lambda} - \left(\dfrac{\psi_b}{\psi_i^{n-1}}\right)^{2+3\lambda}}{\Delta z}\right)\frac{\psi_{i+1}^{n-1} - \psi_i^{n-1}}{\Delta z}\right.$$

$$\left. + K_s\left(\frac{\psi_b}{\psi_i^{n-1}}\right)^{2+3\lambda} \times \frac{\psi_{i+1}^{n-1} - 2\psi_i^{n-1} + \psi_{i-1}^{n-1}}{\left(\Delta z\right)^2} - K_s\left(\frac{\left(\dfrac{\psi_b}{\psi_{i+1}^{n-1}}\right)^{2+3\lambda} - \left(\dfrac{\psi_b}{\psi_i^{n-1}}\right)^{2+3\lambda}}{\Delta z}\right)\right) \tag{3.56}$$

We substitute the above Equations (3.55) and (3.56) into Equation (3.54) to obtain the final numerical solution for the saturated groundwater flow equation:

$$\psi_i^{n+1} = \psi_i^n + \frac{1-\alpha}{M(\alpha)}\left[\frac{1}{-\left(\theta_s - \theta_0\right)\psi_b^\lambda \dfrac{\lambda}{\psi_i^{\lambda+1}}}\left(K_s\frac{\left(\dfrac{\psi_b}{\psi_{i+1}^n}\right)^{2+3\lambda} - \left(\dfrac{\psi_b}{\psi_i^n}\right)^{2+3\lambda}}{\Delta z}\frac{\psi_{i+1}^n - \psi_i^n}{\Delta z}\right.\right.$$

$$\left. + K_s\left(\frac{\psi_b}{\psi_i^n}\right)^{2+3\lambda} \times \frac{\psi_{i+1}^n - 2\psi_i^n + \psi_{i-1}^n}{\left(\Delta z\right)^2} - K_s\frac{\left(\dfrac{\psi_b}{\psi_{i+1}^n}\right)^{2+3\lambda} - \left(\dfrac{\psi_b}{\psi_i^n}\right)^{2+3\lambda}}{\Delta z}\right)$$

$$-\frac{1}{-\left(\theta_s - \theta_0\right)\psi_b^\lambda \dfrac{\lambda}{\psi_i^{\lambda+1}}}\left(K_s\frac{\left(\dfrac{\psi_b}{\psi_{i+1}^{n-1}}\right)^{2+3\lambda} - \left(\dfrac{\psi_b}{\psi_i^{n-1}}\right)^{2+3\lambda}}{\Delta z}\frac{\psi_{i+1}^{n-1} - \psi_i^{n-1}}{\Delta z}\right.$$

$$\left.\left. + K_s\left(\frac{\psi_b}{\psi_i^{n-1}}\right)^{2+3\lambda} \times \frac{\psi_{i+1}^{n-1} - 2\psi_i^{n-1} + \psi_{i-1}^{n-1}}{\left(\Delta z\right)^2} - K_s\frac{\left(\dfrac{\psi_b}{\psi_{i+1}^{n-1}}\right)^{2+3\lambda} - \left(\dfrac{\psi_b}{\psi_i^{j-1}}\right)^{2+3\lambda}}{\Delta z}\right)\right]$$

$$+\frac{\alpha}{M(\alpha)}\left[\frac{3\Delta t}{2}\frac{1}{-\left(\theta_s - \theta_0\right)\psi_b^\lambda \dfrac{\lambda}{\psi_i^{\lambda+1}}}\left(K_s\frac{\left(\dfrac{\psi_b}{\psi_{i+1}^n}\right)^{2+3\lambda} - \left(\dfrac{\psi_b}{\psi_i^n}\right)^{2+3\lambda}}{\Delta z}\frac{\psi_{i+1}^n - \psi_i^n}{\Delta z}\right.\right.$$

$$\left. + K_s\left(\frac{\psi_b}{\psi_i^n}\right)^{2+3\lambda} \times \frac{\psi_{i+1}^n - 2\psi_i^n + \psi_{i-1}^n}{\left(\Delta z\right)^2} - K_s\frac{\left(\dfrac{\psi_b}{\psi_{i+1}^n}\right)^{2+3\lambda} - \left(\dfrac{\psi_b}{\psi_i^n}\right)^{2+3\lambda}}{\Delta z}\right)$$

$$-\frac{\Delta t}{2}\frac{1}{-\left(\theta_s - \theta_0\right)\psi_b^\lambda \dfrac{\lambda}{\psi_i^{\lambda+1}}}\left(K_s\frac{\left(\dfrac{\psi_b}{\psi_{i+1}^{n-1}}\right)^{2+3\lambda} - \left(\dfrac{\psi_b}{\psi_i^{n-1}}\right)^{2+3\lambda}}{\Delta z}\frac{\psi_{i+1}^{n-1} - \psi_i^{n-1}}{\Delta z}\right.$$

$$\left.\left. + K_s\left(\frac{\psi_b}{\psi_i^{n-1}}\right)^{2+3\lambda} \times \frac{\psi_{i+1}^{n-1} - 2\psi_i^{n-1} + \psi_{i-1}^{n-1}}{\left(\Delta z\right)^2} - K_s\frac{\left(\dfrac{\psi_b}{\psi_{i+1}^{n-1}}\right)^{2+3\lambda} - \left(\dfrac{\psi_b}{\psi_i^{n-1}}\right)^{2+3\lambda}}{\Delta z}\right)\right] \tag{3.57}$$

For simplicity, let's use the following parameters:

$$\wp = \left(-(\theta_s - \theta_0)\psi_b{}^\lambda \frac{\lambda}{\psi_l^{\lambda+1}} \right),$$

$$a = \frac{1-\alpha}{M(\alpha)},$$

$$b = \frac{\alpha}{M(\alpha)},$$

$$c = \frac{K_s}{\Delta z},$$

$$d = \frac{1}{\Delta z},$$

$$e = \frac{1}{(\Delta z)^2},$$

$$f = \frac{3\Delta t}{2},$$

$$g = \frac{\Delta t}{2}$$

Now substituting the above parameters we obtain the following:

$$\psi_i^{n+1} = \psi_i^n + a \left[\begin{array}{l} \dfrac{1}{\wp} \left(\begin{array}{l} c\left(\left(\dfrac{\psi_b}{\psi_{i+1}^n}\right)^{2+3\lambda} - \left(\dfrac{\psi_b}{\psi_i^n}\right)^{2+3\lambda}\right) d\left(\psi_{i+1}^n - \psi_i^n\right) + K_s\left(\dfrac{\psi_b}{\psi_i^n}\right)^{2+3\lambda} \\ \times e\left(\psi_{i+1}^n - 2\psi_i^n + \psi_{i-1}^n\right) - c\left(\left(\dfrac{\psi_b}{\psi_{i+1}^n}\right)^{2+3\lambda} - \left(\dfrac{\psi_b}{\psi_i^n}\right)^{2+3\lambda}\right) \end{array} \right) \\ -\dfrac{1}{\wp} \left(\begin{array}{l} c\left(\left(\dfrac{\psi_b}{\psi_{i+1}^{n-1}}\right)^{2+3\lambda} - \left(\dfrac{\psi_b}{\psi_i^{n-1}}\right)^{2+3\lambda}\right) d\left(\psi_{i+1}^{n-1} - \psi_i^{n-1}\right) + K_s\left(\dfrac{\psi_b}{\psi_i^{n-1}}\right)^{2+3\lambda} \\ \times e\left(\psi_{i+1}^{n-1} - 2\psi_i^{n-1} + \psi_{i-1}^{n-1}\right) - c\left(\left(\dfrac{\psi_b}{\psi_{i+1}^{j-1}}\right)^{2+3\lambda} - \left(\dfrac{\psi_b}{\psi_i^{j-1}}\right)^{2+3\lambda}\right) \end{array} \right) \end{array} \right]$$

$$+b \left[\begin{array}{l} f\dfrac{1}{\wp} \left(\begin{array}{l} c\left(\left(\dfrac{\psi_b}{\psi_{i+1}^n}\right)^{2+3\lambda} - \left(\dfrac{\psi_b}{\psi_i^n}\right)^{2+3\lambda}\right) d\left(\psi_{i+1}^n - \psi_i^n\right) + K_s\left(\dfrac{\psi_b}{\psi_i^n}\right)^{2+3\lambda} \\ \times e\left(\psi_{i+1}^n - 2\psi_i^n + \psi_{i-1}^n\right) - c\left(\left(\dfrac{\psi_b}{\psi_{i+1}^n}\right)^{2+3\lambda} - \left(\dfrac{\psi_b}{\psi_i^n}\right)^{2+3\lambda}\right) \end{array} \right) \\ -g\dfrac{1}{\wp} \left(\begin{array}{l} c\left(\left(\dfrac{\psi_b}{\psi_{i+1}^{n-1}}\right)^{2+3\lambda} - \left(\dfrac{\psi_b}{\psi_i^{n-1}}\right)^{2+3\lambda}\right) d\left(\psi_{i+1}^{n-1} - \psi_i^{n-1}\right) + \hat{K}_s\left(\dfrac{\psi_b}{\psi_i^{n-1}}\right)^{2+3\lambda} \\ \times e\left(\psi_{i+1}^{n-1} - 2\psi_i^{n-1} + \psi_{i-1}^{n-1}\right) - c\left(\left(\dfrac{\psi_b}{\psi_{i+1}^{n-1}}\right)^{2+3\lambda} - \left(\dfrac{\psi_b}{\psi_i^{n-1}}\right)^{2+3\lambda}\right) \end{array} \right) \end{array} \right] \quad (3.58)$$

The above numerical solution for the unsaturated zone is highly non-linear; therefore, we cannot apply the stability analysis

3.5 CONCLUSION

The aim of this study was to develop a new 1-d saturated–unsaturated groundwater flow equation by replacing the classical derivative with the Caputo–Fabrizio derivative with non-singular kernel. The stability of the new model is also detailed. Based on the numerical solutions presented above, it can be concluded that the fractional derivatives with exponential decay law are suitable mathematical operators to depict processes with fading memory. This situation is similar to the flow taking place in elastic media, where the geological formation has slow transmissivity and high storativity.

REFERENCES

Alkahtani, B.S.T., (2016). Chua's circuit model with Atangana–Baleanu derivative with fractional order. *Chaos, Solutions and Fractals*, 89: 547–551.

Allepalli, P. K., & Govindaraju, R. S., (2009). *Modeling fate and transport of atrazine in the saturated–unsaturated zone of soil*. Kansas State University, Department of Civil Engineering.

Atanackovic T.M., Pilipovic S., & Zorica D., (2018). Properties of the Caputo–Fabrizio fractional derivative and its distributional settings. *Fractional Calculus and Applied Analysis*, 21, 1: 29–44.

Atangana, A., & Alkahtani, B.S.T., (2015). Analysis of the Keller-Segel model with a fractional derivative without singular kernel. *Entropy*, 17(6): 4439–4453.

Atangana, A., & Baleanu, D., (2016). New fractional derivatives with non-local and non-singular kernel: Theory and application to heat transfer model. *Thermal Science*, 20(2): 763–769.

Brooks, R.H., & Corey, A.T., (1966). Properties of porous media affecting fluid flow. *Journal of the Irrigation and Drainage Division*, ASCE, 92, (IR2): 61–88.

Caputo, M., & Fabrizio, M., (2015). A new definition of fractional derivative without singular kernel. *Progress in Fractional Differentiation and Applications*, 1(2): 87–92.

Cooper, J.C.B., (2005). The Poisson and Exponential distributions. *Applied Probability Trust*, 37(3): 123–125.

Gomez-Aguilar, J. F., Lopez-Lopez, M.G., Alvarado-Martinez, V.M., Baleanu, D., & Khan, H., (2017). Chaos in a cancer model via fractional derivatives with exponential decay and Mittag–Leffler law. *Entropy*, 19(681): 19.

List, F., & Radu, F.A., (2015). A study on iterative methods for solving Richards' equation. *Computer Geoscience*, 20: 341–353.

Papagianakis, A.T., & Fredlund, D.G. (1984). A steady state model for flow in saturated–unsaturated soils. *Canadian Geotechnical Journal*, 21: 419–430.

Pelka, W., (1983). Heat and mass transport in saturated–unsaturated groundwater flow. *Relation of Groundwater Quantity and Quality*. IAHS Publ.no. 146: 160–166.

Rice, J.A., (2006). Mathematical statistics and data analysis. *Cengage Learning*.

Sheikh, N.A., Ali, F., Saqib, M., Khan, I., Jan, S.A.A., Alshomrani, A.S., & Alghamdi, M.S., (2017). Comparison and analysis of the Atangana–Baleanu and Caputo–Fabrizio fractional derivatives for generalized Casson fluid model with heat generation and chemical reaction. *Results in Physics*. 7: 789–800.

4 A New Model of the 1-d Unsaturated–Saturated Groundwater Flow with Crossover from Usual to Sub-Flow Mean Square Displacement

Rendani Vele Makahane and Abdon Atangana
University of the Free State, Bloemfontein, South Africa

CONTENTS

4.1 INTRODUCTION

In 1903, Gosta Mittag-Leffler, a Swedish mathematician, introduced the Mittag-Leffler function, which became prominent in the last two decades as a result of its application to natural problems: for example, in engineering, earth science, and biology. This function responds to the conventional question of complex analysis, in particular, to portray the procedure of the analytic continuation of power law series outside the disk of their convergence (Alqahtani, 2016; Gomez-Aguilar & Atangana, 2019). The ordinary and generalized Mittag-Leffler functions interpolate between a purely exponential and power law (Gomez-Aguilar & Atangana, 2019). In diffusion, Mittag-Leffler is essential because it can capture non-locality and avoids singularity. It is also crucial for the description of long-time behavior. It can be used as a waiting time distribution as well as the first passage time distribution for renewal processes. Properties of Mittag-Leffler include usual differentiation and integration, Euler transforms, Laplace transforms, Whittaker transforms, generalized and hypergeometric series, and others. Recently, a non-local fractional derivative with non-singularity based on the Mittag-Leffler function was suggested by Atangana and Baleanu (2016), 2017; Djida et al., 2016;

Sheikh et al., 2017). This new kernel has an advantage that a normal initial condition can be obtained when using the Laplace transform, unlike with the Riemann–Liouville integral. Furthermore, since this kernel has no singularity it can depict a full memory (Alkahtani, 2016). The kernel is more natural and it is the first of its kind to show the cross-over behavior between the stretched exponential law for earlier time and power-law for latter time (Gomez-Aguilar & Atangana, 2019).The aim of this chapter is to present a new model for groundwater flow through the saturated–unsaturated zone using the Atangana–Baleanu fractional derivative and the Ghanbari–Atangana numerical scheme. Below, we recall the definitions of the new kernel.

4.2   A-B DERIVATIVE WITH FRACTIONAL ORDER

In this section, we illustrate definitions of the fractional derivatives with no singular and non-local kernel. Atangana and Baleanu introduced these definitions in their work published in a thermal science journal.

Definition 1

Let $f \in H'(a, b)$, $b > a$, $\alpha \in [0, 1]$ and not necessarily differentiable then, the definition of the new fractional derivative (A-B fractional derivative in R-L sense) is specified as:

$$^{ABR}_{a}D_t^\alpha \left(f(t) \right) = \frac{B(\alpha)}{1-\alpha} \frac{d}{dt} \int_a^t f(\tau) E_\alpha \left[-\alpha \frac{(t-\tau)^\alpha}{1-\alpha} \right] d\tau \qquad (4.1)$$

Definition 2

Let $f \in H'(a, b)$, $b > a$, $\alpha \in [0, 1]$ then, the definition of the new fractional derivative (A-B fractional derivative in Caputo sense) is given as:

$$^{ABC}_{a}D_t^\alpha \left(f(t) \right) = \frac{B(\alpha)}{1-\alpha} \int_a^t f'(\tau) E_\alpha \left[-\alpha \frac{(t-\tau)^\alpha}{1-\alpha} \right] d\tau \qquad (4.2)$$

Definition 3

The fractional integral associate to the new fractional derivative with the non-local kernel (A-B fractional integral) is defined as:

$$^{AB}_{a}I_t^\alpha \left(f(t) \right) = \frac{1-\alpha}{B(\alpha)} f(t) + \frac{\alpha}{B(\alpha)\Gamma(\alpha)} \int_a^t f(\tau)(t-\tau)^{\alpha-1} d\tau \qquad (4.3)$$

The numerical solution of the integral is:

$$^{AB}_{a}I_t^\alpha \left(\psi(z,t) \right) = \frac{1-\alpha}{B(\alpha)} \psi(z,t) + \frac{\alpha}{B(\alpha)\Gamma(\alpha)} \int_a^t \psi(z,\tau)(t-\tau)^{\alpha-1} d\tau \qquad (4.4)$$

$$^{AB}_{a}I_t^\alpha \left(\psi(z_i,t_n) \right) = \frac{1-\alpha}{B(\alpha)} \psi(z_i,t_n) + \frac{\alpha}{B(\alpha)\Gamma(\alpha)} \int_a^{t_n} \psi(z_i,\tau)(t_n-\tau)^{\alpha-1} d\tau \qquad (4.5)$$

$$
{}^{AB}_{a}I^{\alpha}_{t}\left(\psi\left(z_{i},t_{n}\right)\right)=\frac{1-\alpha}{B(\alpha)}\psi^{n}_{i}+\frac{\alpha}{B(\alpha)\Gamma(\alpha)}\int_{a}^{t_{n}}\psi^{j}_{i}\left(t_{n}-\tau\right)^{\alpha-1}d\tau \tag{4.6}
$$

$$
{}^{AB}_{a}I^{\alpha}_{t}\left(\psi\left(z_{i},t_{n}\right)\right)=\frac{1-\alpha}{B(\alpha)}\psi^{n}_{i}+\frac{\alpha}{B(\alpha)\Gamma(\alpha)}\sum_{j=0}^{n-1}\psi^{j}_{i}\int_{t_{j}}^{t_{j+1}}\left(t_{n}-\tau\right)^{\alpha-1}d\tau \tag{4.7}
$$

Let $y=t_{n}-\tau$, $y=t_{n}-t_{j}$, $y=t_{n}-t_{j+1}$ and $dy=-d\tau$, therefore:

$$
{}^{AB}_{a}I^{\alpha}_{t}\left(\psi\left(z_{i},t_{n}\right)\right)=\frac{1-\alpha}{B(\alpha)}\psi^{n}_{i}+\frac{\alpha}{B(\alpha)\Gamma(\alpha)}\sum_{j=0}^{n-1}\psi^{j}_{i}\int_{t_{n}-t_{j}}^{t_{n}-t_{j+1}}y^{\alpha-1}\left(-dy\right) \tag{4.8}
$$

$$
{}^{AB}_{a}I^{\alpha}_{t}\left(\psi\left(z_{i},t_{n}\right)\right)=\frac{1-\alpha}{B(\alpha)}\psi^{n}_{i}+\frac{\alpha}{B(\alpha)\Gamma(\alpha)}\sum_{j=0}^{n-1}\psi^{j}_{i}\int_{t_{n}-t_{j+1}}^{t_{n}-t_{j}}y^{\alpha-1}dy \tag{4.9}
$$

$$
{}^{AB}_{a}I^{\alpha}_{t}\left(\psi\left(z_{i},t_{n}\right)\right)=\frac{1-\alpha}{B(\alpha)}\psi^{n}_{i}+\frac{\alpha}{B(\alpha)\Gamma(\alpha)}\sum_{j=0}^{n-1}\psi^{j}_{i}\left[y^{\alpha-1}dy\right] \tag{4.10}
$$

$$
{}^{AB}_{a}I^{\alpha}_{t}\left(\psi\left(z_{i},t_{n}\right)\right)=\frac{1-\alpha}{B(\alpha)}\psi^{n}_{i}+\frac{\alpha}{B(\alpha)\Gamma(\alpha)}\sum_{j=0}^{n-1}\psi^{j}_{i}\left[\left(t_{n}-t_{j}\right)^{\alpha-1}-\left(t_{n}-t_{j+1}\right)^{\alpha-1}\right] \tag{4.11}
$$

Using $t_{n}=n\Delta t$ and $t_{j+1}=(j+1)\Delta t$ we can write the above equation as:

$$
{}^{AB}_{a}I^{\alpha}_{t}\left(\psi\left(z_{i},t_{n}\right)\right)=\frac{1-\alpha}{B(\alpha)}\psi^{n}_{i}+\frac{\alpha}{B(\alpha)\Gamma(\alpha)}\sum_{j=0}^{n-1}\psi^{j}_{i}\Delta t^{\alpha-1}\left[\left(n-j\right)^{\alpha-1}-\left(n-j-1\right)^{\alpha-1}\right] \tag{4.12}
$$

The final equation can be written as:

$$
{}^{AB}_{a}I^{\alpha}_{t}\left(\psi\left(z_{i},t_{n}\right)\right)=\frac{1-\alpha}{B(\alpha)}\psi^{n}_{i}+\frac{\alpha\Delta t^{\alpha}}{B(\alpha)\Gamma(\alpha)}\sum_{j=0}^{n-1}\psi^{j}_{i}\left[\left(n-j\right)^{\alpha}-\left(n-j-1\right)^{\alpha}\right] \tag{4.13}
$$

4.3 NUMERICAL SOLUTION OF THE SATURATED–UNSATURATED GROUNDWATER FLOW EQUATION USING THE A-B FRACTIONAL DERIVATIVE

In this section, the model is considered using the non-singular and non-local kernel. To achieve this, the time derivative of the original model is replaced with the time-fractional derivative built on the Mittag-Leffler function to derive the numerical solution of the new model. The 1-d unsaturated–saturated groundwater flow equation is given as:

$$
\left[S_{S}S_{a}\left(\psi\right)+C\left(\psi\right)\right]\frac{\partial\psi}{\partial t}=\frac{\partial}{\partial z}\left[K_{z}\left(\psi\right)\left(\frac{\partial\psi}{\partial z}+1\right)\right] \tag{4.14}
$$

where:
ψ = pressure head
n = porosity

S_S = specific storage of the soil
S_a = saturation of the aqueous phase
$C(\psi)$ = capillary capacity of soil = $d\theta/d\psi$
K_z = hydraulic conductivity
z = vertical coordinate.

The above equation (4.14), consists of the soil hydraulic functions ($K(\psi)$ and $\theta(\psi)$). As a result, empirical relationships are necessary to model conductivity and calculate flow (Allepalli & Govindaraju, 2009). The empirical solution scheme used for this study is that of Brooks and Corey (1966):

$$K(\psi) = K_s \left(\frac{\psi_b}{\psi}\right)^{2+3\lambda} \qquad\qquad \psi \leq 0 \qquad\qquad (4.15)$$

$$K(\psi) = K_s \qquad\qquad \psi > 0 \qquad\qquad (4.16)$$

$$\theta(\psi) = \theta_0 + (\theta_s - \theta_0)\left(\frac{\psi_b}{\psi}\right)^{\lambda} \qquad\qquad \psi \leq 0 \qquad\qquad (4.17)$$

The unsaturated flow is indicated by $\psi \leq 0$ and the saturated by $\psi > 0$. Using the above empirical relationship by Brooks and Corey, we generate the following equations for unsaturated flow and saturated flow, respectively:

$$\frac{\partial \psi}{\partial t} = \frac{K_s}{S_s}\frac{\partial^2 \psi}{\partial z^2} \qquad\qquad (4.18)$$

and,

$$\frac{\partial \theta(\psi)}{\partial t} = \frac{\partial}{\partial z}\left[K_z(\psi)\left(\frac{\partial \psi}{\partial z} - 1\right)\right] \qquad\qquad (4.19)$$

Using the Brooks and Corey equations, the above equation (4.19) become:

$$-(\theta_s - \theta_0)\psi_b^\lambda \frac{\partial \psi}{\partial t}\frac{\lambda}{\psi^{\lambda+1}} = K_s \frac{\partial}{\partial z}\left(\frac{\psi_b}{\psi}\right)^{2+3\lambda}\frac{\partial \psi}{\partial z} + K_s\left(\frac{\psi_b}{\psi}\right)^{2+3\lambda}\frac{\partial^2 \psi}{\partial z^2} - K_s\frac{\partial}{\partial z}\left(\frac{\psi_b}{\psi}\right)^{2+3\lambda} \qquad (4.20)$$

4.3.1 NUMERICAL SOLUTION OF THE SATURATED ZONE USING THE A-B FRACTIONAL DERIVATIVE

By replacing the time derivative in the saturated equation by the A-B fractional time derivative we obtain the following:

$$^{AB}_0 D^\alpha \psi(z,t) = \frac{K_s}{S_s}\frac{\partial^2 \psi}{\partial z^2} \qquad\qquad (4.21)$$

where

$$^{AB}_0 D^\alpha \psi(z,t) = f(z,t,\psi(z,t)) \qquad\qquad (4.22)$$

Then the AB integral is written as:

$$\psi(z_i,t)-\psi(z_i,0)=\frac{(1-\alpha)}{B(\alpha)}f\big(z,t,\psi(z,t)\big)+\frac{\alpha}{B(\alpha)\Gamma(\alpha)}\int_0^t f\big(z_i,\tau,\psi(z_i,\tau)\big)(t-\tau)^{\alpha-1}\,d\tau \tag{4.23}$$

At point t_{n+1}:

$$\psi(z_i,t_{n+1})-\psi(z_i,0)=\frac{(1-\alpha)}{B(\alpha)}f\big(z,t,\psi(z,t)\big)+\frac{\alpha}{B(\alpha)\times\Gamma(\alpha)}\int_0^{t_{n+1}} f\big(z_i,\tau,\psi(z_i,\tau)\big)(t_{n+1}-\tau)^{\alpha-1}\,d\tau \tag{4.24}$$

$$\psi(z_i,t_{n+1})-\psi(z_i,0)=\frac{(1-\alpha)}{B(\alpha)}f\big(z,t,\psi(z,t)\big)+\frac{\alpha(\Delta t)^{\alpha}}{B(\alpha)\times\Gamma(2+\alpha)}$$
$$\times\sum_{j=0}^{n}\left[\begin{array}{l} f\big(z_i,t_j,\psi_i^{j}\big)\left(\begin{array}{l}(n-j+1)^{\alpha}\,(n-j+2+\alpha)\\ -(n-j)^{\alpha}\,(n-j+2+2\alpha)\end{array}\right)\\ -f\big(z_i,t_{j-1},\psi_i^{j-1}\big)\left(\begin{array}{l}(n-j+1)^{\alpha+1}\\ -(n-j)^{\alpha}\,(n-j+1+\alpha)\end{array}\right)\end{array}\right] \tag{4.25}$$

where:

$$F\big(z_i,t_j,\psi_i^{j}\big)=\frac{K_s}{S_s}\frac{\psi_{i+1}^{j}-2\psi_i^{j}+\psi_{i-1}^{j}}{(\Delta z)^2} \tag{4.26}$$

$$F\big(z_i,t_{j-1},\psi_i^{j-1}\big)=\frac{K_s}{S_s}\frac{\psi_{i+1}^{j-1}-2\psi_i^{j-1}+\psi_{i-1}^{j-1}}{(\Delta z)^2} \tag{4.27}$$

The final numerical solution for the saturated groundwater flow equation is given as:

$$\psi(z_i,t_{n+1})-\psi(z_i,0)=\frac{(1-\alpha)}{B(\alpha)}\left(\frac{K_s}{S_s}\frac{\psi_{i+1}^{n}-2\psi_i^{n}+\psi_{i-1}^{n}}{(\Delta z)^2}\right)+\frac{\alpha(\Delta t)^{\alpha}}{B(\alpha)\Gamma(2+\alpha)}$$
$$\times\sum_{j=0}^{n}\left[\begin{array}{l}\dfrac{K_s}{S_s}\dfrac{\psi_{i+1}^{j}-2\psi_i^{j}+\psi_{i-1}^{j}}{(\Delta z)^2}\left(\begin{array}{l}(n-j+1)^{\alpha}\,(n-j+2+\alpha)\\ -(n-j)^{\alpha}\,(n-j+2+2\alpha)\end{array}\right)\\ -\dfrac{K_s}{S_s}\dfrac{\psi_{i+1}^{j-1}-2\psi_i^{j-1}+\psi_{i-1}^{j-1}}{(\Delta z)^2}\left(\begin{array}{l}(n-j+1)^{\alpha+1}\\ -(n-j)^{\alpha}\,(n-j+1+\alpha)\end{array}\right)\end{array}\right] \tag{4.28}$$

We also assume the following parameters:

$$a=\frac{(1-\alpha)}{B(\alpha)}$$

$$b=\frac{K_s}{S_s(\Delta z)^2}$$

$$c = \frac{\alpha \left(\Delta t\right)^{\alpha}}{B\left(\alpha\right)\Gamma\left(2+\alpha\right)}$$

$$R_{n.j}^{\alpha} = \left(\left(n-j+1\right)^{\alpha}\left(n-j+2+\alpha\right)-\left(n-j\right)^{\alpha}\left(n-j+2+2\alpha\right)\right)$$

$$\delta_{n,j}^{\alpha} = \left(\left(n-j+1\right)^{\alpha+1}-\left(n-j\right)^{\alpha}\left(n-j+1+\alpha\right)\right)$$

Then our equation is written as:

$$\psi_i^{n+1} - \psi_i^n = a\left[b\left(\psi_{i+1}^n - 2\psi_i^n + \psi_{i-1}^n\right)\right] + c\sum_{j=0}^{n}\left[\begin{array}{l} b\left(\psi_{i+1}^j - 2\psi_i^j + \psi_{i-1}^j\right)R_{n,j}^{\alpha} \\ -b\left(\psi_{i+1}^{j-1} - 2\psi_i^{j-1} + \psi_{i-1}^{j-1}\right)\delta_{n,j}^{\alpha} \end{array}\right] \quad (4.29)$$

$$\psi_i^{n+1} - \psi_i^n = a\left[b\left(\psi_{i+1}^n - 2\psi_i^n + \psi_{i-1}^n\right)\right] + c\left[\begin{array}{l} b\left(\psi_{i+1}^n - 2\psi_i^n + \psi_{i-1}^n\right)R_{n,n}^{\alpha} \\ -b\left(\psi_{i+1}^{n-1} - 2\psi_i^{n-1} + \psi_{i-1}^{n-1}\right)\delta_{n,n}^{\alpha} \end{array}\right]$$

$$+ c\sum_{j=0}^{n-1}\left[b\left(\psi_{i+1}^j - 2\psi_i^j + \psi_{i-1}^j\right)R_{n,j}^{\alpha} - b\left(\psi_{i+1}^{j-1} - 2\psi_i^{j-1} + \psi_{i-1}^{j-1}\right)\delta_{n,j}^{\alpha}\right] \quad (4.30)$$

To check the stability of the above numerical solution, we assume a Fourier expansion in space of:

$$\psi_i^n = \sigma_n e^{ik_m z}$$

Using the above assumption, Equation (4.29) becomes:

$$\sigma_{n+1}e^{ik_m z} - \sigma_n e^{ik_m z} = a\left[b\left(\sigma_n e^{ik_m\left(z+\Delta z\right)} - 2\sigma_n e^{ik_m z} + \sigma_n e^{ik_m\left(z-\Delta z\right)}\right)\right]$$

$$+ c\left[\begin{array}{l} b\left(\sigma_n e^{ik_m\left(z+\Delta z\right)} - 2\sigma_n e^{ik_m z} + \sigma_n e^{ik_m\left(z-\Delta z\right)}\right)R_{n,n}^{\alpha} \\ -b\left(\sigma_{n-1}e^{ik_m\left(z+\Delta z\right)} - 2\sigma_{n-1}e^{ik_m z} + \sigma_{n-1}e^{ik_m\left(z-\Delta z\right)}\right)\delta_{n,n}^{\alpha} \end{array}\right]$$

$$+ c\sum_{j=0}^{n-1}\left[\begin{array}{l} b\left(\sigma_j e^{ik_m\left(z+\Delta z\right)} - 2\sigma_j e^{ik_m z} + \sigma_j e^{ik_m\left(z-\Delta z\right)}\right)R_{n,j}^{\alpha} \\ -b\left(\sigma_{j-1}e^{ik_m\left(z+\Delta z\right)} - 2\sigma_{j-1}e^{ik_m z} + \sigma_{j-1}e^{ik_m\left(z-\Delta z\right)}\right)\delta_{n,j}^{\alpha} \end{array}\right] \quad (4.31)$$

$$e^{ik_m z}\left(\sigma_{n+1} - \sigma_n\right) = a\left[b\sigma_n e^{ik_m z}\left(e^{ik_m\Delta z} - 2 + e^{-ik_m\Delta z}\right)\right] + c\left[\begin{array}{l} R_{n,n}^{\alpha}b\sigma_n e^{ik_m z}\left(e^{ik_m\Delta z} - 2 + e^{-ik_m\Delta z}\right) \\ -\delta_{n,n}^{\alpha}b\sigma_{n-1}e^{ik_m z}\left(e^{ik_m\Delta z} - 2 + e^{-ik_m\Delta z}\right) \end{array}\right]$$

$$+ c\sum_{j=0}^{n-1}\left[R_{n,j}^{\alpha}b\sigma_j e^{ik_m z}\left(e^{ik_m\Delta z} - 2 + e^{-ik_m\Delta z}\right) - \delta_{n,j}^{\alpha}b\sigma_{j-1}e^{ik_m z}\left(e^{ik_m\Delta z} - 2 + e^{-ik_m\Delta z}\right)\right] \quad (4.32)$$

$$\sigma_{n+1} - \sigma_n = a\left[b\sigma_n\left(e^{ik_m\Delta z} - 2 + e^{-ik_m\Delta z}\right)\right] + c\left[\begin{array}{l} R_{n,n}^{\alpha}b\sigma_n\left(e^{ik_m\Delta z} - 2 + e^{-ik_m\Delta z}\right) \\ -\delta_{n,n}^{\alpha}b\sigma_{n-1}\left(e^{ik_m\Delta z} - 2 + e^{-ik_m\Delta z}\right) \end{array}\right]$$

$$+ c\sum_{j=0}^{n-1}\left[R_{n,j}^{\alpha}b\sigma_j\left(e^{ik_m\Delta z} - 2 + e^{-ik_m\Delta z}\right) - \delta_{n,j}^{\alpha}b\sigma_{j-1}\left(e^{ik_m\Delta z} - 2 + e^{-ik_m\Delta z}\right)\right] \quad (4.33)$$

$$\sigma_{n+1}-\sigma_n = 2ab\sigma_n\left(\frac{e^{ik_m\Delta z}+e^{-ik_m\Delta z}}{2}-1\right)+c\left[\begin{array}{c}2R_{n,n}^{\alpha}b\sigma_n\left(\dfrac{e^{ik_m\Delta z}+e^{-ik_m\Delta z}}{2}-1\right)\\ -2\delta_{n,n}^{\alpha}b\sigma_{n-1}\left(\dfrac{e^{ik_m\Delta z}+e^{-ik_m\Delta z}}{2}-1\right)\end{array}\right]$$

$$+c\sum_{j=0}^{n-1}\left[2R_{n,j}^{\alpha}b\sigma_j\left(\frac{e^{ik_m\Delta z}+e^{-ik_m\Delta z}}{2}-1\right)-2\delta_{n,j}^{\alpha}b\sigma_{j-1}\left(\frac{e^{ik_m\Delta z}+e^{-ik_m\Delta z}}{2}-1\right)\right] \tag{4.34}$$

Considering the definition for a hyperbolic cosine, we can rewrite the above equation (4.34) as:

$$\sigma_{n+1}-\sigma_n = 2ab\sigma_n\left(\cos\left(k_m\Delta z\right)-1\right)+c\left[2R_{n,n}^{\alpha}b\sigma_n\left(\cos\left(k_m\Delta z\right)-1\right)-2\delta_{n,n}^{\alpha}b\sigma_{n-1}\left(\cos\left(k_m\Delta z\right)-1\right)\right]$$

$$+c\sum_{j=0}^{n-1}\left[2R_{n,j}^{\alpha}b\sigma_j\left(\cos\left(k_m\Delta z\right)-1\right)-2\delta_{n,j}^{\alpha}b\sigma_{j-1}\left(\cos\left(k_m\Delta z\right)-1\right)\right] \tag{4.35}$$

If n = 0 the equation above becomes:

$$\sigma_1-\sigma_0 = 2ab\sigma_0\left(\cos\left(k_m\Delta z\right)-1\right)+c\left[2R_{n,n}^{\alpha}b\sigma_0\left(\cos\left(k_m\Delta z\right)-1\right)\right] \tag{4.36}$$

$$\sigma_1 = \sigma_0\left[1+2ab\left(\cos\left(k_m\Delta z\right)-1\right)+R_{n,n}^{\alpha}cb\left(\cos\left(k_m\Delta z\right)-1\right)\right] \tag{4.37}$$

$$\left|\frac{\sigma_1}{\sigma_0}\right|<1$$

$$\left|\frac{1}{1+2ab\left(\cos\left(k_m\Delta z\right)-1\right)+R_{n,n}^{\alpha}cb\left(\cos\left(k_m\Delta z\right)-1\right)}\right|<1 \tag{4.38}$$

From the above results, it is clear that the denominator of the equation will always be greater than the numerator, therefore the A-B fractional derivative for the 1-d saturated groundwater flow equation is unconditionally stable. Now, we assume that $\forall n>0$:

$$\left|\frac{\sigma_n}{\sigma_0}\right|<1$$

Using the above assumption, we will proof that:

$$\left|\frac{\sigma_{n+1}}{\sigma_0}\right|<1$$

The equation then becomes

$$\sigma_{n+1} = \sigma_n\left[1+2ab\left(\cos\left(k_m\Delta z\right)-1\right)+R_{n,n}^{\alpha}cb\left(\cos\left(k_m\Delta z\right)-1\right)\right]-2\delta_{n,n}^{\alpha}cb\sigma_{n-1}\left(\cos\left(k_m\Delta z\right)-1\right)$$

$$+c\sum_{j=0}^{n-1}\left[2R_{n,j}^{\alpha}b\sigma_j\left(\cos\left(k_m\Delta z\right)-1\right)-2\delta_{n,j}^{\alpha}b\sigma_{j-1}\left(\cos\left(k_m\Delta z\right)-1\right)\right] \tag{4.39}$$

We let:

$$A = \left[1 + 2ab\left(\cos\left(k_m\Delta z\right) - 1\right) + R_{n,n}^{\alpha} cb\left(\cos\left(k_m\Delta z\right) - 1\right)\right]$$

$$B = 2\delta_{n,n}^{\alpha} cb\left(\cos\left(k_m\Delta z\right) - 1\right)$$

$$A_1 = 2R_{n,j}^{\alpha} b\left(\cos\left(k_m\Delta z\right) - 1\right)$$

$$A_2 = 2\delta_{n,j}^{\alpha} b\left(\cos\left(k_m\Delta z\right) - 1\right)$$

Then Equation (4.39) becomes:

$$\sigma_{n+1} = \sigma_n A - \sigma_{n-1} B + c\sum_{j=0}^{n-1}\left[\sigma_j A_1 - \sigma_{j-1} A_2\right] \tag{4.40}$$

$$\left|\sigma_{n+1}\right| \leq \left|\sigma_n\right|\left|A\right| - \left|\sigma_{n-1}\right|\left|B\right| + c\sum_{j=0}^{n-1}\left[\left|\sigma_j\right|\left|A_1\right| - \left|\sigma_{j-1}\right|\left|A_2\right|\right] \tag{4.41}$$

$$\left|\sigma_{n+1}\right| \leq \left|\sigma_0\right|\left(\left|A\right| - \left|B\right| + c\sum_{j=0}^{n-1}\left(\left|A_1\right| - \left|A_2\right|\right)\right) \tag{4.42}$$

$$\frac{\left|\sigma_{n+1}\right|}{\left|\sigma_0\right|} \leq \left|A\right| - \left|B\right| + c\sum_{j=0}^{n-1}\left(\left|A_1\right| - \left|A_2\right|\right) \tag{4.43}$$

The stability condition is therefore given by:

$$\left|\frac{1}{\left|A\right| - \left|B\right| + c\sum_{j=0}^{n-1}\left(\left|A_1\right| - \left|A_2\right|\right)}\right| \leq 1 \tag{4.44}$$

4.3.2 Numerical Solution of the Unsaturated Zone Using the A-B Fractional Derivative

Now, using the A-B fractional derivative and the first and second-order derivative, the new model for the unsaturated zone is given by the following equation:

$$-\left(\theta_s - \theta_0\right)\psi_b^{\lambda}\frac{\lambda}{\psi_l^{\lambda+1}}\,{}_0^{ABC}D^{\alpha}\psi\left(z,t\right) = K_s\frac{\partial}{\partial z}\left(\frac{\psi_b}{\psi}\right)^{2+3\lambda}\frac{\partial\psi}{\partial z} + K_s\left(\frac{\psi_b}{\psi}\right)^{2+3\lambda}\frac{\partial^2\psi}{\partial z^2} - K_s\frac{\partial}{\partial z}\left(\frac{\psi_b}{\psi}\right)^{2+3\lambda} \tag{4.45}$$

Now, rearranging Equation (4.45) such that:

$$^{ABC}_0 D^\alpha \psi(z,t) = \frac{1}{-(\theta_s - \theta_0)\psi_b^\lambda \dfrac{\lambda}{\psi_l^{\lambda+1}}} \left(\begin{array}{l} K_s \dfrac{\partial}{\partial z}\left(\dfrac{\psi_b}{\psi}\right)^{2+3\lambda} \dfrac{\partial \psi}{\partial z} + K_s \left(\dfrac{\psi_b}{\psi}\right)^{2+3\lambda} \dfrac{\partial^2 \psi}{\partial z^2} \\ -K_s \dfrac{\partial}{\partial z}\left(\dfrac{\psi_b}{\psi}\right)^{2+3\lambda} \end{array} \right) \tag{4.46}$$

where

$$^{ABC}_0 D^\alpha \psi(z,t) = F(z,t,\psi(z,\alpha)) \tag{4.47}$$

Then the A-B integral is given as:

$$\psi(z_i,t) - \psi(z_i,0) = \frac{(1-\alpha)}{B(\alpha)} f(z,t,\psi(z,t)) + \frac{\alpha}{B(\alpha)\Gamma(\alpha)} \int_0^t f(z_i,\tau,\psi(z_i,\tau))(t-\tau)^{\alpha-1} d\tau \tag{4.48}$$

At point t_{n+1}

$$\psi(z_i,t_{n+1}) - \psi(z_i,0) = \frac{(1-\alpha)}{B(\alpha)} f(z,t,\psi(z,t)) + \frac{\alpha}{B(\alpha)\Gamma(\alpha)} \int_0^{t_{n+1}} F(z_i,\tau,\psi(z_i,\tau))(t_{n+1}-\tau)^{\alpha-1} d\tau \tag{4.49}$$

$$\psi(z_i,t_{n+1}) - \psi(z_i,0) = \frac{(1-\alpha)}{B(\alpha)} f(z,t,\psi(z,t)) + \frac{(\Delta t)^\alpha}{B(\alpha)\Gamma(2+\alpha)}$$
$$\times \sum_{j=0}^n \left[\begin{array}{l} F(z_i,t_j,\psi_i^j)\left(\begin{array}{l}(n-j+1)^\alpha(n-j+2+\alpha) \\ -(n-j)^\alpha(n-j+2+2\alpha)\end{array}\right) \\ -F(z_i,t_{j-1},\psi_i^{j-1})\left(\begin{array}{l}(n-j+1)^{\alpha+1} \\ -(n-j)^\alpha(n-j+1+\alpha)\end{array}\right) \end{array} \right] \tag{4.50}$$

where:

$$F(z_i,t_j,\psi_i^j) = \frac{1}{-(\theta_s-\theta_0)\psi_b^\lambda \dfrac{\lambda}{\psi_i^{\lambda+1}}} \left(\begin{array}{l} K_s \left(\dfrac{\left(\dfrac{\psi_b}{\psi_{i+1}^j}\right)^{2+3\lambda} - \left(\dfrac{\psi_b}{\psi_i^j}\right)^{2+3\lambda}}{\Delta z}\right) \dfrac{\psi_{i+1}^j - \psi_i^j}{\Delta z} + K_s \left(\dfrac{\psi_b}{\psi_i^j}\right)^{2+3\lambda} \\ \times \dfrac{\psi_{i+1}^j - 2\psi_i^j + \psi_{i-1}^j}{(\Delta z)^2} - K_s \left(\dfrac{\left(\dfrac{\psi_b}{\psi_{i+1}^j}\right)^{2+3\lambda} - \left(\dfrac{\psi_b}{\psi_i^j}\right)^{2+3\lambda}}{\Delta z}\right) \end{array} \right) \tag{4.51}$$

$$F\left(z_i, t_{j-1}, \psi_i^{j-1}\right) = \frac{1}{-\left(\theta_s - \theta_0\right)\psi_b^{\lambda}\dfrac{\lambda}{\psi_i^{\lambda+1}}} \left(\begin{array}{l} K_s \left(\dfrac{\left(\dfrac{\psi_b}{\psi_{i+1}^{j-1}}\right)^{2+3\lambda} - \left(\dfrac{\psi_b}{\psi_i^{j-1}}\right)^{2+3\lambda}}{\Delta z} \right) \dfrac{\psi_{i+1}^{j-1} - \psi_i^{j-1}}{\Delta z} + K_s \left(\dfrac{\psi_b}{\psi_i^{j-1}}\right)^{2+3\lambda} \\[4mm] \times \dfrac{\psi_{i+1}^{j-1} - 2\psi_i^{j-1} + \psi_{i-1}^{j-1}}{\left(\Delta z\right)^2} - K_s \left(\dfrac{\left(\dfrac{\psi_b}{\psi_{i+1}^{j-1}}\right)^{2+3\lambda} - \left(\dfrac{\psi_b}{\psi_i^{j-1}}\right)^{2+3\lambda}}{\Delta z} \right) \end{array} \right)$$

$$(4.52)$$

The final numerical solution for the unsaturated groundwater flow equation is given as:

$$\psi_i^{n+1} - \psi_i^0 = \frac{(1-\alpha)}{B(\alpha)} \frac{1}{-\left(\theta_s - \theta_0\right)\psi_b^{\lambda}\dfrac{\lambda}{\psi_i^{\lambda+1}}} \left(\begin{array}{l} K_s \left(\dfrac{\left(\dfrac{\psi_b}{\psi_{i+1}^{n}}\right)^{2+3\lambda} - \left(\dfrac{\psi_b}{\psi_i^{n}}\right)^{2+3\lambda}}{\Delta z} \right) \dfrac{\psi_{i+1}^{n} - \psi_i^{n}}{\Delta z} + K_s \left(\dfrac{\psi_b}{\psi_i^{n}}\right)^{2+3\lambda} \\[4mm] \times \dfrac{\psi_{i+1}^{n} - 2\psi_i^{n} + \psi_{i-1}^{n}}{\left(\Delta z\right)^2} - K_s \left(\dfrac{\left(\dfrac{\psi_b}{\psi_{i+1}^{n}}\right)^{2+3\lambda} - \left(\dfrac{\psi_b}{\psi_i^{n}}\right)^{2+3\lambda}}{\Delta z} \right) \end{array} \right)$$

$$+ \frac{(\Delta t)^{\alpha}}{\Gamma(2+\alpha)} \sum_{j=0}^{n} \left[\begin{array}{l} \dfrac{1}{-\left(\theta_s - \theta_0\right)\psi_b^{\lambda}\dfrac{\lambda}{\psi_i^{\lambda+1}}} \left(\begin{array}{l} K_s \left(\dfrac{\left(\dfrac{\psi_b}{\psi_{i+1}^{j}}\right)^{2+3\lambda} - \left(\dfrac{\psi_b}{\psi_i^{j}}\right)^{2+3\lambda}}{\Delta z} \right) \dfrac{\psi_{i+1}^{j} - \psi_i^{j}}{\Delta z} + K_s \left(\dfrac{\psi_b}{\psi_i^{j}}\right)^{2+3\lambda} \\[4mm] \times \dfrac{\psi_{i+1}^{j} - 2\psi_i^{j} + \psi_{i-1}^{j}}{\left(\Delta z\right)^2} - K_s \left(\dfrac{\left(\dfrac{\psi_b}{\psi_{i+1}^{j}}\right)^{2+3\lambda} - \left(\dfrac{\psi_b}{\psi_i^{j}}\right)^{2+3\lambda}}{\Delta z} \right) \end{array} \right) \\[8mm] \times \left((n-j+1)^{\alpha}(n-j+2+\alpha) - (n-j)^{\alpha}(n-j+2+2\alpha) \right) \\[4mm] - \dfrac{1}{-\left(\theta_s - \theta_0\right)\psi_b^{\lambda}\dfrac{\lambda}{\psi_i^{\lambda+1}}} \left(\begin{array}{l} K_s \left(\dfrac{\left(\dfrac{\psi_b}{\psi_{i+1}^{j-1}}\right)^{2+3\lambda} - \left(\dfrac{\psi_b}{\psi_i^{j-1}}\right)^{2+3\lambda}}{\Delta z} \right) \dfrac{\psi_{i+1}^{j-1} - \psi_i^{j-1}}{\Delta z} + K_s \left(\dfrac{\psi_b}{\psi_i^{j-1}}\right)^{2+3\lambda} \\[4mm] \times \dfrac{\psi_{i+1}^{j-1} - 2\psi_i^{j-1} + \psi_{i-1}^{j-1}}{\left(\Delta z\right)^2} - K_s \left(\dfrac{\left(\dfrac{\psi_b}{\psi_{i+1}^{j-1}}\right)^{2+3\lambda} - \left(\dfrac{\psi_b}{\psi_i^{j-1}}\right)^{2+3\lambda}}{\Delta z} \right) \end{array} \right) \\[8mm] \times \left((n-j+1)^{\alpha+1} - (n-j)^{\alpha}(n-j+1+\alpha) \right) \end{array} \right]$$

$$(4.53)$$

For simplicity of the above equation let:

$$\wp = -(\theta_s - \theta_0)\psi_b^\lambda \frac{\lambda}{\psi_i^{\lambda+1}}$$

$$a = \frac{(\Delta t)^\alpha}{\Gamma(2-\alpha)}$$

$$b = \frac{K_s}{\Delta z}$$

$$c = K_s$$

$$d = \frac{1}{(\Delta z)^2}$$

$$e = \frac{(1-\alpha)}{B(\alpha)}$$

$$f = \frac{1}{\Delta z}$$

$$R_{n,j}^{1,\alpha} = (n-j+1)^\alpha (n-j+2+\alpha) - (n-j)^\alpha (n-j+2+2\alpha)$$

$$R_{n,j}^{2,\alpha} = (n-j+1)^{\alpha+1} - (n-j)^\alpha (n-j+1+\alpha)$$

Now, the equation becomes:

$$\psi_i^{n+1} - \psi_i^0 = e\frac{1}{\wp}\left[\begin{array}{l} b\left(\left(\frac{\psi_b}{\psi_{i+1}^n}\right)^{2+3\lambda} - \left(\frac{\psi_b}{\psi_i^n}\right)^{2+3\lambda} f\left(\psi_{i+1}^n - \psi_i^n\right)\right) + c\left(\frac{\psi_b}{\psi_i^n}\right)^{2+3\lambda} \\ \times d\left(\psi_{i+1}^n - 2\psi_i^n + \psi_{i-1}^n\right) - b\left(\left(\frac{\psi_b}{\psi_{i+1}^n}\right)^{2+3\lambda} - \left(\frac{\psi_b}{\psi_i^n}\right)^{2+3\lambda}\right) \end{array}\right]$$

$$+ a\sum_{j=0}^n \left[\begin{array}{l} \frac{1}{\wp}\left(\begin{array}{l} b\left(\left(\frac{\psi_b}{\psi_{i+1}^j}\right)^{2+3\lambda} - \left(\frac{\psi_b}{\psi_i^j}\right)^{2+3\lambda}\right) f\left(\psi_{i+1}^j - \psi_i^j\right) + c\left(\frac{\psi_b}{\psi_i^j}\right)^{2+3\lambda} \\ \times d\left(\psi_{i+1}^j - 2\psi_i^j + \psi_{i-1}^j\right) - b\left(\frac{\psi_b}{\psi_{i+1}^j}\right)^{2+3\lambda} - \left(\frac{\psi_b}{\psi_i^j}\right)^{2+3\lambda} \end{array}\right) R_{n,j}^{1,\alpha} \\ -\frac{1}{\wp}\left(\begin{array}{l} b\left(\left(\frac{\psi_b}{\psi_{i+1}^{j-1}}\right)^{2+3\lambda} - \left(\frac{\psi_b}{\psi_i^{j-1}}\right)^{2+3\lambda}\right) f\left(\psi_{i+1}^{j-1} - \psi_i^{j-1}\right) + c\left(\frac{\psi_b}{\psi_i^{j-1}}\right)^{2+3\lambda} \\ \times d\left(\psi_{i+1}^{j-1} - 2\psi_i^{j-1} + \psi_{i-1}^{j-1}\right) - b\left(\left(\frac{\psi_b}{\psi_{i+1}^{j-1}}\right)^{2+3\lambda} - \left(\frac{\psi_b}{\psi_i^{j-1}}\right)^{2+3\lambda}\right) \end{array}\right) R_{n,j}^{2,\alpha} \end{array}\right] \tag{4.54}$$

4.3.3 NUMERICAL SOLUTION OF THE SATURATED–UNSATURATED GROUNDWATER FLOW EQUATION USING THE GHANBARI–ATANGANA NUMERICAL SCHEME

Numerical methods such as Adams–Bashforth, the two-step Laplace transform, and Atangana–Toufik including others have been applied successfully when solving complexities associated with non-linear differential and integral equations (Toufik & Atangana, 2017). While these methods have been applied intensively to the case of power law differential and integral operators, one needs to mention that many of them have not really been applied when dealing with differential and integral equations generated by the generalised Mittag-Leffler function (Ghanbari & Atangana, 2019). In this section, we are going to apply the newly introduced numerical scheme by Ghanbari and Atangana to the saturated–unsaturated equation. This newly established numerical scheme is based on the product-integral (PI) rule to solve a functional initial-value problem. Let us consider an A-B fractional derivative equation as:

$$ {}^{ABC}_{t_0}D^\alpha_t = F\big(t, \psi(t)\big) \tag{4.55} $$

where $F(t, \psi(t))$ are continuous functions and the initial condition $\psi(t_0) = \psi_0$. Applying the integral operator on both sides of Equation (4.55) above we provide the following Volterra integral equation:

$$ \psi(t) - \psi(t_0) = \frac{(1-\alpha)}{B(\alpha)}F\big(t, \psi(t)\big) + \frac{\alpha}{B(\alpha)\Gamma(\alpha)}\int_{t_0}^{t} F\big(\tau, \psi(\tau)\big)(t-\tau)^{\alpha-1}\,d\tau \tag{4.56} $$

Taking $t = t_n = t_0 + n\hbar$ in the above equation, where $\hbar = \dfrac{t_f - t_0}{N}$ is a constant step-size, we achieve:

$$ \psi(t_n) - \psi(t_0) = \frac{(1-\alpha)}{B(\alpha)}F\big(t_n, \psi(t_n)\big) + \frac{\alpha}{B(\alpha)\times\Gamma(\alpha)}\sum_{i=0}^{n-1}\int_{t_j}^{t_{j+1}} F\big(\tau, \psi(\tau)\big)(t_n - \tau)^{\alpha-1}\,d\tau\ 1 \le n \le N \tag{4.57} $$

Now, we can approximate the function $F(\tau, \psi(\tau))$ by the first-order Lagrange interpolation:

$$ F\big(\tau, \psi(\tau)\big) \approx F\big(t_{j+1}, \psi_{j+1}\big) + \frac{\tau - t_{j+1}}{\hbar}\big(F(t_{j+1}, \psi_{j+1}) - F(t_j, \psi_j)\big)\tau \in [t_j, t_{j+1}] \tag{4.58} $$

where the notation $\psi_j = \psi(t_j)$ is used. Substituting Equation (4.57) into (4.58) along with carrying out some algebraic manipulations, the following implicit ABC-PI rule is achieved:

$$ \psi^n - \psi^0 = \frac{(1-\alpha)}{B(\alpha)}F\big(t_n, \psi_n\big) + \frac{\alpha\hbar^\alpha}{B(\alpha)}\left(\gamma_n F\big(t_n, \psi_n\big) + \sum_{J=0}^{N}\beta_{n-j}F\big(t_J, \psi_J\big)\right) \tag{4.59} $$

where:

$$ \gamma_n = \frac{(n-1)^{\alpha+1} - n^\alpha(n-\alpha-1)}{\Gamma(\alpha+1)} $$

$$B_J \begin{cases} \dfrac{1}{\Gamma(\alpha+2)}, j=0 \\[3mm] \dfrac{(j-1)^{\alpha+1}-2j^{\alpha+1}+(j+1)^{\alpha+1}}{\Gamma(\alpha+2)}, j=1,2,3.\dots.n-1 \end{cases}$$

Introducing the above equation (4.59) into our 1-d saturated–unsaturated groundwater flow equation we obtain the following equation for the saturated zone:

$$\psi^n - \psi^0 = \frac{(1-\alpha)}{B(\alpha)}\frac{K_s}{S_s}\frac{\psi_{i+1}^n-2\psi_i^n+\psi_{i-1}^n}{(\Delta z)^2} + \frac{\alpha\hbar^\alpha}{B(\alpha)}\left(\gamma_n\frac{K_s}{S_s}\frac{\psi_{i+1}^n-2\psi_i^n+\psi_{i-1}^n}{(\Delta z)^2} + \sum_{J=0}^{N}\beta_{n-j}\frac{K_s}{S_s}\frac{\psi_{i+1}^j-2\psi_i^j+\psi_{i-1}^j}{(\Delta z)^2}\right) \quad (4.60)$$

And the following for the unsaturated zone:

$$\psi^n - \psi^0 = \frac{(1-\alpha)}{B(\alpha)}\frac{1}{-(\theta_s-\theta_0)\psi_b^\lambda\dfrac{\lambda}{\psi_i^{\lambda+1}}}\left[K_s\frac{\left(\dfrac{\psi_b}{\psi_{i+1}^n}\right)^{2+3\lambda}-\left(\dfrac{\psi_b}{\psi_i^n}\right)^{2+3\lambda}}{\Delta z}\frac{\psi_{i+1}^n-\psi_i^n}{\Delta z}+K_s\left(\dfrac{\psi_b}{\psi_i^n}\right)^{2+3\lambda} \atop \times\frac{\psi_{i+1}^n-2\psi_i^n+\psi_{i-1}^n}{(\Delta z)^2}-K_s\frac{\left(\dfrac{\psi_b}{\psi_{i+1}^n}\right)^{2+3\lambda}-\left(\dfrac{\psi_b}{\psi_i^n}\right)^{2+3\lambda}}{\Delta z}\right]$$

$$+\frac{\alpha\hbar^\alpha}{B(\alpha)}\left\{\gamma_n\frac{1}{-(\theta_s-\theta_0)\psi_b^\lambda\dfrac{\lambda}{\psi_i^{\lambda+1}}}\left[K_s\frac{\left(\dfrac{\psi_b}{\psi_{i+1}^n}\right)^{2+3\lambda}-\left(\dfrac{\psi_b}{\psi_i^n}\right)^{2+3\lambda}}{\Delta z}\frac{\psi_{i+1}^n-\psi_i^n}{\Delta z}+K_s\left(\dfrac{\psi_b}{\psi_i^n}\right)^{2+3\lambda} \atop \times\frac{\psi_{i+1}^n-2\psi_i^n+\psi_{i-1}^n}{(\Delta z)^2}-K_s\frac{\left(\dfrac{\psi_b}{\psi_{i+1}^n}\right)^{2+3\lambda}-\left(\dfrac{\psi_b}{\psi_i^n}\right)^{2+3\lambda}}{\Delta z}\right] \right.$$

$$\left.+\sum_{J=0}^{N}\beta_{n-j}\frac{1}{-(\theta_s-\theta_0)\psi_b^\lambda\dfrac{\lambda}{\psi_i^{\lambda+1}}}\left[K_s\frac{\left(\dfrac{\psi_b}{\psi_{i+1}^j}\right)^{2+3\lambda}-\left(\dfrac{\psi_b}{\psi_i^j}\right)^{2+3\lambda}}{\Delta z}\frac{\psi_{i+1}^j-\psi_i^j}{\Delta z}+K_s\left(\dfrac{\psi_b}{\psi_i^j}\right)^{2+3\lambda} \atop \times\frac{\psi_{i+1}^j-2\psi_i^j+\psi_{i-1}^j}{(\Delta z)^2}-K_s\frac{\left(\dfrac{\psi_b}{\psi_{i+1}^j}\right)^{2+3\lambda}-\left(\dfrac{\psi_b}{\psi_i^j}\right)^{2+3\lambda}}{\Delta z}\right]\right\}$$

$$(4.61)$$

To make a further simplification to the above equation, let us say:

$$\wp = -(\theta_s-\theta_0)\psi_b^\lambda\frac{\lambda}{\psi_i^{\lambda+1}}$$

$$a = \frac{(1-\alpha)}{B(\alpha)}$$

$$b = \frac{K_s}{\Delta z}$$

$$c = \frac{1}{\Delta z}$$

$$d = \frac{1}{(\Delta z)^2}$$

$$e = \frac{\alpha \hbar^\alpha}{B(\alpha)}$$

Now we rewrite the equation

$$
\psi^n - \psi^0 = a\frac{1}{\wp}\left(\begin{array}{l} b\left(\left(\frac{\psi_b}{\psi_{i+1}^n}\right)^{2+3\lambda} - \left(\frac{\psi_b}{\psi_i^n}\right)^{2+3\lambda}\right)c\left(\psi_{i+1}^n - \psi_i^n\right) + K_s\left(\frac{\psi_b}{\psi_i^n}\right)^{2+3\lambda} \\ \times d\left(\psi_{i+1}^n - 2\psi_i^n + \psi_{i-1}^n\right) - b\left(\left(\frac{\psi_b}{\psi_{i+1}^n}\right)^{2+3\lambda} - \left(\frac{\psi_b}{\psi_i^n}\right)^{2+3\lambda}\right) \end{array}\right)
$$

$$
+ e\left(\begin{array}{l} \gamma_n \frac{1}{\wp}\left(\begin{array}{l} b\left(\left(\frac{\psi_b}{\psi_{i+1}^n}\right)^{2+3\lambda} - \left(\frac{\psi_b}{\psi_i^n}\right)^{2+3\lambda}\right)c\left(\psi_{i+1}^n - \psi_i^n\right) + K_s\left(\frac{\psi_b}{\psi_i^n}\right)^{2+3\lambda} \\ \times d\left(\psi_{i+1}^n - 2\psi_i^n + \psi_{i-1}^n\right) - b\left(\left(\frac{\psi_b}{\psi_{i+1}^n}\right)^{2+3\lambda} - \left(\frac{\psi_b}{\psi_i^n}\right)^{2+3\lambda}\right) \end{array}\right) \\ + \sum_{J=0}^{N}\beta_{n-j}\frac{1}{\wp}\left(\begin{array}{l} b\left(\left(\frac{\psi_b}{\psi_{i+1}^j}\right)^{2+3\lambda} - \left(\frac{\psi_b}{\psi_i^j}\right)^{2+3\lambda}\right)c\left(\psi_{i+1}^j - \psi_i^j\right) + K_s\left(\frac{\psi_b}{\psi_i^j}\right)^{2+3\lambda} \\ \times d\left(\psi_{i+1}^j - 2\psi_i^j + \psi_{i-1}^j\right) - b\left(\left(\frac{\psi_b}{\psi_{i+1}^j}\right)^{2+3\lambda} - \left(\frac{\psi_b}{\psi_i^j}\right)^{2+3\lambda}\right) \end{array}\right) \end{array}\right)
$$

(4.62)

4.4   CONCLUSION

In this chapter, differential operators with Mittag-Leffler function were used to model groundwater flow in the saturated–unsaturated zone. The main advantage of using a derivative with Mittag-Leffler function is that it can replicate crossover behavior; for example, the operator is able to replicate a passage from a random walk to a power law process.

REFERENCES

Allepalli, P. K., & Govindaraju, R. S., (2009). *Modeling Fate and Transport of Atrazine in the Saturated–Unsaturated Zone of Soil*. Kansas State University, Department of Civil Engineering, Manhattan, Kansas.

Alqahtani, R.T., (2016). Atangana–Baleanu derivative with fractional order applied to the model of groundwater within an unconfined aquifer. *Journal of Nonlinear Sciences and Applications* 9:3647–3654.

Atangana, A., & Baleanu, D., (2016). New fractional derivatives with non-local and non-singular kernel: Theory and application to heat transfer model. *Thermal Science*. 20(2): 763–769.

Brooks, R.H., & Corey, A.T., (1966). Properties of porous media affecting fluid flow. *Journal of the Irrigation and Drainage Division, ASCE*, 92(IR2): 61–88.

Djida, J.D., Area, I., & Atangana, A., 2016. New numerical scheme of Atangana–Baleanu fractional integral: An application to groundwater flow within leaky aquifer. arXiv preprint arXiv:1610.08681

Ghanbari, B., & Atangana, A., 2019. An effective implicit numerical method to solve chaotic systems involving the fractional operator with nonsingular kernel. *Chaos. An Interdisciplinary Journal of Nonlinear Science*, 29(9): 093111.

Gomez-Aguilar, J.F., & Atangana, A., 2019. Time-fractional variable-order telegraph equation involving operators with Mittag-Leffler kernel. *Journal of Electromagnetic Waves and Applications*. 33(2): 165–177.

Sheikh, N.A., Ali, F., Saqib, M., Khan, I., Jan, S.A.A., Alshomrani, A.S., & Alghamdi, M.S., (2017). Comparison and analysis of the Atangana–Baleanu and Caputo–Fabrizio fractional derivatives for generalized Casson fluid model with heat generation and chemical reaction. *Results in Physics* 7: 789–800.

Toufik, M., & Atangana, A., 2017. New numerical approximation of fractional derivative with non-local and non-singular kernel: Application to chaotic models. *European Physical Journal - Plus*, 132: 444.

5 New Model of the 1-d Saturated–Unsaturated Groundwater Flow Using the Fractal-Fractional Derivative

Rendani Vele Makahane and Abdon Atangana
University of the Free State, Bloemfontein, South Africa

CONTENTS

5.1 INTRODUCTION

The Euclidean model is usually satisfactory when describing patterns with a simple and pure structure (Gaddis & Zyda, 1986). However, many patterns in nature are so irregular and fragmented. The existence of such complexity challenges us to study these patterns, since Euclidian geometry cannot explain them (Mandelbrot, 1982). Responding to this challenge, there exists a non-standard derivative in the field of applied mathematics and mathematics known as a fractal derivative where a variable is scaled according to t^a. This derivative was suggested to model real-world problems such as heterogeneity and fracture network in a system that is not appropriate for classical physical law such as Darcy's, Fick's and Fourier's law (Atangana, 2017; Allwright & Atangana, 2018). Unlike the usual Euclidean structures, magnifying fractal results in the resolution of more details. Fractal structures are found everywhere in nature, such as clouds, coastlines, trees, etc. Their properties include scale invariant, an infinite amount of details, self-similarity, etc. (Liu et al., 2003). As our dependence on groundwater resources increases, modeling and prediction of sub-surface flow processes continue to be an important topic for researchers in hydrogeology (Sivakumar et al., 2005). Moreover, it is vital to note that no model can completely describe a natural process; however, at a given scale, if the model is accurate it can describe the process with enough precision (Gaddis & Zyda, 1986) and much success. However, these fractional derivatives are local operators and therefore are not very appropriate for modeling complex natural problems – for example, natural problems that display fractal behaviors (Atangana, 2017). For the past few years, the idea of non-local operators has attracted more researchers because of its ability to model more complex natural problems using mathematical equations. In this chapter we are going to use the new concept of

DOI: 10.1201/9781003266266-5

differentiation that was introduced by Atangana (2017), a combination of the concept of fractional differentiation and that of fractal derivative, which takes into account the fractal effect, memory, non-locality and elasticity. The governing equation considered in this work is one with a 2-phase flow: the flow of water in the saturated zone and flow in the region near the surface where pores are filled with both air and water; this zone is called the unsaturated zone. If we consider that the pressure of air remains constant, water flow through the saturated–unsaturated porous media is described mathematically by a combination of Darcy's law and the continuity law (Pelka, 1983; Allepalli and Govindaraju, 2009; List & Radu, 2015):

$$\left[S_S S_a\left(\psi\right) + C\left(\psi\right)\right]\frac{\partial\psi}{\partial t} = \frac{\partial}{\partial z}\left[K_z\left(\psi\right)\left(\frac{\partial\psi}{\partial z}+1\right)\right] \tag{5.1}$$

where:

ψ = pressure head

n = porosity

S_S = specific storage of the soil

S_a = saturation of the aqueous phase

$\dfrac{\theta}{n}$, $C(\psi)$ = capillary capacity of the soil, $d\theta/d\psi$

$q(t)$ = groundwater flux $= -K_z\left(\psi\right)\left(\dfrac{\partial\psi}{\partial z}-1\right)$

$K_z(\psi)$ = hydraulic conductivity

z = vertical co-ordinate.

Equation (5.1), involves the hydraulic properties of the soil that are a function of the suction head of the soil and therefore are non-linear (Allepalli and Govindaraju, 2009). Many functions have been proposed to describe the soil hydraulic properties empirically. One of the popular models is based on the equations of Brooks and Corey (1966):

$$K\left(\psi\right) = K_s\left(\frac{\psi_b}{\psi}\right)^{2+3\lambda} \qquad\qquad \psi \leq 0 \tag{5.2}$$

$$K\left(\psi\right) = K_s \qquad\qquad \psi > 0 \tag{5.3}$$

$$\theta\left(\psi\right) = \theta_0 + \left(\theta_s - \theta_0\right)\left(\frac{\psi_b}{\psi}\right)^{\lambda} \qquad\qquad \psi \leq 0 \tag{5.4}$$

where: $\psi \leq 0$ represent the unsaturated flow and $\psi > 0$ the saturated flow, ψ = pressure head, ψ_b = air entry suction pressure head, $\theta(\psi)$ = soil water content, θ_s = saturated water content, θ_0 = residual water content, λ = pore size distribution index, K_s = saturated hydraulic conductivity. Using the above soil hydraulic property, we obtain the following equations for flow in the saturated zone and flow in the unsaturated zone:

$$\frac{\partial\psi}{\partial t} = \frac{K_s}{S_s}\frac{\partial^2\psi}{\partial z^2} \qquad\qquad \psi > 0 \tag{5.5}$$

and,

$$\frac{\partial\theta\left(\psi\right)}{\partial t} = \frac{\partial}{\partial z}\left[K_z\left(\psi\right)\left(\frac{\partial\psi}{\partial z}-1\right)\right] \tag{5.6}$$

Using the Brooks and Corey equations, the above equation (5.9) become:

$$\frac{\partial \psi}{\partial t} = \frac{1}{-(\theta_s - \theta_0)\psi_b^{\lambda} \dfrac{\lambda}{\psi_i^{\lambda+1}}} \left(K_s \frac{\partial}{\partial z}\left(\frac{\psi_b}{\psi}\right)^{2+3\lambda} \frac{\partial \psi}{\partial z} + K_s \left(\frac{\psi_b}{\psi}\right)^{2+3\lambda} \frac{\partial^2 \psi}{\partial z^2} - K_s \frac{\partial}{\partial z}\left(\frac{\psi_b}{\psi}\right)^{2+3\lambda} \right) \psi \leq 0 \cdot \quad (5.7)$$

5.2 NUMERICAL SOLUTION OF THE NEW SATURATED–UNSATURATED GROUNDWATER FLOW MODEL USING THE FRACTAL DERIVATIVE

In this section, we introduce the fractal-fractional derivative into our 1-d saturated–unsaturated classical equation by replacing the time derivative of equation with the fractal derivative. The fractal-fractional derivative in the Riemann–Liouville sense is given by:

$$^{FFR}_{0}D_t^{\alpha,\beta}\psi(z,t) = \frac{1}{\Gamma(1-\alpha)}\frac{d}{dt}\int_0^t \psi(z,\tau)(t-\tau)^{-\alpha}\,d\tau \cdot \frac{1}{\beta t^{\beta-1}} \quad (5.8)$$

For us to make use of the above fractional derivative we will first have to discretize it. Using the first order approximation:

$$\frac{dF}{dt} = \frac{F(t+h)-F(t)}{\Delta t} \quad (5.9)$$

$$\frac{dF}{dt} = \frac{F(t_{n+1})-F(t_n)}{\Delta t} \quad (5.10)$$

The solution for $F(t_{n+1})$ is given in detail below:

$$F(t_{n+1}) = \frac{1}{(1-\alpha)} \int_0^{t_{n+1}} \psi(z_i,\tau)(t_{n+1}-\tau)^{-\alpha}\,d\tau\,\frac{1}{\beta t_{n+1}^{\beta-1}} \quad (5.11)$$

$$F(t_{n+1}) = \frac{1}{(1-\alpha)} \sum_{j=0}^{n} \psi_i^j \int_{t_j}^{t_{j+1}} (t_{n+1}-\tau)^{-\alpha}\,d\tau\,\frac{1}{\beta t_{n+1}^{\beta-1}} \quad (5.12)$$

For us to easily integrate the equation, the following substitutions will be done:

$$d\tau = -dy$$

$$t_{n+1}-\tau = y$$

$$\text{Lower boundary} = t_{n+1}-(t_{j+1})$$

$$\text{Upper boundary} = t_{n+1}-t_j$$

$$F\left(t_{n+1}\right) = \frac{1}{(1-\alpha)} \sum_{j=0}^{n} \psi_i^{\,j} \int_{t_j}^{t_{j+1}} (y)^{-\alpha}\, d\tau \, \frac{1}{\beta t_{n+1}^{\beta-1}}$$

$$F\left(t_{n+1}\right) = \frac{1}{(1-\alpha)} \sum_{j=0}^{n} \psi_i^{\,j} \int_{t_{j+1}}^{t_j} \frac{(y)^{1-\alpha}}{1-\alpha}\, d\tau \, \frac{1}{\beta t_{n+1}^{\beta-1}}$$

Now that the equation has been integrated, we can substitute $(\Delta t_{n+1} - \Delta t_j)$ and $(\Delta t_{n+1} - \Delta t_{(j+1)})$ back into the equation to yield the following:

$$F\left(t_{n+1}\right) = \frac{1}{(1-\alpha)} \sum_{j=0}^{n} \psi_i^{\,j} \frac{\left[\left(\Delta t_{n+1} - \Delta t_j\right) - \left(\Delta t_{n+1} - \Delta t_{(j+1)}\right)\right]^{1-\alpha}}{1-\alpha} \frac{1}{\beta t_{n+1}^{\beta-1}} \tag{5.13}$$

$$F\left(t_{n+1}\right) = \frac{1}{(1-\alpha)} \sum_{j=0}^{n} \psi_i^{\,j} \left(\Delta t\right)^{1-\alpha} \frac{\left(\left(n+1-j\right)^{1-\alpha} - \left(n-j\right)^{1-\alpha}\right)}{1-\alpha} \frac{1}{\beta t_{n+1}^{\beta-1}} \tag{5.14}$$

$$F\left(t_{n+1}\right) = \frac{\left(\Delta t\right)^{1-\alpha}}{(2-\alpha)} \sum_{j=0}^{n} \psi_i^{\,j} \left(\left(n+1-j\right)^{1-\alpha} - \left(n-j\right)^{1-\alpha}\right) \frac{1}{\beta t_{n+1}^{\beta-1}} \tag{5.15}$$

The solution for $F(t_n)$ is given in detail below:

$$F\left(t_n\right) = \frac{1}{(1-\alpha)} \int_{0}^{t_n} \psi\left(z_i, \tau\right)\left(t_n - \tau\right)^{-\alpha}\, d\tau \, \frac{1}{\beta t_n^{\beta-1}} \tag{5.16}$$

$$F\left(t_n\right) = \frac{1}{(1-\alpha)} \sum_{j=0}^{n-1} \psi_i^{\,j} \int_{t_j}^{t_{j+1}} \left(t_n - \tau\right)^{-\alpha}\, d\tau \, \frac{1}{\beta t_n^{\beta-1}} \tag{5.17}$$

For us to easily integrate the equation, the following substitutions will be done:

$$d\tau = -dy$$

$$t_n - \tau = y$$

$$\text{Lower boundary} = t_n - \left(t_{j+1}\right)$$

$$\text{Upper boundary} = t_n - t_j$$

$$F\left(t_n\right) = \frac{1}{(1-\alpha)} \sum_{j=0}^{n-1} \psi_i^{\,j} \int_{t_j}^{t_{j+1}} (y)^{-\alpha}\, d\tau \, \frac{1}{\beta t_n^{\beta-1}}$$

$$F\left(t_n\right)=\frac{1}{\left(1-\alpha\right)}\sum_{j=0}^{n-1}\psi_i^j\int_{t_{j+1}}^{t_j}\frac{\left(y\right)^{1-\alpha}}{1-\alpha}d\tau\frac{1}{\beta t_n^{\beta-1}}$$

Now that the equation has been integrated, we can substitute $(\Delta t_n - \Delta t_j)$ and $(\Delta t_n - \Delta t_{(j+1)})$ back into the equation to yield the following:

$$F\left(t_n\right)=\frac{1}{\left(1-\alpha\right)}\sum_{j=0}^{n-1}\psi_i^j\frac{\left[\left(\Delta t_n-\Delta t_j\right)-\left(\Delta t_n-\Delta t_{(j+1)}\right)\right]^{1-\alpha}}{1-\alpha}\frac{1}{\beta t_n^{\beta-1}} \tag{5.18}$$

$$F\left(t_n\right)=\frac{1}{\left(1-\alpha\right)}\sum_{j=0}^{n-1}\psi_i^j\left(\Delta t\right)^{1-\alpha}\frac{\left(\left(n+1-j\right)^{1-\alpha}-\left(n-j\right)^{1-\alpha}\right)}{1-\alpha}\frac{1}{\beta t_n^{\beta-1}} \tag{5.19}$$

$$F\left(t_n\right)=\frac{\left(\Delta t\right)^{1-\alpha}}{\left(2-\alpha\right)}\sum_{j=0}^{n-1}\psi_i^j\left(\left(n-j\right)^{1-\alpha}-\left(n-j-1\right)^{1-\alpha}\right)\frac{1}{\beta t_n^{\beta-1}} \tag{5.20}$$

Now we can substitute the solutions for $F(t_{n+1})$ and $F(t_n)$ back into Equation (5.8), such that:

$$_{0}^{FFR}D_t^{\alpha,\beta}\psi\left(z,t\right)=\frac{\left(\Delta t\right)^{-\alpha}}{\left(2-\alpha\right)}\left[\begin{array}{l}\sum_{j=0}^{n}\psi_i^j\left(\left(n+1-j\right)^{1-\alpha}-\left(n-j\right)^{1-\alpha}\right)\frac{1}{\beta t_{n+1}^{\beta-1}}\\-\sum_{j=0}^{n-1}\psi_i^j\left(\left(n-j\right)^{1-\alpha}-\left(n-j-1\right)^{1-\alpha}\right)\frac{1}{\beta t_{n-1}^{\beta-1}}\end{array}\right] \tag{5.21}$$

5.2.1 NUMERICAL SOLUTION FOR THE 1-D SATURATED ZONE USING THE FRACTAL-FRACTIONAL DERIVATIVE

We can now introduce the derivative into our classical saturated equation (5.5), such that:

$$_{0}^{RL}D_t^{\alpha}\psi\left(z,t\right)=\frac{K_s}{S_s}\beta t^{\beta-1}\frac{\partial^2\psi}{\partial z^2} \tag{5.22}$$

Appling the explicit forward Euler method on the Equation (5.22), we obtain the following numerical solution:

$$\frac{\left(\Delta t\right)^{-\alpha}}{\left(2-\alpha\right)}\left[\begin{array}{l}\sum_{j=0}^{n}\psi_i^j\left(\left(n+1-j\right)^{1-\alpha}-\left(n-j\right)^{1-\alpha}\right)\\-\sum_{j=0}^{n-1}\psi_i^j\left(\left(n-j\right)^{1-\alpha}-\left(n-j-1\right)^{1-\alpha}\right)\end{array}\right]=\frac{K_s}{S_s}\beta t_n^{\beta-1}\left[\frac{\psi_{i+1}^n-2\psi_i^n+\psi_{i-1}^n}{\left(\Delta z\right)^2}\right] \tag{5.23}$$

The above Equation can be simplified using the following parameters:

$$a=\frac{\left(\Delta t\right)^{-\alpha}}{\left(2-\alpha\right)}$$

$$b = \frac{K_s}{\left(\Delta z\right)^2 S_s}\, \beta t_n^{\beta-1}$$

$$\delta_{\alpha,j}^{1,n} = \left(\left(n+1-j\right)^{1-\alpha} - \left(n-j\right)^{1-\alpha}\right)$$

$$\delta_{\alpha,j}^{2,n} = \left(\left(n-j\right)^{1-\alpha} - \left(n-j-1\right)^{1-\alpha}\right)$$

Now we can rewrite the equation with parameters:

$$a\sum_{j=0}^{n}\psi_i^j \delta_{\alpha,j}^{1,n} - a\sum_{j=0}^{n-1}\psi_i^j \delta_{\alpha,j}^{2,n} = b\left(\psi_{i+1}{}^n - 2\psi_i{}^n + \psi_{i-1}{}^n\right) \tag{5.24}$$

$$a\sum_{j=0}^{n}\psi_i^j \delta_{\alpha,j}^{1,n} = b\left(\psi_{i+1}{}^n - 2\psi_i{}^n + \psi_{i-1}{}^n\right) + a\sum_{j=0}^{n-1}\psi_i^j \delta_{\alpha,j}^{2,n} \tag{5.25}$$

We obtain another numerical solution by applying the implicit backward Euler method on Equation (5.22):

$$\frac{\left(\Delta t\right)^{-\alpha}}{\left(2-\alpha\right)}\left[\begin{array}{l}\displaystyle\sum_{j=0}^{n}\psi_i^j \left(\left(n+1-j\right)^{1-\alpha} - \left(n-j\right)^{1-\alpha}\right)\\[2mm] \displaystyle-\sum_{j=0}^{n-1}\psi_i^j \left(\left(n-j\right)^{1-\alpha} - \left(n-j-1\right)^{1-\alpha}\right)\end{array}\right] = \frac{K_s}{S_s}\beta t_n^{\beta-1}\left[\frac{\psi_{i+1}{}^{n+1} - 2\psi_i{}^{n+1} + \psi_{i-1}{}^{n+1}}{\left(\Delta z\right)^2}\right] \tag{5.26}$$

To simplify the above equation, we make use of the following parameters:

$$a = \frac{\left(\Delta t\right)^{-\alpha}}{\left(2-\alpha\right)}$$

$$b = \frac{K_s}{S_s\left(\Delta z\right)^2}\, \beta t_n^{\beta-1}$$

$$\delta_{\alpha,j}^{1,n} = \sum_{j=0}^{n}\psi_i^j \left(\left(n+1-j\right)^{1-\alpha} - \left(n-j\right)^{1-\alpha}\right)$$

$$\delta_{\alpha,j}^{2,n} = \sum_{j=0}^{n-1}\psi_i^j \left(\left(n-j\right)^{1-\alpha} - \left(n-j-1\right)^{1-\alpha}\right)$$

Now we can rewrite the equation with parameters:

$$a \sum_{j=0}^{n} \psi_i^j \delta_{\alpha,j}^{1,n} - a \sum_{j=0}^{n-1} \psi_i^j \delta_{\alpha,j}^{2,n} = b \left(\psi_{i+1}^{n+1} - 2\psi_i^{n+1} + \psi_{i-1}^{n+1} \right) \tag{5.27}$$

$$a \sum_{j=0}^{n} \psi_i^j \delta_{\alpha,j}^{1,n} = b \left(\psi_{i+1}^{n+1} - 2\psi_i^{n+1} + \psi_{i-1}^{n+1} \right) + a \sum_{j=0}^{n-1} \psi_i^j \delta_{\alpha,j}^{2,n} \tag{5.28}$$

Applying the finite difference spatial discretization to Equation (5.28) the implicit Crank–Nicolson discretization is as follows:

$$\frac{(\Delta t)^{1-\alpha}}{(2-\alpha)} \sum_{j=0}^{n} \psi_i^j \left[\begin{array}{l} \left((n+1-j)^{1-\alpha} - (n-j)^{1-\alpha} \right) \\ - \left((n-j)^{1-\alpha} - (n-j-1)^{1-\alpha} \right) \end{array} \right] = \frac{K_s}{S_s} \beta t^{\beta-1} \left[\begin{array}{l} \dfrac{\psi_{i+1}^{n+1} - 2\psi_i^{n+1} + \psi_{i-1}^{n+1}}{2(\Delta z)^2} \\ + \dfrac{\psi_{i+1}^{n} - 2\psi_i^{n} + \psi_{i-1}^{n}}{2(\Delta z)^2} \end{array} \right] \tag{5.29}$$

$$\frac{(\Delta t)^{-\alpha}}{(2-\alpha)} \left[\begin{array}{l} \displaystyle\sum_{j=0}^{n} \psi_i^j \left((n+1-j)^{1-\alpha} - (n-j)^{1-\alpha} \right) \\ - \displaystyle\sum_{j=0}^{n-1} \psi_i^j \left((n-j)^{1-\alpha} - (n-j-1)^{1-\alpha} \right) \end{array} \right] = \frac{K_s}{2(\Delta z)^2 S_s} \beta t_n^{\beta-1} \left[\begin{array}{l} \left(\psi_{i+1}^{n+1} - 2\psi_i^{n+1} + \psi_{i-1}^{n+1} \right) \\ + \left(\psi_{i+1}^{n} - 2\psi_i^{n} + \psi_{i-1}^{n} \right) \end{array} \right] \tag{5.30}$$

To simplify the above equation, we will make use the following parameters:

$$a = \frac{(\Delta t)^{-\alpha}}{(2-\alpha)}$$

$$b = \frac{K_s}{2(\Delta z)^2 S_s} \beta t_n^{\beta-1}$$

$$\delta_{\alpha,j}^{1,n} = \left((n+1-j)^{1-\alpha} - (n-j)^{1-\alpha} \right)$$

$$\delta_{\alpha,j}^{2,n} = \left((n-j)^{1-\alpha} - (n-j-1)^{1-\alpha} \right)$$

Such that:

$$a \sum_{j=0}^{n} \psi_i^j \delta_{\alpha,j}^{1,n} - a \sum_{j=0}^{n-1} \psi_i^j \delta_{\alpha,j}^{2,n} = b \left[\left(\psi_{i+1}^{n+1} - 2\psi_i^{n+1} + \psi_{i-1}^{n+1} \right) + \left(\psi_{i+1}^{n} - 2\psi_i^{n} + \psi_{i-1}^{n} \right) \right] \tag{5.31}$$

$$\sum_{j=0}^{n} \psi_i^j \delta_{\alpha,j}^{1,n} = b \left[\left(\psi_{i+1}^{n+1} - 2\psi_i^{n+1} + \psi_{i-1}^{n+1} \right) + \left(\psi_{i+1}^{n} - 2\psi_i^{n} + \psi_{i-1}^{n} \right) \right] + a \sum_{j=0}^{n-1} \psi_i^j \delta_{\alpha,j}^{2,n} \tag{5.32}$$

5.2.2 NUMERICAL SOLUTION OF THE 1-D UNSATURATED ZONE USING THE FRACTAL-FRACTIONAL DERIVATIVE

Now, using the fractal-fractional derivative, we provide the solution for the unsaturated zone. By replacing the time derivative in the classical equation (5.7) derived for modeling the unsaturated zone we obtain the following equation:

$$
\frac{(\Delta t)^{-\alpha}}{(2-\alpha)}\left[
\begin{array}{l}
\sum_{j=0}^{n}\psi_i^j\left((n+1-j)^{1-\alpha}-(n-j)^{1-\alpha}\right) \\
-\sum_{j=0}^{n-1}\psi_i^j\left((n-j)^{1-\alpha}-(n-j-1)^{1-\alpha}\right)
\end{array}
\right]
$$

$$
=\frac{1}{-(\theta_s-\theta_0)\psi_b^\lambda\dfrac{\lambda}{\psi_i^{\lambda+1}}}K_s\beta t_n^{\beta-1}\left(
\begin{array}{l}
\dfrac{\left(\dfrac{\psi_b}{\psi_{i+1}^{n+1}}\right)^{2+3\lambda}-\left(\dfrac{\psi_b}{\psi_i^{n+1}}\right)^{2+3\lambda}}{\Delta z}\dfrac{\psi_{i+1}^{n+1}-\psi_i^{n+1}}{\Delta z}+\left(\dfrac{\psi_b}{\psi_i^{n+1}}\right)^{2+3\lambda} \\[2em]
\times\dfrac{\psi_{i+1}^{n+1}-2\psi_i^{n+1}+\psi_{i-1}^{n+1}}{(\Delta z)^2}-\dfrac{\left(\dfrac{\psi_b}{\psi_{i+1}^{n+1}}\right)^{2+3\lambda}-\left(\dfrac{\psi_b}{\psi_i^{n+1}}\right)^{2+3\lambda}}{\Delta z}
\end{array}
\right)
$$

$$(5.33)$$

To make further simplification on the above equation, let us say:

$$
\wp=-(\theta_s-\theta_0)\psi_b^\lambda\frac{\lambda}{\psi_i^{\lambda+1}},\ a=\frac{(\Delta t)^{-\alpha}}{\Gamma(2-\alpha)},\ b=K_s\beta t_n^{\beta-1},\ c=\frac{1}{(\Delta z)^2},\ d=\frac{1}{\Delta z}
$$

$$
R_{n,j}^\alpha=\left((n+1-j)^{1-\alpha}-(n-j)^{1-\alpha}\right)-\left((n-j)^{1-\alpha}-(n-j-1)^{1-\alpha}\right)
$$

Now we rewrite the equation as:

$$
aR_{n,j}^\alpha=\frac{1}{\wp}b\left(
\begin{array}{l}
d\left(\left(\dfrac{\psi_b}{\psi_{i+1}^{n+1}}\right)^{2+3\lambda}-\left(\dfrac{\psi_b}{\psi_i^{n+1}}\right)^{2+3\lambda}\times\left(\psi_{i+1}^{n+1}-\psi_i^{n+1}\right)\right)+\left(\dfrac{\psi_b}{\psi_i^{n+1}}\right)^{2+3\lambda} \\[2em]
\times c\left(\psi_{i+1}^{n+1}-2\psi_i^{n+1}+\psi_{i-1}^{n+1}\right)-d\left(\left(\dfrac{\psi_b}{\psi_{i+1}^{n+1}}\right)^{2+3\lambda}-\left(\dfrac{\psi_b}{\psi_i^{n+1}}\right)^{2+3\lambda}\right)
\end{array}
\right)
$$

$$(5.34)$$

The above numerical scheme is highly non-linear; therefore it is not possible to perform a stability analysis.

5.3 NUMERICAL SIMULATIONS, RESULTS AND DISCUSSION

To see the efficiency and the accuracy of the suggested new models with the fractal-fractional differential operator, we present in this section the numerical simulation of the new models with self-similarities. We chose the following theoretical parameters: $K_s=2.5 cm/hr$, $S_s=0.023 cm$, $\Delta t=0.001$,

h (*space step*) = 0.1. For the saturated zone let us consider the equation with the following boundary conditions:

$$\psi(0,t) = 0, \quad \text{and} \quad \psi(1,t) = 0 \, t > 0$$

And the following initial condition:

$$\psi(z,0) = \sin(n\pi z), z \in [0,L]$$

The flow of sub-surface water within a saturated and unsaturated media has attracted the attention of some researchers, while some mathematical models were developed to capture such physical problems, it has to be pointed out that heterogeneity of the geological formation was not really taken into account as many models use classical operators. In order to include in the mathematical formulation some complexity of nature, for example, a flow taking place in a fracture, a new concept was introduced called fractional calculus. One of the great properties of a fractional derivative with power law kernel is the ability to replicate the long-tailed, which can be assimilated with the flow within a fracture. This property can be used to depict the flow from the matrices soil to fracture. Nevertheless, it was revealed recently that even these new operators were unable to depict some complicated flow. For example, flow within a geological formation where the system of network has self-similar properties. Self-similar problems with power law have been recognized as complex real-world problems that could not be represented with classical and fractional differential operators. Thus, fractal-fractional differential operators have been used in this chapter. Numerical simulation of the saturated model with classical and fractal-fractional derivatives are depicted in Figures 5.1 to 5.5

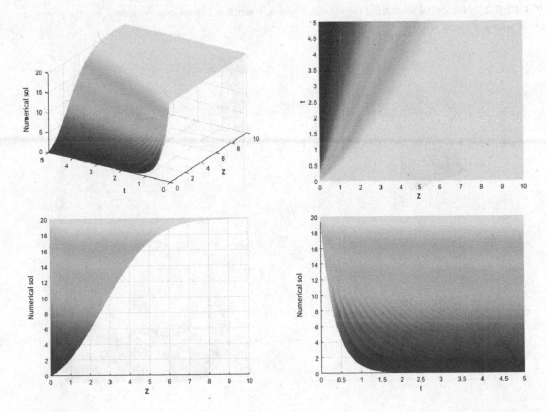

FIGURE 5.1 Numerical solution of the saturated groundwater flow equation using fractal and power law kernel with $\beta = 0.9$ and $\alpha = 1$.

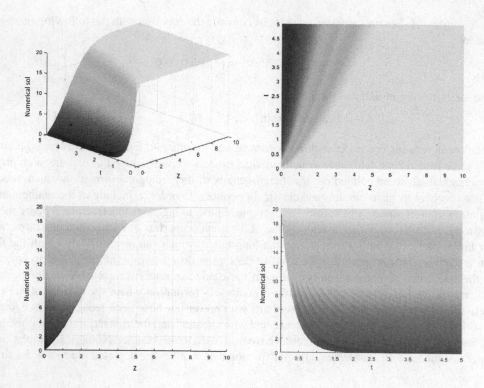

FIGURE 5.2 Numerical solution of the saturated groundwater flow equation using fractal and the power law kernel with $\beta = 1$ and $\alpha = 0.9$.

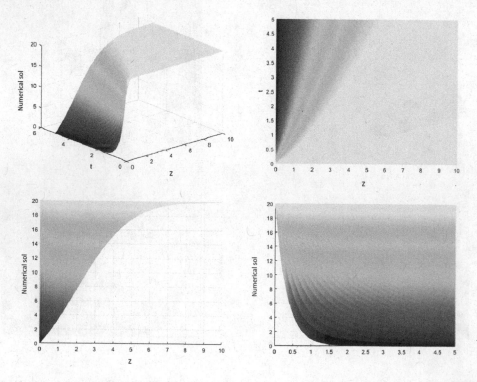

FIGURE 5.3 Numerical solution of the saturated groundwater flow equation using fractal and power law kernel with $\beta = 0.9$ and $\alpha = 0.9$.

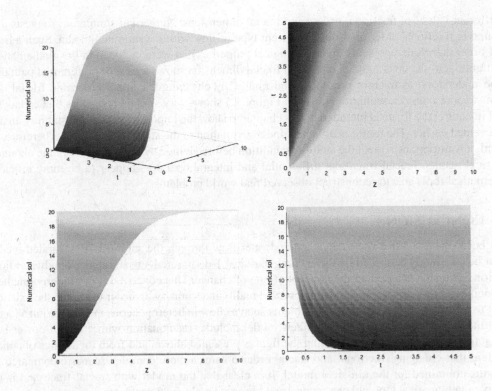

FIGURE 5.4 Numerical solution of the saturated groundwater flow equation using fractal and power law kernel with $\beta = 0.95$ and $\alpha = 0.95$.

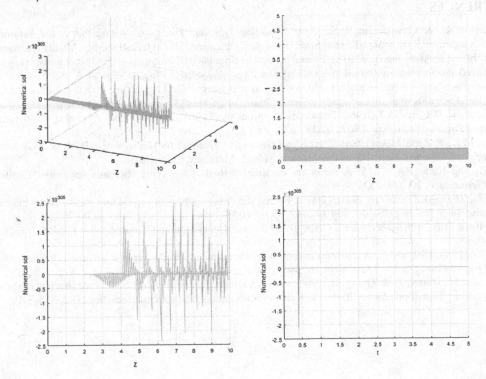

FIGURE 5.5 Numerical solution of the saturated groundwater flow equation using fractal and power law kernel with $\beta = 0.45$ and $\alpha = 1$.

for different values of fractional orders and fractal dimensions. Numerical simulation suggests that a change in fractional order provides a different type of flow within a saturated media. Such a fractional order therefore represents given geological properties that were included in the mathematical formulation. On the other hand, a change in fractal dimension shows a case of preferential path that can be understood as fracture presence. And finally, by changing fractional order and fractal one obtains a more complex scenario. However, Figure 5.5 shows very strange behaviors that cannot be found in nature; the figure is included in this chapter to show the important role of the stability analysis presented earlier. The figure is obtained under a violation of the stability condition. Therefore, in general, it is important to meet the stability condition before giving any interpretation of the obtained numerical results. Fractal-fractional differential and integral operators appear to be more efficient mathematical tools able to reconstruct observed real world problems.

5.4 CONCLUSION

It has been discussed that the vertical groundwater flow through the saturated–unsaturated media cannot be accurately described by the existing classical 1-d saturated–unsaturated groundwater flow equation because it is based on the concept of rate of change. This concept doesn't incorporate heterogeneity and therefore represents the observed reality inaccurately. In order to extend the existing model to capture complex real-world problems such as flow in heterogeneous media, we introduced new reliable and effective models. One new model includes the equation with a fractal-power law approach. The model is illustrated using the figures presented above, and from the figures obtained it is clear that changing fractional and fractal order obtains a more complex scenario compared to the figures obtained for the classical model. It is clear that the model with fractal-fractional is an effective mathematical tool and can reconstruct observed real-world problems.

REFERENCES

Allepalli, P.K., & Govindaraju, R.S., 2009. *Modeling Fate and Transport of Atrazine in the Saturated–Unsaturated Zone of Soil.* Kansas State University, Department of Civil Engineering, Manhattan, Kansas.

Allwright, A., & Atangana, A., (2018). Fractal advection-dispersion equation for groundwater transport in fractured aquifers with self-similarities. *European Physical Journal - Plus*, 133(2): 48.

Atangana, A., 2017. Fractal-fractional differentiation and integration: Connecting fractal calculus and fractional calculus to predict complex system. *Chaos, Solutions and Fractals*, 102: 396–406.

Brooks, R.H., & Corey, A.T., 1966. Properties of porous media affecting fluid flow. *Journal of the Irrigation and Drainage Division, ASCE*, 92(IR2): 61–88.

Gaddis, M.E., & Zyda, M.J., (1986). *The Fractal Geometry of Nature: Its Mathematical Basis and Application of Computer Graphics.* Naval Postgraduate school, Monterey, California.

List, F., and Radu F.A., 2015. A study on iterative methods for solving Richards' equation, *Computer Geoscience*, 20: 341–353.

Liu, J.Z., Zhang, L.D., & Yue, G.H., (2003). Fractal dimension in human cerebellum measured by magnetic resonance imaging. *Biophysical Journal*, 85(6): 4041–4046.

Mandelbrot, B.B., 1982. *The Fractal Geometry of Nature.* Henry Holt and Company, 3rd Edition. New York. 468.

Pelka, W., 1983. Heat and mass transport in saturated–unsaturated groundwater flow. *Relation of Groundwater Quantity and Quality.* IAHS Publ.no. 146: 160–166.

Sivakumar, B., Harter, T., & Zhang, H., (2005). A fractal investigation of solute travel time in a heterogeneous aquifer: Transition probability/Markov chain representation. *Ecological Modelling*, 182: 355–370.

6 Application of the Fractional-Stochastic Approach to a Saturated–Unsaturated Zone Model

Rendani Vele Makahane and Abdon Atangana
University of the Free State, Bloemfontein, South Africa

CONTENTS

6.1 INTRODUCTION

Recently, a new concept that brings us much closer to the reality of complex systems was suggested by Atangana and Bonyah. This new approach is a combination of fractional differentiation and stochastic modeling. These two important approaches have been used independently in previous years and have achieved great success. The differential operators have been used to capture the non-Markovian process; meanwhile, the stochastic approach has been used to capture heterogeneity in a closed system for a Markovian process (Atangana & Bonyah, 2019). The concept of differentiation is used to construct mathematical formulas called differential equations with constant coefficients, where the differential operators can be local, non-local with the singular kernel, local with non-singular and non-local with the non-singular kernel (Atangana & Araz, 2019; Atangana & Bonyah, 2019). However, these approaches have failed to capture some statistical settings of nature for larger space and time. Contrarily, the stochastic approach has been used in many situations to capture statistical setting by converting all parameter inputs to distributions which allow for random variation in all input parameters over a given period (Atangana & Alqahtani, 2019). None

DOI: 10.1201/9781003266266-6

the less, this approach is unable to capture long-range behavior, fading memory and heterogeneity linked with fracture network. Combining these two approaches can help us to capture physical problems appearing in nature, with long-range, fading memory and statistical setting for larger space and time (Atangana & Bonyah, 2019). In this chapter, we aim to apply this approach to the 1-d saturated–unsaturated sub-surface flow equation. We shall illustrate this approach by combining the statistical settings with the Caputo fractional derivative, Caputo–Fabrizio fractional derivative, and Atangana–Baleanu fractional derivative.

6.2   APPLICATION OF THE STOCHASTIC APPROACH

The equation is the combination of Darcy's law and the continuity law where it describes the continuous flow through the saturated–unsaturated medium (Pelka, 1983; List and Radu, 2015). The 1-d saturated–unsaturated groundwater flow equation is given by:

$$\left[S_S S_a\left(\psi\right) + C\left(\psi\right)\right]\frac{\partial\psi}{\partial t} = \frac{\partial}{\partial z}\left[K_z\left(\psi\right)\left(\frac{\partial\psi}{\partial z} - 1\right)\right] \tag{6.1}$$

where:

ψ = pressure head

n = porosity

S_S = specific storage of the soil

S_a = saturation of the aqueous phase $= \dfrac{\theta}{n}$

$C(\psi)$ = capillary capacity of soil $= d\theta/d\psi$

$q(t)$ = groundwater flux $= -K_z\left(\psi\right)\left(\dfrac{\partial\psi}{\partial z} - 1\right)$

K_z = hydraulic conductivity

z = vertical coordinate.

The above equation has the soil hydraulic conductivity function, $K(\psi)$ and the soil water content function, $\theta(\psi)$ needed to solve for water movement. One of the popular models is based on the equations of Brooks and Corey (Allepalli & Govindaraju, 2009):

$$K\left(\psi\right) = K_s\left(\frac{\psi_b}{\psi}\right)^{2+3\lambda} \qquad\qquad \psi \leq 0 \tag{6.2}$$

$$K\left(\psi\right) = K_s \qquad\qquad \psi > 0 \tag{6.3}$$

$$\theta\left(\psi\right) = \theta_0 + \left(\theta_s - \theta_0\right)\left(\frac{\psi_b}{\psi}\right)^{\lambda} \qquad\qquad \psi \leq 0 \tag{6.4}$$

where:

ψ = pressure head

ψ_b = air entry suction pressure head

$\theta(\psi)$ = soil water content

θ_s = saturated water content

θ_0 = residual water content

λ = pore size distribution index

K_s = saturated hydraulic conductivity.

By substituting the above equations into Equation (6.10), we obtain the following classical equations for the saturated and unsaturated zones, respectively:

$$\frac{\partial \psi}{\partial t} = \frac{K_s}{S_s} \frac{\partial^2 \psi}{\partial z^2} \tag{6.5}$$

and

$$\frac{\partial \theta (\psi)}{\partial t} = \frac{\partial}{\partial z}\left[K_z(\psi)\left(\frac{\partial \psi}{\partial z} - 1\right)\right] \tag{6.6a}$$

Using the Brooks and Corey equations, the above equation (6.6a) becomes:

$$-(\theta_s - \theta_0)\psi_b^{\lambda} \frac{\partial \psi}{\partial t}\frac{\lambda}{\psi_i^{\lambda+1}} = K_s \frac{\partial}{\partial z}\left(\frac{\psi_b}{\psi}\right)^{2+3\lambda}\frac{\partial \psi}{\partial z} + K_s\left(\frac{\psi_b}{\psi}\right)^{2+3\lambda}\frac{\partial^2 \psi}{\partial z^2} - K_s\frac{\partial}{\partial z}\left(\frac{\psi_b}{\psi}\right)^{2+3\lambda} \tag{6.6b}$$

Equation (6.1) has two constant coefficients: the hydraulic conductivity K_s and the specific storage S_s of the porous media. To convert these constants into distributions, we need to find the mean and the variance of the two constants.

6.2.1 THE MEAN AND VARIANCE OF THE HYDRAULIC CONDUCTIVITY

The mean for the hydraulic conductivity K_s is given by:

$$\bar{K}_s = \frac{\sum_{j=0}^{n} K_{sj}}{n}. \tag{6.7}$$

The variance is given by:

$$\sigma^2 = \frac{\sum_{j=0}^{n}\left(K_{sj} - \bar{K}_s\right)^2}{(n-1)} \tag{6.8}$$

Now we make use of the log-normal distribution to convert the constant-coefficient K_s into distribution, we convert the parameter $\hat{K}_s$ to:

$$\hat{K}_s = \bar{K}_{sj} + \gamma \log N\left(\bar{K}_{sj}, \sigma^2\right) \tag{6.9}$$

where, n to the number of data points, K_{sj} is each of the values of the data, σ^2 is the variance, and $\overline{K}_s$ is the mean of K_{sj}, γ the stochastic constant, and $\log N$ is the log-normal distribution given as:

$$fx(\sigma) = \frac{1}{x}\frac{1}{\sigma\sqrt{2\pi}}\exp\left[-\frac{\left(lnx - \bar{a}_j\right)}{2\sqrt{2}}\right]. \tag{6.10}$$

6.2.2 THE MEAN AND VARIANCE OF THE SPECIFIC STORAGE

The mean for the specific storage S_s is given by:

$$\overline{S}_s = \frac{\sum_{j=0}^{n} S_{sj}}{n} \tag{6.11}$$

The variance is given by:

$$\sigma^2 = \frac{\sum_{j=0}^{n} \left(S_{sj} - \overline{S}_s\right)^2}{(n-1)} \tag{6.12}$$

where, n = to the number of data points, S_{sj} = each of the values of the data, and $\overline{S}_s$ = the mean of S_{sj}. Now we make use of the log-normal distribution to convert the constant-coefficient S_s into distribution, we convert the parameter $\hat{S}_s$ to:

$$\hat{S}_s = \overline{S}_{sj} + \gamma \log N\left(\overline{S}_{sj}, \sigma^2\right) \tag{6.13}$$

where, n to the number of data points, K_{sj} is each of the values of the data, σ^2 is the variance, and K_s is the mean of K_{sj}, γ is the stochastic constant, and $\log N$ is the log-normal distribution given as:

$$fx(\sigma) = \frac{1}{x} \frac{1}{\sigma\sqrt{2\pi}} \exp\left[-\frac{\left(lnx - \overline{a}_j\right)}{2\sqrt{2}}\right] \tag{6.14}$$

6.2.3 The Stochastic 1-D Saturated–Unsaturated Groundwater Flow Equation

Now introducing the stochastic approach to the 1-D unsaturated–saturated zone equation (6.1), the saturated equation becomes:

$$\frac{\partial \psi}{\partial t} = \frac{\hat{K}_s}{\hat{S}_s} \frac{\partial^2 \psi}{\partial z^2} \tag{6.15}$$

The stochastic unsaturated zone equation is given by:

$$\frac{\partial \psi}{\partial t} = \frac{1}{-(\theta_s - \theta_0)\psi_b^\lambda \frac{\lambda}{\psi^{\lambda+1}}} \left(\hat{K}_s \frac{\partial}{\partial z}\left(\frac{\psi_b}{\psi}\right)^{2+3\lambda} \frac{\partial \psi}{\partial z} + \hat{K}_s\left(\frac{\psi_b}{\psi}\right)^{2+3\lambda} \frac{\partial^2 \psi}{\partial z^2} - \hat{K}_s \frac{\partial}{\partial z}\left(\frac{\psi_b}{\psi}\right)^{2+3\lambda}\right) \tag{6.16}$$

6.3 APPLICATION OF THE FRACTIONAL-STOCHASTIC APPROACH

With the benefits and application of the power law, exponential law, the Mittag-Leffler function, and the stochastic approach, we convert the above classical models into a stochastic differential equation, then perform a stability check for each equation where applicable.

6.3.1 Stochastic Differential Equation Using the Caputo Fractional Derivative

In this section, numerical solutions for the 1-d saturated–unsaturated groundwater flow equation are obtained using three different discretization methods and replacing the time derivative of Equation (6.15) with the Caputo fractional derivative, respectively. The Caputo fractional derivative is based on the power law. The power law has attracted a lot of attention because of its ability to describe material moving in singular media and its long-range memory (Ghoshdastidar and Dukkipati, 2013). The following equation gives the stochastic differential equation in the Caputo sense:

$$f(x, \sigma) = \frac{\hat{K}_s}{\hat{S}_s} \frac{\partial^2 \psi}{\partial z^2} \tag{6.17}$$

and

$$
{}_0^C D^\alpha \psi(z,t) = \frac{1}{-(\theta_s - \theta_0)\psi_b^\lambda \frac{\lambda}{\psi^{\lambda+1}}} \left(\hat{K}_s \frac{\partial}{\partial z}\left(\frac{\psi_b}{\psi}\right)^{2+3\lambda} \frac{\partial \psi}{\partial z} + \hat{K}_s \left(\frac{\psi_b}{\psi}\right)^{2+3\lambda} \cdot \frac{\partial^2 \psi}{\partial z^2} - \hat{K}_s \frac{\partial}{\partial z}\left(\frac{\psi_b}{\psi}\right)^{2+3\lambda} \right)
$$

(6.18)

6.3.1.1 Explicit Forward Euler Method

Applying the explicit forward Euler method on the right-hand side of Equation (6.17), we obtain the following stochastic differential equation for the saturated zone:

$$
\frac{(\Delta t)^{-\alpha}}{\Gamma(2-\alpha)} \sum_{j=0}^{n} \left(\psi_i^{j+1} - \psi_i^j\right)\left((n-j)^\alpha - (n-j+1)^\alpha\right) = \frac{\hat{K}_s}{\hat{S}_s}\left[\frac{\psi_{i+1}^n - 2\psi_i^n + \psi_{i-1}^n}{(\Delta z)^2}\right]
$$

(6.19)

Before we can perform stability analysis for the above numerical solution, we need to simplify the equation. To do that we make use of the following parameters:

$$
a = \frac{(\Delta t)^{-\alpha}}{\Gamma(2-\alpha)},
$$

$$
b = \frac{\hat{K}_s}{\hat{S}_s (\Delta z)^2}
$$

$$
R_{n,j}^\alpha = \left((n-j)^\alpha - (n-j-1)^\alpha\right)
$$

Now we can rewrite the equation with parameters:

$$
\sum_{j=0}^{n} \left(\psi_i^{j+1} - \psi_i^j\right) a R_{n,j}^\alpha = b\left(\psi_{i+1}^n - 2\psi_i^n + \psi_{i-1}^n\right)
$$

(6.20)

The above equation is simplified into:

$$
\left(\psi_i^{n+1} - \psi_i^n\right) a R_{n,n}^\alpha + \sum_{j=0}^{n-1} \left(\psi_i^{j+1} - \psi_i^j\right) a R_{n,j}^\alpha = b\left(\psi_{i+1}^n - 2\psi_i^n + \psi_{i-1}^n\right)
$$

(6.21)

Now we have this equation for a numerical scheme:

$$
\psi_i^{n+1} a R_{n,n}^\alpha = \psi_i^n \left(a R_{n,n}^\alpha - 2b\right) + b\left(\psi_{i+1}^n + \psi_{i-1}^n\right) - \sum_{j=0}^{n-1} \left(\psi_i^{j+1} - \psi_i^j\right) a R_{n,j}^\alpha
$$

(6.22)

For stability analysis, we assume that:

$$
\psi_i^n = \sigma_n e^{ik_m z}
$$

(6.23)

Substituting the above assumption into Equation (6.23) we obtain the following:

$$\sigma^{n+1}e^{ik_mz}aR_{n,n}^{\alpha} = \sigma_n e^{ik_mz}\left(aR_{n,n}^{\alpha} - 2b\right) + be^{ik_mz}\left(\sigma_n e^{ik_m\Delta z} + \sigma_n e^{ik_m\Delta z}\right) - \sum_{j=0}^{n-1}\left(\sigma_{j+1} - \sigma_j\right)e^{ik_mz}aR_{n,j}^{\alpha} \quad (6.24)$$

If we simplify the above equation, we obtain the following:

$$\sigma_{n+1}aR_{n,n}^{\alpha} = \sigma_n\left[aR_{n,n}^{\alpha} - 2b + b\left(e^{ik_m\Delta z} + e^{-ik_m\Delta z}\right)\right] - \sum_{j=0}^{n-1}\left(\sigma_{j+1} - \sigma_j\right)aR_{n,j}^{\alpha} \quad (6.25)$$

$$\sigma_{n+1}aR_{n,n}^{\alpha} = \sigma_n\left[aR_{n,n}^{\alpha} + 2b\left(\frac{e^{ik_m\Delta z} + e^{-ik_m\Delta z}}{2} - 1\right)\right] - \sum_{j=0}^{n-1}\left(\sigma_{j+1} - \sigma_j\right)aR_{n,j}^{\alpha} \quad (6.26)$$

$$\sigma_{n+1}aR_{n,n}^{\alpha} = \sigma_n\left[aR_{n,n}^{\alpha} + 2b\left(\cos\left(k_m\Delta z\right) - 1\right)\right] - \sum_{j=0}^{n-1}\left(\sigma_{j+1} - \sigma_j\right)aR_{n,j}^{\alpha} \quad (6.27)$$

Using the double angle for cos we can simplify the equation into:

$$\sigma_{n+1}aR_{n,n}^{\alpha} = \sigma_n\left[aR_{n,n}^{\alpha} + 2b\left(1 - 2\sin^2\left(\frac{k_m\Delta z}{2}\right) - 1\right)\right] - \sum_{j=0}^{n-1}\left(\sigma_{j+1} - \sigma_j\right)aR_{n,j}^{\alpha} \quad (6.28)$$

$$\sigma_{n+1}aR_{n,n}^{\alpha} = \sigma_n\left[aR_{n,n}^{\alpha} - 4b\sin^2\left(\frac{k_m\Delta z}{2}\right)\right] - \sum_{j=0}^{n-1}\left(\sigma_{j+1} - \sigma_j\right)aR_{n,j}^{\alpha} \quad (6.29)$$

If $n = 0$ the above equation becomes:

$$a\sigma_1 R_{0,n}^{\alpha} = \sigma_0\left[aR_{0,n}^{\alpha} - 4b\sin^2\left(\frac{k_m\Delta z}{2}\right)\right] \quad (6.30)$$

$$\left|\frac{\sigma_1}{\sigma_0}\right| < 1 \rightarrow \left|\frac{\sigma_1}{\sigma_0}\right| = \left|\frac{aR_{0,n}^{\alpha} - 4b\sin^2\left(\frac{k_m\Delta z}{2}\right)}{aR_{0,n}^{\alpha}}\right| < 1 \quad (6.31)$$

$$aR_{0,n}^{\alpha} - 4b\sin^2\left(\frac{k_m\Delta z}{2}\right) < aR_{0,n}^{\alpha} \quad (6.32)$$

$$0 < 4b\sin^2\left(\frac{k_m\Delta z}{2}\right) \quad (6.33)$$

$$0 < 4\frac{\hat{K}_s}{\hat{S}_s\left(\Delta z\right)^2}\sin^2\left(\frac{k_m\Delta z}{2}\right) \quad (6.34)$$

We assume that $\forall n > 0$:

$$\left| \frac{\sigma_n}{\sigma_0} \right| < 1$$

Using the above assumption, we shall prove that:

$$\left| \frac{\sigma_{n+1}}{\sigma_0} \right| < 1$$

$$\sigma_{n+1} a R_{n,n}^\alpha = \sigma_n \left[a R_{n,n}^\alpha \quad 4b \sin^2 \left(\frac{k_m \Delta z}{2} \right) \right] - \sum_{j=0}^{n-1} \left(\sigma_{j+1} - \sigma_j \right) a R_{n,j}^\alpha \tag{6.35}$$

$$\sigma_{n+1} = \frac{\sigma_n \left[a R_{n,n}^\alpha - 4b \sin^2 \left(\frac{k_m \Delta z}{2} \right) \right]}{a R_{n,n}^\alpha} - \frac{\sum_{j=0}^{n-1} \sigma_{n-j} a R_{n,j}^\alpha}{a R_{n,n}^\alpha} \tag{6.36}$$

$$\left| \sigma_{n+1} \right| \le \left| \sigma_n \right| \left| \frac{\left[a R_{n,n}^\alpha - 4b \sin^2 \left(\frac{k_m \Delta z}{2} \right) \right]}{a R_{n,n}^\alpha} \right| - \frac{\sum_{j=0}^{n-1} \left| \sigma_{n-j} \right| a R_{n,j}^\alpha}{a R_{n,n}^\alpha} \tag{6.37}$$

Using the recursive hypothesis:

$$\left| \sigma_{n+1} \right| < \left| \sigma_0 \right| \left[\frac{a R_{n,n}^\alpha - 4b \sin^2 \left(\frac{k_m \Delta z}{2} \right)}{a R_{n,n}^\alpha} \right] + \frac{\sum_{j=0}^{n-1} \left| \sigma_0 \right| a R_{n,j}^\alpha}{a R_{n,n}^\alpha} \tag{6.38}$$

$$\left| \sigma_{n+1} \right| < \left| \sigma_0 \right| \left[\frac{\left[a R_{n,n}^\alpha - 4b \sin^2 \left(\frac{k_m \Delta z}{2} \right) \right]}{a R_{n,n}^\alpha} + \frac{\sum_{j=0}^{n-1} a R_{n,j}^\alpha}{a R_{n,n}^\alpha} \right] \tag{6.39}$$

$$\frac{\left| \sigma_{n+1} \right|}{\left| \sigma_0 \right|} < 1$$

$$\left| \frac{a R_{n,n}^\alpha - 4b \sin^2 \left(\frac{k_m \Delta z}{2} \right) + \sum_{j=0}^{n-1} a R_{n,j}^\alpha}{a R_{n,n}^\alpha} \right| < 1 \tag{6.40}$$

$$a R_{n,n}^\alpha - 4b \sin^2 \left(\frac{k_m \Delta z}{2} \right) + \sum_{j=0}^{n-1} a R_{n,j}^\alpha < a R_{n,n}^\alpha \tag{6.41}$$

The stability condition is given as:

$$\sum_{j=0}^{n-1} aR_{n,j}^{\alpha} < 4b\sin^2\left(\frac{k_m\Delta z}{2}\right) \tag{6.42}$$

6.3.1.2 Implicit Backward Euler Method

Applying the implicit backward Euler method on the right-hand side of Equation (6.17), we obtain the following stochastic differential equation for the saturated zone:

$$\frac{(\Delta t)^{-\alpha}}{\Gamma(2-\alpha)}\sum_{j=0}^{n}\left(\psi_i^{j+1}-\psi_i^j\right)\left((n-j)^{\alpha}-(n-j-1)^{\alpha}\right) = \frac{\hat{K}_s}{\hat{S}_s}\left[\frac{\psi_{i+1}^{n+1}-2\psi_i^{n+1}+\psi_{i-1}^{n+1}}{(\Delta z)^2}\right] \tag{6.43}$$

Before we can perform stability analysis for the above implicit backward Euler method, we need to simplify the equation. To do that, we use the following parameters:

$$a = \frac{(\Delta t)^{-\alpha}}{\Gamma(2-\alpha)},$$

$$b = \frac{\hat{K}_s}{\hat{S}_s(\Delta z)^2}$$

$$R_{n,j}^{\alpha} = \left((n-j)^{\alpha}-(n-j-1)^{\alpha}\right)$$

Now we can rewrite the equation with parameters:

$$\sum_{j=0}^{n}\left(\psi_i^{j+1}-\psi_i^j\right)aR_{n,j}^{\alpha} = b\left(\psi_{i+1}^{n+1}-2\psi_i^{n+1}+\psi_{i-1}^{n+1}\right) \tag{6.44}$$

$$\left(\psi_i^{n+1}-\psi_i^n\right)aR_{n,n}^{\alpha} + \sum_{j=0}^{n-1}\left(\psi_i^{j+1}-\psi_i^j\right)aR_{n,j}^{\alpha} = b\left(\psi_{i+1}^{n+1}-2\psi_i^{n+1}+\psi_{i-1}^{n+1}\right) \tag{6.45}$$

Now we have the above equation as a numerical scheme:

$$\left(\psi_i^{n+1}-\psi_i^n\right)aR_{n,n}^{\alpha} = b\left(\psi_{i+1}^{n+1}-2\psi_i^{n+1}+\psi_{i-1}^{n+1}\right) - \sum_{j=0}^{n-1}\left(\psi_i^{j+1}-\psi_i^j\right)aR_{n,j}^{\alpha} \tag{6.46}$$

Which can be further simplified:

$$-\psi_i^n aR_{n,n}^{\alpha} = -\psi_i^{n+1}\left(aR_{n,j}^{\alpha}+2b\right) + b\left(\psi_{i+1}^{n+1}+\psi_{i-1}^{n+1}\right) - \sum_{j=0}^{n-1}\left(\psi_i^{j+1}-\psi_i^j\right)aR_{n,j}^{\alpha} \tag{6.47}$$

For stability analysis, we assume that:

$$\psi_i^n = \sigma_n e^{ik_m z} \tag{6.48}$$

Substituting the above into Equation (6.48) we obtain the following:

$$-\sigma_n e^{ik_m z} aR_{n,n}^{\alpha} = -\sigma_{n+1} e^{ik_m z} \left(aR_{n,n}^{\alpha} + 2b \right) + b \left(\sigma_{n+1} e^{ik_m(z+\Delta z)} + \sigma_{n+1} e^{ik_m(z-\Delta z)} \right) - \sum_{j=0}^{n-1} \left(\sigma_{j+1} - \sigma_j \right) e^{ik_m z} aR_{n,j}^{\alpha}$$

(6.49)

$$-\sigma_n e^{ik_m z} aR_{n,n}^{\alpha} = -\sigma_{n+1} e^{ik_m z} \left(aR_{n,n}^{\alpha} + 2b \right) + b\sigma_{n+1} e^{ik_m z} \left(e^{ik_m \Delta z} + e^{-ik_m \Delta z} \right) - \sum_{j=0}^{n-1} \left(\sigma_{j+1} - \sigma_j \right) e^{ik_m z} aR_{n,j}^{\alpha}$$

(6.50)

$$-\sigma_n e^{ik_m z} aR_{n,n}^{\alpha} = -\sigma_{n+1} e^{ik_m z} \left(aR_{n,n}^{\alpha} + 2b - b \left(e^{ik_m \Delta z} + e^{-ik_m \Delta z} \right) \right) - \sum_{j=0}^{n-1} \left(\sigma_{j+1} - \sigma_j \right) e^{ik_m z} aR_{n,j}^{\alpha} \quad (6.51)$$

$$-\sigma_n e^{ik_m z} aR_{n,n}^{\alpha} = -\sigma_{n+1} e^{ik_m z} \left(aR_{n,n}^{\alpha} + 2b - 2b \left(\frac{e^{ik_m \Delta z} + e^{-ik_m \Delta z}}{2} \right) \right) - \sum_{j=0}^{n-1} \left(\sigma_{j+1} - \sigma_j \right) e^{ik_m z} aR_{n,j}^{\alpha} \quad (6.52)$$

$$-\sigma_n e^{ik_m z} aR_{n,n}^{\alpha} = -\sigma_{n+1} e^{ik_m z} \left(aR_{n,n}^{\alpha} + 2b - 2b \left(\cos(k_m \Delta z) \right) \right) - \sum_{j=0}^{n-1} \left(\sigma_{j+1} - \sigma_j \right) e^{ik_m z} aR_{n,j}^{\alpha} \quad (6.53)$$

$$-\sigma_n e^{ik_m z} aR_{n,n}^{\alpha} = -\sigma_{n+1} e^{ik_m z} \left(aR_{n,n}^{\alpha} + 2b - 2b \left(1 - 2\sin^2\left(\frac{k_m \Delta z}{2} \right) \right) \right) - \sum_{j=0}^{n-1} \left(\sigma_{j+1} - \sigma_j \right) e^{ik_m z} aR_{n,j}^{\alpha} \quad (6.54)$$

$$-\sigma_n aR_{n,n}^{u} = -\sigma_{n+1} \left(aR_{n,n}^{u} + 4b\sin^2\left(\frac{k_m \Lambda z}{2} \right) \right) - \sum_{j=0}^{n-1} \left(\sigma_{j+1} - \sigma_j \right) aR_{n,j}^{\alpha} \quad (6.55)$$

If $n = 0$ then the above equation becomes:

$$\sigma_1 \left(aR_{0,n}^{\alpha} + 4b\sin^2\left(\frac{k_m \Delta z}{2} \right) \right) = \sigma_0 aR_{0,n}^{\alpha} \quad (6.56)$$

$$\left| \frac{\sigma_1}{\sigma_0} \right| < 1 \rightarrow \left| \frac{\sigma_1}{\sigma_0} \right| = \left| \frac{aR_{0,n}^{\alpha}}{aR_{0,n}^{\alpha} + 4b\sin^2\left(\frac{k_m \Delta z}{2} \right)} \right| < 1 \quad (6.57)$$

$$aR_{0,n}^{\alpha} < aR_{0,n}^{\alpha} + 4b\sin^2\left(\frac{k_m \Delta z}{2} \right) \quad (6.58)$$

$$0 < 4b\sin^2\left(\frac{k_m \Delta z}{2} \right) \quad (6.59)$$

We assume that $\forall n > 0$:

$$\left| \frac{\sigma_n}{\sigma_0} \right| < 1$$

Using the above assumption, we shall prove that:

$$\left| \frac{\sigma_{n+1}}{\sigma_0} \right| < 1$$

$$\sigma_{n+1}\left(aR_{n,n}^{\alpha} + 4b\sin^2\left(\frac{k_m \Delta z}{2} \right) \right) + \sum_{j=0}^{n-1}\left(\sigma_{j+1} - \sigma_j \right)aR_{n,j}^{\alpha} = \sigma_n aR_{n,n}^{\alpha} \tag{6.60}$$

$$\sigma_{n+1} = \frac{\sigma_n aR_{n,n}^{\alpha}}{\left[aR_{n,n}^{\alpha} + 4b\sin^2\left(\frac{k_m \Delta z}{2} \right) \right]} - \frac{\sum_{j=0}^{n-1}\sigma_{n-j}aR_{n,j}^{\alpha}}{aR_{n,n}^{\alpha} + 4b\sin^2\left(\frac{k_m \Delta z}{2} \right)} \tag{6.61}$$

$$\left| \sigma_{n+1} \right| \leq \left| \sigma_n \right|\left| \frac{aR_{n,n}^{\alpha}}{\left[aR_{n,n}^{\alpha} + 4b\sin^2\left(\frac{k_m \Delta z}{2} \right) \right]} \right| - \frac{\sum_{j=0}^{n-1}\left| \sigma_{n-j} \right|aR_{n,j}^{\alpha}}{aR_{n,n}^{\alpha} + 4b\sin^2\left(\frac{k_m \Delta z}{2} \right)} \tag{6.62}$$

Using the recursive hypothesis:

$$\left| \sigma_{n+1} \right| < \left| \sigma_0 \right|\left[\frac{aR_{n,n}^{\alpha}}{aR_{n,n}^{\alpha} + 4b\sin^2\left(\frac{k_m \Delta z}{2} \right)} \right] + \frac{\sum_{j=0}^{n-1}\left| \sigma_0 \right|aR_{n,j}^{\alpha}}{aR_{n,n}^{\alpha} + 4b\sin^2\left(\frac{k_m \Delta z}{2} \right)} \tag{6.63}$$

$$\left| \sigma_{n+1} \right| < \left| \sigma_0 \right|\left[\frac{aR_{n,n}^{\alpha}}{aR_{n,n}^{\alpha} + 4b\sin^2\left(\frac{k_m \Delta z}{2} \right)} + \frac{\sum_{j=0}^{n-1}aR_{n,j}^{\alpha}}{aR_{n,n}^{\alpha} + 4b\sin^2\left(\frac{k_m \Delta z}{2} \right)} \right] \tag{6.64}$$

$$\frac{\left| \sigma_{n+1} \right|}{\left| \sigma_0 \right|} < 1$$

$$\left| \frac{aR_{n,n}^{\alpha} + \sum_{j=0}^{n-1}aR_{n,j}^{\alpha}}{aR_{n,n}^{\alpha} + 4b\sin^2\left(\frac{k_m \Delta z}{2} \right)} \right| < 1 \tag{6.65}$$

$$aR_{n,n}^{\alpha} + \sum_{j=0}^{n-1} aR_{n,j}^{\alpha} < aR_{n,n}^{\alpha} + 4b\sin^2\left(\frac{k_m\Delta z}{2}\right) \tag{6.66}$$

The stability condition is given as:

$$\sum_{j=0}^{n-1} aR_{n,j}^{\alpha} < 4b\sin^2\left(\frac{k_m\Delta z}{2}\right) \tag{6.67}$$

6.3.1.3 Implicit Crank–Nicolson Method

We obtain other numerical solutions by applying the implicit Crank–Nicolson discretization on the right-hand side of Equation (6.17):

$$\frac{(\Delta t)^{-\alpha}}{\Gamma(2-\alpha)} \sum_{j=0}^{n} \left(\psi_i^{j+1} - \psi_i^{j}\right)\left((n-j)^{\alpha} - (n-j-1)^{\alpha}\right) = \frac{\hat{K}_s}{\hat{S}_s}\left[\frac{\psi_{i+1}^{n+1} - 2\psi_i^{n+1} + \psi_{i-1}^{n+1}}{2(\Delta z)^2} + \frac{\psi_{i+1}^{n} - 2\psi_i^{n} + \psi_{i-1}^{n}}{2(\Delta z)^2}\right]$$

$$\tag{6.68}$$

Before we can perform stability analysis for the above equation, we need to simplify the above equation by using the following parameters:

$$a = \frac{(\Delta t)^{-\alpha}}{\Gamma(2-\alpha)},$$

$$c = \frac{\hat{K}_s}{\hat{S}_s 2(\Delta z)^2}$$

$$R_{n,j}^{\alpha} = \left((n-j)^{\alpha} - (n-j-1)^{\alpha}\right)$$

Now we can rewrite the equation as:

$$aR_{n,j}^{\alpha} = c\left[\left(\psi_{i+1}^{n+1} - 2\psi_i^{n+1} + \psi_{i-1}^{n+1}\right) + \left(\psi_{i+1}^{n} - 2\psi_i^{n} + \psi_{i-1}^{n}\right)\right] \tag{6.69}$$

$$\sum_{j=0}^{n} \left(\psi_i^{j+1} - \psi_i^{j}\right) aR_{n,j}^{\alpha} = c\left[\left(\psi_{i+1}^{n+1} - 2\psi_i^{n+1} + \psi_{i-1}^{n+1}\right) + \left(\psi_{i+1}^{n} - 2\psi_i^{n} + \psi_{i-1}^{n}\right)\right] \tag{6.70}$$

$$\left(\psi_i^{n+1} - \psi_i^{n}\right) aR_{n,n}^{\alpha} + \sum_{j=0}^{n-1} \left(\psi_i^{j+1} - \psi_i^{j}\right) aR_{n,j}^{\alpha} = c\left[\left(\psi_{i+1}^{n+1} - 2\psi_i^{n+1} + \psi_{i-1}^{n+1}\right) + \left(\psi_{i+1}^{n} - 2\psi_i^{n} + \psi_{i-1}^{n}\right)\right] \tag{6.71}$$

$$\psi_i^{n+1} aR_{n,n}^{\alpha} = c\left(\psi_{i+1}^{n+1} - 2\psi_i^{n+1} + \psi_{i-1}^{n+1}\right) + c\left(\psi_{i+1}^{n} - 2\psi_i^{n} + \psi_{i-1}^{n}\right) + \psi_i^{n} aR_{n,n}^{\alpha} - \sum_{j=0}^{n-1} \left(\psi_i^{j+1} - \psi_i^{j}\right) aR_{n,j}^{\alpha} \tag{6.72}$$

For stability analysis, we assume that:

$$\psi_i^n = \sigma_n e^{ik_m z} \tag{6.73}$$

$$\begin{aligned}
\sigma_{n+1} e^{ik_m z} a R_{n,n}^{\alpha} &= c\left(\sigma_{n+1} e^{ik_m(z+\Delta z)} - 2\sigma_{n+1} e^{ik_m z} + \sigma_{n+1} e^{ik_m(z-\Delta z)}\right) \\
&\quad + c\left(\sigma_n e^{ik_m(z+\Delta z)} - 2\sigma_n e^{ik_m z} + \sigma_n e^{ik_m(z-\Delta z)}\right) \\
&\quad + \sigma_n e^{ik_m z} a R_{n,n}^{\alpha} - \sum_{j=0}^{n-1}\left(\sigma_{j+1} - \sigma_j\right) e^{ik_m z} a R_{n,j}^{\alpha}
\end{aligned} \tag{6.74}$$

$$\begin{aligned}
\sigma_{n+1} e^{ik_m z} a R_{n,n}^{\alpha} - c\sigma_{n+1} e^{ik_m z}\left(e^{ik_m \Delta z} - 2 + e^{-ik_m \Delta z}\right) &= c\sigma_n e^{ik_m z}\left(e^{ik_m \Delta z} + \sigma_n e^{-ik_m \Delta z}\right) \\
&\quad + \sigma_n e^{ik_m z}\left(a R_{n,n}^{\alpha} - 2c\right) - \sum_{j=0}^{n-1}\left(\sigma_{j+1} - \sigma_j\right) e^{ik_m z} a R_{n,j}^{\alpha}
\end{aligned} \tag{6.75}$$

$$\sigma_{n+1} a R_{n,n}^{\alpha} - c\sigma_{n+1}\left(e^{ik_m \Delta z} - 2 + e^{-ik_m \Delta z}\right) = c\sigma_n\left(e^{ik_m \Delta z} + e^{-ik_m \Delta z}\right) + \sigma_n\left(a R_{n,n}^{\alpha} - 2c\right) - \sum_{j=0}^{n-1}\left(\sigma_{j+1} - \sigma_j\right) a R_{n,j}^{\alpha} \tag{6.76}$$

$$\sigma_{n+1}\left(a R_{n,n}^{\alpha} - 2c\left(\frac{e^{ik_m \Delta z} + e^{-ik_m \Delta z}}{2} - 1\right)\right) = \sigma_n\left(a R_{n,n}^{\alpha} + 2c\left(\frac{e^{ik_m \Delta z} + e^{-ik_m \Delta z}}{2} - 1\right)\right) - \sum_{j=0}^{n-1}\left(\sigma_{j+1} - \sigma_j\right) a R_{n,j}^{\alpha} \tag{6.77}$$

$$\sigma_{n+1}\left(a R_{n,n}^{\alpha} - 2c\left(\cos\left(k_m \Delta z\right) - 1\right)\right) = \sigma_n\left(a R_{n,n}^{\alpha} + 2c\left(\cos\left(k_m \Delta z\right) - 1\right)\right) - \sum_{j=0}^{n-1}\left(\sigma_{j+1} - \sigma_j\right) a R_{n,j}^{\alpha} \tag{6.78}$$

If $n = 0$ the above equation becomes:

$$\sigma_1\left(a R_{n,n}^{\alpha} - 2c\left(\cos\left(k_m \Delta z\right) - 1\right)\right) = \sigma_0\left(a R_{n,n}^{\alpha} + 2c\left(\cos\left(k_m \Delta z\right) - 1\right)\right) \tag{6.79}$$

$$\left|\frac{\sigma_1}{\sigma_0}\right| < 1 \rightarrow \left|\frac{\sigma_1}{\sigma_0}\right| = \left|\frac{a R_{n,n}^{\alpha} + 2c\left(\cos\left(k_m \Delta z\right) - 1\right)}{a R_{n,n}^{\alpha} - 2c\left(\cos\left(k_m \Delta z\right) - 1\right)}\right| < 1 \tag{6.80}$$

$$a R_{n,n}^{\alpha} + 2c\left(\cos\left(ik_m \Delta z\right) - 1\right) < a R_{n,n}^{\alpha} - 2c\left(\cos\left(k_m \Delta z\right) - 1\right) \tag{6.81}$$

$$0 < -4c\left(\cos\left(k_m \Delta z\right) - 1\right) \tag{6.82}$$

$$4c\cos\left(k_m \Delta z\right) < 4c \tag{6.83}$$

$$\cos\left(k_m \Delta z\right) < 1 \tag{6.84}$$

We can write 1 as $\cos 2\pi$ such that:

$$\cos(k_m\Delta z) < \cos 2\pi \tag{6.85}$$

$$ik_m\Delta z < 2\pi \tag{6.86}$$

Therefore:

$$k_m < \frac{2\pi}{\Delta z} \tag{6.87}$$

We assume that $\forall n > 0$:

$$\left|\frac{\sigma_n}{\sigma_0}\right| < 1$$

Using the above assumption, we will prove that:

$$\left|\frac{\sigma_{n+1}}{\sigma_0}\right| < 1$$

$$\sigma_{n+1}\left(aR_{n,n}^\alpha - 2c\left(\cos(k_m\Delta z)-1\right)\right) = \sigma_n\left(aR_{n,n}^\alpha + 2c\left(\cos(k_m\Delta z)-1\right)\right) - \sum_{j=0}^{n-1}\sigma_{n-j}aR_{n,j}^\alpha \tag{6.88}$$

$$\sigma_{n+1} = \sigma_n\left[\frac{aR_{n,n}^\alpha + 2c\left(\cos(ik_m\Delta z)-1\right)}{aR_{n,n}^\alpha - 2c\left(\cos(k_m\Delta z)-1\right)}\right] - \frac{\sum_{j=0}^{n-1}\sigma_{n-j}aR_{n,j}^\alpha}{aR_{n,n}^\alpha - 2c\left(\cos(k_m\Delta z)-1\right)} \tag{6.89}$$

$$|\sigma_{n+1}| \leq |\sigma_n|\left|\frac{aR_{n,n}^\alpha + 2c\left(\cos(k_m\Delta z)-1\right)}{aR_{n,n}^\alpha - 2c\left(\cos(k_m\Delta z)-1\right)}\right| + \frac{\sum_{j=0}^{n-1}|\sigma_{n-j}|aR_{n,j}^\alpha}{aR_{n,n}^\alpha - 2c\left(\cos(k_m\Delta z)-1\right)} \tag{6.90}$$

Using the recursive hypothesis:

$$|\sigma_{n+1}| < |\sigma_0|\left|\frac{aR_{n,n}^\alpha + 2c\left(\cos(k_m\Delta z)-1\right)}{aR_{n,n}^\alpha - 2c\left(\cos(k_m\Delta z)-1\right)}\right| + \frac{\sum_{j=0}^{n-1}|\sigma_0|aR_{n,j}^\alpha}{aR_{n,n}^\alpha - 2c\left(\cos(k_m\Delta z)-1\right)} \tag{6.91}$$

$$|\sigma_{n+1}| < |\sigma_0|\left[\frac{aR_{n,n}^\alpha + 2c\left(\cos(k_m\Delta z)-1\right)}{aR_{n,n}^\alpha - 2c\left(\cos(k_m\Delta z)-1\right)} + \frac{\sum_{j=0}^{n-1}aR_{n,j}^\alpha}{aR_{n,n}^\alpha - 2c\left(\cos(k_m\Delta z)-1\right)}\right] \tag{6.92}$$

$$\left| \frac{aR_{n,n}^{\alpha} + 2c\left(\cos\left(k_m\Delta z\right)-1\right) + \sum_{j=0}^{n-1} aR_{n,j}^{\alpha}}{aR_{n,n}^{\alpha} - 2c\left(\cos\left(k_m\Delta z\right)-1\right)} \right| < 1 \tag{6.93}$$

$$aR_{n,n}^{\alpha} + 2c\left(\cos\left(km\Delta z\right)-1\right) + \sum_{j=0}^{n-1} aR_{n,j}^{\alpha} < aR_{n,n}^{\alpha} - 2c\left(\cos\left(k_m\Delta z\right)-1\right) \tag{6.94}$$

$$\sum_{j=0}^{n-1} aR_{n,j}^{\alpha} < -4c\left(\cos\left(k_m\Delta z\right)-1\right) \tag{6.95}$$

The stability condition is given as:

$$4c\cos\left(k_m\Delta z\right) + \sum_{j=0}^{n-1} aR_{n,j}^{\alpha} < 4c \tag{6.96}$$

The above stability condition is simplified into:

$$4c\cos\left(k_m\Delta z\right) < 4c - \sum_{j=0}^{n-1} aR_{n,j}^{\alpha} \tag{6.97}$$

$$k_m\Delta z < \cos^{-1}\left(1 - \frac{\sum_{j=0}^{n-1} aR_{n,j}^{\alpha}}{4c}\right) \tag{6.98}$$

6.3.1.4 New Model of the Unsaturated Zone in the Caputo Sense

Now we consider Equation (6.18) to provide the fractional-stochastic numerical solution for the unsaturated zone in the Caputo sense:

$$\frac{(\Delta t)^{-\alpha}}{\Gamma(2-\alpha)} \sum_{j=0}^{n} \left(\psi_i^{j+1} - \psi_i^{j}\right)\left((n-j)^{\alpha} - (n-j-1)^{\alpha}\right) = \frac{1}{-(\theta_s - \theta_0)\psi_b^{\lambda}\dfrac{\lambda}{\psi_i^{\lambda+1}}}$$

$$\times \left(\begin{array}{c} K_s \dfrac{\left(\dfrac{\psi_b}{\psi_{i+1}^{n+1}}\right)^{2+3\lambda} - \left(\dfrac{\psi_b}{\psi_i^{n+1}}\right)^{2+3\lambda}}{\Delta z} \dfrac{\psi_{i+1}^{n+1} - \psi_i^{n+1}}{\Delta z} + K_s\left(\dfrac{\psi_b}{\psi_i^{n+1}}\right)^{2+3\lambda} \\ \times \dfrac{\psi_{i+1}^{n+1} - 2\psi_i^{n+1} + \psi_{i-1}^{n+1}}{(\Delta z)^2} - K_s\dfrac{\left(\dfrac{\psi_b}{\psi_{i+1}^{n+1}}\right)^{2+3\lambda} - \left(\dfrac{\psi_b}{\psi_i^{n+1}}\right)^{2+3\lambda}}{\Delta z} \end{array} \right) \tag{6.99}$$

To make further simplification on the above equation, let us say:

$$\wp = -\left(\theta_s - \theta_0\right)\psi_b^\lambda \frac{\lambda}{\psi_i^{\lambda+1}}$$

$$a = \frac{\left(\Delta t\right)^{-\alpha}}{\Gamma\left(2-\alpha\right)}$$

$$b = \frac{K_s}{\Delta z}$$

$$c = \frac{1}{\Delta z}$$

$$d = \frac{1}{\left(\Delta z\right)^2}$$

$$R_{n,j}^\alpha = \left(\left(n-j\right)^\alpha - \left(n-j-1\right)^\alpha\right)$$

Now we rewrite the equation as:

$$\sum_{j=0}^{n}\left(\psi_i^{j+1} - \psi_i^{j}\right)aR_{n,j}^\alpha = \frac{1}{\wp}\left(\begin{array}{l} b\left(\left(\dfrac{\psi_b}{\psi_{i+1}^{n+1}}\right)^{2+3\lambda} - \left(\dfrac{\psi_b}{\psi_i^{n+1}}\right)^{2+3\lambda}\right)c\left(\psi_{i+1}^{n+1} - \psi_i^{n+1}\right) + K_s\left(\dfrac{\psi_b}{\psi_i^{n+1}}\right)^{2+3\lambda} \\ \times d\left(\psi_{i+1}^{n+1} - 2\psi_i^{n+1} + \psi_{i-1}^{n+1}\right) - b\left(\left(\dfrac{\psi_b}{\psi_{i+1}^{n+1}}\right)^{2+3\lambda} - \left(\dfrac{\psi_b}{\psi_i^{n+1}}\right)^{2+3\lambda}\right) \end{array}\right) \quad (6.100)$$

The above numerical scheme is highly non-linear; therefore it is not possible to perform a stability analysis.

6.3.2 STOCHASTIC DIFFERENTIAL EQUATION USING THE CAPUTO–FABRIZIO FRACTIONAL DERIVATIVE

For another numerical solution, the constant coefficients are converted to distributions. To introduce the fading memory and exponential decay law without singular kernel, the time derivative on our saturated equation (6.15) is replaced by the Caputo–Fabrizio fractional derivative, such that:

$$^{CF}_0 D^\alpha \psi\left(z,t\right) = \frac{\hat{K}_s}{\hat{S}_s}\frac{\partial^2\psi}{\partial z^2} \quad (6.101)$$

where:

$$^{CF}_0 D^\alpha \psi\left(z,t\right) = f\left(z,t,\psi\left(z,t\right)\right) \quad (6.102)$$

We apply the Caputo–Fabrizio integral to the above equation:

$$\psi(z,t) = \psi(z,0) + \frac{1-\alpha}{M(\alpha)} f\big(z,t,\psi(z,t)\big) + \frac{\alpha}{M(\alpha)} \int_0^t f\big(z,\tau\,\psi(z,\tau)\big)\,d\tau \qquad (6.103)$$

When $t = t_{n+1}$ we have the following equation:

$$\psi(z_i,t_{n+1}) = \psi(z_i,0) + \frac{1-\alpha}{M(\alpha)} f\big(z_i,t_n,\psi(z_i,t_n)\big) + \frac{\alpha}{M(\alpha)} \int_0^{t_{n+1}} f\big(z_i,\tau\,\psi(z_i,\tau)\big)\,d\tau \qquad (6.104)$$

When $t = t_n$:

$$\psi(z_i,t_n) = \psi(z_i,0) + \frac{1-\alpha}{M(\alpha)} f\big(z_i,t_{n-1},\psi(z_i,t_{n-1})\big) + \frac{\alpha}{M(\alpha)} \int_0^{t_n} f\big(z_i,\tau\,\psi(z_i,\tau)\big)\,d\tau \qquad (6.105)$$

Subtracting the above solution for t_n, from t_{n+1} we obtain the following:

$$\psi_i^{n+1} - \psi_i^n = \frac{1-\alpha}{M(\alpha)}\Big[f\big(z_i,t_n,\psi_i^n\big) - f\big(z_i,t_{n-1},\psi_i^{n-1}\big) \Big] + \frac{\alpha}{M(\alpha)} \int_{t_n}^{t_{n+1}} f\big(z_i,\tau,\psi(x_i,\tau)\big)\,d\tau \quad (6.106)$$

$$\psi_i^{n+1} = \psi_i^n + \frac{1-\alpha}{M(\alpha)}\Big[f\big(z_i,t_n,\psi_i^n\big) - f\big(z_i,t_{n-1},\psi_i^{n-1}\big) \Big] + \frac{\alpha}{M(\alpha)}\Big[\frac{3}{2}\Delta t f\big(z_i,t_n,\psi_i^n\big) - \frac{\Delta t}{2} f\big(z_i,t_{n-1},\psi_i^{n-1}\big) \Big]$$

$$(6.107)$$

where:

$$F\big(z_i,t_j,\psi_i^n\big) = \frac{\hat{K}_s}{\hat{S}_s} \frac{\psi_{i+1}^n - 2\psi_i^n + \psi_{i-1}^n}{(\Delta z)^2} \qquad (6.108)$$

$$F\big(z_i,t_{j-1},\psi_i^{n-1}\big) = \frac{\hat{K}_s}{\hat{S}_s} \frac{\psi_{i+1}^{n-1} - 2\psi_i^{n-1} + \psi_{i-1}^{n-1}}{(\Delta z)^2} \qquad (6.109)$$

We substitute the above Equations (6.108) and (6.109) into Equation (6.107) to obtain the final numerical solution for the saturated groundwater flow equation. The stochastic differential equation for the saturated zone in Caputo–Fabrizio sense is given by:

$$\psi_i^{n+1} = \psi_i^n + \frac{1-\alpha}{M(\alpha)}\left[\frac{\hat{K}_s}{\hat{S}_s} \frac{\psi_{i+1}^n - 2\psi_i^n + \psi_{i-1}^n}{(\Delta z)^2} - \frac{\hat{K}_s}{\hat{S}_s} \frac{\psi_{i+1}^{n-1} - 2\psi_i^{n-1} + \psi_{i-1}^{n-1}}{(\Delta z)^2} \right]$$

$$+ \frac{\alpha}{M(\alpha)}\left[\frac{3\Delta t}{2}\left(\frac{\hat{K}_s}{\hat{S}_s} \frac{\psi_{i+1}^n - 2\psi_i^n + \psi_{i-1}^n}{(\Delta z)^2} \right) - \frac{\Delta t}{2}\left(\frac{\hat{K}_s}{\hat{S}_s} \frac{\psi_{i+1}^{n-1} - 2\psi_i^{n-1} + \psi_{i-1}^{n-1}}{(\Delta z)^2} \right) \right] \qquad (6.110)$$

To check the stability of the above numerical solution, we assume a Fourier expansion in space of:

$$\psi_i^n = \sigma_n e^{ik_m z} \qquad (6.111)$$

We also assume the following parameters:

$$a = \frac{1-\alpha}{M(\alpha)},$$

$$b = \frac{\alpha}{M(\alpha)},$$

$$c = \frac{\hat{K}_s}{\hat{S}_s (\Delta z)^2}$$

$$d = \frac{\Delta t}{2}.$$

Then our equation becomes:

$$\sigma_{n+1} e^{ik_m z} = \sigma_n e^{ik_m z} + a \left[\begin{array}{l} c\left(\sigma^n e^{ik_m(z+\Delta z)} - 2\sigma^n e^{ik_m z} + \sigma^n e^{ik_m(z-\Delta z)}\right) \\ -c\left(\sigma_{n-1} e^{ik_m(z+\Delta z)} - 2\sigma_{n-1} e^{ik_m z} + \sigma_{n-1} e^{ik_m(z-\Delta z)}\right) \end{array} \right]$$
$$+ b \left[\begin{array}{l} 3dc\left(\sigma^n e^{ik_m(z+\Delta z)} - 2\sigma^n e^{ik_m z} + \sigma^n e^{ik_m(z-\Delta z)}\right) \\ -dc\left(\sigma^{n-1} e^{ik_m(z+\Delta z)} - 2\sigma^{n-1} e^{ik_m z} + \sigma^{n-1} e^{ik_m(z-\Delta z)}\right) \end{array} \right] \tag{6.112}$$

$$\sigma^{n+1} e^{ik_m z} = \sigma^n e^{ik_m z} + a \left[\sigma^n e^{ik_m z} c\left(e^{ik_m \Delta z} - 2 + e^{-ik_m \Delta z}\right) - \sigma^{n-1} e^{ik_m z} c\left(e^{ik_m \Delta z} - 2 + e^{-ik_m \Delta z}\right) \right]$$
$$+ b \left[3\sigma^n e^{ik_m z} dc\left(e^{ik_m \Delta z} - 2 + e^{-ik_m \Delta z}\right) - \sigma^{n-1} e^{ik_m z} c\left(e^{ik_m \Delta z} - 2 + e^{-ik_m \Delta z}\right) \right] \tag{6.113}$$

$$\sigma^{n+1} = \sigma^n + a \left[\sigma^n c\left(2\cos(k_m \Delta z) - 2\right) - \sigma^{n-1} c\left(2\cos(k_m \Delta z) - 2\right) \right]$$
$$+ b \left[3\sigma^n dc\left(2\cos(k_m \Delta z) - 2\right) - \sigma^{n-1} dc\left(2\cos(k_m \Delta z) - 2\right) \right] \tag{6.114}$$

$$\frac{\sigma^{n+1}}{\sigma^n} = a \left[2c\left(\cos(k_m \Delta z) - 1\right) - 2c\frac{\sigma^{n-1}}{\sigma^n}\left(\cos(k_m \Delta z) - 1\right) \right]$$
$$+ b \left[3dc\left(\cos(k_m \Delta z) - 1\right) - 2dc\frac{\sigma^{n-1}}{\sigma^n}\left(\cos(k_m \Delta z) - 1\right) \right] \tag{6.115}$$

$$\frac{\sigma^{n+1}}{\sigma^n} = 2ac\left(\cos(k_m \Delta z) - 1\right)\left[1 - \frac{\sigma^{n-1}}{\sigma^n}\right] + bdc\left(2\cos(k_m \Delta z) - 2\right)\left[3 - 2\frac{\sigma^{n-1}}{\sigma^n}\right] \tag{6.116}$$

For $n = 0$, the equation becomes:

$$\left|\frac{\sigma^1}{\sigma^0}\right| = \left|2ac\left(\cos(k_m \Delta z) - 1\right) + 3bc\left(2\cos(k_m \Delta z) - 2\right)\right| \tag{6.117}$$

The stability is given by:

$$\left| 2ac\left(\cos\left(k_m\Delta z\right)-1\right)+3bc\left(2\cos\left(k_m\Delta z\right)-2\right)\right| < 1 \qquad (6.118)$$

Now we assume that $\forall n > 0$:

$$\left|\frac{\sigma_n}{\sigma_0}\right| < 1$$

Using the above assumption, we will proof that:

$$\left|\frac{\sigma_{n+1}}{\sigma_0}\right| < 1$$

The equation then becomes:

$$\sigma^{n+1} = \sigma^n + a\left[\sigma^n 2c\left(\cos\left(k_m\Delta z\right)-1\right)-\sigma^{n-1}2c\left(\cos\left(k_m\Delta z\right)-1\right)\right]$$
$$+ b\left[\sigma^n 3dc\left(2\cos\left(k_m\Delta z\right)-2\right)-\sigma^{n-1}dc\left(2\cos\left(k_m\Delta z\right)-2\right)\right] \qquad (6.119)$$

Which implies:

$$\left|\sigma^{n+1}\right| = \left|\sigma^n\right|\left(1+2ac\left(\cos\left(k_m\Delta z\right)-1\right)\right)-\left|\sigma^{n-1}\right|2ac\left(\cos\left(k_m\Delta z\right)-1\right)$$
$$+ \left|\sigma^n\right|3bdc\left(2\cos\left(k_m\Delta z\right)-2\right)-\left|\sigma^{n-1}\right|bdc\left(2\cos\left(k_m\Delta z\right)-2\right) \qquad (6.120)$$

Using recursive hypothesis, the equation becomes:

$$\left|\sigma^{n+1}\right| < \left|\sigma^0\right|\left(1+2ac\left(\cos\left(k_m\Delta z\right)-1\right)\right)-\left|\sigma^0\right|2ac\left(\cos\left(k_m\Delta z\right)-1\right)$$
$$+ \left|\sigma^0\right|3bdc\left(2\cos\left(k_m\Delta z\right)-2\right)-\left|\sigma^0\right|bdc\left(2\cos\left(k_m\Delta z\right)-2\right) \qquad (6.121)$$

$$\left|\sigma^{n+1}\right| < \left|\sigma^0\right|\left[\begin{array}{c}\left(1+2ac\left(\cos\left(k_m\Delta z\right)-1\right)\right)-2ac\left(\cos\left(k_m\Delta z\right)-1\right)\\ +3bdc\left(2\cos\left(k_m\Delta z\right)-2\right)-bdc\left(2\cos\left(k_m\Delta z\right)-2\right)\end{array}\right] \qquad (6.122)$$

$$\left|\sigma^{n+1}\right| < \left|\sigma^0\right|\left[1+2bdc\left(2\cos\left(k_m\Delta z\right)-2\right)\right] \qquad (6.123)$$

$$1+4bdc\left(\cos\left(k_m\Delta z\right)-1\right) < 1 \qquad (6.124)$$

$$4bdc\cos\left(k_m\Delta z\right) < 4bdc \qquad (6.125)$$

The stability condition is given by:

$$\cos\left(k_m\Delta z\right) < 1 \qquad (6.126)$$

Now, we replace the time derivative on our stochastic unsaturated equation (6.16) with the Caputo–Fabrizio fractional derivative, such that:

$$
{}^{CF}_0 D^\alpha \psi\left(z,t\right) = \frac{1}{-\left(\theta_s - \theta_0\right)\psi_b^\lambda \dfrac{\lambda}{\psi_i^{\lambda+1}}} \left(
\begin{array}{l}
\hat{K}_s \dfrac{\left(\dfrac{\psi_b}{\psi_{i+1}^{n+1}}\right)^{2+3\lambda} - \left(\dfrac{\psi_b}{\psi_i^{n+1}}\right)^{2+3\lambda}}{\Delta z} \dfrac{\psi_{i+1}^{n+1} - \psi_i^{n+1}}{\Delta z} + \hat{K}_s \left(\dfrac{\psi_b}{\psi_i^{n+1}}\right)^{2+3\lambda} \\[4ex]
\times \dfrac{\psi_{i+1}^{n+1} - 2\psi_i^{n+1} + \psi_{i-1}^{n+1}}{\left(\Delta z\right)^2} - \hat{K}_s \dfrac{\left(\dfrac{\psi_b}{\psi_{i+1}^{n+1}}\right)^{2+3\lambda} - \left(\dfrac{\psi_b}{\psi_i^{n+1}}\right)^{2+3\lambda}}{\Delta z}
\end{array}
\right)
$$

$$(6.127)$$

where:

$$
{}^{CF}_0 D^\alpha \psi\left(z,t\right) = f\left(z,t,\psi\left(z,t\right)\right)
$$

$$(6.128)$$

We apply the Caputo–Fabrizio integral to the above equation:

$$
\psi\left(z,t\right) = \psi\left(z,0\right) + \frac{1-\alpha}{M\left(\alpha\right)} f\left(z,t,\psi\left(z,t\right)\right) + \frac{\alpha}{M\left(\alpha\right)} \int_0^t f\left(z,\tau,\psi\left(z,\tau\right)\right) d\tau
$$

$$(6.129)$$

When $t = t_{n+1}$ we now have the following equation:

$$
\psi\left(z_i,t_{n+1}\right) = \psi\left(z_i,0\right) + \frac{1-\alpha}{M\left(\alpha\right)} f\left(z_i,t_j,\psi\left(z_i,t_j\right)\right) + \frac{\alpha}{M\left(\alpha\right)} \int_0^{t_{n+1}} f\left(z_i,\tau,\psi\left(z_i,\tau\right)\right) d\tau
$$

$$(6.130)$$

When $= t_n$:

$$
\psi\left(z_i,t_n\right) = \psi\left(z_i,0\right) + \frac{1-\alpha}{M\left(\alpha\right)} f\left(z_i,t_{n-1},\psi\left(z_i,t_{n-1}\right)\right) + \frac{\alpha}{M\left(\alpha\right)} \int_0^{t_n} f\left(z_i,\tau,\psi\left(z_i,\tau\right)\right) d\tau
$$

$$(6.131)$$

Subtracting the above solution for t_n from t_{n+1}, we obtain the following:

$$
\psi_i^{n+1} - \psi_i^n = \frac{1-\alpha}{M\left(\alpha\right)} \left[f\left(z_i,t_n,\psi_i^n\right) - f\left(z_i,t_{n-1},\psi_i^{n-1}\right) \right] + \frac{\alpha}{M\left(\alpha\right)} \int_{t_n}^{t_{n+1}} f\left(z_i,\tau,\psi\left(z_i,\tau\right)\right) d\tau
$$

$$(6.132)$$

$$
\psi_i^{n+1} = \psi_i^n + \frac{1-\alpha}{M\left(\alpha\right)} \left[f\left(z_i,t_n,\psi_i^n\right) - f\left(z_i,t_{n-1},\psi_i^{n-1}\right) \right] + \frac{\alpha}{M\left(\alpha\right)} \left[\frac{3}{2}\Delta t f\left(z_i,t_n,\psi_i^n\right) - \frac{\Delta t}{2} f\left(z_i,t_{n-1},\psi_i^{n-1}\right) \right]
$$

$$(6.133)$$

where:

$$
f\left(z_i,t_n,\psi_i^n\right) = \frac{1}{-\left(\theta_s - \theta_0\right)\psi_b^\lambda \dfrac{\lambda}{\psi_i^{\lambda+1}}} \left(
\begin{array}{l}
\hat{K}_s \dfrac{\left(\dfrac{\psi_b}{\psi_{i+1}^n}\right)^{2+3\lambda} - \left(\dfrac{\psi_b}{\psi_i^n}\right)^{2+3\lambda}}{\Delta z} \dfrac{\psi_{i+1}^n - \psi_i^n}{\Delta z} + \hat{K}_s \left(\dfrac{\psi_b}{\psi_i^n}\right)^{2+3\lambda} \\[4ex]
\times \dfrac{\psi_{i+1}^n - 2\psi_i^n + \psi_{i-1}^n}{\left(\Delta z\right)^2} - \hat{K}_s \dfrac{\left(\dfrac{\psi_b}{\psi_{i+1}^n}\right)^{2+3\lambda} - \left(\dfrac{\psi_b}{\psi_i^n}\right)^{2+3\lambda}}{\Delta z}
\end{array}
\right)
$$

$$(6.134)$$

$$F\left(z_i, t_{n-1}, \psi_i^{n-1}\right) = \frac{1}{-(\theta_s - \theta_0)\psi_b^\lambda \dfrac{\lambda}{\psi_i^{\lambda+1}}} \left(\begin{array}{l} \hat{K}_s \dfrac{\left(\dfrac{\psi_b}{\psi_{i+1}^{n-1}}\right)^{2+3\lambda} - \left(\dfrac{\psi_b}{\psi_i^{n-1}}\right)^{2+3\lambda}}{\Delta z} \dfrac{\psi_{i+1}^{n-1} - \psi_i^{n-1}}{\Delta z} + \hat{K}_s \left(\dfrac{\psi_b}{\psi_i^{n-1}}\right)^{2+3\lambda} \\[4ex] \times \dfrac{\psi_{i+1}^{n-1} - 2\psi_i^{n-1} + \psi_{i-1}^{n-1}}{(\Delta z)^2} - \hat{K}_s \dfrac{\left(\dfrac{\psi_b}{\psi_{i+1}^{n-1}}\right)^{2+3\lambda} - \left(\dfrac{\psi_b}{\psi_i^{n-1}}\right)^{2+3\lambda}}{\Delta z} \end{array} \right)$$

$$(6.135)$$

We substitute the above Equations (6.134) and (6.135) into Equation (6.133) to obtain the final numerical solution for the saturated groundwater flow equation

$$\psi_i^{n+1} = \psi_i^n + \frac{1-\alpha}{M(\alpha)} \left[\begin{array}{l} \dfrac{1}{-(\theta_s - \theta_0)\psi_b^\lambda \dfrac{\lambda}{\psi_i^{\lambda+1}}} \left(\begin{array}{l} \hat{K}_s \dfrac{\left(\dfrac{\psi_b}{\psi_{i+1}^n}\right)^{2+3\lambda} - \left(\dfrac{\psi_b}{\psi_i^n}\right)^{2+3\lambda}}{\Delta z} \dfrac{\psi_{i+1}^n - \psi_i^n}{\Delta z} + \hat{K}_s \left(\dfrac{\psi_b}{\psi_i^n}\right)^{2+3\lambda} \\[4ex] \times \dfrac{\psi_{i+1}^n - 2\psi_i^n + \psi_{i-1}^n}{(\Delta z)^2} - \hat{K}_s \dfrac{\left(\dfrac{\psi_b}{\psi_{i+1}^n}\right)^{2+3\lambda} - \left(\dfrac{\psi_b}{\psi_i^n}\right)^{2+3\lambda}}{\Delta z} \end{array} \right) \\[10ex] - \dfrac{1}{-(\theta_s - \theta_0)\psi_b^\lambda \dfrac{\lambda}{\psi_i^{\lambda+1}}} \left(\begin{array}{l} \hat{K}_s \dfrac{\left(\dfrac{\psi_b}{\psi_{i+1}^{n-1}}\right)^{2+3\lambda} - \left(\dfrac{\psi_b}{\psi_i^{n-1}}\right)^{2+3\lambda}}{\Delta z} \dfrac{\psi_{i+1}^{n-1} - \psi_i^{n-1}}{\Delta z} + \hat{K}_s \left(\dfrac{\psi_b}{\psi_i^{n-1}}\right)^{2+3\lambda} \\[4ex] \times \dfrac{\psi_{i+1}^{n-1} - 2\psi_i^{n-1} + \psi_{i-1}^{n-1}}{(\Delta z)^2} - \hat{K}_s \dfrac{\left(\dfrac{\psi_b}{\psi_{i+1}^{j-1}}\right)^{2+3\lambda} - \left(\dfrac{\psi_b}{\psi_i^{j-1}}\right)^{2+3\lambda}}{\Delta z} \end{array} \right) \end{array} \right]$$

$$+ \frac{\alpha}{M(\alpha)} \left[\begin{array}{l} \dfrac{3\Delta t}{2} \dfrac{1}{-(\theta_s - \theta_0)\psi_b^\lambda \dfrac{\lambda}{\psi_i^{\lambda+1}}} \left(\begin{array}{l} \hat{K}_s \dfrac{\left(\dfrac{\psi_b}{\psi_{i+1}^n}\right)^{2+3\lambda} - \left(\dfrac{\psi_b}{\psi_i^n}\right)^{2+3\lambda}}{\Delta z} \dfrac{\psi_{i+1}^n - \psi_i^n}{\Delta z} + \hat{K}_s \left(\dfrac{\psi_b}{\psi_i^n}\right)^{2+3\lambda} \\[4ex] \times \dfrac{\psi_{i+1}^n - 2\psi_i^n + \psi_{i-1}^n}{(\Delta z)^2} - \hat{K}_s \dfrac{\left(\dfrac{\psi_b}{\psi_{i+1}^n}\right)^{2+3\lambda} - \left(\dfrac{\psi_b}{\psi_i^n}\right)^{2+3\lambda}}{\Delta z} \end{array} \right) \\[10ex] - \dfrac{\Delta t}{2} \dfrac{1}{-(\theta_s - \theta_0)\psi_b^\lambda \dfrac{\lambda}{\psi_i^{\lambda+1}}} \left(\begin{array}{l} \hat{K}_s \dfrac{\left(\dfrac{\psi_b}{\psi_{i+1}^{n-1}}\right)^{2+3\lambda} - \left(\dfrac{\psi_b}{\psi_i^{n-1}}\right)^{2+3\lambda}}{\Delta z} \dfrac{\psi_{i+1}^{n-1} - \psi_i^{n-1}}{\Delta z} + \hat{K}_s \left(\dfrac{\psi_b}{\psi_i^{n-1}}\right)^{2+3\lambda} \\[4ex] \times \dfrac{\psi_{i+1}^{n-1} - 2\psi_i^{n-1} + \psi_{i-1}^{n-1}}{(\Delta z)^2} - \hat{K}_s \dfrac{\left(\dfrac{\psi_b}{\psi_{i+1}^{n-1}}\right)^{2+3\lambda} - \left(\dfrac{\psi_b}{\psi_i^{n-1}}\right)^{2+3\lambda}}{\Delta z} \end{array} \right) \end{array} \right]$$

$$(6.136)$$

Now, let:

$$\wp = -(\theta_s - \theta_0)\psi_b^\lambda \frac{\lambda}{\psi_i^{\lambda+1}},$$

$$a = \frac{1-\alpha}{M(\alpha)}$$

$$b = \frac{\alpha}{M(\alpha)}$$

$$c = \frac{\hat{K}_s}{\Delta z}$$

$$d = \frac{\hat{K}_s}{(\Delta z)^2}$$

$$e = \frac{1}{\Delta z}$$

$$d = \frac{1}{(\Delta z)^2}$$

$$f = \frac{3\Delta t}{2}$$

$$g = \frac{\Delta t}{2}$$

Using the above substitutions, Equation (6.136) becomes:

$$
\psi_i^{n+1} = \psi_i^n + a\left[
\begin{array}{l}
\frac{1}{\wp}\left(
\begin{array}{l}
\left(c\left(\left(\frac{\psi_b}{\psi_{i+1}^n}\right)^{2+3\lambda} - \left(\frac{\psi_b}{\psi_i^n}\right)^{2+3\lambda}\right)d\left(\psi_{i+1}^n - \psi_i^n\right) + \hat{K}_s\left(\frac{\psi_b}{\psi_i^n}\right)^{2+3\lambda}\right) \\
\times e\left(\psi_{i+1}^n - 2\psi_i^n + \psi_{i-1}^n\right) - c\left(\left(\frac{\psi_b}{\psi_{i+1}^n}\right)^{2+3\lambda} - \left(\frac{\psi_b}{\psi_i^n}\right)^{2+3\lambda}\right)
\end{array}
\right) \\
-\frac{1}{\wp}\left(
\begin{array}{l}
\left(c\left(\left(\frac{\psi_b}{\psi_{i+1}^{n-1}}\right)^{2+3\lambda} - \left(\frac{\psi_b}{\psi_i^{n-1}}\right)^{2+3\lambda}\right)d\left(\psi_{i+1}^{n-1} - \psi_i^{n-1}\right) + \hat{K}_s\left(\frac{\psi_b}{\psi_i^{n-1}}\right)^{2+3\lambda}\right) \\
\times e\left(\psi_{i+1}^{n-1} - 2\psi_i^{n-1} + \psi_{i-1}^{n-1}\right) - c\left(\left(\frac{\psi_b}{\psi_{i+1}^{j-1}}\right)^{2+3\lambda} - \left(\frac{\psi_b}{\psi_i^{j-1}}\right)^{2+3\lambda}\right)
\end{array}
\right)
\end{array}
\right]
$$

$$
+ b\left[
\begin{array}{l}
f\frac{1}{\wp}\left(
\begin{array}{l}
\left(c\left(\left(\frac{\psi_b}{\psi_{i+1}^n}\right)^{2+3\lambda} - \left(\frac{\psi_b}{\psi_i^n}\right)^{2+3\lambda}\right)d\left(\psi_{i+1}^n - \psi_i^n\right) + \hat{K}_s\left(\frac{\psi_b}{\psi_i^n}\right)^{2+3\lambda}\right) \\
\times e\left(\psi_{i+1}^n - 2\psi_i^n + \psi_{i-1}^n\right) - c\left(\left(\frac{\psi_b}{\psi_{i+1}^n}\right)^{2+3\lambda} - \left(\frac{\psi_b}{\psi_i^n}\right)^{2+3\lambda}\right)
\end{array}
\right) \\
-g\frac{1}{\wp}\left(
\begin{array}{l}
\left(c\left(\left(\frac{\psi_b}{\psi_{i+1}^{n-1}}\right)^{2+3\lambda} - \left(\frac{\psi_b}{\psi_i^{n-1}}\right)^{2+3\lambda}\right)d\left(\psi_{i+1}^{n-1} - \psi_i^{n-1}\right) + \hat{K}_s\left(\frac{\psi_b}{\psi_i^{n-1}}\right)^{2+3\lambda}\right) \\
\times e\left(\psi_{i+1}^{n-1} - 2\psi_i^{n-1} + \psi_{i-1}^{n-1}\right) - c\left(\left(\frac{\psi_b}{\psi_{i+1}^{n-1}}\right)^{2+3\lambda} - \left(\frac{\psi_b}{\psi_i^{n-1}}\right)^{2+3\lambda}\right)
\end{array}
\right)
\end{array}
\right].
$$

(6.137)

6.3.3 STOCHASTIC DIFFERENTIAL EQUATION USING THE ATANGANA-BALEANU FRACTIONAL DERIVATIVE

In this section, the model is considered using the non-local and non-singular kernel. To do this, the time derivative of the stochastic model is replaced with the time-fractional derivative based on the Mittag-Leffler function to derive the numerical solution of the new model. Therefore, the fractional-stochastic saturated equation is written as follows:

$$
{}_0^{AB}D^\alpha \psi\left(z,t\right) = \frac{\hat{K}_s}{\hat{S}_s}\frac{\partial^2\psi}{\partial z^2}
\tag{6.138}
$$

Where:

$$
{}_0^{AB}D^\alpha \psi\left(z,t\right) = f\left(z,t,\psi\left(z,t\right)\right)
\tag{6.139}
$$

Then the AB integral is written as:

$$
\psi\left(z_i,t\right)-\psi\left(z_i,0\right) = \frac{\left(1-\alpha\right)}{B\left(\alpha\right)}f\left(z,t,\psi\left(z,t\right)\right)+\frac{\alpha}{B\left(\alpha\right)\Gamma\left(\alpha\right)}\int_0^t f\left(z_i,\tau,\psi\left(z_i,\tau\right)\right)\left(t-\tau\right)^{\alpha-1}d\tau
\tag{6.140}
$$

At point t_{n+1}

$$
\psi\left(z_i,t_{n+1}\right)-\psi\left(z_i,0\right) = \frac{\left(1-\alpha\right)}{\mathrm{B}\left(\alpha\right)}f\left(z,t,\psi\left(z,t\right)\right)+\frac{\alpha}{\mathrm{B}\left(\alpha\right)\times\Gamma\left(\alpha\right)}\int_0^{t_{n+1}} f\left(z_i,\tau,\psi\left(z_i,\tau\right)\right)\left(t_{n+1}-\tau\right)^{\alpha-1}d\tau
$$

$$
\tag{6.141}
$$

$$
\psi\left(z_i,t_{n+1}\right)-\psi\left(z_i,0\right) = \frac{\left(1-\alpha\right)}{\mathrm{B}\left(\alpha\right)}f\left(z,t,\psi\left(z,t\right)\right)+\frac{\alpha\left(\Delta t\right)^\alpha}{\mathrm{B}\left(\alpha\right)\times\Gamma\left(2+\alpha\right)}
$$

$$
\times\sum_{j=0}^n\left[\begin{array}{l}f\left(z_i,t_j,\psi_i^j\right)\left(\left(n-j+1\right)^\alpha\left(n-j+2+\alpha\right)-\left(n-j\right)^\alpha\left(n-j+2+2\alpha\right)\right)\\ -f\left(z_i,t_{j-1},\psi_i^{j-1}\right)\left(\left(n-j+1\right)^{\alpha+1}-\left(n-j\right)^\alpha\left(n-j+1+\alpha\right)\right)\end{array}\right]
\tag{6.142}
$$

where:

$$
F\left(z_i,t_j,\psi_i^j\right) = \frac{\hat{K}_s}{\hat{S}_s}\frac{\psi_{i+1}^j-2\psi_i^j+\psi_{i-1}^j}{\left(\Delta z\right)^2}
\tag{6.143}
$$

$$
F\left(z_i,t_{j-1},\psi_i^{j-1}\right) = \frac{\hat{K}_s}{\hat{S}_s}\frac{\psi_{i+1}^{j-1}-2\psi_i^{j-1}+\psi_{i-1}^{j-1}}{\left(\Delta z\right)^2}
\tag{6.144}
$$

The final numerical solution for the saturated groundwater flow equation is given as:

$$\psi\left(z_i,t_{n+1}\right)-\psi\left(z_i,0\right)=\frac{(1-\alpha)}{B(\alpha)}\left(\frac{\hat{K}_s}{\hat{S}_s}\frac{\psi_{i+1}^n-2\psi_i^n+\psi_{i-1}^n}{(\Delta z)^2}\right)+\frac{\alpha(\Delta t)^\alpha}{B(\alpha)\Gamma(2+\alpha)}$$

$$\times\sum_{j=0}^n\left[\begin{array}{l}\dfrac{\hat{K}_s}{\hat{S}_s}\dfrac{\psi_{i+1}^j-2\psi_i^j+\psi_{i-1}^j}{(\Delta z)^2}\left((n-j+1)^\alpha(n-j+2+\alpha)-(n-j)^\alpha(n-j+2+2\alpha)\right)\\[4mm]-\dfrac{\hat{K}_s}{\hat{S}_s}\dfrac{\psi_{i+1}^{j-1}-2\psi_i^{j-1}+\psi_{i-1}^{j-1}}{(\Delta z)^2}\left((n-j+1)^{\alpha+1}-(n-j)^\alpha(n-j+1+\alpha)\right)\end{array}\right]$$

$$(6.145)$$

We also assume the following parameters:

$$a=\frac{(1-\alpha)}{B(\alpha)}$$

$$b=\frac{\hat{K}_s}{\hat{S}_s(\Delta z)^2}$$

$$c=\frac{\alpha(\Delta t)^\alpha}{B(\alpha)\Gamma(2+\alpha)}$$

$$R_{n,j}^\alpha=\left((n-j+1)^\alpha(n-j+2+\alpha)-(n-j)^\alpha(n-j+2+2\alpha)\right)$$

$$\delta_{n,j}^\alpha=\left((n-j+1)^{\alpha+1}-(n-j)^\alpha(n-j+1+\alpha)\right)$$

Then our equation is written as:

$$\psi_i^{n+1}-\psi_i^n=a\left[b\left(\psi_{i+1}^n-2\psi_i^n+\psi_{i-1}^n\right)\right]+c\sum_{j=0}^n\left[b\left(\psi_{i+1}^j-2\psi_i^j+\psi_{i-1}^j\right)R_{n,j}^\alpha-b\left(\psi_{i+1}^{j-1}-2\psi_i^{j-1}+\psi_{i-1}^{j-1}\right)\delta_{n,j}^\alpha\right]$$

$$(6.146)$$

$$\psi_i^{n+1}-\psi_i^n=a\left[b\left(\psi_{i+1}^n-2\psi_i^n+\psi_{i-1}^n\right)\right]+c\left[b\left(\psi_{i+1}^n-2\psi_i^n+\psi_{i-1}^n\right)R_{n,n}^\alpha-b\left(\psi_{i+1}^{n-1}-2\psi_i^{n-1}+\psi_{i-1}^{n-1}\right)\delta_{n,n}^\alpha\right]$$

$$+c\sum_{j=0}^{n-1}\left[b\left(\psi_{i+1}^j-2\psi_i^j+\psi_{i-1}^j\right)R_{n,j}^\alpha-b\left(\psi_{i+1}^{j-1}-2\psi_i^{j-1}+\psi_{i-1}^{j-1}\right)\delta_{n,j}^\alpha\right]$$

$$(6.147)$$

To check the stability of the above numerical solution, we assume a Fourier expansion in space of:

$$\psi_i^n=\sigma_n e^{ik_m z}$$

$$(6.148)$$

Using the above assumption, Equation (6.148) becomes:

$$\sigma_{n+1}e^{ik_m z} - \sigma_n e^{ik_m z} = a\left[b\left(\sigma_n e^{ik_m(z+\Delta z)} - 2\sigma_n e^{ik_m z} + \sigma_n e^{ik_m(z-\Delta z)}\right)\right]$$
$$+ c\begin{bmatrix} b\left(\sigma_n e^{ik_m(z+\Delta z)} - 2\sigma_n e^{ik_m z} + \sigma_n e^{ik_m(z-\Delta z)}\right)R_{n,n}^{\alpha} \\ -b\left(\sigma_{n-1}e^{ik_m(z+\Delta z)} - 2\sigma_{n-1}e^{ik_m z} + \sigma_{n-1}e^{ik_m(z-\Delta z)}\right)\delta_{n,n}^{\alpha} \end{bmatrix}$$
$$+ c\sum_{j=0}^{n-1}\begin{bmatrix} b\left(\sigma_j e^{ik_m(z+\Delta z)} - 2\sigma_j e^{ik_m z} + \sigma_j e^{ik_m(z-\Delta z)}\right)R_{n,j}^{\alpha} \\ -b\left(\sigma_{j-1}e^{ik_m(z+\Delta z)} - 2\sigma_{j-1}e^{ik_m z} + \sigma_{j-1}e^{ik_m(z-\Delta z)}\right)\delta_{n,j}^{\alpha} \end{bmatrix} \quad (6.149)$$

$$e^{ik_m z}\left(\sigma_{n+1} - \sigma_n\right) = a\left[b\sigma_n e^{ik_m z}\left(e^{ik_m \Delta z} - 2 + e^{-ik_m \Delta z}\right)\right]$$
$$+ c\left[R_{n,n}^{\alpha}b\sigma_n e^{ik_m z}\left(e^{ik_m \Delta z} - 2 + e^{-ik_m \Delta z}\right) - \delta_{n,n}^{\alpha}b\sigma_{n-1}e^{ik_m z}\left(e^{ik_m \Delta z} - 2 + e^{-ik_m \Delta z}\right)\right]$$
$$+ c\sum_{j=0}^{n-1}\left[R_{n,j}^{\alpha}b\sigma_j e^{ik_m z}\left(e^{ik_m \Delta z} - 2 + e^{-ik_m \Delta z}\right) - \delta_{n,j}^{\alpha}b\sigma_{j-1}e^{ik_m z}\left(e^{ik_m \Delta z} - 2 + e^{-ik_m \Delta z}\right)\right]$$

$$(6.150)$$

$$\sigma_{n+1} - \sigma_n = a\left[b\sigma_n\left(e^{ik_m \Delta z} - 2 + e^{-ik_m \Delta z}\right)\right]$$
$$+ c\left[R_{n,n}^{\alpha}b\sigma_n\left(e^{ik_m \Delta z} - 2 + e^{-ik_m \Delta z}\right) - \delta_{n,n}^{\alpha}b\sigma_{n-1}\left(e^{ik_m \Delta z} - 2 + e^{-ik_m \Delta z}\right)\right]$$
$$+ c\sum_{j=0}^{n-1}\left[R_{n,j}^{\alpha}b\sigma_j\left(e^{ik_m \Delta z} - 2 + e^{-ik_m \Delta z}\right) - \delta_{n,j}^{\alpha}b\sigma_{j-1}\left(e^{ik_m \Delta z} - 2 + e^{-ik_m \Delta z}\right)\right] \quad (6.151)$$

$$\sigma_{n+1} - \sigma_n = 2ab\sigma_n\left(\frac{e^{ik_m \Delta z} + e^{-ik_m \Delta z}}{2} - 1\right) + c\begin{bmatrix} 2R_{n,n}^{\alpha}b\sigma_n\left(\dfrac{e^{ik_m \Delta z} + e^{-ik_m \Delta z}}{2} - 1\right) \\ -2\delta_{n,n}^{\alpha}b\sigma_{n-1}\left(\dfrac{e^{ik_m \Delta z} + e^{-ik_m \Delta z}}{2} - 1\right) \end{bmatrix}$$
$$+ c\sum_{j=0}^{n-1}\left[2R_{n,j}^{\alpha}b\sigma_j\left(\frac{e^{ik_m \Delta z} + e^{-ik_m \Delta z}}{2} - 1\right) - 2\delta_{n,j}^{\alpha}b\sigma_{j-1}\left(\frac{e^{ik_m \Delta z} + e^{-ik_m \Delta z}}{2} - 1\right)\right] \quad (6.152)$$

Considering the definition for a hyperbolic cosine, we can rewrite the above equation (6.152) as:

$$\sigma_{n+1} - \sigma_n = 2ab\sigma_n\left(\cos\left(k_m \Delta z\right) - 1\right) + c\left[2R_{n,n}^{\alpha}b\sigma_n\left(\cos\left(k_m \Delta z\right) - 1\right) - 2\delta_{n,n}^{\alpha}b\sigma_{n-1}\left(\cos\left(k_m \Delta z\right) - 1\right)\right]$$
$$+ c\sum_{j=0}^{n-1}\left[2R_{n,j}^{\alpha}b\sigma_j\left(\cos\left(k_m \Delta z\right) - 1\right) - 2\delta_{n,j}^{\alpha}b\sigma_{j-1}\left(\cos\left(k_m \Delta z\right) - 1\right)\right]$$

$$(6.153)$$

If n = 0, the equation above becomes:

$$\sigma_1 - \sigma_0 = 2ab\sigma_0\left(\cos\left(k_m \Delta z\right) - 1\right) + c\left[2R_{n,n}^{\alpha}b\sigma_0\left(\cos\left(k_m \Delta z\right) - 1\right)\right] \quad (6.154)$$

$$\sigma_1 = \sigma_0 \left[1 + 2ab\left(\cos\left(k_m\Delta z\right) - 1\right) + R_{n,n}^{\alpha}cb\left(\cos\left(k_m\Delta z\right) - 1\right) \right]$$ (6.155)

$$\left| \frac{\sigma_1}{\sigma_0} \right| < 1$$

$$\left| \frac{1}{1 + 2ab\left(\cos\left(k_m\Delta z\right) - 1\right) + R_{n,n}^{\alpha}cb\left(\cos\left(k_m\Delta z\right) - 1\right)} \right| < 1$$ (6.156)

From the above results, it is clear that the denominator of the equation will always be greater than the numerator; therefore, the A-B fractional derivative for the 1-d saturated groundwater flow equation is unconditionally stable. Now we assume $\forall n > 0$ to prove that:

$$\left| \frac{\sigma_{n+1}}{\sigma_0} \right| < 1$$

$$\sigma_{n+1} = \sigma_n \left[1 + 2ab\left(\cos\left(k_m\Delta z\right) - 1\right) + R_{n,n}^{\alpha}cb\left(\cos\left(k_m\Delta z\right) - 1\right) \right] - 2\delta_{n,n}^{\alpha}cb\sigma_{n-1}\left(\cos\left(k_m\Delta z\right) - 1\right)$$
$$+ c\sum_{j=0}^{n-1} \left[2R_{n,j}^{\alpha}b\sigma_j\left(\cos\left(k_m\Delta z\right) - 1\right) - 2\delta_{n,j}^{\alpha}b\sigma_{j-1}\left(\cos\left(k_m\Delta z\right) - 1\right) \right]$$ (6.157)

We let:

$$A = \left[1 + 2ab\left(\cos\left(k_m\Delta z\right) - 1\right) + R_{n,n}^{\alpha}cb\left(\cos\left(k_m\Delta z\right) - 1\right) \right]$$

$$B = 2\delta_{n,n}^{\alpha}cb\left(\cos\left(k_m\Delta z\right) - 1\right)$$

$$A_1 = 2R_{n,j}^{\alpha}b\left(\cos\left(k_m\Delta z\right) - 1\right)$$

$$A_2 = 2\delta_{n,j}^{\alpha}b\left(\cos\left(k_m\Delta z\right) - 1\right)$$

Then Equation (6.23) becomes:

$$\sigma_{n+1} = \sigma_n A - \sigma_{n-1} B + c\sum_{j=0}^{n-1} \left[\sigma_j A_1 - \sigma_{j-1}A_2 \right]$$ (6.158)

$$|\sigma_{n+1}| \leq |\sigma_n||A| - |\sigma_{n-1}||B| + c\sum_{j=0}^{n-1} \left[|\sigma_j||A_1| - |\sigma_{j-1}||A_2| \right]$$ (6.159)

$$|\sigma_{n+1}| \leq |\sigma_0| \left(|A| - |B| + c\sum_{j=0}^{n-1} \left(|A_1| - |A_2| \right) \right)$$ (6.160)

$$\frac{|\sigma_{n+1}|}{|\sigma_0|} \leq |A| - |B| + c \sum_{j=0}^{n-1} \left(|A_1| - |A_2| \right) \tag{6.161}$$

The stability condition is therefore given by:

$$\left| \frac{1}{|A| - |B| + c \sum_{j=0}^{n-1} \left(|A_1| - |A_2| \right)} \right| \leq 1 \tag{6.162}$$

Now, using the A-B fractional derivative and stochastic new model for the unsaturated zone is given by the following equation:

$$^{ABC}_{0}D^{\alpha}\psi(z,t) = \frac{1}{-(\theta_s - \theta_0)\psi_b^{\lambda} \frac{\lambda}{\psi_i^{\lambda+1}}} \left(\begin{array}{l} \hat{K}_s \dfrac{\left(\dfrac{\psi_b}{\psi_{i+1}^{n+1}}\right)^{2+3\lambda} - \left(\dfrac{\psi_b}{\psi_i^{n+1}}\right)^{2+3\lambda}}{\Delta z} \dfrac{\psi_{i+1}^{n+1} - \psi_i^{n+1}}{\Delta z} \\[4mm] + \hat{K}_s \left(\dfrac{\psi_b}{\psi_i^{n+1}}\right)^{2+3\lambda} \times \dfrac{\psi_{i+1}^{n+1} - 2\psi_i^{n+1} + \psi_{i-1}^{n+1}}{(\Delta z)^2} \\[4mm] - \hat{K}_s \dfrac{\left(\dfrac{\psi_b}{\psi_{i+1}^{n+1}}\right)^{2+3\lambda} - \left(\dfrac{\psi_b}{\psi_i^{n+1}}\right)^{2+3\lambda}}{\Delta z} \end{array} \right) \tag{6.163}$$

where:

$$^{ABC}_{0}D^{\alpha}\psi(z,t) = F\left(z,t,\psi(z,t)\right) \tag{6.164}$$

Then the A-B integral is given as:

$$\psi(z_i,t) - \psi(z_i,0) = \frac{(1-\alpha)}{B(\alpha)} f\left(z,t,\psi(z,t)\right) + \frac{\alpha}{B(\alpha)\Gamma(\alpha)} \int_0^t f\left(z_i,\tau,\psi(z_i,\tau)\right)(t-\tau)^{\alpha-1} d\tau \tag{6.165}$$

At point t_{n+1}

$$\psi(z_i,t_{n+1}) - \psi(z_i,0) = \frac{(1-\alpha)}{B(\alpha)} f\left(z,t,\psi(z,t)\right) + \frac{\alpha}{B(\alpha)\Gamma(\alpha)} \int_0^{t_{n+1}} F\left(z_i,\tau,\psi(z_i,\tau)\right)(t_{n+1}-\tau)^{\alpha-1} d\tau$$

$$\tag{6.166}$$

$$\psi(z_i,t_{n+1}) - \psi(z_i,0) = \frac{(1-\alpha)}{B(\alpha)} f\left(z,t,\psi(z,t)\right) + \frac{(\Delta t)^{\alpha}}{B(\alpha)\Gamma(2+\alpha)}$$

$$\sum_{j=0}^{n} \left[\begin{array}{l} F\left(z_i,t_j,\psi_i^j\right)\left((n-j+1)^{\alpha}(n-j+2+\alpha) - (n-j)^{\alpha}(n-j+2+2\alpha)\right) \\[2mm] -F\left(z_i,t_{j-1},\psi_i^{j-1}\right)\left((n-j+1)^{\alpha+1} - (n-j)^{\alpha}(n-j+1+\alpha)\right) \end{array} \right]$$

$$\tag{6.167}$$

Where:

$$F\left(z_i,t_j,\psi_i^j\right)=\cfrac{1}{-(\theta_s-\theta_0)\psi_b^\lambda\,\cfrac{\lambda}{\psi_i^{\lambda+1}}}\left(\begin{array}{l}\hat{K}_s\cfrac{\left(\cfrac{\psi_b}{\psi_{i+1}^j}\right)^{2+3\lambda}-\left(\cfrac{\psi_b}{\psi_i^j}\right)^{2+3\lambda}}{\Delta z}\cfrac{\psi_{i+1}^j-\psi_i^j}{\Delta z}+\hat{K}_s\left(\cfrac{\psi_b}{\psi_i^j}\right)^{2+3\lambda}\\[2em]\times\cfrac{\psi_{i+1}^j-2\psi_i^j+\psi_{i-1}^j}{(\Delta z)^2}-\hat{K}_s\cfrac{\left(\cfrac{\psi_b}{\psi_{i+1}^j}\right)^{2+3\lambda}-\left(\cfrac{\psi_b}{\psi_i^j}\right)^{2+3\lambda}}{\Delta z}\end{array}\right)$$

(6.168)

$$F\left(z_i,t_{j-1},\psi_i^{j-1}\right)=\cfrac{1}{-(\theta_s-\theta_0)\psi_b^\lambda\,\cfrac{\lambda}{\psi_i^{\lambda+1}}}\left(\begin{array}{l}\hat{K}_s\cfrac{\left(\cfrac{\psi_b}{\psi_{i+1}^{j-1}}\right)^{2+3\lambda}-\left(\cfrac{\psi_b}{\psi_i^{j-1}}\right)^{2+3\lambda}}{\Delta z}\cfrac{\psi_{i+1}^{j-1}-\psi_i^{j-1}}{\Delta z}+\hat{K}_s\left(\cfrac{\psi_b}{\psi_i^{j-1}}\right)^{2+3\lambda}\\[2em]\times\cfrac{\psi_{i+1}^{j-1}-2\psi_i^{j-1}+\psi_{i-1}^{j-1}}{(\Delta z)^2}-\hat{K}_s\cfrac{\left(\cfrac{\psi_b}{\psi_{i+1}^{j-1}}\right)^{2+3\lambda}-\left(\cfrac{\psi_b}{\psi_i^{j-1}}\right)^{2+3\lambda}}{\Delta z}\end{array}\right)$$

(6.169)

The final numerical solution for the unsaturated groundwater flow equation is given as:

$$\psi_i^{n+1}-\psi_i^0=\cfrac{(1-\alpha)}{B(\alpha)}\cfrac{1}{-(\theta_s-\theta_0)\psi_b^\lambda\,\cfrac{\lambda}{\psi_i^{\lambda+1}}}\left(\begin{array}{l}\hat{K}_s\cfrac{\left(\cfrac{\psi_b}{\psi_{i+1}^n}\right)^{2+3\lambda}-\left(\cfrac{\psi_b}{\psi_i^n}\right)^{2+3\lambda}}{\Delta z}\cfrac{\psi_{i+1}^n-\psi_i^n}{\Delta z}+\hat{K}_s\left(\cfrac{\psi_b}{\psi_i^n}\right)^{2+3\lambda}\\[2em]\times\cfrac{\psi_{i+1}^n-2\psi_i^n+\psi_{i-1}^n}{(\Delta z)^2}-\hat{K}_s\cfrac{\left(\cfrac{\psi_b}{\psi_{i+1}^n}\right)^{2+3\lambda}-\left(\cfrac{\psi_b}{\psi_i^n}\right)^{2+3\lambda}}{\Delta z}\end{array}\right)$$

$$+\cfrac{(\Delta t)^\alpha}{\Gamma(2+\alpha)}\sum_{j=0}^n\left[\begin{array}{l}\cfrac{1}{-(\theta_s-\theta_0)\psi_b^\lambda\,\cfrac{\lambda}{\psi_i^{\lambda+1}}}\left(\begin{array}{l}\hat{K}_s\cfrac{\left(\cfrac{\psi_b}{\psi_{i+1}^j}\right)^{2+3\lambda}-\left(\cfrac{\psi_b}{\psi_i^j}\right)^{2+3\lambda}}{\Delta z}\cfrac{\psi_{i+1}^j-\psi_i^j}{\Delta z}+\hat{K}_s\left(\cfrac{\psi_b}{\psi_i^j}\right)^{2+3\lambda}\\[2em]\times\cfrac{\psi_{i+1}^j-2\psi_i^j+\psi_{i-1}^j}{(\Delta z)^2}-\hat{K}_s\cfrac{\left(\cfrac{\psi_b}{\psi_{i+1}^j}\right)^{2+3\lambda}-\left(\cfrac{\psi_b}{\psi_i^j}\right)^{2+3\lambda}}{\Delta z}\end{array}\right)\\[3em]\Big((n-j+1)^\alpha(n-j+2+\alpha)-(n-j)^\alpha(n-j+2+2\alpha)\Big)\\[2em]-\cfrac{1}{-(\theta_s-\theta_0)\psi_b^\lambda\,\cfrac{\lambda}{\psi_i^{\lambda+1}}}\left(\begin{array}{l}\hat{K}_s\cfrac{\left(\cfrac{\psi_b}{\psi_{i+1}^{j-1}}\right)^{2+3\lambda}-\left(\cfrac{\psi_b}{\psi_i^{j-1}}\right)^{2+3\lambda}}{\Delta z}\cfrac{\psi_{i+1}^{j-1}-\psi_i^{j-1}}{\Delta z}+\hat{K}_s\left(\cfrac{\psi_b}{\psi_i^{j-1}}\right)^{2+3\lambda}\\[2em]\times\cfrac{\psi_{i+1}^{j-1}-2\psi_i^{j-1}+\psi_{i-1}^{j-1}}{(\Delta z)^2}-\hat{K}_s\cfrac{\left(\cfrac{\psi_b}{\psi_{i+1}^{j-1}}\right)^{2+3\lambda}-\left(\cfrac{\psi_b}{\psi_i^{j-1}}\right)^{2+3\lambda}}{\Delta z}\end{array}\right)\\[3em]\Big((n-j+1)^{\alpha+1}-(n-j)^\alpha(n-j+1+\alpha)\Big)\end{array}\right]$$

(6.170)

For simplicity of the above equation let:

$$\wp = -\left(\theta_s - \theta_0\right)\psi_b^\lambda \frac{\lambda}{\psi_i^{\lambda+1}}, \quad a = \frac{(\Delta t)^\alpha}{\Gamma(2-\alpha)}, \quad b = \frac{\hat{K}_s}{\Delta z}, \quad c = \hat{K}_s, \quad d = \frac{1}{(\Delta z)^2}, \quad e = \frac{1}{\Delta z}, \quad f = \frac{(1-\alpha)}{B(\alpha)}$$

$$R_{n.j}^\alpha = \left(\left(n-j+1\right)^\alpha \left(n-j+2+\alpha\right) - \left(n-j\right)^\alpha \left(n-j+2+2\alpha\right)\right)$$

$$\delta_{n.j}^\alpha = \left(\left(n-j+1\right)^{\alpha+1} - \left(n-j\right)^\alpha \left(n-j+1+\alpha\right)\right)$$

$$\psi_i^{n+1} - \psi_i^0 = f\frac{1}{\wp}\begin{pmatrix} b\left(\left(\frac{\psi_b}{\psi_{i+1}^n}\right)^{2+3\lambda} - \left(\frac{\psi_b}{\psi_i^n}\right)^{2+3\lambda}\right)e\left(\psi_{i+1}^n - \psi_i^n\right) + c\left(\frac{\psi_b}{\psi_i^n}\right)^{2+3\lambda} \\ \times d\left(\psi_{i+1}^n - 2\psi_i^n + \psi_{i-1}^n\right) - b\left(\left(\frac{\psi_b}{\psi_{i+1}^n}\right)^{2+3\lambda} - \left(\frac{\psi_b}{\psi_i^n}\right)^{2+3\lambda}\right) \end{pmatrix}$$

$$+ a\sum_{j=0}^n \begin{bmatrix} \frac{1}{\wp}\begin{pmatrix} b\left(\left(\frac{\psi_b}{\psi_{i+1}^j}\right)^{2+3\lambda} - \left(\frac{\psi_b}{\psi_i^j}\right)^{2+3\lambda}\right)e\left(\psi_{i+1}^j - \psi_i^j\right) + c\left(\frac{\psi_b}{\psi_i^j}\right)^{2+3\lambda} \\ \times d\left(\psi_{i+1}^j - 2\psi_i^j + \psi_{i-1}^j\right) - b\left(\left(\frac{\psi_b}{\psi_{i+1}^j}\right)^{2+3\lambda} - \left(\frac{\psi_b}{\psi_i^j}\right)^{2+3\lambda}\right) \end{pmatrix}R_{n.j}^\alpha \\ -\frac{1}{\wp}\begin{pmatrix} b\left(\left(\frac{\psi_b}{\psi_{i+1}^{j-1}}\right)^{2+3\lambda} - \left(\frac{\psi_b}{\psi_i^{j-1}}\right)^{2+3\lambda}\right)e\left(\psi_{i+1}^{j-1} - \psi_i^{j-1}\right) + c\left(\frac{\psi_b}{\psi_i^{j-1}}\right)^{2+3\lambda} \\ \times d\left(\psi_{i+1}^{j-1} - 2\psi_i^{j-1} + \psi_{i-1}^{j-1}\right) - b\left(\left(\frac{\psi_b}{\psi_{i+1}^{j-1}}\right)^{2+3\lambda} - \left(\frac{\psi_b}{\psi_i^{j-1}}\right)^{2+3\lambda}\right) \end{pmatrix}\delta_{n.j}^\alpha \end{bmatrix}$$

$$(6.171)$$

6.4 CONCLUSION

For many years, the stochastic approach and fractional differentiation and integration approach were used separately to describe problems found in nature. These methods have been proven to be powerful tools. However, limitations have been pointed out against these approaches. This is why in this chapter we used the new approach, which is a combination of the two concepts. Combining these two approaches helps us to capture physical problems appearing in nature, with long-range, fading memory and statistical setting for larger space and time. It is clear that the proposed method can be useful in many complex real-world problems, and therefore can bring us much closer to reality.

REFERENCES

Allepalli, P.K., & Govindaraju, R.S., 2009. Modeling Fate and Transport of Atrazine in the Saturated–Unsaturated Zone of Soil. Kansas State University, Department of Civil Engineering.

Atangana, A., & Alqahtani, R.T., 2019. A new approach to capture heterogeneity in groundwater problem: An illustration with an earth equation. *Mathematical Modelling of Natural Phenomena* 14(3): 313–323.

Atangana, A., & Araz, S.I., 2019. Fractional stochastic modelling illustration with modified Chua attractor. *The European Physical Journal Plus*, 134(160): 23.

Atangana, A., & Bonyah, E., 2019. Fractional stochastic modelling: New approach to capture more heterogeneity. *Chaos*, 29: 013118.

Brooks, R.H., & Corey, A.T., 1966. Properties of porous media affecting fluid flow. *Journal of the Irrigation and Drainage Division, ASCE*, 92(IR2): 61–88.

Caputo, M., & Fabrizio, M., (2015). A new definition of fractional derivative without singular kernel. *Progress in Fractional Differentiation and Applications*, 1(2): 87–92.

Ghoshdastidar, D., & Dukkipati, A., 2013. *On Power-law Kernels, Corresponding Reproducing Kernel Hilbert Space and Application*. Department of Computer Science and Automation. Indian Institute of Science, Bangalore. 7pp.

List, F., & Radu, F.A., 2015. A study on iterative methods for solving Richards' equation, *Computational Geosciences*, 20: 341–353.

Pelka, W., 1983. Heat and mass transport in saturated–unsaturated groundwater flow. *Relation of Groundwater Quantity and Quality*. IAHS Publ.no. 146: 160–166.

7 Transfer Function of the Sumudu, Laplace Transforms and Their Application to Groundwater

Rendani Vele Makahane and Abdon Atangana
University of the Free State, Bloemfontein, South Africa

CONTENTS

7.1 INTRODUCTION

Historically, the genesis of the integral transforms can be traced back to the 1970s from the work of P.S. Laplace (Maitama & Zhao, 2019). The Laplace transform is a potent tool for solving partial and ordinary differential equations, it transforms the ordinary differential equation into a simple algebraic expression. This can then be transformed back into the original solution (Schiff, 1999). A fascinating mathematical modeling technique called the transfer function uses the Laplace transform to make sense of complex processes. A transfer function is a mathematical relationship existing between the numerical input and the subsequent output in a dynamic system. For a linear time-invariant system, the transfer function is defined as a ratio of the Laplace transform of the output variable $Y(s) = \mathcal{L}\{y(t)\}$, to the Laplace transform of the input variable $F(s) = \mathcal{L}\{f(t)\}$ with all zero initial conditions (Zwart, 2004). Together with the Bode plots, the transfer function can show us the frequency response of a linear time-invariant system. This system could be any system that experiences a change in behavior causing a change in frequency; an example of such a system is the flow of groundwater. The theory surrounding the transfer functions of linear time-invariant systems has been available for many years and was used until recently predominantly in connection with electrical and mechanical systems described in continuous time (Pollock, 2011; Cook, 1979). There exists another integral transform called the Sumudu transform. This transform was introduced by G.K. Watugala (1993) to provide solutions to control engineering problems and differential equations. It is theoretically a dual of the Laplace transform (Panchal et al., 2016; Atangana & Kilicman, 2013). This new transform rivals the Laplace transform in solving problems. The main advantage of the Sumudu transform is that it can provide solutions to problems resorting to a new frequency domain, through its ability to preserve the same scales and physical unit properties (Panchal et al., 2016; Atangana & Kilicman, 2013). Other advantages of the Sumudu transform over the Laplace transform includes the resemblance between the function $f(x,t)$ in the (x,t) domain and the subsequent function $\mathcal{S}\{f(x,t)\}$ in the (u,v) domain, the equality of $f(x,t)$ and $\mathcal{S}\{f(x,t)\}$ for constant functions and the

DOI: 10.1201/9781003266266-7

limit of $f(t)$ as t tends to zero is equal to the limit of $F(u)$ as u tends to zero (Jarad & Tas, 2012). For these set of functions, the Sumudu transform is defined as (Kilicman & Eltayeb, 2012):

$$\mathcal{A} = \left\{ f(t) : \exists M, \tau_1, \tau_2 > 0, |f(t)| < Me^{t/\tau_j}, \text{ if } t \in (-1)^j \times [0, \infty) \right\}$$

By:

$$F(u) = \mathcal{S}[f(t)] = \int_0^\infty e^{-t} f(ut) dt, u \in (-\tau_1, \tau_2) \tag{7.1}$$

If $F(u)$ is the Sumudu transform of $f(t)$, then the Sumudu transform of the nth a derivative is as follows:

$$\mathcal{S}\left[\frac{d^n f(t)}{dt^n} \right] = u^{-n} \left[F(u) - \sum_{k=0}^{n-1} u^k \frac{d^n f(t)}{dt^n} \bigg|_{t=0} \right] \tag{7.2}$$

And the Sumudu of the first derivative is given by:

$$\mathcal{S}\left[\frac{df(t)}{dt} \right] = \frac{1}{u}\left[F(u) - f(0) - u \frac{df(t)}{dt} \bigg|_{t=0} \right] \tag{7.3}$$

And the Sumudu of the second derivative is given by:

$$\mathcal{S}\left[\frac{d^2 f(t)}{dt^2} \right] = \frac{1}{u^2}\left[F(u) - f(0) - u \frac{df(t)}{dt} \bigg|_{t=0} \right] \tag{7.4}$$

To illustrate the application of the Laplace and the Sumudu transform in groundwater, we consider the 1-d saturated classical groundwater flow equation. We also consider this equation using the Caputo, Caputo–Fabrizio and Atangana–Baleanu fractional derivatives. The aim of this analysis is to find out if we could get better results when using the Sumudu transform than the Laplace transform. We view our partial differential equation as a transfer function from time to space. One thing of importance in a transfer function is that once you have the transfer function of a system you can estimate the output of that system for any given input. Solutions for each case are detailed below. However, we shall note that the theory under analysis in this section is not yet fully developed. This will open doors for new investigations within the field of partial differential equations and their applications in real-world problems. The equation under consideration is given below:

$$(S_s) \frac{\partial \psi}{\partial t} = \frac{\partial}{\partial z}\left[K_s \left(\frac{\partial \psi}{\partial z} - 1 \right) \right] \tag{7.5}$$

Since the hydraulic conductivity of the saturated zone is constant, the above equation can be further simplified into:

$$\frac{\partial \psi}{\partial t} = \frac{K_s}{S_s} \frac{\partial^2 \psi}{\partial z^2} \tag{7.6}$$

7.2   APPLICATION OF THE LAPLACE TRANSFORM TO THE SATURATED GROUNDWATER EQUATION

Let us recall the 1-d saturated groundwater flow equation (7.6). Applying the Laplace transform in time, we find:

$$\mathcal{L}\left(\frac{\partial \psi}{\partial t}\right) = \mathcal{L}\left(\frac{K_s}{S_s}\frac{\partial^2 \psi}{\partial z^2}\right) \tag{7.7}$$

$$s\tilde{\psi} - \psi(0) = \mathcal{L}\left(\frac{K_s}{S_s}\frac{\partial^2 \psi}{\partial z^2}\right) \tag{7.8}$$

$$s\tilde{\psi} - \psi(0) = \frac{K_s}{S_s}\frac{\partial^2 \tilde{\psi}}{\partial z^2} \tag{7.9}$$

$$\frac{\tilde{\psi} - \psi(0)}{\dfrac{K_s}{S_s}\dfrac{\partial^2 \tilde{\psi}}{\partial z^2}} = \frac{1}{s} = R(s) \tag{7.10}$$

Now we find the Laplace transform in space:

$$\mathcal{L}\left(\frac{\partial \psi}{\partial t}\right) = \frac{K_s}{S_s}\left(s^2\psi(z,s) - s\psi(z,0) + \psi'(z,0)\right) \tag{7.11}$$

$$\frac{\dfrac{\partial \tilde{\psi}}{\partial t}\dfrac{S_s}{K_s} + s\tilde{\psi}(z,s) + \psi'(z,0)}{\psi(z,s)} = s^2 \tag{7.12}$$

$$\frac{\psi(z,s)}{\dfrac{\partial \tilde{\psi}}{\partial t}\dfrac{S_s}{K_s} + s\tilde{\psi}(z,0) + \psi'(z,0)} = \frac{1}{s^2} = R(s) \tag{7.13}$$

For the Caputo case we have:

$$_0^C D_t^\alpha \psi(z,t) = \frac{K_s}{S_s}\frac{\partial^2 \psi(z,t)}{\partial z^2} \tag{7.14}$$

$$\frac{1}{\Gamma(\alpha)}\int_0^t \frac{\partial \psi(z,t)}{\partial t}(t-\tau)^{-\alpha}\, d\tau = \frac{K_s}{S_s}\frac{\partial^2 \psi(z,t)}{\partial z^2} \tag{7.15}$$

Applying the Laplace transform on both sides with respect to t we have:

$$s^\alpha\tilde{\psi}(z,s) - s^{\alpha-1}\psi(z,s) = \frac{K_s}{S_s}\frac{\partial^2 \tilde{\psi}(z,s)}{\partial z^2} \tag{7.16}$$

$$\frac{\tilde{\psi}(z,s) - s^{-1}\psi(z,s)}{\dfrac{K_s}{S_s}\dfrac{\partial^2 \tilde{\psi}(z,s)}{\partial z^2}} = \frac{1}{s^\alpha} = R^\alpha(s) \tag{7.17}$$

For the Laplace transform with respect to space, we get a solution like the one above.

Now, with Caputo–Fabrizio we have:

$$
{}_{0}^{CF}D_{t}^{\alpha}\psi\left(z,t\right) = \frac{K_{s}}{S_{s}}\frac{\partial^{2}\psi\left(z,t\right)}{\partial z^{2}} \tag{7.18}
$$

$$
\frac{M\left(\alpha\right)}{1-\alpha}\int_{0}^{t}\frac{d\psi\left(z,\tau\right)}{d\tau}\exp\left[-\frac{\alpha}{1-\alpha}\left(t-\tau\right)\right]d\tau = \frac{K_{s}}{S_{s}}\frac{\partial^{2}\psi\left(z,t\right)}{\partial z^{2}} \tag{7.19}
$$

$$
\frac{M\left(\alpha\right)}{1-\alpha}\frac{d\psi\left(z,t\right)}{dt}*\exp\left[-\frac{\alpha}{1-\alpha}t\right] = \frac{K_{s}}{S_{s}}\frac{\partial^{2}\psi\left(z,t\right)}{\partial z^{2}} \tag{7.20}
$$

Applying the Laplace transform on both sides of Equation (7.20) with respect to time we get:

$$
\frac{M\left(\alpha\right)}{1-\alpha}\left(s\tilde{\psi}\left(z,s\right)-\psi\left(z,0\right)\right)\frac{1}{s+\dfrac{\alpha}{1-\alpha}} = \frac{K_{s}}{S_{s}}\frac{\partial^{2}\tilde{\psi}\left(z,s\right)}{\partial z^{2}} \tag{7.21}
$$

$$
s\tilde{\psi}\left(z,s\right)-\psi\left(z,0\right) = \left(s+\frac{\alpha}{1-\alpha}\right)\frac{1-\alpha}{M\left(\alpha\right)}\frac{K_{s}}{S_{s}}\frac{\partial^{2}\tilde{\psi}\left(z,s\right)}{\partial z^{2}} \tag{7.22}
$$

$$
\tilde{\psi}\left(z,s\right) = \frac{\psi\left(z,0\right)}{s}+\frac{\left(s+\dfrac{\alpha}{1-\alpha}\right)}{s}\frac{1-\alpha}{sM\left(\alpha\right)}\frac{K_{s}}{S_{s}}\frac{\partial^{2}\tilde{\psi}\left(z,s\right)}{\partial z^{2}} \tag{7.23}
$$

We find the same solution for the Laplace transform in space.
Now, with the Atangana–Baleanu derivative, we have:

$$
{}_{0}^{ABC}D_{t}^{\alpha}\psi\left(z,t\right) = \frac{K_{s}}{S_{s}}\frac{\partial^{2}\psi\left(z,t\right)}{\partial z^{2}} \tag{7.24}
$$

$$
\frac{AB\left(\alpha\right)}{1-\alpha}\int_{0}^{t}\frac{d\psi\left(z,\tau\right)}{d\tau}E_{\alpha}\left[-\frac{\alpha}{1-\alpha}\left(t-\tau\right)^{\alpha}\right]d\tau = \frac{K_{s}}{S_{s}}\frac{\partial^{2}\psi\left(z,t\right)}{\partial z^{2}} \tag{7.25}
$$

Applying the Laplace transform, we obtain:

$$
\frac{AB\left(\alpha\right)}{1-\alpha}\left(s\tilde{\psi}\left(z,s\right)-\psi\left(z,0\right)\right)\frac{s^{\alpha-1}}{s^{\alpha}+\dfrac{\alpha}{1-\alpha}} = \frac{K_{s}}{S_{s}}\frac{\partial^{2}\tilde{\psi}\left(z,s\right)}{\partial z^{2}} \tag{7.26}
$$

$$
\frac{s\tilde{\psi}\left(z,s\right)-\psi\left(z,0\right)}{\dfrac{K_{s}}{S_{s}}\dfrac{\partial^{2}\tilde{\psi}\left(z,s\right)}{\partial z^{2}}} = \frac{1-\alpha}{AB\left(\alpha\right)}\frac{s^{\alpha}+\dfrac{\alpha}{1-\alpha}}{s^{\alpha-1}} = R\left(s\right) \tag{7.27}
$$

We obtain the same solution for the Laplace transform in space.

7.3 APPLICATION OF THE SUMUDU TRANSFORM TO THE SATURATED GROUNDWATER EQUATION

Let us recall the 1-d saturated groundwater flow equation (7.6). Applying the Sumudu transform in time, we find:

$$S\left(\frac{\partial \psi}{\partial t}\right) = S\left(\frac{K_s}{S_s} \frac{\partial^2 \psi}{\partial z^2}\right) \tag{7.28}$$

$$\frac{S(\psi) - \psi(0)}{p} = \frac{K_s}{S_s} S\left(\frac{\partial^2 \psi}{\partial z^2}\right) \tag{7.29}$$

$$\frac{S(\psi) - \psi(0)}{\frac{K_s}{S_s} S\left(\frac{\partial^2 \psi}{\partial z^2}\right)} = p = R(p) \tag{7.30}$$

$$\frac{\tilde{\psi}(z,p) - \psi(0)}{\frac{K_s}{S_s} \frac{\partial^2 \tilde{\psi}(z,p)}{\partial z^2}} = p = R(p) \tag{7.31}$$

Now we find the Sumudu transform in space:

$$S_z\left(\frac{\partial \psi}{\partial t}\right) = S_z\left(\frac{K_s}{S_s} \frac{\partial^2 \psi}{\partial z^2}\right) \tag{7.32}$$

$$\frac{\partial \tilde{\psi}}{\partial t} = \frac{K_s}{S_s} S_z\left(\frac{\partial^2 \psi}{\partial z^2}\right) \tag{7.33}$$

$$\frac{\partial \tilde{\psi}}{\partial t} = \frac{K_s}{S_s} \frac{\tilde{\psi}(z,p) - \psi(z,0)}{p^2} - \frac{\psi'(z,0)}{p} \tag{7.34}$$

$$\frac{K_s}{S_s} \frac{\tilde{\psi}(z,p) - \psi(z,0)}{\frac{\partial \tilde{\psi}}{\partial t}} - \psi'(z,0) p = p^2 = R(p) \tag{7.35}$$

For the Caputo case we have:

$$_0^C D_t^\alpha \psi(z,t) = \frac{K_s}{S_s} \frac{\partial^2 \psi(z,t)}{\partial z^2} \tag{7.36}$$

$$\frac{1}{\Gamma(\alpha)} \int_0^t \frac{\partial \psi(z,t)}{\partial t} (t-\tau)^{-\alpha} d\tau = \frac{K_s}{S_s} \frac{\partial^2 \psi(z,t)}{\partial z^2} \tag{7.37}$$

Applying the Sumudu transform on both sides with respect to t we have:

$$\frac{S\psi(z,p) - \psi(z,0)}{p^\alpha} = \frac{K_s}{S_s} S\left(\frac{\partial^2 \psi(z,p)}{\partial z^2}\right) \tag{7.38}$$

$$\frac{\tilde{\psi}(z,p) - \psi(z,0)}{\frac{K_s}{S_s} \frac{\partial^2 \tilde{\psi}(z,p)}{\partial z^2}} = p^\alpha = R(p) \tag{7.39}$$

For the Sumudu transform with respect to space, we obtain the same solution as the one above. Now, with Caputo–Fabrizio we have:

$$
{}^{CF}_{0}D_t^\alpha \psi(z,t) = \frac{K_s}{S_s} \frac{\partial^2 \psi(z,t)}{\partial z^2}
\tag{7.40}
$$

$$
\frac{M(\alpha)}{1-\alpha} \int_0^t \frac{d\psi(z,\tau)}{d\tau} \exp\left[-\frac{\alpha}{1-\alpha}(t-\tau)\right] d\tau = \frac{K_s}{S_s} \frac{\partial^2 \psi(z,t)}{\partial z^2}
\tag{7.41}
$$

Applying the Sumudu transform on both sides of Equation (7.40) with respect to time we get:

$$
\frac{M(\alpha)}{1-\alpha} \mathcal{S}\left(\int_0^t \frac{d\psi(z,\tau)}{d\tau} \exp\left[-\frac{\alpha}{1-\alpha}(t-\tau)\right] d\tau \right) = \frac{K_s}{S_s} \mathcal{S}\left(\frac{\partial^2 \psi(z,p)}{\partial z^2} \right)
\tag{7.42}
$$

$$
\left(\mathcal{S}(\psi) - \psi(z,0) \right) \frac{M(\alpha)}{\alpha p + 1 - \alpha} = \frac{K_s}{S_s} \frac{\partial^2 \tilde{\psi}(z,p)}{\partial z^2}
\tag{7.43}
$$

$$
\frac{\left(\tilde{\psi}(z,p) - \psi(z,0) \right) M(\alpha)}{\alpha p + 1 - \alpha} = \frac{K_s}{S_s} \frac{\partial^2 \tilde{\psi}(z,p)}{\partial z^2}
\tag{7.44}
$$

$$
\frac{\tilde{\psi}(z,p) - \psi(z,0)}{\dfrac{K_s}{S_s} \dfrac{\partial^2 \tilde{\psi}(z,p)}{\partial z^2}} = \frac{\alpha p + 1 - \alpha}{M(\alpha)} = R(p)
\tag{7.45}
$$

We find the same solution for the Sumudu transform in space. Now, with the Atangana–Baleanu derivative, we have:

$$
{}^{ABC}_{0}D_t^\alpha \psi(z,t) = \frac{K_s}{S_s} \frac{\partial^2 \psi(z,t)}{\partial z^2}
\tag{7.46}
$$

$$
\frac{AB(\alpha)}{1-\alpha} \int_0^t \frac{d\psi(z,\tau)}{d\tau} E_\alpha\left[-\frac{\alpha}{1-\alpha}(t-\tau)^\alpha\right] d\tau = \frac{K_s}{S_s} \frac{\partial^2 \psi(z,t)}{\partial z^2}
\tag{7.47}
$$

Applying the Sumudu transform with respect to time, we obtain:

$$
\frac{AB(\alpha)}{1-\alpha} \mathcal{S}\left(\int_0^t \frac{d\psi(z,\tau)}{d\tau} E_\alpha\left[-\frac{\alpha}{1-\alpha}(t-\tau)^\alpha\right] d\tau \right) = \frac{K_s}{S_s} \mathcal{S}\left(\frac{\partial^2 \psi(z,t)}{\partial z^2} \right)
\tag{7.48}
$$

$$
\frac{\left(\tilde{\psi}(z,p) - \psi(z,0) \right) AB(\alpha)}{1 - \alpha + \alpha p^\alpha} = \frac{K_s}{S_s} \frac{\partial^2 \tilde{\psi}(z,p)}{\partial z^2}
\tag{7.49}
$$

$$
\frac{\tilde{\psi}(z,p) - \psi(z,0)}{\dfrac{K_s}{S_s} \dfrac{\partial^2 \tilde{\psi}(z,p)}{\partial z^2}} = \frac{1 - \alpha + \alpha p^\alpha}{AB(\alpha)} = R(s)
\tag{7.50}
$$

We obtain the same solution for the Sumudu transform in space.

7.4   BODE PLOTS OF THE LAPLACE AND SUMUDU TRANSFORM

A Bode plot is a graph that shows the system's frequency response. It was initially considered by Hendrick Wade Bode in the 1930s. It combines two logarithmic plots, one expressing the magnitude, and the other the phase shift of a system with respect to a given input frequency (York, 2009). The x-axis displays frequency, whereas the y axis displays magnitude and phase angle (Summer, 2004). The Bode plot is used to test and analyze the filters of a system. Two types of filters are recognized, a high pass filter allowing passage of signals with frequencies higher than the cut-off frequency, and restricts signals with frequencies lower than the cut-off frequency. The second filter is a low pass filter; it permits signals with a frequency of less than a selected cut-off frequency to pass and restricts signals with frequencies greater than the cut-off frequency. In the next few pages, we shall give the subsequent Bode plots for the Laplace and Sumudu transforms with respect to time and space as well as with Caputo, Caputo–Fabrizio and Atangana–Baleanu derivatives. Analysis and comparison between the Bode plots are also presented. The following plots were produced using MatLab software (Figures 7.1 to 7.6).

From the Bode plots presented above, it is evident that the plots do not represent similar behaviors for the system. Most plots for the Laplace transform indicate that the system is a low pass filter as we observe the most gain at low frequencies, and less gain at higher frequencies. All the Sumudu transform Bode plots and Laplace transform with the Atangana–Baleanu derivative and with respect to time indicates that the system is a high pass filter as it provides less gain at low frequencies and the most gain at high frequencies. Now, the bigger question here is, which transform between the two represents the correct behavior of the system, or which transform could give better results? However, from analyzing the two transform, one could suggest that the Sumudu give better results due to the advantages it has over the Laplace transform.

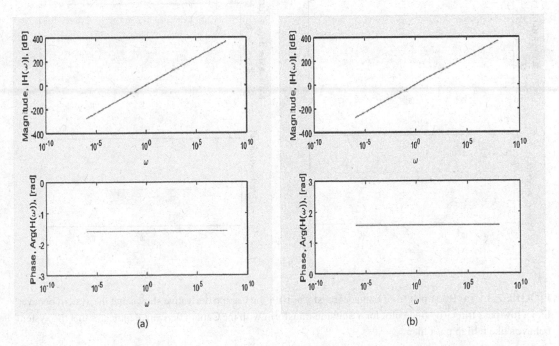

FIGURE 7.1   (a) Bode plot of a Laplace transform in time indicates that the system behaves like a high pass filter. (b) Bode plot of a Sumudu transform in time indicates that the system behaves like a high pass filter.

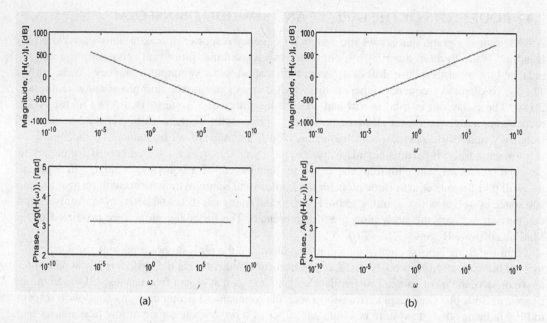

FIGURE 7.2 (a) Bode plot for a Laplace transform in space indicates that the system behaves like a low pass filter. (b) Bode plot for a Sumudu transform in space indicates that the system behaves like a high pass filter.

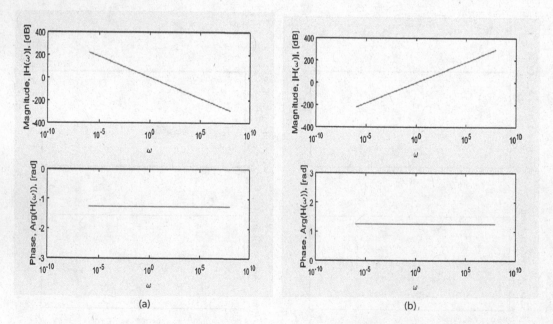

FIGURE 7.3 (a) Bode plot for a Laplace transform with the Caputo derivative shows that the system behaves like a low pass filter. (b) Bode plot for a Sumudu transform with the Caputo derivative indicates that the system behaves like a high pass filter.

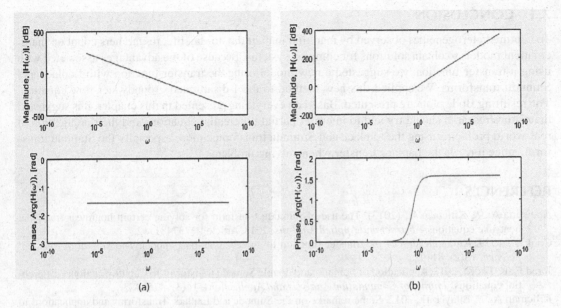

FIGURE 7.4 (a) Bode plot of a Laplace transform with the Caputo-Fabrizio derivative shows that the system behaves like a low pass filter. (b) Bode plot of a Sumudu transform with the Caputo-Fabrizio derivative shows that the system behaves like a high pass filter.

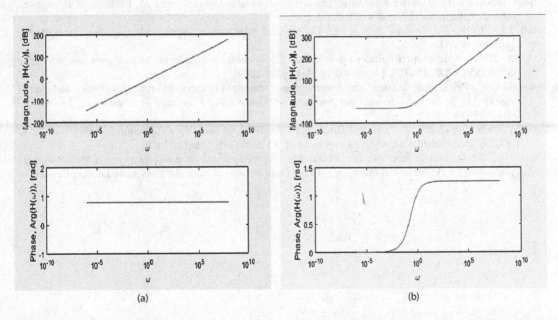

FIGURE 7.5 (a) Bode plot of the Laplace transform with Atangana-Baleanu derivative indicates that the system behaves like a low pass filter. (b) Bode plot of the Sumudu transform with Atangana-Baleanu derivative indicates that the system behaves like a high pass filter.

7.5   CONCLUSION

To capture heterogeneities observed by mankind in their day-to-day life, researchers count on mathematical models to obtain solutions for complex systems. Because of the advantages associated with using a transfer function, we suggested a new model using the transfer function with Laplace and Sumudu transforms. We applied this new method to the 1-d saturated groundwater flow equation. The resulting Bode plots are presented. Based on everything presented in this chapter, it is suggested that future research should try and to interpret partial differential equations and their application to real-world problems using the Laplace and Sumudu transforms; more especially the Sumudu transform, since it rivals the Laplace transfer when solving problems.

REFERENCES

Atangana A., & Kilicman A., (2013). The use of Sumudu transform for solving certain nonlinear fractional heat-like equations. *Abstract and Applied Analysis*, 2013, Article ID737481, 12.

Cook T., 1979. An application of the transfer function to an economic-base model. *The Annual of Regional Science*, 13(2): 81–92.

Jarad F., & Tas K., 2012. Application of Sumudu and double Sumudu transforms to Caputo-fractional differential equations. *Journal of Computation Analysis and Applications*, 14(3): 475–483.

Kilicman A., & Eltayeb H., 2012. Some remarks on the Sumudu and Laplace Transforms and applications to differential equation. *ISRN Applied Mathematics*, 2012: 13.

Maitama S., & Zhao W., 2019. New integral transform: Shehu transforms generalization of Sumudu and Laplace transform for solving differential equations. *International Journal of Analysis and Applications* 17(2): 167–190.

Panchal S.K., & Khandagale A.D., Dole P.V., 2016. Sumudu transform of Hilfer–Prabhakar fractional derivatives and applications. arXiv:1608.08017

Pollock D.S.G., 2011. *Transfer Functions, Discussion Papers in Economics 11/15*, Division of Economics, School of Business, University of Leicester.

Schiff J.L., 1999. *The Laplace Transform: Theory and Applications*. Springer-Verlag, New York/Berlin/Heidelberg. 233.

Summer, 2004. Bode plots. Engineering Sciences 22-Systems. Available at: https://www.academia.edu/38586755/BODE_PLOTS. [Accessed 10 November 2019]

Watugala G.K., 1993. Sumudu transform:A new integral transform to solve differential equations and control engineering problems. *International Journals of Mathematical Education in Science and Technology*, 24(1): 35–43.

York B., 2009. Frequency response and Bode plots. Available at: https://web.njit.edu/~levkov/classes_files/ECE232/Handouts/Frequency%20Response.pdf. [Accessed 08 October 2019]

Zwart, H.J., 2004. Transfer functions for infinite-dimensional systems. *Proceedings of the 16th international symposium on mathematical theory of networks and systems*. Katholieke University Leuven, Leuven.

8 Analyzing the New Generalized Equation of Groundwater Flowing within a Leaky Aquifer Using Power Law, Exponential Decay Law and Mittag–Leffler Law

Amanda Ramotsho and Abdon Atangana
University of the Free State, Bloemfontein, South Africa

CONTENTS

8.1 INTRODUCTION

Nonlocal derivatives have been identified as useful tools in many branches. They are widely used in physics, biology, engineering and other fields of science. Nonlocal operators are applied, or rather used, in modeling. Several models are used to describe or assist in anomalous diffusion processes, viscoelasticity, signal processing, geomorphology, materials sciences, fractals and so on. These derivatives are capable of capturing long-term interactions and to describe memory effect. It has been captured in literature that, if the memory can be described by fractional derivatives, the integral

DOI: 10.1201/9781003266266-8

describes the reverse memory as the inverse operator of the derivatives (Atangana and Gomez, 2018). The three fundamental mathematical laws mainly observed in nature, of which physical scenarios follow, are power law, exponential law, and Mittag–Leffler law.

The presence of power law behavior in nature is such an extremely common phenomenon that considerable lore has now grown up concerning its genesis. The onset of power law behavior actually occurs at a much more primitive level and can be analyzed directly in terms of the underlying differential equations (Visser and Yunes, 2008). Power law has recently been used to model nature and its complexities, but unfortunately the singularity associated with power law disadvantages its use (Atangana and Gomez, 2018). The memory with power law is usually not complete and has no initial conditions or beginning. There are two known power law operators, which are Riemann–Liouville and the Caputo operator, which are explained in the context of this chapter.

Exponential decay law appears in many scenarios. That the law was applied way back without proper definition and use, was already apparent in Schrödinger's solution of the hydrogen atom, but it is only recently that a satisfactory mathematical theory has been developed for the problem. In contrast to the power law, it can capture the initial or beginning condition, and it is complete. The well-known Caputo–Fabrizio operator resemble this specific law. The ordinary and generalized Mittag–Leffler functions interpolate between a purely exponential law and power law (Haubold et al., 2011). Thus Atangana and Baleanu (2017) used the generalized Mittag–Leffler function to construct a derivative with no singular or nonlocal kernel.

8.2 POWER LAW OPERATORS

8.2.1 Riemann–Liouville Fractional Derivative

The Riemann–Liouville operator was the most fractional derivative to be used before the newly proposed different fractional-time operators. The derivative sometimes acts like classical operators, because the memory is not always associative or commutative (Atangana, 2017). This tool has been identified as the only mathematical tool to assist in assessing all physical problems. It is given as:

$$_a D_t^a f(t) = \frac{d^n}{dt^n} \, _a D_t^{-(n-a)} f(t) = \frac{d^n}{dt^n} \, _a I_t^{(n-a)} f(t) \tag{8.1}$$

8.2.2 Caputo Fractional Derivative

Caputo made the first contribution to the Riemann–Liouville operators by transforming a fractional differential operator from a derivative of a convolution, to a convolution of a derivative with power to obtain normal initial condition rather than an initial condition with no physical meaning and difficult to compute (Atangana, 2018). The Caputo fractional-time derivative it is given as,

$$D_t^a f(t) = \frac{1}{\Gamma(1-a)} \int_a^t \frac{f(\tau)}{(t-\tau)^a} \, d\tau \tag{8.2}$$

The Caputo derivative is introduced into the new generalized equation of leaky aquifers: Ramotsho and Atangana (2018)

$$\frac{S}{T} h(r,t) = \frac{\partial}{\partial r}\left(\frac{\partial h}{\partial r}\right)\left(1 + \frac{\Delta r}{r}\right) + \frac{1}{r} h(r,t) + \frac{h(r,t)}{\lambda^2} \tag{8.3}$$

The new generalized equation with the Caputo derivative can be given as follows:

$$\frac{S}{T} {}_0^c D_t^\beta h(r,t) = \frac{\partial}{\partial r}\left(\frac{\partial h}{\partial r}\right)\left(1+\frac{\Delta r}{r}\right) + \frac{1}{r} {}_0^c D_r^\alpha h(r,t) + \frac{h(r,t)}{\lambda^2}$$

(8.4)

8.2.2.1 Applying the Crank–Nicolson Scheme into the Classical New Groundwater Equation of Flow within a Leaky Aquifer

Numerical tools are used to solve PDEs with local and nonlocal operators. The Crank–Nicolson scheme is one of the most stable and reliable numerical schemes. The Crank–Nicolson scheme is a second-order method in time and it is implicit in time. Applying Crank–Nicolson to Equation (8.5), the following is obtained:

We recall that,

$$D_t^a f(t) = \frac{1}{\Gamma(1-a)} \int_0^t \frac{d}{d\tau} h(\tau)(t-\tau)^{-a} d\tau$$

(8.5)

Therefore,

$$D_t^a f(t_{n+1}) = \frac{1}{\Gamma(1-a)} \sum_{j=0}^{n} \int_{t_j}^{t_{j+1}} \frac{h^{j+1}-h^j}{\Delta t}(t_{n+1}-\tau)^{-a} d\tau$$

(8.6)

$$D_t^a f(t_{n+1}) = \frac{1}{\Gamma(1-a)} \sum_{j=0}^{n} \frac{h^{j+1}-h^j}{\Delta t} \int_{t_j}^{t_{j+1}} (t_{n+1}-\tau)^{-a} d\tau$$

(8.7)

Let,

$$\int_{t_j}^{t_{j+1}} (t_{n+1}-\tau)^{-a} d\tau = \sigma_{n,j}^\alpha$$

(8.8)

and,

$$\int_{t_j}^{t_{j+1}} (t_{n+1}-\tau)^{-\beta} d\tau = \sigma_{n,j}^\beta$$

(8.9)

Therefore,

$$\frac{S}{T}\sum_{j=0}^{n}\left(h_i^{j+1}-h_i^j\right)\sigma_{n,j}^\beta = \frac{\partial^2 h}{\partial r^2}\left(1+\frac{\Delta r}{r}\right) + \frac{1}{r}\sum_{l=0}^{m}\left(h_{i+1}^{n+1}-h_{i-1}^{n+1}\right)\sigma_{n,j}^\alpha$$

$$+\frac{h(r,t)}{\lambda^2} = \frac{1}{2}\left(\frac{h_{i+1}^{n+1}-2h_i^{n+1}+h_{i-1}^{n+1}}{(\Delta r)^2} + \frac{h_{i+1}^n-2h_i^n+h_{i-1}^n}{(\Delta r)^2}\right)\left(1+\frac{\Delta r}{r_i}\right)$$

$$+\frac{1}{r_i}\sum_{l=0}^{m}\left(h_{i+1}^{n+1}-h_{i-1}^{n+1}\right)\sigma_{n,j}^\alpha + \frac{h_i^{n+1}}{\lambda^2}$$

(8.10)

In order to simplify the notation, let

$$\lambda_1 = \frac{S}{T}\sigma_{n,j}^{\beta}$$

$$\lambda_2 = \frac{1}{2(\Delta r)^2}\left(1+\frac{\Delta r}{r_i}\right)$$

$$\lambda_3 = \frac{1}{r_i}\sigma_{n,j}^{\alpha}$$

$$\lambda_4 = \frac{1}{\lambda^2}$$

$$\sum_{j=0}^{n-1}\left(h_i^{j+1}-h_i^{j}\right)\lambda_1 + \lambda_1\left(h_i^{n+1}-h_i^{n}\right) = \lambda_2\left(h_{i+1}^{n+1}-2h_i^{n+1}+h_{i-1}^{n+1}+h_{i+1}^{n}-2h_i^{n}+h_{i-1}^{n}\right)$$
$$+ \lambda_3\sum_{l=0}^{m}\left(h_{i+1}^{n+1}-h_{i-1}^{n+1}\right)+\lambda_4 h_i^{n+1} \tag{8.11}$$

$$\sum_{j=0}^{n-1}\left(h_i^{j+1}-h_i^{j}\right)\lambda_1 = \lambda_2\left(h_{i+1}^{n+1}-2h_i^{n+1}+h_{i-1}^{n+1}+h_{i+1}^{n}-2h_i^{n}+h_{i-1}^{n}\right)$$
$$- \lambda_1\left(h_i^{n+1}-h_i^{n}\right)+\lambda_3\sum_{l=0}^{m}\left(h_{i+1}^{n+1}-h_{i-1}^{n+1}\right)+\lambda_4 h_i^{n+1} \tag{8.12}$$

$$h_i^{n+1}\left(\lambda_1+2\lambda_2-\lambda_4\right) = h_i^{n}\left(\lambda_1-2\lambda_2\right)+\lambda_2\left(h_{i+1}^{n+1}+h_{i-1}^{n+1}+h_{i+1}^{n}+h_{i-1}^{n}\right)$$
$$- \sum_{l=0}^{n-1}\left(h_i^{j+1}-h_i^{j}\right)\lambda_1+\lambda_3\sum_{l=0}^{m}\left(h_{i+1}^{n+1}-h_{i-1}^{n+1}\right) \tag{8.13}$$

The final numerical equation with the Crank–Nicolson method is presented.

8.2.2.1.1 Stability Analysis

The stability was verified using the Fourier series, iteration method and recursion method. The Fourier series is a very useful technique used to analyze the stability of the finite difference method. The stability analysis is given below:

$$h_i^{n+1}\left(\lambda_1+2\lambda_2-\lambda_4\right) = h_i^{n}\left(\lambda_1-2\lambda_2\right)+\lambda_2\left(h_{i+1}^{n+1}+h_{i-1}^{n+1}+h_{i+1}^{n}+h_{i-1}^{n}\right)$$
$$- \sum_{j=0}^{n-1}\left(h_i^{j+1}-h_i^{j}\right)\lambda_1+\lambda_3\sum_{l=0}^{m-1}\left(h_{i+1}^{l+1}-h_{i-1}^{l+1}\right) \tag{8.14}$$

$$\delta_{n+1}e^{iklr}\left(\lambda_1+2\lambda_2-\lambda_4\right) = \delta_{n}e^{iklr}\left(\lambda_1-2\lambda_2\right)+\lambda_2\left(\delta_{n+1}e^{ikl(r+\Delta r)}+\delta_{n+1}e^{ikm(r-\Delta r)}+\delta_{n}e^{ikl(r+\Delta r)}+\delta_{n}e^{ikl(r-\Delta r)}\right)$$
$$- \sum_{j=0}^{n-1}\left(\delta_{j+1}e^{iklr}-\delta_{j}e^{iklr}\right)\lambda_1+\lambda_3\sum_{l=0}^{m-1}\left(\delta_{l+1}e^{ikl(r+\Delta r)}-\delta_{l+1}e^{ikl(r-\Delta r)}\right)$$

$$\tag{8.15}$$

Divide both sides by e^{iklr} and group the like terms,

$$\delta_{n+1}\left(\lambda_1 + 2\lambda_2 - \lambda_4\right) = \delta_n\left(\lambda_1 - 2\lambda_2\right) + \lambda_2\delta_{n+1}e^{ikl\Delta r} + \lambda_2\delta_{n+1}e^{-ikl\Delta r}$$

$$+ \lambda_2\delta_n e^{ikl\Delta r} + \lambda_2\delta_n e^{-ikl\Delta r} - \sum_{j=0}^{n-1}\left(\delta_{j+1} - \delta_j\right)\lambda_1 + \lambda_3\sum_{l=0}^{m-1}\left(\delta_{l+1}e^{ikl\Delta r} - \delta_{l+1}e^{-ikl\Delta r}\right)$$

$$(8.16)$$

$$\delta_{n+1}\left(\lambda_1 + 2\lambda_2 - \lambda_4 - \lambda_2 e^{ikl\Delta r} - \lambda_2 e^{-ikl\Delta r}\right) = \delta_n\left(\lambda_1 - 2\lambda_2 + \lambda_2 e^{ikl\Delta r} + \lambda_2 e^{-ikl\Delta r}\right)$$

$$-\sum_{j=0}^{n-1}\left(\delta_{j+1} - \delta_j\right)\lambda_1 + \lambda_3\sum_{l=0}^{m-1}\delta_{l+1}\left(e^{ikl\Delta r} - e^{-ikl\Delta r}\right) \quad (8.17)$$

Therefore,

$$\delta_{n+1}\left(\lambda_1 + 2\lambda_2 - \lambda_4 - \lambda_2\left[2\cos\left(kl\Delta r\right)\right]\right) = \delta_n\left(\lambda_1 - 2\lambda_2 + \lambda_2\left[2\cos\left(kl\Delta r\right)\right]\right)$$

$$-\sum_{j=0}^{n-1}\left(\delta_{n-j}\right)\lambda_1 + \lambda_3\sum_{l=0}^{m-1}\left(\delta_{l+1}\left[2i\sin\left(kl\Delta r\right)\right]\right) \quad (8.18)$$

If,

$$n = 0$$

Therefore,

$$\delta_{n+1}\left(\lambda_1 + 2\lambda_2 - \lambda_4 - \lambda_2\left[2\cos\left(kl\Delta r\right)\right]\right) = \delta_n\left(\lambda_1 - 2\lambda_2 + \lambda_2\left[2\cos\left(kl\Delta r\right)\right]\right)$$

$$-\lambda_3\sum_{l=0}^{m-1}\left(\delta_{l+1}\left[2i\sin\left(kl\Delta r\right)\right]\right) \quad (8.19)$$

Applying the continuous recursion,
If,

$$m = 0$$

Therefore,

$$\delta_{n+1}\left(\lambda_1 + 2\lambda_2 - \lambda_4 - \lambda_2\left[2\cos\left(kl\Delta r\right)\right]\right) = \delta_n\left(\lambda_1 - 2\lambda_2 + \lambda_2\left[2\cos\left(kl\Delta r\right)\right]\right) \quad (8.20)$$

To check the stability of the above equation, iteration and recursion methods will be applied.
To prove that,

$$\left|\frac{\delta_1}{\delta_0}\right| < 1$$

Let,

$$n = 0$$

$$\left| \frac{\left(\lambda_1 - 2\lambda_2 + \lambda_2 \left[2\cos\left(kl\Delta r\right) \right] \right)}{\left(\lambda_1 + 2\lambda_2 - \lambda_4 - \lambda_2 \left[2\cos\left(kl\Delta r\right) \right] \right)} \right| < 1 \tag{8.21}$$

Assuming that the following different conditions were applied:

Condition 1
If,

$$\lambda_1 - 2\lambda_2 + \lambda_2 \left[2\cos\left(kl\Delta r\right) \right] > 0 \tag{8.22}$$

and,

$$\lambda_1 + 2\lambda_2 - \lambda_4 - \lambda_2 \left[2\cos\left(kl\Delta r\right) \right] > 0 \tag{8.23}$$

Therefore

$$\lambda_1 - 2\lambda_2 + \lambda_2 \left[2\cos\left(kl\Delta r\right) \right] < \lambda_1 + 2\lambda_2 - \lambda_4 - \lambda_2 \left[2\cos\left(kl\Delta r\right) \right] \tag{8.24}$$

$$\lambda_2 \left[4\cos\left(kl\Delta r\right) \right] < 4\lambda_2 - \lambda_4 \tag{8.25}$$

$$\cos\left(kl\Delta r\right) < 1 - \frac{\lambda_4}{4\lambda_2} \tag{8.26}$$

Condition 2
If,

$$\lambda_1 - 2\lambda_2 + \lambda_2 \left[2\cos\left(kl\Delta r\right) \right] < 0 \tag{8.27}$$

and,

$$\lambda_1 + 2\lambda_2 - \lambda_4 - \lambda_2 \left[2\cos\left(kl\Delta r\right) \right] < . \tag{8.28}$$

Therefore,

$$-\lambda_1 + 2\lambda_2 - \lambda_2 \left[2\cos\left(kl\Delta r\right) \right] < -\lambda_1 - 2\lambda_2 + \lambda_4 + \lambda_2 \left[2\cos\left(kl\Delta r\right) \right] \tag{8.29}$$

$$4\lambda_2 < \lambda_4 + 4\lambda_2 \left[\cos\left(kl\Delta r\right) \right] \tag{8.30}$$

$$1 < \frac{\lambda_4}{4\lambda_2} + \cos\left(kl\Delta r\right) \tag{8.31}$$

Condition 3
If,

$$\lambda_1 - 2\lambda_2 + \lambda_2 \left[2\cos\left(kl\Delta r\right) \right] > 0 \tag{8.32}$$

and,

$$\lambda_1 + 2\lambda_2 - \lambda_4 - \lambda_2 \left[2\cos\left(kl\Delta r\right) \right] < 0 \tag{8.33}$$

Therefore,

$$\lambda_1 - 2\lambda_2 + \lambda_2 \left[2\cos\left(kl\Delta r\right)\right] < -\lambda_1 - 2\lambda_2 + \lambda_4 + \lambda_2 \left[2\cos\left(kl\Delta r\right)\right] \tag{8.34}$$

$$2\lambda_1 < \lambda_4 \tag{8.35}$$

Condition 4
If,

$$\lambda_1 - 2\lambda_2 + \lambda_2 \left[2\cos\left(kl\Delta r\right)\right] < 0 \tag{8.36}$$

and,

$$\lambda_1 + 2\lambda_2 - \lambda_4 - \lambda_2 \left[2\cos\left(kl\Delta r\right)\right] > 0 \tag{8.37}$$

Therefore,

$$-\lambda_1 + 2\lambda_2 - \lambda_2 \left[2\cos\left(kl\Delta r\right)\right] < \lambda_1 + 2\lambda_2 - \lambda_4 - \lambda_2 \left[2\cos\left(kl\Delta r\right)\right] \tag{8.38}$$

$$-2\lambda_1 < -\lambda_4 \tag{8.39}$$

$$2\lambda_1 > \lambda_4 \tag{8.40}$$

We assume that when all integers are greater than 1,

$$\left|\delta_{l+1}\right| < \left|\delta_0\right|$$

When, n = 0, then at m+1

$$\delta_1 \left(\lambda_1 + 2\lambda_2 - \lambda_4 - \lambda_2 \left[2\cos\left(kl\Delta r\right)\right]\right) = \delta_0 \left(\lambda_1 - 2\lambda_2 + \lambda_2 \left[2\cos\left(kl\Delta r\right)\right]\right)$$
$$- \lambda_3 \sum_{l=0}^{m} \left(\delta_{l+1} \left[2i\sin\left(kl\Delta r\right)\right]\right) \tag{8.41}$$

$$\left|\delta_1(\lambda_1 + 2\lambda_2 - \lambda_4 - 2\lambda_2 \cos\left(kl\Delta r\right)\right| = \begin{vmatrix} \delta_0 \left(\lambda_1 - 2\lambda_2 + 2\lambda_2 \cos\left(kl\Delta r\right)\right) \\ - \lambda_3 \sum_{l=0}^{m} \left(\delta_{l+1} \left[2i\sin\left(kl\Delta r\right)\right]\right) \end{vmatrix} \tag{8.42}$$

$$\left|\delta_1\right|\left|(\lambda_1 + 2\lambda_2 - \lambda_4 - 2\lambda_2 \cos\left(kl\Delta r\right)\right| \le \left|\delta_0\right|\left|\left(\lambda_1 - 2\lambda_2 + 2\lambda_2 \cos\left(kl\Delta r\right)\right)\right|$$
$$+ \left|\lambda_3 \sum_{l=0}^{m} \left(\delta_{l+1} \left[2i\sin\left(kl\Delta r\right)\right]\right)\right| \tag{8.43}$$

$$\left|\delta_1\right|\left|(\lambda_1 + 2\lambda_2 - \lambda_4 - 2\lambda_2 \cos\left(kl\Delta r\right)\right| \le \left|\delta_0\right|\left|\left(\lambda_1 - 2\lambda_2 + 2\lambda_2 \cos\left(kl\Delta r\right)\right)\right|$$
$$+ \sum_{l=0}^{m} \left(\left|\delta_{l+1}\right|\left|\lambda_3\right|\left|2i\sin\left(kl\Delta r\right)\right|\right) \tag{8.44}$$

Hypothetically the above can be written as follows,

$$|\delta_1|\left|(\lambda_1 + 2\lambda_2 - \lambda_4 - 2\lambda_2 \cos(kl\Delta r))\right| < |\delta_0|\left|(\lambda_1 - 2\lambda_2 + 2\lambda_2 \cos(kl\Delta r))\right| + 2|\delta_0||\lambda_3| \sum_{l=0}^{m}(1) \quad (8.45)$$

$$|\delta_1|\left|(\lambda_1 + 2\lambda_2 - \lambda_4 - 2\lambda_2 \cos(kl\Delta r))\right| < |\delta_0|\left|(\lambda_1 - 2\lambda_2 + 2\lambda_2 \cos(kl\Delta r))\right| + 2|\delta_0||\lambda_3|(m+1) \quad (8.46)$$

$$\left|\frac{\delta_1}{\delta_0}\right| < \frac{\left|(\lambda_1 - 2\lambda_2 + 2\lambda_2 \cos(kl\Delta r))\right| + 2|\lambda_3|(m+1)}{\left|(\lambda_1 + 2\lambda_2 - \lambda_4 - 2\lambda_2 \cos(kl\Delta r))\right|} \quad (8.47)$$

$$\frac{\left|(\lambda_1 - 2\lambda_2 + 2\lambda_2 \cos(kl\Delta r))\right| + 2|\lambda_3|(m+1)}{\left|(\lambda_1 + 2\lambda_2 - \lambda_4 - 2\lambda_2 \cos(kl\Delta r))\right|} < 1 \quad (8.48)$$

Considering different conditions of Equation (8.70):

Condition 1
If,

$$\lambda_1 - 2\lambda_2 + 2\lambda_2 \cos(kl\Delta r) + 2\lambda_3(m+1) > 0 \quad (8.49)$$

and,

$$\lambda_1 + 2\lambda_2 - \lambda_4 - 2\lambda_2 \cos(kl\Delta r) > 0 \quad (8.50)$$

Therefore,

$$\lambda_1 - 2\lambda_2 + 2\lambda_2 \cos(kl\Delta r) + 2\lambda_3(m+1) < \lambda_1 + 2\lambda_2 - \lambda_4 - 2\lambda_2 \cos(kl\Delta r) \quad (8.51)$$

$$4\lambda_2 \cos(kl\Delta r) + 2\lambda_3(m+1) < 4\lambda_2 - \lambda_4 \quad (8.52)$$

$$\cos(kl\Delta r) + \frac{\lambda_3(m+1)}{2\lambda_2} + \frac{\lambda_4}{4\lambda_2} < 1 \quad (8.53)$$

Condition 2
If,

$$\lambda_1 - 2\lambda_2 + 2\lambda_2 \cos(kl\Delta r) + 2\lambda_3(m+1) < 0 \quad (8.54)$$

and,

$$\lambda_1 + 2\lambda_2 - \lambda_4 - 2\lambda_2 \cos(kl\Delta r) < 0 \quad (8.55)$$

Therefore,

$$-\lambda_1 + 2\lambda_2 - 2\lambda_2 \cos(kl\Delta r) - 2\lambda_3(m+1) < -\lambda_1 - 2\lambda_2 + \lambda_4 + 2\lambda_2 \cos(kl\Delta r) \quad (8.56)$$

$$4\lambda_2 - 2\lambda_3(m+1) < \lambda_4 + 4\lambda_2 \cos(kl\Delta r) \quad (8.57)$$

$$\cos(kl\Delta r) + \frac{\lambda_4}{4\lambda_2} + \frac{\lambda_3(m+1)}{2\lambda_2} > 1 \quad (8.58)$$

Condition 3

If,

$$\lambda_1 - 2\lambda_2 + 2\lambda_2 \cos\left(kl\Delta r\right) + 2\lambda_3\left(m+1\right) < 0 \tag{8.59}$$

and,

$$\lambda_1 + 2\lambda_2 - \lambda_4 - 2\lambda_2 \cos\left(kl\Delta r\right) > 0 \tag{8.60}$$

Or

If,

$$\lambda_1 - 2\lambda_2 + 2\lambda_2 \cos\left(kl\Delta r\right) + 2\lambda_3\left(m+1\right) > 0 \tag{8.61}$$

and,

$$\lambda_1 + 2\lambda_2 - \lambda_4 - 2\lambda_2 \cos\left(kl\Delta r\right) < 0 \tag{8.62}$$

The resulting solution for both conditions is the same.

$$-\lambda_1 + 2\lambda_2 - 2\lambda_2 \cos\left(kl\Delta r\right) - 2\lambda_3\left(m+1\right) < \lambda_1 + 2\lambda_2 - \lambda_4 - 2\lambda_2 \cos\left(kl\Delta r\right) \tag{8.63}$$

$$-2\lambda_3\left(m+1\right) < -\lambda_4 + 2\lambda_1 \tag{8.64}$$

$$\lambda_1 + \lambda_3\left(m+1\right) < \frac{\lambda_4}{2} \tag{8.65}$$

For all n > 0,

Then,

$$\left|\sum_{l=0}^{m} \delta_{l+1}\right| < \left|\delta_0\right|\left(m+1\right) \tag{8.66}$$

We assume that,

$$\left|\delta_n\right| < \left|\delta_0\right| \tag{8.67}$$

$$\delta_{n+1}\left(\lambda_1 + 2\lambda_2 - \lambda_4 - \lambda_2\left[2\cos\left(kl\Delta r\right)\right]\right) = \delta_n\left(\lambda_1 - 2\lambda_2 + \lambda_2\left[2\cos\left(kl\Delta r\right)\right]\right)$$
$$- \sum_{j=0}^{n-1}\left(\delta_{n-j}\right)\lambda_1 + \lambda_3 \sum_{l=0}^{m-1}\left(\delta_{l+1}\left[2i\sin\left(kl\Delta r\right)\right]\right) \tag{8.68}$$

$$\left|\delta_{n+1}\left(\lambda_1 + 2\lambda_2 - \lambda_4 - \lambda_2\left[2\cos\left(kl\Delta r\right)\right]\right)\right| = \left|\begin{array}{c}\delta_n\left(\lambda_1 - 2\lambda_2 + 2\lambda_2 \cos\left(kl\Delta r\right)\right) \\ - \sum_{j=0}^{n-1}\left(\delta_{n-j}\right)\lambda_1 + \lambda_3 \sum_{l=0}^{m-1}\left(\delta_{l+1}\left[2i\sin\left(kl\Delta r\right)\right]\right)\end{array}\right| \tag{8.69}$$

Considering the above assumptions:

$$\left| \delta_{n+1} \left(\lambda_1 + 2\lambda_2 - \lambda_4 - \lambda_2 \left[2\cos\left(kl\Delta r\right)\right]\right)\right| < \left| \begin{array}{c} \delta_0 \left(\lambda_1 - 2\lambda_2 + 2\lambda_2 \cos\left(kl\Delta r\right)\right) \\ -\sum_{j=0}^{n-1}\left(\delta_0\right)\lambda_1 + 2\lambda_3 \sum_{l=0}^{m-1}\left(\delta_0\right) \end{array}\right| \tag{8.70}$$

$$\left|\delta_{n+1}\right|\left|\left(\lambda_1 + 2\lambda_2 - \lambda_4 - \lambda_2\left[2\cos\left(kl\Delta r\right)\right]\right)\right| < \left|\delta_0\right|\left|\left(\lambda_1 - 2\lambda_2 + 2\lambda_2\cos\left(kl\Delta r\right)\right)\right| \\ -\left|\delta_0\right|\left|\lambda_1\right|\left(n+1\right) + 2\lambda_3\left|\delta_0\right|\left(m+1\right) \tag{8.71}$$

$$\left|\frac{\delta_{n+1}}{\delta_0}\right| < \frac{\left|\left(\lambda_1 - 2\lambda_2 + 2\lambda_2\cos\left(kl\Delta r\right)\right)\right| - \left|\lambda_1\right|\left(n+1\right) + \left|2\lambda_3\right|\left(m+1\right)}{\left|\left(\lambda_1 + 2\lambda_2 - \lambda_4 - \lambda_2\left[2\cos\left(kl\Delta r\right)\right]\right)\right|} \tag{8.72}$$

$$\frac{\left|\left(\lambda_1 - 2\lambda_2 + 2\lambda_2\cos\left(kl\Delta r\right)\right)\right| - \left|\lambda_1\right|\left(n+1\right) + \left|2\lambda_3\right|\left(m+1\right)}{\left|\left(\lambda_1 + 2\lambda_2 - \lambda_4 - \lambda_2\left[2\cos\left(kl\Delta r\right)\right]\right)\right|} < 1 \tag{8.73}$$

Different conditions of Equation (8.73) were applied and the solutions below were obtained.

Condition 1
If,

$$\lambda_1 - 2\lambda_2 + 2\lambda_2\cos\left(kl\Delta r\right) - \lambda_1\left(n+1\right) + 2\lambda_3\left(m+1\right) > 0 \tag{8.74}$$

and,

$$\lambda_1 + 2\lambda_2 - \lambda_4 - 2\lambda_2\cos\left(kl\Delta r\right) > 0 \tag{8.75}$$

Therefore,

$$\lambda_1 - 2\lambda_2 + 2\lambda_2\cos\left(kl\Delta r\right) - \lambda_1\left(n+1\right) + 2\lambda_3\left(m+1\right) < \lambda_1 + 2\lambda_2 - \lambda_4 - 2\lambda_2\cos\left(kl\Delta r\right) \tag{8.76}$$

$$4\lambda_2\cos\left(kl\Delta r\right) - \lambda_1\left(n+1\right) + 2\lambda_3\left(m+1\right) < 4\lambda_2 - \lambda_4 \tag{8.77}$$

$$\cos\left(kl\Delta r\right) - \frac{\lambda_1\left(n+1\right)}{4\lambda_2} + \frac{\lambda_3\left(m+1\right)}{2\lambda_2} + \frac{\lambda_4}{4\lambda_2} < 1 \tag{8.78}$$

Condition 2
If,

$$\lambda_1 - 2\lambda_2 + 2\lambda_2\cos\left(kl\Delta r\right) - \lambda_1\left(n+1\right) + 2\lambda_3\left(m+1\right) < 0 \tag{8.79}$$

and,

$$\lambda_1 + 2\lambda_2 - \lambda_4 - 2\lambda_2\cos\left(kl\Delta r\right) < 0 \tag{8.80}$$

Therefore,

$$-\lambda_1 + 2\lambda_2 - 2\lambda_2\cos\left(kl\Delta r\right) + \lambda_1\left(n+1\right) - 2\lambda_3\left(m+1\right) < -\lambda_1 - 2\lambda_2 + \lambda_4 + 2\lambda_2\cos\left(kl\Delta r\right) \tag{8.81}$$

$$4\lambda_2 + \lambda_1 (n+1) - 2\lambda_3 (m+1) < \lambda_4 + 4\lambda_2 \cos(kl\Delta r) \tag{8.82}$$

$$1 < \cos(kl\Delta r) - \frac{\lambda_1 (n+1)}{4\lambda_2} + \frac{\lambda_3 (m+1)}{2\lambda_2} + \frac{\lambda_4}{4\lambda_2} \tag{8.83}$$

Condition 3

If,

$$\lambda_1 - 2\lambda_2 + 2\lambda_2 \cos(kl\Delta r) - \lambda_1 (n+1) + 2\lambda_3 (m+1) > 0 \tag{8.84}$$

and,

$$\lambda_1 + 2\lambda_2 - \lambda_4 - 2\lambda_2 \cos(kl\Delta r) < 0 \tag{8.85}$$

or

If,

$$\lambda_1 - 2\lambda_2 + 2\lambda_2 \cos(kl\Delta r) - \lambda_1 (n+1) + 2\lambda_3 (m+1) < 0 \tag{8.86}$$

and,

$$\lambda_1 + 2\lambda_2 - \lambda_4 - 2\lambda_2 \cos(kl\Delta r) > 0 \tag{8.87}$$

The resulting solution for both conditions is the same.

$$-\lambda_1 + 2\lambda_2 - 2\lambda_2 \cos(kl\Delta r) + \lambda_1 (n+1) - 2\lambda_3 (m+1) < \lambda_1 + 2\lambda_2 - \lambda_4 - 2\lambda_2 \cos(kl\Delta r)$$

$$\lambda_1 (n+1) - 2\lambda_3 (m+1) < 2\lambda_1 \tag{8.88}$$

$$\lambda_1 (n+1) - 2\lambda_1 < 2\lambda_3 (m+1) - \lambda_4 \tag{8.89}$$

$$\lambda_1 (n+1-2) < 2\lambda_3 (m+1) - \lambda_4 \tag{8.90}$$

8.2.2.2 Applying the New Numerical Approximation Compiled by Atangana and Toufik

The concept of fractional derivative has come into wider use recently, and several studies have been conducted. The concept of nonlocal and non-singular kernel was introduced to extend the limitations of the two known power law methods; namely, the Riemann–Liouville and Caputo fractional derivatives. However, it has recently identified that the current numerical schemes cannot be used to solve the fractional derivative mentioned above. Around the same time, a new scheme was developed to analyze PDEs with nonlocal and non-singular kernel fractional derivatives. The scheme incorporates the two well-known concepts: the fundamental theorem of fractional calculus, and the two-step Lagrange polynomial (Toufik and Atangana, 2017). The following numerical scheme was obtained:

$$y_{n+1} = y(0) + \frac{1-\alpha}{ABC(\alpha)} f(t_n, y(t_n))$$

$$+ \frac{\alpha}{ABC(\alpha)} \sum_{k=0}^{n} \left(\begin{array}{l} \frac{h^\alpha f(t_k, y_k)}{\Gamma(\alpha+2)} (n+1-k)^\alpha (n-k+2+\alpha) \\ -(n-k)^\alpha (n-k+2+2\alpha) \end{array} \right)$$

$$- \frac{h^\alpha f(t_{k-1}, y_{k-1})}{\Gamma(\alpha+2)} \left((n+1-k)^{\alpha+1} - (n-k)^\alpha (n-k+1+\alpha) \right) \tag{8.91}$$

The above version of numerical scheme will be used to analyze the equation below.

$$
{}_{0}^{C}D_{t}^{\beta}h(r,t) = \frac{T}{S}\left[\frac{\partial^{2}h}{\partial r^{2}}\left(1+\frac{\Delta r}{r}\right)+\frac{1}{r}\,{}_{0}^{C}D_{r}^{\alpha}h(r,t)+\frac{h(r,t)}{\lambda^{2}}\right]
\tag{8.92}
$$

Let the function above be represented as follows,

$$
{}_{0}^{C}D_{t}^{\beta}h(r,t) = f\big(r,t,h(r,t)\big)
\tag{8.93}
$$

at interval r_{i}, t_{j+1} $F\big(r_{i},t_{j},h_{i}^{j}\big)$ will be given as follows,

$$
F\big(r_{i},t_{j},h_{i}^{j}\big) = \frac{T}{S}\left[\frac{h_{i+1}^{j}-2h_{i}^{j}+h_{i-1}^{j}}{(\Delta r)^{2}}\left(1+\frac{\Delta r}{r_{i}}\right)+\frac{h_{i}^{j}}{\lambda^{2}}+\frac{1}{r_{i}}\,{}_{0}^{C}D_{r}^{\alpha}h(r_{i},t_{j})\right]
\tag{8.94}
$$

However ${}_{0}^{C}D_{r}^{\alpha}h(r_{i},t_{j})$ can be written as,

$$
\begin{aligned}
{}_{0}^{C}D_{r}^{\alpha}h(r_{i},t_{j}) &= \frac{1}{\Gamma(1-\alpha)}\int_{0}^{r_{i}}\frac{\partial}{\partial\tau}h(\tau,t_{j})(r_{i}-\tau)^{-\alpha}\,d\tau \\[2mm]
&= \frac{1}{\Gamma(1-\alpha)}\sum_{k=0}^{i-1}\left(\int_{r_{k}}^{r_{k+1}}\frac{h(r_{i+1},t_{j})-h(r_{i-1},t_{j})}{\Delta r}(r_{i}-\tau)^{-\alpha}\,d\tau\right) \\[2mm]
&= \frac{1}{\Gamma(1-\alpha)}\sum_{k=0}^{i-1}\left(\frac{h_{i+1}^{j}-h_{i-1}^{j}}{\Delta r}\right)\int_{r_{k}}^{r_{k+1}}(r_{i}-\tau)^{-\alpha}\,d\tau \\[2mm]
&= \frac{1}{\Gamma(1-\alpha)}\sum_{k=0}^{i-1}\left(\frac{h_{i+1}^{j}-h_{i-1}^{j}}{\Delta r}\right)\int_{r_{i}-r_{k+1}}^{r_{i}-r_{k}}y^{-\alpha}\,dy \\[2mm]
&= \frac{1}{\Gamma(1-\alpha)}\sum_{k=0}^{i-1}\left(\frac{h_{i+1}^{j}-h_{i-1}^{j}}{\Delta r}\left[\frac{(r_{i}-r_{k})^{1-\alpha}}{1-\alpha}-\frac{(r_{i}-r_{k+1})^{1-\alpha}}{1-\alpha}\right]\right) \\[2mm]
&= \frac{1}{\Gamma(2-\alpha)}\sum_{k=0}^{i-1}\left(\frac{h_{i+1}^{j}-h_{i-1}^{j}}{\Delta r}\left[(r_{i}-r_{k})^{1-\alpha}-(r_{i}-r_{k+1})^{1-\alpha}\right]\right)
\end{aligned}
\tag{8.95}
$$

Let

$$
r_{i}=\Delta r\,i,\, r_{k}=\Delta r\,k \text{ and } r_{k+1}=\Delta r\,(k+1)
$$

Therefore,

$$
{}_{0}^{C}D_{r}^{\alpha}h(r_{i},t_{j}) = \frac{(\Delta r)^{-\alpha}}{\Gamma(2-\alpha)}\sum_{k=0}^{i-1}\left(\frac{h_{i+1}^{j}-h_{i-1}^{j}}{\Delta r}\left[(i-k)^{1-\alpha}-(i-k-1)^{1-\alpha}\right]\right)
\tag{8.96}
$$

From the solution above, $F\big(r_{i},t_{j},h_{i}^{j}\big)$ and $F\big(r_{i},t_{j-1},h_{i}^{j-1}\big)$ can be written as follows,

$$
\begin{aligned}
F\big(r_{i},t_{j},h_{i}^{j}\big) &= \frac{T}{S}\left[\frac{h_{i+1}^{j}-2h_{i}^{j}+h_{i-1}^{j}}{(\Delta r)^{2}}\left(1+\frac{\Delta r}{r_{i}}\right)+\frac{h_{i}^{j}}{\lambda^{2}}\right] \\[2mm]
&\quad + \frac{T}{S}\left[\frac{1}{r_{i}}\frac{(\Delta r)^{-\alpha}}{\Gamma(2-\alpha)}\sum_{k=0}^{i-1}\left(\frac{h_{i+1}^{j}-h_{i-1}^{j}}{\Delta r}\left[(i-k)^{1-\alpha}-(i-k-1)^{1-\alpha}\right]\right)\right]
\end{aligned}
\tag{8.97}
$$

and,

$$F\left(r_i,t_{j-1},h_i^{j-1}\right)=\frac{T}{S}\left[\frac{h_{i+1}^{j-1}-2h_i^{j-1}+h_{i-1}^{j-1}}{\left(\Delta r\right)^2}\left(1+\frac{\Delta r}{r_i}\right)+\frac{h_i^{j-1}}{\lambda^2}\right]$$
$$+\frac{T}{S}\left[\frac{1}{r_i}\frac{\left(\Delta r\right)^{-\alpha}}{\Gamma\left(2-\alpha\right)}\sum_{k=0}^{i-1}\left(\frac{h_{i+1}^{j-1}-h_{i-1}^{j-1}}{\Delta r}\left[\left(i-k\right)^{1-\alpha}-\left(i-k-1\right)^{1-\alpha}\right]\right)\right]$$

(8.98)

Discretizing $\,_0^C D_t^\beta h\left(r,t\right)$ the following is obtained,

$$h\left(r,t\right)-h\left(r,0\right)=\frac{1}{\Gamma\left(\beta\right)}\int_0^t\left(t-\tau\right)^{\beta-1}f\left(r,\tau,h\left(r,\tau\right)\right)d\tau$$

(8.99)

at interval r_i, t_{n+1}

$$h_i^{n+1}=h_i^0+\sum_{j=0}^n\left[\frac{\left(\Delta t\right)^\beta}{\Gamma\left(\beta+2\right)}f\left(r_i,t_j,h_i^j\right)\left\{\left(n+1-j\right)^\beta\left(n-j+2+\beta\right)-\left(n-j\right)^\beta\left(n-j+2+2\beta\right)\right\}\right]$$
$$-\sum_{j=o}^n\left[\frac{\left(\Delta t\right)^\beta}{\Gamma\left(\beta+2\right)}f\left(r_i,t_{j-1},h_i^{j-1}\right)\left\{\left(n+1-j\right)^{\beta+1}-\left(n-j\right)^\beta\left(n-j+1+\beta\right)\right\}\right]$$

(8.100)

Therefore, the solution can be given as,

$$h_i^{n+1}=h_i^0+\sum_{j=0}^n\left[\begin{array}{l}\dfrac{\left(\Delta t\right)^\beta}{\Gamma\left(\beta+2\right)}\dfrac{T}{S}\left\{\dfrac{h_{i+1}^j-2h_i^j+h_{i-1}^j}{\left(\Delta r\right)^2}\left(1+\dfrac{\Delta r}{r_i}\right)+\dfrac{h_i^j}{\lambda^2}\right\}\\\times\left\{\left(n+1-j\right)^\beta\left(n-j+2+\beta\right)-\left(n-j\right)^\beta\left(n-j+2+2\beta\right)\right\}\end{array}\right]$$
$$+\sum_{j=0}^n\left[\begin{array}{l}\dfrac{\left(\Delta t\right)^\beta}{\Gamma\left(\beta+2\right)}\dfrac{T}{S}\left\{\dfrac{1}{r_i}\dfrac{\left(\Delta r\right)^{-\alpha}}{\Gamma\left(2-\alpha\right)}\sum_{k=0}^{i-1}\dfrac{h_{i+1}^j-h_{i-1}^j}{\Delta r}\left[\left(i-k\right)^{1-\alpha}-\left(i-k-1\right)^{1-\alpha}\right]\right\}\\\times\left\{\left(n+1-j\right)^\beta\left(n-j+2+\beta\right)-\left(n-j\right)^\beta\left(n-j+2+2\beta\right)\right\}\end{array}\right]$$
$$-\sum_{j=o}^n\left[\begin{array}{l}\dfrac{\left(\Delta t\right)^\beta}{\Gamma\left(\beta+2\right)}\dfrac{T}{S}\left\{\dfrac{h_{i+1}^{j-1}-2h_i^{j-1}+h_{i-1}^{j-1}}{\left(\Delta r\right)^2}\left(1+\dfrac{\Delta r}{r_i}\right)+\dfrac{h_i^{j-1}}{\lambda^2}\right\}\\\times\left\{\left(n+1-j\right)^{\beta+1}-\left(n-j\right)^\beta\left(n-j+1+\beta\right)\right\}\end{array}\right]$$
$$-\sum_{j=o}^n\left[\begin{array}{l}\dfrac{\left(\Delta t\right)^\beta}{\Gamma\left(\beta+2\right)}\dfrac{T}{S}\left\{\dfrac{1}{r_i}\dfrac{\left(\Delta r\right)^{-\alpha}}{\Gamma\left(2-\alpha\right)}\sum_{k=0}^{i-1}\dfrac{h_{i+1}^{j-1}-h_{i-1}^{j-1}}{\Delta r}\left[\left(i-k\right)^{1-\alpha}-\left(i-k-1\right)^{1-\alpha}\right]\right\}\\\times\left\{\left(n+1-j\right)^{\beta+1}-\left(n-j\right)^\beta\left(n-j+1+\beta\right)\right\}\end{array}\right]$$

(8.101)

8.3   EXPONENTIAL DECAY LAW

8.3.1   Caputo–Fabrizio Fractional Derivative

Caputo and Fabrizio were the first researchers to indicate that the commonly used fractional differential operators have some limitations and also produced misleading results while modeling real world scenarios. They were mainly concerned about the singularity of the power law. The Caputo–Fabrizio fractional derivative, because of certain of its factors, does not satisfy the index law for the fractional order, but the solution of its associate evolution equation satisfies all the properties of a strongly continuous one-parameter semi-group and is a generalization of the exponential function (Atangana, 2018). The Caputo–Fabrizio derivative is defined as:

$$
{}^{CF}_{0}D_t^{\alpha}u(t) = \frac{1}{1-\alpha}\int_0^t u'(\tau)\exp\left[-\frac{\alpha}{1-\alpha}(t-\tau)\right]d\tau \tag{8.102}
$$

The law of exponential decay will be applied on the modified or new groundwater equation for water flowing within a leaky aquifer. The new generalized equation (Equation 8.3) will be modified and given as,

$$
{}^{CF}_{0}D_t^{\beta}h(r,t) = \frac{T}{S}\left[\frac{\partial^2 h}{\partial r^2}\left(1+\frac{\Delta r}{r}\right) + \frac{1}{r}\,{}^{CF}_{0}D_r^{\alpha}h(r,t) + \frac{h(r,t)}{\lambda^2}\right] \tag{8.103}
$$

Where the ${}^{CF}_{0}D_t^{\beta}$ is the Caputo–Fabrizio derivative is defined as,

$$
{}^{CF}_{0}D_t^{\beta}f(t) = \frac{M(\beta)}{1-\beta}\int_0^t f'(x)\exp\left[-\frac{\beta(t-x)}{1-\beta}dx\right] \tag{8.104}
$$

Where M (β) is the normalization function,
   If n ≥ 1 then

$$
{}^{CF}_{0}D_t^{\beta}f(t_n) = \frac{M(\beta)}{1-\beta}\int_0^{t_n} f'(\tau)\exp\left[-\frac{\beta(t_n-\tau)}{1-\beta}d\tau\right] \tag{8.105}
$$

$$
= \frac{M(\beta)}{1-\beta}\sum_{j=0}^{n}\left(\frac{f(t_{j+1})-f(t_j)}{\Delta t}\right)\int_{t_n-t_{j+1}}^{t_n-t_j} f'(\tau)\exp\left[-\frac{\beta(t_n-\tau)}{1-\beta}d\tau\right] \tag{8.106}
$$

$$
= \frac{M(\beta)}{1-\beta}\sum_{j=0}^{n}\left(\frac{f(t_{j+1})-f(t_j)}{\Delta t}\right)\frac{1-\beta}{\beta}\exp\Big|_{t_j-t_{j+1}}^{t_n-t_j}. \tag{8.107}
$$

$$
= \frac{M(\beta)}{1-\beta}\sum_{j=0}^{n}\left(\frac{f(t_{j+1})-f(t_j)}{\Delta t}\right)\left\{\exp\left[\frac{\beta}{1-\beta}(t_n-t_j)\right] - \exp\left[\frac{\beta}{1-\beta}(t_n-t_{j+1})\right]\right\} \tag{8.108}
$$

Let,

$$
\exp\left[\frac{\beta}{1-\beta}(t_n-t_j)\right] - \exp\left[\frac{\beta}{1-\beta}(t_n-t_{j+1})\right] = A_j^{\beta} \tag{8.109}
$$

$$
{}^{CF}_{0}D_t^{\beta}f(t_n) = \frac{M(\beta)}{\beta}\sum_{j=0}^{n}\left(\frac{f(t_{j+1})-f(t_j)}{\Delta r}\right)A_j^{\alpha} \tag{8.110}
$$

8.3.1.1 Numerical Approximation Using the Adam–Bashforth Method

The Adam–Bashforth method will be used to numerically solve the equation with the Caputo–Fabrizio fractional derivative. Introducing the Caputo–Fabrizio fractional derivative and solving the equation with the Adam–Bashforth method:

$$
{}^{CF}_{0}D_t^\beta h(r,t) = \frac{T}{S}\left[\frac{\partial^2 h(r,t)}{\partial r^2}\left(1+\frac{\Delta r}{r}\right)+\frac{1}{r}\frac{\partial h(r,t)}{\partial r}+\frac{h(r,t)}{\lambda^2}\right]
\tag{8.111}
$$

$$
u(x,t)-u(x,0) = \frac{1-\beta}{M(\beta)} f(x,t,u(x,t))+\frac{\beta}{M(\beta)}\int_0^{t_{n+1}} f(x,\tau,u(x,\tau))\,d\tau
\tag{8.112}
$$

At

$$
t_{n+1}, u(x,t_{n+1})-u(x,0) = \frac{1-\beta}{M(\beta)} f(x,t_n,u(x,t_n))+\frac{\beta}{M(\beta)}\int_0^{t_{n+1}} f(x,\tau,u(x,\tau))\,d\tau
\tag{8.113}
$$

At

$$
t_n, u(x,t_n)-u(x,0) - \frac{1-\beta}{M(\beta)} f(x,t_{n-1},u(x,t_{n-1}))+\frac{\beta}{M(\beta)}\int_0^{t_{n+1}} f(x,\tau,u(x,\tau))\,d\tau
\tag{8.114}
$$

Taking the difference $t_{n+1}-t_n$,

$$
u(x,t_{n+1})-u(x,t_n) = \frac{1-\beta}{M(\beta)}\left[f(x,t_n,u(x,t_n))-f(x,t_{n-1},u(x,t_{n-1}))\right]+\frac{\beta}{M(\beta)}\int_0^{t_{n+1}} f(x,\tau,u(x,\tau))\,d\tau
\tag{8.115}
$$

At

$$
x_i, u_i^{n+1}-u_i^n = \frac{1-\beta}{M(\beta)}\left[f\left(x_i,t_n,u\left(x_i u_i^n\right)\right)-f\left(x_i,t_{n-1},u\left(x_i,u_i^{n-1}\right)\right)\right]
$$
$$
+\frac{\beta}{M(\beta)}\left[\frac{3\Delta t}{2} f\left(x_i,t_n,u_i^n\right)-\frac{\Delta t}{2} f\left(x_i,t_{n-1},u_i^{n-1}\right)\right]
\tag{8.116}
$$

Therefore the function can be given as,

$$
u_i^{n+1}-u_i^n = \frac{1-\beta}{M(\beta)}\frac{T}{S}\left[\left(\frac{u_{i+1}^n-2u_i^n+u_{i-1}^n}{(\Delta r)^2}\right)\left(1+\frac{r_{i+1}-r_i}{r_i}\right)+\frac{1}{r_i}\frac{u_{i+1}^n-u_i^n}{\Delta r}+\frac{u_i^n}{\lambda^2}\right]
$$
$$
-\frac{1-\beta}{M(\beta)}\frac{T}{S}\left[\left(\frac{u_{i+1}^{n-1}-2u_i^{n-1}+u_{i-1}^{n-1}}{(\Delta r)^2}\right)\left(1+\frac{r_{i+1}-r_i}{r_i}\right)+\frac{1}{r_i}\frac{u_{i+1}^{n-1}-u_i^{n-1}}{\Delta r}+\frac{u_i^{n-1}}{\lambda^2}\right]
$$
$$
+\frac{\beta}{M(\beta)}\frac{3T\Delta t}{2S}\left[\left(\frac{u_{i+1}^n-2u_i^n+u_{i-1}^n}{(\Delta r)^2}\right)\left(1+\frac{r_{i+1}-r_i}{r_i}\right)+\frac{1}{r_i}\frac{u_{i+1}^n-u_i^n}{\Delta r}+\frac{u_i^n}{\lambda^2}\right]
$$
$$
-\frac{\beta}{M(\beta)}\frac{T\Delta t}{2S}\left[\left(\frac{u_{i+1}^{n-1}-2u_i^{n-1}+u_{i-1}^{n-1}}{(\Delta r)^2}\right)\left(1+\frac{r_{i+1}-r_i}{r_i}\right)++\frac{1}{r_i}\frac{u_{i+1}^{n-1}-u_i^{n-1}}{\Delta r}+\frac{u_i^{n-1}}{\lambda^2}\right]
\tag{8.117}
$$

Grouping the like terms,

$$
\begin{aligned}
u_i^{n+1} = u_i^n &\left(
\begin{array}{c}
1 + \dfrac{1-\beta}{M(\beta)}\dfrac{T}{\lambda^2 S} - \dfrac{1-\beta}{M(\beta)}\dfrac{2T}{S(\Delta r)^2}\left(1 + \dfrac{r_{i+1}-r_i}{r_i}\right) - \dfrac{1-\beta}{M(\beta)}\dfrac{T}{Sr_i\Delta r} + \dfrac{1-\beta}{M(\beta)}\dfrac{3T\Delta t}{2\lambda^2 S} \\[2mm]
- \dfrac{1-\beta}{M(\beta)}\dfrac{3T\Delta t}{S(\Delta r)^2}\left(1 + \dfrac{r_{i+1}-r_i}{r_i}\right) - \dfrac{1-\beta}{M(\beta)}\dfrac{3T\Delta t}{2Sr_i\Delta r}
\end{array}
\right) \\[3mm]
+ u_{i+1}^n &\left(
\begin{array}{c}
\dfrac{1-\beta}{M(\beta)}\dfrac{T}{S(\Delta r)^2}\left(1 + \dfrac{r_{i+1}-r_i}{r_i}\right) + \dfrac{1-\beta}{M(\beta)}\dfrac{T}{Sr_i\Delta r} + \dfrac{1-\beta}{M(\beta)}\dfrac{3T\Delta t}{2S(\Delta r)^2}\left(1 + \dfrac{r_{i+1}-r_i}{r_i}\right) \\[2mm]
+ \dfrac{1-\beta}{M(\beta)}\dfrac{3T\Delta t}{2Sr_i\Delta r}
\end{array}
\right) \\[3mm]
+ u_{i-1}^n &\left(
\dfrac{1-\beta}{M(\beta)}\dfrac{T}{S(\Delta r)^2}\left(1 + \dfrac{r_{i+1}-r_i}{r_i}\right) + \dfrac{1-\beta}{M(\beta)}\dfrac{3T\Delta t}{2S(\Delta r)^2}\left(1 + \dfrac{r_{i+1}-r_i}{r_i}\right)
\right) \\[3mm]
- u_{i+1}^{n-1} &\left(
\begin{array}{c}
\left(\dfrac{1-\beta}{M(\beta)}\dfrac{T}{S(\Delta r)^2}\right)\left(1 + \dfrac{r_{i+1}-r_i}{r_i}\right) + \dfrac{1-\beta}{M(\beta)}\dfrac{T}{Sr_i\Delta r} + \left(\dfrac{1-\beta}{M(\beta)}\dfrac{T\Delta t}{2S(\Delta r)^2}\right)\left(1 + \dfrac{r_{i+1}-r_i}{r_i}\right) \\[2mm]
+ \dfrac{1-\beta}{M(\beta)}\dfrac{T\Delta t}{2Sr_i\Delta r}
\end{array}
\right) \\[3mm]
- u_i^{n-1} &\left(
\begin{array}{c}
\left(\dfrac{1-\beta}{M(\beta)}\dfrac{2T}{S(\Delta r)^2}\right)\left(1 + \dfrac{r_{i+1}-r_i}{r_i}\right) - \dfrac{1-\beta}{M(\beta)}\dfrac{T}{Sr_i\Delta r} + \dfrac{1-\beta}{M(\beta)}\dfrac{T}{\lambda^2 S} \\[2mm]
- \left(\dfrac{1-\beta}{M(\beta)}\dfrac{T\Delta t}{S(\Delta r)^2}\right)\left(1 + \dfrac{r_{i+1}-r_i}{r_i}\right) - \dfrac{1-\beta}{M(\beta)}\dfrac{T\Delta t}{2Sr_i\Delta r} + \dfrac{1-\beta}{M(\beta)}\dfrac{T\Delta t}{2\lambda^2 S}
\end{array}
\right) \\[3mm]
- u_{i-1}^{n-1} &\left(
\left(\dfrac{1-\beta}{M(\beta)}\dfrac{T}{S(\Delta r)^2}\right)\left(1 + \dfrac{r_{i+1}-r_i}{r_i}\right) + \left(\dfrac{1-\beta}{M(\beta)}\dfrac{T\Delta t}{2S(\Delta r)^2}\right)\left(1 + \dfrac{r_{i+1}-r_i}{r_i}\right)
\right)
\end{aligned}
$$

$$(8.118)$$

Therefore,

Let,

$$
a_1 = \left(
\begin{array}{c}
1 + \dfrac{1-\beta}{M(\beta)}\dfrac{T}{\lambda^2 S} - \dfrac{1-\beta}{M(\beta)}\dfrac{2T}{S(\Delta r)^2}\left(1 + \dfrac{r_{i+1}-r_i}{r_i}\right) - \dfrac{1-\beta}{M(\beta)}\dfrac{T}{Sr_i\Delta r} + \dfrac{1-\beta}{M(\beta)}\dfrac{3T\Delta t}{2\lambda^2 S} \\[2mm]
- \dfrac{1-\beta}{M(\beta)}\dfrac{3T\Delta t}{S(\Delta r)^2}\left(1 + \dfrac{r_{i+1}-r_i}{r_i}\right) - \dfrac{1-\beta}{M(\beta)}\dfrac{3T\Delta t}{2Sr_i\Delta r}
\end{array}
\right)
$$

$$(8.119)$$

$$a_2 = \left(\begin{array}{l} \dfrac{1-\beta}{M(\beta)} \dfrac{T}{S(\Delta r)^2}\left(1+\dfrac{r_{i+1}-r_i}{r_i}\right) + \dfrac{1-\beta}{M(\beta)} \dfrac{T}{Sr_i\Delta r} + \dfrac{1-\beta}{M(\beta)} \dfrac{3T\Delta t}{2S(\Delta r)^2}\left(1+\dfrac{r_{i+1}-r_i}{r_i}\right) \\ \qquad\qquad + \dfrac{1-\beta}{M(\beta)} \dfrac{3T\Delta t}{2Sr_i\Delta r} \end{array} \right) \tag{8.120}$$

$$a_3 = \left(\dfrac{1-\beta}{M(\beta)} \dfrac{T}{S(\Delta r)^2}\left(1+\dfrac{r_{i+1}-r_i}{r_i}\right) + \dfrac{1-\beta}{M(\beta)} \dfrac{3T\Delta t}{2S(\Delta r)^2}\left(1+\dfrac{r_{i+1}-r_i}{r_i}\right) \right) \tag{8.121}$$

$$a_4 = \left(\begin{array}{l} \left(\dfrac{1-\beta}{M(\beta)} \dfrac{T}{S(\Delta r)^2}\right)\left(1+\dfrac{r_{i+1}-r_i}{r_i}\right) + \dfrac{1-\beta}{M(\beta)} \dfrac{T}{Sr_i\Delta r} \\[2mm] \quad + \left(\dfrac{1-\beta}{M(\beta)} \dfrac{T\Delta t}{2S(\Delta r)^2}\right)\left(1+\dfrac{r_{i+1}-r_i}{r_i}\right) \\[2mm] \qquad + \dfrac{1-\beta}{M(\beta)} \dfrac{T\Delta t}{2Sr_i\Delta r} \end{array} \right) \tag{8.122}$$

$$a_5 = \left(\begin{array}{l} \left(\dfrac{1-\beta}{M(\beta)} \dfrac{2T}{S(\Delta r)^2}\right)\left(1+\dfrac{r_{i+1}-r_i}{r_i}\right) - \dfrac{1-\beta}{M(\beta)} \dfrac{T}{Sr_i\Delta r} + \dfrac{1-\beta}{M(\beta)} \dfrac{T}{\lambda^2 S} \\[2mm] - \left(\dfrac{1-\beta}{M(\beta)} \dfrac{T\Delta t}{S(\Delta r)^2}\right)\left(1+\dfrac{r_{i+1}-r_i}{r_i}\right) - \dfrac{1-\beta}{M(\beta)} \dfrac{T\Delta t}{2Sr_i\Delta r} + \dfrac{1-\beta}{M(\beta)} \dfrac{T\Delta t}{2\lambda^2 S} \end{array} \right) \tag{8.123}$$

and,

$$a_6 = \left(\left(\dfrac{1-\beta}{M(\beta)} \dfrac{T}{S(\Delta r)^2}\right)\left(1+\dfrac{r_{i+1}-r_i}{r_i}\right) + \left(\dfrac{1-\beta}{M(\beta)} \dfrac{T\Delta t}{2S(\Delta r)^2}\right)\left(1+\dfrac{r_{i+1}-r_i}{r_i}\right) \right) \tag{8.124}$$

$$u_i^{n+1} = u_i^n a_1 + u_{i+1}^n a_2 + u_{i-1}^n a_3 - u_{i+1}^{n-1} a_4 - u_i^{n-1} a_5 - u_{i-1}^{n-1} a_6 \tag{8.125}$$

8.3.1.1.1 Stability Analysis Using the Von Neumann Method

The Von Neumann stability method incorporates the Fourier's series and the Euler formulas. This method was developed in Los Alamos during World War II by Von Neumann and was considered classified until its brief description in Crank and Nicolson (1947) and in a publication in 1950 (Charney et al., 1950). At present this is the most widely applied technique for stability analysis.

$$\rho(x,t) = \sum_f \left(\hat{\rho}(t)e^{i\Delta x k_x}\right) \tag{8.126}$$

Equation (8.96) will be written as follows,

$$\hat{\rho}_{n+1}e^{i\Delta xk_x} = a_1\hat{\rho}_n e^{i\Delta xk_x} + a_2\hat{\rho}_n e^{i(x+\Delta r)k_x} + a_3\hat{\rho}_n e^{i(\Delta r-x)k_x}$$
$$+ a_4\hat{\rho}_{n-1}e^{i(x+\Delta r)k_x} + a_5\hat{\rho}_{n-1}e^{i\Delta xk_x} + a_6\hat{\rho}_{n-1}e^{i(\Delta r-x)k_x} \tag{8.127}$$

Divide by both sides by $e^{i\Delta xk_x}$ and grouping like terms,

$$\hat{\rho}_{n+1} = \hat{\rho}_n a_1 + \hat{\rho}_n a_2 e^{ixk_x} + \hat{\rho}_n a_3 e^{-ixk_x} + \hat{\rho}_{n-1}a_4 e^{ixk_x} + \hat{\rho}_{n-1}a_5 + \hat{\rho}_{n-1}a_6 e^{-ixk_x} \tag{8.128}$$

$$\hat{\rho}_{n+1} = \left(a_1 + a_2 e^{ixk_x} + a_3 e^{-ixk_x}\right)\hat{\rho}_n + \left(a_4 e^{ixk_x} + a_5 + a_6 e^{-ixk_x}\right)\hat{\rho}_{n-1} \tag{8.129}$$

When n = 0 and $\hat{\rho}_{-1}$ will not be applicable,

$$\hat{\rho}_1 = \left(a_1 + a_2 e^{ixk_x} + a_3 e^{-xik_x}\right)\hat{\rho}_0 + \left(e^{ixk_x}a_4 + a_5 + e^{-xik_x}a_6\right)\hat{\rho}_{-1} \tag{8.130}$$

$$\hat{\rho}_1 = \left(a_1 + a_2 e^{ixk_x} + a_3 e^{-xik_x}\right)\hat{\rho}_0 \tag{8.131}$$

$$\frac{\hat{\rho}_1}{\hat{\rho}_0} = \left(a_1 + a_2 e^{ixk_x} + a_3 e^{-ixk_x}\right) \tag{8.132}$$

$$\left|\frac{\hat{\rho}_1}{\hat{\rho}_0}\right| \le \left|\left(a_1 + a_2 e^{ixk_x} + a_3 e^{-ixk_x}\right)\right| < 1 \tag{8.133}$$

Applying Euler's formulas for simplification,

$$\left|\frac{\hat{\rho}_1}{\hat{\rho}_0}\right| \le \left|\left(a_1 + a_2\left[\cos\left(xk_x\right) + i\sin\left(xk_x\right)\right] + a_3\left[\cos\left(xk_x\right) - i\sin\left(xk_x\right)\right]\right)\right| < 1 \tag{8.134}$$

$$\left|\frac{\hat{\rho}_1}{\hat{\rho}_0}\right| \le \left|\left[a_1 + \cos\left(xk_x\right)\left(a_2 + a_3\right) + i\sin\left(xk_x\right)\left(a_2 - a_3\right)\right]\right| < 1 \tag{8.135}$$

$$|a_1| + \left|\left(a_2 + a_3\right)\right|\left|\cos\left(xk_x\right)\right| + \left|\left(a_2 - a_3\right)\right|\left|i\sin\left(xk_x\right)\right| < 1 \tag{8.136}$$

a_1, a_2 and a_3 are given, we can substitute them in the equation above,

$$\left(a_2 + a_3\right) = \begin{pmatrix} \dfrac{1-\beta}{M\left(\beta\right)}\dfrac{T}{S\left(\Delta r\right)^2}\left(1 + \dfrac{r_{i+1}-r_i}{r_i}\right) + \dfrac{1-\beta}{M\left(\beta\right)}\dfrac{T}{Sr_i\Delta r} \\[4mm] + \dfrac{1-\beta}{M\left(\beta\right)}\dfrac{3T\Delta t}{2S\left(\Delta r\right)^2}\left(1 + \dfrac{r_{i+1}-r_i}{r_i}\right) + \dfrac{1-\beta}{M\left(\beta\right)}\dfrac{3T\Delta t}{2Sr_i\Delta r} \\[4mm] + \left(\dfrac{1-\beta}{M\left(\beta\right)}\dfrac{T}{S\left(\Delta r\right)^2}\left(1 + \dfrac{r_{i+1}-r_i}{r_i}\right) + \dfrac{1-\beta}{M\left(\beta\right)}\dfrac{3T\Delta t}{2S\left(\Delta r\right)^2}\left(1 + \dfrac{r_{i+1}-r_i}{r_i}\right)\right) \end{pmatrix} \tag{8.137a}$$

$$\left(a_2 + a_3\right) = \frac{1-\beta}{M(\beta)} \frac{T}{S(\Delta r)^2} \left(\frac{2r_{i+1}}{r_i}\right) + \frac{1-\beta}{M(\beta)} \frac{T}{Sr_i\Delta r} \left(\frac{2+3\Delta t}{2}\right) + \frac{1-\beta}{M(\beta)} \frac{3T\Delta t}{S(\Delta r)^2} \left(\frac{r_{i+1}}{r_i}\right)$$

(8.137b)

$$\left(a_2 + a_3\right) = \frac{1-\beta}{M(\beta)} \frac{T}{S(\Delta r)^2} \left(\frac{r_{i+1}(2+3\Delta t)}{r_i}\right) + \frac{1-\beta}{M(\beta)} \frac{T}{Sr_i\Delta r} \left(\frac{2+3\Delta t}{2}\right) \tag{8.138}$$

$$\left(a_2 - a_3\right) = \begin{pmatrix} \dfrac{1-\beta}{M(\beta)} \dfrac{T}{S(\Delta r)^2} \left(1 + \dfrac{r_{i+1}-r_i}{r_i}\right) \\[2mm] + \dfrac{1-\beta}{M(\beta)} \dfrac{T}{Sr_i\Delta r} \\[2mm] + \dfrac{1-\beta}{M(\beta)} \dfrac{3T\Delta t}{2S(\Delta r)^2} \left(1 + \dfrac{r_{i+1}-r_i}{r_i}\right) \\[2mm] + \dfrac{1-\beta}{M(\beta)} \dfrac{3T\Delta t}{2Sr_i\Delta r} \end{pmatrix} - \begin{pmatrix} \dfrac{1-\beta}{M(\beta)} \dfrac{T}{S(\Delta r)^2} \left(1 + \dfrac{r_{i+1}-r_i}{r_i}\right) \\[2mm] + \dfrac{1-\beta}{M(\beta)} \dfrac{3T\Delta t}{2S(\Delta r)^2} \left(1 + \dfrac{r_{i+1}-r_i}{r_i}\right) \end{pmatrix}$$

$$= \frac{1-\beta}{M(\beta)} \frac{T}{Sr_i\Delta r} \left(\frac{2+3\Delta t}{2}\right) \tag{8.139}$$

Therefore,

$$\left\| \left(1 + \frac{1-\beta}{M(\beta)} \frac{T}{\lambda^2 S} - \frac{1-\beta}{M(\beta)} \frac{2T}{S(\Delta r)^2} \left(1 + \frac{r_{i+1}-r_i}{r_i}\right) - \frac{1-\beta}{M(\beta)} \frac{T}{Sr_i\Delta r} + \frac{1-\beta}{M(\beta)} \frac{3T\Delta t}{2\lambda^2 S} \right. \right.$$
$$\left. \left. - \frac{1-\beta}{M(\beta)} \frac{3T\Delta t}{S(\Delta r)^2} \left(1 + \frac{r_{i+1}-r_i}{r_i}\right) - \frac{1-\beta}{M(\beta)} \frac{3T\Delta t}{2Sr_i\Delta r} \right) \right\|$$
$$+ \left| \frac{1-\beta}{M(\beta)} \frac{T}{S(\Delta r)^2} \left(\frac{r_{i+1}(2+3\Delta t)}{r_i}\right) + \frac{1-\beta}{M(\beta)} \frac{T}{Sr_i\Delta r} \left(\frac{2+3\Delta t}{2}\right) \right| \left\| \cos(xk_x) \right\|$$
$$+ \left| \frac{1-\beta}{M(\beta)} \frac{T}{Sr_i\Delta r} \left(\frac{2+3\Delta t}{2}\right) \right| \left\| i\sin(xk_x) \right\| < 1 \tag{8.140}$$

If $a_1 \langle 0$ and $\left(a_2 + a_3\right)$ and $\left(a_2 - a_3\right) \rangle 0$,

$$-1 - \frac{1-\beta}{M(\beta)} \frac{T}{\lambda^2 S} + \frac{1-\beta}{M(\beta)} \frac{2T}{S(\Delta r)^2} \left(1 + \frac{r_{i+1}-r_i}{r_i}\right)$$
$$+ \frac{1-\beta}{M(\beta)} \frac{T}{Sr_i\Delta r} - \frac{1-\beta}{M(\beta)} \frac{3T\Delta t}{2\lambda^2 S} + \frac{1-\beta}{M(\beta)} \frac{3T\Delta t}{S(\Delta r)^2} \left(1 + \frac{r_{i+1}-r_i}{r_i}\right)$$
$$+ \frac{1-\beta}{M(\beta)} \frac{3T\Delta t}{2Sr_i\Delta r} + \left[\begin{array}{c} \dfrac{1-\beta}{M(\beta)} \dfrac{T}{S(\Delta r)^2} \left(\dfrac{r_{i+1}(2+3\Delta t)}{r_i}\right) \\[2mm] + \dfrac{1-\beta}{M(\beta)} \dfrac{T}{Sr_i\Delta r} \left(\dfrac{2+3\Delta t}{2}\right) \end{array} \right] \cos(xk_x)$$
$$+ \frac{1-\beta}{M(\beta)} \frac{T}{Sr_i\Delta r} \left(\frac{2+3\Delta t}{2}\right) < 1 \tag{8.141}$$

Adding the like terms,

$$-1 - \frac{1-\beta}{M(\beta)} \frac{T}{\lambda^2 S} \left(1 + \frac{3\Delta t}{2}\right) + \frac{1-\beta}{M(\beta)} \frac{T}{S(\Delta r)^2} \left(\frac{r_{i+1}(2+3\Delta t)}{r_i}\right)$$

$$+ \frac{1-\beta}{M(\beta)} \frac{T}{Sr_i \Delta r} \left(\frac{4+6\Delta t}{2}\right)$$

$$+ \left[\frac{1-\beta}{M(\beta)} \frac{T}{S(\Delta r)^2} \left(\frac{r_{i+1}(2+3\Delta t)}{r_i}\right) + \frac{1-\beta}{M(\beta)} \frac{T}{Sr_i \Delta r} \left(\frac{2+3\Delta t}{2}\right)\right] \cos(xk_x) < 1$$

(8.142)

Simplifying the equations,

$$-\frac{1-\beta}{M(\beta)} \frac{T}{\lambda^2 S} \left(1 + \frac{3\Delta t}{2}\right) + \frac{1-\beta}{M(\beta)} \frac{T}{S(\Delta r)^2} \left(\frac{r_{i+1}(2+3\Delta t)}{r_i}\right)(\cos(xk_x)+1)$$

$$+ \frac{1-\beta}{M(\beta)} \frac{T}{Sr_i \Delta r} \left(\frac{4+6\Delta t}{2} + \left(\frac{2+3\Delta t}{2}\right)\cos(xk_x)\right) < 2$$

(8.143)

If $a_1, (a_2 + a_3)$ and $(a_2 - a_3) > 0$,

$$1 + \frac{1-\beta}{M(\beta)} \frac{T}{\lambda^2 S} \left(1 + \frac{3\Delta t}{2}\right) - \frac{1-\beta}{M(\beta)} \frac{T}{S(\Delta r)^2} \left(\frac{r_{i+1}(2+3\Delta t)}{r_i}\right) - \frac{1-\beta}{M(\beta)} \frac{T}{Sr_i \Delta r} \left(\frac{2+3\Delta t}{2}\right)$$

$$+ \left[\frac{1-\beta}{M(\beta)} \frac{T}{S(\Delta r)^2} \left(\frac{r_{i+1}(2+3\Delta t)}{r_i}\right) + \frac{1-\beta}{M(\beta)} \frac{T}{Sr_i \Delta r} \left(\frac{2+3\Delta t}{2}\right)\right] \cos(xk_x)$$

$$+ \frac{1-\beta}{M(\beta)} \frac{T}{Sr_i \Delta r} \left(\frac{2+3\Delta t}{2}\right) < 1$$

(8.144)

Simplifying the equations,

$$\frac{1-\beta}{M(\beta)} \frac{T}{\lambda^2 S} \left(1 + \frac{3\Delta t}{2}\right) + \frac{1-\beta}{M(\beta)} \frac{T}{S(\Delta r)^2} \left(\frac{r_{i+1}(2+3\Delta t)}{r_i}\right)(\cos(xk_x)-1)$$

$$+ \frac{1-\beta}{M(\beta)} \frac{T}{Sr_i \Delta r} \left(\frac{2+3\Delta t}{2}\right) \cos(xk_x) < 0.$$

(8.145)

8.4 MITTAG–LEFFLER

8.4.1 Mittag–Leffler Special Function and Its General Form

The Mittag–Leffler function is named after the great Swedish mathematician Gosta Magnus Mittag–Leffler (1846–1927). He has worked on the general theory of functions, studying the relationship between independent and dependent variables. The Mittag–Leffler function arises naturally in the solution of fractional order integral equations or fractional order differential equations, and in particular in the investigations of the fractional generalization of the kinetic equation, random walks, Lévy flight, super diffusive transport, and in the study of complex systems (Haubold et al., 2011).

The special function,

$$E_{\alpha,\beta}\left(\lambda z^{\alpha}\right) \tag{8.146}$$

Its general form,

$$E_{\alpha,\beta}\left(z\right) = \sum_{k=0}^{\infty}\left(\frac{\left(\lambda z^{\alpha}\right)^{k}}{\Gamma\left(\beta+\alpha k\right)}\right)\alpha > 0, \beta > 0, \lambda < \infty \text{ and } \lambda = -\alpha\left(1-\alpha\right)^{-1} \tag{8.147}$$

8.4.1.1 Applying the Atangana–Baleanu (A–B) Fractional Derivative

Applying the A–B fractional derivative,

$${}_{a}^{AB}D_{t}^{\beta}h\left(x,t\right) = \frac{T}{S}\left[\frac{\partial}{\partial r}\left(\frac{\partial h}{\partial r}\right)\left(1+\frac{\Delta r}{r}\right) + \frac{1}{r}\,{}_{a}^{AB}D_{r}^{\alpha}h\left(r,t\right) + \frac{h\left(r,t\right)}{\lambda^{2}}\right] \tag{8.148}$$

However ${}_{a}^{AB}D_{r}^{\alpha}h\left(r_{i},t_{j}\right)$ can be written as,

$${}_{a}^{AB}D_{r}^{\alpha}h\left(r_{i},t_{j}\right) = \frac{AB\left(\alpha\right)}{\left(1-\alpha\right)}\int_{0}^{r_{i}}\frac{\partial h\left(\tau,t_{j}\right)}{\partial\tau}E_{\alpha}\left[-\frac{\alpha}{1-\alpha}\left(r_{i}-\tau\right)^{\alpha}\right]d\tau \tag{8.149}$$

$${}_{a}^{AB}D_{r}^{\alpha}h\left(r_{i},t_{j}\right) = \frac{AB\left(\alpha\right)}{\left(1-\alpha\right)}\sum_{k=0}^{i-1}\left(\int_{r_{k}}^{r_{k+1}}\frac{h\left(r_{i+1},t_{j}\right)-h\left(r_{i-1},t_{j}\right)}{\Delta r}E_{\alpha}\left[-\frac{\alpha}{1-\alpha}\left(r_{i}-\tau\right)^{\alpha}\right]\right)d\tau\,(121)\,{}_{a}^{AB}D_{r}^{\alpha}h\left(r_{i},t_{j}\right)$$

$$= \frac{AB\left(\alpha\right)}{\left(1-\alpha\right)}\sum_{k=0}^{i-1}\left(\frac{h_{i+1}^{j}-h_{i-1}^{j}}{\Delta r}\right)\int_{r_{k}}^{r_{k+1}}E_{\alpha}\left[-\frac{\alpha}{1-\alpha}\left(r_{i}-\tau\right)^{\alpha}\right]d\tau \tag{8.150}$$

We recall that the Mittag–Leffler in general form can be given as follows,

$$E_{\alpha}\left[-\frac{\alpha}{1-\alpha}\left(r_{i}-\tau\right)^{\alpha}\right] = \sum_{l=0}^{\infty}\left(\frac{\left(\frac{-\alpha}{1-\alpha}\right)^{l}\left(t_{n}-\tau\right)^{\alpha l}}{\Gamma\left(\alpha l+1\right)}\right) \tag{8.151}$$

Therefore,

$$\int_{r_{k}}^{r_{k+1}}E_{\alpha}\left[-\frac{\alpha}{1-\alpha}\left(r_{i}-\tau\right)^{\alpha}\right]d\tau = \int_{r_{k}}^{r_{k+1}}\sum_{l=0}^{\infty}\left(\frac{\left(\frac{-\alpha}{1-\alpha}\right)^{l}\left(t_{n}-\tau\right)^{\alpha l}}{\Gamma\left(\alpha l+1\right)}\right)d\tau \tag{8.152}$$

$$\int_{r_k}^{r_{k+1}} E_\alpha\left[-\frac{\alpha}{1-\alpha}\left(r_i-\tau\right)^\alpha\right]d\tau = \sum_{l=0}^{\infty}\left(\frac{\left(\frac{-\alpha}{1-\alpha}\right)^l}{\Gamma(\alpha l+1)}\right)\int_{r_k}^{r_{k+1}}\left(t_n-\tau\right)^{\alpha l}d\tau \tag{8.153}$$

Let

$$Y = t_n-\tau, \int_{r_k}^{r_{k+1}} E_\alpha\left[-\frac{\alpha}{1-\alpha}\left(r_i-\tau\right)^\alpha\right]d\tau = \sum_{l=0}^{\infty}\left(\frac{\left(\frac{-\alpha}{1-\alpha}\right)^l}{\Gamma(\alpha l+1)}\right)\int_{r_j-r_k}^{r_j-r_{k+1}} Y^{\alpha l}\left(-dy\right)$$

$$= \sum_{l=0}^{\infty}\left(\frac{\left(\frac{-\alpha}{1-\alpha}\right)^l}{\Gamma(\alpha l+1)}\right)\frac{Y^{\alpha l+1}}{\alpha l+1}\bigg|_{r_j-r_{k+1}}^{r_j-r_k}\int_{r_k}^{r_{k+1}} E_\alpha\left[-\frac{\alpha}{1-\alpha}\left(r_i-\tau\right)^\alpha\right]d\tau$$

$$= \sum_{l=0}^{\infty}\left(\frac{\left(\frac{-\alpha}{1-\alpha}\right)^l}{\Gamma(\alpha l+1)}\right)\left[\frac{\left(r_j-r_k\right)^{\alpha l+1}}{\alpha l+1}-\frac{\left(r_j-r_{k+1}\right)^{\alpha l+1}}{\alpha l+1}\right] \tag{8.154}$$

$$= \sum_{l=0}^{\infty}\left[\frac{\left(\frac{-\alpha}{1-\alpha}\right)^l\left(r_j-r_k\right)^{\alpha l+1}}{\Gamma(\alpha l+1)(\alpha l+1)}-\frac{\left(\frac{-\alpha}{1-\alpha}\right)^l\left(r_j-r_{k+1}\right)^{\alpha l+1}}{\Gamma(\alpha l+1)(\alpha l+1)}\right]$$

$$= r_j-r_k\sum_{l=0}^{\infty}\frac{\left(\frac{-\alpha}{1-\alpha}\left(r_j-r_k\right)^\alpha\right)^l}{\Gamma(\alpha l+2)}-r_j-r_{k+1}\sum_{l=0}^{\infty}\frac{\left(\frac{-\alpha}{1-\alpha}\left(r_j-r_{k+1}\right)^\alpha\right)^l}{\Gamma(\alpha l+2)}$$

$$= \left(r_j-r_k\right)E_{\alpha,2}\left[\frac{-\alpha}{1-\alpha}\left(r_j-r_k\right)^\alpha\right]-\left(r_j-r_{k+1}\right)E_{\alpha,2}\left[\frac{-\alpha}{1-\alpha}\left(r_j-r_{k+1}\right)^\alpha\right]$$

Let $r_j = j\Delta r$, $r_k = k\Delta r$ and $r_{k+1} = (k+1)\Delta r$,
   Therefore,

$$\int_{r_k}^{r_{k+1}} E_\alpha\left[-\frac{\alpha}{1-\alpha}\left(r_i-\tau\right)^\alpha\right]d\tau = \left(j\Delta r-k\Delta r\right)E_{\alpha,2}\left[\frac{-\alpha}{1-\alpha}\left(j\Delta r-k\Delta r\right)^\alpha\right]$$

$$-\left(j\Delta r-(k+1)\Delta r\right)E_{\alpha,2}\left[\frac{-\alpha}{1-\alpha}\left(j\Delta r-(k+1)\Delta r\right)^\alpha\right] \tag{8.155}$$

$$\int_{r_k}^{r_{k+1}} E_\alpha\left[-\frac{\alpha}{1-\alpha}\left(r_i-\tau\right)^\alpha\right]d\tau = \Delta r(j-k)E_{\alpha,2}\left[\frac{-\alpha}{1-\alpha}\Delta r^\alpha\left(j-k\right)^\alpha\right]$$

$$-\Delta r\left(j-(k+1)\right)E_{\alpha,2}\left[\frac{-\alpha}{1-\alpha}\Delta r^\alpha\left(j-(k+1)\right)^\alpha\right] \tag{8.156}$$

$${}_a^{AB}D_r^\alpha h\left(r_i,t_j\right) = \frac{AB(\alpha)}{(1-\alpha)}\sum_{k=0}^{i-1}h_{i+1}^j-h_{i-1}^j\left\{\begin{array}{l}\left(j-k\right)E_{\alpha,2}\left[\frac{-\alpha}{1-\alpha}\Delta r^\alpha\left(j-k\right)^\alpha\right]\\[2mm]-\left(j-(k+1)\right)E_{\alpha,2}\left[\frac{-\alpha}{1-\alpha}\Delta r^\alpha\left(j-(k+1)\right)^\alpha\right]\end{array}\right\} \tag{8.157}$$

We represent $^{AB}_a D^\beta_t h(x,t)$ as $f(x,t,h(x,t))$ and the fractional integral associated to the new fractional derivative with nonlocal kernel is defined as,

$$^{AB}_a I^\beta_t h(x,t) = \frac{1-\beta}{AB(\beta)} h(t) + \frac{\beta}{AB(\beta)\Gamma(\beta)} \int_a^t h(\tau)(t-\tau)^{\beta-1}\, d\tau \tag{8.158}$$

Therefore $^{AB}_a D^\beta_t h(x,t)$ can be written as,

$$^{AB}_a D^\beta_t h(x,t) = \frac{1-\beta}{AB(\beta)} f\big(x,t,h(x,t)\big) + \frac{\beta}{AB(\beta)\Gamma(\beta)} \int_0^t f\big(x,\tau,h(x,\tau)\big)(t-\tau)^{\beta-1}\, d\tau \tag{8.159}$$

Discretizing,

$$h(x,t) - h(x,0) = \frac{1-\beta}{AB(\beta)} f\big(x,t,h(x,t)\big) + \frac{\beta}{AB(\beta)\Gamma(\beta)} \int_0^t f\big(x,\tau,h(x,\tau)\big)(t-\tau)^{\beta-1}\, d\tau \tag{8.160}$$

at interval $x_i,\, t_{j+1}$,

$$h(x_i,t_{j+1}) - h(x_i,0) = \frac{1-\beta}{AB(\beta)} f\big(x_i,t_{j+1},h(x_i,t_{j+1})\big)$$
$$+ \frac{\beta}{AB(\beta)\Gamma(\beta)} \sum_{l=0}^{j} \int_{t_j}^{t_{j+1}} f\big(x_i,\tau,h(x_i,\tau)\big)(t_{j+1}-\tau)^{\beta-1}\, d\tau \tag{8.161}$$

We recall that,

$$\beta_j = \begin{cases} \dfrac{1}{\Gamma(\alpha+2)} & j=0 \\[2ex] \dfrac{(j-1)^{\alpha+1} - 2j^{\alpha+1} + (j+1)^{\alpha+1}}{\Gamma(\alpha+l)} & j=1,2,\ldots,n-1 \end{cases} \tag{8.162}$$

The final equation can be given as,

$$
\begin{aligned}
h_i^{j+1} = h_i^0 &+ \frac{1-\beta}{AB(\beta)}\frac{T}{S}\left[\frac{h_{i+1}^j - 2h_i^j + h_{i-1}^j}{(\Delta r)^2}\left(1+\frac{\Delta r}{r_i}\right) + \frac{h_i^j}{\lambda^2} \right. \\
&\left. + \frac{1}{r_i}\frac{AB(\alpha)}{(1-\alpha)}\sum_{k=0}^{i-1}\big(h_{i+1}^j - h_{i-1}^j\big)\left\{ \begin{array}{l} (j-k)E_{\alpha,2}\left[\frac{-\alpha}{1-\alpha}\Delta r^\alpha (j-k)^\alpha\right] \\ -(j-(k+1))E_{\alpha,2}\left[\frac{-\alpha}{1-\alpha}\Delta r^\alpha (j-(k+1))^\alpha\right] \end{array}\right\} \right] \\
&+ \sum_{j=\alpha}^{n-1}(\beta_{n-j})\frac{T}{S}\left[\begin{array}{l} \frac{h_{i+1}^j - 2h_i^j + h_{i-1}^j}{(\Delta r)^2}\left(1+\frac{\Delta r}{r_i}\right) + \frac{h_i^j}{\lambda^2} + \frac{1}{r_i}\frac{AB(\alpha)}{(1-\alpha)}\sum_{k=0}^{i-1}\big(h_{i+1}^j - h_{i-1}^j\big) \\ \left[(j-k)E_{\alpha,2}\left[\frac{-\alpha}{1-\alpha}\Delta r^\alpha (j-k)^\alpha\right]\right. \\ \left. -(j-(k+1))E_{\alpha,2}\left[\frac{-\alpha}{1-\alpha}\Delta r^\alpha (j-(k+1))^\alpha\right]\right] \end{array} \right]
\end{aligned}
\tag{8.163}
$$

Let,

$$\frac{1-\beta}{AB(\beta)}\frac{T}{S(\Delta r)^2}\left(1+\frac{\Delta r}{r_i}\right)=\alpha_1 \tag{8.164}$$

$$\frac{1-\beta}{AB(\beta)}\frac{T}{S\lambda^2}=\alpha_2 \tag{8.165}$$

$$\left\{\begin{array}{l}\dfrac{1-\beta}{AB(\beta)}\dfrac{T}{S}\dfrac{1}{r_i}\dfrac{AB(\alpha)}{(1-\alpha)}\\[2mm]\left\{\begin{array}{l}(j-k)E_{\alpha,2}\left[\dfrac{-\alpha}{1-\alpha}\Delta r^\alpha\left(j-k\right)^\alpha\right]\\[2mm]-(j-(k+1))E_{\alpha,2}\left[\dfrac{-\alpha}{1-\alpha}\Delta r^\alpha\left(j-(k+1)\right)^\alpha\right]\end{array}\right\}\end{array}\right\}=\alpha_3 \tag{8.166}$$

$$\left(1+\frac{\Delta r}{r_i}\right)\frac{T}{S}\sum_{j=\alpha}^{n-1}\left(\beta_{n-j}\right)=\alpha_4 \tag{8.167}$$

$$\frac{T}{S\lambda^2}\sum_{j=\alpha}^{n-1}\left(\beta_{n-j}\right)=\alpha_5 \tag{8.168}$$

$$\left\{\begin{array}{l}\left\{\begin{array}{l}(j-k)E_{\alpha,2}\left[\dfrac{-\alpha}{1-\alpha}\Delta r^\alpha\left(j-k\right)^\alpha\right]\\[2mm]-(j-(k+1))E_{\alpha,2}\left[\dfrac{-\alpha}{1-\alpha}\Delta r^\alpha\left(j-(k+1)\right)^\alpha\right]\end{array}\right\}\\[4mm]\sum_{j=\alpha}^{n-1}\left(\beta_{n-j}\right)\dfrac{T}{S}\dfrac{1}{r_i}\dfrac{AB(\alpha)}{(1-\alpha)}\end{array}\right\}=\alpha_6 \tag{8.169}$$

Therefore,

$$h_i^{j+1}=h_i^0+\alpha_1\left(h_{i+1}^j-2h_i^j+h_{i-1}^j\right)+\alpha_2 h_i^j+\alpha_3\sum_{k=0}^{i-1}\left(h_{i+1}^j-h_{i-1}^j\right)$$
$$+\alpha_4+\left(h_{i+1}^j-2h_i^j+h_{i-1}^j\right)+\alpha_5 h_i^j+\alpha_6\sum_{k=0}^{i-1}\left(h_{i+1}^j-h_{i-1}^j\right) \tag{8.170}$$

8.4.1.1.1 *Stability Analysis*

To analyze the stability of the above equation, the Von Neumann stability analysis method is used. The method was fully explained in the previous subsection.

$$h_i^{j+1}=h_i^0+\alpha_1\left(h_{i+1}^j-2h_i^j+h_{i-1}^j\right)+\alpha_2 h_i^j+\alpha_3\sum_{k=0}^{i-1}\left(h_{i+1}^j-h_{i-1}^j\right)$$
$$+\alpha_4\left(h_{i+1}^j-2h_i^j+h_{i-1}^j\right)+\alpha_5 h_i^j+\alpha_6\sum_{k=0}^{i-1}\left(h_{i+1}^j-h_{i-1}^j\right) \tag{8.171}$$

We have $h_i^j = \delta_j e^{ikmx}$,

Therefore,

$$\delta_{j+1}e^{ikmx} = \delta_0 e^{ikmx} + \alpha_1\left(\delta_j e^{ikm(x+\Delta x)} - 2\delta_j e^{ikmx} + \delta_j e^{ikm(x-\Delta x)}\right)$$
$$+ \alpha_2\delta_j e^{ikmx} + \alpha_3\sum_{k=0}^{i-1}\left(\delta_j e^{ikm(x+\Delta x)} - \delta_j e^{ikm(x-\Delta x)}\right)$$
$$+ \alpha_4\left(\delta_j e^{ikm(x+\Delta x)} - 2\delta_j e^{ikmx} + \delta_j e^{ikm(x-\Delta x)}\right) \tag{8.172}$$
$$+ \alpha_5\delta_j e^{ikmx} + \alpha_6\sum_{k=0}^{i-1}\left(\delta_j e^{ikm(x+\Delta x)} - \delta_j e^{ikm(x-\Delta x)}\right)$$

Dividing both sides by e^{ikmx},

$$\delta_{j+1} = \delta_0 + \alpha_1\left(\delta_j e^{ikm\Delta x} - 2\delta_j + \delta_j e^{-ikm\Delta x}\right) + \alpha_2\delta_j + \alpha_3\sum_{k=0}^{i-1}\left(\delta_j e^{ikm\Delta x} - \delta_j e^{-ikm\Delta x}\right)$$
$$+ \alpha_4\left(\delta_j e^{ikm\Delta x} - 2\delta_j + \delta_j e^{-ikm\Delta x}\right) + \alpha_5\delta_j + \alpha_6\sum_{k=0}^{i-1}\left(\delta_j e^{ikm\Delta x} - \delta_j e^{-ikm\Delta x}\right) \tag{8.173}$$

Taking out the common factors,

$$\delta_{j+1} = \delta_0 + \alpha_1\delta_j\left(e^{ikm\Delta x} - 2 + e^{-ikm\Delta x}\right) + \alpha_2\delta_j + \alpha_3\delta_j\sum_{k=0}^{i-1}\left(e^{ikm\Delta x} - e^{-ikm\Delta x}\right)$$
$$+ \alpha_4\delta_j\left(e^{ikm\Delta x} - 2 + e^{-ikm\Delta x}\right) + \alpha_5\delta_j + \alpha_6\delta_j\sum_{k=0}^{i-1}\left(e^{ikm\Delta x} - e^{-ikm\Delta x}\right) \tag{8.174}$$

We recall that,

$$\frac{e^{i\theta} + e^{i\theta}}{2i} = \cos\theta \text{ and } \frac{e^{i\theta} - e^{i\theta}}{2i} = \sin\theta$$

$$\cos 2\theta = \cos\theta^2 - \sin\theta^2 = 1 - 2\sin\theta^2 \text{ and } \cos\theta = 1 - 2\sin\frac{\theta^2}{2}$$

Therefore,

$$\delta_{j+1} = \delta_0 - \alpha_1 4\sin\left(\frac{km\Delta x}{2}\right)^2\delta_j + \alpha_2\delta_j + 2i\alpha_3\sum_{k=0}^{i-1}\left(\sin\left(km\Delta x\right)\right)\delta_j$$
$$- \alpha_4 4\sin\left(\frac{km\Delta x}{2}\right)^2\delta_j + \alpha_5\delta_j + 2i\alpha_6\sum_{k=0}^{i-1}\left(\sin\left(km\Delta x\right)\right)\delta_j \tag{8.175}$$

We assume that for all j = 0 $|\delta_j| < |\delta_0|$,

$$\delta_{j+1} = \delta_0 - \alpha_1 4\sin\left(\frac{km\Delta x}{2}\right)^2\delta_0 + \alpha_2\delta_0 + 2i\alpha_3\sum_{k=0}^{i-1}\left(\sin\left(km\Delta x\right)\right)\delta_0$$
$$- \alpha_4 4\sin\left(\frac{km\Delta x}{2}\right)^2\delta_0 + \alpha_5\delta_0 + 2i\alpha_6\sum_{k=0}^{i-1}\left(\sin\left(km\Delta x\right)\right)\delta_0 \tag{8.176}$$

Let,

$$\alpha_1 4 \sin\left(\frac{km\Delta x}{2}\right)^2 + \alpha_2 - \alpha_4 4 \sin\left(\frac{km\Delta x}{2}\right)^2 + \alpha_5 = a \tag{8.177}$$

$$2i\alpha_3 \sum_{k=0}^{i-1} \left(\sin\left(km\Delta x\right)\right) + 2i\alpha_6 \sum_{k=0}^{i-1} \left(\sin\left(km\Delta x\right)\right) = ib \tag{8.178}$$

Therefore,

$$\delta_1 = \delta_0 \left(a + ib\right) \tag{8.179}$$

$$\left|\frac{\delta_1}{\delta_0}\right| = |a + ib| = \sqrt{a^2 + b^2} \tag{8.180}$$

We know that,

$$\left|\frac{\delta_1}{\delta_0}\right| < 1 \tag{8.181}$$

Therefore,

$$\sqrt{a^2 + b^2} < 1 \tag{8.182}$$

We assume that for all j>0 $|\delta_j| < |\delta_0|$,

$$\delta_{j+1} = \delta_0 - \delta_j \left[\begin{array}{c} \alpha_1 4 \sin\left(\frac{km\Delta x}{2}\right)^2 + \alpha_2 + 2i\alpha_3 \sum_{k=0}^{i-1} \left(\sin\left(km\Delta x\right)\right) \\ -\alpha_4 4 \sin\left(\frac{km\Delta x}{2}\right)^2 + \alpha_5 + 2i\alpha_6 \sum_{k=0}^{i-1} \left(\sin\left(km\Delta x\right)\right) \end{array}\right] \tag{8.183}$$

Let,

$$\alpha_1 4 \sin\left(\frac{km\Delta x}{2}\right)^2 + \alpha_2 - \alpha_4 4 \sin\left(\frac{km\Delta x}{2}\right)^2 + \alpha_5 = a \tag{8.184}$$

$$2i\alpha_3 \sum_{k=0}^{i-1} \left(\sin\left(km\Delta x\right)\right) + 2i\alpha_6 \sum_{k=0}^{i-1} \left(\sin\left(km\Delta x\right)\right) = ib \tag{8.185}$$

Therefore,

$$\delta_{j+1} = \delta_0 - \delta_j \left(a + ib\right) \tag{8.186}$$

$$\left|\delta_{j+1}\right| \le \left|\delta_0\right| + \left|\delta_j\right|\left|a + ib\right| \tag{8.187}$$

And we recall that $|\delta_0| = |\delta_j|$,

$$\left|\delta_{j+1}\right| < \left|\delta_0\right| + \left|\delta_0\right|\left|a + ib\right| \tag{8.188}$$

$$\left|\delta_{j+1}\right| < \left|\delta_0\right|\left(1 + \left|a + ib\right|\right). \tag{8.189}$$

8.5 SIMULATIONS

8.5.1 Caputo Numerical Figures and Interpretation

Figures 8.1 to 8.5 represents the Caputo derivative. The cone of depression decreases as the time increases, and decreases as the space decreases. From the simulations it is depicted that the lower

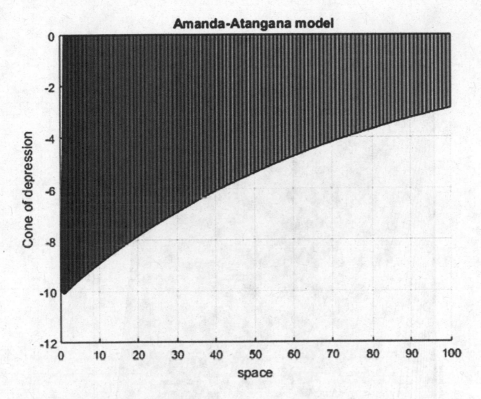

FIGURE 8.1 Cone of depression at order 0.4 in space.

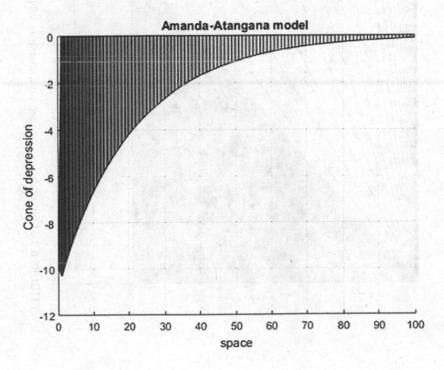

FIGURE 8.2 Cone of depression at order 0.7 in space.

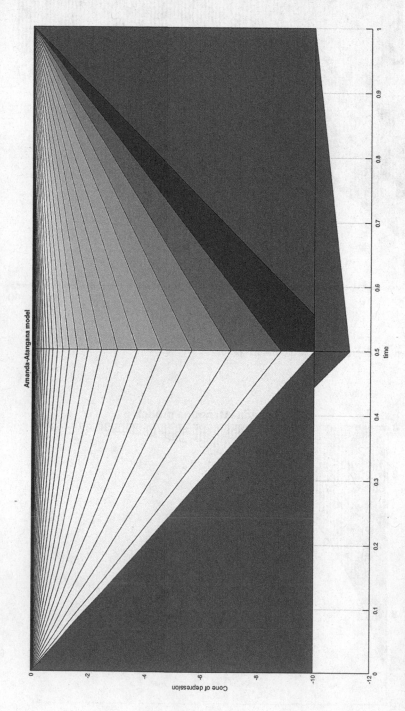

FIGURE 8.3 Cone of depression at order 0.9 time space solution.

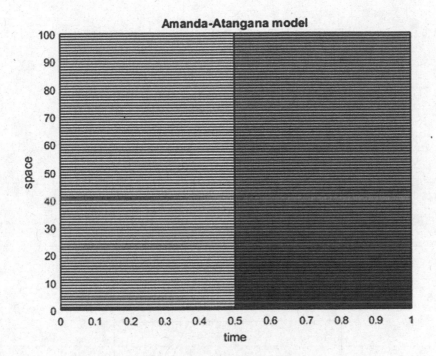

FIGURE 8.4   Amanda–Atangana model for order 0.4 time space solution.

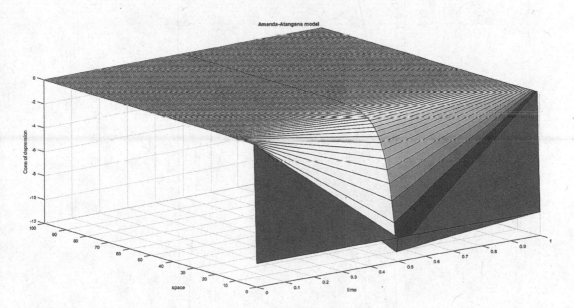

FIGURE 8.5   Cone of depression for order 0.9 time space solution.

the order, the higher the cone of depression. When the order is 0.04, the cone of depression is high, compared to when the order is 0.4. When the order increases to 0.7 and 0.9, the cone of depression is declining. At the order of 1 the cone of depression is normal. We can simply summarize that at orders 0.04 and 0.4, the water is flowing through a sand or silt–sand type of soil. At order 0.7 the type of soil can be silt and at 0.9 the type of soil would be expected to be clay. At order 1 the water flow is normal, as it is depicted in the figures. This indicates that the water flow is occurring through

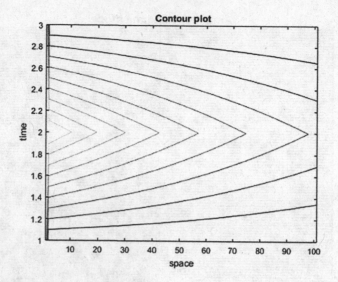

FIGURE 8.6   Contour plot at 0.4 time space solution with Atangana–Baleanu derivative.

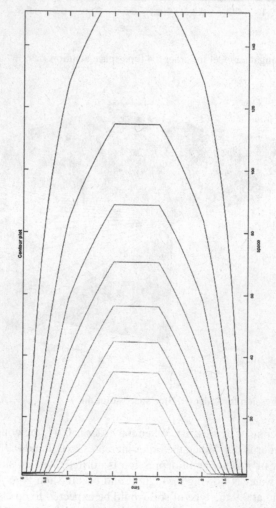

FIGURE 8.7   Contour plot at 0.4 time space solution with Caputo—Fabrizio derivative.

the fractures at order 1. The contour plots indicated that the memory is fully captured at order 0.7. This interpretation is presented in literature that a permeable material is capable of releasing more water and the cone of depression will be high. As groundwater flows within a geological formation, it encounters different formations with different porosity and permeability. We recall that the higher the porosity, the large the extent of water to spread out, and the flow velocity or permeability will be very low (Figures 8.6 and 8.7).

8.6 CONCLUSION

One of the huge responsibilities of humankind is to protect nature and the environment. However, to reach the position where which humankind can claim to be in control of their environment, they must consider three major steps: they must observe, predict and analyze. In the recent decades, researchers have intensively used the concept of differentiation and integration to model world problems based on the rate of change. These operators have been used in many situations to model the flow of sub-surface water within the geological formations called leaky aquifers. While these models have been applied intensively to determine the aquifer's parameters, it was clear that the model used was applicable only in the case of a non-memory process where the geological formation is considered homogeneous. Nature, on the other hand, suggests that the geological formation is not homogeneous, therefore modeling such flow using classical differential and integral operators lead to inadequate results. Therefore, either the aquifers' parameters are highly estimated or underestimated. Additional to this limitation, some simplifications were made while deriving the groundwater flow model within a leaky aquifer. Ramatsho and Atangana solved this problem; their mathematical equation was able to include the scale factor of the aquifer. However their equation was based on local differentiation operators, therefore a mathematical equation with fractal operators was proposed. To further capture more complexities of the geological formation, the concept of nonlocal operators is adopted with three different laws, one based on power law, one on exponential decay law, and the generalized Mittag–Leffler function. An additional extension was done as the classical nonlocal fractional differential and integral operators were replaced by fractal-fractional operators.

REFERENCES

Atangana A., (2017). Fractal-fractional differentiation and integration: Connecting fractal calculus and fractional calculus to predict complex system. *Chaos, Solitons and Fractals* 102: 399–406.

Atangana A., (2018). Blind in a commutative world: Simple illustrations with functions and chaotic attractors. *Chaos, Solitons and Fractals* 144: 347–363.

Atangana A., & Baleanu D., (2017). New fractional derivatives with nonlocal and non-singular kernel: Theory and application to heat transfer model. *Thermal Science* 20(2): 763–769.

Atangana A., & Gomez-Aguilar J.F., (2018). Decolonisation of fractional calculus rules: Breaking commutativity and associativity to capture more natural phenomena. *The European Physical Journal Plus* 133(4): 1–22.

Charney J. G., Fjortoft R., & Von Neumann J., (1950). Numerical integration of the barotropic vorticity equation. *Tellus* 2: 237–254.

Crank J., & Nicolson P., (1947). A practical method for numerical evaluation of solutions of partial differential equations of the heat conduction type. *Mathematical Proceedings of the Cambridge Philosophical Society.* 43(1): 50–67.

Haubold H.J., Mathai A.M., & Saxena R.K., (2011). Mittag-Leffler functions and their applications: Hindawi Publishing Corporation. *Journal of Applied Mathematics.*

Ramotsho A., & Atangana A., (2018). Derivation of a groundwater flow model within leaky and self-similar aquifers: Beyond Hantush model. *Chaos, Solitons and Fractals* 116: 414–423.

Toufik M., & Atangana A., (2017). New numerical approximation of fractional derivative with non-local and non-singular kernel: Application to chaotic models. *The European Physical Journal Plus* 132: 444.

Visser M., & Yunes N., (2008). Power laws, scale invariance, and generalized Frobenius series: Applications to Newtonian and TOV stars near criticality. *International Journal of Modern Physics* 18(20): 3433–3468.

9 Application of the New Numerical Method with Caputo Fractal-Fractional Derivative on the Self-Similar Leaky Aquifer Equations

Amanda Ramotsho and Abdon Atangana
University of the Free State, Bloemfontein, South Africa

CONTENTS

9.1 INTRODUCTION

It has been clearly indicated that there is a lack of implementation of documented strategies and policies for the conservation and management of water. The water demand, management, purification to drinkable standards and supply do not balance. The 2030 visions can be accomplished if sustainable and equitable water is secured. This can only be achieved if the information that is available is used accordingly and well interpreted. However there is a need for expanding the monitoring programs and points, and data sharing between stakeholders and custodians.

Through Section 24 of the constitution we recognize our environmental rights, which guarantee everyone the right to an environment that is not harmful, and it can only be applicable or achievable if we conserve, manage and protect our natural resources. For example, the mining sector must compile an effective and accurate water and salt balance for their water demand and conservation, farmers must put in place measuring devices, in overall water quality must be taken as the most crucial and important aspect for management of water. Groundwater researches and innovations provide effective and efficient water management solutions that must be taken into account. Water management is complex, considering that it is used for different objectives, economic and ecological activities. Climate change contributes a lot to economy and social development. Given the limited water resources available, this chapter covers the concept of self-similar leaky aquifers which are governed by fractures.

DOI: 10.1201/9781003266266-9

Self-similarity can be elaborated as an object which a part of the whole replicates itself. For example, the Romanesco broccoli vegetable is a natural example of replication. The vegetable exemplifies the Fibonacci sequence in its appearance, with spirals upon spirals of the vegetable pattern repeating themselves at different levels of scale. The other example can be the deciduous tree leaves, where the branching veins from the smallest to the biggest simplify the shape of the leaf and so on. Other examples can be feathers of a peacock, salt flats, snowflakes, mountains, and etc. In a perfect mathematical fractal, such as the famous Mandelbrot set, represent the self-similarity pattern.

The characteristics and physical properties of water flowing within an aquifer are determined using the well-known groundwater flow equations, which are derived from classical equations. This results in mathematical models; for example, partial differential equations with the first partition conditions, which are solved using analytical and numerical solutions. Analytical and numerical solutions can be seen complementary rather than competing, as numerical methods approximate solutions of mathematical equations. This can only be done when the algebraic methods cannot determine the exact solutions (Remani, 2013).

There are different numerical schemes that have been developed, but the most widely used is the Adams–Bashforth method, based on the well-known Lagrange polynomial. The new numerical schemes were developed using a different polynomial known as Newton polynomial because of its effectiveness and accuracy. Two calculus definitions regarding the Caputo operator and the power law are presented below (Atangana and Araz, 2021).

9.2   DEFINITIONS IN TERMS OF DIFFERENTIATION

Definition 1

Accede that f(t) is continuous and fractal differentiable on (a, b) with order β, then the fractal-fractional derivative of $f(t)$ with order α in the Caputo sense having a power law type kernel has been defined as follows:

$$_{0}^{FFP}D_{t}^{\alpha,\beta}f\left(t\right)=\frac{1}{\Gamma\left(1-\alpha\right)}\int_{0}^{\tau}\frac{df\left(\tau\right)}{d\tau^{\beta}}\left(t-\tau\right)^{n-\alpha-1}d\tau \tag{9.1}$$

When

$$n-1<\alpha\leq n,\,0<n-1<\beta\leq n$$

$$_{0}^{FFP}D_{t}^{\alpha,\beta}f\left(t\right)=\frac{1}{\Gamma\left(1-\alpha\right)}\frac{d}{dt}\int_{0}^{\tau}\frac{df\left(\tau\right)}{d\tau^{\beta}}\left(t-\tau\right)^{-\alpha}d\tau\frac{t^{1-\alpha}}{\alpha} \tag{9.2}$$

The more generalized version is given as:

$$_{0}^{FFP}D_{t}^{\alpha,\beta\lambda}f\left(t\right)=\frac{1}{\Gamma\left(1-\alpha\right)}\int_{0}^{\tau}\frac{df^{\lambda}}{d\tau^{\beta}}\left(t-\tau\right)^{n-\alpha-1}d\tau \tag{9.3}$$

When

$$n-1<\alpha\leq n,\,0<n-1<\lambda,\,\beta\leq n$$

$$^{FFP}_0 D_t^{\alpha,\beta} f(t) = \frac{1}{\Gamma(1-\alpha)} \int_0^t \frac{df}{d\tau} \frac{\tau^{1-\beta}}{\beta} (t-\tau)^{-\alpha} \, d\tau. \tag{9.4}$$

Definition 2

Accede that $f(t)$ be continuous on an open interval (a, b) then the fractal-fractional integral of $f(t)$ with order α having power law type kernel is defined as follows:

$$^{FFP}_0 J_t^{\alpha,\beta} f(t) = \frac{\beta}{\Gamma(\alpha)} \int_0^t (t-\tau)^{-\alpha} \tau^{1-\beta} f(\tau) \, d\tau. \tag{9.5}$$

9.3　NEW NUMERICAL METHOD WITH CAPUTO FRACTAL-FRACTIONAL DERIVATIVE BY ATANGANA AND ARAZ

Fractional calculus is the branch of mathematics that has been used to derive derivatives that are widely applied in many fields of science, engineering, technology and other fields. There are three types of fractional differential and integral operators including the first, namely Riemann–Liouville and Caputo derivatives: this version is a convolution of power law with first derivative, while the second is the Caputo–Fabrizio fractional derivative which is the convolution of first derivative and exponential decay law with the Dirac-delta property, and finally the third is known as the Atangana–Baleanu fractional derivative which is the convolution of the generalized Mittag-Leffler function and the first derivative (Atangana and Qureshi, 2019).

Fractal and fractional are classified or elaborated as characteristics and fundamental problems which arise in all fields of science and technology. They are widely applied in different fields recently. These problems are characterized by having some non-integer order features. Fractals are geometrical objects with non-integer dimension, while fractional is the non-integer order of differential operators. The following numerical scheme was derived with Caputo fractal-fractional derivative.

The method was initiated or developed using the Cauchy problem

$$^{FFP}_0 D_t^{\alpha,\beta} h(t) = f(t, h(t)) \tag{9.6}$$

Integrating the equation above

$$h(t) - h(0) = \frac{\alpha\beta}{\Gamma(\alpha)} \int_0^t \tau^{\beta-1} f(\tau, h(\tau)) (t-\tau)^{\alpha-1} \, d\tau \tag{9.7}$$

$F(t, h(t))$ can be represented as $t^{-1} f(t, h(t))$, thus we have

$$h(t) - h(0) = \frac{1}{\Gamma(\alpha)} \int_0^t F(\tau, h(\tau)) (t-\tau)^{\alpha-1} \, d\tau \tag{9.8}$$

At the point $t_{n+1} = (n + 1)\Delta t$, where the function F is nonlinear, we have

$$h(t_{n+1}) - h(0) = \frac{1}{\Gamma(\alpha)} \int_0^{t_{n+1}} F(\tau, h(\tau)) (t_{n+1}-\tau)^{\alpha-1} \, d\tau \tag{9.9}$$

Therefore

$$h(t_{n+1}) - h(0) = \frac{1}{\Gamma(\alpha)} \sum_{j=2}^{n} \int_{t_j}^{t_{j+1}} F(\tau, h(\tau))(t_{n+1} - \tau)^{\alpha-1} d\tau \qquad (9.10)$$

The Newton polynomial is given as:

$$P_n(\tau) = f(t_{n-2}, h(t_{n-2})) + \left[\frac{f(t_{n-1}, h(t_{n-1})) - f(t_{n-2}, h(t_{n-2}))}{\Delta t} \right](\tau - t_{n-2})$$

$$+ \left[\frac{f(t_n, h(t_n)) - 2f(t_{n-1}, h(t_{n-1})) + f(t_{n-2}, h(t_{n-2}))}{2(\Delta t)^2} \right](\tau - t_{n-2})(\tau - t_{n-1}) \qquad (9.11)$$

By replacing the Newton polynomial in the discretized equation the following is obtained

$$h^{n+1} = h^0 + \frac{1}{\Gamma(\alpha)} \sum_{j=2}^{n} \int_{t_j}^{t_{j+1}} \left\{ \begin{array}{c} F(t_{j-2}, h^{j-2}) \\ + \dfrac{F(t_{j-1}, h^{j-1}) - F(t_{j-2}, h^{j-2})}{\Delta t}(\tau - t_{j-2}) \\ + \dfrac{F(t_j, h^j) - 2F(t_{j-1}, h^{j-1}) + F(t_{j-2}, h^{j-2})}{2(\Delta t)^2} \\ \times (\tau - t_{j-2})(\tau - t_{j-1}) \end{array} \right\} (t_{n+1} - \tau)^{\alpha-1} d\tau \qquad (9.12)$$

The following equality is obtained:

$$h^{n+1} = h^0 + \frac{1}{\Gamma(\alpha)} \sum_{j=2}^{n} F(t_{j-2}, h^{j-2}) \int_{t_j}^{t_{j+1}} (t_{n+1} - \tau)^{\alpha-1} d\tau$$

$$+ \frac{1}{\Gamma(\alpha)} \sum_{j=2}^{n} \frac{F(t_{j-1}, h^{j-1}) - F(t_{j-2}, h^{j-2})}{\Delta t} \int_{t_j}^{t_{j+1}} (\tau - t_{j-2})(t_{n+1} - \tau)^{\alpha-1} d\tau$$

$$+ \frac{1}{\Gamma(\alpha)} \sum_{j=2}^{n} \frac{F(t_j, h^j) - 2F(t_{j-1}, h^{j-1}) + F(t_{j-2}, h^{j-2})}{2(\Delta t)^2}$$

$$\times \int_{t_j}^{t_{j+1}} (\tau - t_{j-2})(\tau - t_{j-1})(t_{n+1} - \tau)^{\alpha-1} d\tau \qquad (9.13)$$

The above integrals can be calculated and grouped as:

$$\int_{t_j}^{t_{j+1}} (t_{n+1} - \tau)^{\alpha-1} d\tau = \frac{(\Delta t)^\alpha}{\alpha} \left[(n - j + 1)^\alpha - (n - j)^\alpha \right] \qquad (9.14)$$

$$\int_{t_j}^{t_{j+1}} (\tau - t_{j-2})(t_{n+1} - \tau)^{\alpha-1} d\tau = \frac{(\Delta t)^{\alpha+1}}{\alpha(\alpha+1)} \left[\begin{array}{l} (n-j+1)^{\alpha}(n-j+3+2\alpha) \\ -(n-j)^{\alpha}(n-j+3+3\alpha) \end{array} \right] \quad (9.15)$$

$$\int_{t_j}^{t_{j+1}} (\tau - t_{j-2})(\tau - t_{j-1})(t_{n+1} - \tau)^{\alpha-1} d\tau = \frac{(\Delta t)^{\alpha+2}}{\alpha(\alpha+1)(\alpha+2)} \times$$

$$\left[\begin{array}{l} (n-j+1)^{\alpha} \left[\begin{array}{l} 2(n-j)^2 + (3\alpha+10)(n-j) \\ + (2\alpha^2 + 9\alpha + 12) \end{array} \right] \\ -(n-j)^{\alpha} \left[\begin{array}{l} 2(n-j)^2 + (5\alpha+10)(n-j) \\ + (6\alpha^2 + 18\alpha + 12) \end{array} \right] \end{array} \right] \quad (9.16)$$

Therefore substituting the above integrals in the equality, the following is obtained:

$$h^{n+1} - h^0 + \frac{(\Delta t)^{\alpha}}{\Gamma(\alpha+1)} \sum_{j=2}^{n} F(t_{j-2}, h^{j-2}) \left[(n-j+1)^{\alpha} - (n-j)^{\alpha} \right] +$$

$$\frac{(\Delta t)^{\alpha}}{\Gamma(\alpha+2)} \sum_{j=2}^{n} \left(F(t_{j-1}, h^{j-1}) - F(t_{j-2}, h^{j-2}) \right) \left[\begin{array}{l} (n-j+1)^{\alpha}(n-j+3+2\alpha) \\ -(n-j)^{\alpha}(n-j+3+3\alpha) \end{array} \right] +$$

$$\frac{(\Delta t)^{\alpha}}{2\Gamma(\alpha+3)} \sum_{j=2}^{n} \left(F(t_j, h^j) - 2F(t_{j-1}, h^{j-1}) + F(t_{j-2}, h^{j-2}) \right) \times$$

$$\left[\begin{array}{l} (n-j+1)^{\alpha} \left[2(n-j)^2 + (3\alpha+10)(n-j) + (2\alpha^2+9\alpha+12) \right] \\ -(n-j)^{\alpha} \left[-2(n-j)^2 + (5\alpha+10)(n-j) + (6\alpha^2+18\alpha+12) \right] \end{array} \right] \quad (9.17)$$

Therefore:

$$h^{n+1} = h^0 + \frac{\beta(\Delta t)^{\alpha}}{\Gamma(\alpha+1)} \sum_{j=2}^{n} t_{j-2}^{\beta-1} f(t_{j-2}, h^{j-2}) \left[(n-j+1)^{\alpha} - (n-j)^{\alpha} \right] +$$

$$\frac{\beta(\Delta t)^{\alpha}}{\Gamma(\alpha+2)} \sum_{j=2}^{n} t_{j-1}^{\beta-1} f(t_{j-1}, h^{j-1}) - t_{j-2}^{\beta-1} f(t_{j-2}, h^{j-2}) \left[\begin{array}{l} (n-j+1)^{\alpha}(n-j+3+2\alpha) \\ -(n-j)^{\alpha}(n-j+3+3\alpha) \end{array} \right] +$$

$$\frac{\beta(\Delta t)^{\alpha}}{2\Gamma(\alpha+3)} \sum_{j=2}^{n} t_j^{\beta-1} f(t_j, h^j) - 2t_{j-1}^{\beta-1} f(t_{j-1}, h^{j-1}) + t_{j-2}^{\beta-1} f(t_{j-2}, h^{j-2}) \times$$

$$\left[\begin{array}{l} (n-j+1)^{\alpha} \left[2(n-j)^2 + (3\alpha+10)(n-j) + (2\alpha^2+9\alpha+12) \right] \\ -(n-j)^{\alpha} \left[-2(n-j)^2 + (5\alpha+10)(n-j) + (6\alpha^2+18\alpha+12) \right] \end{array} \right]. \quad (9.18)$$

9.3.1 APPLICATION OF THE NEW CAPUTO FRACTAL-FRACTIONAL DERIVATIVE ON THE SELF-SIMILAR LEAKY AQUIFER EQUATION: SCENARIO 1

The mathematical model below was incorporated to assess self-similar leaky aquifers, the inflow and outflow is through a self-similar leaky aquifer.

$$\frac{S}{T}\frac{\partial h}{\partial t} = \frac{\partial}{\partial r^{\alpha}}\left(\frac{\partial h}{\partial r^{\alpha}}\right)\left(1+\frac{dr}{r}\right)+\frac{1}{r}\frac{\partial h}{\partial r^{\alpha}}+\frac{h(r,t)}{\lambda^2} \tag{9.19}$$

Applying the Caputo fractal-fractional derivative to the function

$$\frac{S}{T}\,{}_0^c D_t^{\beta,\alpha} h(r,t) = \frac{\partial}{\partial r^{\alpha}}\left(\frac{\partial h(r,t)}{\partial r^{\alpha}}\right)\left(1+\frac{dr}{r}\right)+\frac{1}{r}\frac{\partial h(r,t)}{\partial r^{\alpha}}+\frac{h(r,t)}{\lambda^2} \tag{9.20}$$

Differentiating the single and double fractal derivative $\dfrac{\partial h(r,t)}{\partial r^{\alpha}}+\dfrac{\partial}{\partial r^{\alpha}}\left(\dfrac{\partial h(r,t)}{\partial r^{\alpha}}\right)$

$$\frac{S}{T}\,{}_0^c D_t^{\beta,\alpha} h(r,t) = \frac{\partial}{\partial r^{\alpha}}\left(\frac{\partial h(r,t)}{\partial r^{\alpha}}\right)\left(1+\frac{dr}{r}\right)+\frac{1}{r}\frac{\partial h(r,t)}{\partial r^{\alpha}}\frac{r^{1-\alpha}}{\alpha}+\frac{h(r,t)}{\lambda^2} \tag{9.21}$$

$$\begin{aligned}{}_0^c D_t^{\beta,\alpha} h(r,t) &= \frac{T}{S}\frac{r^{1-\alpha}}{\alpha}\left[\frac{\partial r^2 h(r,t)}{\partial r^2}\left(\frac{r^{1-\alpha}}{\alpha}\right)+\left(\frac{1-\alpha}{\alpha}\right)r^{-\alpha}\frac{\partial h(r,t)}{\partial r}\right]\left(1+\frac{dr}{r}\right)\\ &\quad+\frac{T}{S}\frac{1}{r}\frac{\partial h(r,t)}{\partial r}\frac{r^{1-\alpha}}{\alpha}+\frac{T}{S}\frac{h(r,t)}{\lambda^2}\end{aligned} \tag{9.22}$$

The right-hand side of the above equation can be simplified by replacing $f(\alpha,r,t,h(r,\tau))$ such that:

$$\begin{aligned}f\big(\alpha,r,t,h(r,t)\big) &= \frac{T}{S}\frac{r^{1-\alpha}}{\alpha}\left[\frac{\partial r^2 h(r,t)}{\partial r^2}\left(\frac{r^{1-\alpha}}{\alpha}\right)+\left(\frac{1-\alpha}{\alpha}\right)r^{-\alpha}\frac{\partial h(r,t)}{\partial r}\right]\left(1+\frac{dr}{r}\right)\\ &\quad+\frac{T}{S}\frac{1}{r}\frac{\partial h(r,t)}{\partial r}\frac{r^{1-\alpha}}{\alpha}+\frac{T}{S}\frac{h(r,t)}{\lambda^2}\end{aligned} \tag{9.23}$$

Such that:

$$\,{}_0^C D_t^{\beta,\alpha} h(r,t) = f\big(\alpha,r,t,h(r,t)\big) \tag{9.24}$$

Discretizing at (x_i, t_{n+1})

$$h(r_i,t_{n+1})-h(r_i,0) = \frac{\alpha}{\Gamma(\beta)}\int_0^{t_{n+1}}\tau^{\alpha-1}\left(t_{n+1}-\tau\right)^{\beta-1}f\big(\alpha,r_i,t_{n+1},h(r_i,\tau)\big)d\tau \tag{9.25}$$

Therefore the numerical scheme will be given as

$$
\begin{aligned}
h_i^{n+1} = h_i^0 &+ \frac{\alpha(\Delta t)^\beta}{\Gamma(\beta+1)} \sum_{j=2}^n t_{j-2}^{\alpha-1} \left\{ f\left(\alpha, r_i, t_{j-2}, h\left(r_i, t_{j-2}\right)\right) \right\} \\
&+ \frac{\alpha(\Delta t)^\beta}{\Gamma(\beta+2)} \sum_{j=2}^n \left\{ \begin{array}{l} t_{j-1}^{\alpha-1} f\left(\alpha, r_i, t_{j-1}, h\left(r_i, t_{j-1}\right)\right) \\ -t_{j-2}^{\alpha-1} f\left(\alpha, r_i, t_{j-2}, h\left(r_i, t_{j-2}\right)\right) \end{array} \right\} \left[\begin{array}{l} (n-j+1)^\beta \left(n-j+3+2\beta\right) \\ -(n-j)^\beta \left(n-j+3+3\beta\right) \end{array} \right] \\
&+ \frac{\alpha(\Delta t)^\beta}{2\Gamma(\beta+3)} \sum_{j=2}^n \left(\begin{array}{l} t_j^{\alpha-1} f\left(\alpha, r_i, t_j, h\left(r_i, t_j\right)\right) \\ -2t_{j-1}^{\alpha-1} f\left(\alpha, r_i, t_{j-1}, h\left(r_i, t_{j-1}\right)\right) \\ +t_{j-2}^{\alpha-1} f\left(\alpha, r_i, t_{j-2}, h\left(r_i, t_{j-2}\right)\right) \end{array} \right) \left[\begin{array}{l} (n-j+1)^\beta \left[\begin{array}{l} 2(n-j)^2 \\ +(3\beta+10)(n-j) \\ +(2\beta^2+9\beta+12) \end{array} \right] \\ -(n-j)^\beta \left[\begin{array}{l} -2(n-j)^2 \\ +(5\beta+10)(n-j) \\ +(6\beta^2+18\beta+12) \end{array} \right] \end{array} \right]
\end{aligned}
\tag{9.26}
$$

Let the functions be represented as:

$$
\begin{aligned}
f\left(\alpha, r_i, t_{j-2}, h\left(r_i, t_{j-2}\right)\right) = \frac{T}{S} \frac{r_i^{1-\alpha}}{\alpha} &\left[\frac{\partial r^2 h\left(r_i, t_{j-2}\right)}{\partial r^2} \left(\frac{r_i^{1-\alpha}}{\alpha} \right) + \left(\frac{1-\alpha}{\alpha} \right) r_i^{-\alpha} \frac{\partial h\left(r_i, t_{j-2}\right)}{\partial r} \right] \left(1 + \frac{dr}{r_i} \right) \\
&+ \frac{T}{S} \frac{1}{r} \frac{\partial h\left(r_i, t_{j-2}\right)}{\partial r} \frac{r_i^{1-\alpha}}{\alpha} + \frac{T}{S} \frac{h\left(r_i, t_{j-2}\right)}{\lambda^2}
\end{aligned}
\tag{9.27}
$$

$$
\begin{aligned}
f\left(\alpha, r_i, t_{j-1}, h\left(r_i, t_{j-1}\right)\right) = \frac{T}{S} \frac{r_i^{1-\alpha}}{\alpha} &\left[\frac{\partial r^2 h\left(r_i, t_{j-1}\right)}{\partial r^2} \left(\frac{r_i^{1-\alpha}}{\alpha} \right) + \left(\frac{1-\alpha}{\alpha} \right) r_i^{-\alpha} \frac{\partial h\left(r_i, t_{j-1}\right)}{\partial r} \right] \left(1 + \frac{dr}{r_i} \right) \\
&+ \frac{T}{S} \frac{1}{r} \frac{\partial h\left(r_i, t_{j-1}\right)}{\partial r} \frac{r_i^{1-\alpha}}{\alpha} + \frac{T}{S} \frac{h\left(r_i, t_{j-1}\right)}{\lambda^2}
\end{aligned}
\tag{9.28}
$$

$$
\begin{aligned}
f\left(\alpha, r_i, t_j, h\left(r_i, t_j\right)\right) = \frac{T}{S} \frac{r_i^{1-\alpha}}{\alpha} &\left[\frac{\partial r^2 h\left(r_i, t_j\right)}{\partial r^2} \left(\frac{r_i^{1-\alpha}}{\alpha} \right) + \left(\frac{1-\alpha}{\alpha} \right) r_i^{-\alpha} \frac{\partial h\left(r_i, t_j\right)}{\partial r} \right] \left(1 + \frac{dr}{r_i} \right) \\
&+ \frac{T}{S} \frac{1}{r} \frac{\partial h\left(r_i, t_j\right)}{\partial r} \frac{r_i^{1-\alpha}}{\alpha} + \frac{T}{S} \frac{h\left(r_i, t_j\right)}{\lambda^2}
\end{aligned}
\tag{9.29}
$$

Therefore the final numerical solution can be given as:

$$
\begin{aligned}
h_i^{n+1} h_i^0 &+ \frac{\alpha(\Delta t)^\beta}{\Gamma(\beta+1)} \sum_{j=2}^n t_{j-2}^{\alpha-1} \left\{ \frac{T}{S} \frac{r_i^{1-\alpha}}{\alpha} \left[\frac{\frac{\partial r^2 h(r_i,t_{j-2})}{\partial r^2}\left(\frac{r_i^{1-\alpha}}{\alpha}\right)}{+\left(\frac{1-\alpha}{\alpha}\right)r_i^{-\alpha}\frac{\partial h(r_i,t_{j-2})}{\partial r}} \right]\left(1+\frac{dr}{r_i}\right) \right. \\
&\left. + \frac{T}{S}\frac{1}{r}\frac{\partial h(r_i,t_{j-2})}{\partial r}\frac{r_i^{1-\alpha}}{\alpha} + \frac{T}{S}\frac{h(r_i,t_{j-2})}{\lambda^2} \right\}
\end{aligned}
$$

$$
\begin{aligned}
&+ \frac{\alpha(\Delta t)^\beta}{\Gamma(\beta+2)} \sum_{j=2}^n \left\{ t_{j-1}^{\alpha-1}\left(\frac{T}{S}\frac{r_i^{1-\alpha}}{\alpha}\left[\frac{\partial r^2 h(r_i,t_{j-1})}{\partial r^2}\left(\frac{r_i^{1-\alpha}}{\alpha}\right) + \left(\frac{1-\alpha}{\alpha}\right)r_i^{-\alpha}\frac{\partial h(r_i,t_{j-1})}{\partial r} \right]\left(1+\frac{dr}{r_i}\right) \right. \right. \\
&\left. \left. \qquad + \frac{T}{S}\frac{1}{r}\frac{\partial h(r_i,t_{j-1})}{\partial r}\frac{r_i^{1-\alpha}}{\alpha} + \frac{T}{S}\frac{h(r_i,t_{j-1})}{\lambda^2} \right) \right. \\
&\left. \quad -t_{j-2}^{\alpha-1}\left(\frac{T}{S}\frac{r_i^{1-\alpha}}{\alpha}\left[\frac{\partial r^2 h(r_i,t_{j-2})}{\partial r^2}\left(\frac{r_i^{1-\alpha}}{\alpha}\right) + \left(\frac{1-\alpha}{\alpha}\right)r_i^{-\alpha}\frac{\partial h(r_i,t_{j-2})}{\partial r} \right]\left(1+\frac{dr}{r_i}\right) \right. \right. \\
&\left. \left. \qquad + \frac{T}{S}\frac{1}{r}\frac{\partial h(r_i,t_{j-2})}{\partial r}\frac{r_i^{1-\alpha}}{\alpha} + \frac{T}{S}\frac{h(r_i,t_{j-2})}{\lambda^2} \right) \right\}
\end{aligned}
$$

$$
\times \left[\begin{array}{l} (n-j+1)^\beta(n-j+3+2\beta) \\ -(n-j)^\beta(n-j+3+3\beta) \end{array} \right]
$$

$$
\begin{aligned}
&+ \frac{\alpha(\Delta t)^\beta}{2\Gamma(\beta+3)} \sum_{j=2}^n \left\{ t_j^{\alpha-1}\left(\frac{T}{S}\frac{r_i^{1-\alpha}}{\alpha}\left[\frac{\partial r^2 h(r_i,t_j)}{\partial r^2}\left(\frac{r_i^{1-\alpha}}{\alpha}\right) + \left(\frac{1-\alpha}{\alpha}\right)r_i^{-\alpha}\frac{\partial h(r_i,t_j)}{\partial r} \right]\left(1+\frac{dr}{r_i}\right) \right. \right. \\
&\left. \left. \qquad + \frac{T}{S}\frac{1}{r}\frac{\partial h(r_i,t_j)}{\partial r}\frac{r_i^{1-\alpha}}{\alpha} + \frac{T}{S}\frac{h(r_i,t_j)}{\lambda^2} \right) \right. \\
&\left. \quad -2t_{j-1}^{\alpha-1}\left(\frac{T}{S}\frac{r_i^{1-\alpha}}{\alpha}\left[\frac{\partial r^2 h(r_i,t_{j-1})}{\partial r^2}\left(\frac{r_i^{1-\alpha}}{\alpha}\right) + \left(\frac{1-\alpha}{\alpha}\right)r_i^{-\alpha}\frac{\partial h(r_i,t_{j-1})}{\partial r} \right]\left(1+\frac{dr}{r_i}\right) \right. \right. \\
&\left. \left. \qquad + \frac{T}{S}\frac{1}{r}\frac{\partial h(r_i,t_{j-1})}{\partial r}\frac{r_i^{1-\alpha}}{\alpha} + \frac{T}{S}\frac{h(r_i,t_{j-1})}{\lambda^2} \right) \right. \\
&\left. \quad +t_{j-2}^{\alpha-1}\left(\frac{T}{S}\frac{r_i^{1-\alpha}}{\alpha}\left[\frac{\partial r^2 h(r_i,t_{j-2})}{\partial r^2}\left(\frac{r_i^{1-\alpha}}{\alpha}\right) + \left(\frac{1-\alpha}{\alpha}\right)r_i^{-\alpha}\frac{\partial h(r_i,t_{j-2})}{\partial r} \right]\left(1+\frac{dr}{r_i}\right) \right. \right. \\
&\left. \left. \qquad + \frac{T}{S}\frac{1}{r}\frac{\partial h(r_i,t_{j-2})}{\partial r}\frac{r_i^{1-\alpha}}{\alpha} + \frac{T}{S}\frac{h(r_i,t_{j-2})}{\lambda^2} \right) \right\}
\end{aligned}
$$

$$
\times \left[\begin{array}{l} (n-j+1)^\beta\left[2(n-j)^2+(3\beta+10)(n-j)+(2\beta^2+9\beta+12)\right] \\ -(n-j)^\beta\left[-2(n-j)^2+(5\beta+10)(n-j)+(6\beta^2+18\beta+12)\right] \end{array} \right].
$$

$$(9.30)$$

9.3.2 APPLICATION OF THE NEW CAPUTO FRACTAL-FRACTIONAL DERIVATIVE ON THE SELF-SIMILAR LEAKY AQUIFER EQUATION: SCENARIO 2

Another scenario was considered where the water is not flowing out of a small portion of a self-similar leaky aquifer but a normal leaky aquifer and the equation below was derived.

$$\frac{S}{T}\frac{\partial h}{\partial t} = \frac{\partial}{\partial r^\alpha}\left(\frac{\partial h}{\partial r^\alpha}\right)\left(1+\frac{dr}{r}\right) + \frac{\partial h}{\partial r^\alpha}\left(\frac{1}{dr}-\frac{1}{r}\right) + \frac{h(r,t)}{\lambda^2} - \frac{1}{dr}\left[\frac{\partial h}{\partial r}\right] \tag{9.31}$$

Introducing the Caputo fractal-fractional derivative

$$\frac{S}{T}{}_0^C D_t^{\beta,\alpha}h(r,t) = \frac{\partial}{\partial r^\alpha}\left(\frac{\partial h}{\partial r^\alpha}\right)\left(1+\frac{dr}{r}\right) + \frac{\partial h}{\partial r^\alpha}\left(\frac{1}{dr}-\frac{1}{r}\right) + \frac{h(r,t)}{\lambda^2} - \frac{1}{dr}\left[\frac{\partial h}{\partial r}\right] \tag{9.32}$$

Differentiating the single and double fractal derivative $\dfrac{\partial h(r,t)}{\partial r^\alpha} + \dfrac{\partial}{\partial r^\alpha}\left(\dfrac{\partial h(r,t)}{\partial r^\alpha}\right)$

$$\frac{\partial h(r,t)}{\partial r^\alpha} = \frac{\partial h(r,t)}{\partial r}\frac{r^{1-\alpha}}{\alpha} \tag{9.33}$$

$$\frac{\partial}{\partial r^\alpha}\left(\frac{\partial h(r,t)}{\partial r^\alpha}\right) = \frac{\partial}{\partial r}\left(\frac{\partial h(r,t)}{\partial r}\frac{r^{1-\alpha}}{\alpha}\right)\frac{r^{1-\alpha}}{\alpha}$$
$$= \frac{\partial r^2 h(r,t)}{\partial r^2}\left(\frac{r^{2-2\alpha}}{\alpha^2}\right) + \frac{(1-\alpha)}{\alpha^2}\frac{\partial h(r,t)}{\partial r}r^{1-2\alpha} \tag{9.34}$$

The right-hand side of the above equation can be simplified by replacing

$$f(\alpha,r,t,h(r,t)) = \frac{T}{S}\left[\frac{\partial r^2 h(r,t)}{\partial r^2}\left(\frac{r^{2-2\alpha}}{\alpha^2}\right) + \frac{(1-\alpha)}{\alpha^2}\frac{\partial h(r,t)}{\partial r}r^{1-2\alpha}\right]$$
$$\left(1+\frac{dr}{r}\right) + \frac{T}{S}\left[\frac{\partial h(r,t)}{\partial r}\frac{r^{1-\alpha}}{\alpha}\right]\left(\frac{1}{dr}-\frac{1}{r}\right)$$
$$-\frac{T}{S}\left[\frac{\partial h(r,t)}{\partial r}\right]\left(\frac{1}{dr}\right) + \frac{T}{S}\left[\frac{h(r,t)}{\lambda^2}\right] \tag{9.35}$$

Such that

$$_0^C D_t^{\beta,\alpha}h(r,t) = f(\alpha,r,t,h(r,t)) \tag{9.36}$$

$$
{}_0^C D_t^{\beta,\alpha} h(r,t) = \frac{T}{S}\left[\frac{\partial r^2 h(r,t)}{\partial r^2}\left(\frac{r^{2-2\alpha}}{\alpha^2}\right) + \frac{(1-\alpha)}{\alpha^2}\frac{\partial h(r,t)}{\partial r}r^{1-2\alpha}\right]
$$

$$
\times\left(1+\frac{dr}{r}\right) + \frac{T}{S}\left[\frac{\partial h(r,t)}{\partial r}\frac{r^{1-\alpha}}{\alpha}\right]\left(\frac{1}{dr}-\frac{1}{r}\right)
$$

$$
-\frac{T}{S}\left[\frac{\partial h(r,t)}{\partial r}\right]\left(\frac{1}{dr}\right) + \frac{T}{S}\left[\frac{h(r,t)}{\lambda^2}\right] \tag{9.37}
$$

The functions can be represented as:

$$
f\left(\alpha, r_i, t_{j-2}, h\left(r_i, t_{j-2}\right)\right) = \frac{T}{S}\left[\frac{\partial r^2 h\left(r_i, t_{j-2}\right)}{\partial r^2}\left(\frac{r_i^{2-2\alpha}}{\alpha^2}\right) + \frac{(1-\alpha)}{\alpha^2}\frac{\partial h\left(r_i, t_{j-2}\right)}{\partial r}r_i^{1-2\alpha}\right]
$$

$$
\times\left(1+\frac{dr}{r_i}\right) + \frac{T}{S}\left[\frac{\partial h\left(r_i, t_{j-2}\right)}{\partial r}\frac{r^{1-\alpha}}{\alpha}\right]\left(\frac{1}{dr}-\frac{1}{r_i}\right)
$$

$$
-\frac{T}{S}\left[\frac{\partial h\left(r_i, t_{j-2}\right)}{\partial r}\right]\left(\frac{1}{dr}\right) + \frac{T}{S}\left[\frac{h\left(r_i, t_{j-2}\right)}{\lambda^2}\right] \tag{9.38}
$$

$$
f\left(\alpha, r_i, t_{j-1}, h\left(r_i, t_{j-1}\right)\right) = \frac{T}{S}\left[\begin{array}{c}\dfrac{\partial r^2 h\left(r_i, t_{j-1}\right)}{\partial r^2}\left(\dfrac{r_i^{2-2\alpha}}{\alpha^2}\right)\\[2mm]+\dfrac{(1-\alpha)}{\alpha^2}\dfrac{\partial h\left(r_i, t_{j-1}\right)}{\partial r}r_i^{1-2\alpha}\end{array}\right]\left(1+\frac{dr}{r_i}\right)
$$

$$
+\frac{T}{S}\left[\frac{\partial h\left(r_i, t_{j-1}\right)}{\partial r}\frac{r^{1-\alpha}}{\alpha}\right]\left(\frac{1}{dr}-\frac{1}{r_i}\right)
$$

$$
-\frac{T}{S}\left[\frac{\partial h\left(r_i, t_{j-1}\right)}{\partial r}\right]\left(\frac{1}{dr}\right) + \frac{T}{S}\left[\frac{h\left(r_i, t_{j-1}\right)}{\lambda^2}\right] \tag{9.39}
$$

$$
f\left(\alpha, r_i, t_j, h\left(r_i, t_j\right)\right) = \frac{T}{S}\left[\frac{\partial r^2 h\left(r_i, t_j\right)}{\partial r^2}\left(\frac{r_i^{2-2\alpha}}{\alpha^2}\right) + \frac{(1-\alpha)}{\alpha^2}\frac{\partial h\left(r_i, t_j\right)}{\partial r}r_i^{1-2\alpha}\right]
$$

$$
\left(1+\frac{dr}{r_i}\right) + \frac{T}{S}\left[\frac{\partial h\left(r_i, t_j\right)}{\partial r}\frac{r^{1-\alpha}}{\alpha}\right]\left(\frac{1}{dr}-\frac{1}{r_i}\right)
$$

$$
-\frac{T}{S}\left[\frac{\partial h\left(r_i, t_j\right)}{\partial r}\right]\left(\frac{1}{dr}\right) + \frac{T}{S}\left[\frac{h\left(r_i, t_j\right)}{\lambda^2}\right] \tag{9.40}
$$

Therefore the final numerical solution can be given as:

$$
\begin{aligned}
h_i^{n+1} = h_i^0 &+ \frac{\alpha(\Delta t)^\beta}{\Gamma(\beta+1)} \sum_{j=2}^n t_{j-2}^{\alpha-1} \left\{ \frac{T}{S} \left[\begin{array}{l} \frac{\partial r^2 h(r_i,t_{j-2})}{\partial r^2}\left(\frac{r_i^{2-2\alpha}}{\alpha^2}\right) \\ + \frac{(1-\alpha)}{\alpha^2}\frac{\partial h(r_i,t_{j-2})}{\partial r} r_i^{1-2\alpha} \end{array} \right] \left(1+\frac{dr}{r_i}\right) - \frac{T}{S}\left[\frac{\partial h(r_i,t_{j-2})}{\partial r}\right]\left(\frac{1}{dr}\right) + \frac{T}{S}\left[\frac{h(r_i,t_{j-2})}{\lambda^2}\right] \right\} \\
&+ \frac{\alpha(\Delta t)^\beta}{\Gamma(\beta+2)} \sum_{j=2}^n \left\{ \begin{array}{l} t_{j-1}^{\alpha-1}\left(\frac{T}{S}\left[\frac{\partial r^2 h(r_i,t_{j-1}r,t)}{\partial r^2}\left(\frac{r_i^{2-2\alpha}}{\alpha^2}\right) + \frac{(1-\alpha)}{\alpha^2}\frac{\partial h(r_i,t_{j-1}r,t)}{\partial r}r_i^{1-2\alpha}\right]\left(1+\frac{dr}{r_i}\right) \right. \\ \left. + \frac{T}{S}\left[\frac{\partial h(r_i,t_{j-1})}{\partial r}\frac{r^{1-\alpha}}{\alpha}\right]\left(\frac{1}{dr}-\frac{1}{r_i}\right) - \frac{T}{S}\left[\frac{\partial h(r_i,t_{j-1})}{\partial r}\right]\left(\frac{1}{dr}\right) + \frac{T}{S}\left[\frac{h(r_i,t_{j-1})}{\lambda^2}\right]\right) \\ -t_{j-2}^{\alpha-1}\left(\frac{T}{S}\left[\frac{\partial r^2 h(r_i,t_{j-2})}{\partial r^2}\left(\frac{r_i^{2-2\alpha}}{\alpha^2}\right) + \frac{(1-\alpha)}{\alpha^2}\frac{\partial h(r_i,t_{j-2})}{\partial r}r_i^{1-2\alpha}\right]\left(1+\frac{dr}{r_i}\right) \right. \\ \left. + \frac{T}{S}\left[\frac{\partial h(r_i,t_{j-2})}{\partial r}\frac{r^{1-\alpha}}{\alpha}\right]\left(\frac{1}{dr}-\frac{1}{r_i}\right) - \frac{T}{S}\left[\frac{\partial h(r_i,t_{j-2})}{\partial r}\right]\left(\frac{1}{dr}\right) + \frac{T}{S}\left[\frac{h(r_i,t_{j-2})}{\lambda^2}\right]\right) \end{array} \right\} \\
&\times \left[(n-j+1)^\beta (n-j+3+2\beta) - (n-j)^\beta(n-j+3+3\beta) \right] \\
&+ \frac{\alpha(\Delta t)^\beta}{2\Gamma(\beta+3)} \sum_{j=2}^n \left\{ \begin{array}{l} t_j^{\alpha-1}\left(\frac{T}{S}\left[\frac{\partial r^2 h(r_i,t_j)}{\partial r^2}\left(\frac{r_i^{2-2\alpha}}{\alpha^2}\right) + \frac{(1-\alpha)}{\alpha^2}\frac{\partial h(r_i,t_j)}{\partial r}r_i^{1-2\alpha}\right]\left(1+\frac{dr}{r_i}\right) \right. \\ \left. + \frac{T}{S}\left[\frac{\partial h(r_i,t_j)}{\partial r}\frac{r^{1-\alpha}}{\alpha}\right]\left(\frac{1}{dr}-\frac{1}{r_i}\right) - \frac{T}{S}\left[\frac{\partial h(r_i,t_j)}{\partial r}\right]\left(\frac{1}{dr}\right) + \frac{T}{S}\left[\frac{h(r_i,t_j)}{\lambda^2}\right]\right) \\ -2t_{j-1}^{\alpha-1}\left(\frac{T}{S}\left[\frac{\partial r^2 h(r_i,t_{j-1}r,t)}{\partial r^2}\left(\frac{r_i^{2-2\alpha}}{\alpha^2}\right) + \frac{(1-\alpha)}{\alpha^2}\frac{\partial h(r_i,t_{j-1}r,t)}{\partial r}r_i^{1-2\alpha}\right]\left(1+\frac{dr}{r_i}\right) \right. \\ \left. + \frac{T}{S}\left[\frac{\partial h(r_i,t_{j-1})}{\partial r}\frac{r^{1-\alpha}}{\alpha}\right]\left(\frac{1}{dr}-\frac{1}{r_i}\right) - \frac{T}{S}\left[\frac{\partial h(r_i,t_{j-1})}{\partial r}\right]\left(\frac{1}{dr}\right) + \frac{T}{S}\left[\frac{h(r_i,t_{j-1})}{\lambda^2}\right]\right) \\ +t_{j-2}^{\alpha-1}\left(\frac{T}{S}\left[\frac{\partial r^2 h(r_i,t_{j-2})}{\partial r^2}\left(\frac{r_i^{2-2\alpha}}{\alpha^2}\right) + \frac{(1-\alpha)}{\alpha^2}\frac{\partial h(r_i,t_{j-2})}{\partial r}r_i^{1-2\alpha}\right]\left(1+\frac{dr}{r_i}\right) \right. \\ \left. + \frac{T}{S}\left[\frac{\partial h(r_i,t_{j-2})}{\partial r}\frac{r^{1-\alpha}}{\alpha}\right]\left(\frac{1}{dr}-\frac{1}{r_i}\right) - \frac{T}{S}\left[\frac{\partial h(r_i,t_{j-2})}{\partial r}\right]\left(\frac{1}{dr}\right) + \frac{T}{S}\left[\frac{h(r_i,t_{j-2})}{\lambda^2}\right]\right) \end{array} \right\} \\
&\times \left[\begin{array}{l} (n-j+1)^\beta\left[2(n-j)^2 + (3\beta+10)(n-j) + (2\beta^2+9\beta+12)\right] \\ -(n-j)^\beta\left[-2(n-j)^2 + (5\beta+10)(n-j) + (6\beta^2+18\beta+12)\right] \end{array} \right].
\end{aligned}
\tag{9.41}
$$

9.4 SIMULATION

In this section, we present a numerical solution. This is depicted in Figures 9.1 to 9.5 for different values of fractal and fractional order. In this case we chose the fractal order to be 1.

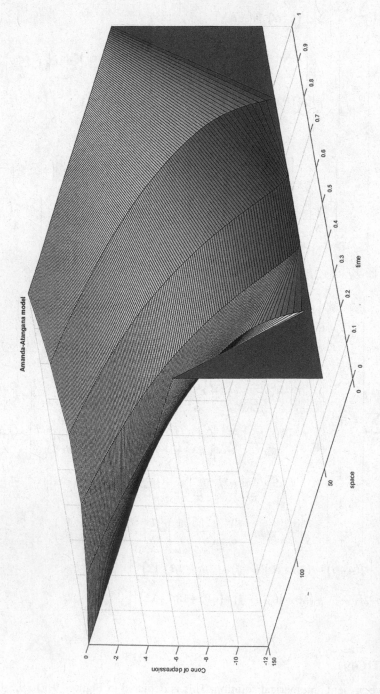

FIGURE 9.1 Cone of depression for order (1, 0.5) time space solution.

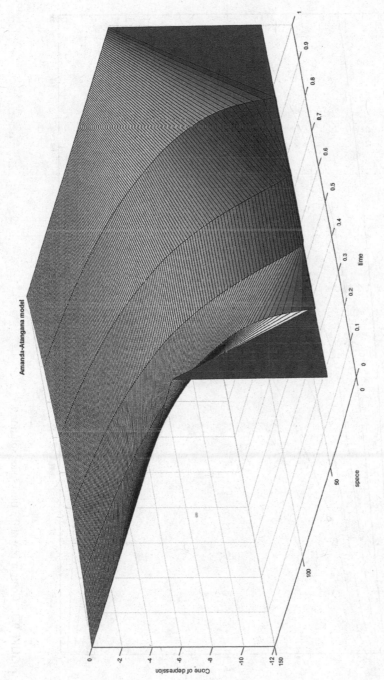

FIGURE 9.2 Cone of depression for order (1, 0.7) time space solution.

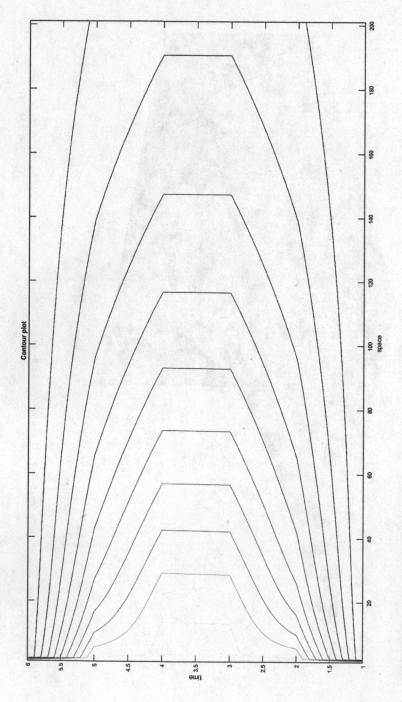

FIGURE 9.3 Contour plot at 0.3 time space solution.

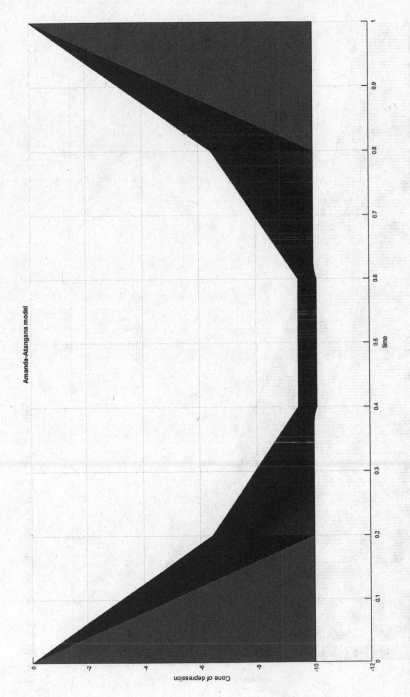

FIGURE 9.4 Cone of depression at order 0.3 in time.

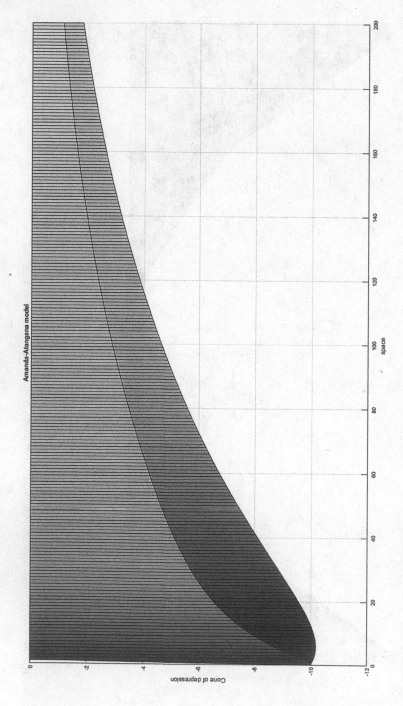

FIGURE 9.5   Cone of depression at order (1, 0.3) in space.

A model of groundwater flowing within a leaky aquifer has been considered in this chapter. The model took into account the effect of self-similarities, thus a differential operator with fractal dimension was used. A detailed stability analysis was presented and numerical simulations depicted for different values of fractional and fractal dimension 1.

REFERENCES

Atangana, A., & Araz, A.I., (2021). New numerical scheme with newton polynomial: theory, methods, and applications. *Academic Press*.

Atangana, A., & Qureshi, S., (2019). Modeling attractors of chaotic dynamic systems with fractal-fractional operators. *Chaos, Solitons and Fractals*. 320–337.

Ramotsho, A., & Atangana, A., (2018). Derivation of a groundwater flow model within leaky and self-similar aquifers: Beyond Hantush model. *Chaos, Solitons and Fractals*, 116 (2018) 414–423.

Remani, C., (2013). *Numerical Methods for Solving Systems of Nonlinear Equations*. Lakehead University Thunder Bay, Ontario, Canada.

10 Application of the New Numerical Method with Caputo–Fabrizio Fractal-Fractional Derivative on the Self-Similar Leaky Aquifer Equations

Amanda Ramotsho and Abdon Atangana
University of the Free State, Bloemfontein, South Africa

CONTENTS

10.1 INTRODUCTION

The Mandelbrot set is a set of complex numbers c for which the iteration $z \to z^2 + c$, starting from $z = 0$ does not diverge to infinity when iteration methods are used. Julia sets are either connected (one piece) or a dust of infinitely many points. The Mandelbrot set is that c for which the Julia set is connected. The Julia set produces similarity complex fractal shapes (Mandelbrot, 2010). Though it is well known that the mathematician Mandelbrot invented fractals in order to relate the objects and mathematics – we shall recall that the idea was already well-developed in Africa, where, when the first European arrived and saw African's constructions and structures concluded that they built with disorganized structures. Fractals change our mind set on how to elaborate and interpret field data. From the past, we know that when non-fractal objects are enlarged we can obtain details and be able to understand, but with fractals we need to enlarge further in order to accumulate the information needed. Fractals can be found in nature patterns and are the new approach used in mathematics and art research. Fractals as used for solving certain anomalous physical behaviors, they are related to fractional derivative, levy statistics, fractional Brownian motion and empirical power law scaling (Chen, 2005). Fractals can be explained in simpler terms according to a spring article as a geometric object whose fractal dimension is larger than its topological dimension. Fractals are made up of the three most important properties, namely self-similarity, scaling, and statistics. A fractal is a never-ending pattern that repeats itself at different scales. Self-similarity occurs when a system replicate itself and shows iteration throughout. Previously scientists only studied the movement of groundwater within one leaky aquifer system based on assumptions outlined by (Hantush, 1959) without considering that the aquifer can actually replicate itself and form stacks. Fractures are subdivided into joints, fissures and faults, which are formed by the fracturing of rocks. The majority of high yielding aquifers are associated with fracture zones. They resemble a self-similar leaky aquifer in nature. Self-similar leaky aquifers cannot be described or solved mathematically using the known classical physical laws that are used in groundwater such as Darcy's law and other relevant laws used because they are not similar to the regular geometries defined in Euclidean geometry. Fractal and fractional derivatives are used in this situation, as they can assist in solving porous or heterogeneous aquifers. Fractured rocks often have two flow regimes – fast flow in the fractures and slow flow in rock matrices with long residence time. Darcy's law can describe the flow in the matrix, but it is not valid in the fractures, conduits and caves, where the flow may be turbulent (Karay and Hajnal, 2015). There are different approaches that were previously used for complexity of discontinuity in terms of fractures. Approaches such as the continuum approach, which is used for the classical study of flow through porous media, understanding flow through the fractures based on the assumption that the rock matrix surrounding the fractures is practically impermeable (Sekar, 1994). A new concept of differentiation was recently suggested where the operator has two orders, the first is fractional order, and the second fractal dimension. The concept is looming its way up and it has been applied in few articles. Referencing the article that was submitted in 2019 and accepted in 2020, the new generalization of integer order diffusion equations to a two-order fractional diffusion problem was considered (Atangana et al, 2020). This type of equation has been applied to model chaotic systems and also in biology. Fractional order models are improved and they in turn improve our knowledge of differentiability, incorporate non-local and system memory effects through space and time fractional order derivatives. These features can be used when modeling self-similarity without partitioning the problem into increasingly small compartments. It has been a very essential tool and many problems can be illustrated more conveniently and more accurately with differential equations having arbitrary order (Veeresha and Prakasha, 2020).

10.2 DEFINITIONS: FRACTAL-FRACTIONAL DERIVATIVE IN CAPUTO–FABRIZIO SENSE

Following the two articles that were accepted in 2017 and 2019, the definitions given below were outlined (Atangana, 2017 and Atangana and Qureshi, 2019).

Definition 1

Accede that $f(t)$ is continuous on (a, b) and also fractal differentiable with order β then the fractal-fractional derivative of $f(t)$ with order α in the Caputo–Fabrizio sense having an exponential law type kernel has been written as:

$$^{FFE}_{\ 0}D_t^{\alpha,\beta} f(t) = \frac{M(\alpha)}{\Gamma(1-\alpha)} \int_0^t \frac{df(\tau)}{d\tau^\beta} \exp\left[\frac{-\alpha}{1-\alpha}(t-\tau)\right] d\tau \tag{10.1}$$

When $0 < \alpha, \beta \leq n, M(0) = M(1) = 1$

$$^{FFE}_{\ 0}D_t^{\alpha,\beta} f(t) = \frac{M(\alpha)}{\Gamma(1-\alpha)} \frac{d}{dt} \int_0^t \frac{df(\tau)}{d\tau} \exp\left[\frac{-\alpha}{1-\alpha}(t-\tau)\right] d\tau \frac{t^{1-\alpha}}{\beta} \tag{10.2}$$

The more generalized version is given as:

$$
{}^{FFE}_{0}D_t^{\alpha,\beta,\lambda}f(t) = \frac{M(\alpha)}{\Gamma(1-\alpha)}\frac{d}{dt}\int_0^t \frac{d^\lambda f(\tau)}{dt^\beta}\exp\left[\frac{-\alpha}{1-\alpha}(t-\tau)\right]d\tau \tag{10.3}
$$

When $0 < \alpha, \lambda, \beta \le 1$

$$
{}^{FFE}_{0}D_t^{\alpha,\beta,\lambda}f(t) = \frac{M(\alpha)}{\Gamma(1-\alpha)}\int_0^t \frac{d^\lambda f(\tau)}{d\beta}\exp\left[\frac{-\alpha}{1-\alpha}(t-\tau)\right]d\tau\, \frac{t^{1-\alpha}}{\beta}. \tag{10.4}
$$

Definition 2

Accede that $f(t)$ be continuous on (a, b) which are open intervals, then the fractal-fractional integral of $f(t)$ with order α having exponential decay law type kernel is defined as follows:

$$
{}^{FFE}J_{0,t}^{\alpha,\beta}f(t) = \frac{\alpha\beta}{M(\alpha)}\int_0^t \tau^{\alpha-1}f(\tau)d\tau + \frac{\beta(1-\alpha)t^{\beta-1}f(t)}{M(\alpha)}. \tag{10.5}
$$

10.3 THE NEW NUMERICAL SCHEME FOR ORDINARY DIFFERENTIAL EQUATIONS AND PARTIAL DIFFERENTIAL EQUATIONS WITH CAPUTO–FABRIZIO FRACTIONAL DERIVATIVE BY ATANGANA AND ARAZ

The method was developed using the Cauchy problem and implementing or introducing the Caputo–Fabrizio fractional derivative

Let

$$
{}^{CF}_{0}D_t^{\alpha}y(t) = f(t, y(t)) \tag{10.6}
$$

The equation is reformulated as follows

$$
y(t) - y(0) = \frac{1-\alpha}{M(\alpha)}f(t, y(t)) + \frac{\alpha}{M(\alpha)}\int_0^t f(\tau, y(\tau))d\tau \tag{10.7}
$$

At the point $t_{n+1} = (n+1)\Delta t$, we have

$$
y(t_{n+1}) - y(0) = \frac{1-\alpha}{M(\alpha)}f(t_n, y(t_n)) + \frac{\alpha}{M(\alpha)}\int_0^{t_{n+1}} f(\tau, y(\tau))d\tau \tag{10.8}
$$

At the point $t_n = n\Delta t$, we have

$$
y(t_n) - y(0) = \frac{1-\alpha}{M(\alpha)}f(t_{n-1}, y(t_{n-1})) + \frac{\alpha}{M(\alpha)}\int_0^{t_n} f(\tau, y(\tau))d\tau \tag{10.9}
$$

and at $t_{n+1} - t_n$, we have

$$y(t_{n+1}) - y(t_n) = \frac{1-\alpha}{M(\alpha)}\left[f(t_n, y(t_n)) - f(t_{n-1}, y(t_{n-1}))\right] + \frac{\alpha}{M(\alpha)}\int_{t_n}^{t_{n+1}} f(\tau, y(\tau))d\tau \qquad (10.10)$$

and

$$y(t_{n+1}) - y(t_n) = \frac{1-\alpha}{M(\alpha)}\left[f(t_n, y(t_n)) - f(t_{n-1}, y(t_{n-1}))\right] + \frac{\alpha}{M(\alpha)}\int_{t_n}^{t_{n+1}} f(\tau, y(\tau))d\tau \qquad (10.11)$$

$f(t, y(t))$ was represented using the Newton Polynomial therefore

$$y^{n+1} - y^n = \frac{1-\alpha}{M(\alpha)}\left[f(t_n, y(t_n)) - f(t_{n-1}, y(t_{n-1}))\right]$$
$$+ \frac{\alpha}{M(\alpha)}\int_{j}^{t_{j+1}} \left\{ \begin{array}{c} f(t_{j-2}, y^{j-2}) \\ + \dfrac{f(t_{j-1}, y^{j-1}) - f(t_{j-2}, y^{j-2})}{\Delta t}(\tau - t_{j-2}) \\ + \dfrac{f(t_j, y^j) - 2f(t_{j-1}, y^{j-1}) + f(t_{j-2}, y^{j-2})}{2(\Delta t)^2} \\ \times (\tau - t_{-2})(\tau - t_{j-1}) \end{array} \right\} d\tau \qquad (10.12)$$

The following is obtained

$$h^{n+1} - h^n = \frac{1-\alpha}{M(\alpha)}\left[f(t_n, y(t_n)) - f(t_{n-1}, y(t_{n-1}))\right]$$
$$+ \frac{\alpha}{M(\alpha)} \left\{ \begin{array}{c} f(t_{j-2}, y^{j-2})\Delta t \\ + \dfrac{f(t_{j-1}, y^{j-1}) - f(t_{j-2}, y^{j-2})}{\Delta t}\int_{t_n}^{t_{n+1}}(\tau - t_{j-2})d\tau \\ + \dfrac{f(t_j, y^j) - 2f(t_{j-1}, y^{j-1}) + f(t_{j-2}, y^{j-2})}{2(\Delta t)^2} \\ \times \int_{t_j}^{t_{n+1}}(\tau - t_{j-2})(\tau - t_{j-1})d\tau \end{array} \right\} \qquad (10.13)$$

Therefore, the following equalities can be written as

$$\int_{t_j}^{t_{j+1}} (\tau - t_{j-2})d\tau = \frac{5}{2}(\Delta t)^2 \qquad (10.14)$$

$$\int_{t_j}^{t_{j+1}} (\tau - t_{j-2})(\tau - t_{j-1})d\tau = \frac{23}{6}(\Delta t)^3 \qquad (10.15)$$

The above scheme can be given as

$$
y^{n+1} = y^n + \frac{1-\alpha}{M(\alpha)}\Big[f\big(t_n,y^n\big)-f\big(t_{n-1},y\big(t_{n-1}\big)\big)\Big]
$$

$$
+ \frac{\alpha}{M(\alpha)}\sum_{j=2}^{n}\left\{
\begin{array}{l}
f\big(t_{j-2},y^{j-2}\big)\Delta t + \Big[f\big(t_{j-1},y^{j-1}\big)-f\big(t_{j-2},y^{j-2}\big)\Big]\dfrac{5}{2}\Delta t \\[2mm]
+\Big[f\big(t_j,y^j\big)-2f\big(t_{j-1},y^{j-1}\big)+f\big(t_{j-2},y^{j-2}\big)\Big]\dfrac{23\Delta t}{12}
\end{array}
\right\}. \tag{10.16}
$$

10.4 DISCRETIZING USING THE CAPUTO–FABRIZIO DERIVATIVE AND APPLYING THE NUMERICAL SCHEME GIVEN ABOVE ON THE SELF-SIMILAR LEAKY AQUIFER EQUATION SCENARIO 1

The mathematical model below was incorporated to assess self-similar leaky aquifers, the equation below represents the scenario where the in- and out-flow is within a self-similar leaky aquifer,

$$
\frac{S}{T}\frac{\partial h}{\partial t} = \frac{\partial}{\partial r^\alpha}\left(\frac{\partial h}{\partial r^\alpha}\right)\left(1+\frac{dr}{r}\right)+\frac{1}{r}\frac{\partial h}{\partial r^\alpha}+\frac{h(r,t)}{\lambda^2} \tag{10.17}
$$

The discretized equation with Caputo–Fabrizio the resulting equation is given as:

$$
{}^{CF}_0D_t^{\alpha,\beta}h(r,t) = \frac{T}{S}\frac{r^{1-\alpha}}{\alpha}\left[\frac{\partial r^2 h(r,t)}{\partial r^2}\left(\frac{r^{1-\alpha}}{\alpha}\right)+\frac{\partial h(r,t)}{\partial r}\left(\frac{1-\alpha}{\alpha}\right)r^{-\alpha}\right]
$$

$$
\left(1+\frac{dr}{r}\right)+\frac{T}{S}\frac{1}{r}\frac{\partial h(r,t)}{\partial r}\frac{r^{1-\alpha}}{\alpha}+\frac{T}{S}\frac{h(r,t)}{\lambda^2} \tag{10.18}
$$

The equation can be simplified as follows

$$
f\big(\alpha,r,t,h(r,t)\big) = \frac{T}{S}\frac{r^{1-\alpha}}{\alpha}\left[\frac{\partial r^2 h(r,t)}{\partial r^2}\left(\frac{r^{1-\alpha}}{\alpha}\right)+\left(\frac{1-\alpha}{\alpha}\right)r^{-\alpha}\frac{\partial h(r,t)}{\partial r}\right]
$$

$$
\left(1+\frac{dr}{r}\right)+\frac{T}{S}\frac{1}{r}\frac{\partial h(r,t)}{\partial r}\frac{r^{1-\alpha}}{\alpha}+\frac{T}{S}\frac{h(r,t)}{\lambda^2} \tag{10.19}
$$

Such that

$$
{}^{CF}_0D_t^{\beta,\alpha}h(r,t) = f\big(\alpha,r,t,h(r,t)\big) \tag{10.20}
$$

We recall that the Caputo–Fabrizio fractional derivative can be given as

$$
\frac{M(\alpha)}{1-\alpha}\frac{d}{dt^\beta}\int_0^t h(r,t)\exp\left[\frac{\alpha}{1-\alpha}(t-\tau)\right]d\tau = F\big(\alpha,r,t,h(r,t)\big) \tag{10.21}
$$

Differentiating in terms of β and multiplying by $\beta t\beta^{-1}$

$$\frac{M(\alpha)}{1-\alpha}\frac{d}{dt}\int_0^t h(r,\tau)\exp\left[\frac{\alpha}{1-\alpha}(t-\tau)\right]d\tau\,\frac{1}{\beta t^{\beta-1}} = F\left(\alpha,r,t,h(r,t)\right) \qquad (10.22)$$

$$\frac{M(\alpha)}{1-\alpha}\frac{d}{dt}\int_0^t h(r,\tau)\exp\left[\frac{\alpha}{1-\alpha}(t-\tau)\right]d\tau = F\left(\alpha,r,t,h(r,t)\right)\beta t^{\beta-1} \qquad (10.23)$$

Integrating

$$h(r,t) = \frac{1-\alpha}{M(\alpha)}\beta t^{\beta-1}F\left(\alpha,r,t,h(r,t)\right) + \frac{\beta\alpha}{M(\alpha)}\int_0^t \tau^{\beta-1}F\left(\alpha,r,t,h(r,t)\right)d\tau \qquad (10.24)$$

At (r_i, t_{n+1}) we have

$$h_i^{n+1} = \frac{1-\alpha}{M(\alpha)}\beta t_{n+1}^{\beta-1}F\left(\alpha,r_i,t_{n+1},h(r_i,t_{n+1})\right) + \frac{\beta\alpha}{M(\alpha)}\int_0^{t_{n+1}}\tau^{\beta-1}F\left(\alpha,r_i,\tau,h(r_i,\tau)\right)d\tau \qquad (10.25)$$

At (r_i, t_n) we have

$$h_i^n = \frac{1-\alpha}{M(\alpha)}\beta t_{n-1}^{\beta-1}F\left(\alpha,r_i,t_{n-1},h(r_i,t_{n-1})\right) + \frac{\beta\alpha}{M(\alpha)}\int_0^{t_n}\tau^{\beta-1}F\left(\alpha,r_i,\tau,h(r_i,\tau)\right)d\tau \qquad (10.26)$$

At $h(r_i, t_{n+1}) - h(r_i, t_n)$ we have

$$h_i^{n+1} - h_i^n = \frac{1-\alpha}{M(\alpha)}\beta\left[t_{n+1}^{\beta-1}F\left(\alpha,r_i,t_n,h_i^n\right) - t_{n-1}^{\beta-1}F\left(\alpha,r_i,t_{n-1},h_i^{n-1}\right)\right]$$
$$+ \frac{\beta\alpha}{M(\alpha)}\int_{t_n}^{t_{n+1}}\tau^{\beta-1}F\left(\alpha,r_i,\tau,h(r,\tau)\right)d\tau \qquad (10.27)$$

Therefore the final numerical scheme is denoted as,

$$y^{n+1} = y^n + \frac{(1-\alpha)\beta}{M(\alpha)}\left[f\left(t_n,y^n\right) - f\left(t_{n-1},y(t_{n-1})\right)\right]$$
$$+ \frac{\alpha}{M(\alpha)}\sum_{j=2}^n\left\{\begin{array}{l}-\dfrac{4}{3}f\left(t_{j-1},y^{j-1}\right)\Delta t + \dfrac{5}{12}f\left(t_{j-2},y^{j-2}\right)\Delta t \\ +\dfrac{23}{12}f\left(t_j,y^j\right)\Delta t\end{array}\right\} \qquad (10.28)$$

$$h_i^{n+1} - h_i^n = \frac{1-\alpha}{M(\alpha)}\beta\left[t_{n+1}^{\beta-1}F\left(\alpha,r_i,t_n,h_i^n\right) - t_{n-1}^{\beta-1}F\left(\alpha,r_i,t_{n-1},h_i^{n-1}\right)\right]$$
$$+ \frac{\alpha\beta}{M(\alpha)}\left\{\begin{array}{l}-\dfrac{4}{3}t_{n-1}^{\beta-1}F\left(\alpha,r_i,t_{n-1},h(r_i,t_{n-1})\right)\Delta t \\ +\dfrac{5}{12}t_{n-2}^{\beta-1}F\left(\alpha,r_i,t_{n-2},h(r_i,t_{n-2})\right)\Delta t \\ +\dfrac{23}{12}t_n^{\beta-1}F\left(\alpha,r_i,t_n,h(r_i,t_n)\right)\Delta t\end{array}\right\} \qquad (10.29)$$

$$
h_i^{n+1} - h_i^n = \frac{1-\alpha}{M(\alpha)} \beta t_{n-1}^{\beta-1} \left\{ \frac{T}{S} \frac{r^{1-\alpha}}{\alpha} \left[\frac{\frac{\partial r^2 h(r_i, t_n)}{\partial r^2} \left(\frac{r^{1-\alpha}}{\alpha} \right)}{+ \frac{\partial h(r_i, t_n)}{\partial r} \left(\frac{1-\alpha}{\alpha} \right) r^{-\alpha}} \right] \left(1 + \frac{dr}{r_i} \right) \right.
$$

$$
\left. + \frac{T}{S} \frac{1}{r_i} \frac{\partial h(r, t_n)}{\partial r} \frac{r^{1-\alpha}}{\alpha} + \frac{T}{S} \frac{h(r_i, t_n)}{\lambda^2} \right\}
$$

$$
- \frac{1-\alpha}{M(\alpha)} \beta t_{n-1}^{\beta-1} \left\{ \frac{T}{S} \frac{r^{1-\alpha}}{\alpha} \left[\frac{\frac{\partial r^2 h(r_i, t_{n-1})}{\partial r^2} \left(\frac{r^{1-\alpha}}{\alpha} \right)}{+ \frac{\partial h(r_i, t_{n-1})}{\partial r} \left(\frac{1}{\alpha} \frac{\alpha}{\alpha} \right) r^{-\alpha}} \right] \left(1 + \frac{dr}{r_i} \right) \right.
$$

$$
\left. + \frac{T}{S} \frac{1}{r_i} \frac{\partial h(r, t_{n-1})}{\partial r} \frac{r^{1-\alpha}}{\alpha} + \frac{T}{S} \frac{h(r_i, t_{n-1})}{\lambda^2} \right\}
$$

$$
+ \frac{\alpha\beta}{M(\alpha)} \left\{ - \frac{4\Delta t}{3} t_{n-1}^{\beta-1} \left[\frac{T}{S} \frac{r^{1-\alpha}}{\alpha} \left[\frac{\frac{\partial r^2 h(r_i, t_{n-1})}{\partial r^2} \left(\frac{r^{1-\alpha}}{\alpha} \right)}{+ \frac{\partial h(r_i, t_{n-1})}{\partial r} \left(\frac{1-\alpha}{\alpha} \right) r^{-\alpha}} \right] \left(1 + \frac{dr}{r_i} \right) \right. \right.
$$

$$
\left. + \frac{T}{S} \frac{1}{r_i} \frac{\partial h(r, t_{n-1})}{\partial r} \frac{r^{1-\alpha}}{\alpha} + \frac{T}{S} \frac{h(r_i, t_{n-1})}{\lambda^2} \right]
$$

$$
+ \frac{5\Delta t}{12} t_{n-2}^{\beta-1} \left[\frac{T}{S} \frac{r^{1-\alpha}}{\alpha} \left[\frac{\frac{\partial r^2 h(r_i, t_{n-2})}{\partial r^2} \left(\frac{r^{1-\alpha}}{\alpha} \right)}{+ \frac{\partial h(r_i, t_{n-2})}{\partial r} \left(\frac{1-\alpha}{\alpha} \right) r^{-\alpha}} \right] \left(1 + \frac{dr}{r_i} \right) \right.
$$

$$
\left. + \frac{T}{S} \frac{1}{r_i} \frac{\partial h(r, t_{n-2})}{\partial r} \frac{r^{1-\alpha}}{\alpha} + \frac{T}{S} \frac{h(r_i, t_{n-2})}{\lambda^2} \right]
$$

$$
+ \frac{23\Delta t}{12} t_n^{\beta-1} \left[\frac{T}{S} \frac{r^{1-\alpha}}{\alpha} \left[\frac{\frac{\partial r^2 h(r_i, t_n)}{\partial r^2} \left(\frac{r^{1-\alpha}}{\alpha} \right)}{+ \frac{\partial h(r_i, t_n)}{\partial r} \left(\frac{1-\alpha}{\alpha} \right) r^{-\alpha}} \right] \left(1 + \frac{dr}{r_i} \right) \right.
$$

$$
\left. \left. + \frac{T}{S} \frac{1}{r_i} \frac{\partial h(r, t_n)}{\partial r} \frac{r^{1-\alpha}}{\alpha} + \frac{T}{S} \frac{h(r_i, t_n)}{\lambda^2} \right] \right\} \tag{10.30}
$$

10.5 IMPLEMENTATION OF CAPUTO–FABRIZIO FRACTAL-FRACTIONAL DERIVATIVE ON THE SELF-SIMILAR LEAKY AQUIFER EQUATION SCENARIO 2

Another scenario was considered where the water is not flowing out of a small portion of a self-similar leaky aquifer but a normal leaky aquifer, and the equation below was derived.

$$
\frac{S}{T} \frac{\partial h}{\partial t} = \frac{\partial}{\partial r^\alpha} \left(\frac{\partial h}{\partial r^\alpha} \right) \left(1 + \frac{dr}{r} \right) + \frac{\partial h}{\partial r^\alpha} \left(\frac{1}{dr} - \frac{1}{r} \right) + \frac{h(r,t)}{\lambda^2} - \frac{1}{dr} \left[\frac{\partial h}{\partial r} \right] \tag{10.31}
$$

The second discretized equation in terms of the Caputo–Fabrizio derivative is given as

$$
\begin{aligned}
{}^{CF}_0 D_t^{\alpha,\beta} h(r,t) &= \frac{T}{S}\left[\frac{\partial r^2 h(r,t)}{\partial r^2}\left(\frac{r^{2-2\alpha}}{\alpha^2}\right) + \frac{(1-\alpha)}{\alpha^2}\frac{\partial h(r,t)}{\partial r}r^{1-2\alpha}\right] \\
&\quad \left(1+\frac{dr}{r}\right) + \frac{T}{S}\left[\frac{\partial h(r,t)}{\partial r}\frac{r^{1-\alpha}}{\alpha}\right]\left(\frac{1}{dr}-\frac{1}{r}\right) \\
&\quad -\frac{T}{S}\left[\frac{\partial h(r,t)}{\partial r}\right]\left(\frac{1}{dr}\right) + \frac{T}{S}\left[\frac{h(r,t)}{\lambda^2}\right]
\end{aligned}
\tag{10.32}
$$

Differentiating the single and double fractal derivative $\dfrac{\partial h(r,t)}{\partial r^\alpha} + \dfrac{\partial}{\partial r^\alpha}\left(\dfrac{\partial h(r,t)}{\partial r^\alpha}\right)$

$$
\frac{S}{T}\,{}^{CF}_0 D_t^{\alpha,\beta} h(r,t) = \frac{\partial}{\partial r^\alpha}\left(\frac{\partial h(r,t)}{\partial r^\alpha}\right)\left(1+\frac{dr}{r}\right) + \frac{1}{r}\frac{\partial h(r,t)}{\partial r^\alpha}\frac{r^{1-\alpha}}{\alpha} + \frac{h(r,t)}{\lambda^2}
\tag{10.33}
$$

$$
\begin{aligned}
{}^{CF}_0 D_t^{\alpha,\beta} h(r,t) &= \frac{T}{S}\frac{r^{1-\alpha}}{\alpha}\left[\frac{\partial r^2 h(r,t)}{\partial r^2}\left(\frac{r^{1-\alpha}}{\alpha}\right) + \left(\frac{1-\alpha}{\alpha}\right)r^{-\alpha}\frac{\partial h(r,t)}{\partial r}\right] \\
&\quad \left(1+\frac{dr}{r}\right) + \frac{T}{S}\frac{1}{r}\frac{\partial h(r,t)}{\partial r}\frac{r^{1-\alpha}}{\alpha} + \frac{T}{S}\frac{h(r,t)}{\lambda^2}
\end{aligned}
\tag{10.34}
$$

The right-hand side of the above equation can be simplified by replacing

$$
\begin{aligned}
f\big(\alpha,r,t,h(r,\tau)\big) &= \frac{T}{S}\frac{r^{1-\alpha}}{\alpha}\left[\frac{\partial r^2 h(r,t)}{\partial r^2}\left(\frac{r^{1-\alpha}}{\alpha}\right) + \left(\frac{1-\alpha}{\alpha}\right)r^{-\alpha}\frac{\partial h(r,t)}{\partial r}\right] \\
&\quad \left(1+\frac{dr}{r}\right) + \frac{T}{S}\frac{1}{r}\frac{\partial h(r,t)}{\partial r}\frac{r^{1-\alpha}}{\alpha} + \frac{T}{S}\frac{h(r,t)}{\lambda^2}
\end{aligned}
\tag{10.35}
$$

Such that

$$
{}^{CF}_0 D_t^{\beta,\alpha} h(r,t) = f\big(\alpha,r,t,h(r,\tau)\big)
\tag{10.36}
$$

Applying the numerical Caputo–Fabrizio fractal-fractional scheme, the resulting numerical solution is given as

$$
h_i^{n+1} - h_i^n = \frac{1-\alpha}{M(\alpha)} \beta t_{n-1}^{\beta-1} \left[\begin{array}{c} \dfrac{T}{S}\left[\dfrac{\partial r^2 h(r_i,t_n)}{\partial r^2}\left(\dfrac{r_i^{2-2\alpha}}{\alpha^2}\right) + \dfrac{(1-\alpha)}{\alpha^2}\dfrac{\partial h(r_i,t_n)}{\partial r} r_i^{1-2\alpha}\right]\left(1+\dfrac{dr}{r_i}\right) \\[2mm] + \dfrac{T}{S}\left[\dfrac{\partial h(r_i,t_n)}{\partial r}\dfrac{r^{1-\alpha}}{\alpha}\right]\left(\dfrac{1}{dr}-\dfrac{1}{r_i}\right) - \dfrac{T}{S}\left[\dfrac{\partial h(r_i,t_n)}{\partial r}\right]\left(\dfrac{1}{dr}\right) \\[2mm] + \dfrac{T}{S}\left[\dfrac{h(r_i,t_n)}{\lambda^2}\right] \end{array}\right]
$$

$$
- \frac{1-\alpha}{M(\alpha)}\beta t_{n-1}^{\beta-1}\left[\begin{array}{c} \dfrac{T}{S}\begin{bmatrix}\dfrac{\partial r^2 h(r_i,t_{n-1})}{\partial r^2}\left(\dfrac{r_i^{2-2\alpha}}{\alpha^2}\right)\\[2mm] + \dfrac{(1-\alpha)}{\alpha^2}\dfrac{\partial h(r_i,t_{n-1})}{\partial r} r_i^{1-2\alpha}\end{bmatrix}\left(1+\dfrac{dr}{r_i}\right) \\[2mm] + \dfrac{T}{S}\left[\dfrac{\partial h(r_i,t_{n-1})}{\partial r}\dfrac{r^{1-\alpha}}{\alpha}\right]\left(\dfrac{1}{dr}-\dfrac{1}{r_i}\right) \\[2mm] - \dfrac{T}{S}\left[\dfrac{\partial h(r_i,t_{n-1})}{\partial r}\right]\left(\dfrac{1}{dr}\right) + \dfrac{T}{S}\left[\dfrac{h(r_i,t_{n-1})}{\lambda^2}\right] \end{array}\right]
$$

$$
+\frac{\alpha\beta}{M(\alpha)}\left\{ \begin{array}{c} -\dfrac{4\Delta t}{3}t_{n-1}^{\beta-1}\left[\begin{array}{c}\dfrac{T}{S}\begin{bmatrix}\dfrac{\partial r^2 h(r_i,t_{n-1})}{\partial r^2}\left(\dfrac{r_i^{2-2\alpha}}{\alpha^2}\right)\\[2mm]+\dfrac{(1-\alpha)}{\alpha^2}\dfrac{\partial h(r_i,t_{n-1})}{\partial r}r_i^{1-2\alpha}\end{bmatrix}\left(1+\dfrac{dr}{r_i}\right)\\[2mm]+\dfrac{T}{S}\left[\dfrac{\partial h(r_i,t_{n-1})}{\partial r}\dfrac{r^{1-\alpha}}{\alpha}\right]\left(\dfrac{1}{dr}-\dfrac{1}{r_i}\right)\\[2mm]-\dfrac{T}{S}\left[\dfrac{\partial h(r_i,t_{n-1})}{\partial r}\right]\left(\dfrac{1}{dr}\right)+\dfrac{T}{S}\left[\dfrac{h(r_i,t_{n-1})}{\lambda^2}\right]\end{array}\right] \\[6mm] +\dfrac{5\Delta t}{12}t_{n-2}^{\beta-1}\left[\begin{array}{c}\dfrac{T}{S}\begin{bmatrix}\dfrac{\partial r^2 h(r_i,t_{n-2})}{\partial r^2}\left(\dfrac{r_i^{2-2\alpha}}{\alpha^2}\right)\\[2mm]+\dfrac{(1-\alpha)}{\alpha^2}\dfrac{\partial h(r_i,t_{n-2})}{\partial r}r_i^{1-2\alpha}\end{bmatrix}\left(1+\dfrac{dr}{r_i}\right)\\[2mm]+\dfrac{T}{S}\left[\dfrac{\partial h(r_i,t_{n-2})}{\partial r}\dfrac{r^{1-\alpha}}{\alpha}\right]\left(\dfrac{1}{dr}-\dfrac{1}{r_i}\right)\\[2mm]-\dfrac{T}{S}\left[\dfrac{\partial h(r_i,t_{n-2})}{\partial r}\right]\left(\dfrac{1}{dr}\right)+\dfrac{T}{S}\left[\dfrac{h(r_i,t_{n-2})}{\lambda^2}\right]\end{array}\right] \\[6mm] +\dfrac{23\Delta t}{12}t_n^{\beta-1}\left[\begin{array}{c}\dfrac{T}{S}\left[\dfrac{\partial r^2 h(r_i,t_n)}{\partial r^2}\left(\dfrac{r_i^{2-2\alpha}}{\alpha^2}\right)+\dfrac{(1-\alpha)}{\alpha^2}\dfrac{\partial h(r_i,t_n)}{\partial r}r_i^{1-2\alpha}\right]\left(1+\dfrac{dr}{r_i}\right)\\[2mm]+\dfrac{T}{S}\left[\dfrac{\partial h(r_i,t_n)}{\partial r}\dfrac{r^{1-\alpha}}{\alpha}\right]\left(\dfrac{1}{dr}-\dfrac{1}{r_i}\right)-\dfrac{T}{S}\left[\dfrac{\partial h(r_i,t_n)}{\partial r}\right]\left(\dfrac{1}{dr}\right)+\dfrac{T}{S}\left[\dfrac{h(r_i,t_n)}{\lambda^2}\right]\end{array}\right] \end{array}\right\}
$$

$$(10.37)$$

10.6 SIMULATIONS AND INTERPRETATION

In this section, we present numerical solutions. They are depicted in Figures 10.1 to 10.5 for different values of fractal and fractional order. In this case we chose the fractal order to be 1 (Figures 10.6 and 10.7).

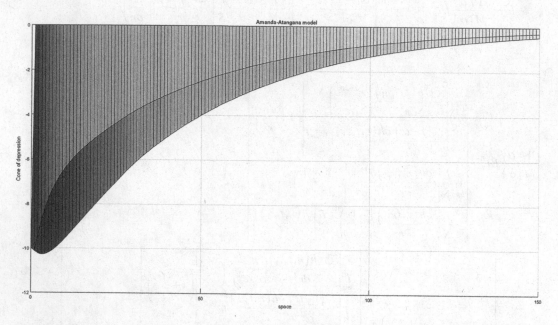

FIGURE 10.1 Cone of depression at order (1, 0.7) in space.

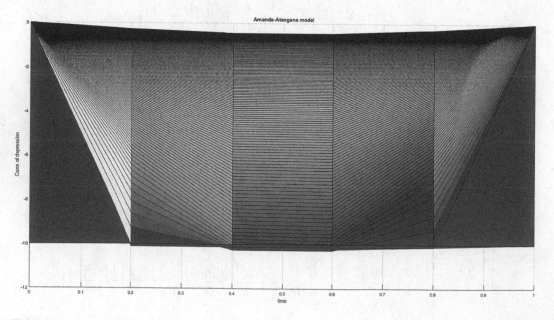

FIGURE 10.2 Cone of depression at order (1, 0.7) in time.

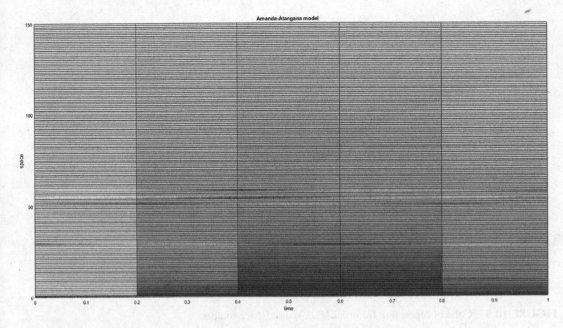

FIGURE 10.3   Amanda–Atangana model for order (1, 0.7) time/space solution.

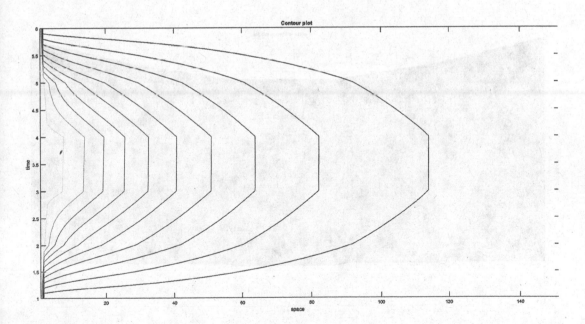

FIGURE 10.4   Contour plot at (1, 0.7) time/space solution.

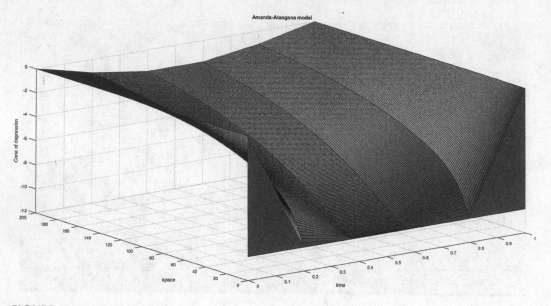

FIGURE 10.5   Cone of depression for order (1, 0.3) time/space solution.

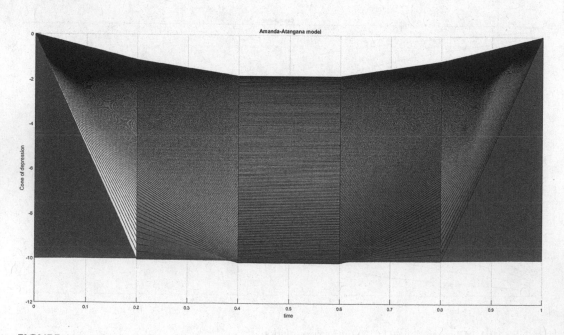

FIGURE 10.6   Cone of depression at order (1, 0.03) in time.

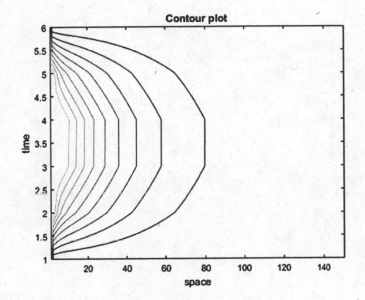

FIGURE 10.7 Contour plot at $(1, 1)$ time/space solution.

10.7 CONCLUSION

As opposed to Caputo, the cone of depression is too high. The time against cone of depression figure at order 0.03 indicates a huge abstraction of water as the cone of depression is too high. The Caputo–Fabrizio figures indicate that the pores within the aquifer are larger and the friction is less so that is able to flow freely within the aquifer. The contour plot at order 1 represents a fractured geological formation and the normal flow within an aquifer. At order 0.7 the flow is too low; this indicates that the aquitard is clay as it does not allow a significant amount of water to be abstracted from the aquifer.

REFERENCES

Atangana, A., (2017). Fractal-fractional differentiation and integration: connecting fractal calculus and fractional calculus to predict complex system. *Chaos, Solitons and Fractals*. pp. 399–406.

Atangana, A., Akgul, A., & Owolabi, M. K., (2020). Analysis of fractal fractional differential equations. *Alexandria Engineering Journal*. 59(3), pp. 1117–1134.

Atangana, A., & Araz, S.I., (2021). *New Numerical Scheme with Newton Polynomial: Theory, Methods, and Applications*. Academic Press.

Atangana, A., & Qureshi, S., (2019). Modeling attractors of chaotic dynamic systems with fractal-fractional operators. *Chaos, Solitons and Fractals*. pp. 320–337.

Chen, W. (2005). Time–Space fabric underlying anomalous diffusion. *Chaos, Soliton and Fractals*, 923–929.

Hantush, M.S., (1959). N*onsteady flow to flowing wells in leaky aquifers. Journal of Geophysical Research*, 64(8), 1043–1052.

Karay, G., & Hajnal, G., (2015). Modelling of groundwater flow in fractured rocks. *Procedia Environmental Sciences*, 142–149.

Mandelbrot, B.B., (2010). *The Fractalist, Memoir of a Scientific Maverick*. New York: Vintage Books, Division of Random House.

Sekar, M., Mohan Kumar, M., & Sridharan, K., (1994). A leaky aquifer model for hard rock aquifers. *Applied Hydrogeology*, 32–39.

Veeresha, P., & Prakasha, D.G., (2020). Solution for fractional generalized Zakharov equations with Mittag-Leffler function. *Results in Engineering*, 5, p. 100085.

11 Application of the New Numerical Method with Atangana–Baleanu Fractal-Fractional Derivative on the Self-Similar Leaky Aquifer Equations

Amanda Ramotsho and Abdon Atangana
University of the Free State, Bloemfontein, South Africa

CONTENTS

11.1 INTRODUCTION

Fractional calculus is considered to be the bible of mathematics, but recently it has been widely used in many fields to model real-world problems. It has been applied recently to model complex biological systems with non-linear behavior and long-term memory. Fractional calculus is also applied in physical components and motions (Tarasov, 2016). Fractional derivatives can be used to model real problems in ecology, diffusion processes, viscoelastic systems, control processing, fractions stochastic systems, signal processing, and in biology (Changpin et al., 2011). A fractional derivative is defined by Marchaud, Riemann, Grünwald, Letnikov, Liouville, Riesz and Caputo–Fabrizio. Scaling variables such as (x) according to x^α it has been described as a fractal derivative in mathematics and applied mathematics. Porous media, aquifer and turbulence usually exhibit fractal properties. Local fractional derivative has been proposed in a sense that it can assist in describing, or rather understanding and studying properties of fractal objects and processes in relation to them (Tarasov, 2016). Local operators are classified or defined as local products of fields and their space–time derivative, in an interacting quantum field of theory (Zimmermann, 1970). The classical physical laws such as Darcy's law, Fick's law of diffusion and Fourier's law cannot be used to assess heterogeneous media. Nonlocal derivatives have been identified as a useful tool in many branches.

DOI: 10.1201/9781003266266-11

Nonlocal operators are applied, or rather used in, modeling. Several models are used to describe or assist in anomalous diffusion processes, viscoelasticity, signal processing, geomorphology, materials sciences, fractals and so on. These derivatives are capable of capturing long-term interactions. Fractional differentiation has the ability to describe memory effects. It has been captured in literature that the reverse memory is described by the integral as the inverse operator of the derivatives if and only if the fractional derivative is capable of describing the memory (Atangana and Gomez-Aguilar, 2018). Nonlocal behavior, introduced by the following integral operators: the fractional Laplacian, the Caputo and Marchaud fractional derivatives has been investigated by Bucur and proved that the nonlocal character given by Caputo and Marchaud induces some properties similar to those of the fractional Laplacian (Bucur, 2017).

11.2   MITTAG-LEFFLER LAW TYPE

The new conceptualization of differentiation was introduced, which combines the concept of fractional differentiation and fractal derivative.

Definition 1

Accede that $f(t)$ is continuous on (a, b) and also fractal differentiable with order β then the fractal-fractional derivative of $f(t)$ with order α having Mittag-Leffler law type kernel has been defined as follows:

$$^{FFM}_{\quad 0}D_t^{\alpha,\beta} f(t) = \frac{AB(\alpha)}{\Gamma(1-\alpha)} \frac{d}{dt^\beta} \int_0^t f(\tau) E_\alpha\left[\frac{-\alpha}{1-\alpha}(t-\tau)^\alpha\right] d\tau \tag{11.1}$$

where $AB(\alpha) = 1 - \alpha + \dfrac{\alpha}{\Gamma(\alpha)}$

$$^{FFM}_{\quad 0}D_t^{\alpha,\beta} f(t) = \frac{AB(\alpha)}{\Gamma(1-\alpha)} \frac{d}{dt} \int_0^t f(\tau) E_\alpha\left[\frac{-\alpha}{1-\alpha}(t-\tau)^\alpha\right] d\tau \frac{t^{1-\alpha}}{\beta} \tag{11.2}$$

The more generalized version is given as:

$$^{FFM}_{\quad 0}D_t^{\alpha,\beta,\lambda} f(t) = \frac{AB(\alpha)}{\Gamma(1-\alpha)} \frac{d^\lambda}{dt^\beta} \int_0^t f(\tau) E_\alpha\left[\frac{-\alpha}{1-\alpha}(t-\tau)^\alpha\right] d\tau \tag{11.3}$$

When $0 < \alpha, \beta, \lambda \le 1$,

$$^{FFM}_{\quad 0}D_t^{\alpha,\beta} f(t) = \frac{AB(\alpha)}{\Gamma(1-\alpha)} \frac{d^\lambda}{dt} \int_0^t f(\tau) E_\alpha\left[\frac{-\alpha}{1-\alpha}(t-\tau)^\alpha\right] d\tau \frac{t^{1-\alpha}}{\beta} \tag{11.4}$$

Definition 2

Accede that $f(t)$ is continuous and fractal differentiable on (a, b) with order β then the fractal-fractional derivative of $f(t)$ with order α in the Caputo sense having Mittag-Leffler law type kernel has been defined as follows:

$$^{FFM}_{\quad 0}D_t^{\alpha,\beta} f(t) = \frac{AB(\alpha)}{\Gamma(1-\alpha)} \int_0^t \frac{df(\tau)}{d\tau^\beta} E_\alpha\left[\frac{-\alpha}{1-\alpha}(t-\tau)^\alpha\right] d\tau \tag{11.5}$$

where $AB(\alpha) = 1 - \alpha + \dfrac{\alpha}{\Gamma(\alpha)}$

$$^{FFM}_0 D_t^{\alpha,\beta} f(t) = \frac{AB(\alpha)}{1-\alpha} \int_0^t \frac{df(\tau)}{d\tau^\beta} \frac{\tau^{1-\beta}}{\beta} E_\alpha \left[-\frac{\alpha}{1-\alpha} (t-\tau)^\alpha \right] dz \qquad (11.6)$$

The more generalized version is given as:

$$^{FFM}_0 D_t^{\alpha,\beta,\lambda} f(t) = \frac{AB(\alpha)}{\Gamma(1-\alpha)} \int_0^t \frac{d^\lambda f(\tau)}{d\tau^\beta} E_\alpha \left[\frac{-\alpha}{1-\alpha} (t-\tau)^\alpha \right] d\tau \qquad (11.7)$$

When $0 < \alpha, \lambda, \beta \leq 1$

$$^{FFM}_0 D_t^{\alpha,\beta,\lambda} f(t) = \frac{AB(\alpha)}{1-\alpha} \int_0^t \frac{d^\lambda f(\tau)}{d\tau} \frac{\tau^{1-\beta}}{\beta} E_\alpha \left[-\frac{\alpha}{1-\alpha} (t-\tau)^\alpha \right] dz \cdot \qquad (11.8)$$

Definition 3

Accede that $f(t)$ be continuous on (a, b) at an open interval then the fractal-fractional integral of $f(t)$ with order α having generalized Mittag-Leffler type is given as:

$$^{FFM}_0 J_t^{\alpha,\beta} f(t) = \frac{\alpha\beta}{AB(\alpha)} \int_0^t (t-\tau)^{\alpha-1} \tau^{\alpha-1} f(\tau) d\tau + \frac{\beta(1-\alpha)t^{\beta-1} f(t)}{AB(\alpha)} \cdot \qquad (11.9)$$

11.3 NUMERICAL SCHEME: USING ATANGANA–BALEANU FRACTAL-FRACTIONAL DERIVATIVE

The Cauchy problem was analyzed

$$^{FFM}_0 D_t^{\alpha,\beta} h(t) = f\big(t,\big(h(t)\big)\big) \qquad (11.10)$$

Applying the Atangana–Baleanu fractal-fractional derivative on the equation above and integrating the equation

$$h(t) - h(0) = \frac{1-\alpha}{AB(\alpha)} \beta t^{\beta-1} f\big(t\big(h(t)\big)\big) + \frac{\alpha\beta}{AB(\alpha)\Gamma(\alpha)} \int_0^t \tau^{\beta-1} f\big(\tau, h(\tau)\big)(t-\tau)^{\alpha-1} d\tau \qquad (11.11)$$

Let

$$\begin{aligned} f\big(t, h(t)\big) &= \beta t^{\beta-1} f\big(t\big(h(t)\big)\big) h(t) - h(0) \\ &= \frac{1-\alpha}{AB(\alpha)} F\big(t, h(t)\big) + \frac{\alpha\beta}{AB(\alpha)\Gamma(\alpha)} \int_0^t F\big(\tau, h(\tau)\big)(t-\tau)^{\alpha-1} d\tau \end{aligned} \qquad (11.12)$$

At the point $t_{n+1} = (n+1)\Delta t$, we have

$$h(t_{n+1}) - h(0) = \frac{1-\alpha}{AB(\alpha)} F\big(t_n, h(t_n)\big) + \frac{\alpha\beta}{AB(\alpha)\Gamma(\alpha)} \int_0^{t_{n+1}} F\big(\tau, h(\tau)\big)(t_{n+1}-\tau)^{\alpha-1} d\tau \qquad (11.13)$$

This can also be given as,

$$h(t_{n+1}) = h(0) + \frac{1-\alpha}{AB(\alpha)} F(t_n, h(t_n)) + \frac{\alpha}{AB(\alpha)\Gamma(\alpha)} \sum_{j=2}^{n} \int_{t_j}^{t_{j+1}} F(\tau, h(\tau))(t_{n+1} - \tau)^{\alpha-1} d\tau \quad (11.14)$$

We can now consider the approximation of $f(\tau, y(t))$ as a Newton polynomial given below

$$
\begin{aligned}
P_n(\tau) = {}& f(t_{n-2}, h(t_{n-2})) + \left[\frac{f(t_{n-1}, h(t_{n-1})) - f(t_{n-2}, h(t_{n-2}))}{\Delta t} \right](\tau - t_{n-2}) \\
& + \left[\frac{f(t_n, h(t_n)) - 2f(t_{n-1}, h(t_{n-1})) + f(t_{n-2}, h(t_{n-2}))}{2(\Delta t)^2} \right](\tau - t_{n-2})(\tau - t_{n-1})
\end{aligned}
\quad (11.15)
$$

Substituting Equation (11.15) into Equation (11.14)

$$
\begin{aligned}
h^{n+1} = {}& h^0 + \frac{1-\alpha}{AB(\alpha)} F(t_n, h(t_n)) \\
& + \frac{\alpha}{AB(\alpha)\Gamma(\alpha)} \sum_{j=2}^{n} \int_{t_j}^{t_{j+1}} \left\{
\begin{array}{l}
F(t_{j-2}, h^{j-2}) \\[4pt]
+ \dfrac{F(t_{j-1}, h^{j-1}) - F(t_{j-2}, h^{j-2})}{\Delta t}(\tau - t_{j-2}) \\[10pt]
+ \dfrac{F(t_j, h^j) - 2F(t_{j-1}, h^{j-1}) + F(t_{j-2}, h^{j-2})}{2(\Delta t)^2} \\[10pt]
\times (\tau - t_{j-2})(\tau - t_{j-1})
\end{array}
\right\}(t_{n+1} - \tau)^{\alpha-1} d\tau
\end{aligned}
\quad (11.16)
$$

This can be represented or simplified as follows:

$$
\begin{aligned}
h^{n+1} = {}& h^0 + \frac{1-\alpha}{AB(\alpha)} F(t_n, h(t_n)) \\
& + \frac{\alpha}{AB(\alpha)\Gamma(\alpha)} \sum_{j=2}^{n} \left\{
\begin{array}{l}
\displaystyle\int_{t_j}^{t_{j+1}} F(t_{j-2}, h^{j-2})(t_{n+1} - \tau)^{\alpha-1} d\tau \\[12pt]
+ \displaystyle\int_{t_j}^{t_{j+1}} \dfrac{F(t_{j-1}, h^{j-1}) - F(t_{j-2}, h^{j-2})}{\Delta t}(\tau - t_{j-2})(t_{n+1} - \tau)^{\alpha-1} d\tau \\[12pt]
+ \displaystyle\int_{t_j}^{t_{j+1}} \dfrac{F(t_j, h^j) - 2F(t_{j-1}, h^{j-1}) + F(t_{j-2}, h^{j-2})}{2(\Delta t)^2} \\[10pt]
\times (\tau - t_{j-2})(\tau - t_{j-1})(t_{n+1} - \tau)^{\alpha-1} d\tau
\end{array}
\right\}
\end{aligned}
\quad (11.17)
$$

Therefore, the summations can be given as:

$$h^{n+1} = h^0 + \frac{1-\alpha}{AB(\alpha)} F\left(t_n, h(t_n)\right) + \frac{\alpha}{AB(\alpha)\Gamma(\alpha)} \sum_{j=2}^{n} F\left(t_{j-2}, h^{j-2}\right) \int_{t_j}^{t_{j+1}} \left(t_{n+1} - \tau\right)^{\alpha-1} d\tau$$

$$+ \frac{\alpha}{AB(\alpha)\Gamma(\alpha)} \sum_{j=2}^{n} \frac{F\left(t_{j-1}, h^{j-1}\right) - F\left(t_{j-2}, h^{j-2}\right)}{\Delta t} \int_{t_j}^{t_{j+1}} \left(\tau - t_{j-2}\right)\left(t_{n+1} - \tau\right)^{\alpha-1} d\tau$$

$$+ \frac{\alpha}{AB(\alpha)\Gamma(\alpha)} \sum_{j=2}^{n} \frac{F\left(t_j, h^j\right) - 2F\left(t_{j-1}, h^{j-1}\right) + F\left(t_{j-2}, h^{j-2}\right)}{2(\Delta t)^2}$$

$$\times \int_{t_j}^{t_{j+1}} \times \left(\tau - t_{j-2}\right)\left(\tau - t_{j-1}\right)\left(t_{n+1} - \tau\right)^{\alpha-1} d\tau$$

(11.18)

The integrals can be calculated and given as:

$$\int_{t_j}^{t_{j+1}} \left(t_{n+1} - \tau\right)^{\alpha-1} d\tau = \frac{(\Delta t)^\alpha}{\alpha} \left[\left(n-j+1\right)^\alpha - \left(n-j\right)^\alpha\right]$$

(11.19)

$$\int_{t_j}^{t_{j+1}} \left(\tau - t_{j-2}\right)\left(t_{n+1} - \tau\right)^{\alpha-1} d\tau = \frac{(\Delta t)^{\alpha+1}}{\alpha(\alpha+1)} \begin{bmatrix} \left(n-j+1\right)^\alpha \left(n-j+3+2\alpha\right) \\ -\left(n-j+1\right)^\alpha \left(n-j+3+3\alpha\right) \end{bmatrix}$$

(11.20)

$$\int_{t_j}^{t_{j+1}} \left(\tau - t_{j-2}\right)\left(\tau - t_{j-1}\right)\left(t_{n+1} - \tau\right)^{\alpha-1} d\tau = \frac{(\Delta t)^{\alpha+2}}{\alpha(\alpha+1)(\alpha+2)}$$

$$\times \begin{bmatrix} \left(n-j+1\right)^\alpha \begin{bmatrix} 2(n-j)^2 + (3\alpha+10)(n-j) \\ +(2\alpha^2+9\alpha+12) \end{bmatrix} \\ -(n-j)^\alpha \begin{bmatrix} 2(n-j)^2 + (5\alpha+10)(n-j) \\ +(6\alpha^2+18\alpha+12) \end{bmatrix} \end{bmatrix}$$

(11.21)

Replacing the calculated integrals in the equality above:

$$h^{n+1} = h^0 + \frac{1-\alpha}{AB(\alpha)} F\left(t_n, h(t_n)\right)$$

$$+ \frac{\alpha(\Delta t)^\alpha}{AB(\alpha)\Gamma(\alpha+1)} \sum_{j=2}^{n} F\left(t_{j-2}, h^{j-2}\right)\left[\left(n-j+1\right)^\alpha - \left(n-j\right)^\alpha\right]$$

$$+ \frac{\alpha(\Delta t)^\alpha}{AB(\alpha)\Gamma(\alpha+2)} \sum_{j=2}^{n} \left[F\left(t_{j-1}, h^{j-1}\right) - F\left(t_{J-2}\, h^{j-2}\right)\right]$$

$$\times \begin{bmatrix} \left(n-j+1\right)^\alpha \left(n-j+3+2\alpha\right) \\ -\left(n-j\right)^\alpha \left(n-j+3+3\alpha\right) \end{bmatrix} + \frac{\alpha(\Delta t)^\alpha}{2AB(\alpha)\Gamma(\alpha+3)} \sum_{j=2}^{n} F\left(t_j,\, h^j\right) - 2F\left(t_{j-1}, h^{j-1}\right) + F\left(t_{j-2}, h^{j-2}\right)$$

$$\times \begin{bmatrix} \left(n-j+1\right)^\alpha \left[2(n-j)^2 + (3\alpha+10)(n-j) + (2\alpha^2+9\alpha+12)\right] \\ -\left(n-j\right)^\alpha \left[2(n-j)^2 + (5\alpha+10)(n-j) + (6\alpha^2+18\alpha+12)\right] \end{bmatrix}$$

(11.22)

Therefore, the final numerical scheme can be given as:

$$h^{n+1} = h^0 + \frac{1-\alpha}{AB(\alpha)} \beta t_n^{\beta-1} f\left(t_n, h(t_n)\right) + \frac{\alpha\beta(\Delta t)^\alpha}{AB(\alpha)\Gamma(\alpha+1)} \sum_{j=2}^n t_{j-2}^{\beta-1} f\left(t_{j-2}, h^{j-2}\right)\left[(n-j+1)^\alpha - (n-j)^\alpha\right]$$

$$+ \frac{\alpha\beta(\Delta t)^\alpha}{AB(\alpha)\Gamma(\alpha+2)} \sum_{j=2}^n \left[t_{j-1}^{\beta-1} f\left(t_{j-1}, h^{j-1}\right) - t_{j-2}^{\beta-1} f\left(t_{J-2}, h^{j-2}\right)\right] \times \begin{bmatrix}(n-j+1)^\alpha(n-j+3+2\alpha)\\ -(n-j)^\alpha(n-j+3+3\alpha)\end{bmatrix}$$

$$+ \frac{\alpha\beta(\Delta t)^\alpha}{2AB(\alpha)\Gamma(\alpha+3)} \sum_{j=2}^n \left[t_j^{\beta-1} f\left(t_j, h^j\right) - 2t_{j-1}^{\beta-1} f\left(t_{j-1}, h^{j-1}\right) + t_{j-2}^{\beta-1} f\left(t_{j-2}, h^{j-2}\right)\right]$$

$$\times \begin{bmatrix}(n-j+1)^\alpha\left[2(n-j)^2 + (3\alpha+10)(n-j) + (2\alpha^2+9\alpha+12)\right]\\ -(n-j)^\alpha\left[2(n-j)^2 + (5\alpha+10)(n-j) + (6\alpha^2+18\alpha+12)\right]\end{bmatrix}.$$

$$(11.23)$$

11.4 IMPLEMENTATION OF ATANGANA–BALEANU FRACTAL-FRACTIONAL DERIVATIVE ON THE SELF-SIMILAR LEAKY AQUIFER EQUATION SCENARIO 1

The mathematical model below was incorporated to assess self-similar leaky aquifers, where the in and out flow is within a self-similar leaky aquifer

$$\frac{S}{T}\frac{\partial h}{\partial t} = \frac{\partial}{\partial r^\alpha}\left(\frac{\partial h}{\partial r^\alpha}\right)\left(1 + \frac{dr}{r}\right) + \frac{1}{r}\frac{\partial h}{\partial r^\alpha} + \frac{h(r,t)}{\lambda^2} \tag{11.24}$$

The discretized equation with Atangana–Baleanu the resulting equation is given as:

$$_0^{AB}D_t^{\beta,\alpha}h(r,t) = \frac{T}{S}\frac{r^{1-\alpha}}{\alpha}\begin{bmatrix}\dfrac{\partial r^2 h(r,t)}{\partial r^2}\left(\dfrac{r^{1-\alpha}}{\alpha}\right)\\ +\left(\dfrac{1-\alpha}{\alpha}\right)r^{-\alpha}\dfrac{\partial h(r,t)}{\partial r}\end{bmatrix}\left(1+\frac{dr}{r}\right) + \frac{T}{S}\frac{1}{r}\frac{\partial h(r,t)}{\partial r}\frac{r^{1-\alpha}}{\alpha} + \frac{T}{S}\frac{h(r,t)}{\lambda^2} \tag{11.25}$$

The equation can be simplified and given as

$$f\left(\alpha,r,t,h(r,\tau)\right) = \frac{T}{S}\frac{r^{1-\alpha}}{\alpha}\begin{bmatrix}\dfrac{\partial r^2 h(r,t)}{\partial r^2}\left(\dfrac{r^{1-\alpha}}{\alpha}\right)\\ +\left(\dfrac{1-\alpha}{\alpha}\right)r^{-\alpha}\dfrac{\partial h(r,t)}{\partial r}\end{bmatrix}\left(1+\frac{dr}{r}\right) + \frac{T}{S}\frac{1}{r}\frac{\partial h(r,t)}{\partial r}\frac{r^{1-\alpha}}{\alpha} + \frac{T}{S}\frac{h(r,t)}{\lambda^2}$$

$$(11.26)$$

Such that

$$\,_{0}^{AB}D_t^{\beta,\alpha}h(r,t) = f\left(\alpha,r,t,h(r,\tau)\right)$$ (11.27)

As presented before, discretization, we convert the above equation into an integral equation by applying the Atangana–Baleanu fractional integral:

$$h(r,t)-h(r,0)=\left\{\begin{array}{c}\dfrac{1-\alpha}{\mathrm{AB}(\alpha)}f\left(\alpha,r,t,h(r,t)\right)\\[3mm]+\dfrac{\alpha}{\mathrm{AB}(\alpha)\Gamma(\alpha)}\displaystyle\int_0^t(t-l)^{\alpha-1}f\left(\alpha,r,l,h(r,l)\right)dl\end{array}\right\}$$ (11.28)

We evaluate the above equation at the point r_i, t_{n+1}, such that the function $h(r,t)$ will be represented as:

$$h(r_i,t_{n+1})=h(r_i,0)+\frac{1-\alpha}{\mathrm{AB}(\alpha)}f\left(r_i,t_n,h(r_i,t_n)\right)+\frac{\alpha}{\mathrm{AB}(\alpha)\Gamma(\alpha)}\sum_{j=2}^{n}\int_{t_j}^{t_{j+1}}f(\alpha,r_i,\tau,h(r_i,\tau)(t_{n+1}-\tau)^{\alpha-1}\,d\tau$$ (11.29)

Following the scheme presented earlier, which was derived using the Newton polynomials and the Cauchy problem the numerical solution, will be given as:

$$h^{n+1}=h^0+\frac{1-\alpha}{\mathrm{AB}(\alpha)}\beta t_n^{\beta-1}f\left(\alpha,r_i,t_n,h_i^n\right)$$

$$+\frac{\alpha\beta(\Delta t)^\alpha}{\mathrm{AB}(\alpha)\Gamma(\alpha+1)}\sum_{j=2}^{n}t_{j-2}^{\beta-1}f\left(\alpha,r_i,t_{j-2},h(r,t_{j-2})\right)\left[(n-j+1)^\alpha-(n-j)^\alpha\right]$$

$$+\frac{\alpha\beta(\Delta t)^\alpha}{\mathrm{AB}(\alpha)\Gamma(\alpha+2)}\sum_{j=2}^{n}\left[\begin{array}{c}t_{j-1}^{\beta-1}f\left(\alpha,r_i,t_{j-1},h(r,t_{j-1})\right)\\-t_{j-2}^{\beta-1}f\left(\alpha,r_i,t_{j-2},h(r,t_{j-2})\right)\end{array}\right]\left[\begin{array}{c}(n-j+1)^\alpha(n-j+3+2\alpha)\\-(n-j)^\alpha(n-j+3+3\alpha)\end{array}\right]$$

$$+\frac{\alpha\beta(\Delta t)^\alpha}{2\mathrm{AB}(\alpha)\Gamma(\alpha+3)}\sum_{j=2}^{n}\left[\begin{array}{c}t_j^{\beta-1}f\left(\alpha,r_i,t_j,h(r,t_j)\right)\\-2t_{j-1}^{\beta-1}f\left(\alpha,r_i,t_{j-1},h(r,t_{j-1})\right)\\+t_{j-2}^{\beta-1}f\left(\alpha,r_i,t_{j-2},h(r,t_{j-2})\right)\end{array}\right]\times\left[\begin{array}{c}(n-j+1)^\alpha\left[\begin{array}{c}2(n-j)^2+(3\alpha+10)(n-j)\\+(2\alpha^2+9\alpha+12)\end{array}\right]\\-(n-j)^\alpha\left[\begin{array}{c}2(n-j)^2+(5\alpha+10)(n-j)\\+(6\alpha^2+18\alpha+12)\end{array}\right]\end{array}\right]$$ (11.30)

Therefore, the final numerical solution can be given as:

$$
\begin{aligned}
h^{n+1} = h^0 &+ \frac{1-\alpha}{AB(\alpha)}\beta t_n^{\beta-1}\left[\frac{\dfrac{T}{S}\dfrac{r_i^{1-\alpha}}{\alpha}\left[\dfrac{\partial r^2 h(r_i,t_j)}{\partial r^2}\left(\dfrac{r_i^{1-\alpha}}{\alpha}\right)+\left(\dfrac{1-\alpha}{\alpha}\right)r_i^{-\alpha}\dfrac{\partial h(r_i,t_j)}{\partial r}\right]}{\left(1+\dfrac{dr}{r_i}\right)+\dfrac{T}{S}\dfrac{1}{r}\dfrac{\partial h(r_i,t_j)}{\partial r}\dfrac{r_i^{1-\alpha}}{\alpha}+\dfrac{T}{S}\dfrac{h(r_i,t_j)}{\lambda^2}}\right] \\[10pt]
&+ \frac{\alpha\beta(\Delta t)^\alpha}{AB(\alpha)\Gamma(\alpha+1)}\sum_{j=2}^{n}t_{j-2}^{\beta-1}\left\{\frac{\dfrac{T}{S}\dfrac{r_i^{1-\alpha}}{\alpha}\left[\begin{array}{c}\dfrac{\partial r^2 h(r_i,t_{j-2})}{\partial r^2}\left(\dfrac{r_i^{1-\alpha}}{\alpha}\right)\\[6pt]+\left(\dfrac{1-\alpha}{\alpha}\right)r_i^{-\alpha}\dfrac{\partial h(r_i,t_{j-2})}{\partial r}\end{array}\right]}{\left(1+\dfrac{dr}{r_i}\right)+\dfrac{T}{S}\dfrac{1}{r}\dfrac{\partial h(r_i,t_{j-2})}{\partial r}\dfrac{r_i^{1-\alpha}}{\alpha}+\dfrac{T}{S}\dfrac{h(r_i,t_{j-2})}{\lambda^2}}\right\}\left(\left[(n-j+1)^\alpha-(n-j)^\alpha\right]\right) \\[10pt]
&+ \frac{\alpha\beta(\Delta t)^\alpha}{AB(\alpha)\Gamma(\alpha+2)}\sum_{j=2}^{n}\left[\begin{array}{c}t_{j-1}^{\beta-1}\left\{\dfrac{\dfrac{T}{S}\dfrac{r_i^{1-\alpha}}{\alpha}\left[\dfrac{\partial r^2 h(r_i,t_{j-1})}{\partial r^2}\left(\dfrac{r_i^{1-\alpha}}{\alpha}\right)+\left(\dfrac{1-\alpha}{\alpha}\right)r_i^{-\alpha}\dfrac{\partial h(r_i,t_{j-1})}{\partial r}\right]}{\left(1+\dfrac{dr}{r_i}\right)+\dfrac{T}{S}\dfrac{1}{r}\dfrac{\partial h(r_i,t_{j-1})}{\partial r}\dfrac{r_i^{1-\alpha}}{\alpha}+\dfrac{T}{S}\dfrac{h(r_i,t_{j-1})}{\lambda^2}}\right\} \\[10pt] -t_{j-2}^{\beta-1}\left\{\dfrac{\dfrac{T}{S}\dfrac{r_i^{1-\alpha}}{\alpha}\left[\dfrac{\partial r^2 h(r_i,t_{j-2})}{\partial r^2}\left(\dfrac{r_i^{1-\alpha}}{\alpha}\right)+\left(\dfrac{1-\alpha}{\alpha}\right)r_i^{-\alpha}\dfrac{\partial h(r_i,t_{j-2})}{\partial r}\right]}{\left(1+\dfrac{dr}{r_i}\right)+\dfrac{T}{S}\dfrac{1}{r}\dfrac{\partial h(r_i,t_{j-2})}{\partial r}\dfrac{r_i^{1-\alpha}}{\alpha}+\dfrac{T}{S}\dfrac{h(r_i,t_{j-2})}{\lambda^2}}\right\}\end{array}\right] \\
&\qquad\times\left[\begin{array}{c}(n-j+1)^\alpha(n-j+3+2\alpha)\\-(n-j)^\alpha(n-j+3+3\alpha)\end{array}\right] \\[10pt]
&+ \frac{\alpha\beta(\Delta t)^\alpha}{2AB(\alpha)\Gamma(\alpha+3)}\sum_{j=2}^{n}\left[\begin{array}{c}t_j^{\beta-1}\left\{\dfrac{\dfrac{T}{S}\dfrac{r_i^{1-\alpha}}{\alpha}\left[\dfrac{\partial r^2 h(r_i,t_j)}{\partial r^2}\left(\dfrac{r_i^{1-\alpha}}{\alpha}\right)+\left(\dfrac{1-\alpha}{\alpha}\right)r_i^{-\alpha}\dfrac{\partial h(r_i,t_j)}{\partial r}\right]}{\left(1+\dfrac{dr}{r_i}\right)+\dfrac{T}{S}\dfrac{1}{r}\dfrac{\partial h(r_i,t_j)}{\partial r}\dfrac{r_i^{1-\alpha}}{\alpha}+\dfrac{T}{S}\dfrac{h(r_i,t_j)}{\lambda^2}}\right\} \\[10pt] -2t_{j-1}^{\beta-1}\left\{\dfrac{\dfrac{T}{S}\dfrac{r_i^{1-\alpha}}{\alpha}\left[\dfrac{\partial r^2 h(r_i,t_{j-1})}{\partial r^2}\left(\dfrac{r_i^{1-\alpha}}{\alpha}\right)+\left(\dfrac{1-\alpha}{\alpha}\right)r_i^{-\alpha}\dfrac{\partial h(r_i,t_{j-1})}{\partial r}\right]}{\left(1+\dfrac{dr}{r_i}\right)+\dfrac{T}{S}\dfrac{1}{r}\dfrac{\partial h(r_i,t_{j-1})}{\partial r}\dfrac{r_i^{1-\alpha}}{\alpha}+\dfrac{T}{S}\dfrac{h(r_i,t_{j-1})}{\lambda^2}}\right\} \\[10pt] +t_{j-2}^{\beta-1}\left\{\dfrac{\dfrac{T}{S}\dfrac{r_i^{1-\alpha}}{\alpha}\left[\dfrac{\partial r^2 h(r_i,t_{j-2})}{\partial r^2}\left(\dfrac{r_i^{1-\alpha}}{\alpha}\right)+\left(\dfrac{1-\alpha}{\alpha}\right)r_i^{-\alpha}\dfrac{\partial h(r_i,t_{j-2})}{\partial r}\right]}{\left(1+\dfrac{dr}{r_i}\right)+\dfrac{T}{S}\dfrac{1}{r}\dfrac{\partial h(r_i,t_{j-2})}{\partial r}\dfrac{r_i^{1-\alpha}}{\alpha}+\dfrac{T}{S}\dfrac{h(r_i,t_{j-2})}{\lambda^2}}\right\}\end{array}\right] \\
&\qquad\times\left[\begin{array}{c}(n-j+1)^\alpha\left[2(n-j)^2+(3\alpha+10)(n-j)+(2\alpha^2+9\alpha+12)\right]\\-(n-j)^\alpha\left[2(n-j)^2+(5\alpha+10)(n-j)+(6\alpha^2+18\alpha+12)\right]\end{array}\right]
\end{aligned}
\tag{11.31}
$$

11.5 IMPLEMENTATION OF ATANGANA–BALEANU FRACTAL-FRACTIONAL DERIVATIVE ON THE SELF-SIMILAR LEAKY AQUIFER EQUATION SCENARIO 2

Another scenario was considered where the water is not flowing out of a small portion of a self-similar leaky aquifer but a normal leaky aquifer and the equation below was derived.

$$\frac{S}{T}\frac{\partial h}{\partial t} = \frac{\partial}{\partial r^\alpha}\left(\frac{\partial h}{\partial r^\alpha}\right)\left(1+\frac{dr}{r}\right) + \frac{\partial h}{\partial r^\alpha}\left(\frac{1}{dr}-\frac{1}{r}\right) + \frac{h(r,t)}{\lambda^2} - \frac{1}{dr}\left[\frac{\partial h}{\partial r}\right] \tag{11.32}$$

The second discretized equation in terms of Atangana–Baleanu is given as

$$
{}_{0}^{c}D_{t}^{\beta,\alpha}h(r,t) = \frac{T}{S}\left[\frac{\partial r^2 h(r,t)}{\partial r^2}\left(\frac{r^{2-2\alpha}}{\alpha^2}\right) + \frac{(1-\alpha)}{\alpha^2}\frac{\partial h(r,t)}{\partial r}r^{1-2\alpha}\right]\left(1+\frac{dr}{r}\right)
$$
$$
+ \frac{T}{S}\left[\frac{\partial h(r,t)}{\partial r}\frac{r^{1-\alpha}}{\alpha}\right]\left(\frac{1}{dr}-\frac{1}{r}\right) - \frac{T}{S}\left[\frac{\partial h(r,t)}{\partial r}\right]\left(\frac{1}{dr}\right) + \frac{T}{S}\left[\frac{h(r,t)}{\lambda^2}\right] \tag{11.33}
$$

The right-hand side of the above equation can be simplified by replacing

$$
f\big(\alpha,r,t,h(r,\tau)\big) = \frac{T}{S}\left[\frac{\partial r^2 h(r,t)}{\partial r^2}\left(\frac{r^{2-2\alpha}}{\alpha^2}\right) + \frac{(1-\alpha)}{\alpha^2}\frac{\partial h(r,t)}{\partial r}r^{1-2\alpha}\right]\left(1+\frac{dr}{r}\right)
$$
$$
+ \frac{T}{S}\left[\frac{\partial h(r,t)}{\partial r}\frac{r^{1-\alpha}}{\alpha}\right]\left(\frac{1}{dr}-\frac{1}{r}\right) - \frac{T}{S}\left[\frac{\partial h(r,t)}{\partial r}\right]\left(\frac{1}{dr}\right) \, \frac{T}{S}\left[\frac{h(r,t)}{\lambda^2}\right] \tag{11.34}
$$

Such that:

$$
{}_{0}^{AB}D_{t}^{\beta,\alpha}h(r,t) = f\big(\alpha,r,t,h(r,\tau)\big) \tag{11.35}
$$

As presented before discretization, we convert the above equation into an integral equation by applying the Atangana–Baleanu fractional integral

$$
h(r,t) - h(r,0) = \left\{ \begin{array}{l} \dfrac{1-\alpha}{AB(\alpha)}f\big(\alpha,r,t,h(r,t)\big) \\[2mm] + \dfrac{\alpha}{AB(\alpha)\Gamma(\alpha)}\displaystyle\int_{0}^{t}(t-l)^{\alpha-1}f\big(\alpha,r,l,h(r,l)\big)dl \end{array} \right\} \tag{11.36}
$$

We evaluate the above equation at the point r_i, t_{n+1}, such that the function $h(r,t)$ will be represented as

$$h(r_i,t_{n+1}) = h(r_i,0) + \frac{1-\alpha}{AB(\alpha)} f(r_i,t_n,h(r_i,t_n))$$

$$+ \frac{\alpha}{AB(\alpha)\Gamma(\alpha)} \sum_{j=2}^{n} \int_{t_j}^{t_{j+1}} f(\alpha,r_i,\tau,h(r_i,\tau))(t_{n+1}-\tau)^{\alpha-1} d\tau \qquad (11.37)$$

Therefore

$$h^{n+1} = h^0 + \frac{1-\alpha}{AB(\alpha)}\beta t_n^{\beta-1} \left[\begin{array}{l} \dfrac{T}{S}\left[\dfrac{\partial r^2 h(r_i,t_j)}{\partial r^2}\left(\dfrac{r_i^{2-2\alpha}}{\alpha^2}\right) + \dfrac{(1-\alpha)}{\alpha^2}\dfrac{\partial h(r_i,t_j)}{\partial r} r_i^{1-2\alpha}\right]\left(1+\dfrac{dr}{r_i}\right) \\[3mm] +\dfrac{T}{S}\left[\dfrac{\partial h(r_i,t_j)}{\partial r}\dfrac{r^{1-\alpha}}{\alpha}\right]\left(\dfrac{1}{dr}-\dfrac{1}{r_i}\right)-\dfrac{T}{S}\left[\dfrac{\partial h(r_i,t_j)}{\partial r}\right]\left(\dfrac{1}{dr}\right)+\dfrac{T}{S}\left[\dfrac{h(r_i,t_j)}{\lambda^2}\right] \end{array}\right]$$

$$+ \frac{\alpha\beta(\Delta t)^\alpha}{AB(\alpha)\Gamma(\alpha+1)} \sum_{j=2}^{n} t_{j-2}^{\beta-1} \left\{ \begin{array}{l} \dfrac{T}{S}\left[\dfrac{\partial r^2 h(r_i,t_{j-2})}{\partial r^2}\left(\dfrac{r_i^{2-2\alpha}}{\alpha^2}\right) + \dfrac{(1-\alpha)}{\alpha^2}\dfrac{\partial h(r_i,t_{j-2})}{\partial r} r_i^{1-2\alpha}\right]\left(1+\dfrac{dr}{r_i}\right) \\[3mm] +\dfrac{T}{S}\left[\dfrac{\partial h(r_i,t_{j-2})}{\partial r}\dfrac{r^{1-\alpha}}{\alpha}\right]\left(\dfrac{1}{dr}-\dfrac{1}{r_i}\right)-\dfrac{T}{S}\left[\dfrac{\partial h(r_i,t_{j-2})}{\partial r}\right]\left(\dfrac{1}{dr}\right) \\[3mm] +\dfrac{T}{S}\left[\dfrac{h(r_i,t_{j-2})}{\lambda^2}\right] \end{array}\right\}$$

$$\times\left[(n-j+1)^\alpha-(n-j)^\alpha\right] + \frac{\alpha\beta(\Delta t)^\alpha}{AB(\alpha)\Gamma(\alpha+2)}\sum_{j=2}^{n} \left[\begin{array}{l} t_{j-1}^{\beta-1}\left\{ \begin{array}{l} \dfrac{T}{S}\left[\begin{array}{l}\dfrac{\partial r^2 h(r_i,t_{j-1})}{\partial r^2}\left(\dfrac{r_i^{2-2\alpha}}{\alpha^2}\right)\\[2mm]+\dfrac{(1-\alpha)}{\alpha^2}\dfrac{\partial h(r_i,t_{j-1})}{\partial r}r_i^{1-2\alpha}\end{array}\right]\left(1+\dfrac{dr}{r_i}\right) \\[5mm] +\dfrac{T}{S}\left[\dfrac{\partial h(r_i,t_{j-1})}{\partial r}\dfrac{r^{1-\alpha}}{\alpha}\right]\left(\dfrac{1}{dr}-\dfrac{1}{r_i}\right) \\[3mm] -\dfrac{T}{S}\left[\dfrac{\partial h(r_i,t_{j-1})}{\partial r}\right]\left(\dfrac{1}{dr}\right)+\dfrac{T}{S}\left[\dfrac{h(r_i,t_{j-1})}{\lambda^2}\right] \end{array}\right\} \\[10mm] -t_{j-2}^{\beta-1}\left\{ \begin{array}{l} \dfrac{T}{S}\left[\begin{array}{l}\dfrac{\partial r^2 h(r_i,t_{j-2})}{\partial r^2}\left(\dfrac{r_i^{2-2\alpha}}{\alpha^2}\right)\\[2mm]+\dfrac{(1-\alpha)}{\alpha^2}\dfrac{\partial h(r_i,t_{j-2})}{\partial r}r_i^{1-2\alpha}\end{array}\right]\left(1+\dfrac{dr}{r_i}\right) \\[5mm] +\dfrac{T}{S}\left[\dfrac{\partial h(r_i,t_{j-2})}{\partial r}\dfrac{r^{1-\alpha}}{\alpha}\right]\left(\dfrac{1}{dr}-\dfrac{1}{r_i}\right) \\[3mm] -\dfrac{T}{S}\left[\dfrac{\partial h(r_i,t_{j-2})}{\partial r}\right]\left(\dfrac{1}{dr}\right)+\dfrac{T}{S}\left[\dfrac{h(r_i,t_{j-2})}{\lambda^2}\right] \end{array}\right\} \end{array}\right]$$

$$
\times \left[\begin{array}{l} (n-j+1)^{\alpha} \left(n-j+3+2\alpha\right) \\ -(n-j)^{\alpha} \left(n-j+3+3\alpha\right) \end{array} \right]
$$

$$
+ \frac{\alpha\beta\left(\Delta t\right)^{\alpha}}{2AB(\alpha)\Gamma(\alpha+3)} \sum_{j=2}^{n} \left[\begin{array}{l} t_j^{\beta-1} \left\{ \begin{array}{l} \dfrac{T}{S}\left[\dfrac{\partial r^2 h\left(r_i,t_j\right)}{\partial r^2}\left(\dfrac{r_i^{2-2\alpha}}{\alpha^2} \right) + \dfrac{(1-\alpha)}{\alpha^2}\dfrac{\partial h\left(r_i,t_j\right)}{\partial r} r_i^{1-2\alpha} \right]\left(1+\dfrac{dr}{r_i}\right) \\[3mm] + \dfrac{T}{S}\left[\dfrac{\partial h\left(r_i,t_j\right)}{\partial r}\dfrac{r^{1-\alpha}}{\alpha} \right]\left(\dfrac{1}{dr}-\dfrac{1}{r_i}\right) - \dfrac{T}{S}\left[\dfrac{\partial h\left(r_i,t_j\right)}{\partial r} \right]\left(\dfrac{1}{dr}\right) \\[3mm] + \dfrac{T}{S}\left[\dfrac{h\left(r_i,t_j\right)}{\lambda^2} \right] \end{array} \right\} \\[14mm] -2t_{j-1}^{\beta-1} \left\{ \begin{array}{l} \dfrac{T}{S}\left[\dfrac{\partial r^2 h\left(r_i,t_{j-1}\right)}{\partial r^2}\left(\dfrac{r_i^{2-2\alpha}}{\alpha^2} \right) + \dfrac{(1-\alpha)}{\alpha^2}\dfrac{\partial h\left(r_i,t_{j-1}\right)}{\partial r} r_i^{1-2\alpha} \right]\left(1+\dfrac{dr}{r_i}\right) \\[3mm] + \dfrac{T}{S}\left[\dfrac{\partial h\left(r_i,t_{j-1}\right)}{\partial r}\dfrac{r^{1-\alpha}}{\alpha} \right]\left(\dfrac{1}{dr}-\dfrac{1}{r_i}\right) - \dfrac{T}{S}\left[\dfrac{\partial h\left(r_i,t_{j-1}\right)}{\partial r} \right]\left(\dfrac{1}{dr}\right) \\[3mm] + \dfrac{T}{S}\left[\dfrac{h\left(r_i,t_{j-1}\right)}{\lambda^2} \right] \end{array} \right\} \\[14mm] +t_{j-2}^{\beta-1} \left\{ \begin{array}{l} \dfrac{T}{S}\left[\dfrac{\partial r^2 h\left(r_i,t_{j-2}\right)}{\partial r^2}\left(\dfrac{r_i^{2-2\alpha}}{\alpha^2} \right) + \dfrac{(1-\alpha)}{\alpha^2}\dfrac{\partial h\left(r_i,t_{j-2}\right)}{\partial r} r_i^{1-2\alpha} \right]\left(1+\dfrac{dr}{r_i}\right) \\[3mm] + \dfrac{T}{S}\left[\dfrac{\partial h\left(r_i,t_{j-2}\right)}{\partial r}\dfrac{r^{1-\alpha}}{\alpha} \right]\left(\dfrac{1}{dr}-\dfrac{1}{r_i}\right) - \dfrac{T}{S}\left[\dfrac{\partial h\left(r_i,t_{j-2}\right)}{\partial r} \right]\left(\dfrac{1}{dr}\right) \\[3mm] + \dfrac{T}{S}\left[\dfrac{h\left(r_i,t_{j-2}\right)}{\lambda^2} \right] \end{array} \right\} \end{array} \right]
$$

$$
\times \left[\begin{array}{l} (n-j+1)^{\alpha} \left[2(n-j)^2 + (3\alpha+10)(n-j) + \left(2\alpha^2 + 9\alpha + 12\right) \right] \\ -(n-j)^{\alpha} \left[2(n-j)^2 + (5\alpha+10)(n-j) + \left(6\alpha^2 + 18\alpha + 12\right) \right] \end{array} \right]
$$

$$\tag{11.38}$$

11.6   SIMULATIONS AND INTERPRETATION

In this section, we present numerical simulation for different values of fractional orders. The solutions are depicted in Figures 11.1 to 11.12.

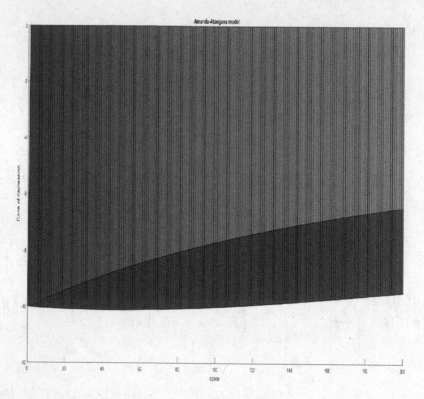

FIGURE 11.1   Cone of depression at order 0.03 in space.

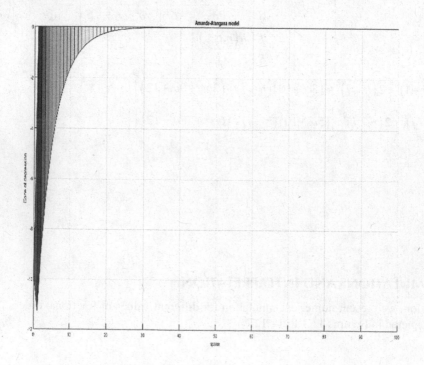

FIGURE 11.2   Cone of depression at order 1 in space.

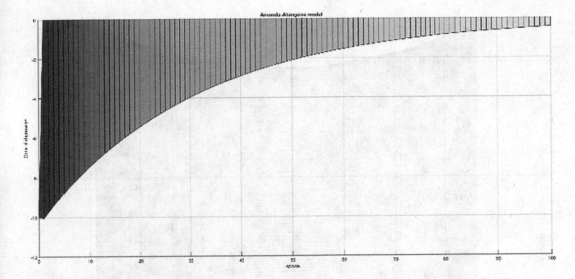

FIGURE 11.3 Cone of depression at order 0.9 in space.

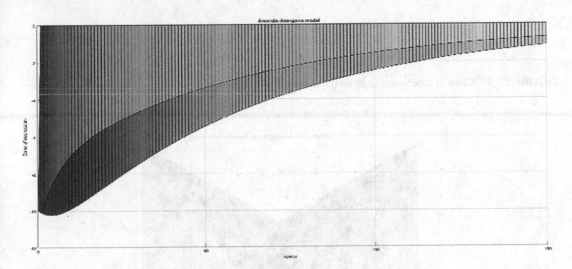

FIGURE 11.4 Cone of depression at order 0.5 in space.

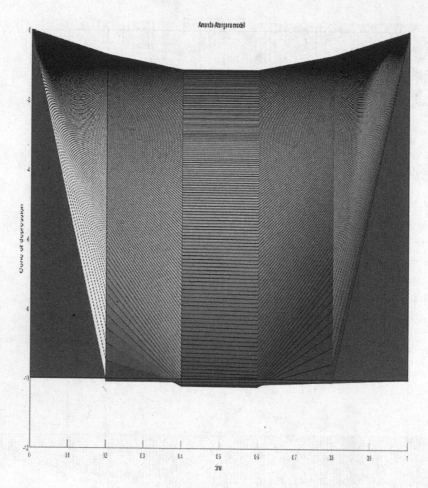

FIGURE 11.5   Cone of depression at order 0.5 in time.

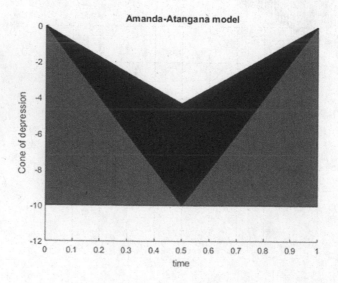

FIGURE 11.6   Cone of depression for order 0.04 in time.

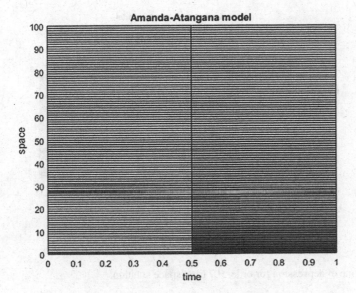

FIGURE 11.7   Amanda-Atangana model for order 0.7 time space solution.

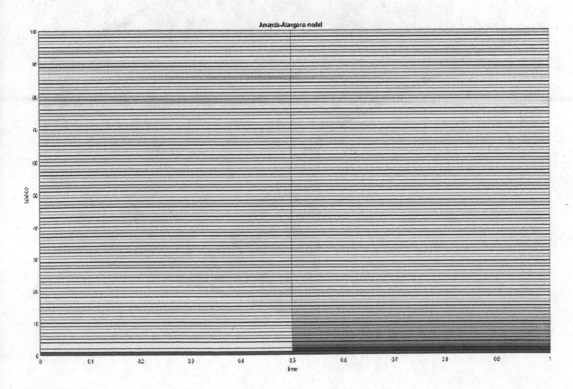

FIGURE 11.8   Amanda-Atangana model for order 0.9 time space solution.

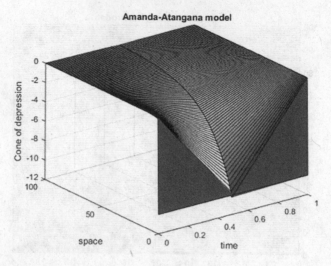

FIGURE 11.9   Cone of depression for order 0.7 time space solution.

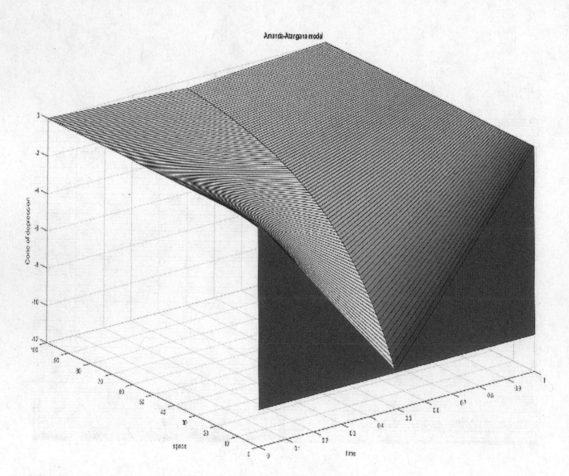

FIGURE 11.10   Cone of depression for order 1 time space solution.

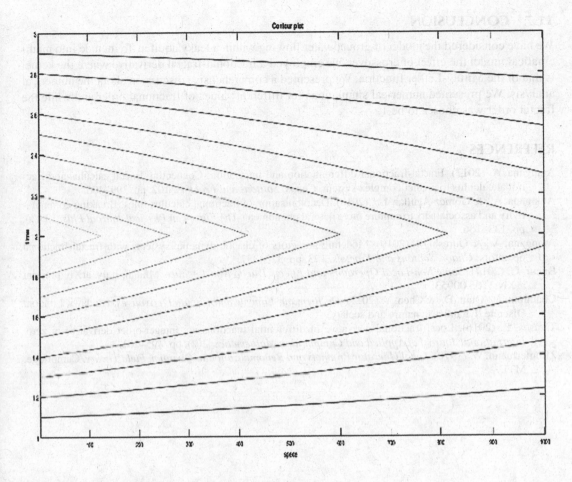

FIGURE 11.11   Contour plot at 0.04 time space solution.

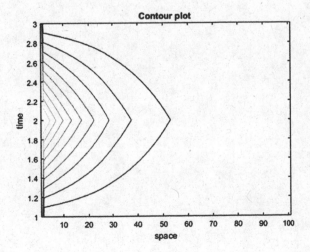

FIGURE 11.12   Contour plot at 0.7 time space solution.

11.7   CONCLUSION

We have considered the model of groundwater flowing within a leaky aquifer. To include into mathematical model the effect of crossover, we employed a fractional-fractal derivative where the kernel is that of the Mittag-Leffler function. We presented a comprehensive discussion about the numerical analysis. We presented numerical simulations for different values of fractional order and while the fractal order was chosen to be 1.

REFERENCES

Atangana, A. (2017). Fractal-fractional differentiation and integration: Connecting fractal calculus and fractional calculus to predict complex system. *Chaos, Solitons and Fractals.* 102, pp. 399–406.

Atangana, A., & Gomez-Aguilar, J.F. (2018). Decolonisation of fractional calculus rules: Breaking commutativity and associativity to capture more natural phenomena. *The European Physical Journal Plus.* 133(4), pp. 133–166.

Atangana, A., & Qureshi, S. (2019). Modeling attractors of chaotic dynamic systems with fractal-fractional operators. *Chaos, Solitons and Fractals.* 123, pp. 320–337.

Bucur, C. (2017). *Some Non-Local Operators and Effects Due to Non-Locality.* Milano, Italy. arXiv preprint arXiv:1705.00953.

Changpin, L., Qian, D., & Chen, Y. (2011). *On Riemann-Liouvile and Caputo Derivatives.* Hindawi, London: Discrete dynamics in nature and society.

Tarasov, E. (2016). Local fractional derivative of differential functions are integer-order derivative or zero. *International Journal of Applied and Computation Mathematics.* 2(2), pp. 195–201.

Zimmermann, W. (1970). *Local Operator Products and Renomalization in Quantum Field Theory.* Cambridge: MTI.

12 Analysis of General Groundwater Flow Equation within a Confined Aquifer Using Caputo Fractional Derivative and Caputo–Fabrizio Fractional Derivative

Mashudu Mathobo and Abdon Atangana
University of Free State, Bloemfontein, South Africa

CONTENTS

12.1 INTRODUCTION

Recently, Mathobo and Atangana (2018) derived the exact groundwater flow equation in a confined aquifer. In their research, they argued that the Theis groundwater flow model approximates the formulation of the model as he removed some components of the equation to simply the problem. In their equation, they included all high order terms that had been removed by Theis, and to consider the assumptions that were used during the derivation of the groundwater flow by Theis. Thereafter, it was proved that the new groundwater flow equation in confined aquifers has a unique solution. The derivation of the exact solution was derived and shown using the Boltzmann transform. However, their research has limitations. The exact solution is limited in terms of derivatives. It does not consider heterogeneity, random walk and memory effect etc. Therefore, it can only explain Markovian processes but not non-Markovian.

This research aims at using differentiation operators that are nonlocal, meaning that they can introduce the mathematical formulation of memory and random walk. A differential operator (which is generally discontinuous, unbounded and nonlinear on its domain) is an operator defined by some differential expression, and acting on space (usually vector valued) functions

DOI: 10.1201/9781003266266-12

(or sections of a differentiable vector bundle) on differentiable manifolds. Examples include the Atangana–Baleanu fractional derivative, Caputo fractional derivative and Caputo–Fabrizio fractional derivative.

12.2 ANALYSIS OF GENERAL GROUNDWATER FLOW WITH CAPUTO FRACTIONAL DERIVATIVE

In this section, we are going to present analysis of general groundwater flow equation using Caputo fractional derivative. Discretization was carried out using novel method called Newton polynomial. This method was developed by Atangana and Araz (2019). The method has attracted worldwide interest. A numerical scheme was developed, numerical models created and discussed under numerical simulations and the discussion section.

Using the Caputo fractional derivative:

$$
{}_{0}^{c}D_{t}^{\alpha} y(t) = f(t, y(t))
\tag{12.1}
$$

A numerical scheme is developed to solve the above equation.
The Caputo equation was transformed into:

$$
y(t_{n+1}) - y(0) = \frac{1}{\Gamma(\alpha)} \int_{0}^{t_{n+1}} f(\tau, y(\tau))(t_{n+1} - \tau)^{\alpha-1} d\tau
\tag{12.2}
$$

at point $t_{n+1} = (n+1)\Delta t$, we get:

$$
y(t_{n+1}) - y(0) = \frac{1}{\Gamma(\alpha)} \int_{0}^{t_{n+1}} f(\tau, y(\tau))(t_{n+1} - \tau)^{\alpha-1} d\tau
\tag{12.3}
$$

We also have:

$$
y(t_{n+1}) = y(0) + \frac{1}{\Gamma(\alpha)} \sum_{l=2}^{n} \int_{0}^{t_{n+1}} f(\tau, y(\tau))(t_{n+1} - \tau)^{\alpha-1} d\tau
\tag{12.4}
$$

We replace the Newton polynomial into Equation (12.4) to get:

$$
y^{n+1} = y^{0} + \frac{1}{\Gamma(\alpha)} \sum_{l=2}^{n} \int_{t_{l}}^{t_{l+1}} \left\{ \begin{array}{l} f(t_{l-2}, y^{l-2}) \\[4pt] + \dfrac{f(t_{l-1}, y^{l-1}) - f(t_{l-2}, y^{l-2})}{\Delta t}(\tau - t_{l-2}) \\[8pt] \dfrac{f(t_{l}, y^{l}) - 2f(t_{l-1}, y^{l-1}) + f(t_{l-2}, y^{l-2})}{2(\Delta t)^{2}} \\[8pt] \times (\tau - t_{l-2})(\tau - t_{l-1}) \end{array} \right\} (t_{n+1} - \tau)^{\alpha-1} d\tau
\tag{12.5}
$$

We then get:

$$y^{n+1} = y^0 + \frac{1}{\Gamma(\alpha)} \sum_{l=2}^{n} \left\{ \begin{array}{l} \displaystyle\int_{t_l}^{t_{l+1}} f\left(t_{l-2}, y^{l-2}\right)\left(t_{n+1} - \tau\right)^{\alpha-1} \\[2mm] + \displaystyle\int_{t_l}^{t_{l+1}} \frac{f\left(t_{l-1}, y^{l-1}\right) - f\left(t_{l-2}, y^{l-2}\right)}{\Delta t}\left(\tau - t_{l-2}\right) \\[2mm] + \displaystyle\int_{l}^{t_{l+1}} \frac{f\left(t_l, y^l\right) - 2f\left(t_{l-1}, y^{l-1}\right) + f\left(t_{l-2}, y^{l-2}\right)}{2\left(\Delta t\right)^2} \\[2mm] \times \displaystyle\int_{t_l}^{t_{l+1}} \left(\tau - t_{l-2}\right)\left(\tau - t_{l-1}\right)\left(t_{n+1} - \tau\right)^{\alpha-1} d\tau \end{array} \right\} \tag{12.6}$$

For the above integrals, we can get:

$$\int_{l}^{t_{l+1}} \left(t_{n+1} - \tau\right)^{\alpha-1} d\tau = \frac{\left(\Delta t\right)^\alpha}{\alpha}\left[\left(n-l+1\right)^\alpha - \left(n-l\right)^\alpha\right] \tag{12.7}$$

$$\int_{t_l}^{t_{l+1}} \left(\tau - t_{l-2}\right)\left(t_{n+1} - \tau\right)^{\alpha-1} d\tau = \frac{\left(\Delta t\right)^{\alpha+1}}{\alpha\left(\alpha+1\right)}\left[\left(n-l+1\right)^\alpha\left(n-l+3+2\alpha\right) - \left(n-l\right)^\alpha\left(n-l+3+3\alpha\right)\right] \tag{12.8}$$

$$\int_{t_l}^{t_{l+1}} \left(\tau - t_{l-2}\right)\left(\tau - t_{l-1}\right)\left(t_{n+1} - \tau\right)^{\alpha-1} d\tau = \frac{\left(\Delta t\right)^{\alpha+2}}{\alpha\left(\alpha+1\right)\left(\alpha+2\right)} \left[\begin{array}{l} \left(n-l+1\right)^\alpha\left[2\left(n-l\right)^2 + \left(3\alpha+10\right)\left(n-l\right)\right] \\[2mm] -\left(n-l\right)^\alpha \begin{bmatrix} 2\left(n-l\right)^2 \\ +\left(5\alpha+10\right)\left(n-l\right) \\ +6\alpha^2 + 18_\alpha + 12 \end{bmatrix} \end{array} \right] \tag{12.9}$$

By putting equality into above, the following scheme was obtained,

$$y^{n+1} = y^0 + \frac{\left(\Delta t\right)^\alpha}{\alpha\left(\alpha+1\right)} \sum_{l=2}^{n} f\left(t_{l-2}, y^{l-2}\right)\left(n-l+1\right)^\alpha - \left(n-l\right)^\alpha$$

$$+ \frac{\left(\Delta t\right)^\alpha}{\alpha\left(\alpha+2\right)} \sum_{l=2}^{n} f\left[\left(t_{l-1}, y^{l-1}\right) - f\left(t_{l-2}, y^{l-2}\right)\right]$$

$$\times \left[\left(n-l+1\right)^\alpha\left(n-l+3+2\alpha\right) - \left(n-l+1\right)^\alpha\left(n-l+3+3\alpha\right)\right]$$

$$+ \frac{\left(\Delta t\right)^2}{2\Gamma\left(\alpha+3\right)} \sum_{l-2}^{n} f\left(t_l, y^l\right) - 2f\left(t_{l-1}, y^{l-1}\right)$$

$$+ f\left(t_{l-2}, y^{l-2}\right) \left[\begin{array}{l} \left(n-l+1\right)^\alpha\left[2\left(n-l\right)^2 + \left(3\alpha+10\right)\left(n-l\right)\right] \\[2mm] -\left(n-l\right)^\alpha\left[2\left(n-l\right)^2 + \left(5\alpha+10\right)\left(n-l\right) + 6\alpha^2 + 18\alpha + 12\right] \end{array} \right]. \tag{12.10}$$

12.3 ANALYSIS OF GENERAL GROUNDWATER FLOW EQUATION WITH CAPUTO–FABRIZIO FRACTIONAL DERIVATIVE

The section aims to show the analysis of the general groundwater flow equation using Caputo–Fabrizio fractional derivative. We shall also discuss the properties of the Caputo–Fabrizio derivative and applications of the derivative. A graph illustrating this type of derivative is also shown. Stability analysis of the solution was presented using the Von Neumann method.

12.3.1 PROPERTIES AND APPLICATIONS OF CAPUTO–FABRIZIO FRACTIONAL DERIVATIVE

The Caputo–Fabrizio fractional derivative has been applied in many branches of science. This derivative has been used in real-life problems as it is much more suitable than the classical derivative. It is noted that this new derivative has advantageous properties such as that its able to show material heterogeneities and configurations with different scales (Tateishi et al., 2017).

This is something that cannot be achieved with previously used popular derivatives. The following is a case where the new Caputo–Fabrizio derivative was applied (Tateishi et al., 2017). Some numerical simulations were perfumed to display the efficiency of the method. Goufo (2016) applied the newly developed Caputo–Fabrizio fractional derivative to explore the possibility of extending the analysis of Korteweg-de-Vries–Burgers equation with two perturbation levels.

By the use of Caputo–Fabrizio derivative, Goufo (2016) wanted to gain an understanding of the unusual irregularities observed in wave motions and movements of liquids. Previously, the Korteweg-de-Vries–Burgers equation was applied to describe and analyze physical contexts related to liquids and wave dynamics, including the investigation of the propagation of waves in an elastic tube with viscous fluid, etc. Alqahtani and Atangana (2016) examined the Vallis model for El Nino using fractional derivatives, namely the Caputo derivative and Caputo–Fabrizio fractional derivative. Using fractional derivatives, they demonstrated the existence and uniqueness of the solution for the model. These were compared with solutions obtained by local derivatives. Numerical simulations of both local derivatives and fractional derivatives were created and compared. Based on the simulations, Alqahtani and Atangana (2016) observed that the Caputo–Fabrizio model shows and gives more information when compared to the local derivative model. Chaotic behavior was also observed on the Caputo–Fabrizio model.

Tateishi et al. 2017 carried out a research on the role of fractional time-derivative operators on anomalous diffusion and their results showed that fractional operators may be a simple and efficient way for incorporating different structural aspects into the system. While investigating general solutions and processes related to the diffusion equation when considering various choices for the kernel $\mathcal{K}(t)$, they observed that with Caputo–Fabrizio operator, its connection with continuous time random walk is complex, therefore not compatible with its standard interpretation. The diffusion equation associated with this operator is connected to a diffusive process with stochastic resetting, where the waiting time distribution is exponential (Tateishi et al., 2017; Shkilev, 2017; Méndez and Campos, 2016).

Based on the equation,

$$\frac{\partial}{\partial t} p(x,t) = {}_0 D_t^{1-\alpha} \left(\{ p(x,t) \} \right) \qquad (12.11)$$

Tateishi et al. (2017) derived solutions of the above equation using fractional operators.

The solutions were acquired by applying inverse of Fourier and Laplace transforms. For Caputo–Fabrizio operator, they observed that the distribution is the same as a Gaussian for small times, while showing a tent shape for longer times. For $t \to \infty$, a stationery behavior was observed, and this was interpreted to correspond with confined diffusion.

This type of behavior is also apparent for the mean square displacement which showed linear behavior in time for small times and saturates for long times. (Evans and Majumdar, 2011) noted that Caputo–Fabrizio fractional operator may be related to a diffusion with stochastic resetting. This was also proved on the work done by Tateishi, Ribeiro and Lenzi in 2015 and Hristov (2016). Tateishi et al. (2017) established an acceptable continuous time random walk formulation by considering a density of particle $\mathcal{J}(x,t)$ whose dynamics are governed by,

$$
\begin{aligned}
\mathcal{J}(x,t) = \delta(t)\delta(x) + r\delta(x)\int_0^t dt'\omega(t')\mathcal{J}(x,t-t') \\
+ (1-r)\int_0^t dt' \, \Psi(x',t')\mathcal{J}(x-x',t-t')
\end{aligned}
\tag{12.12}
$$

where r is a resetting rate, $\Psi(x,t)$ is joint distribution of jump length and waiting time, $\lambda x = \int_0^\infty \Psi(x,t)dt$ is the jump length distribution, and $\omega(t) = \int_{-\infty}^\infty \Psi(x,t)dx$ is waiting time distribution. They also concluded that the diffusion equation with Caputo–Fabrizio operator leads to the same waiting time distribution of the usual diffusion, which is an exponential.

12.3.2 ANALYSIS OF GENERAL GROUNDWATER FLOW WITH CAPUTO–FABRIZIO FRACTIONAL DERIVATIVE

In this section, we are going to present an analysis of general groundwater flow equation for confined aquifer using the Caputo–Fabrizio fractional derivative.

Discretization will be done using a novel method called the Newton polynomial. Stability analysis was also conducted on the discretized formula.

$$
\frac{S}{T}\cdot\frac{\partial h}{\partial t} = \frac{1}{r}\frac{\partial h}{\partial r} + \frac{\partial h^2}{\partial r^2}\left[1 + \frac{\Delta r}{r}\right]
\tag{12.13}
$$

$$
{}_0^{cf}D_*^\alpha f(t) = \frac{1}{1-\alpha}\int_0^t \exp\left(-\frac{\alpha}{1-\alpha}(t-s)\right)f'(s)ds, t \geq 0
\tag{12.14}
$$

When alpha is 1, we consider a general Cauchy problem as in Equation (12.15); we then discretize the equation using a numerical scheme based on the Newton polynomial as follows;

$$
\frac{dy(t)}{dt} = f(t, y(t))
\tag{12.15}
$$

$$
y(t) - y(0) = \int_0^t f(t, y(\tau))d\tau
\tag{12.16}
$$

at $t = t_{n+1}$

$$
y(t_{n+1}) - y(0) = \int_{t_n}^{t_{n+1}} f(\tau, y(\tau))d\tau
\tag{12.17}
$$

at $t = t_n$

$$y(t_{n+1}) - y(0) = \int_{t_n}^{t_{n+1}} f(\tau, y(\tau)) d\tau \qquad (12.18)$$

$$y(t_{n+1}) - y(t_n) = \int_{t_n}^{t_{n+1}} f(\tau, y(\tau)) d\tau \qquad (12.19)$$

$$P_n(\tau) = F(t_{n-2}, y(t_{n-2})) + \frac{F(t_{n-1}, y(t_{n-1})) - F(t_{n-2}, y_{n-2})(\tau - t_{n-2})}{\Delta t}$$
$$+ \frac{F(t_n, y(t_n)) - 2F(t_{n-1}, y(t_{n-1})) + F(t_{n-2}, y(t_{n-2}))(\tau - t_{n-1})(\tau - t_{n-2})}{2(\Delta t)^2} \qquad (12.20)$$

$$y(t_{n+1}) - y(t_n) = \int_{t_n}^{t_{n+1}} P_n(\tau) d\tau \qquad (12.21)$$

$$y(t_{n+1}) - y(t_n) = \frac{5}{12} F(t_{n-2}, y(t_{n-2})) \Delta t - \frac{4}{3} \Delta t \left[F(t_{n-1}, y(t_{n-1})) \Delta t + \frac{23}{12} F(t_n, y(t_n)) \right] \qquad (12.22)$$

thus

$$y(t_{n+1}) - y(t_n) = \frac{1-\alpha}{M(\alpha)} \left[F(t_n, y(t_n)) - F(t_{n-1}, y(t_{n-1})) \right]$$
$$+ \frac{\alpha}{M(\alpha)} \left[\frac{5}{12} F(t_{n-2}, y(t_{n-2})) \Delta t - \frac{4}{3} \Delta t \left[F(t_{n-1}, y(t_{n-1})) \Delta t + \frac{23}{12} F(t_n, y(t_n)) \right] \right] \qquad (12.23)$$

If we want to apply it in our equation, then;

$$_0^{cf} D_t^\alpha h(r, t) = \frac{T}{S} \left[\frac{1}{r} \frac{\partial h(r, t)}{\partial r} + \frac{\partial^2 h(r, t)}{\partial r^2} \left[1 + \frac{\Delta r}{r} \right] \right] \qquad (12.24)$$

$$_0^{CF} D_t^\alpha h(r, t) = F(t, r, h(r, t)) \qquad (12.25)$$

Applying the Caputo–Fabrizio integral, we have

$$h(r, t) - h(r, 0) = \frac{1-\alpha}{M(\alpha)} F(t, r, h(r, t)) + \frac{\alpha}{M(\alpha)} \int_0^t F(r, \tau, h(r, \tau)) d\tau \qquad (12.26)$$

at $t = t_{n+1}$ and $t = t_n$, following the derivation presented earlier, we have

$$h(r, t_{n+1}) - h(r, t_n) = \frac{1-\alpha}{M(\alpha)} \left[F(r, t_n, h(r, t_n)) - F(r, t_{n-1}, h(r, t_{n-1})) \right]$$
$$+ \frac{\alpha}{M(\alpha)} \left[\begin{array}{l} \frac{5}{12} \Delta t\, F(r, t_{n-2}, h(r, t_{n-2})) - \frac{4}{3} F(r, t_{n-1}, h(r, t_{n-1})) \\ + \frac{23}{12} F(r, t_n, h(r, t_n)) \end{array} \right] \qquad (12.27)$$

Now at $r = r_i$, we have

$$h(r_i, t_{n+1}) - h(r_i, t_n) = \frac{1-\alpha}{M(\alpha)} \Big[F(r_i, t_n, h(r_i, t_n)) - F(r_i, t_{n-1}, h(r_i, t_{n-1})) \Big]$$

$$+ \frac{\alpha}{M(\alpha)} \begin{bmatrix} \dfrac{5}{12} \Delta t \, F(r_i, t_{n-2}, h(r_i, t_{n-2})) \\[2mm] - \dfrac{4}{3} F(r_i, t_{n-1}, h(r_i, t_{n-1})) + \dfrac{23}{12} F(r_i, t_n, h(r_i, t_n)) \end{bmatrix} \qquad (12.28)$$

$$F(r_i, t_n, h(r_i, t_n)) = \frac{T}{S} \left[\frac{1}{r_i} \frac{h(r_{i+1}, t_n) - h(r_{i-1}, t_n)}{2\Delta r} \right.$$

$$+ \left[1 + \frac{\Delta r}{r_i} \right] \frac{h(r_{i+1}, t_n) - 2h(r_i, t_n) + h(r_{i-1}, t_n)}{\Delta r^2} \qquad (12.29)$$

$$F(r_i, t_{n-2}, h(r_i, t_{n-2})) = \frac{T}{S} \left[\frac{1}{r_i} \frac{h(r_{i+1}, t_{n-2}) - h(r_{i-1}, t_{n-2})}{2\Delta r} \right.$$

$$+ \left[1 + \frac{\Delta r}{r_i} \right] \frac{h(r_{i+1}, t_{n-2}) - 2h(r_i, t_{n-2}) + h(r_{i-1}, t_{n-2})}{\Delta r^2} \qquad (12.30)$$

We now replace into our equation:

$$h(r_i, t_{n+1}) - h(r_i, t_n) = \frac{1-\alpha}{M(\alpha)} \frac{T}{S} \left[\frac{1}{r_i} \frac{h(r_{i+1}, t_n) - h(r_{i-1}, t_n)}{2\Delta r} \right.$$

$$+ \left[1 + \frac{\Delta r}{r_i} \right] \frac{h(r_{i+1}, t_n) - 2h(r_i, t_n) + h(r_{i-1}, t_n)}{\Delta r^2}$$

$$- \left(\frac{T}{S} \left[\frac{1}{r_i} \frac{h(r_{i+1}, t_{n-1}) - h(r_{i-1}, t_{n-1})}{2\Delta r} \right. \right.$$

$$+ \left[1 + \frac{\Delta r}{r_i} \right] \frac{h(r_{i+1}, t_{n-1}) - 2h(r_i, t_{n-1}) + h(r_{i-1}, t_{n-1})}{\Delta r^2} \right)$$

$$+ \frac{\alpha}{M(\alpha)} \left[\frac{5}{12} \Delta t \frac{T}{S} \left(\frac{1}{r_i} \frac{h(r_{i+1}, t_{n-2}) - h(r_{i-1}, t_{n-2})}{2\Delta r} \right) \right]$$

$$+ \left[1 + \frac{\Delta r}{r_i} \right] \frac{h(r_{i+1}, t_{n-2}) - 2h(r_i, t_{n-2}) + h(r_{i-1}, t_{n-2})}{\Delta r^2} \qquad (12.31)$$

$$- \frac{4}{3} \left(\frac{T}{S} \left[\frac{1}{r_i} \frac{h(r_{i+1}, t_{n-1}) - h(r_{i-1}, t_{n-1})}{2\Delta r} \right] \right)$$

$$+ \left[1 + \frac{\Delta r}{r_i} \right] \frac{h(r_{i+1}, t_{n-1}) - 2h(r_i, t_{n-1}) + h(r_{i-1}, t_{n-1})}{\Delta r^2}$$

$$+ \frac{23}{12} \frac{T}{S} \left[\frac{1}{r_i} \frac{h(r_{i+1}, t_n) - h(r_{i-1}, t_n)}{2\Delta r} \right.$$

$$+ \left[1 + \frac{\Delta r}{r_i} \right] \frac{h(r_{i+1}, t_n) - 2h(r_i, t_n) + h(r_{i-1}, t_n)}{\Delta r^2}$$

by factorization, we now put like terms together

$$h(r_i, t_{n+1}) = h(r_i, t_n)\left(1 - \frac{2h}{\Delta r^2}\left(\frac{\Delta r}{r_i}\right)\frac{1-\alpha}{M(\alpha)}\frac{T}{S} - \frac{2h}{\Delta r^2}\frac{\Delta r}{r_i}\right)$$

$$+ h(r_i + 1, t_n)\left(\frac{1}{2\Delta r} + \frac{1}{\Delta r^2}\left(\frac{\Delta r}{r_i}\right)\frac{1-\alpha}{M(\alpha)}\frac{T}{S} + \frac{1}{2\Delta r} + \frac{1}{\Delta r^2}\left(\frac{\Delta r}{r_i}\right)\frac{23}{12}\frac{T}{S}\right)$$

$$- h(r_{i-1}, t_n)\left(\frac{1}{2\Delta r} + \frac{1}{\Delta r^2}\left(\frac{\Delta r}{r_i}\right)\frac{1-\alpha}{M(\alpha)} - \frac{1}{2\Delta r} + \frac{1}{\Delta r^2}\left(\frac{\Delta r}{r_i}\right)\frac{23}{12}\frac{T}{S}\right)$$

$$- h(r_{i+1}, t_{n-1})\left(\frac{1}{2\Delta r} + \frac{1}{\Delta r^2}\frac{T}{S} + \frac{1}{\Delta r^2}\frac{\Delta r}{r_i}\frac{\alpha}{M(\alpha)} - \frac{1}{2\Delta r} + \frac{1}{\Delta r^2}\frac{\Delta r}{r_i}\frac{4}{3}\frac{T}{S}\right) \quad (12.32)$$

$$- h(r_{i-1}, t_{n-1})\left(\frac{1}{2\Delta r} + \frac{1}{\Delta r^2}\left(\frac{\Delta r}{r_i}\right)\frac{T}{S} - \frac{1}{2\Delta r} + \frac{1}{\Delta r^2}\left(\frac{\Delta r}{r_i}\frac{4}{3}\frac{T}{S}\right)\right)$$

$$- 2h(r_i, t_{n-1})\left(\frac{1}{\Delta r^2}\left(\frac{\Delta r}{r_i}\right)\frac{T}{S} - \frac{1}{\Delta r^2}\left(\frac{\Delta r}{r}\right)\frac{4}{3}\frac{T}{S}\right)$$

$$+ h(r_{i+1}, t_{n-2})\left(\frac{1}{2\Delta r}\left(\frac{\alpha}{M(\alpha)}\right)\right)$$

$$- h(r_{i-1}, t_{n-2})\left(\frac{1}{2\Delta r} + \frac{1}{\Delta r^2}\left(\frac{\Delta r}{r_i}\right)\frac{\alpha}{M(\alpha)}\right)$$

Simplification in terms of alpha yields:

$$\alpha_1 h(r_i, t_{n+1}) = \alpha_2 h(r_i, t_n) + \alpha_3 h(r_{i+1}, t_n) - \alpha_3 h(r_{i-1}, t_n) \quad (12.33)$$
$$- \alpha_4 h(r_{i+1}, t_{n-1}) - \alpha_5 h(r_{i-1}, t_{n-1}) - \alpha_5 2h(r_i, t_{n-1}) + \alpha_6 h(r_{i+1}, t_{n-2}) - \alpha_7 h(r_{i-1}, t_{n-2})$$

Hence,

$$h_i^n = \delta_n e^{ik_m x} \quad (12.34)$$

$$h_{i-1}^{n-2} = \delta_{n-2} e^{ik_m(x - \Delta r)} \quad (12.35)$$

$$h_i^{n-2} = \delta_{n-2} e^{ik_m x} \quad (12.36)$$

$$h_{i+1}^n = \delta_n e^{ik_m(x + \Delta r)} \quad (12.37)$$

$$h_{i-1}^n = \delta_n e^{-ik_m x} \quad (12.38)$$

By replacing the above in our solution, we get:

$$\alpha_1 \delta_{n+1} e^{ik_m x} = \alpha_2 \delta_n e^{ik_m x} + \alpha_3 \delta_n e^{ik_m(x + \Delta r)} - \alpha_3 \delta_n e^{-ik_m(x - \Delta r)} - \alpha_4 \delta_{n-1} e^{ik_m(x + \Delta r)}$$
$$- \alpha_5 \delta_{n-1} e^{-ik_m(x - \Delta r)} - \alpha_5 2\delta_{n-1} e^{ik_m x} + \alpha_6 \delta_{n-2} e^{ik_m(x + \Delta r)} - \alpha_7 \delta_{n-2} e^{-ik_m(x - \Delta r)} \quad (12.39)$$

Simplification yields:

$$\alpha_1 \delta_{n+1} = \alpha_2 \delta_n + \alpha_3 \delta_n e^{ik_m \Delta r} - \alpha_3 \delta_n e^{-ik_m \Delta r} - \alpha_4 \delta_{n-1} e^{ik_m \Delta r}$$
$$- \alpha_5 \delta_{n-1} e^{-ik_m \Delta r} - \alpha_5 2\delta_{n-1} + \alpha_6 \delta_{n-2} e^{ik_m \Delta r} - \alpha_7 \delta_{n-2} e^{-ik_m \Delta r}$$
$$\quad (12.40)$$

Putting like terms together and factorizing, we get:

$$\delta_{n+1}(\alpha_1) = \delta_n\left(\alpha_2 + \alpha_3 e^{ik_m\Delta r} - \alpha_3 e^{-ik_m\Delta r}\right) - \delta_{n-1}\left(\alpha_4 e^{ik_m\Delta r} - \alpha_5 e^{-ik_m\Delta r} - 2\alpha_5\right) \qquad (12.41)$$
$$+ \delta_{n-2}\left(\alpha_6 e^{ik_m\Delta r} - \alpha_7 e^{-ik_m\Delta r}\right)$$

Therefore,

$$\alpha_1\delta_{n+1} = \delta_n\left(\alpha_2 + \alpha_3\left(2i\sin k_m\Delta r\right)\right)$$
$$- \delta_{n-1}\left(\alpha_4\left(\cos k_m\Delta r + i\sin k_m\Delta r\right) - \alpha_5\left(\cos k_m\Delta r + i\sin k_m\Delta r\right) - 2\alpha_5\right)$$
$$+ \delta_{n-2}\left[\alpha_6\left(\cos k_m\Delta r + i\sin k_m\Delta r\right) - \alpha_7\left(\cos\left(k_m\Delta r\right) \quad i\sin\left(k_m\Delta r\right)\right)\right] \qquad (12.42)$$

Since,

$$e^{ik_m\Delta r} = \cos\left(k_m\Delta r\right) + \sin\left(k_m\Delta r\right)e^{-ik_m\Delta r}$$
$$= \cos\left(k_m\Delta r\right) - i\sin\left(k_m\Delta r\right)\alpha_1\delta_{n+1}$$
$$= \delta_n\left(\alpha_2 + 2i\alpha_3\sin\left(k_m\Delta r\right)\right)$$
$$- \delta_{n-1}\left[\left(\alpha_4 - \alpha_5\right)\cos k_m\Delta r - 2\alpha_5 + i\left(\alpha_4 - \alpha_5\right)\sin\left(k_m\Delta r\right)\right]$$
$$+ \delta_{n-2}\left[\left(\alpha_6 - \alpha_7\right)\cos\left(k_m\Delta r\right) + i\left(\alpha_6 - \alpha_7\right)\sin\left(k_m\Delta r\right)\right] \qquad (12.43)$$

When $n = 0$, we have:

$$\alpha_1\delta_1 = \delta_0\left(\alpha_2 + 2i\alpha_3\right)\sin\left(k_m\Delta r\right) \qquad (12.44)$$

Hence,

$$\left|\frac{\delta_1}{\delta_0}\right| = \left|\frac{\alpha_2 + 2i\alpha_3\sin\left(k_m\Delta r\right)}{\alpha_1}\right| \qquad (12.45)$$

$$= \frac{\sqrt{\alpha_2^2 + \left(2\alpha_3\sin k_m\Delta r\right)^2}}{|\alpha_1|} \qquad (12.46)$$

$$\left|\frac{\delta_1}{\delta_0}\right| < 1 \Rightarrow \sqrt{\alpha^2 + \left(2\alpha_3\sin k_m\Delta r\right)^2} < |\alpha_1| \qquad (12.47)$$

We assume that $\forall_n \geq 1$

If $\left|\dfrac{\delta_n}{\delta_0}\right| < 1$ then, we want to prove that $\left|\dfrac{\delta_{n+1}}{\delta_0}\right| < 1$

However,

$$\left|\delta_{n+1}\right| \leq \left|\delta_n\right|\left|\alpha_2 + 2i\alpha_3\sin\left(k_m\Delta r\right)\right|$$
$$+ \left|\delta_{n-1}\right|\left|\left(\alpha_4 - \alpha_5\right)\cos k_m\Delta r - 2\alpha_5 + i\left(\alpha_4 + \alpha_5\right)\sin\left(k_m\Delta r\right)\right|$$
$$+ \left|\delta_{n-2}\right|\left|\left(\alpha_6 - \alpha_7\right)\cos\left(k_m\Delta r\right) + i\left(\alpha_6 - \alpha_7\right)\sin\left(k_m\Delta r\right)\right| \qquad (12.48)$$

However, according to the inductive formula,

$$\left|\delta_n\right| < \left|\delta_0\right|, \left|\delta_{n-1}\right| < \left|\delta_0\right| \text{ and } \left|\delta_{n-2}\right| < \left|\delta_0\right|$$

Therefore,

$$|\delta_{n+1}| < |\delta_0| \left[\begin{array}{c} \sqrt{\alpha^2 + \left(2\alpha_3 \sin k_m \Delta r\right)^2} \\ + \sqrt{\left((\alpha_4 - \alpha_5)\cos(k_m \Delta r) - 2\alpha_5\right)^2 + (\alpha_4 + \alpha_5)^2 \sin^2(k_m \Delta r)} \\ + \sqrt{(\alpha_6 - \alpha_7)\cos(k_m \Delta r)^2 + \left((\alpha_6 - \alpha_7)\sin(k_m \Delta r)\right)^2} \end{array} \right] \tag{12.49}$$

$$\frac{|\delta_{n+1}|}{|\delta_0|} < 1 \tag{12.50}$$

The scheme is stable if

$$\text{Max} \left\{ \frac{\sqrt{\alpha^2 + \left(2\alpha_3 \sin k_m \Delta r\right)^2}}{|\alpha_1|} ; \left(\begin{array}{c} \sqrt{\alpha^2 + \left(2\alpha_3 \sin k_m \Delta r\right)^2} \\ + \sqrt{\left((\alpha_4 - \alpha_5)\cos(k_m \Delta r) - 2\alpha_5\right)^2 + (\alpha_4 + \alpha_5)^2 \sin^2(k_m \Delta r)} \\ + \sqrt{(\alpha_6 - \alpha_7)\cos(k_m \Delta r)^2 + \left((\alpha_6 - \alpha_7)\sin(k_m \Delta r)\right)^2} \end{array} \right) \right\} \tag{12.51}$$

The above completes our stability analysis.

12.4 NUMERICAL SIMULATIONS AND DISCUSSION

In this section, numerical simulations are depicted in the Figures below. The Figures suggest three different types of flow according to the fractional orders.

For fractional orders from 0.99 to 0.51, one can see the depiction of slow flow, which mean the geological formation, does not really connect, helping water to pass through easily. This situation can be observed in real-world situations, but sometime researchers change storativity or transmissivity to match the mathematical formula to the observed facts, so such results could be misleading, however, here the fractional order helps the mathematical formula to capture such physical behavior. When the fractional order is below 0.5, we can observe a fast flow; and finally when the fractional order is 1, we recover normal flow, where the geological formation is assumed to be homogenous. Below, figure with T = time, S = space and H = hydraulic head (Figures 12.1 to 12.13).

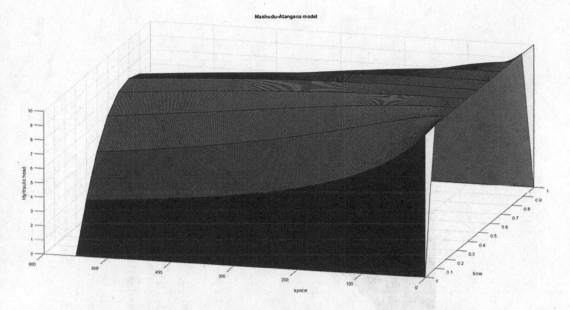

FIGURE 12.1   Numerical simulation of Caputo with theoretical value of 1 showing H, S and T.

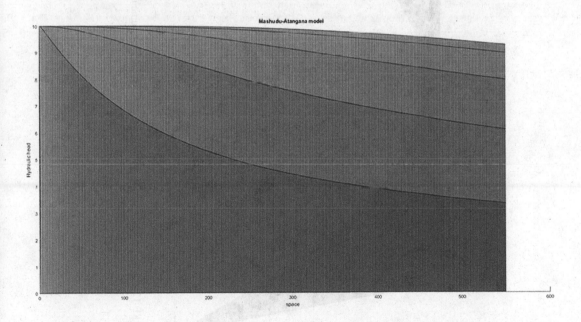

FIGURE 12.2   Numerical simulation of Caputo with theoretical value of 1 showing H and S.

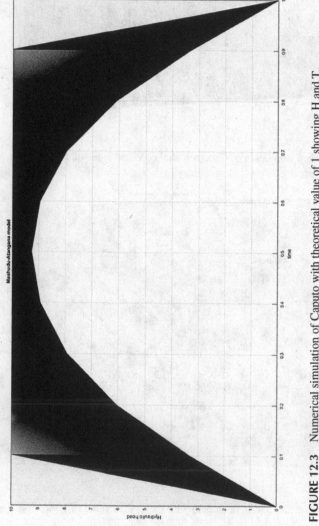

FIGURE 12.3   Numerical simulation of Caputo with theoretical value of 1 showing H and T.

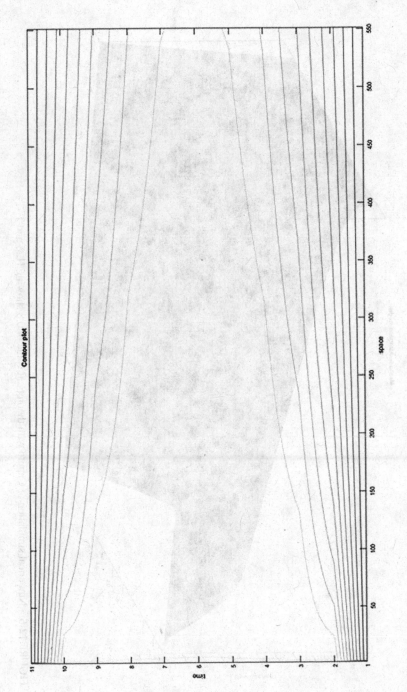

FIGURE 12.4 Contour slot of Caputo with theoretical value of 1.

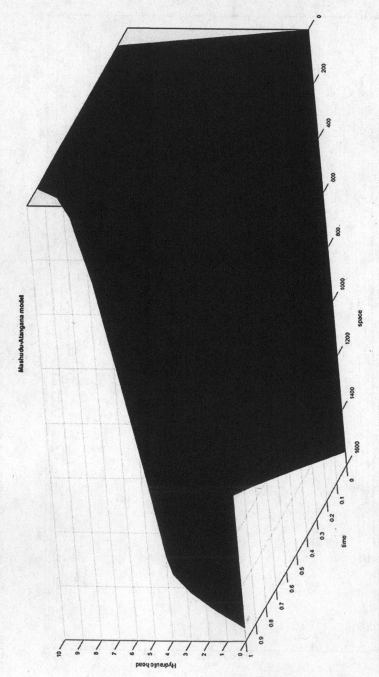

FIGURE 12.5　Numerical simulation of Caputo with theoretical value of 0.8 showing H, S and T.

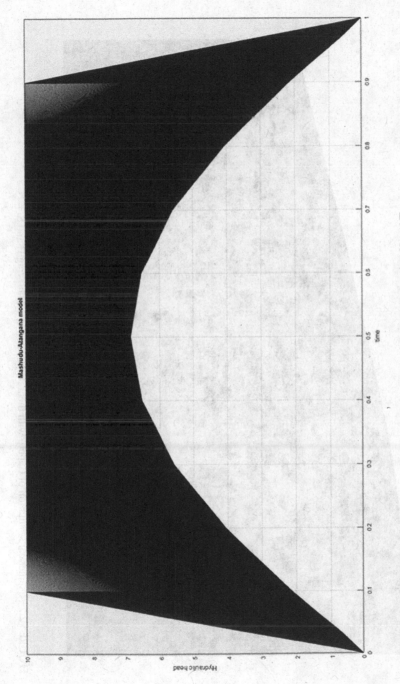

FIGURE 12.6 Numerical simulation of Caputo with theoretical value of 0.8 showing H and T.

FIGURE 12.7 Numerical simulation of Caputo with theoretical value of 0.8 showing H and S.

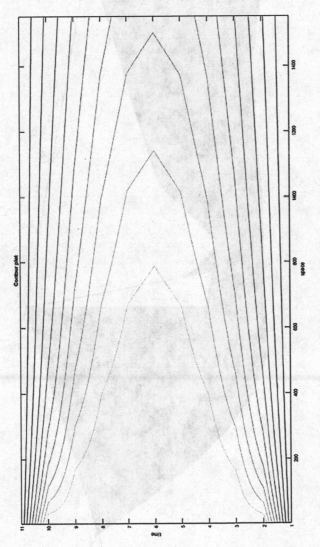

FIGURE 12.8 Contour slot of Caputo with theoretical value of 0.

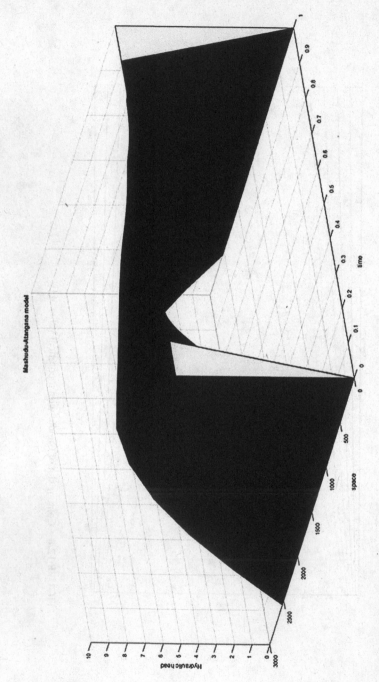

FIGURE 12.9 Numerical simulation of Caputo with theoretical value of 0.4 showing H, S and T.

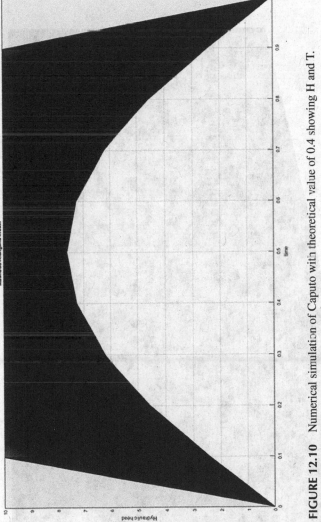

FIGURE 12.10 Numerical simulation of Caputo with theoretical value of 0.4 showing H and T.

FIGURE 12.11   Numerical simulation of Caputo with theoretical value of 0.4 showing H and S.

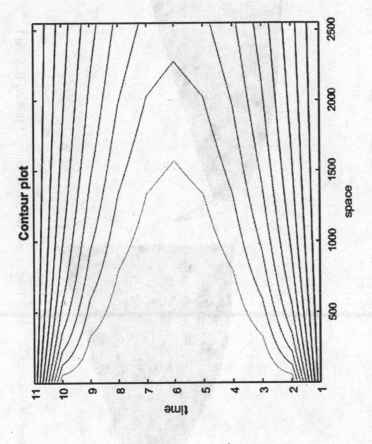

FIGURE 12.12   Contour slot of Caputo with theoretical value of 0.4.

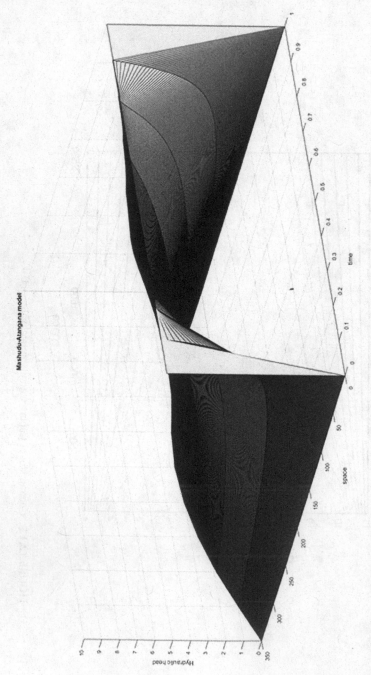

FIGURE 12.13   Numerical simulation of Caputo–Fabrizio with theoretical value of 1 showing H, S and T.

12.5 CONCLUSION

The main aim of the research was to analyze the general groundwater flow equation for a confined aquifer with nonlocal operators that include the Caputo fractional derivative, the Caputo–Fabrizio fractional derivative, Atangana–Baleanu fractional derivative and fractal-fractional differential operators. For the sole purpose of this chapter, only the Caputo fractional and Caputo–Fabrizio fractional derivative were presented. In addressing this aim, a newly developed general groundwater flow equation for a confined aquifer, developed by Mashudu and Atangana (2017) was adopted. The equation was then analyzed using the fractional derivatives mentioned above. New numerical schemes were developed using the classical Adams–Bashforth method and the newly developed Newton polynomial, developed by Atangana and Araz (2019). Numerical simulations were developed for the Caputo fractional derivative and Caputo–Fabrizio fractional derivative. Based on the simulations, it can be observed that for a fractional order from 0.99 to 0.51, there is slow flow, meaning the geological formation is not really connected and as such, does not help water to pass through easily.

When the fractional order is below 0.5, we observe a fast flow, and when the fractional order is 1, normal flow is observed and thus the geology is assumed to be homogenous. In conclusion, this research helped to capture complexities of the geological formation through which the sub-surface water flows.

REFERENCES

Atangana, A. & Araz, S.I. (2019). Fractional stochastic modelling illustration with modified Chua attractor. *European Physical Journal*, 134: 160.

Caputo, M. & Fabrizio, M. (2015). A new definition of fractional derivative without singular kernel. *Progress in Fractional Differentiation and Applications. An International Journal*, 1(2): 1–13.

Evans, M.R. & Majumdar, S.H. (2011). Diffusion with stochastic resetting. *Physical Review Letters*, 106(16): 160601.

Goufo, D.E.F. (2016). Application of the Caputo–Fabrizio fractional derivative without singular kernel to Korteweg-de Vries–Burgers equation. *Mathematical Modelling and Analysis*, 21(2): 188–198.

Hristov, J. (2016). Transient heat diffusion with a non-singular fading memory: From the Cattaneo constitutive equation with Jeffrey's Kernel to the Caputo–Fabrizio time-fractional derivative. *Thermal Science*, 20: 757–762. doi: 10.2298/TSCI160112019H.

Mathobo, M. & Atangana, A. (2018). Analysis of exact groundwater model within a confined aquifer: New proposed model beyond the Theis equation. *The European Journal Plus*, 133: 415.

Méndez, V. & Campos, D. (2016). Characterization of stationary states in random walks with stochastic resetting. *Physical Review E* 93: 022106. doi: 10.1103/PhysRevE.93.022106.

Shkilev, V.P. (2017). Continuous-time random walk under time-dependent resetting. *Physical Review E*, 96: 012126. doi: 10.1103/PhysRevE.96.012126.

Tateishi, A.A., Ribeiro, H.V., & Lenzi, E.K., (2017). The role of fractional Time-Derivative operators on Anomalous Diffusion. *Frontiers in Physics*, 5: 52. doi: 10.3389/fphy.2017.00052.

13 Analysis of General Groundwater Flow Equation with Fractal Derivative

Mashudu Mathobo and Abdon Atangana
University of Free State, Bloemfontein, South Africa

CONTENTS

13.1 INTRODUCTION

In applied mathematics and mathematical analysis, the fractal derivative is a nonstandard type of derivative in which the variable, such as t, has been scaled according to t^α. (Yang, 2012) argued that the fractal derivative can be categorized as a special local fractional derivative. This type of derivative can be applied to solve problems associated with a discontinuous media and equations with fractal derivatives can easily be solved. Fractal derivatives have also been employed as an alternative modeling approach to the classical Fick's second law, where it is used to derive the linear anomalous transport-diffusion equation underlying the diffusion process (Chen et al., 2010).

Recently Ramotsho and Atangana (2017) used fractal derivatives to derive the exact numerical solution of diffusion within a leaky aquifer. In their work, they used one of the fractals properties, self-similarity. Self-similarity occurs when a system replicate itself and shows iteration throughout. Ramotsho and Atangana (2017) derived the new groundwater flow equations within a self-similar leaky aquifer, with one showing a scenario of abstracting water from a self-similar leaky aquifer, and another showing a scenario of abstracting water from a leaky aquifer. Numerical solutions were derived using the newly developed Adams–Bashforth method by Atangana and Batogna. Groundwater models were also created. Their models took into account the geological formation of the system compared to models developed by classical formulas. Allwright and Atangana (2018) conducted research on groundwater transport in fractured aquifers with self-similarities. In their work, they highlighted that groundwater transport within a fractured aquifer with a fractal nature exhibiting self-similarity cannot be simulated accurately using the Fickian advection-dispersion transport equation. In order to close the gap in knowledge, Allwright and Atangana (2018) developed a fractal advection-dispersion groundwater transport equation and its integral with theorem and proof. Numerical simulations were developed and it was shown in their studies that by incorporating a fractal dimension, anomalous diffusion may be modeled in an effective and efficient way.

This research aims to analyze the general groundwater flow equation with fractal derivatives. We aim to introduce a differential operator that will be able to account for groundwater flow in fractures as well as the heterogeneity of the system. This differential operator is known as the fractal operator, which will be integrated into the general groundwater flow equation.

DOI: 10.1201/9781003266266-13

13.2 PROPERTIES OF FRACTALS

Initially, fractals were only used in mathematics to describe self-similarity in structures. Through continuous research and new developments in mathematics, fractals are being used in various areas of science as they aid in describing structures in nature that conventional mathematical tools cannot solve. Fractals are an easy and convenient way of solving and creating things that may otherwise be difficult or impossible.

Fractals are generally self-similar and independent of scale. Fractals have the following characteristics:

- Self-similarity;
- Their form is extremely irregular or fragmented, and remains so whatever the scale or examination;
- They contain distinct elements whose scales are very varied and cover a large range;
- Formation by iteration; and
- A fractional dimension.

13.3 ANALYSIS OF GENERAL GROUNDWATER FLOW WITH FRACTAL DERIVATIVE

This section aims at analyzing the general groundwater flow equation using a fractal derivative. We are going to derive and propose a new derivative operator that is capable of describing groundwater flow in fractures, and takes into account heterogeneity of the system. This could not be accounted for by the general groundwater flow equation. We are going to use the Forward, Backward and Crank–Nicolson scheme to develop a numerical approximation of the operator and show the stability analysis.

$$\frac{df(z)}{dz^{\alpha}} = \lim_{s \to x} \frac{f(z) - f(s)}{z^{\alpha} - s^{\alpha}} \tag{13.1}$$

$$\frac{d}{dt}(f) = \lim_{h \to 0} \frac{f(x+h) - f(x)}{h} \tag{13.2}$$

By expressing the fractal differentiable operator with Equation (13.2) we get,

$$\frac{d}{dt^{\beta}} f(x) = \lim_{x \to s} \frac{f(x) - f(s)}{x^{\alpha} - s^{\alpha}} = \lim_{x \to s} \frac{f(x) - f(s)}{x - s} \left[\frac{x - s}{x^{\alpha} - s^{\alpha}} \right] \tag{13.3}$$

Therefore,

$$\lim_{x \to s} \frac{f(x) - f(s)}{x - s} \cdot \left(\frac{\frac{1}{x^{\alpha} - s^{\alpha}}}{x - s} \right) \tag{13.4}$$

$$\lim_{x \to s} \frac{f(x) - f(s)}{x - s} = f'(s) \tag{13.5}$$

and,

$$\lim_{x \to s} \frac{f(x) - f(s)}{x - s} \cdot \left(\frac{\frac{1}{x^\alpha - s^\alpha}}{x - s} \right) = f'(s) \frac{1}{\alpha s^{\alpha - 1}} \tag{13.6}$$

$$\lim_{x \to s} \frac{f(x) - f(s)}{x - s} \cdot \left(\frac{\frac{1}{x^\alpha - s^\alpha}}{x - s} \right) = \frac{f'(s)}{\alpha} s^{1-\alpha} \tag{13.7}$$

In order to further extend the work done by Mathobo and Atangana (2018) we replace the general groundwater flow derivative with a fractal derivative. The initial phase of our analysis starts from the general groundwater flow equation,

$$\frac{S}{T} \cdot \frac{\partial h}{\partial t} = \frac{\partial h}{r \partial r} + \frac{\partial^2 h}{\partial r^2} \left[1 + \frac{\Delta r}{r} \right] \tag{13.8}$$

The above equation is local and does not account for groundwater flow within fractures in a geological environment, heterogeneity of the aquifer, random walk, diffusion and memory effect. We aim to apply a differential operator that will be able to account for groundwater flow in fractures as well as the heterogeneity of the system. This differential operator is known as the fractal differential operator, which will be integrated into the general groundwater flow equation.

$$\frac{S}{T} \cdot \frac{\partial h}{\partial t} = \frac{1}{r} \frac{\partial h}{\partial r^\alpha} + \frac{\partial^2 h}{\partial r^2} \left[1 + \frac{\Delta r}{r} \right] \tag{13.9}$$

Because of the physical problem under investigation, advection in groundwater flow exists, therefore the function may act as a differentiable, thus the general groundwater flow equation can be reformulated as,

$$\frac{S}{T} \cdot \frac{\partial h}{\partial t} = \frac{1}{r} \frac{\partial h}{\partial r} \cdot \frac{1}{\alpha r^{\alpha - 1}} + \frac{\partial^2 h}{\partial r^2} \left[1 + \frac{\Delta r}{r} \right] \tag{13.10}$$

By canceling out the like terms, we get;

$$\frac{S}{T} \frac{\partial h}{\partial t} = \frac{1}{\alpha r^{\alpha - 1}} \frac{\partial h}{\partial r} + \frac{\partial^2 h}{\partial r^2} \left[1 + \frac{\Delta r}{r} \right] \tag{13.11}$$

By discretization of the first and second order derivatives, we get;

$$\frac{d}{dt} f(t_n) = \frac{f^{n+1} - f(t_n)}{\Delta t} \tag{13.12}$$

For 2nd order,

$$\frac{\partial^2}{\partial x^2} f(x_i, t_n) = \frac{f_{i+1}^n - 2 f_i^n + f_{i-1}^n}{(\Delta x)^2} \tag{13.13}$$

$$\frac{\partial^2}{\partial x^2} f\left(x_i, t_{n+1}\right) = \frac{f_{i+1}^{n+1} - 2f_i^n + f_{i-1}^n}{\left(\Delta x\right)^2} \tag{13.14}$$

We then discretize the forward, backward and Crank–Nicolson scheme into our equation, which was reformulated as follows,

$$\frac{S}{T} \frac{\partial h}{\partial t} = \frac{1}{\alpha r^\alpha} \frac{\partial h}{\partial r} + \frac{\partial^2 h}{\partial r^2}\left[1 + \frac{\Delta r}{r}\right].$$

To represent our numerical approximation using forward Euler, we get,

$$\frac{\partial}{\partial x} f\left(x_i, t_n\right) = \frac{f_{i+1}^n - f_i^n}{\Delta x} \tag{13.15}$$

Backward Euler gives us,

$$\frac{\partial}{\partial x} f\left(x_i, t_{n+1}\right) = \frac{f_{i+1}^{n+1} - f_i^{n+1}}{\Delta x} \tag{13.16}$$

Substituting Equation (13.15) and (13.16) into Equation (13.11), the following is obtained,

$$\frac{S}{T} \frac{h_i^{n+1} - h_i^n}{\Delta t} = \frac{1}{\alpha r_i^\alpha} \frac{h_{i+1}^n - h_i^n}{\Delta r} + \frac{h_{i+1}^n - 2h_i^n + h_{i-1}^n}{\left(\Delta r\right)^2}\left[1 + \frac{r_{i+1} - r_i}{r_i}\right] \tag{13.17}$$

$$\frac{S}{T} \frac{h_i^{n+1} - h_i^n}{\Delta t} = \frac{1}{\alpha r_i^\alpha} \frac{h_{i+1}^{n+1} - h_i^{n+1}}{\Delta r} + \frac{h_{i+1}^{n+1} - 2h_i^{n+1} + h_{i-1}^{n+1}}{\left(\Delta r\right)^2}\left[1 + \frac{r_{i+1} - r_i}{r_i}\right] \tag{13.18}$$

$$\frac{S}{T} \frac{h_i^{n+1} - h_i^n}{\Delta t} = \frac{1}{\alpha r_i^\alpha} \frac{h_{i+1}^n - h_i^n}{\Delta r} + \left\{ \frac{h_{i+1}^{n+1} - 2h_i^{n+1} + h_{i-1}^{n+1}}{2\left(\Delta r\right)^2} + \frac{h_{i+1}^n - 2h_i^n + h_{i-1}^n}{\left(\Delta r\right)^2}\left[1 + \frac{r_{i+1} - r_i}{r_i}\right] \right\} \tag{13.19}$$

By putting like terms together, we get:

$$\frac{S}{T\Delta t} + \frac{1}{\left(r\right)^2}\left[\left(1 + \frac{r_{i+1} - r_i}{r_i}\right)\right]h_i^{n+1} = \left(\frac{S}{T\Delta t} - \frac{1}{\alpha r_i^\alpha \Delta r} - \left[1 + \frac{r_{i+1} - r_i}{r_i}\right]\right)h_i^n + \frac{1}{2\left(\Delta r\right)^2}\left[1 + \frac{r_{i+1} - r_i}{r_i}\right]h_{i+1}^{n+1}$$

$$+ h_{i+1}^n\left(\frac{1}{\alpha r^\alpha \Delta r} + \frac{1}{2\left(\Delta r\right)^2}\left[1 + \frac{r_{i+1} - r_i}{r_i}\right]\right) + h_{i-1}^{n+1}\left(\frac{1}{2\left(\Delta r\right)^2}\left[1 + \frac{r_{i+1} - r_i}{r_i}\right]\right)$$

$$+ h_{i-1}^n\left(\left[1 + \frac{r_{i+1} - r_i}{r_i}\right]\frac{1}{2\left(\Delta r\right)^2}\right)$$

$$\tag{13.20}$$

By simplifying the above in terms of alpha, we get:

$$a_1 h_i^{n+1} = a_2 h_i^n + a_3 h_{i+1}^{n+1} + a_4 h_{i+1}^n + a_3 h_{i-1}^{n+1} + a_3 h_{i-1}^n \tag{13.21}$$

Above is the solution derived by analyzing the general groundwater flow equation using a fractal derivative. We then replace the above with an epsilon variant to get the following,

$$a_1 e^{a(t+\Delta t)} e^{ik_m x} = a_2 e^{at} e^{ik_m x} + a_3 e^{a(t+\Delta t)} e^{ik_m (x+\Delta x)} + a_4 e^{at} e^{ik_m (x+\Delta x)} + a_3 e^{a(t+\Delta t)} e^{ik_m (x-\Delta x)} + a_3 e^{at} e^{ik_m (x-\Delta x)} \tag{13.22}$$

by simplification, we get:

$$a_1 e^{a\Delta t} = a_2 + a_3 e^{a\Delta t} e^{ik_m \Delta x} + a_4 e^{ik_m \Delta x} + a_3 e^{a\Delta t} e^{-ik_m \Delta x} + a_3 e^{-ik_m \Delta x} \tag{13.23}$$

by putting like terms together, we get:

$$e^{a\Delta t} \left(a_1 - a_3 e^{ik_m \Delta x} - a_3 e^{-ik_m \Delta x} \right) = a_2 + a_4 e^{ik_m \Delta x} + a_3 e^{-ik_m \Delta x} \tag{13.24}$$

Thus,

$$e^{a\Delta t} = \frac{a_2 + a_4 e^{ik_m \Delta x} + a_3 e^{-ik_m \Delta x}}{a_1 - a_3 e^{ik_m \Delta x} - a_3 e^{-k_m \Delta x}} \tag{13.25}$$

We now have to prove that,

$$\left| e^{a\Delta t} \right| < 1$$

Therefore,

$$e^{a\Delta t} = \frac{a_2 + a_4 e^{ik_m \Delta x} + a_3 e^{-ik_m \Delta x}}{a_1 - a_3 \left(e^{ik_m \Delta x} + e^{ik_m \Delta x} \right)} \tag{13.26}$$

$$e^{a\Delta t} = \frac{a_2 + a_4 \left(\cos\left(k_m \Delta x\right) + i\sin\left(k_m \Delta x\right) \right) + a_3 \left(\cos\left(k_m \Delta x\right) - i\sin\left(k_m \Delta x\right) \right)}{a_1 - 2a_3 \cos\left(k_m \Delta x\right)} \tag{13.27}$$

by simplification,

$$e^{a\Delta t} = \frac{a_2 + \left(a_4 + a_3\right)\cos\left(k_m \Delta x\right) + i\left(a_4 - a_3\right)\sin\left(k_m \Delta x\right)}{a_1 - 2a_3 \cos\left(k_m \Delta x\right)} \tag{13.28}$$

$$\left| e^{a\Delta t} \right| < 1 \Rightarrow \frac{\left| a_2 + \left(a_4 + a_3\right)\cos\left(k_m \Delta x\right) + i\left(a_4 - a_3\right)\sin\left(k_m \Delta x\right) \right|}{\left| a_1 - 2a_3 \cos\left(k_m \Delta x\right) \right|} < 1 \tag{13.29}$$

Based on the above, Equation (13.29) can be rewritten as:

$$\left|e^{a\Delta t}\right|^2 < 1 \Rightarrow \frac{\left|a_2+\left(a_4+a_3\right)\cos\left(k_m\Delta x\right)+i\left(a_4-a_3\right)\sin\left(k_m\Delta x\right)\right|^2}{\left|a_1-2a_3\cos\left(k_m\Delta x\right)\right|^2} < 1 \tag{13.30}$$

$$\Rightarrow \frac{\left(a_2+\left(a_4+a_3\right)\cos\left(k_m\Delta x\right)\right)^2+\left(\sin\left(k_m\Delta x\right)\left(a_4-a_3\right)\right)^2}{\left(a_1-2a_3\cos\left(k_m\Delta x\right)\right)^2} < 1 \tag{13.31}$$

For conditions of stability, the following must be satisfied:

$$\frac{\begin{array}{c} a_2^2+2a_2\left(a_4+a_3\right)\cos\left(k_m\Delta x\right)+\left(a_4+a_3\right)^2\cos^2\left(k_m\Delta x\right) \\ +\left(a_4-a_3\right)^2\sin^2\left(k_m\Delta x\right) \end{array}}{a_1^2-4a_1a_3\cos\left(k_m\Delta x\right)+4a_3^2\cos\left(k_m\Delta x\right)} < 1 \tag{13.32}$$

By simplification and putting like terms together, we get:

$$\begin{array}{c} a_1^2-a_1^2+\left[2a_2\left(a_4+a_3+4a_1a_3\right)\right]\cos\left(k_m\Delta x\right) \\ +\left[\left(a_4+a_3\right)^2-4a_3^2\right]\cos^2\left(k_m\Delta x\right)+\left(a_4-a_3\right)^2\sin^2\left(k_m\Delta x\right) \end{array} < 0 \tag{13.33}$$

The above is conditionally stable only when the above conditions are satisfied.

We are now going to apply the Atangana–Batogna numerical scheme to the general groundwater flow equation and then present the stability analysis using the iterative method. First, we shall present how the Atangana–Batogna method was developed. The method derives a two-step Adams–Bashforth numerical scheme in Laplace space and the solution is taken back into real space via Inverse Laplace transform form (Atangana & Batogna, 2018).

By considering the following partial differential equation,

$$\frac{\partial u\left(x,t\right)}{\partial t} = Lu\left(x,t\right)+Nu\left(x,t\right) \tag{13.34}$$

where L is a linear operator and N is a non-linear operator.

By applying a Laplace transform on both sides of Equation (13.34) we get,

$$\mathcal{L}\left(\frac{\partial u\left(x,t\right)}{\partial t}\right) = \mathcal{L}\left(Lu\left(x,t\right)+Nu\left(x,t\right)\right) \tag{13.35}$$

$$\frac{\partial u\left(p,t\right)}{\partial t} = \mathcal{L}\left(Lu\left(x,t\right)+Nu\left(x,t\right)\right) \tag{13.36}$$

$$v\left(t\right) = u\left(p,t\right) \tag{13.37}$$

$$\frac{d}{dt}\left(v\left(t\right)\right) = F\left(t,v\left(t\right)\right) \tag{13.38}$$

where $v(t) = u(p,t)$ and $F\left(t,v\left(t\right)\right) = \mathcal{L}\left(Lu\left(x,t\right)+Nu\left(x,t\right)\right)$.

By applying principles of calculus, we get;

$$v(t) - v(0) = \int_0^t F(v,\tau) d\tau \tag{13.39}$$

and

$$v(t) = v(0) + \int_0^t F(v,\tau) d\tau \tag{13.40}$$

When $t = t_{n+1}$

$$v(t_{n+1}) = v(0) + \int_0^{t_{n+1}} F(v,\tau) d\tau \tag{13.41}$$

$$v(0) = C \tag{13.42}$$

C is a constant
When $t = t_n$

$$v(t_n) = v(0) + \int_0^{t_n} F(v,\tau) d\tau \tag{13.43}$$

It follows that,

$$v(t_{n+1}) - v(t_n) = \int_0^{t_{n+1}} F(v,\tau) d\tau - \int_0^{t_n} F(v,\tau) d\tau \tag{13.44}$$

$$v(t_{n+1}) - v(t_n) = \int_0^{t_{n+1}} F(v,\tau) d\tau + \int_{t_n}^0 F(v,\tau) d\tau \tag{13.45}$$

$$v_{n+1} - v_n = \int_{t_n}^{t_{n+1}} F(v,\tau) d\tau \tag{13.46}$$

By making use of the Lagrange polynomial, we approximate $F(t, v(t))$

$$p(t) = \frac{t - t_{n-1}}{t_n - t_{n-1}} F(v,t_n) + \frac{t - t_n}{t_{n-1} - t_n} F(v,t_{n-1}) \tag{13.47}$$

Therefore,

$$v_{n+1} - v_n = \int_{t_n}^{t_{n+1}} \frac{t - t_{n-1}}{t_n - t_{n-1}} \left(F(v,\tau_n) + \frac{t - t_n}{t_{n-1} - t_n} F(v,t_{n-1}) dt \right) \tag{13.48}$$

$$= \int_{t_n}^{t_{n+1}} \frac{t-t_{n-1}}{\Delta t} F(v,t_n) - \frac{t-t_n}{\Delta t} F(v,t_{n-1}) dt \tag{13.49}$$

$$\frac{1}{\Delta t} F(v,t_n) \int_{t_n}^{t_{n+1}} (t-t_{n-1}) dt - F\left(\frac{v,t_{n-1}}{\Delta t}\right) \int_{t_n}^{t_{n+1}} (t-t_n) \tag{13.50}$$

$$\frac{1}{\Delta t} F(v,t_n) \left[\frac{t^2}{2} - tt_{n-1} \right]_{t_n}^{t_{n+1}} - \frac{F(v,t_{n+1})}{\Delta t} \left[\frac{t^2}{2} - tt_n \right]_{t_n}^{t_{n+1}} \tag{13.51}$$

Replacing in terms of $v(t) = u(p,t)$, we get;

$$u(p,t_{n+1}) = u(p,t_n) + \frac{3}{2} \Delta t F(u(P,t_n),t_n) - \frac{\Delta t}{2} F(u(p,t_{n-1}),t_{n-1}) \tag{13.52}$$

The above has been discretized using the Adams–Bashforth method in terms of time. We applied the Inverse Laplace transform in order to return to real space, we get

$$u(x,t_{n+1}) = u(x,t_n) + \frac{3}{2} \Delta t F(u(x,t_n),t_n) - \frac{\Delta t}{2} F(u(x,t_{n-1}),t_{n-1}) \tag{13.53}$$

$$u_i^{n+1} = u_i^n + \frac{3}{2} \Delta t F_i^n - \frac{\Delta t}{2} F_i^{n-1} \tag{13.54}$$

We are going to apply the above directly into the general groundwater flow equation and present stability analysis.

$$\frac{S}{T} \cdot \frac{\partial h}{\partial t} = \frac{\partial h}{r \partial r} + \frac{\partial^2 h}{\partial r^2} \left[1 + \frac{\Delta r}{r} \right]$$

The general groundwater flow equation is rearranged as follows:

$$\frac{\partial h}{\partial t} = \frac{S}{T} \left[\frac{\partial h}{r \partial r} + \frac{\partial^2 h}{\partial r^2} \left(1 + \frac{\Delta r}{r} \right) \right] \tag{13.55}$$

Therefore,

$$h(r,t) - h(r,o) = \int_0^t F(h,\tau) d\tau \tag{13.56}$$

$$F(h,t) = \frac{T}{S} \left[\frac{\partial h}{r \partial r} + \frac{\partial^2 h}{\partial r^2} \left(1 + \frac{\Delta r}{r} \right) \right] \tag{13.57}$$

We then replace u in terms of h,

$$h_i^{n+1} = h_i^n + \frac{3}{2}\Delta t F_i^n - \frac{\Delta t}{2} F_i^{n-1}$$ (13.58)

By discretization, we get

$$F_i^n = \left[\frac{1}{r_i}\frac{h_i^n - h_{i-1}^n}{\Delta r} + \frac{h_{i+1}^n - 2h_i^n + h_{i-1}^n}{(\Delta r)^2}\left(1 + \frac{r_i - r_{i-1}}{r_i}\right)\right]\frac{T}{S}$$ (13.59)

$$F_i^n = \frac{T}{S}\left[\frac{1}{r_i}\frac{h_i^{n-1} - h_{i-1}^{n-1}}{\Delta r} + \frac{h_{i+1}^{n-1} - 2h_i^{n-1} + h_{i-1}^{n-1}}{(\Delta r)^2}\left(1 + \frac{r_i - r_{i-1}}{r_i}\right)\right]$$ (13.60)

We substitute Equation (13.60) and (13.59) into Equation (13.58),

$$h_i^{n+1} = h_i^n + \frac{3}{2}\Delta t\left[\frac{1}{r_i}\frac{h_i^n - h_{i-1}^n}{\Delta r} + \frac{h_{i+1}^n - 2h_i^n + h_{i-1}^n}{(\Delta r)^2}\left(1 + \frac{r_i - r_{i-1}}{r_i}\right)\right]\frac{T}{S}$$
$$-\frac{\Delta t}{2}\left[\frac{1}{r_i}\frac{h_i^{n-1} - h_{i-1}^{n-1}}{\Delta r} + \frac{h_{i+1}^{n-1} - 2h_i^{n-1} + h_{i-1}^{n-1}}{(\Delta r)^2}\left(1 + \frac{r_i - r_{i-1}}{r_i}\right)\right]\frac{T}{S}$$ (13.61)

We are now going to present stability analysis of the above solution using a method used for solving classical and the non-local derivatives.

The round off error is defined as follows:

$$\epsilon_i^n = L_i^n - h_i^n$$ (13.62)

With u_i^n, as the solution of the discretized equation (13.61). The difference equation error is linear, thus,

$$\epsilon_i^n = \exp[at]\exp[jk_m x]$$ (13.63)

By replacing the above into Equation (13.61), we get;

$$\exp[a(t+\Delta t)]\exp[jk_m x]$$
$$= \exp[at]\exp[jk_m x]\frac{3}{2}\Delta t\left[\frac{1}{r_i}\frac{\exp[at]\exp[jk_m x] - \exp[at]\exp[jk_m(x-\Delta x)]}{\Delta r}\right]$$
$$\frac{\exp[at]\exp[jk_m x+\Delta x] - 2\exp[at]\exp[jk_m x] + \exp[at]\exp[jk_m(x-\Delta x)]}{(\Delta r)^2}$$
$$\left(1 + \frac{r_i - r_{i-1}}{r_i}\right)\frac{T}{S}\frac{\Delta t}{2}\left[\frac{1}{r_i}\frac{\exp[a(t-\Delta t)]\exp[jk_m x] - \exp[a(t-\Delta t)]\exp[jk_m x - \Delta x]}{\Delta r}\right]$$
$$+ \exp[a(t-\Delta t)]\exp[jk_m(x+\Delta x)] - 2\exp[a(t-\Delta t)]\exp[jk_m x]$$
$$\frac{+\exp[a(t-\Delta t)]\exp[jk_m x - \Delta x]}{(\Delta r)^2}\left(1 + \frac{r_i - r_{i-1}}{r_i}\right)\frac{T}{S}$$ (13.64)

Simplification yields,

$$\exp\left[a\Delta t\right] = 1 + \frac{3}{2}\frac{\Delta t}{r_i \Delta t} - \frac{3}{2}\exp\left[-jk_m\Delta x\right] + \exp\left[jk_m\Delta x\right] - \frac{2T}{(\Delta r)^2 S}\left(1 + \frac{\Delta r}{r_i}\right) - \frac{2T}{S(\Delta r)^2}\left(1 + \frac{\Delta r}{r_i}\right)$$

$$\times \exp\left[-jk_m\Delta x\right] - \frac{\Delta t}{(\Delta r)^2}\exp\left[-a\Delta t\right]\frac{\Delta t}{2(\Delta r)r_i}\exp\left[-a\Delta t\right]\exp\left[-jk_m\Delta x\right]$$

$$+ \frac{\exp\left[-a\Delta t\right]\exp\left[jk_m\Delta x\right]}{(\Delta r)^2}\left(1 + \frac{\Delta r}{r_i}\right)\frac{T}{S} - \frac{2T}{S(\Delta r)^2}\left(1 + \frac{\Delta r}{r_i}\right)\exp\left[-a\Delta t\right]$$

$$+ \frac{\exp\left[-a\Delta t\right]\exp\left[-jk_m\Delta x\right]}{(\Delta r)^2}\left(1 + \frac{\Delta r}{r_i}\right) \tag{13.65}$$

We have,

$$\exp\left[a\Delta t\right] = \frac{\epsilon_i^{n+1}}{\epsilon_i^n} \tag{13.66}$$

The numerical scheme will be stable if,

$$\|\exp\left[a\Delta t\right]\|\| = \left\|\frac{\epsilon_i^{n+1}}{\epsilon_i^n}\right\| < 1 \tag{13.67}$$

Therefore Equation (13.65) can be transformed to become,

$$\exp\left[a\Delta t\right] = 1 + \frac{3}{2}\frac{\Delta t}{r_i \Delta t} - \frac{3}{2}\left[\cos(k_m\Delta x) - i\sin(k_m\Delta x)\right] + \cos(k_m\Delta x) + i\sin(k_m\Delta x)$$

$$- \frac{2T}{(\Delta r)^2 S}\left(1 + \frac{\Delta r}{r_i}\right) - \frac{2T}{S(\Delta r)^2}\left(1 + \frac{\Delta r}{r_i}\right)\left[\cos(k_m\Delta x) - i\sin(k_m\Delta x)\right]$$

$$- \frac{\Delta t}{(\Delta r)^2}\exp\left[-a\Delta t\right] + \frac{\Delta t}{2(\Delta r)r_i}\exp\left[-a\Delta t\right]\left[\cos(k_m\Delta x) - i\sin(k_m\Delta x)\right]$$

$$+ \frac{\exp\left[-a\Delta t\right]\left[\cos(k_m\Delta x) + i\sin(k_m\Delta x)\right]}{(\Delta r)^2}\left(1 + \frac{\Delta r}{r_i}\right)\frac{T}{S} - \frac{2T}{S(\Delta r)^2}\left(1 + \frac{\Delta r}{r_i}\right)$$

$$\times \exp\left[-a\Delta t\right] + \frac{\exp\left[-a\Delta t\right]\left[\cos(k_m\Delta x) - i\sin(k_m\Delta x)\right]}{(\Delta r)^2}\left(1 + \frac{\Delta r}{r_i}\right) \tag{13.68}$$

By simplification,

$$\exp\left[a\Delta t\right] = 1 + \frac{3}{2}\frac{\Delta t}{r_i \Delta t} - \frac{3}{2}\cos(k_m\Delta x) + \cos(k_m\Delta x) - \frac{2T}{(\Delta r)^2 S}\left(1 + \frac{\Delta r}{r_i}\right)$$

$$- \frac{2T}{S(\Delta r)^2}\left(1 + \frac{\Delta r}{r_i}\right)\left[\cos(k_m\Delta x)\right] - \frac{\Delta t}{(\Delta r)^2}\exp\left[-a\Delta t\right]$$

$$+ \frac{\Delta t}{2(\Delta r)r_i}\exp\left[-a\Delta t\right]\left[\cos(k_m\Delta x)\right] + \frac{\exp\left[-a\Delta t\right]\left[\cos(k_m\Delta x)\right]}{(\Delta r)^2}\left(1 + \frac{\Delta r}{r_i}\right)\frac{T}{S}$$

$$- \frac{2T}{S(\Delta r)^2}\left(1 + \frac{\Delta r}{r_i}\right)\exp\left[-a\Delta t\right] + \frac{\exp\left[-a\Delta t\right]\left[\cos(k_m\Delta x)\right]}{(\Delta r)^2}\left(1 + \frac{\Delta r}{r_i}\right)$$

$$+ i \left\{ \begin{array}{l} -\dfrac{3}{2}\left[\sin\left(k_m \Delta x\right)\right]+\sin\left(k_m \Delta x\right)-\dfrac{2T}{S\left(\Delta r\right)^2}\left(1+\dfrac{\Delta r}{r_i}\right) \\[3mm] -\dfrac{3}{2}\sin\left(k_m \Delta x\right)+\sin\left(k_m \Delta x\right)-\dfrac{2T}{S\left(\Delta r\right)^2}\left(1+\dfrac{\Delta r}{r_i}\right)\left[-\sin\left(k_m \Delta x\right)\right] \\[3mm] +\dfrac{\Delta t}{2\left(\Delta r\right)r_i}\exp\left[-a\Delta t\right]\left[-\sin\left(k_m \Delta x\right)+\dfrac{\exp\left[-a\Delta t\right]\left[\sin\left(k_m \Delta x\right)\right]}{\left(\Delta r\right)^2}\right]\left(1+\dfrac{\Delta r}{r}\right)\dfrac{T}{S} \\[3mm] +\dfrac{\exp\left[-a\Delta t\right]\left[-\sin\left(k_m \Delta x\right)\right]}{\left(\Delta r\right)^2}\left(1+\dfrac{\Delta r}{r_i}\right) \end{array} \right\} \quad (13.69)$$

The above can be further simplified as

$$\exp\left[a\Delta t\right]=1+\dfrac{3}{2}\dfrac{\Delta t}{r_i \Delta t}-\dfrac{3}{2}\cos\left(k_m \Delta x\right)+\cos\left(k_m \Delta x\right)-\dfrac{2T}{\left(\Delta r\right)^2 S}\left(1+\dfrac{\Delta r}{r_i}\right)$$

$$-\dfrac{2T}{S\left(\Delta r\right)^2}\left(1+\dfrac{\Delta r}{r_i}\right)\left[\cos\left(k_m \Delta x\right)\right]-\dfrac{\Delta t}{\left(\Delta r\right)^2}\exp\left[-a\Delta t\right]$$

$$+\dfrac{\Delta t}{2\left(\Delta r\right)r_i}\exp\left[-a\Delta t\right]\left[\cos\left(k_m \Delta x\right)\right]+\dfrac{\exp\left[-a\Delta t\right]\left[\cos\left(k_m \Delta x\right)\right]}{\left(\Delta r\right)^2}\left(1+\dfrac{\Delta r}{r_i}\right)\dfrac{T}{S}$$

$$-\dfrac{2T}{S\left(\Delta r\right)^2}\left(1+\dfrac{\Delta r}{r_i}\right)\exp\left[-a\Delta t\right]+\dfrac{\exp\left[-a\Delta t\right]\left[\cos\left(k_m \Delta x\right)\right]}{\left(\Delta r\right)^2}\left(1+\dfrac{\Delta r}{r_i}\right)$$

$$+ i \left\{ \begin{array}{l} -\sin\left(k_m \Delta x\right)-\dfrac{2T}{S\left(\Delta r\right)^2}\left(1+\dfrac{\Delta r}{r_i}\right)-\dfrac{2T}{S\left(\Delta r\right)^2}\left(1+\dfrac{\Delta r}{r_i}\right)\left[-\sin\left(k_m \Delta x\right)\right] \\[3mm] +\dfrac{\Delta t}{2\left(\Delta r\right)r_i}\exp\left[-a\Delta t\right]\left[-\sin\left(k_m \Delta x\right)+\dfrac{\exp\left[-a\Delta t\right]\left[\sin\left(k_m \Delta x\right)\right]}{\left(\Delta r\right)^2}\right]\left(1+\dfrac{\Delta r}{r}\right)\dfrac{T}{S} \\[3mm] +\dfrac{\exp\left[-a\Delta t\right]\left[-\sin\left(k_m \Delta x\right)\right]}{\left(\Delta r\right)^2}\left(1+\dfrac{\Delta r}{r_i}\right) \end{array} \right\}$$

$$(13.70)$$

Based on the above, the equation cannot be further simplified or analyzed, and hence does not give us conditions for stability. It is in this regard that we propose an alternative method.

We are now going to use the iterative method to determine the conditions for stability.

Analysis starts from Equation (13.61) as follows,

$$h_i^{n+1}=h_i^n+\dfrac{3}{2}\Delta t\left[\dfrac{1}{r_i}\dfrac{h_i^n-h_{i-1}^n}{\Delta r}+\dfrac{h_{i+1}^n-2h_i^n+h_{i-1}^n}{\left(\Delta r\right)^2}\left(1+\dfrac{r_i-r_{i-1}}{r_i}\right)\right]\dfrac{T}{S}$$

$$-\dfrac{\Delta t}{2}\left[\dfrac{1}{r_i}\dfrac{h_i^{n-1}-h_{i-1}^{n-1}}{\Delta r}+\dfrac{h_{i+1}^{n-1}-2h_i^{n-1}+h_{i-1}^{n-1}}{\left(\Delta r\right)^2}\left(1+\dfrac{r_i-r_{i-1}}{r_i}\right)\right]\dfrac{T}{S}$$

By applying the iterative method, we get;

$$\delta_{n+1}e^{ik_mx} = \delta_ne^{ik_mx} + \frac{3}{2}\Delta t\left[\frac{1}{r_i}\frac{\delta_ne^{ik_mx}-\delta_ne^{ik_m(x-\Delta x)}}{\Delta r} + \frac{\delta_ne^{ik_m(x+\Delta x)}-2\delta_ne^{ik_mx}+\delta_ne^{ik_m(x-x)}}{(\Delta r)^2}\left(1+\frac{r_i-r_{i-1}}{r_i}\right)\right]\frac{T}{S}$$

$$-\frac{\Delta t}{2}\left[\frac{1}{r_i}\frac{\delta_{n-1}e^{ik_mx}-\delta_{n-1}e^{ik_m(x-\Delta x)}}{\Delta r} + \frac{\delta_{n+1}e^{ik_m(x-\Delta x)}-2\delta_{n-1}e^{ik_mx}+\delta_{n-1}e^{ik_m(x-\Delta x)}}{(\Delta r)^2}\left(1+\frac{r_i-r_{i-1}}{r_i}\right)\right]\frac{T}{S}$$

(13.71)

Therefore,

if $$\qquad a = \frac{3}{2}\Delta t + \frac{1}{r_i} - \frac{1}{\Delta r}, b = \left(1+\frac{r_i-r_{i-1}}{r_i}\right)\frac{T}{S}\frac{1}{(\Delta r)^2}, C = \frac{\Delta t}{2} + \frac{1}{r_i}\frac{1}{\Delta r}$$

Simplification using the above yields,

$$\delta_{n+1}e^{ik_mx} = \delta_ne^{ik_mx} + \delta_nae^{ik_mx} - \delta_nae^{ik_m(x-\Delta x)} + \delta_nbe^{ik_m(x+\Delta x)} - 2\delta_nbe^{ik_mx} + \delta_nbe^{ik_m(x-\Delta x)}$$
$$-\delta_{n-1}ce^{ik_mx} - \delta_{n-1}ce^{ik_m(x-\Delta x)} + \delta_{n-1}be^{ik_m(x-\Delta x)} - 2\delta_{n-1}be^{ik_mx} + 2\delta_{n-1}be^{ik_m(x-\Delta x)}$$ (13.72)

By removing e^{ik_mx} we get,

$$\delta_{n+1} = \delta_n + \delta_na - \delta_nae^{ik_mx}.e^{-ik_m\Delta x} + \delta_nbe^{ik_mx}.e^{ik_m\Delta x} - 2\delta_nb + \delta_nbe^{ik_mx}.e^{ik_m\Delta x} - \delta_{n-1}c$$
$$-\delta_{n-1}ce^{ik_mx}.e^{-ik_m\Delta x} + \delta_{n+1}be^{ik_mx}.e^{-ik_m\Delta x} - 2\delta_{n-1}b + \delta_{n-1}be^{ik_mx}.e^{-ik_m\Delta x}$$ (13.73)

Further simplification yields,

$$\delta_{n+1} = \delta_n + \delta_na - \delta_nae^{-ik_m\Delta x} + \delta_{n-1}be^{ik_m\Delta x} - 2\delta_{n-1}b + \delta_nbe^{ik_m\Delta x} - \delta_{n-1}c - \delta_{n-1}ce^{-ik_m\Delta x}$$
$$+\delta_{n+1}be^{-ik_m\Delta x} - 2\delta_{n-1}b + \delta_{n-1}be^{-ik_m\Delta x}$$ (13.74)

By factorization, we get

$$\delta_{n+1} = \delta_n\left(1+a-ae^{-ik_m\Delta x}\right) + \delta_{n-1}\left(be^{ik_m\Delta x} - 4b + 3be^{-ik_m\Delta x} - c\right)$$ (13.75)

Equation (13.75) can be written in terms of cos and sin,

$$\delta_{n+1} = \delta_n\left(1+a-a\cos(k_m\Delta x) - i\sin(k_m\Delta x)\right) + \delta_{n-1}b\begin{pmatrix}\cos(k_m\Delta x) + i\sin(k_m\Delta x) - 4b\\ +3b(\cos(k_m\Delta x) - i\sin(k_m\Delta x)) - c\end{pmatrix}$$ (13.76)

By induction when $n = 0$

$$\delta_1 = \delta_0\left(1+a-a\cos(k_m\Delta x) - ai\sin(k_m\Delta x)\right)$$ (13.77)

$$\left|\frac{\delta_1}{\delta_0}\right| < 1 = \left|1-a-a\cos(k_m\Delta x) - ai\sin(k_m\Delta x)\right|$$ (13.78)

$$|a+ib| = \sqrt{a^2+b^2}$$ (13.79)

$$= \sqrt{1-a-a\cos(k_m\Delta x)^2 + (a\sin(k_m\Delta x))} < 1 \tag{13.80}$$

$$\left(1-a-a\cos(k_m\Delta x)\right)^2 + \left(a\sin(k_m\Delta x)\right)^2 \tag{13.81}$$

$$(1-a)^2 - 2(1-a)a\cos(k_m\Delta x) + \left(a\cos(k_m\Delta x)\right)^2 + \left(a\sin(k_m\Delta x)\right)^2 < 1 \tag{13.82}$$

$$(1-a)^2 - 2(1-a)a\cos(k_m\Delta x) + a^2\left(\cos^2(k_m\Delta x) + \sin^2(k_m\Delta x)\right) < 1 \tag{13.83}$$

$$(1-a)^2 - 2(1-a)a\cos(k_m\Delta x) + a < 1 \tag{13.84}$$

$$(1-a)^2 - 2(1-a)\cos(k_m\Delta x) < 1-a \tag{13.85}$$

If $1 - a < 1$

$$(1-a) - 2\cos(k_m\Delta x) < 1 \tag{13.86}$$

$$-a - 2\cos(k_m\Delta x) < 0 \tag{13.87}$$

$$a + 2\cos(k_m\Delta x) > 0 \tag{13.88}$$

$$\cos(k_m\Delta x) > \frac{-a}{2} \tag{13.89}$$

This is one of the conditions for stability.

$$\delta_{n+1} = \delta_n\left(1+a-a\cos(k_m\Delta x) - i\sin(k_m\Delta x)\right) + \delta_{n-1}b\left(\begin{array}{c}\cos(k_m\Delta x) + i\sin(k_m\Delta x) - 4b \\ +3b\left(\cos(k_m\Delta x) - i\sin(k_m\Delta x)\right) - c\end{array}\right)$$

for simplification, we let

$$z_1 = \left(1+a-a(\cos(k_m\Delta x) - i\sin(k_m\Delta x)\right) \tag{13.90}$$

$$z_2 = b\left(\cos(k_m\Delta x) + i\sin(k_m\Delta x) - 4b + 3b\left(\cos(k_m\Delta x) - i\sin(k_m\Delta x)\right) - c\right) \tag{13.91}$$

$$\delta_{n+1} = \delta_n z_1 + \delta_{n-1} z_2 \tag{13.92}$$

$\forall_n \geq 0$ we assume that $\|\delta_n\| \leq \|\delta_0\|$

We then want to find the conditions for which $\|\delta_{n+1}\| < \|\delta_0\|$
By definition we have,

$$\delta_{n+1} = \delta_n z_1 + \delta_{n-1} z_2$$

$$\|\delta_{n+1}\| = \|\delta_n z_1 + \delta_{n-1} z_2\| \tag{13.93}$$

$$\leq \|\delta_n z_1\| + \|\delta_{n-1} z_2\| \tag{13.94}$$

$$\leq \|\delta_n\| \ \|z_1\| + \|\delta_{n-1}\| \|z_2\| \tag{13.95}$$

By induction, we have $\|\delta_n\| < \|\delta_0\|$ and $\|\delta_{n-1}\| < \|\delta_0\|$.
Therefore,

$$\|\delta_{n+1}\| < \|\delta_0\| \ \ \|z_1\| + \|\delta_0\|\|z_2\| \tag{13.96}$$

$$< \|\delta_0\| \big(\|z_1\| + \|z_2\| \big) \tag{13.97}$$

$$\|\delta_{n+1}\| < \|\delta_0\| \quad \text{if} \quad \|z_1\| + \|z_2\| < 1$$

$$z_1 = \big(1 + a - a(\cos(k_m \Delta x) - i\sin(k_m \Delta x) \big)$$

$$z_2 = b\big(\cos(k_m \Delta x) + i\sin(k_m \Delta x) - 4b + 3b\big(\cos(k_m \Delta x) - i\sin(k_m \Delta x)\big) - c \big)$$

Therefore,

$$z_1 = \big(1 + a - a\cos(k_m \Delta x) \big) - ia\sin(k_m \Delta x) \tag{13.98}$$

$$z_2 = \big(4b\cos(k_m \Delta x) - 4b - c \big) - i2b\sin(k_m \Delta x) \tag{13.99}$$

$$\|z_1\| = \sqrt{\big(1 + a - a\cos(k_m \Delta x)\big)^2 + \big(a\sin(k_m \Delta x)\big)^2} \tag{13.100}$$

$$\|z_2\| = \sqrt{\big(4b\cos(k_m \Delta x) - 4b - c\big)^2 + \big(2b\sin(k_m \Delta x)\big)^2} \tag{13.101}$$

$$\|z_1\| + \|z_2\| < 1 \Rightarrow$$

$$\sqrt{\big(1 + a - a\cos(k_m \Delta x)\big)^2 + \big(a\sin(k_m \Delta x)\big)^2} + \sqrt{\big(4b\cos(k_m \Delta x) - 4b - c\big)^2 + \big(2b\sin(k_m \Delta x)\big)^2} < 1 \tag{13.102}$$

Simplification yields,

$$\sqrt{\big(1 + a\big)^2 - 2\big(1 + a\big)a\cos(k_m \Delta x)} + \sqrt{\big(4b\cos(k_m \Delta x) - 4b - c\big)^2 + \big(2b\sin(k_m \Delta x)\big)^2} < 1 \tag{13.103}$$

The above satisfy the condition for stability.
We are now going to determine self-similarity with respect to time.

$$\frac{T}{S}\frac{\partial h(r,t)}{\partial t} = \frac{1}{\alpha^{\alpha-1}}\frac{\partial h(r,t)}{\alpha r} + \left(1+\frac{\Delta r}{r}\right)\frac{\partial^2 h}{\partial r^2} \qquad (13.104)$$

$$\frac{T}{S}\frac{\partial h(r,t)}{\partial t^\beta} = \frac{1}{\alpha r^{\alpha-1}}\frac{\partial h(r,t)}{\partial r} + \left(1+\frac{\Delta r}{r}\right)\frac{\partial^2 h}{\partial r^2} \qquad (13.105)$$

Since $h(r,t)$ is differentiable,

$$\frac{T}{S}\frac{\partial h(r,t)}{\partial t}\frac{1}{\beta t^{\beta-1}} = \frac{1}{\alpha r^{\alpha-1}}\frac{\partial h(r,t)}{\partial r} + \left(1+\frac{\Delta r}{r}\right)\frac{\partial^2 h}{\partial r^2} \qquad (13.106)$$

$$\frac{T}{S}\frac{\partial h(r,t)}{\partial t} = \frac{\beta t^{\beta-1}}{\alpha r^{\alpha-1}}\frac{\partial h(r,t)}{\alpha r} + \beta t^{\beta-1}\left(1+\frac{\Delta r}{r}\right)\frac{\partial^2 h}{\partial r^2} \qquad (13.107)$$

$$\frac{T}{S}\frac{\partial h(r,t)}{\partial t} = V_\beta^\alpha(r,t)\frac{\partial h(r,t)}{\alpha r} + D^\beta(r,t)\frac{\partial^2 h}{\partial r^2} \qquad (13.108)$$

We now discretize the above equation,

$$\frac{T}{S}\frac{h_i^{n+1}-h_i^n}{\Delta t} = V_\beta^\alpha(r_i,t_n)\frac{h_{i+1}^n-h_i^n}{\Delta r} + D^\beta(r_i,t_n)\frac{h_{i+1}^n-2h_i^n+h_{i-1}^n}{(\Delta r)^2}\left[1+\frac{r_{i+1}-r_i}{r_i}\right] \qquad (13.109)$$

$$\frac{T}{S}\frac{h_i^{n+1}-h_i^n}{\Delta t} = V_\beta^\alpha(r_i,t_n)\frac{h_{i+1}^{n+1}-h_i^{n+1}}{\Delta r} + D^\beta(r_i,t_n)\frac{h_{i+1}^{n+1}-2h_i^{n+1}+h_{i-1}^{n+1}}{(\Delta r)^2}\left[1+\frac{r_{i+1}-r_i}{r_i}\right] \qquad (13.110)$$

$$\frac{T}{S}\frac{h_i^{n+1}-h_i^n}{\Delta t} = V_\beta^\alpha(r_i,t_n)\frac{h_{i+1}^n-h_i^n}{\Delta r} + \left[\frac{h_{i+1}^{n+1}-2h_i^{n+1}+h_{i-1}^{n+1}}{2(\Delta r)^2} + \frac{h_{i+1}^n-2h_i^n+h_{i-1}^n}{2(\Delta r)^2}\right]\left[1+\frac{r_{i+1}-r_i}{r_i}\right] \qquad (13.111)$$

Simplification yields,

$$h_i^{n+1}\left[\frac{T}{S\Delta t}+\frac{1}{(\Delta r)^2}\left[1+\frac{r_{i+1}-r_i}{r_i}\right]\right] = h_i^n\left[\frac{1}{(\Delta r)^2}\left(1+\frac{r_{i+1}-r_i}{r_i}\right)\right] + h_{i+1}^n\left[V_\beta^\alpha\frac{(r_i t_n)}{\Delta r}+\left(\frac{1+\frac{r_{i+1}-r_i}{r_i}}{2(\Delta r)^2}\right)\right]$$

$$+h_{i-1}^n\left[V_\beta^\alpha\frac{(r_i t_n)}{\Delta r}+\left(\frac{1+\frac{r_{i+1}-r_i}{r_i}}{2(\Delta r)^2}\right)\right] + h_{i-1}^{n+1}\left(\frac{1+\frac{r_{i+1}-r_i}{r_i}}{2(\Delta r)^2}\right)$$

$$(13.112)$$

Simplification in terms of alpha gives,

$$a_1 h_i^{n+1} = a_2 h_i^n + a_3 h_{i+1}^n + a_3 h_{i-1}^n + a_4 h_{i-1}^{n+1} \tag{13.113}$$

The above solution is derived by analyzing self-similarity with respect to time. Hence,

$$h_i^{n+1} = \delta_{n+1} e^{ik_m x}$$

$$h_i^n = \delta_n e^{ik_m x}$$

$$h_{i+1}^n = \delta_n e^{ik_m (x+\Delta x)}$$

$$h_{i-1}^n = \delta_n e^{-ik_m (x-\Delta x)}$$

$$h_{i-1}^{n+1} = \delta_{n+1} e^{ik_m (x-\Delta x)}$$

By replacing the above in our solution, we get

$$a_1 \delta_{n+1} e^{ik_m \Delta x} = a_2 \delta_n e^{ik_m \Delta x} + a_3 \delta_n e^{ik_m (x+\Delta x)} + a_3 \delta_n e^{-ik_m (x-\Delta x)} + a_4 \delta_{n+1} e^{ik_m (x-\Delta x)} \tag{13.114}$$

Simplification yields,

$$a_1 \delta_{n+1} = a_2 \delta_n + a_3 \delta_n e^{ik_m \Delta x} + a_3 \delta_n e^{-ik_m \Delta x} + a_4 \delta_{n+1} e^{-ik_m \Delta x} \tag{13.115}$$

Putting like terms together and factorization, we get

$$\delta_{n+1} \left(a_1 - a_4 e^{ik_m \Delta x} \right) = \delta_n \left(a_2 + a_3 e^{ik_m \Delta x} + a_3 e^{-ik_m \Delta x} \right) \tag{13.116}$$

Therefore,

$$\frac{\delta_{n+1}}{\delta_n} = \frac{a_2 + a_3 \left(e^{ik_m \Delta x} - e^{-ik_m \Delta x} \right)}{a_1 - a_4 e^{ik_m \Delta x}} \tag{13.117}$$

We now have to prove that,

$$\left| \frac{\delta_{n+1}}{\delta_n} \right| < 1$$

Therefore,

$$\left|\frac{\delta_{n+1}}{\delta_n}\right| < 1 = \frac{\left|a_2 + 2a_3 i \sin\left(k_m \Delta x\right)\right|}{\left|a_1 - a_4 \cos\left(k_m \Delta x\right) + i \sin\left(k_m \Delta x\right)\right|} < 1 \qquad (13.118)$$

Since $\left|a + ib\right| = \sqrt{a^2 + b^2}$ we get,

$$\sqrt{\frac{a_2^2 + \left(2a_3 \sin\left(k_m \Delta x\right)\right)^2}{\left(a_1 - a_4 \cos\left(k_m \Delta x\right)\right)^2 + \sin^2\left(k_m \Delta x\right)}} < 1 \qquad (13.119)$$

Removal of square root gives us,

$$\frac{a_2^2 + \left(2a_3 \sin\left(k_m \Delta x\right)\right)^2}{\left(a_1 - a_4 \cos\left(k_m \Delta x\right)\right)^2 + \sin^2\left(k_m \Delta x\right)} < 1 \qquad (13.120)$$

For conditions of stability, the above conditions must be satisfied.

13.4 NUMERICAL SIMULATIONS AND DISCUSSION

In this section, using the already presented numerical solution, we present some numerical simulations for different values of fractal dimension. The simulations are depicted in Figures 13.1 to 13.3.

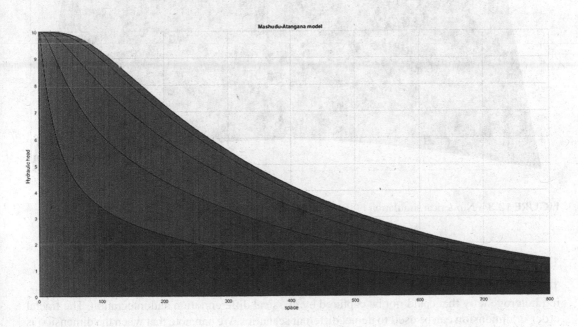

FIGURE 13.1 Numerical simulation with theoretical value of 0.7 showing H and S.

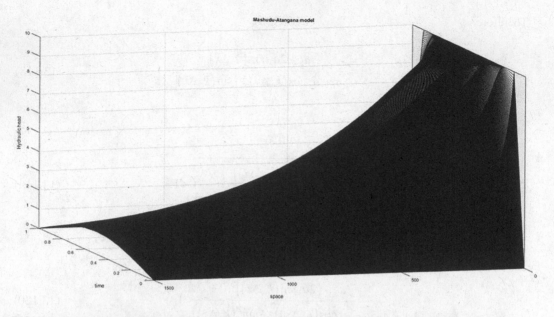

FIGURE 13.2   Numerical simulation with theoretical value of 0.4 showing H, S and T.

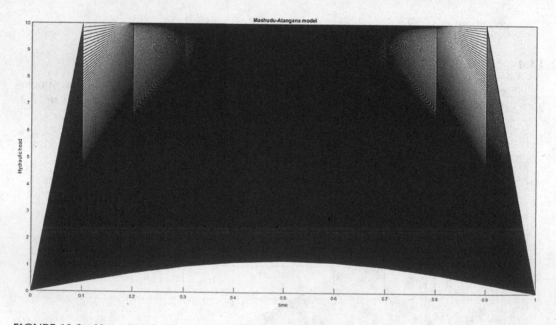

FIGURE 13.3   Numerical simulation of theoretical value of 0.4 showing H and T.

13.5   CONCLUSION

The concept of fractal differential and integral operators have been suggested with the aim of capturing heterogeneity that could not be captured by classical differentiation and integration. The fractal order or dimension can be used to depict different scenarios. We can note that when the dimension is 1 we capture Markovian process. This work considered the possible application of fractal derivatives to model a flow of water within a fractured confined aquifer.

REFERENCES

Allwright, A. & Atangana, A. (2018). Fractal advection-dispersion equation for groundwater transport in fractured aquifers with self-similarities. *The European Physical Journal Plus*, 133: 48.

Amanda, R. & Atangana, A. (2018). Derivation of a groundwater flow model within leaky and self-similar aquifers: Beyond Hantush model. *Chaos, Solitons & Fractals*, 116: 414–423.

Chen, W., Sun, H., Zhang, X., & Korosak, D. (2010). Anomalous diffusion modelling by fractal and fractional derivatives. *Computers and Mathematics with Applications*, 59(2010): 1754–1758.

Mathobo, M. & Atangana, A. (2018). Analysis of exact groundwater model within a confined aquifer: New proposed model beyond the Theis equation. *The European Physical Journal Plus*, 133(10): 415.

14 Analysis of General Groundwater Flow Equation with Fractal-Fractional Differential Operators

Mashudu Mathobo and Abdon Atangana
University of Free State, Bloemfontein, South Africa

CONTENTS

14.1 INTRODUCTION

The fractal derivative is a non-Newtonian type of derivative in which the variable such as t has been scaled according to t^α. This derivative was defined in fractal geometry. Porous media, aquifers, turbulence and other media exhibit fractal properties. Physical concepts such as distance and velocity in fractal media are required to be redefined; the scales for space and time should be transformed according to (x^β, t^μ) (Chen, 2006). The elementary physical concepts such as velocity in a fractal spacetime (x^β, t^α) can be redefined by:

$$V' = \frac{dx'}{dt'} = \frac{dx^\beta}{dt^\alpha}, \alpha, \beta > 0 \tag{14.1}$$

where $S^{\alpha, \beta}$ represents the fractal spacetime with scaling indices α and β (Chen, 2006).

The concept of the fractal derivative of a function $u(t)$, with respect to a fractal measure t was developed as follows;

$$\frac{\partial f(t)}{\partial (t)^\alpha} = \lim_{t_1 \to t} \frac{f(t_1) - f(t)}{t_1^\alpha - t^\alpha}, \alpha > 0 \tag{14.2}$$

A more generalized definition was given by,

$$\frac{\partial^\beta f(t)}{\alpha t^\alpha} = \lim_{t_1 \to t} \frac{f^\beta(t_1) - f^\beta(t)}{t_1^\alpha - t^\alpha}, \alpha > 0, \beta > 0 \tag{14.3}$$

DOI: 10.1201/9781003266266-14

According to Matlob and Jamali (2017) fractional calculus is a generalization of ordinary differentiation and integration to arbitrary non-integer order. Fractional calculus is a field of mathematics that grows out of the traditional definitions of calculus integral and derivative operators in much the same way that fractional exponents is an outgrowth of exponents with integer values (Dalir and Bashour, 2010). Different scholars defined fractional derivatives as follows:

1. L. Euler (1730):
   Euler generalized the formula,

   $$\frac{d^n x^m}{dx^n} m(m-1)\ldots(m-n+1)x^{m-n} \tag{14.4}$$

   By use of the following property of gamma function,

   $$\Gamma(m+1) = m(m-1)\ldots(m-n+1)x^{m-n} \tag{14.5}$$

   To obtain,

   $$\frac{d^n x^m}{dx^n} = \frac{\Gamma(m+1)}{\Gamma(m-n+1)} x^{m-n} \tag{14.6}$$

   Gamma function is defined as follows,

   $$\Gamma(z) = \int_0^\infty e^{-t} t^{z-1} dt, \operatorname{Re}(z) > 0 \tag{14.7}$$

2. J.J. Fourier (1820–1822):
   By means of integral representation, he wrote

   $$\frac{d^n f(x)}{dx^n} = \frac{1}{2\pi} \int_{-\infty}^\infty f(z) dz \int_{-\infty}^\infty \cos\left(px - pz + n\frac{\pi}{2}\right) dp \tag{14.8}$$

3. N.H. Abel (1823–1826):
   Abel considered the integration representation $\int_0^\infty \frac{s'(\eta) d\eta}{(x-\eta)^\alpha} = \psi(x)$ for arbitrary α and then wrote,

   $$s(x) = \frac{1}{\Gamma(1-\alpha)} \frac{d^{-\alpha} \psi(x)}{dx^{-\alpha}} \tag{14.9}$$

4. L. Liouville (1832–1855):
   · I. In his first definition, according to exponential representation of a function $f(x) = \sum_n{}^\infty = 0 c_n e^{\alpha_n x}$, he generalized the formula $\frac{d^m e^{\alpha x}}{dx^n} = a^m e^{ax}$ as,

   $$\frac{d^v f(x)}{dx^v} = \sum_{n=0}^\infty c_n a_n^v e^{a_n x} \tag{14.10}$$

II. His second type of definition was fractional integral

$$\int^{\mu} \Phi(x) dx^{\mu} = \frac{1}{(-1)^{\mu} \Gamma(\mu)} \int_{0}^{\infty} \Phi(x+\alpha) \alpha^{\mu-1} d\alpha \qquad (14.11)$$

$$\int^{\mu} \Phi(x) dx^{\mu} = \frac{1}{\Gamma(\mu)} \int_{0}^{\infty} \Phi(x-\alpha) \alpha^{\mu-1} d\alpha \qquad (14.12)$$

By substituting of $r = x + \alpha$ and $r = x - \alpha$ in the above, the following was obtained,

$$\int^{\mu} \Phi(x) dx^{\mu} = \frac{1}{(-1)^{\mu} \Gamma(\mu)} \int_{0}^{\infty} \Phi(r-\alpha)^{\mu-1} \Phi(r) dr \qquad (14.13)$$

$$\int^{\mu} \Phi(x) dx^{\mu} = \frac{1}{\Gamma(\mu)} \int_{-\infty}^{\infty} \Phi(x-r)^{\mu-1} \Phi(r) dr \qquad (14.14)$$

III. His third definition includes fractional derivatives,

$$\frac{d^{\mu} f(x)}{dx^{\mu}} = \frac{1}{h^{\mu}} \left(f(x) \frac{\mu}{1} f(x+h) + + \frac{\mu(\mu-1)}{1.2} f(x+2h) - \cdots \right) \qquad (14.15)$$

$$\frac{d^{\mu} f(x)}{dx^{\mu}} \frac{1}{h^{\mu}} \left(f(x) \frac{\mu}{1} f(x-h) + + \frac{\mu(\mu-1)}{1.2} f(x-2h) - \cdots \right) \qquad (14.16)$$

5. F.B. Riemann (1847–1876)
   His definition of fractal integral is,

$$D^{-\nu} f(x) = \frac{1}{\Gamma(\nu)} \int_{c}^{x} (x-t)^{\nu-1} f(t) dt + \psi(t) \qquad (14.17)$$

6. N.Y.A. Sonin (1869), A.V. Letnikov (1872), H. Laurent (1884), N. Nekrasove (1888), K. Nishimoto (1987):
   Using the Cauchy Integral formula

$$f^{(n)}(z) = \frac{n!}{2\pi i} \int_{c} \frac{f(t)}{(t-2)^{n+1}} dt \qquad (14.18)$$

7. Riemann–Liouville
   The popular definition of fractional calculus shows the joining of two previous definitions,

$$_{a}D_{t}^{\alpha} f(t) = \frac{1}{\Gamma(n-\alpha)} \left(\frac{d}{dt} \right)^{n} \int_{a}^{t} \frac{f(r) dr}{(t-r)^{\alpha-n+1}}, (n-1 \leq \alpha < n) \qquad (14.19)$$

8. Grunwald-Letnikov

Below is another joined definition which may be useful at times,

$$_aD_t^\alpha f(t) = \lim_{h \to 0} h^{-\alpha} \sum_{j=0}^{\left[\frac{t-a}{h}\right]} (-1)^j \binom{\alpha}{j} f(t-jh) \tag{14.20}$$

9. M. Caputo (1967)

This is the second most popular definition,

$$_a^C D_t^\alpha f(t) = \frac{1}{\Gamma(\alpha-n)} \int_a^t \frac{f^n(r)dr}{(t-r)^{\alpha+1-n}}, (n-1 \le \alpha < n) \tag{14.21}$$

10. K.S. Miller, B. Ross (1993)

Differential operator D was used as

$$D^{\bar{\alpha}} f(t) = D^{\alpha_1} D^{\alpha_2} \dots D^{\alpha_n} f(t), \bar{\alpha} = (\alpha_1, \alpha_2, \dots \alpha_n) \tag{14.22}$$

Which $D\alpha_i$ is Riemann–Liouville or Caputo definitions.

14.2 APPLICATION OF FRACTAL-FRACTIONAL DERIVATIVE

Atangana and Araz (2019) developed a new numerical method for ordinary and partial differential equations, including those with non-integers called Newton polynomials. In their work, Atangana and Araz mentioned that as much as the Adams–Bashforth method have been recognized as a very efficient numerical method to solve linear and nonlinear orders, the method was developed using the Lagrange interpolation, which is less accurate as compared to their newly developed Newton polynomial. The new method was further used with the Caputo–Fabrizio fractal-fractional derivative, the Atangana–Baleanu fractal-fractional derivative, and with the Caputo fractal-fractional derivative.

14.2.1 ANALYSIS WITH ATANGANA–BALEANU FRACTAL-FRACTIONAL DERIVATIVE

The general groundwater flow equation for a confined aquifer was analyzed using the Atangana–Baleanu Fractal-Fractional derivative. Discretization of the equation conducted using the Newton polynomial. Stability analyses were shown. Numerical simulations developed using the new scheme.

By considering the following Cauchy problem,

$$_{0}^{FFM}D_t^{\alpha,\beta} y(t) = f(t, y(t)) \tag{14.23}$$

By integrating the Atangana–Baleanu fractal-fractional derivative, we get

$$y(t) - y(0) = \frac{1-\alpha}{AB(\alpha)} \beta t^{\beta-1} f(t, y(t)) + \frac{\alpha\beta}{AB(\alpha)\Gamma(\alpha)} \int_0^t \tau^{\beta-1} f(\tau, y(\tau))(t-\tau)^{\alpha-1} d\tau \tag{14.24}$$

by taking $F(t, y(t)) = \beta t^{\beta-1} f(t, y(t))$, the above equation can be reformulated as

$$y(t) - y(0) = \frac{1-\alpha}{AB(\alpha)} F(t, y(t)) + \frac{\alpha\beta}{AB(\alpha)\Gamma(\alpha)} \int_0^t F(\tau, y(\tau))(t-\tau)^{\alpha-1} d\tau \tag{14.25}$$

at the point $t_{n+1} = (n+1)\Delta t$, we have

$$y(t_{n+1}) - y(0) = \frac{1-\alpha}{AB(\alpha)} F(t_n, y(t_n)) + \frac{\alpha}{AB(\alpha)\Gamma(\alpha)} \int_0^{t_{n+1}} F(\tau, y(\tau))(t_{n+1} - \tau)^{\alpha-1} d\tau \quad (14.26)$$

We also get,

$$y(t_{n+1}) = y(0) + \frac{1-\alpha}{AB(\alpha)} F(t_n, y(t_n)) + \frac{\alpha}{AB(\alpha)\Gamma(\alpha)} \sum_{l=2}^{n} \int_{t_l}^{t_{l+1}} F(\tau, y(\tau))(t_{n+1} - \tau)^{\alpha-1} d\tau \quad (14.27)$$

The Newton polynomial was used to approximate the function $f(\tau, y(\tau))$ as follows,

$$p_n(\tau) = f(t_{n-2}, y(t_{n-2})) + \frac{f(t_{n-1}, y(t_{n-1})) - f(t_{n-2}, y(t_{n-2}))}{\Delta t}(\tau - t_{n-2})$$
$$+ \frac{f(t_n, y(t_n)) - 2f(t_{n-1}, y(t_{n-1})) + f(t_{n-2}, y(t_{n-2}))}{2(\Delta t)^2} \times (\tau - t_{n-2})(\tau - t_{n-1}) \quad (14.28)$$

By putting polynomial into Equation (14.6) we derive it as

$$y^{n+1} = y^0 + \frac{1-\alpha}{AB(\alpha)} F(t_n, y(t_n)) + \frac{\alpha}{AB(\alpha)\Gamma(\alpha)}$$
$$\sum_{l=2}^{n} \int_{t_l}^{t_{l+1}} \left\{ \begin{matrix} F(t_{l-2}, y^{l-2}) + \frac{F(t_{j-1}, y^{l-1}) - F(t_{l-2}, y^{l-2})}{\Delta t}(\tau - t_{l-2}) \\ + \frac{F(t_l, y^l) - 2F(t_{l-1}, y^{l-1}) + F(t_{l-2}, y^{l-2})}{2(\Delta t)^2} \\ \times (\tau - t_{l-2})(\tau - t_{l-1}). \end{matrix} \right\} (t_{n+1} - \tau)^{\alpha-1} d\tau \quad (14.29)$$

If we order the above equality, we obtain:

$$y^{n+1} = y^0 + \frac{1-\alpha}{AB(\alpha)} F(t_n, y(t_n))$$
$$+ \frac{\alpha}{AB(\alpha)\Gamma(\alpha)} \sum_{l=2}^{n} \int_{t_l}^{t_{l+1}} \left\{ \begin{matrix} F(t_{l-2}, y^{l-2})(t_{n+1} - \tau)^{\alpha-1} d\tau \\ + \int_{t_l}^{t_{l+1}} \frac{F(t_{l-1}, y^{l-1}) - F(t_{l-2}, y^{l-2})}{\Delta t} \\ \times (\tau - t_{l-2})(t_{n+1} - \tau)^{\alpha-1} d\tau \\ + \int_{t_l}^{t_{l+1}} \frac{F(t_l, y^l) - 2F(t_{l-1}, y^{l-1}) + F(t_{l-2}, y^{l-2})}{2(\Delta t)^2} \\ \times (\tau - t_{l-2})(\tau - t_{l-1})(t_{n+1} - \tau)^{\alpha-1} d\tau \end{matrix} \right\}$$

$$(14.30)$$

and write the following,

$$y^{n+1} = y^0 + \frac{1-\alpha}{AB(\alpha)} F\left(t_n, y(t_n)\right) + \frac{\alpha}{AB(\alpha)\Gamma(\alpha)} \sum_{l=2}^{n} F\left(t_{l-2}, y^{l-2}\right) \int_{t_l}^{t_{l+1}} \left(t_{n+1} - \tau\right)^{\alpha-1} d\tau$$

$$+ \frac{\alpha}{AB(\alpha)\Gamma(\alpha)} \sum_{l=2}^{n} \frac{F\left(t_{l-1}, y^{l-1}\right) - F\left(t_{l-2}, y^{l-2}\right)}{\Delta t} \times \int_{t_l}^{t_{l+1}} \left(\tau - t_{l-2}\right)\left(t_{n+1} - \tau\right)^{\alpha-1} d\tau$$

$$+ \frac{\alpha}{AB(\alpha)\Gamma(\alpha)} \sum_{l=2}^{n} \frac{F\left(t_l, y^l\right) - 2F\left(t_{l-1}, y^{l-1}\right) + F\left(t_{l-2}, y^{l-2}\right)}{2(\Delta t)^2}$$

$$\times \int_{t_l}^{t_{l+1}} \left(\tau - t_{l-2}\right)\left(\tau - t_{l-1}\right)\left(t_{n+1} - \tau\right)^{\alpha-1} d\tau \tag{14.31}$$

The above integrals were calculated as,

$$\int_{t_l}^{t_{l+1}} \left(t_{n+1} - \tau\right)^{\alpha-1} d\tau = \frac{(\Delta t)^\alpha}{\alpha} \left[\left(n-l+1\right)^\alpha - \left(n-l\right)^\alpha\right] \tag{14.32}$$

$$\int_{t_l}^{t_{l+1}} \left(\tau - t_{l-2}\right)\left(t_{n+1} - \tau\right)^{\alpha-1} d\tau = \frac{(\Delta t)^{\alpha+1}}{\alpha(\alpha+1)} \left[\begin{array}{l} \left(n-l+1\right)^\alpha \left(n-l+3+2\alpha\right) \\ -\left(n-l+1\right)^\alpha \left(n-l+3+3\alpha\right) \end{array}\right] \tag{14.33}$$

$$\int_{t_l}^{t_{l+1}} \left(\tau - t_{l-2}\right)\left(\tau - t_{l-1}\right)\left(t_{n+1} - \tau\right)^{\alpha-1} d\tau = \frac{(\Delta t)^{\alpha+2}}{\alpha(\alpha+1)(\alpha+2)}$$

$$\times \left[\begin{array}{l} \left(n-l+1\right)^\alpha \left[2\left(n-l\right)^2 + \left(3\alpha+10\right)\left(n-l\right) + 2\alpha^2 + 9\alpha + 12\right] \\ -\left(n-l\right)^\alpha \left[2\left(n-l\right)^2 + \left(5\alpha+10\right)\left(n-l\right) + 6\alpha^2 + 18\alpha + 12\right] \end{array}\right] \tag{14.34}$$

By replacing into the above scheme, we get:

$$y^{n+1} = y^0 + \frac{1-\alpha}{AB(\alpha)} F\left(t_n, y(t_n)\right) + \frac{\alpha(\Delta t)^\alpha}{AB(\alpha)\Gamma(\alpha+1)} \sum_{l=2}^{n} F\left(t_{l-2}, y^{l-2}\right)\left(n-l+1\right)^\alpha - \left(n-l\right)^\alpha$$

$$+ \frac{\alpha(\Delta t)^\alpha}{AB(\alpha)\Gamma(\alpha+2)} \sum_{l=2}^{n} \left[F\left(t_{l-1}, y^{l-1}\right) - F\left(t_{l-2}, y^{l-2}\right) \times \left[\begin{array}{l} \left(n-l+1\right)^\alpha \left(n-l+3+2\alpha\right) \\ -\left(n-l\right)^\alpha \left(n-l+3+3\alpha\right) \end{array}\right]\right]$$

$$+ \frac{\alpha(\Delta t)^\alpha}{2AB(\alpha)\Gamma(\alpha+3)} \sum_{l=2}^{n} \left[F\left(\left(t_l, y^l\right)\right) - 2F\left(t_{l-1}, y^{l-1}\right) + F\left(t_{l-2}, y^{l-2}\right)\right]$$

$$\times \left[\begin{array}{l} \left(n-l+1\right)^\alpha \left[2\left(n-l\right)^2 + \left(3\alpha+10\right)\left(n-l\right) + 2\alpha^2 + 9\alpha + 12\right] \\ -\left(n-l\right)^\alpha \left[2\left(n-l\right)^2 + \left(5\alpha+10\right)\left(n-l\right) + 6\alpha^2 + 18\alpha + 12\right] \end{array}\right] \tag{14.35}$$

The following approximation is thus obtained:

$$y^{n+1} = y^0 + \frac{1-\alpha}{AB(\alpha)}\beta t_n^{\beta-1} f\left(t_n, y(t_n)\right) + \frac{\alpha(\Delta t)^\alpha}{AB(\alpha)\Gamma(\alpha+1)}\sum_{l=2}^{n} t_{l-2}^{\beta-1} F\left(t_{l-2}, y^{l-2}\right)$$

$$\left[(n-l+1)^\alpha - (n-l)^\alpha\right] + \frac{\alpha(\Delta t)^\alpha}{AB(\alpha)\Gamma(\alpha+2)}\sum_{l=2}^{n} F\left(t_{l-2}^{\beta-1}\left(t_{l-1}, y^{l-1}\right) - ft_{l-2}^{\beta-1}\left(t_{l-2}, y^{l-2}\right)\right)$$

$$\times\begin{bmatrix}(n-l+1)^\alpha(n-l+3+2\alpha)-\\(n-l)^\alpha(n-l+3+3\alpha)\end{bmatrix} + \frac{\alpha\beta(\Delta t)^\alpha}{AB(\alpha)\Gamma(\alpha+3)}\sum_{l=2}^{n} t_i^{\beta-1} f\left(t_n, y(t_n)\right)$$

$$-2t_{l-2}^{\beta-1} f\left(t_{l-1}, y^{l-1}\right) + t_{l-2}^{\beta-1}\left(t_{l-2}, y^{l-2}\right)$$

$$\begin{bmatrix}(n-l+1)^\alpha\left[2(n-l)^2+(3\alpha+10)(n-l)+2\alpha^2+9\alpha+12\right]\\-(n-l)^\alpha\left[2(n-l)^2+(5\alpha+10)(n-l)+6\alpha^2+18\alpha+12\right]\end{bmatrix}$$

(14.36)

We now discretize in space,

$$f\left(r_i, t_n, h(r_i, t_n)\right) = \frac{T}{S}\left[\begin{array}{c}\dfrac{1}{r_i}\dfrac{h(r_{i+1}, t_n)-h(r_{i-1}, t_n)}{2\Delta r}\\+\dfrac{h(r_{i+1}, t_n)-2h(r_i, t_n)+h(r_{i-1}, t_n)}{(\Delta r)^2}\left(1+\dfrac{\Delta r}{r_i}\right)\end{array}\right],$$

$$f\left(r_i, t_{l-2}, h(r_i, t_{l-2})\right) = \frac{T}{S}\left[\begin{array}{c}\dfrac{1}{r_i}\dfrac{h(r_{i+1}, t_{l-2})-h(r_{i-1}, t_{l-2})}{2\Delta r}\\+\dfrac{h(r_{i+1}, t_{l-2})-2h(r_i, t_{l-2})+h(r_{i-1}, t_{l-2})}{(\Delta r)^2}\left(1+\dfrac{\Delta r}{r_i}\right)\end{array}\right],$$

$$f\left(r_i, t_{l-1}, h(r_i, t_{l-1})\right) = \frac{T}{S}\left[\begin{array}{c}\dfrac{1}{r_i}\dfrac{h(r_{i+1}, t_{l-1})-h(r_{i-1}, t_{l-1})}{2\Delta r}\\+\dfrac{h(r_{i+1}, t_{l-1})-2h(r_i, t_{l-1})+h(r_{i-1}, t_{l-1})}{(\Delta r)^2}\left(1+\dfrac{\Delta r}{r_i}\right)\end{array}\right].$$

14.2.2 ANALYSIS WITH CAPUTO FRACTAL-FRACTIONAL DERIVATIVES

We are now going to analyze general groundwater flow equation for a confined aquifer using the Caputo fractal-fractional derivative. Similar to previous section, we will apply the Newton polynomial to discretize our formula. This method has been developed by Atangana and Araz (2019) as an addition to Adams–Bashforth that has been used by many scholars in their respective researches.

Analyses begins by handling the following Cauchy problem,

$$^{FFP}_{0}D_t^{\alpha,\beta} y(t) = f\left(t, y(t)\right)$$

(14.37)

Integrating Equation (14.37), we get:

$$y(t) - y(0) = \frac{\alpha\beta}{\Gamma(\alpha)} \int_0^t \tau^{\beta-1} f(\tau, y(\tau))(t-\tau)^{\alpha-1} d\tau \tag{14.38}$$

By taking $F(t, y(t)) = t^{\beta-1} f(t, y(t))$.
Thus, we get

$$y(t) - y(0) = \frac{1}{\Gamma(\alpha)} \int_0^t F(\tau, y(\tau))(t-\tau)^{\alpha-1} d\tau \tag{14.39}$$

where the function F is nonlinear. A numerical scheme to solve the above was provided. At the point $t_{n+1} = (n+1)\Delta t$, we have:

$$y(t_{n+1}) - y(0) = \frac{1}{\Gamma(\alpha)} \int_0^{t_{n+1}} F(\tau, y(\tau))(t_{n+1} - \tau)^{\alpha-1} d\tau \tag{14.40}$$

Hence,

$$y(t_{n+1}) = y(0) + \frac{1}{\Gamma(\alpha)} \sum_{l=2}^n \int_{t_l}^{t_{l+1}} F(\tau, y(\tau))(t_{n+1} - \tau)^{\alpha-1} d\tau \tag{14.41}$$

The Newton polynomial was then substituted to the Caputo fractal-fractional derivative, we get:

$$y(t_{n+1}) = y(0) + \frac{1}{\Gamma(\alpha)} \sum_{l=2}^n \int_{t_l}^{t_{l+1}} \left\{ \begin{array}{c} F(t_{l-2}, y^{l-2}) \\ + \dfrac{F(t_{l-1}, y^{l-1}) - F(t_{l-2}, y^{l-2})}{\Delta t}(\tau - t_{l-2}) \\ + \dfrac{F(t_l, y^l) - 2F(t_{l-1}, y^{l-1}) + F(t_{l-2}, y^{l-2})}{2(\Delta t)^2} \\ \times (\tau - t_{l-2})(\tau - t_{l-1}) \\ (t_{n+1} - \tau)^{\alpha-1} d\tau \end{array} \right\} \tag{14.42}$$

The above equation can also be written as,

$$y(t_{n+1}) = y(0) + \frac{1}{\Gamma(\alpha)} \sum_{l=2}^n \left\{ \begin{array}{c} \displaystyle\int_{t_l}^{t_{l+1}} F(t_{l-2}, y^{l-2}) \\ + \displaystyle\int_{t_l}^{t_{l+1}} \dfrac{F(t_{l-1}, y^{l-1}) - F(t_{l-2}, y^{l-2})}{\Delta t}(\tau - t_{l-2})(t_{n+1} - \tau)^{\alpha-1} d\tau \\ + \displaystyle\int_{t_l}^{t_{l+1}} \dfrac{F(t_l, y^l) - 2F(t_{l-1}, y^{l-1}) + F(t_{l-2}, y^{l-2})}{2(\Delta t)^2} \\ \times (\tau - t_{l-2})(\tau - t_{l-1})(t_{n+1} - \tau)^{\alpha-1} d\tau \end{array} \right\} \tag{14.43}$$

Thus the following equality is obtained

$$y(t_{n+1}) = y(0) + \frac{1}{\Gamma(\alpha)} \sum_{l=2}^{n} F(t_{l-2}, y^{l-2}) \int_{t_l}^{t_{l+1}} (t_{n+1} - \tau)^{\alpha-1} d\tau$$

$$+ \frac{1}{\Gamma(\alpha)} \sum_{l=2}^{n} \frac{F(t_{l-1}, y^{l-1}) - F(t_{l-2}, y^{l-2})}{\Delta t} \int_{t_l}^{t_{l+1}} (\tau - t_{l-2})(t_{n+1} - \tau)^{\alpha-1} d\tau$$

$$+ \frac{1}{\Gamma(\alpha)} \sum_{l=2}^{n} \frac{F(t_l, y^l) - 2F(t_{l-1}, y^{l-1}) + F(t_{l-2}, y^{l-2})}{2(\Delta t)^2}$$

$$\times \int_{t_l}^{t_{l+1}} (\tau - t_{l-2})(\tau - t_{l-1})(t_{n+1} - \tau)^{\alpha-1} d\tau \tag{14.44}$$

The above integrals are then calculated as follows:

$$\int_{t_l}^{t_{l+1}} (t_{n+1} - \tau)^{\alpha-1} d\tau = \frac{(\Delta t)^\alpha}{\alpha} \left[(n-l+1)^\alpha - (n-l)^\alpha \right] + \int_{t_l}^{t_{l+1}} (\tau - t_{l-2})(t_{n+1} - \tau)^{\alpha-1} d\tau$$

$$= \frac{(\Delta t)^{\alpha+1}}{\alpha(\alpha+1)} \left[\begin{array}{c} (n-l+1)^\alpha (n-l+3+2\alpha) \\ -(n-l)^\alpha (n-l+3+3\alpha) \end{array} \right]$$

$$+ \int_{t_l}^{t_{l+1}} (\tau - t_{l-2})(\tau - t_{l-1})(t_{n+1} - \tau)^{\alpha-1} d\tau = \frac{(\Delta t)^{\alpha+2}}{\alpha(\alpha+2)}$$

$$\left[(n-l+1)^\alpha \left[\begin{array}{c} 2(n-l)^2 + (3\alpha+10)(n-l) + 2\alpha^2 + 9\alpha + 12 \\ -(n-l)^\alpha \left[2(n-l)^2 + (5\alpha+10)(n-l)^\alpha + 6\alpha^2 + 18\alpha + 12 \right] \end{array} \right] \right] \tag{14.45}$$

The above completes the analysis.
We now discretize in space,

$$x_i = \Delta x i$$

$$f(r_i, t_n, h(r_i, t_n)) = \frac{T}{S} \left[\begin{array}{c} \dfrac{1}{r_i} \dfrac{h(r_{i+1}, t_n) - h(r_{i-1}, t_n)}{2\Delta r} \\ + \dfrac{h(r_{i+1}, t_n) - 2h(r_i, t_n) + h(r_{i-1}, t_n)}{(\Delta r)^2} \left(1 + \dfrac{\Delta r}{r_i} \right) \end{array} \right]$$

$$f(r_i, t_{l-2}, h(r_i, t_{l-2})) = \frac{T}{S} \left[\begin{array}{c} \dfrac{1}{r_i} \dfrac{h(r_{i+1}, t_{l-2}) - h(r_{i-1}, t_{l-2})}{2\Delta r} \\ + \dfrac{h(r_{i+1}, t_{l-2}) - 2h(r_i, t_{l-2}) + h(r_{i-1}, t_{l-2})}{(\Delta r)^2} \left(1 + \dfrac{\Delta r}{r_i} \right) \end{array} \right]$$

$$f(r_i, t_{l-1}, h(r_i, t_{l-1})) = \frac{T}{S} \left[\begin{array}{c} \dfrac{1}{r_i} \dfrac{h(r_{i+1}, t_{l-2}) - h(r_{i-1}, t_{l-2})}{2\Delta r} \\ + \dfrac{h(r_{i+1}, t_{l-1}) - 2h(r_i, t_{l-1}) + h(r_{i-1}, t_{l-1})}{(\Delta r)^2} \left(1 + \dfrac{\Delta r}{r_i} \right) \end{array} \right]$$

14.3   NUMERICAL SIMULATION AND DISCUSSION

Using the above obtained numerical scheme, numerical simulations are depicted in Figures 14.1 to 14.8 below for different values of fractal dimension and fractional orders.

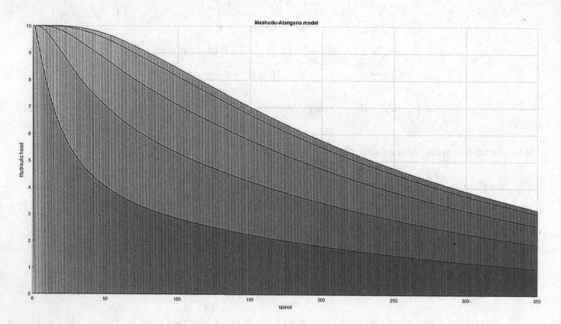

FIGURE 14.1   Numerical simulation with theoretical value of (1, 1) showing H and S.

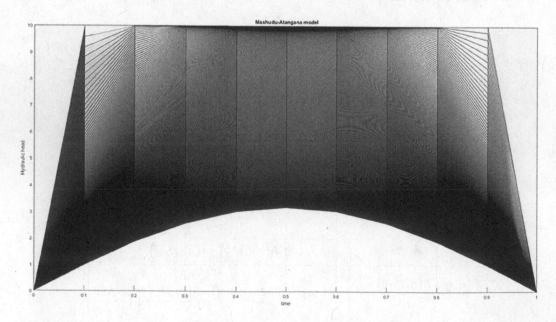

FIGURE 14.2   Numerical simulation with theoretical value of 1 showing H and T.

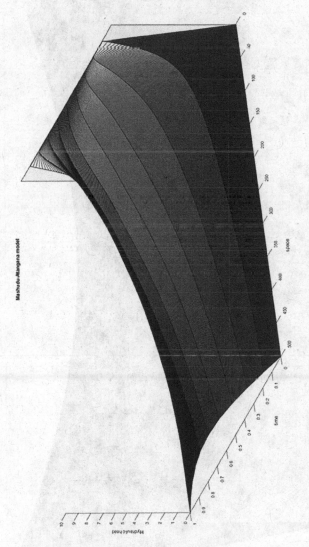

FIGURE 14.3 Numerical simulation with theoretical value of 0.9 showing H, S and T.

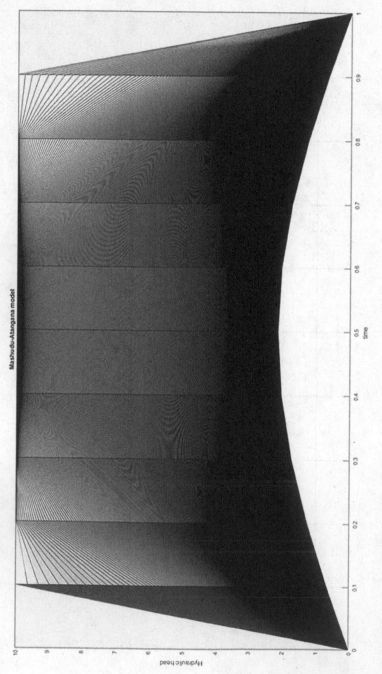

FIGURE 14.4 Numerical simulation with theoretical value of 0.9 showing H and T.

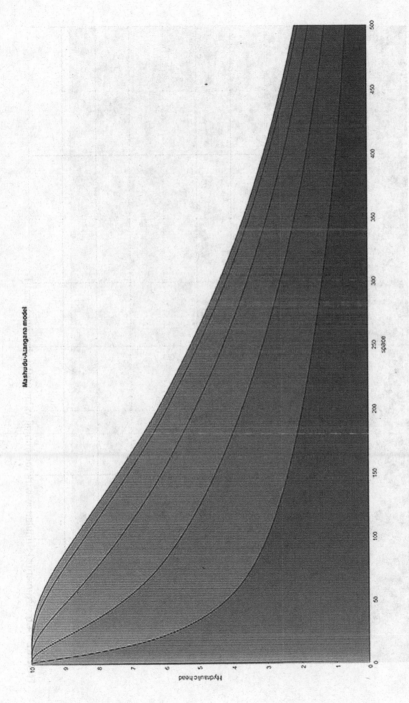

FIGURE 14.5 Numerical simulation with theoretical value of 0.9 showing H and S.

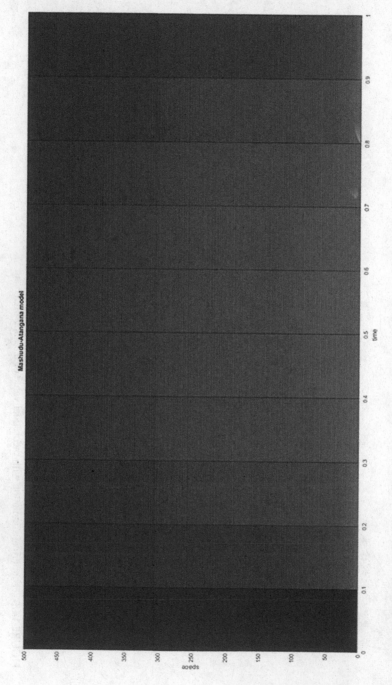

FIGURE 14.6   Numerical Simulation with theoretical value of 0.9 showing S and T.

Mashudu-Atangana model

FIGURE 14.7 Numerical Simulation with theoretical value of 0.7 showing H, S and T.

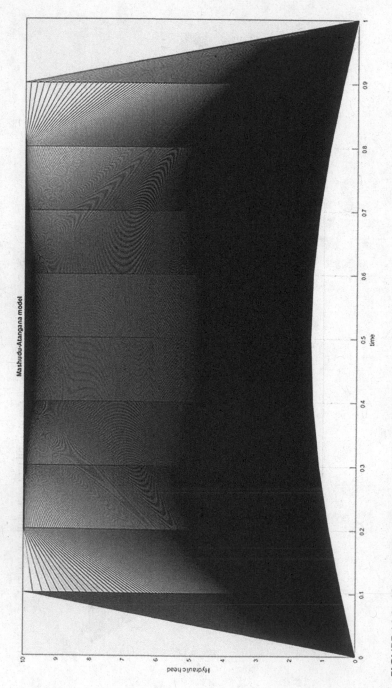

FIGURE 14.8   Numerical Simulation with theoretical value of 0.7 showing H and T.

14.4 CONCLUSION

To include into mathematical formulation the effect of self-similarities together with power processes, we have made use of a differential operator able to depict at the same time power law processes and self-similarities. The power law processes can be captured by fractional orders while self-similarities can be capture by fractal dimension. A new numerical scheme, based on the Newton interpolation, was used to provide numerical solutions. Stability analysis and numerical simulations were performed for different values of fractal-fractional couple orders.

REFERENCES

Atangana, A. & Araz, S.I. (2019). Fractional stochastic modelling illustration with modified Chua attractor. *European Physical Journal*, 134(4): 1–23.

Caputo, M. & Fabrizio, M. (2015). A new Definition of Fractional Derivative without Singular Kernel. *Progress in Fractional Differentiation and Applications*, 1(2): 73–85.

Chen, W. (2006). Time–space fabric underlying anomalous diffusion. *Chaos, Solitons and Fractals*, 28(4): 923–929.

Dalir, M. & Bashour, M. (2010). Applications of fractional calculus. *Applied Mathematical Sciences*, 4: 1021–1032.

Matlob, M.A. & Jamali, Y. (2017). The concepts and applications of fractional order differential calculus in modelling of viscoelastic systems: A primer. *Critical Reviews in Biomedical Engineering*, 47(4): 249–276.

Samko, S.G., Kilbas, A.A., & Marichev, O.I. (1993). Fractional integrals and derivatives. Translated from the 1987 *Russian Original*. Gordon and Breach, Yverdon.

15 A New Model for Groundwater Contamination Transport in Dual Media

Mpafane Deyi
University of the Free State, Bloemfontein, South Africa

Abdon Atangana
China Medical University, Taichung, Taiwan
University of the Free State, Bloemfontein, South Africa

CONTENTS

15.1 INTRODUCTION

Groundwater has been considered to be, and used as, a reliable source of drinking water and irrigation in many places across the globe. In Egypt, wells were already being used in 3000 BC (Katko, 1997, in Sharma, 2001). Groundwater is still regarded nowadays, as a reliable source of water for many people in different countries and for different water uses ranging from domestic, and irrigation to industrial. According to the study done by Margat & van der Gun (2013): the worldwide groundwater use distribution based on the data collected from different countries is presented below:

- Groundwater abstraction for purposes of irrigation is reported to be at 70% across the world;
- Approximately 21% of groundwater abstraction is for domestic use purposes; and
- Approximately 9% of groundwater abstracted is intended/used for industrial and for mining activities.

By and large, groundwater is used the most for agricultural (irrigation) activities in many parts of the world. Contaminated groundwater may translate to a health risk for millions of people consuming

DOI: 10.1201/9781003266266-15

produce from farmlands. On the other hand, irrigation activities may mobilize the transport of contaminants (chemicals and fertilizers) into the aquifers.

Heavy metal contamination in food produce is regarded as one of the major contaminating agents (Gholizadeh et al., 2009). Research has shown that the most toxic and abundant metals in food are lead and cadmium. Chronic health conditions such as cardiovascular, bone, and kidney diseases, to name but a few, are in the main caused as a result of an excessive accumulation of heavy metals in human bodies (Khan et al., 2009).

Incidents of arsenic in food have been studied and found to be in excessive concentrations in certain countries. Arsenic is well-known to be a highly toxic element, and the consumption of food containing arsenic is a cause for concern for the well-being of human beings and animals (Al Rmalli et al., 2005). Arsenic presence in groundwater has been reported as one of serious environmental health hazards across the globe. The consumption of food grown in fields irrigated with arsenic water is one serious health risk to human beings (Das et al., 2004).

15.2 GROUNDWATER CONTAMINATION

Groundwater is deemed to have been stored and filtered through the soil matrix over long periods and therefore in most cases regarded as 'pure'. Furthermore, groundwater is often available closer to where the demand is located and requires less water treatment as compared to water sourced from a river.

Against these common advantages, it should be noted that groundwater is not immune to contamination by various anthropogenic activities such as agricultural, domestic and industrial. Contrary to the popular impression and believe that the waters from the springs and wells are 'pure', patterns of pervasive pollution of groundwater are being uncovered (Sharma, 2001).

Groundwater contamination incidents have occurred in different places in the world, and a wide range of materials identified as contaminants are found in groundwater. These include organic hydrocarbons, chemicals (organic and inorganic), inorganic cations, inorganic anions, pathogens, and radionuclides. Most of these materials are dissolved in water at various concentrations. Some of the organic compounds exist in both a dissolved form and as an insoluble non-aqueous phase, which can also migrate through the ground (Fetter et al., 2017).

One of the worst breakouts of groundwater pollution is perhaps the Love Canal scenario that led to the loss of many lives of humans and animals. The Love Canal became the first human-made disaster, and is the worst chemical disposal and pollution disaster ever recorded in America. It is estimated that there are well over 50,000 known landfill sites containing toxic waste that may pose a threat to public health and groundwater contamination (Worthley & Torkelson, 1981).

Other emerging contaminants found in groundwater include pharmaceutical active compounds, personal care products, industrial chemicals, and hormones such as endocrine disrupting compounds (Fetter et al., 2017; Maeng, 2010).

15.3 CONTAMINATION TRANSPORT IN DUAL MEDIA

A fair amount of effort has been invested in the development of an understanding of groundwater flow in dual media; however, this can't be said for contamination transport in dual media, and less research has been done on this topic thus far.

Over the years we have seen great interest in groundwater transport in dual media; for example, Beranblatt et al. (1960) introduced a concept of modeling exchange between a fracture network and the matrix. They assumed that the flow occurs at steady state between fracture and matrix. Furthermore, in their model they assumed that a porous medium is made up of two separate but connected continua.

- Double-priority type models to estimate the flow of water in a fractured reservoir are also reported in the literature (Warren and Root, 1963; Duguid and Lee, 1977; Moench, 1984);

- Kazemi et al. (1976) introduced an extension of the dual porosity model proposed by Warren and Root (1963) to a two-phase flow that could account for fluid movement, the effects of gravity, and variation in formation properties; and
- Thomas et al. (1983) developed a three-dimensional and a three-phase model for simulating the flow of water, oil and gas in fractured systems.

In this chapter we attempt to introduce a system of equations to model contamination transport in dual media, and we test the uniqueness of the proposed equations. The derived equations are evaluated and solved numerically using different methods proposed in earlier literature.

15.4   DERIVATION OF EQUATIONS AND NUMERICAL ANALYSIS

Mathematical equations have been used extensively to explain natural, physical, chemical, biological and mechanical processes taking place during groundwater transport. Edelman (1972) summarized his thoughts into 'we have no everyday experience with groundwater, as we have with mechanical phenomena; the best way to get acquainted with the nature of the phenomena is to solve, as an exercise, a series of elementary problems'.

We begin the derivation of the equation from the ancient Darcy's law equation.

$$Q = AK\frac{dh}{dL} \tag{15.1}$$

where:

Q = discharge [m^3/hr]
A = cross-sectional area [m^2]
K = hydraulic conductivity [m/hr]
$\dfrac{dh}{dL}$ = hydraulic gradient or change in hydraulic head (h) per change in distance (L)

Now deriving a Darcian velocity or flux [m^3/hr/m^2], where we divide the flow rate by cross-sectional area, therefore the volumetric discharge reduces to a 1-dimensional discharge (q).

$$\frac{Q}{A} - K\frac{dh}{dL} \tag{15.2}$$

Therefore the Darcian velocity (q) is:

$$q = K\frac{dh}{dL} \tag{15.3}$$

It must be noted that mentioned above is the flow rate and Darcian velocity of water. When dealing with different types of liquids with varying densities or viscosity, an intrinsic permeability (k) must be determined. k is a value that characterizes the ease with which any fluid flows through porous media.

$$Q = Ak\frac{\rho_{fluid}g}{\mu_{fluid}}\frac{dh}{dL} \tag{15.4}$$

where:

Q = discharge [m^3/]
A = cross-sectional area [m^2]
k = intrinsic permeability [m^2]
ρ_{fluid} = density of the fluid [kg/m^3]

μ_{fluid} = viscosity of the fluid [kg/m*hr]

g = gravitational constant [m/t^2]

$\dfrac{dh}{dL}$ = hydraulic gradient or change in hydraulic head (h) per change in distance (L).

15.5 RELATIONSHIP BETWEEN HYDRAULIC CONDUCTIVITY AND INTRINSIC PERMEABILITY

A relationship between hydraulic conductivity (K) and intrinsic permeability (k) is demonstrated below:

$$k = K\,\frac{\mu_{fluid}}{\rho_{fluid}\,g} \tag{15.5}$$

Deriving an advection equation, advection is nothing but the movement of a volume of a chemical or biological solute or dissolved over an area at a particular time. It is considered to be the main process that drives the transport mechanism called advective transport.

Deriving an advection equation in one dimension of Darcy's law, it is basically the flux, or the advection flux and can be written as:

$$\frac{\partial C}{\partial t} = -v\,\frac{\partial C}{\partial x} \tag{15.6}$$

The above equation can also be written as

$$\frac{\partial C}{\partial t} = -K\,\frac{dh}{dL}\left(\frac{\partial C}{\partial x}\right) \tag{15.7}$$

We must note that due to the heterogeneity of geologic materials, advective transport in different strata can result in solute fronts spreading at different rates in each stratum (Fetter et al., 2017).

15.6 HYDRODYNAMIC DISPERSION

Dutta (2013) defines the hydrodynamic dispersion as the stretching of a solute band in the flow direction during its transport by an advecting fluid. Hydrodynamic dispersion coefficient is:

$$D_{hydrodynamic} = D_{mechanical} + D_{molecular}D_h = D_{mech.} + D_{molec} \tag{15.8}$$

Molecular diffusion is the spread of the solute molecules because of thermal motion.
Given by Fick's first law:

$$F_{diff} = -D_{molec}\,\frac{\partial C}{\partial x} \tag{15.9}$$

The negative sign indicates that the movement is from areas of greater concentration to those of lesser concentration (Fetter et al., 2017).

where $\dfrac{dC}{dx}$ = the concentration gradient.

Therefore, the diffusion Flux is Fick's first law

$$\text{Diffusion}_{flux} = -D_{molec}\,\frac{dC}{dx} \tag{15.10}$$

$$\frac{\partial C}{\partial t} = -D_{molec}\frac{\partial^2 C}{\partial x^2}$$ (15.11)

Therefore the total flux (F) is:

1. Dimension Flux:

$$\frac{\partial C}{\partial t} = -v\frac{\partial C}{\partial x} + D_h\left(\frac{\partial^2 C}{\partial x^2}\right)$$ (15.12)

2. Dimension Flux:

$$\frac{\partial C}{\partial t} = -v\frac{\partial C}{\partial x} + D_L\left(\frac{\partial^2 C}{\partial x^2}\right) + D_T\left(\frac{\partial^2 C}{\partial y^2}\right)$$ (15.13)

D_L = is the longitudinal dispersivity or diffusion
D_T = is the transverse dispersivity or diffusion.

15.7 RETARDATION FACTOR

According to Berkowitz (2002), sorption reactions in groundwater systems involve many kinds of chemical constituents and different physical and chemical processes. Chemical sorption into colloids is usually governed by a slow kinetic process. Highly mobile colloids can propagate quickly through an interconnected network of fractures and thus enhance significantly the rate of contaminant migration. The added difficulty associated with measuring colloid movement (and estimating, e.g., colloid filtration coefficients) and colloid-chemical interactions at the field scale serves to further increase model and prediction uncertainty.

According to Fetter et al. (2017) sorption processes include adsorption, chemisorption, absorption, and ion exchange. Adsorption includes the processes by which a solute clings to a solid surface. Chemisorption occurs when the solute is incorporated on a sediment, soil, or rock surface by a chemical reaction. Absorption occurs when the aquifer particles are porous so that the solute can diffuse into the particle and be sorbet on to interior surfaces.

Equilibrium surface reaction isotherm is basically a relationship that is not a function of time, showing the concentration in solution (C) versus that absorbed (S) on the solid surface.

Shown below is a transport equation has two variables, C and S.

$$\frac{\partial C}{\partial t} + \beta\frac{\partial S}{\partial t} = D_h\frac{\partial^2 C}{\partial x^2} - v\frac{\partial C}{\partial x}$$ (15.14)

where:
β = is the maximum amount of solute that can be sorbed.
Replacing S with $K_D C$

$$\frac{\partial C}{\partial t} + \beta\frac{\partial k_D C}{\partial t} = D_h\frac{\partial^2 C}{\partial x^2} - v\frac{\partial C}{\partial x}$$ (15.15)

Now simplifying the above equation:

$$\left(1+\beta k_D\right)\frac{\partial C}{\partial t}=D_h\frac{\partial^2 C}{\partial x^2}-v\frac{\partial C}{\partial x} \tag{15.16}$$

where:

$(1+\beta k_D)$ = is the retardation factor

Therefore the equation to model transport in the matrix is:

$$\left(1+\beta k_D\right)\frac{\partial C}{\partial t}=D_h\frac{\partial^2 C}{\partial x^2}-v\frac{\partial C}{\partial x}. \tag{15.17}$$

15.8   GROUNDWATER TRANSPORT IN FRACTURE

Now deriving an equation for groundwater transport in fracture, we first bring an understanding of the movement of water through soil as a steady flow and can be best described with Darcy's equation:

$$q=K\frac{dh}{dL} \tag{15.18}$$

Now defining the effective K as:

$$K_{effective}=-\delta^3\lambda\frac{\gamma}{12\,\mu} \tag{15.19}$$

where:

δ = aperture [L]

λ = fracture density [L^2/L^3] or [L/L^2] or [1/L]

γ = unit weight of water

μ = dynamic viscosity

v = velocity

ρ = density [M/L^3]

n = porosity of a fracture [dimension less]

Now replacing K in Equation (15.18) we get:

$$q=-\delta^3\lambda\frac{\gamma}{12\,\mu}\frac{dh}{dL} \tag{15.20}$$

The above equation represents Darcy's law with the effective hydraulic conductivity $K_{effective}$ and q being the effective flux.

15.9   SOLVING FOR AN APERTURE

Now solving for an effective aperture we get:

$$\delta=\left(\frac{K}{\lambda}\frac{12\mu}{\gamma}\right)^{1/3} \tag{15.21}$$

Please note that the above equation does not consider things like infilling and fluid pressure, as both have an impact on an aperture. The average velocity through fracture:

$$v_x = \frac{q}{n} = \frac{-K}{n} \frac{dh}{dx} \tag{15.22}$$

The average porosity in a fracture is expressed as:

$$\text{porosity} = n = \delta\lambda \tag{15.23}$$

Now substituting n by $\delta\lambda$ we get:

$$v_x = \frac{q}{n} = \frac{-K}{\delta\lambda} \frac{dh}{dx} \tag{15.24}$$

Substituting with the effective aperture in the equation as expressed in Equation (15.22) above.

$$v_x = -K^{2/3} \left(12\lambda^2 \frac{\mu}{\gamma}\right)^{-1/3} \frac{dh}{dx} \tag{15.25}$$

Going back to the one dimension advection diffusion/dispersion equation derived above for the matrix.

$$\frac{\partial C}{\partial t} = -v_x \frac{\partial C}{\partial x} + D_L \left(\frac{\partial^2 C}{\partial x^2}\right) + D_T \left(\frac{\partial^2 C}{\partial y^2}\right) \tag{15.26}$$

Now replacing v_x in Equation (15.26) to make it an advection diffusion/dispersion equation for a fracture:

$$\frac{\partial C_f}{\partial t} = \left(-K^{2/3} \left(12\lambda^2 \frac{\mu}{\gamma}\right)^{-1/3} \frac{dh}{dx}\right) \frac{\partial C_f}{\partial x} + D_L \left(\frac{\partial^2 C_f}{\partial x^2}\right) + D_T \left(\frac{\partial^2 C_f}{\partial y^2}\right) \tag{15.27}$$

Transverse diffusion (D_T) is negligible because it is expected to be very small, we can therefore regard that term as zero, and therefore the advection diffusion equation for a fracture is

$$\frac{\partial C_f}{\partial t} = \left(-K^{2/3} \left(12\lambda^2 \frac{\mu}{\gamma}\right)^{-1/3} \frac{dh}{dx}\right) \frac{\partial C_f}{\partial x} + D_L \left(\frac{\partial^2 C_f}{\partial x^2}\right) \tag{15.28}$$

Now we have two sets of equations, for matrix and fracture:
Matrix:

$$(1 + \beta k_D) \frac{\partial C}{\partial t} = D_h \frac{\partial^2 C}{\partial x^2} - v \frac{\partial C}{\partial x} \tag{15.29}$$

Fracture when considering retardation effect:

$$(1 + \beta k_D) \frac{\partial C_f}{\partial t} = \left(-K^{2/3} \left(12\lambda^2 \frac{\mu}{\gamma}\right)^{-1/3} \frac{dh}{dx}\right) \frac{\partial C_f}{\partial x} + D_h \left(\frac{\partial^2 C_f}{\partial x^2}\right). \tag{15.30}$$

15.10   UNIQUENESS OF THE PROPOSED EQUATIONS

The Picard–Lindelöf theorem is used to prove the uniqueness of the proposed system of equations for the matrix and fracture network. The Picard–Lindelöf theorem is a well-known and important theorem on existence and uniqueness of solutions to first-order differential equations with given initial conditions (Nevanlinna, 1989).

In this section we are considering the proposed equations:

For the soil matrix:

$$(1+\beta k_D)\frac{\partial Cm}{\partial t} = D_h\frac{\partial^2 C_m}{\partial x^2} - v_m\frac{\partial C_m}{\partial x} \tag{15.31}$$

For the fracture:

$$(1+\beta k_D)\frac{\partial C_f}{\partial t} = \left(-K^{2/3}\left(12\lambda^2\frac{\mu}{\gamma}\right)^{-1/3}\frac{dh}{dx}\right)\frac{\partial C_f}{\partial x} + D_h\left(\frac{\partial^2 C_f}{\partial x^2}\right) \tag{15.32}$$

$$(1+\beta k_D)\frac{\partial C_f}{\partial t} = D_h\left(\frac{\partial^2 C_f}{\partial x^2}\right) - (v_f)\frac{\partial C_f}{\partial x} \tag{15.33}$$

Now replacing the retardation factor $(1 + \beta k_D)$ with R in both equations we get the following:

For matrix:

$$R\frac{\partial C_m}{\partial t} = D_h\frac{\partial^2 C_m}{\partial x^2} - v_m\frac{\partial C_m}{\partial x} \tag{15.34}$$

For fracture:

$$R\frac{\partial C_f}{\partial t} = D_h\left(\frac{\partial^2 C_f}{\partial x^2}\right) - (v_f)\frac{\partial C_f}{\partial x} \tag{15.35}$$

Taking into account the transfer term, we add it. A transfer term determines the communication between fracture and matrix domains. The transfer term is the most relevant feature in a dual porosity model as it governs the flow between matrix and fracture.

Adopting the definition of Warren & Root (1963) and change the pressure differential to between the matrix and fracture to concentration difference.

$$R\frac{\partial C_f}{\partial t} = D_h\left(\frac{\partial^2 C_f}{\partial x^2}\right) - (v_f)\frac{\partial C_f}{\partial x} + \sigma\frac{k_m}{\mu}(C_m - C_f) \tag{15.36}$$

To simplify the equations we divide the above set of equations by the retardation factor.

$$\frac{\partial C_m}{\partial t} = \frac{D_h}{R}\frac{\partial^2 C_m}{\partial x^2} - v_m\frac{v_m}{R}\frac{\partial C_m}{\partial x} \tag{15.37}$$

$$\frac{\partial C_f}{\partial t} = \frac{D_h}{R}\left(\frac{\partial^2 C_f}{\partial x^2}\right) - \frac{v_f}{R}\frac{\partial C_f}{\partial x} + \frac{\sigma}{R}\frac{k_m}{\mu}(C_m - C_f) \tag{15.38}$$

For simplicity, we let:

$$\frac{\partial C_m}{\partial t} = F\left(x, t, C_m, C_f\right) \tag{15.39}$$

$$\frac{\partial C_f}{\partial t} = H\left(x, t, C_m, C_f\right) \tag{15.40}$$

then applying the integral on both sides we get:
For matrix, we have

$$C_m\left(x, t\right) - C_m\left(x, 0\right) = \int_0^t F\left(x, \tau, C_m\left(x, \tau\right), C_f\left(x, \tau\right)\right) d\tau \tag{15.41}$$

$$C_m\left(x, t\right) = C_m\left(x, 0\right) + \int_0^t F\left(x, \tau, C_m\left(x, \tau\right), C_f\left(x, \tau\right)\right) d\tau \tag{15.42}$$

For fracture, we have

$$C_f\left(x, t\right) - C_f\left(x, 0\right) = \int_0^t H\left(x, \tau, C_m\left(x, \tau\right), C_f(x, \tau)\right) d\tau \tag{15.43}$$

$$C_f\left(x, t\right) = C_f\left(x, 0\right) + \int_0^t H\left(x, \tau, C_m\left(x, \tau\right), C_f(x, \tau)\right) d\tau \tag{15.44}$$

Then we define:

$$C\left(x, t\right) - C\left(x, 0\right) + \int_0^t \pi\left(x, \tau, C\left(x, \tau\right)\right) d\tau \tag{15.45}$$

$$C\left(x, t\right) = \left[C_m\left(x, t\right), C_f\left(x, t\right)\right] \tag{15.46}$$

$$C\left(x, 0\right) = \left[C_m\left(x, 0\right), C_f\left(x, 0\right)\right] \tag{15.47}$$

where:

$$\pi\left(x, t, C\left(x, t\right)\right) = F\left(x, t, C_m\left(x, t\right), C_f\left(x, t\right)\right) H\left(x, t, C_m\left(x, t\right), C_f(x, t)\right) \tag{15.48}$$

Now defining Aquifer parameter (A_p)

$$A_p = T_{t_{\max}}\left(t_0\right) X_{C_0}\left(b\right) \tag{15.49}$$

$$T_{t_{\max}}\left(t_0\right) = \left[t_{\max} - t_0, t_0 + t_{\max}\right] \tag{15.50}$$

$$X_{C_0}(b) = [C_0 - b, C_0 + b] \tag{15.51}$$

$$\left\| F(x,t,C_{m1},C_f) \right\| < M_1 \tag{15.52}$$

$$\left\| H(x,t,C_{m2},C_f) \right\| < M_2 \tag{15.53}$$

$$\left\| \pi(x,t,C_m,C_f) \right\| < M_\pi \tag{15.54}$$

We define a contraction mapping (Υ) as:

$$\Upsilon \vee (x,t) = C(x,0) + \int_0^t \pi(x,\tau,C(x,\tau)) d\tau \tag{15.55}$$

The supremum infimum is defined below as:

$$\|\vee\| = \mathop{\sup}_{t \in T_{t_{max}}(t_0)} |\vee(x,t)| \tag{15.56}$$

$$\|\vee\| = \mathop{\sup}_{x \in X_{C_0}(b)} |\vee(x,t)| \tag{15.57}$$

$$\left\| \Upsilon \vee (x,t) - C(x,0) \right\| = \left\| \int_0^t \pi(x,\tau,C(x,\tau)) d\tau \right\| \tag{15.58}$$

$$\left\| \Upsilon \vee (x,t) - C(x,o) \right\| = \int_0^t \pi(x,\tau,C(x,\tau)) d\tau \tag{15.59}$$

$$\leq \int_0^t \left\| \pi(x,\tau,C(x,\tau)) \right\| d\tau$$

$$< \int_0^t M_\pi d\tau$$

$$< M_\pi t_{max} < b$$

$$t_{max} < \frac{b}{M_\pi}$$

$$\left\| \Upsilon \vee (x,t) - \Upsilon \cup (x,t) \right\| < K \left\| \vee(x,t) - \cup(x,t) \right\| \tag{15.60}$$

To prove the above we must first check that:

$$\left\| F\left(x,t,C_m,C_f\right) - F\left(x,t,C'_m,C'_f\right) \right\| < K \left\| C_m - C'_m \right\| \tag{15.61}$$

$$F\left(x,t,C_m,C_f\right) - F\left(x,t,C'_m,C'_f\right) = \frac{D_h}{R} \frac{\partial^2}{\partial x^2}\left[\left(C_m - C'_m\right)\right] - v\frac{\partial}{\partial x}\left[\left(C_m - C'_m\right)\right] \tag{15.62}$$

$$\left\| F\left(x,t,C_m,C_f\right) - F\left(x,t,C'_m,C'_f\right) \right\| \le \frac{D_h}{R} \tag{15.63}$$

$$\left\| \frac{\partial^2}{\partial x^2}\left[C_m - C'_m\right] \right\| + \frac{v}{R}\left\| \frac{\partial}{\partial x}\left[C_m - C'_m\right] \right\| \le \frac{D_h}{R}\theta_2^2 \left\| C_m - C'_m \right\| + \frac{v}{R}\theta_1 \left\| C_m - C'_m \right\| \tag{15.64}$$

$$\le \left(\frac{D_h}{R}\theta_2^2 + \frac{v}{R}\theta_1\right)\left\| C_m - C'_m \right\|$$

$$\le L \left\| C_m - C'_m \right\|$$

In a similar way, we show that:

$$\left\| H\left(x,t,C_m,C_f\right) - H\left(x,t,C'_m,C'_f\right) \right\| \le L \left\| C_m - C'_m \right\| \tag{15.65}$$

and

$$\left\| H\left(x,t,C_m,C_f\right) - H\left(x,t,C'_m,C'_f\right) \right\| \le L_2 \left\| C_f - C'_f \right\|. \tag{15.66}$$

15.11 NUMERICAL ANALYSIS OF SYSTEM OF EQUATIONS

The derived equations above are evaluated and solved numerically using different methods proposed in earlier literature. The first method considered is the Crank–Nicolson method, it is a finite difference method used for numerically solving the heat equation and similar partial differential equations (Cebeci & Bradshaw, 2012). The Crank–Nicolson method is considered to be an unconditionally stable, implicit numerical scheme with second-order accuracy in both time and space (Sun & Trueman, 2003).

The second method considered to do the stability analysis on the proposed equations is von Neumann's. The von Neumann's stability analysis method has been applicable to non-stationary convection–diffusion equations, and on the investigation into the transient behavior of viscoelastic pipelines (Wesseling, 1996; Zecchin et al., 2008).

15.11.1 SOLVING 1-D DIFFUSION WITH ADVECTION FOR STEADY FLOW

The previously derived equations considered include one for the matrix equation (15.17) and one for the fracture equation (15.29).

Equation for the soil matrix:

$$\left(1 + \beta k_D\right)\frac{\partial Cm}{\partial t} = D_h \frac{\partial^2 C_m}{\partial x^2} - v_m \frac{\partial C_m}{\partial x} \tag{15.67}$$

Concentration defined on a spatial interval is shown below:

$$\frac{C_{im}^{j+1} - C_{im}^{j}}{\Delta t} = D_h \frac{1}{2(\Delta x)^2} \left(\left(C_{im}^{j+1} - 2C_{im}^{j+1} + C_{im-1}^{j+1} \right) + \left(C_{im+1}^{j} - 2C_{im}^{j} + C_{im-1}^{j} \right) \right)$$

$$-v_m \frac{1}{2} \left(\left(\frac{C_{im+1}^{j+1} - C_{im-1}^{j+1}}{2\Delta x} \right) + \left(\frac{C_{im+1}^{j} + C_{im-1}^{j}}{2\Delta x} \right) \right) \tag{15.68}$$

Now looking at the equation for fracture:

$$(1 + \beta k_D) \frac{\partial C_f}{\partial t} = \left(-K^{2/3} \left(12\lambda^2 \frac{\mu}{\gamma} \right)^{-1/3} \frac{dh}{dx} \right) \frac{\partial C_f}{\partial x} + D_h \left(\frac{\partial^2 C_f}{\partial x^2} \right) \tag{15.69}$$

Now replacing the term $-K^{2/3} \left(12\lambda^2 \frac{\mu}{\gamma} \right)^{-1/3} \frac{dh}{dx}$ with v_f we get:

$$(1 + \beta k_D) \frac{\partial C_f}{\partial t} = D_h \left(\frac{\partial^2 C_f}{\partial x^2} \right) - (v_f) \frac{\partial C_f}{\partial x} \tag{15.70}$$

Applying and solving the first equation (15.67) numerically using the Crank–Nicolson method where i represents the position (in space) and j represent time in space.
For the soil matrix:

$$(1 + \beta k_D) \frac{\partial C_m}{\partial t} = D_h \frac{\partial^2 C_m}{\partial x^2} - v_m \frac{\partial C_m}{\partial x} \tag{15.71}$$

For simplicity on the analysis, we consider splitting the equation into three terms for the ease of analysis and presentation, we have:

$$(1 + \beta k_D) \frac{\partial C_m}{\partial t}; D_h \frac{\partial^2 C_m}{\partial x^2}; -v_m \frac{\partial C_m}{\partial x} \tag{15.72}$$

Then we present and analyze each term as follow:

$$(1 + \beta k_D) \frac{\partial C_m}{\partial t} = (1 + \beta k_D) \frac{C_{i,m}^{j+1} - C_{i,m}^{j}}{\Delta t} \tag{15.73}$$

$$D_h \frac{\partial^2 C_m}{\partial x^2} = D_h \frac{1}{2(\Delta x)^2} \left(\left(C_{i+1,m}^{j+1} - 2C_{i,m}^{j+1} + C_{i-1,m}^{j+1} \right) + \left(C_{i+1,m}^{j} - 2C_{i,m}^{j} + C_{i-1,m}^{j} \right) \right) \tag{15.74}$$

$$-v_m \frac{\partial C_m}{\partial x} = -v_m \frac{1}{2} \left(\left(\frac{C_{i+1,m}^{j+1} - C_{i-1,m}^{j+1}}{2\Delta x} \right) + \left(\frac{C_{i+1,m}^{j} - C_{i-1,m}^{j}}{2\Delta x} \right) \right) \tag{15.75}$$

Therefore the discretized equation for the matrix is shown below:

$$(1 + \beta k_D) \frac{C_{i,m}^{j+1} - C_{i,m}^{j}}{\Delta t} = D_h \frac{1}{2(\Delta x)^2} \left(\left(C_{i+1,m}^{j+1} - 2C_{i,m}^{j+1} + C_{i-1,m}^{j+1} \right) + \left(C_{i+1,m}^{j} - 2C_{i,m}^{j} + C_{i-1,m}^{j} \right) \right)$$

$$-v_m \frac{1}{2} \left(\left(\frac{C_{i+1,m}^{j+1} - C_{i-1,m}^{j+1}}{2\Delta x} \right) + \left(\frac{C_{i+1,m}^{j} - C_{i-1,m}^{j}}{2\Delta x} \right) \right) \tag{15.76}$$

For the fracture:

$$\left(1+\beta k_D\right)\frac{\partial Cf}{\partial t} = D_h \frac{\partial^2 C_f}{\partial x^2} - v_f \frac{\partial C_f}{\partial x} \tag{15.77}$$

Splitting the equation into three terms for the ease of analysis and presentation we have:

$$\left(1+\beta k_D\right)\frac{\partial Cf}{\partial t}; D_h \frac{\partial^2 C_f}{\partial x^2}; -v_f \frac{\partial C_f}{\partial x} \tag{15.78}$$

Then we present and analyze each term as follow:

$$\left(1+\beta k_D\right)\frac{\partial Cf}{\partial t} = \left(1+\beta k_D\right)\frac{C_{i,f}^{j+1} - C_{i,f}^{j}}{\Delta t} \tag{15.79}$$

$$D_h \frac{\partial^2 C_f}{\partial x^2} = D_h \frac{1}{2\left(\Delta x\right)^2}\left(\left(C_{i+1,f}^{j+1} - 2C_{i,f}^{j+1} + C_{i-1,f}^{j+1}\right) + \left(C_{i+1,f}^{j} - 2C_{i,f}^{j} + C_{i-1,f}^{j}\right)\right) \tag{15.80}$$

$$-v_f \frac{\partial Cf}{\partial x} = -v_f \frac{1}{2}\left(\left(\frac{C_{i+1,f}^{j+1} - C_{i-1,f}^{j+1}}{2\Delta x}\right) + \left(\frac{C_{i+1,f}^{j} - C_{i-1,f}^{j}}{2\Delta x}\right)\right) \tag{15.81}$$

Therefore the discretized equation for the fracture is displayed below:

$$\left(1+\beta k_D\right)\frac{\partial C_f}{\partial t} = D_h \left(\frac{\partial^2 C_f}{\partial x^2}\right) - \left(v_f\right)\frac{\partial C_f}{\partial x} \tag{15.82}$$

Then we have:

$$\left(1+\beta k_D\right)\frac{C_{i,f}^{j+1} - C_{i,f}^{j}}{\Delta t} = D_h \frac{1}{2\left(\Delta x\right)^2}\left(\left(C_{i+1,f}^{j+1} - 2C_{i,f}^{j+1} + C_{i-1,f}^{j+1}\right) + \left(C_{i+1,f}^{j} - 2C_{i,f}^{j} + C_{i-1,f}^{j}\right)\right)$$
$$-v_f \frac{1}{2}\left(\left(\frac{C_{i+1,f}^{j+1} - C_{i-1,f}^{j+1}}{2\Delta x}\right) + \left(\frac{C_{i+1,f}^{j} - C_{i-1,f}^{j}}{2\Delta x}\right)\right) \tag{15.83}$$

In the next section we consider the stability analysis using the von Neumann method.

15.12 STABILITY ANALYSIS USING VON NEUMANN'S METHOD

In this section, Equations (15.74) and (15.81) are analyzed numerically using von Neumann's method to establish the stability of the two equations. Von Neumann's method has been applied to groundwater problems, and particularly on the derivation of new models to test the stability of equations numerically. Bethke (1985) developed a new numerical method that allows for the solving of compaction-driven groundwater flow and associated heat transfer in evolving sedimentary basins, and used von Neumann's method to test the stability of the new model.

Assessing the stability of the equation for the matrix:

$$\left(1+\beta k_D\right)\frac{C_{i,m}^{j+1}-C_{i,m}^{j}}{\Delta t} = D_h \frac{1}{2\left(\Delta x\right)^2}\left(\left(C_{i+1,m}^{j+1}-2C_{i,m}^{j+1}+C_{i-1,m}^{j+1}\right)+\left(C_{i+1,m}^{j}-2C_{i,m}^{j}+C_{i-1,m}^{j}\right)\right)$$
$$-v_m \frac{1}{2}\left(\left(\frac{C_{i+1,m}^{j+1}-C_{i-1,m}^{j+1}}{2\Delta x}\right)+\left(\frac{C_{i+1,m}^{j}-C_{i-1,m}^{j}}{2\Delta x}\right)\right)$$

(15.84)

Now replacing $(1 + \beta k_D)$ with R_m then we have:

$$R_m \frac{C_{i,m}^{j+1}-C_{i,m}^{j}}{\Delta t} = D_h \frac{1}{2\left(\Delta x\right)^2}\left(\left(C_{i+1,m}^{j+1}-2C_{i,m}^{j+1}+C_{i-1,m}^{j+1}\right)+\left(C_{i+1,m}^{j}-2C_{i,m}^{j}+C_{i-1,m}^{j}\right)\right)$$
$$-v_m \frac{1}{2}\left(\left(\frac{C_{i+1,m}^{j+1}-C_{i-1,m}^{j+1}}{2\Delta x}\right)+\left(\frac{C_{i+1,m}^{j}-C_{i-1,m}^{j}}{2\Delta x}\right)\right)$$

(15.85)

Multiplying by Δt on both sides and dividing by R_m

$$\left(C_{i,m}^{j+1}-C_{i,m}^{j}\right) = \frac{D_h}{R_m}\frac{\Delta t}{2\left(\Delta x\right)^2}\left(\left(C_{i+1,m}^{j+1}-2C_{i,m}^{j+1}+C_{i-1,m}^{j+1}\right)+\left(C_{i+1,m}^{j}-2C_{i,m}^{j}+C_{i-1,m}^{j}\right)\right)$$
$$-\frac{\Delta t\, v_m}{R_m}\frac{1}{2}\left(\left(\frac{C_{i+1,m}^{j+1}-C_{i-1,m}^{j+1}}{2\Delta x}\right)+\left(\frac{C_{i+1,m}^{j}-C_{i-1,m}^{j}}{2\Delta x}\right)\right)$$

(15.86)

Now let $\dfrac{D_h}{R_m}\dfrac{\Delta t}{2\left(\Delta x\right)^2} = A$ and let $\dfrac{v_m}{R_m}\dfrac{\Delta t}{2*2\Delta x} = B$

Therefore Equation (15.82) becomes:

$$\left(C_{i,m}^{j+1}-C_{i,m}^{j}\right) = A*\left(\left(C_{i+1,m}^{j+1}-2C_{i,m}^{j+1}+C_{i-1,m}^{j+1}\right)+\left(C_{i+1,m}^{j}-2C_{i,m}^{j}+C_{i-1,m}^{j}\right)\right)$$
$$\times B*\left(\left(C_{i+1,m}^{j+1}-C_{i-1,m}^{j+1}\right)+\left(C_{i+1,m}^{j}-C_{i-1,m}^{j}\right)\right)$$

(15.87)

Multiplying it by A and B we get:

$$\left(C_{i,m}^{j+1}-C_{i,m}^{j}\right) = \left(\left(AC_{i+1,m}^{j+1}-2AC_{i,m}^{j+1}+AC_{i-1,m}^{j+1}\right)+\left(AC_{i+1,m}^{j}-2AC_{i,m}^{j}+AC_{i-1,m}^{j}\right)\right)$$
$$-\left(\left(BC_{i+1,m}^{j+1}-BC_{i-1,m}^{j+1}\right)+\left(BC_{i+1,m}^{j}-BC_{i-1,m}^{j}\right)\right)$$

(15.88)

Solving the equation:

$$C_{i,m}^{j+1}-C_{i,m}^{j} = AC_{i+1,m}^{j+1}-2AC_{i,m}^{j+1}+AC_{i-1,m}^{j+1}+AC_{i+1,m}^{j}-2AC_{i,m}^{j}+AC_{i-1,m}^{j}-BC_{i+1,m}^{j+1}$$
$$+BC_{i-1,m}^{j+1}-BC_{i+1,m}^{j}+BC_{i-1,m}^{j}$$

(15.89)

$$C_{i,m}^{j+1}+2AC_{i,m}^{j+1} = C_{i,m}^{j}+AC_{i+1,m}^{j+1}+AC_{i-1,m}^{j+1}+AC_{i+1,m}^{j}-2AC_{i,m}^{j}+AC_{i-1,m}^{j}-BC_{i+1,m}^{j+1}$$
$$+BC_{i-1,m}^{j+1}-BC_{i+1,m}^{j}+BC_{i-1,m}^{j}$$

(15.90)

Simplifying the equation:

$$C_{i,m}^{j+1}\left(1+2A\right)-C_{i,m}^{j}-2AC_{i,m}^{j}+AC_{i+1,m}^{j+1}-BC_{i+1,m}^{j+1}+AC_{i-1,m}^{j+1}+BC_{i-1,m}^{j+1}+AC_{i+1,m}^{j}$$
$$-BC_{i+1,m}^{j}+AC_{i-1,m}^{j}+BC_{i-1,m}^{j} \qquad (15.91)$$

And when simplifying the equation further we get:

$$C_{i,m}^{j+1}\left(1+2A\right)=C_{i,m}^{j}\left(1-2A\right)+C_{i+1,m}^{j+1}\left(A-B\right)+C_{i-1,m}^{j+1}\left(A+B\right)+C_{i+1,m}^{j}\left(A-B\right)+C_{i-1,m}^{j}\left(A+B\right) \qquad (15.92)$$

Furthermore, we get:

$$\zeta_{j+1,m}e^{ibmx}\left(1+2A\right)=\zeta_{j,m}e^{ibmx}\left(1-2A\right)+\zeta_{j+1,m}e^{ibm(x+\Delta x)}\left(A-B\right)+\zeta_{j+1,m}e^{ibm(x-\Delta x)}\left(A+B\right)$$
$$+\zeta_{j,m}e^{ibm(x+\Delta x)}\left(A-B\right)+\zeta_{j,m}e^{ibm(x-\Delta x)}\left(A+B\right) \qquad (15.93)$$

$$\zeta_{j+1,m}\left(1+2A\right)=\zeta_{j,m}\left(1-2A\right)+\zeta_{j+1,m}e^{ibm\Delta x}\left(A-B\right)+\zeta_{j+1,m}e^{-ibm(\Delta x)}\left(A+B\right)$$
$$+\zeta_{j,m}e^{ibm(\Delta x)}\left(A-B\right)+\zeta_{j,m}e^{-ibm(\Delta x)}\left(A+B\right) \qquad (15.94)$$

$$\zeta_{j+1,m}\left(1+2A\right)+\zeta_{j+1,m}e^{ibm(\Delta x)}\left(A-B\right)+\zeta_{j+1,m}e^{-ibm(\Delta x)}\left(A+B\right)$$
$$=\zeta_{j,m}\left(1-2A\right)+\zeta_{j,m}e^{ibm(\Delta x)}\left(A-B\right)+\zeta_{j,m}e^{-ibm(\Delta x)}\left(A+B\right) \qquad (15.95)$$

$$\zeta_{j+1,m}\left(1+2A\right)+\zeta_{j+1,m}e^{ibm\Delta x}\left(A-B\right)+\zeta_{j+1,m}e^{-ibm(\Delta x)}\left(A+B\right)$$
$$=\zeta_{j,m}\left(1-2A\right)+\zeta_{j,m}e^{ibm(\Delta x)}\left(A-B\right)+\zeta_{j,m}e^{-ibm(\Delta x)}\left(A+B\right) \qquad (15.96)$$

$$\left|\frac{\zeta_{j+1,m}}{\zeta_{j,m}}\right|<\left|\frac{\left(1-2A\right)++e^{ibm(\Delta x)}\left(A-B\right)+e^{-ibm(\Delta x)}\left(A+B\right)}{\left(1+2A\right)-e^{ibm\Delta x}\left(A-B\right)-e^{-ibm(\Delta x)}\left(A+B\right)}\right| \qquad (15.97)$$

Since we know that $e^{ibm\Delta x}=\cos\left(km\Delta x\right)+i\sin\left(km\Delta x\right)$, therefore, the above equation become:

$$\left|\frac{\zeta_{j+1,m}}{\zeta_{j,m}}\right|<\left|\frac{\left(1-2A\right)+\left(A-B\right)\left[\cos\left(km\Delta x\right)+i\sin\left(km\Delta x\right)\right]+\left(A+B\right)\left[\cos\left(km\Delta x\right)-i\sin\left(km\Delta x\right)\right]}{\left(1+2A\right)-\left(A-B\right)\left[\cos\left(km\Delta x\right)+i\sin\left(km\Delta x\right)\right]-\left(A+B\right)\left[\cos\left(km\Delta x\right)-i\sin\left(km\Delta x\right)\right]}\right|$$
$$(15.98)$$

Therefore:

$$\left|\frac{\zeta_{j+1,m}}{\zeta_{j,m}}\right|<\left|\frac{\left(1-2A\right)+\left(A-B\right)\left[\cos\left(km\Delta x\right)+i\sin\left(km\Delta x\right)\right]+\left(A+B\right)\left[\cos\left(km\Delta x\right)-i\sin\left(km\Delta x\right)\right]}{\left(1+2A\right)-\left(A-B\right)\left[\cos\left(km\Delta x\right)+i\sin\left(km\Delta x\right)\right]-\left(A+B\right)\left[\cos\left(km\Delta x\right)-i\sin\left(km\Delta x\right)\right]}\right|$$
$$(15.99)$$

Then Equation (15.96) becomes:

$$\left|\frac{\zeta_{j+1,m}}{\zeta_{j,m}}\right|<\left|\frac{\left(1-2A\right)+2A\cos\left(km\Delta x\right)-2Bi\sin\left(km\Delta x\right)}{\left(1+2A\right)-2A\cos\left(km\Delta x\right)+2Bi\sin\left(km\Delta x\right)}\right| \qquad (15.100)$$

Knowing that,

$$|a + ib| = \sqrt{a^2 + b^2} \tag{15.101}$$

We then get:

$$\left| \frac{\zeta_{j+1,m}}{\zeta_{j,m}} \right| < \left| \left(\frac{\left(1 - 2A + 2A\cos(km\Delta x)\right)^2 + \left(2B\sin(km\Delta x)\right)^2}{\left(1 + 2A - 2A\cos(km\Delta x)\right)^2 + \left(2B\sin(km\Delta x)\right)^2} \right) \right| \tag{15.102}$$

$$\frac{\left(1 - 2A + 2A\cos(km\Delta x)\right)^2 + \left(2B\sin(km\Delta x)\right)^2}{\left(1 + 2A - 2A\cos(km\Delta x)\right)^2 + \left(2B\sin(km\Delta x)\right)^2} < 1 \tag{15.103}$$

$$\left(1 - 2A + 2A\cos(km\Delta x)\right)^2 + \left(2B\sin(km\Delta x)\right)^2 < \left(1 + 2A - 2A\cos(km\Delta x)\right)^2 + \left(2B\sin(km\Delta x)\right)^2 \tag{15.104}$$

$$4A\cos(km\Delta x) < 4A \tag{15.105}$$

$$\cos(km\Delta x) < 1 \tag{15.106}$$

Therefore we conclude that the equation is stable. The same condition is expected for fracture as variables are similar.

15.13   CONCLUSION

In this chapter, an attempt to solve and extend the limitations of the classical advection dispersion equation was initiated. We have proposed equations to model the fate of contamination transport between soil matrix and fracture. The new proposed model takes into account the transition of movement from matrix to fracture, which helps to obtain a new mathematical model with variable dispersion and advection. The Picard–Lindelöf theorem was used to prove the uniqueness of the proposed system of equations for the matrix and fracture network, the proposed equations were found to be unique. Crank–Nicolson and von Neumann's methods were used to perform numerical and stability analysis of the proposed system of equations was conducted and equation was found to be stable.

REFERENCES

Al Rmalli, S.W., Haris, P.I., Harrington, C.F., & Ayub, M. (2005). A survey of arsenic in foodstuffs on sale in the United Kingdom and imported from Bangladesh. *Science of The Total Environment, 337*(1), 23–30. doi:10.1016/j.scitotenv.2004.06.008.

Berkowitz, B. (2002). Characterizing flow and transport in fractured geological media: A review. *Advances in Water Resources, 25*(8), 861–884. doi:10.1016/S0309-1708(02)00042-8.

Bethke, C. (1985). A numerical model of compaction-driven groundwater flow and heat transfer and its application to the paleohydrology of intracratonic sedimentary basins. *Journal of Geophysical Research: Solid Earth, 90*(B8). https://agupubs.onlinelibrary.wiley.com/doi/abs/10.1029/JB090iB08p06817.

Cebeci, T., & Bradshaw, P. (2012). *Physical and Computational Aspects of Convective Heat Transfer.* Springer Science & Business Media.

Das, H.K., Mitra, A. K., Sengupta, P.K., Hossain, A., Islam, F., & Rabbani, G.H. (2004). Arsenic concentrations in rice, vegetables, and fish in Bangladesh: A preliminary study. *Environment International, 30*(3), 383–387. doi:10.1016/j.envint.2003.09.005.

Duguid, J.O., & Lee, P.C.Y. (1977). Flow in fractured porous media. *Water Resources Research, 13*(3), 558–566. doi:10.1029/WR013i003p00558.

Dutta, D. (2013). Hydrodynamic Dispersion. In D. Li (ed.), *Encyclopedia of Microfluidics and Nanofluidics* (pp. 1–14). Springer US. doi:10.1007/978-3-642-27758-0_660-3.

Edelman, J.H. (1972). *Ground Water Hydraulics of Extensive Aquifers*. Wageningen, The Netherlands: International Institute for Land Reclamation and Improvement ILI.

Fetter, C.W., Boving, T., & Kreamer, D. (2017). *Contaminant Hydrogeology: Third Edition*. Waveland Press.

Gholizadeh, A., Ardalan, M., Tehrani, M.M., Hosseini, H.M., & Karimian, N. (2009). Solubility test in some phosphate rocks and their potential for direct application in soil. *World Applied Sciences Journal, 6*(2), 182–190.

Kazemi, H., Merrill, L.S.J., Porterfield, K.L., & Zeman, P.R. (1976). Numerical simulation of water–oil flow in naturally fractured reservoirs. *Society of Petroleum Engineers Journal, 16*(06), 317–326. doi:10.2118/5719-PA.

Khan, S., Farooq, R., Shahbaz, S., Khan, M.A., & Sadique, M. (2009). Health risk assessment of heavy metals for population via consumption of vegetables, 6.

Maeng, S.K. (2010). *Multiple Objective Treatment Aspects of Bank Filtration*. Netherlands: CRC Press/ Balkema.

Margat, J., & van der Gun, J. (2013). *Groundwater around the World: A Geographic Synopsis*. CRC Press.

Moench, A.F. (1984). Double-porosity models for a fissured groundwater reservoir with fracture skin. *Water Resources Research, 20*(7), 831–846. doi:10.1029/WR020i007p00831.

Nevanlinna, O., (1989). Remarks on Picard–Lindelöf iteration. *BIT Numerical Mathematics, 29*(2), 328–346. doi:10.1007/BF01952687.

Sharma, S.K. (2001). *Adsorptive Iron Removal from Groundwater*. CRC Press.

Sun, C., & Trueman, C.W. (2003). Unconditionally stable Crank–Nicolson scheme for solving two-dimensional Maxwell's equations. *Electronics Letters, 39*(7), 595–597. doi:10.1049/el:20030416.

Thomas, L.K., Dixon, T.N., & Pierson, R.G. (1983). Fractured Reservoir Simulation. *Society of Petroleum Engineers Journal, 23*(01), 42–54. doi:10.2118/9305-PA.

Warren, J.E., & Root, P.J. (1963). The behavior of naturally fractured reservoirs. *Society of Petroleum Engineers Journal, 3*(03), 245–255. doi:10.2118/426-PA.

Wesseling, P. (1996). Von Neumann stability conditions for the convection–diffusion equation. *IMA Journal of Numerical Analysis*, 16(4), 583–598. doi:10.1093/imanum/16.4.583.

Worthley, J.A., & Torkelson, R. (1981). Managing the toxic waste problem – Lessons from the Love Canal. *Sage Journals Publication, Inc, 13*(2), 145–160.

Zecchin, A., Simpson, A., & Lambert, M. (2008, May 14). Von neumann stability analysis of a method of characteristics visco-elastic pipeline model.

16 Groundwater Contamination Transport Model with Fading Memory Property

Mpafane Deyi
University of the Free State, Bloemfontein, South Africa

Abdon Atangana
China Medical University Hospital, China Medical University, Taichung, Taiwan
University of the Free State, Bloemfontein, South Africa

CONTENTS

16.1 INTRODUCTION

Groundwater contamination can have long-term effects as it may take some time to remediate a contaminated site or aquifer. Remediating a contaminated groundwater site is a difficult challenge which may in some cases not be an achievable goal (Travis and Doty, 2003; MacDonald & Kavanaugh, 2008). Groundwater contamination is further complicated because there are different geological formations underground with different behaviors depending on the contamination load to which they are subjected. Remediating and cleaning a contaminated groundwater site can take a very long time and may require high expenditure. In the year 2011, the China State Council made available a budget to the tune of US$5.5 billion over a period of 10 years to prevent and treat groundwater contamination (Qiu, 2011). The Love Canal project is regarded as one of the worst environmental disasters that ever happened on American soil.

Mathematical models have many advantages compared with other methods of solving complex problems such as the one mentioned above. They have the capability of reflecting the complex physical structures and irregular geometric shapes of an aquifer system; and furthermore, they are convenient and flexible to use and easy to calibrate where there is reliable input data; and they can describe not only the phenomenon of water flow in porous media but also the mass and energy transports and other complex physical–chemical–biological phenomena in porous media (Sun, 1996).

DOI: 10.1201/9781003266266-16

Deyi and Atangana (2020) proposed a system of equations for modeling contamination transport in dual media, a soil–fracture matrix. Contamination transport is deemed to be driven, in the main, by diffusion and advection processes. Diffusion processes have been studied and diffusion equations used to model the movement of many particles in different environments or media. The objects or particles can be as small as basic particles in physics, bacteria, molecules or cells, or very large objects like animals, plants, or certain types of events like epidemics. Thus, diffusion processes are important in many applications in the fields of science and engineering (Ibe, 2013).

Of particular interest to us are diffusion processes during contamination transport in dual media or matrix–fracture contamination transport. In groundwater, molecular diffusion is defined as the concentration difference between two points, a process driven by the random Brownian movement of molecules (Appelo and Postma, 2004). The molecules diffuse away from their original source because molecules always diffuse along their concentration gradient. This means that they diffuse from where they are in high concentration to where they are in low concentration (Ibe, 2013).

16.2 INTRODUCING A CAPUTO–FABRIZIO OPERATOR INTO MATRIX–FRACTURE EQUATIONS

Deyi and Atangana (2020) proposed matrix–fracture equations and showed that Equations (16.1) and (16.2) for matrix and fracture can be discretized, and they proved that the equations are stable after conducting von Neumann's method of analysis in their study.

Equation for the matrix:

$$\left(1 + \beta k_D\right)\frac{\partial Cm}{\partial t} = D_h \frac{\partial^2 C_m}{\partial x^2} - v_m \frac{\partial C_m}{\partial x} \tag{16.1}$$

Equation for the fracture:

$$\left(1 + \beta k_D\right)\frac{\partial Cf}{\partial t} = D_h \frac{\partial^2 C_f}{\partial x^2} - v_m \frac{\partial C_f}{\partial x} \tag{16.2}$$

In this chapter we have considered the introduction of fractional differential operator to the equations because of their advantages. Fractional calculus, as a domain of calculus, is one of the most powerful mathematical tools used to model real-world problems in many fields of science, technology and engineering (Atangana and Gómez-Aguilar, 2018). Furthermore, the fractional calculus provides a suitable framework to deal with complex systems such as fracture networks (Tateishi et al., 2017). The strength of the application of fractional differentiation is that it offers the ability to describe the memory effect (Caputo, 1967; Atangana and Gómez-Aguilar, 2018). The nature of contamination transport in the geological environments of the Earth's subsurface is often observed to be anomalous (Berkowitz and Scher, 1997).

16.3 CAPUTO AND FABRIZIO DERIVATIVE

Caputo (1967) introduced a differential operator as an imperial formula. Some applied mathematicians suggested that the Caputo is suitable for real-world problems because it allows normal initial conditions when playing with integral transform; for example, the Laplace transform (Atangana and Gómez-Aguilar 2018). The Caputo derivative is shown below in Equation (16.3).

$$_0^C D_t^\alpha f(t) = \frac{1}{\Gamma(1-\alpha)} \int_0^t \frac{d}{d\tau} f(\tau)(t-\tau)^{-\alpha} \, d\tau \tag{16.3}$$

Caputo and Fabrizio (2015) presented a new definition of a fractional derivative with a smooth kernel which takes on two different representations for the temporal and spatial variable. The interest for this new approach with a regular kernel was born from the prospect that there is a class of nonlocal systems, which have the ability to describe the material heterogeneities and the fluctuations of different scales, which cannot be well described by classical local theories, or by fractional models with a singular kernel. Properties of Caputo–Fabrizio fractional derivative were analyzed in classical and distributional settings by (Atanacković et al., 2018).

The Caputo–Fabrizio derivate is shown below in Equation (16.4).

$$ {}^{CF}_{0}D^\alpha_t f(t) = \frac{M(\alpha)}{(1-\alpha)} \int_0^t \frac{d}{d\tau} f(\tau) \exp\left[-\frac{\alpha}{1-\alpha}(t-\tau) \right] d\tau \tag{16.4} $$

Some applications of the Caputo–Fabrizio derivative are presented in the work of Caputo and Fabrizio (2016) and Tateishi et al., (2017).

In previous studies it has been established that fractional operators modify the behavior of the waiting time distribution in the context of fractional diffusion. Caputo–Fabrizio can be related to a well-defined physical quantity (resetting rate) (Tateishi et al., 2017).

Atangana and Gómez-Aguilar (2018) show in their paper after running simulations for different values of α, that the memory is consistent from the beginning to the end is fading. This is an improvement by comparison to the previously discussed power law kernel which proved to have an incomplete memory.

An integral of the Caputo–Fabrizio derivative is provided in Equation (16.5):

$$ {}^{CF}_{0}J^\alpha_t f(t) = f(0) + \frac{1-\alpha}{M(\alpha)} f(t) + \frac{\alpha}{M(\alpha)} \int_0^t f(\tau) d\tau. \tag{16.5} $$

16.4 LAPLACE TRANSFORM

The Caputo–Fabrizio derivative from Equation (16.4) is brought forward:

$$ {}^{CF}_{0}D^\alpha_t f(t) = \frac{M(\alpha)}{(1-\alpha)} \int_0^t \frac{d}{d\tau} f(\tau) \exp\left[-\frac{\alpha}{1-\alpha}(t-\tau) \right] d\tau \tag{16.6} $$

Now rewriting the equation, we get:

$$ {}^{CF}_{0}D^\alpha_t f(t) = \int_0^t \frac{d}{d\tau} f(\tau) \exp\left[-\frac{(t-\tau)\alpha}{1-\alpha} \right] \frac{M(\alpha)}{1-\alpha} d\tau \tag{16.7} $$

Then finally we have:

$$ {}^{CF}_{0}D^\alpha_t f(t) = \frac{d}{dt} f(t) * \exp\left[-\frac{\alpha t}{1-\alpha} \right] \tag{16.8} $$

Now finding the Laplace transform for Equation (16.7).

$$ \mathcal{L}\left({}^{CF}_{0}D^\alpha_t f(t) \right) = \mathcal{L}\left(\frac{df(t)}{dt} * \frac{M(\alpha)}{1-\alpha} \exp\left[-\frac{\alpha}{1-\alpha}t \right] \right) \tag{16.9} $$

$$\mathcal{L}\left({}_{0}^{CF}D_t^\alpha f(t) \right) = \mathcal{L}\left(\frac{df(t)}{dt} \right) \frac{M(\alpha)}{1-\alpha} \mathcal{L}\left(\exp\left[-\frac{\alpha}{1-\alpha}t \right] \right) \tag{16.10}$$

$$\mathcal{L}\left({}_{0}^{CF}D_t^\alpha f(t) \right) = \left[S\tilde{f}(s) - f(0) \right] \frac{M(\alpha)}{1-\alpha} \frac{1}{S + \dfrac{\alpha}{1-\alpha}} \tag{16.11}$$

Finally we have:

$$\mathcal{L}\left({}_{0}^{CF}D_t^\alpha f(t) \right) = \frac{S\tilde{f}(s) - f(0)}{S + \dfrac{\alpha}{1-\alpha}} M(\alpha). \tag{16.12}$$

16.5 APPLYING THE LAPLACE TRANSFORM TECHNIQUE TO THE CAPUTO–FABRIZIO INTEGRAL

An integral of Caputo–Fabrizio derivative is brought forward from Equation (16.4).

$$_{0}^{CF}J_t^\alpha f(t) = f(0) + \frac{1-\alpha}{M(\alpha)} f(t) + \frac{\alpha}{M(\alpha)} \int_0^t f(\tau)d\tau \tag{16.13}$$

Now applying the Laplace transform technique:

$$\mathcal{L}\left({}_{0}^{CF}J_t^\alpha f(t) \right) = \mathcal{L}\left(f(0) \right) + \mathcal{L}\left(\frac{1-\alpha}{M(\alpha)} \right) f(t) + \mathcal{L}\left(\frac{\alpha}{M(\alpha)} \int_0^t f(\tau)d\tau \right) \tag{16.14}$$

we have:

$$\mathcal{L}\left({}_{0}^{CF}J_t^\alpha f(t) \right) = f(0) + \frac{1-\alpha}{M(\alpha)} \tilde{f}(s) + \frac{\alpha}{M(\alpha)} \frac{\tilde{f}(s)}{S}. \tag{16.15}$$

16.6 NUMERICAL APPROXIMATION

Numerical approximation of partial differential equations was considered because it is regarded as an important branch of numeral analysis (Quarteroni & Valli, 2008). Furthermore, numerical approximation or schemes have been recognized as powerful mathematical tools for solving nonlinear ordinary differential equations with local and nonlocal operators. In the main, they are normally used when all the applied analytical methods have failed (Toufik and Atangana, 2017).

16.7 NUMERICAL APPROXIMATION OF CAPUTO–FABRIZIO DERIVATIVE

Now the numerical approximation and the discretization of the Caputo–Fabrizio Derivative is done:

$$_{0}^{CF}D_t^\alpha f(t) = \frac{M(\alpha)}{(1-\alpha)} \int_0^t \frac{d}{d\tau} f(\tau) \exp\left[-\frac{\alpha}{1-\alpha}(t-\tau) \right] d\tau \tag{16.16}$$

At (x_i, t_{n+1}) we present the numerical approximation as follows:

$$_{0}^{CF}D_t^\alpha f\left(x_i, t_{n+1}\right) = \frac{M(\alpha)}{(1-\alpha)} \int_0^{t_{n+1}} \frac{df(x_i, r)}{dr} \exp\left[\frac{-\alpha\left(t_{n+1}-r\right)}{1-\alpha}\right] dr \tag{16.17}$$

$$_{0}^{CF}D_t^\alpha f\left(x_i, t_{n+1}\right) = \frac{M(\alpha)}{(1-\alpha)} \int_{t_j}^{t_{j+1}} \frac{df(x_i, r)}{dr} \exp\left[\frac{-\alpha\left(t_{n+1}-r\right)}{1-\alpha}\right] dr \tag{16.18}$$

$$_{0}^{CF}D_t^\alpha f\left(x_i, t_{n+1}\right) = \frac{M(\alpha)}{(1-\alpha)} \sum_{j=0}^{n} \frac{f_i^{j+1} - f_i^j}{\Delta t} \int_{t_j}^{t_{j+1}} \exp\left(\frac{-\alpha\left(t_{n+1}-r\right)}{1-\alpha}\right) dr \tag{16.19}$$

$$_{0}^{CF}D_t^\alpha f\left(x_i, t_{n+1}\right) = \frac{M(\alpha)}{(1-\alpha)} \sum_{j=0}^{n} \frac{f_i^{j+1} - f_i^j}{\Delta t} \int_{t_j}^{t_{j+1}} \left[\exp\frac{\dfrac{-\alpha\left(t_{n+1}-t_{j+1}\right)}{1-\alpha}}{\dfrac{1}{1-\alpha}} - \exp\frac{-\alpha(t_{n+1}-t_j)}{1-\alpha}\right] \tag{16.20}$$

The discretized derivative for Caputo–Fabrizio is shown below:

$$_{0}^{CF}D_t^\alpha f(t) = \frac{M(\alpha)}{\alpha} \sum_{j=0}^{n} \frac{f_i^{j+1} - f_i^j}{\Delta t} \left[\exp\left(\frac{-\alpha(n-j)\Delta t}{1-\alpha}\right) - \exp\left(\frac{-\alpha(n-j+1)\Delta t}{1-\alpha}\right)\right]. \tag{16.21}$$

16.8 NUMERICAL APPROXIMATION OF CAPUTO–FABRIZIO INTEGRAL

Now the numerical approximation for the Caputo–Fabrizio integral:

$$_{0}^{CF}J_t^\alpha f(t) = f(0) + \frac{1-\alpha}{M(\alpha)} f(t) + \frac{\alpha}{M(\alpha)} \int_0^t f(\tau) d\tau \tag{16.22}$$

At (x_i, t_{n+1}) we present the numerical approximation, as follows:

$$_{0}^{CF}J_t^\alpha f(t) = f\left(x_i, 0\right) + \frac{1-\alpha}{M(\alpha)} f\left(x_i, t_{n+1}\right) + \frac{\alpha}{M(\alpha)} \int_0^t f\left(x_i, t_{n+1}\right) dt \tag{16.23}$$

$$_{0}^{CF}J_t^\alpha f(t) = f\left(x_i, 0\right) + \frac{1-\alpha}{M(\alpha)} \left(\frac{f_i^{n+1} + f_i^n}{2}\right) + \frac{\alpha}{M(\alpha)} \sum_{j=0}^{n} \int_{t_j}^{t_{j+1}} f\left(x_i, t_{n+1}\right) dt \tag{16.24}$$

$$_{0}^{CF}J_t^\alpha f(t) = f\left(x_i, 0\right) + \frac{1-\alpha}{M(\alpha)} \left(\frac{f_i^{n+1} + f_i^n}{2}\right) + \frac{\alpha}{M(\alpha)} \sum_{j=0}^{n} \frac{f_i^{j+1} + f_i^j}{2} \left(t\Big|_{t_j}^{t_{j+1}}\right) \tag{16.25}$$

$$_{0}^{CF}J_t^\alpha f(t) = f\left(x_i, 0\right) + \frac{1-\alpha}{M(\alpha)} \left(\frac{f_i^{n+1} + f_i^n}{2}\right) + \frac{\alpha}{M(\alpha)} \sum_{j=0}^{n} \frac{f_i^{j+1} + f_i^j}{2} \left(t_{j+1} - t_j\right) \tag{16.26}$$

Then, finally, we have:

$$
{}^{CF}_{0}J^{\alpha}_{t}f(t) = f(x_i, 0) + \frac{1-\alpha}{M(\alpha)}\left(\frac{f_i^{n+1} + f_i^n}{2}\right) + \frac{\alpha}{M(\alpha)}\sum_{j=0}^{n}\frac{f_i^{j+1} + f_i^j}{2}(\Delta t). \tag{16.27}
$$

16.9 MODEL WITH CAPUTO–FABRIZIO

In this section we introduce the Caputo–Fabrizio derivative into the proposed model. The Caputo–Fabrizio derivate has been applied to solve groundwater flow, free convection flow, fluid dynamics problems, and advection diffusion problems (Atangana & Alkahtani, 2015; Ali et al., 2016).

Tateishi et al. (2017), in their study, showed us that the Caputo–Fabrizio operator is characterized by the following: the waiting time distribution is deemed to be exponential and it recovers a diffusive process with stochastic resetting; and the fractional order exponent is related to the resetting rate.

Again, we bring forward the diffusion equation:

$$
(1 + \beta k_D)\frac{\partial Cm}{\partial t} = D_h\frac{\partial^2 C_m}{\partial x^2} - v_m\frac{\partial C_m}{\partial x} \tag{16.28}
$$

The Caputo–Fabrizio derivative is as follows:

$$
{}^{CF}_{0}D^{\alpha}_{t}f(t) = \frac{M(\alpha)}{(1-\alpha)}\int_{0}^{t}\frac{d}{d\tau}f(\tau)\exp\left[-\frac{\alpha}{1-\alpha}(t-\tau)\right]d\tau \tag{16.29}
$$

Now introducing the derivative to the derived equation for the matrix:

$$
(1 + \beta k_D)\frac{M(\alpha)}{(1-\alpha)}\int_{0}^{t}\frac{\partial}{\partial\tau}C_m(x, \tau)\exp\left[-\frac{\alpha}{1-\alpha}(t-\tau)\right]d\tau = D_h\frac{\partial^2 C_m}{\partial x^2} - v_m\frac{\partial C_m}{\partial x} \tag{16.30}
$$

$$
(1 + \beta k_D)\frac{M(\alpha)}{(1-\alpha)}\int_{0}^{t}\frac{\partial}{\partial\tau}C_m(x, \tau)\exp\left[-\frac{\alpha}{1-\alpha}(t-\tau)\right]d\tau = D_h\frac{\partial^2 C_{m(x,t)}}{\partial x^2} - v_m\frac{\partial C_{m(x,t)}}{\partial x} \tag{16.31}
$$

Dividing with $(1 + \beta k_D)$ we get:

$$
\frac{M(\alpha)}{(1-\alpha)}\int_{0}^{t}\frac{\partial}{\partial\tau}C_m(x, \tau)\exp\left[-\frac{\alpha}{1-\alpha}(t-\tau)\right]d\tau = \frac{1}{1 + \beta k_D}\left[D_h\frac{\partial^2 C_{m(x,t)}}{\partial x^2} - v_m\frac{\partial C_{m(x,t)}}{\partial x}\right] \tag{16.32}
$$

To simplify the equation, we let $\dfrac{1}{1 + \beta k_D}\left[D_h\dfrac{\partial^2 C_{m(x,t)}}{\partial x^2} - v_m\dfrac{\partial C_{m(x,t)}}{\partial x}\right] = F(x, t, C_m(x, t))$

Therefore we can represent the above equation as the following:

$$
\frac{M(\alpha)}{(1-\alpha)}\int_{0}^{t}\frac{\partial}{\partial\tau}C_m(x, \tau)\exp\left[-\frac{\alpha}{1-\alpha}(t-\tau)\right]d\tau = F(x, t, C_m(x, t)) \tag{16.33}
$$

$$
\frac{M(\alpha)}{(1-\alpha)}\int_{0}^{t}\frac{\partial}{\partial\tau}C_m(x, \tau)\exp\left[-\frac{\alpha}{1-\alpha}(t-\tau)\right]d\tau = F(x, t, C_m(x, t)) \tag{16.34}
$$

We apply the associate integral to obtain the following:

$$C_m(x,t) - C_m(x,0) = \frac{1-\alpha}{M(\alpha)} F\big(x,t,C_m(x_i,t_i)\big) + \frac{\alpha}{M(\alpha)} \int_0^t F\big(x,\tau,C_m(x,\tau)\big)d\tau \quad (16.35)$$

Now at (x_i, t_{n+1}) we have:

$$C_m(x_i,t_{n+1}) - C_m(x_i,0) = \frac{1-\alpha}{M(\alpha)} F\big(x_i,t_n,C_m(x_i,t_n)\big) + \frac{\alpha}{M(\alpha)} \int_0^{t_{n+1}} F\big(x_i,\tau,C_m(x_i,\tau)\big)d\tau \quad (16.36)$$

And at (x_i, t_n) we have:

$$C_m(x_i,t_n) - C_m(x_i,0) = \frac{1-\alpha}{M(\alpha)} F\big(x_i,t_{n-1},C_m(x_i,t_{n-1})\big) + \frac{\alpha}{M(\alpha)} \int_0^{t_n} F\big(x_i,\tau,C_m(x_i,\tau)\big)d\tau \quad (16.37)$$

Then Equation (16.36) minus Equation (16.37) gives us:

$$C_m(x_i,t_{n+1}) - C_m(x_i,t_n) = \frac{1-\alpha}{M(\alpha)} \left[F\left(\begin{array}{c} x_i,t_n,C_m(x_i,t_n) - F\big(x_i,t_{n-1},C_m(x_i,t_{n-1})\big) \\ + \dfrac{\alpha}{M(\alpha)} \displaystyle\int_0^{t_{n+1}} F\big(x_i,\alpha,\tau,C_m(x_i,\tau)\big) \end{array} \right) d\tau \\ - \dfrac{\alpha}{M(\alpha)} \displaystyle\int_0^{t_n} F\big(x_i,\tau,C_m(x_i,\tau)\big)d\tau \right] \quad (16.38)$$

Since we know that:

$$\int_a^b f(x)dx + \int_b^c f(x)dx = \int_a^c f(x)dx \quad (16.39)$$

we apply the above principle on to the following:

$$C_m(x_i,t_{n+1}) - C_m(x_i,t_n) = \frac{1-\alpha}{M(\alpha)} \Big[F\big(x_i,t_n,C_m(x_i,t_n)\big) - F\big(x_i,t_{n-1},C_m(x_i,t_{n-1})\big) \Big]$$

$$+ \frac{\alpha}{M(\alpha)} \int_{t_n}^{t_{n+1}} F\big(x_i,\tau,C_m(x_i,\tau)\big)d\tau \quad (16.40)$$

Therefore, within $[t_n, t_{n+1}]$ we approximate the function $F(x_i,\tau,C_m(x_i,\tau))$ using a Lagrange approximation.

$$F\big(x_i,\tau,C_m(x_i,\tau)\big) = P_i(\tau) \quad (16.41)$$

Then we obtain:

$$P_i(\tau) = \frac{\tau - t_{n-1}}{t_n - t_{n-1}} \Big[F\big(x_i,t_n,C_m(x_i,t_n)\big) \Big] + \frac{\tau - t_n}{t_{n-1} - t_n} \Big[F\big(x_i,t_{n-1},C_m(x_i,t_{n-1})\big) \Big] \quad (16.42)$$

And we have:

$$C_m\left(x_i, t_{n+1}\right) - C_m\left(x_i, t_n\right) = \frac{1-\alpha}{M(\alpha)}\left[F\left(x_i, t_n, C_m\left(x_i, t_n\right)\right) - F\left(x_i, t_{n-1}, C_m\left(x_i, t_{n-1}\right)\right)\right]$$

$$+ \frac{\alpha}{M(\alpha)}\int_{t_n}^{t_{n+1}} P_i(\alpha)\,d\tau \tag{16.43}$$

Further, we obtain:

$$C_m\left(x_i, t_{n+1}\right) - C_m\left(x_i, t_n\right) = \frac{1-\alpha}{M(\alpha)}\left[F\left(x_i, t_n, C_m\left(x_i, t_n\right)\right) - F\left(x_i, t_{n-1}, C_m\left(x_i, t_{n-1}\right)\right)\right]$$

$$+ \frac{\alpha}{M(\alpha)}\int_{t_n}^{t_{n+1}}\left[\frac{\tau - t_{n-1}}{t_n - t_{n-1}} F\left(x_i, t_n, C_m\left(x_i, t_n\right)\right)d\tau\right] \tag{16.44}$$

Finally, after integrating, we obtain:

$$C_m\left(x_i, t_{n+1}\right) - C_m\left(x_i, t_n\right) = \frac{1-\alpha}{M(\alpha)}\left[F\left(x_i, t_n, C_m\left(x_i, t_n\right)\right) - F\left(x_i, t_{n-1}, C_m\left(x_i, t_{n-1}\right)\right)\right]$$

$$+ \frac{\alpha}{M(\alpha)}\left[\frac{3}{2}\Delta t F\left(x_i, t_n, C_m\left(x_i, t_n\right)\right) - \frac{\Delta t}{2} F\left(x_i, t_{n-1}, C_m\left(x_i, t_{n-1}\right)\right)\right]. \tag{16.45}$$

16.10 CONCLUSION

In this chapter we have considered the introduction of a fractional differential operator into the equations because of their advantages. The strength of the application of fractional differentiation is that it offers the ability to describe the memory effect. To capture the effect of elasticity of the geological formation, the well-known Caputo–Fabrizio fractional derivative was used to extend the classical model. Furthermore, to include the effect of fracture, a differential operator based on power law kernel was used to extend the classical advection dispersion equation with variable coefficient. Laplace transform technique was applied to the Caputo–Fabrizio derivative and integral. Furthermore, numerical approximation or schemes have been recognized as powerful mathematical tools to solve nonlinear ordinary differential equations with local and nonlocal operators. Numeral approximation techniques were applied to both the Caputo–Fabrizio derivative and integral. Finally, the results were introduced into the proposed equation by Deyi and Atangana (2020). The obtained new equation gives us hope in terms of application.

The proposed model can be introduced into computer programs or software such as FEFLOW and MODFLOW.

REFERENCES

Ali, F., Saqib, M., Khan, I., & Sheikh, N.A. (2016). Application of Caputo–Fabrizio derivatives to MHD free convection flow of generalized Walters'-B fluid model. *The European Physical Journal Plus*, 131(10), 1–10. doi:10.1140/epjp/i2016-16377-x.

Appelo, C.A.J., & Postma, D. (2004). *Geochemistry, Groundwater and Pollution*. CRC Press.

Atanacković, T.M., Pilipović, S., & Zorica, D. (2018). Properties of the Caputo–Fabrizio fractional derivative and its distributional settings. *Fractional Calculus and Applied Analysis*, 21(1), 29–44. doi:10.1515/fca-2018-0003.

Atangana, A., & Alkahtani, B.S.T. (2015). New model of groundwater flowing within a confined aquifer: Application of Caputo–Fabrizio derivative. *Arabian Journal of Geosciences*, 9(1), 8. doi:10.1007/s12517-015-2060-8.

Atangana, A., & Gómez-Aguilar, J.F. (2018). Decolonisation of fractional calculus rules: Breaking commutativity and associativity to capture more natural phenomena. *The European Physical Journal Plus*, 133(4), 166. doi:10.1140/epjp/i2018-12021-3.

Berkowitz, B., & Scher, H. (1997). Anomalous transport in random fracture networks. *Physical Review Letters*, 79, 4038–4041. doi:10.1103/PhysRevLett.79.4038.

Caputo, M. (1967). Linear models of dissipation whose Q is almost Frequency Independent – II. *Geophysical Journal International*, 13(5), 529–539. doi:10.1111/j.1365-246X.1967.tb02303.x.

Caputo, M., & Fabrizio, M. (2015). *A New Definition of Fractional Derivative without Singular Kernel.*

Caputo, M., & Fabrizio, M. (2016). Applications of new time and spatial fractional derivatives with exponential kernels. *Progress in Fractional Differentiation and Applications*, 2(1), 1–11. doi:10.18576/pfda/020101.

Deyi, M., & Atangana, A. (2020). A new model for groundwater contamination transport in dual media. PhD Thesis, University of the Free State, Bloemfontein, South Africa.

Ibe, O.C. (2013). *Elements of Random Walk and Diffusion Processes.* John Wiley & Sons, Inc.

MacDonald, J.A., & Kavanaugh, M.C. (2008). Restoring contaminated groundwater: An achievable goal? [research-article]. doi:10.1021/es00057a001.

Qiu, J. (2011). China to spend billions cleaning up groundwater. *Science*, 334(6057), 745–745. doi:10.1126/science.334.6057.745.

Quarteroni, A., & Valli, A. (2008). *Numerical Approximation of Partial Differential Equations.* Springer Science & Business Media.

Sun, N.Z. (1996). *Mathematical Modeling of Groundwater Pollution* (2nd Edition). Springer Science & Business Media, New York.

Tateishi, A.A., Ribeiro, H.V., & Lenzi, E.K. (2017). The role of fractional time-derivative operators on anomalous diffusion. *Frontiers in Physics*, 5.

Toufik, M. & Atangana, A. 2017. New numerical approximation of fractional derivative with non-local and non-singular kernel: Application to chaotic models. *The European Physical Journal Plus* 132(10), 444.

Travis, C., & Doty, C. (2003). Environmental science & technology views: Can contaminated aquifers at superfund sites be remediated? doi:10.1021/es00080a600.

17 A New Groundwater Transport in Dual Media with Power Law Process

Mpafane Deyi
University of the Free State, Bloemfontein, South Africa

Abdon Atangana
China Medical University Hospital, China Medical University, Taichung, Taiwan

University of the Free State, Bloemfontein, South Africa

CONTENTS

17.1 INTRODUCTION

Fractional calculus is a study with specific focus on fractional order integrals and derivatives, and their diverse applications in various fields (Kilbas et al., 2006). Caputo and Riemann–Liouville are, for example, some of the fractional derivatives. In their form, they generalize the ordinary integral and differential operators. Nonetheless, fractional derivatives have been found to have fewer properties than corresponding classical ones. As a result, it makes these derivatives very useful for describing anomalous phenomena (Abdeljawad, 2011).

The Riemann–Liouville power law differential operator is regarded as the only mathematical tool with the uniqueness to replicate all physical problems. Atangana and Gómez-Aguilar (2018) argue in their study that not all physical problems follow the power law distribution.

Caputo (1967) introduced a differential operator as an imperial formula. Some applied mathematicians suggested that the Caputo operator is suitable for real-world problems because it allows normal initial conditions when dealing with integral transform; for example, the Laplace transform (Atangana and Gómez-Aguilar, 2018). Therefore, in this research the power law process is applied to the proposed groundwater transport model.

DOI: 10.1201/9781003266266-17

17.2 INTRODUCING THE CAPUTO OPERATOR INTO THE MATRIX–FRACTURE EQUATIONS

Deyi and Atangana (2020) proposed matrix–fracture equations and showed that Equations (17.1) and (17.2) for matrix and fracture can be discretized, and proved that the equations are stable after conducting the von Neumann's method of analysis.

Equation for the matrix:

$$\left(1 + \beta k_D\right)\frac{\partial Cm}{\partial t} = D_h \frac{\partial^2 C_m}{\partial x^2} - v_m \frac{\partial C_m}{\partial x} \tag{17.1}$$

Equation for the fracture:

$$\left(1 + \beta k_D\right)\frac{\partial Cf}{\partial t} = D_h \frac{\partial^2 C_f}{\partial x^2} - v_m \frac{\partial C_f}{\partial x}. \tag{17.2}$$

17.3 RIEMANN–LIOUVILLE POWER LAW

The Riemann–Liouville power law differential operator had long been regarded as the only mathematical tool to replicate all physical problems, but Atangana and Gómez-Aguilar (2018) argue that not all physical problems follow the power law distribution.

Because power law cannot be used to model all physical problems, the Caputo and Fabrizio derivative has been suggested as an alternative, the concept of differentiation using exponential decay as kernel rather than power law (Atangana and Gómez-Aguilar, 2018).

Graphs presented in the literature by different scholars show that the power law kernel does not provide or present good memory and has non-Gaussian probability distribution characteristics (Tateishi et al., 2017; Atangana and Gómez-Aguilar, 2018).

17.4 MITTAG-LEFFLER LAW

Gösta Magnus Mittag-Leffler (1846–1927), a Swedish mathematician, discovered and introduced special functions in 1903. The Mittag-Leffler function arises naturally in the solution of fractional order integral equations or fractional order differential equations, and in particular in the investigations of the fractional generalization of kinetic equations, random walks, Levy flights, super-diffusive transport, and in the study of complex systems (Haubold et al., 2011). Shown below is the function named after Mittag-Leffler.

$$E_\alpha(z) = \sum_{k=0}^{\infty} \frac{z^k}{\Gamma(\alpha k + 1)}, \alpha \epsilon \, \mathbb{C}, \mathrm{Re}\, \alpha > 0. \tag{17.3}$$

There has been growing interest in the application of the Mittag-Leffler function in physical, biological and earth sciences, including engineering disciplines (Haubold et al., 2011). The recent interest in this function is mainly because of its close relation to fractional calculus, and in particular to fractional problems coming from applications (Gorenflo et al., 2014).

The Mittag-Leffler function has been applied across many disciplines: for example, anomalous diffusion applications, biochemical transport kinetics, molecular transport, information processing in neural networks, viscoelasticity in polymer networks, heat conduction, scaling behavior in human travel, and signal processing among many others (Owolabi & Atangana, 2019).

Because of the limitations faced by power law, a new derivative with a fractional order base on the concept of the Mittag-Leffler-law was suggested by Atangana and Baleanu (Atangana and Gómez-Aguilar 2018).

Atangana and Gómez-Aguilar (2018) showed in their paper that, after running simulations for different values of α, the Mittag-Leffler kernel demonstrates a partial exponential decay memory as well as power law memory. Further, it proves that the Mittag-Leffler kernel has excellent memory in comparison to the previously discussed kernels, and has properties of both power law and exponential law.

17.5 CAPUTO DERIVATIVE

The Caputo derivative is provided below:

$$ {}_0^C D_t^\alpha f(t) = \frac{1}{\Gamma(1-\alpha)} \int_0^t \frac{d}{d\tau} f(\tau)(t-\tau)^{-\alpha} d\tau \tag{17.4} $$

Rearranging the equation, we obtain:

$$ {}_0^C D_t^\alpha f(t) = \int_0^t \frac{d}{d\tau} f(\tau) \frac{(t-\tau)^{-\alpha}}{\Gamma(1-\alpha)} d\tau \tag{17.5} $$

$$ {}_0^C D_t^\alpha f(t) = \frac{df(t)}{dt} * \frac{(t)^{-\alpha}}{\Gamma(1-\alpha)} \tag{17.6} $$

Now finding the Laplace transform for Equation (17.6), the purpose of using the Laplace transform technique is to establish an equation we could use to find the exact analytical solution from:

$$ \mathcal{L}\left({}_0^C D_t^\alpha f(t) \right) = \mathcal{L}\left(\frac{df(t)}{dt} \right) \mathcal{L}\left(\frac{t^{-\alpha}}{\Gamma(1-\alpha)} \right) \tag{17.7} $$

$$ \mathcal{L}\left({}_0^C D_t^\alpha f(t) \right) = \left(S\tilde{f}(s) - f(0) \right) S^{\alpha-1} \tag{17.8} $$

Then we have:

$$ \mathcal{L}\left({}_0^C D_t^\alpha f(t) \right) = S^\alpha \tilde{f}(s) - S^{\alpha-1} f(0). \tag{17.9} $$

17.6 CAPUTO DERIVATIVE INTEGRAL AND APPLYING THE LAPLACE TRANSFORM

An integral of Caputo derivative is presented below:

$$ {}_0^C J_t^\alpha f(t) = f(0) + \frac{1}{\Gamma(\alpha)} \int_0^t f(\tau)(t-\tau)^{\alpha-1} d\tau \tag{17.10} $$

Rewriting the integral we get:

$$ {}_0^C J_t^\alpha f(t) = f(0) + \int_0^t f(\tau) \frac{(t-\tau)^{\alpha-1}}{\Gamma(\alpha)} d\tau \tag{17.11} $$

Then we have the following:

$$\,_0^C J_t^\alpha f(t) = f(0) + f(\tau) * \frac{t^{\alpha-1}}{\Gamma(\alpha)} \tag{17.12}$$

Now applying the Laplace transform technique:

$$\mathcal{L}\left(\,_0^C J_t^\alpha f(t)\right) = \mathcal{L}\left(f(0)\right) + \mathcal{L}\left(f(t)\right) \mathcal{L} \frac{t^{\alpha-1}}{\Gamma(\alpha)} \tag{17.13}$$

$$\mathcal{L}\left(\,_0^C J_t^\alpha f(t)\right) = \frac{f(0)}{S} + \tilde{f}(s) S^{-\alpha} \tag{17.14}$$

$$\mathcal{L}\left(\,_0^C J_t^\alpha f(t)\right) = \frac{f(0)}{S} + \frac{\tilde{f}(s)}{S^\alpha}. \tag{17.15}$$

17.7   NUMERICAL APPROXIMATION OF THE CAPUTO DERIVATIVES

The purpose of numerical approximation is to calculate an accurate approximation of the derivates. Numerical approximation of derivates have been applied by many scholars when solving numerical schemes, including Eckhoff & Wasberg, 1995; Quarteroni & Valli, 2008; and Toufik and Atangana, 2017.

We start a numerical approximation and the discretization of the Caputo derivative.

$$\,_0^C D_t^\alpha f(t) = \frac{1}{\Gamma(1-\alpha)} \int_0^t \frac{d}{d\tau} f(\tau)(t-\tau)^{-\alpha}\, d\tau \tag{17.16}$$

Now doing a partial differentiation of Equation (17.16):

$$\,_0^C D_t^\alpha f(x,t) = \frac{1}{\Gamma(1-\alpha)} \int_0^t \frac{d}{d\tau} f(x,\tau)(t-\tau)^{-\alpha}\, d\tau \tag{17.17}$$

At (x_i, t_{n+1}) we present the numerical approximation as follows:

$$\,_0^C D_t^\alpha f(x_i, t_{n+1}) = \frac{1}{\Gamma(1-\alpha)} \int_0^{t_{n+1}} \frac{d}{d\tau} f(x_i, \tau)(t_{n+1}-\tau)^{-\alpha}\, d\tau \tag{17.18}$$

$$\,_0^C D_t^\alpha f(x_i, t_{n+1}) = \frac{1}{\Gamma(1-\alpha)} \sum_{j=0}^{n} \int_{t_j}^{t_{n+1}} \frac{df}{dr}(x_i, \tau)(t_{n+1}-\tau)^{-\alpha}\, d\tau \tag{17.19}$$

$$\,_0^C D_t^\alpha f(x_i, t_{n+1}) = \frac{1}{\Gamma(1-\alpha)} \sum_{j=0}^{n} \int_{t_j}^{t_{j+1}} \frac{df}{dr}(x_i, \tau)(t_{n+1}-\tau)^{-\alpha}\, d\tau \tag{17.20}$$

$$\,_0^C D_t^\alpha f(x_i, t_{n+1}) = \frac{1}{\Gamma(1-\alpha)} \sum_{j=0}^{n} \int_{tj}^{t_{j+1}} \left(\frac{f(x_i, t_{j+1}) - f(x_i, t_j)}{\Delta t} \right)(t_{n+1}-\tau)^{-\alpha}\, d\tau \tag{17.21}$$

$$
{}_0^C D_t^\alpha f\left(x_i, t_{n+1}\right) = \frac{1}{\Gamma\left(1-\alpha\right)} \sum_{j=0}^n \frac{f_i^{j+1} - f_i^{j+1}}{\Delta t} \int_{t_i}^{t_{j+1}} \left(t_{n+1} - \tau\right)^{-\alpha} d\tau \tag{17.22}
$$

Now if we let $y = t_{n+1} - \tau$, where $\tau = t_j$, $y = t_{n+1} - t_j$
Now let us let $\tau = t_j + 1$, $y = t_{n+1} - t_{j+1}$ $dy = -d\tau$ therefore $d\tau = -dy$

$$
{}_0^C D_t^\alpha f\left(x_i, t_{n+1}\right) = \frac{1}{\Gamma\left(1-\alpha\right)} \sum_{j=0}^n \frac{f_i^{j+1} - f_i^j}{\Delta t} \int_{t_i}^{t_{n+1}-t_{j+1}} y^{-\alpha} \left(-dy\right) \tag{17.23}
$$

Since we know that $-\int_a^b = \int_b^a$ and that:

$$
\left(1-\alpha\right)\Gamma\left(1-\alpha\right) = \Gamma\left(2-\alpha\right)
$$

Since we also know that:

$$
\int_{t_{n+1}-t_{j+1}}^{t_{n+1}-t_j} y^{-\alpha} dy = \left. \frac{y^{1-\alpha}}{\left(1-\alpha\right)} \right|_{t_{n+1}-t_{j+1}}^{t_{n+1}-t_j} y \tag{17.24}
$$

Therefore, we have:

$$
{}_0^C D_t^\alpha f\left(x_i, t_{n+1}\right) = \frac{1}{\Gamma\left(1-\alpha\right)} \sum_{j=0}^n \frac{f_i^{j+1} - f_i^j}{\Delta t} \left. \frac{y^{1-\alpha}}{\left(1-\alpha\right)} \right|_{t_{n+1}-t_{j+1}}^{t_{n+1}-t_j} y \tag{17.25}
$$

$$
{}_0^C D_t^\alpha f\left(x_i, t_{n+1}\right) = \frac{1}{\Gamma\left(1-\alpha\right)} \sum_{j=0}^n \frac{f_i^{j+1} - f_i^j}{\Delta t} \left\{ \frac{\left(t_{n+1} - t_j\right)^{1-\alpha}}{1-\alpha} - \frac{\left(t_{n+1} - t_{j+1}\right)^{1-\alpha}}{1-\alpha} \right\} \tag{17.26}
$$

$$
= \frac{1}{\Gamma\left(1-\alpha\right)} \sum_{j=0}^n \frac{f_i^{j+1} - f_i^j}{\Delta t} \left\{ \left((n+1)\Delta t - j\Delta t\right)^{1-\alpha} - \left((n+1)\Delta t - (j+1)\Delta t\right)^{1-\alpha} \right\} \tag{17.27}
$$

The discretized derivative for Caputo is shown below:

$$
{}_0^C D_t^\alpha f\left(x_i, t_{n+1}\right) = \frac{\left(\Delta t\right)^{-\alpha}}{\Gamma\left(2-\alpha\right)} \sum_{j=0}^n \left(f_i^{j+1} - f_i^j\right) \left\{ \left(n+1-\alpha\right)^{1-\alpha} - \left(n-j\right)^{1-\alpha} \right\}. \tag{17.28}
$$

17.8 NUMERICAL APPROXIMATION OF INTEGRALS

The numerical approximation of integrals is considered when analytical integrations are difficult to evaluate analytically (Odibat, 2006). According to Kumar et al. (2019), numerical integration for the fractional integral also becomes important in developing the algorithms for solving applied problems defined using fractional derivatives.

Riemann–Liouville

We start the numerical approximation with the Riemann–Liouville approach.

$$J_0^\alpha f(t) = \frac{1}{\Gamma(\alpha)} \int_0^t f(\tau)(t-\tau)^{\alpha-1} \, d\tau \tag{17.29}$$

$$J_0^\alpha f(x,t) = \frac{1}{\Gamma(\alpha)} \int_0^t f(x,\tau)(t-\tau)^{\alpha-1} \, d\tau \tag{17.30}$$

At (x_i, t_{n+1}) we present the numerical approximation as follows:

$$J_0^\alpha f(x_i, t_{n+1}) = \frac{1}{\Gamma(\alpha)} \int_0^{t_{n+1}} f(x_i,\tau)(t_{n+1}-\tau)^{\alpha-1} \, d\tau \tag{17.31}$$

$$J_0^\alpha f(x_i, t_{n+1}) = \frac{1}{\Gamma(\alpha)} \int_{t_j}^{t_{j+1}} f(x_i,\tau)(t_{n+1}-\tau)^{\alpha-1} \, d\tau \tag{17.32}$$

$$J_0^\alpha f(x_i, t_{n+1}) = \frac{1}{\Gamma(\alpha)} \sum_{j=0}^n \frac{f_i^{j+1} + f_i^j}{2} \int_{t_j}^{t_{j+1}} (t_{n+1}-\tau)^{\alpha-1} \, d\tau \tag{17.33}$$

$$J_0^\alpha f(x_i, t_{n+1}) = \frac{1}{\Gamma(\alpha)} \sum_{j=0}^n \frac{f_i^{j+1} + f_i^j}{2} \left(\frac{(t_{n+1}-\tau)^{\alpha-1+1}}{-(\alpha+1-1)} \Bigg|_{t_j}^{t_{j+1}} \right) \tag{17.34}$$

$$J_0^\alpha f(x_i, t_{n+1}) = \frac{1}{\Gamma(\alpha)} \sum_{j=0}^n \frac{f_i^{j+1} + f_i^j}{2} \left(\frac{(t_{n+1}-t_j)^{\alpha}}{\alpha} - \frac{(t_{n+1}-t_{j+1})^{\alpha}}{\alpha} \right) \tag{17.35}$$

$$J_0^\alpha f(x_i, t_{n+1}) = \frac{1}{\Gamma(\alpha)} \sum_{j=0}^n \frac{f_i^{j+1} + f_i^j}{2} \frac{1}{\alpha} \left(((n-j+1)\Delta t)^{\alpha} - ((n-j)\Delta t)^{\alpha} \right) \tag{17.36}$$

Then, finally, we have:

$$J_0^\alpha f(x_i, t_{n+1}) = \frac{1}{\Gamma(\alpha+1)} \sum_{j=0}^n \frac{f_i^{j+1} + f_i^j}{2} (\Delta t)^{\alpha} \left((n-j+1)^{\alpha} - (n-j)^{\alpha} \right) \tag{17.37}$$

Now we do numerical approximation for the **Caputo integral**.

$$_0^C J_t^\alpha f(t) = f(0) + \frac{1}{\Gamma(\alpha)} \int_0^t f(\tau)(t-\tau)^{\alpha-1} \, d\tau \tag{17.38}$$

At $f(x_i, 0)$ we present the numerical approximation as follows:

$$_0^C J_t^\alpha f(t) = f(x_i, 0) + \frac{1}{\Gamma(\alpha)} \int_0^t f(x_i, t)(t-\tau)^{\alpha-1} \, d\tau \tag{17.39}$$

$$\,_0^C J_t^\alpha f(t) = f(x_i, 0) + \frac{1}{\Gamma(\alpha)} \int_0^{t_{n+1}} f(x_i, \tau)(t_{n+1} - \tau)^{\alpha-1} \, d\tau \tag{17.40}$$

$$\,_o^C J_t^\alpha f(t) = f(x_i, 0) + \frac{1}{\Gamma(\alpha)} \sum_{j=0}^{n} \int_0^{t_{j+1}} f(x_i, \tau)(t_{n+1} - \tau)^{\alpha-1} \, d\tau \tag{17.41}$$

$$\,_o^C J_t^\alpha f(t) = f(x_i, 0) + \frac{1}{\Gamma(\alpha)} \sum_{j=0}^{n} \frac{f_i^{j+1} + f_i^{j}}{2} \int_0^{t_{j+1}} (t_{n+1} - \tau)^{\alpha-1} \, d\tau \tag{17.42}$$

$$\,_o^C J_t^\alpha f(t) = f(x_i, 0) + \frac{1}{\Gamma(\alpha)} \sum_{j=0}^{n} \frac{f_i^{j+1} + f_i^{j}}{2} \left(\frac{(t_{n+1} - t_j)^\alpha \, \Delta t^\alpha}{\alpha} - \frac{(t_{n+1} - t_{j+1})^\alpha \, \Delta t^\alpha}{\alpha} \right) \tag{17.43}$$

$$= f(x_i, 0) + \frac{1-\alpha}{\Gamma(\alpha+1)} \left(\frac{f_i^{n+1} - f_i^n}{2} \right) + \frac{\alpha}{AB(\alpha)\Gamma(\alpha)} \sum_{j=0}^{n} \frac{f_i^{j+1} - f_i^{j}}{2} \left[\frac{(t_{n+1} - t_j)^\alpha}{\alpha} - \frac{(t_{n+1} - t_{j+1})^\alpha}{\alpha} \right] \tag{17.44}$$

$$\,_o^C J_t^\alpha f(t) = f(x_i, 0) + \frac{1}{\Gamma(\alpha+1)} \sum_{j=0}^{n} \frac{f_i^{j+1} + f_i^{j}}{2} \Delta t^\alpha \left((n-j+1)^\alpha - (n-j)^\alpha \right) \tag{17.45}$$

Then, finally, we have:

$$\,_o^C J_t^\alpha f(t) = f(x_i, 0) + \frac{1}{\Gamma(\alpha+1)} \sum_{j=0}^{n} \frac{f_i^{j+1} + f_i^{j}}{2} \Delta t^\alpha \left((n-j+1)^\alpha - (n-j)^\alpha \right). \tag{17.46}$$

17.9 LAGRANGE APPROXIMATION

In this section we consider a numerical approximation scheme that combines the fundamental theorem of fractional calculus and the two-step Lagrange polynomial, which has been observed from the works of Mekkaoui and Atangana (2017). The Lagrange method in previous studies has proved to be beneficial in terms of being computationally less complex, yet comparable in accuracy (Yeh & Kwan, 1978).

Using the Lagrange approximation method on the following equation expressed in the Caputo sense:

$$\,_0^C D_t^\alpha f(x, t) = F\big(f(x, t), t\big) \tag{17.47}$$

and using Riemann–Liouville, we obtain:

$$f(x,t) = f(x, 0) + \frac{1}{\Gamma(\alpha)} \int_0^t (t - \tau)^{\alpha-1} F\big(f(x, \tau), \tau\big) d\tau \tag{17.48}$$

At (x_i, t_{n+1}) we present the numerical approximation as follows:

$$f_i^{n+1} - f_i^0 = \frac{1}{\Gamma(\alpha)} \int_0^{t_{n+1}} (t_{n+1} - \tau)^{\alpha-1} F\big(f(x_i, \tau), \tau\big) d\tau \tag{17.49}$$

$$f_i^{n+1} - f_i^0 = \frac{1}{\Gamma(\alpha)} \sum_{j=0}^{n} \int_{t_j}^{t_{j+1}} (t_{n+1} - \tau)^{\alpha-1} F\big(f(x_i, \tau), \tau\big) d\tau \tag{17.50}$$

Then $[t_j, t_{j+1}]$

$$P_j(\tau) = \frac{\tau - t_j}{t_j - t_{j-1}} F\big(f_i^j, t_j\big) + \frac{\tau - t_{j-1}}{t_{j-1} - t_j} F\big(f_i^{j-1}, t_{j-1}\big) \tag{17.51}$$

$$f_i^{n+1} - f_i^0 = \frac{1}{\Gamma(\alpha)} \sum_{j=0}^{n} \int_{t_j}^{t_{j+1}} (t_{n+1} - \tau)^{\alpha-1} \left[\frac{\tau - t_j}{t_j - t_{j-1}} F\big(f_i^j, t_j\big) + \frac{\tau - t_{j-1}}{t_{j-1} - t_j} F\big(f_i^{j-1}, t_{j-1}\big) \right] d\tau \tag{17.52}$$

$$f_i^{n+1} - f_i^0 = \frac{1}{\Gamma(\alpha)} \sum_{j=0}^{n} \left[\begin{array}{l} \dfrac{F\big(f_i^j, t_j\big)}{t_i - t_{j-1}} \displaystyle\int_{t_j}^{t_{j+1}} (t_{n+1} - \tau)^{\alpha-1} (\tau - t_j) d\tau + \\[2mm] F\left(\dfrac{f_i^{j-1}, \tau}{t_{j-1} - t_j} \right) \displaystyle\int_{t_j}^{t_{j+1}} (t_{n+1} - \tau)^{\alpha-1} (\tau - t_{j-1}) d\tau \end{array} \right] \tag{17.53}$$

Therefore we have:

$$
\begin{aligned}
f_i^{n+1} - f_i^0 = {} & \frac{1}{\Gamma(\alpha)} \sum_{j=0}^{n} \frac{F\big(f_{i,}^j\, t_j\big)}{\Delta t} \frac{(\Delta t)^{\alpha+1}}{\alpha} \left[-(n-j)^\alpha - \frac{(n-j)^{\alpha+1}}{\alpha+1} + \frac{(n-j+1)^{\alpha+1}}{\alpha+1} \right] \\
& + \frac{1}{\Gamma(\alpha)} \sum_{j=0}^{n} \frac{F\big(f_i^{j-1}, t_{j-1}\big)}{(-\Delta t)} \frac{(\Delta t)^{\alpha+1}}{\alpha} \left[-(n-j-1)^\alpha - \frac{(n-j-1)^{\alpha+1}}{\alpha+1} \right]
\end{aligned}
\tag{17.54}
$$

$$
f_i^{n+1} - f_i^0 = \frac{1}{\Gamma(\alpha)} \sum_{j=0}^{n} \left[\begin{array}{l} F\big(f_i^j, t_j\big)(\Delta t)^\alpha \left(-(n-j)^\alpha - \dfrac{(n-j)^{\alpha+1}}{\alpha+1} + \dfrac{(n-j+1)^{\alpha+1}}{\alpha+1} \right) \\[4mm] -F\big(f_i^{j-1}, t_{j-1}\big)(\Delta t)^\alpha \left(-(n-j-1)^\alpha - \dfrac{(n-j-1)^{\alpha+1}}{\alpha+1} + \dfrac{(n-j)^{\alpha+1}}{\alpha+1} \right) \end{array} \right] \tag{17.55}
$$

$$
f_i^{n+1} = f_i^0 + \frac{1}{\Gamma(\alpha)} \sum_{j=0}^{n} \left[\frac{\Delta t^\alpha F\big(f_i^j, t_j\big)}{\Gamma(\alpha+2)} \left(\begin{array}{l} (n-j+1)^\alpha (n-j+2+\alpha) \\[2mm] -(n-j)^\alpha (n-j+2+2\alpha) - \dfrac{\Delta t^\alpha F\big(f_{i-1}^j, t_{j-1}\big)}{\Gamma(\alpha+2)} \\[2mm] \big((n+1-j)^{\alpha+1} - (n-j)^\alpha (n-j+1+\alpha)\big) \end{array} \right) \right] \tag{17.56}
$$

17.10 MODEL WITH POWER LAW PROCESS

In this section we discretized the proposed diffusion equation considering the power law function. The power law process has previously been applied to different fractional advection-dispersion equations by different researchers (Benson et al., 2000; Tateishi et al., 2017). It is believed that processes such as Brownian motion are driven by the power law process, in a diffusion process we consider a power law memory kernel.

Detailed analysis of the diffusion equation is considered where the process of power law is applied. The time derivative is converted to a fractional derivative whereby the derivative is the Caputo.

Below we recall and present the diffusion equation in the matrix.

$$\left(1+\beta k_D\right)\frac{\partial Cm}{\partial t} = D_h\frac{\partial^2 C_m}{\partial x^2} - v_m\frac{\partial C_m}{\partial x} \tag{17.57}$$

We also present the Caputo derivative:

$$\,^C_0 D^\alpha_t f(t) = \frac{1}{\Gamma(1-\alpha)}\int_0^t \frac{d}{d\tau} f(\tau)(t-\tau)^{-\alpha}\, d\tau \tag{17.58}$$

Then the numerical approximation of the above is presented here:

$$\left(1+\beta k_D\right)\frac{1}{\Gamma(1-\alpha)}\int_0^t \frac{\partial}{\partial\tau} C_m(x,\tau)(t-\tau)^{-\alpha}\, d\tau = D_h(x)\frac{\partial^2 C_m}{\partial x^2} - v_m\frac{\partial C_m}{\partial x} \tag{17.59}$$

$$\left(1+\beta k_D\right)\frac{1}{\Gamma(1-\alpha)}\int_0^t \frac{\partial}{\partial\tau} C_m(x,\tau)(t-\tau)^{-\alpha}\, d\tau = D_h(x)\frac{\partial^2 C_{m(x,t)}}{\partial x^2} - v_m\frac{\partial C_{m(x,t)}}{\partial x} \tag{17.60}$$

$$\frac{1}{\Gamma(1-\alpha)}\int_0^t \frac{\partial}{\partial\tau} C_m(x,\tau)(t-\tau)^{-\alpha}\, d\tau = \frac{1}{1+\beta k_D}\left[D_h(x)\frac{\partial^2 C_{m(x,t)}}{\partial x^2} - v_m\frac{\partial C_{m(x,t)}}{\partial x}\right] \tag{17.61}$$

For simplicity, we let $\dfrac{1}{1+\beta k_D}\left[D_h(x)\dfrac{\partial^2 C_{m(x,t)}}{\partial x^2} - v_m\dfrac{\partial C_{m(x,t)}}{\partial x}\right]$ be equal to $F(x, t, C_m(x,t))$

Then we have:

$$\frac{1}{\Gamma(1-\alpha)}\int_0^t \frac{\partial}{\partial\tau} C_m(x,\tau)(t-\tau)^{-\alpha}\, d\tau = F\big(x, t, C_m(x,t)\big) \tag{17.62}$$

Now applying the Riemann–Liouville integral we obtain the following:

$$C_m(x,t) - C_m(x,0) = \frac{1}{\Gamma(\alpha)}\int_0^t F\big(x,\tau, C_m(x,\tau)\big)(t-\tau)^{\alpha-1}\, d\tau \tag{17.63}$$

Then at $t = t_n$, we have $t_n = \Delta tn$, $x_i = i\Delta x$, thus at t_{n+1} we discretize.

$$C_m(x_i, t_{n+1}) - C_m(x_i, 0) = \frac{1}{\Gamma(\alpha)}\int_0^{t_{n+1}} F\big(x_i,\tau, C_m(x_i,\tau)\big)(t_{n+1}-\tau)^{\alpha-1}\, d\tau \tag{17.64}$$

Now

$$C_{m,i}^{n+1} - C_{m,i}^0 = \frac{1}{\Gamma(\alpha)} \sum_{j=0}^n \int_{t_j}^{t_{j+1}} F\left(x_i, \tau, C_m(x_i, \tau)\right)(t_{n+1} - \tau)^{\alpha-1} d\tau \tag{17.65}$$

We approximate $F(x_i, \tau, C_m(x_i, \tau)) = P_j(\tau)$ at $[t_{j-1}, t_{j+1}]$

$$P_j(\tau) = \frac{\tau - t_{j-1}}{t_j - t_{j-1}} F\left(x_i, t_j, C_{m,i}^j\right) + \frac{\tau - t_j}{t_{j-1} - t_j} F\left(x_i, t_{j-1}, C_{m,i}^{j-1}\right) \tag{17.66}$$

Thus:

$$C_{m,i}^{n+1} - C_{m,i}^0 = \frac{1}{\Gamma(\alpha)} \sum_{j=0}^n \int_{t_j}^{t_{j+1}} \left[\frac{\tau - t_{j-1}}{\Delta t} F\left(x_i, t_j, C_{m,i}^j\right) - \frac{\tau - t_j}{\Delta t} F\left(x_i, t_{j-1}, C_{m,i}^{j-1}\right) \right] (t_{n+1} - \tau)^{\alpha-1} d\tau \tag{17.67}$$

Now using the integral routine presented above, we obtain the following:

$$C_{m,i}^{n+1} - C_{m,i}^0 = \sum_{j=0}^n \left[\begin{array}{l} \dfrac{(\Delta t)^\alpha}{\Gamma(\alpha+2)} F\left(x_i, t_j, C_{m,i}^j\right) \left\{ \begin{array}{l} (n+1-j)^\alpha (n-j+2+\alpha) \\ -(n-j)^\alpha (n-j+2+2\alpha) \end{array} \right\} \\ - \dfrac{(\Delta t)^\alpha}{\Gamma(\alpha+2)} F\left(x_i, t_{j-1}, C_{m,i}^{j-1}\right) \left\{ \begin{array}{l} (n+2-j)^{\alpha+1} - (n-j)^\alpha \\ (n-j+1+\alpha) \end{array} \right\} \end{array} \right] \tag{17.68}$$

However, we know that we have the following:

$$F\left(x_i, t_j, C_{m,i}^j\right) = \frac{1}{1+\beta k_D} \left[D_h(x_i) \frac{C_{m,i+1}^j - 2C_{m,i}^j + C_{m,i+1}^j}{\Delta x^2} - v_m \frac{C_{m,i+1}^j - C_{m,i-1}^j}{2\Delta x} \right] \tag{17.69}$$

$$F\left(x_i, t_j, C_{m,i}^{j-1}\right) = \frac{1}{1+\beta k_D} \left[D_h(x_i) \frac{C_{m,i+1}^{j-1} - 2C_{m,i}^{j-1} + C_{m,i+1}^{j-1}}{\Delta x^2} - v_m \frac{C_{m,i+1}^{j-1} - C_{m,i-1}^{j-1}}{2\Delta x} \right] \tag{17.70}$$

Then substituting the above two equations into Equation (17.68) we get:

$$C_{m,i}^{n+1} - C_{m,i}^0 = \sum_{j=0}^n \frac{(\Delta t)^\alpha}{\Gamma(\alpha+2)} \left\{ \frac{1}{1+\beta k_D} \left[D_h(x_i) \frac{C_{m,i+1}^j - 2C_{m,i}^j + C_{m,i-1}^j}{\Delta x^2} - v_m \frac{C_{m,i+1}^j - C_{m,i+1}^j}{2\Delta x} \right] \right\}$$

$$\left\{ (n-j+1)^\alpha (n-j+2+\alpha) - (n-j+2+2\alpha)(n-j)^\alpha \right\} - \frac{(\Delta t)^\alpha}{\Gamma(\alpha+2)}$$

$$\left\{ \frac{1}{1+\beta K_D} \left[D_h(x_i) \frac{C_{m,i+1}^{j-1} - 2C_{m,i}^{j-1} + C_{m,i-1}^{j-1}}{\Delta x^2} - v_m \frac{C_{m,i+1}^{j-1} - C_{m,i-1}^{j-1}}{2\Delta x} \right] \right\}$$

$$\left\{ (n-j+1)^{\alpha+1} - (n-j)^\alpha (n-j+1+\alpha) \right\} \tag{17.71}$$

Below we then present the stability analysis of the above numerical scheme. To achieve this, we consider $C_{m,i}^{j}$ to be the exact solution of the equation and $\overline{C_{m,i}^{j}}$ the approximate solution.

Thus:

$$\epsilon_{m,i}^{j} = C_{m,i}^{j} - \overline{C_{m,i}^{j}}, \text{ the error} \tag{17.72}$$

Applying the Fourier transform, we know that:

$$\epsilon_{m,i}^{j} = \delta_{j} e^{ik_{e}x}, \ \epsilon_{m,i-1}^{j-1} = \delta_{j} e^{ik_{e}(x-\Delta x)} \tag{17.73}$$

$$\epsilon_{m,i+1}^{j+1} = \delta_{j+1} e^{ik_{e}(x+\Delta x)}, \epsilon_{m,i+1}^{j+1} = \delta_{j+1} e^{ik_{e}(x-\Delta x)} \tag{17.74}$$

Then, for simplicity, we let:

$$\delta_{n,j}^{\alpha} = (n-j+1)^{\alpha}(n-j+2+\alpha) - (n-j+2+2\alpha)(n-j)^{\alpha} \tag{17.75}$$

and

$$\delta_{n,j-1}^{\alpha} = (n-j+1)^{\alpha+1} - (n-j)^{\alpha}(n-j+\alpha+1) \tag{17.76}$$

and

$$A_{i} = \frac{D_{h}(x_{i})(\Delta t)^{\alpha}}{(1+\beta k_{D})\Delta x^{2}\Gamma(\alpha+2)} \tag{17.77}$$

$$B_{i} = \frac{-v_{m}\Delta t^{\alpha}}{2(1+\beta k_{D})\Delta x\Gamma(\alpha+2)} \tag{17.78}$$

Substituting the above, we have:

$$C_{m,i}^{j+1} - C_{m,i}^{0} = \sum_{j=0}^{n}\left[A_{i}\left(C_{m,i+1}^{j} - 2C_{m,i}^{j} + C_{m,i-1}^{j}\right) + B_{i}\frac{\left(C_{m,i+1}^{j} - C_{m,i-1}^{j}\right)}{2\Delta x}\right]\delta_{n,j}^{\alpha}$$
$$- \sum_{j=0}^{n}\left[A_{i}\left(C_{m,j+1}^{j-1} - 2C_{m,i}^{j-1} + C_{m,i-1}^{j-1}\right) + B_{i}\left(C_{m,i+1}^{j-1} - C_{m,i-1}^{j-1}\right)\right]\delta_{n,j-1}^{\alpha} \tag{17.79}$$

Now introducing an error, we have:

$$\delta_{n+1}e^{ik_{e}(x)} - \delta_{0}e^{ik_{e}x} = \sum_{j=0}^{n}\left[\begin{array}{l}A_{i}\left(\delta_{j}e^{ik_{e}(x+\Delta x)} - 2\delta_{j}e^{ik_{e}x} + \delta_{j}e^{ik_{e}(x-\Delta x)}\right)\\ +B_{i}\left(\delta_{j}e^{ik_{e}(x+\Delta x)} - \delta_{j}e^{ik_{e}(x-\Delta x)}\right)\end{array}\right]\delta_{n,j}^{\alpha}$$
$$- \sum_{j=0}^{n}\left[\begin{array}{l}A_{i}\left(\delta_{j-1}e^{ik_{e}(x+\Delta x)} - 2\delta_{j-1}e^{ik_{e}x} + \delta_{j-1}e^{ik_{e}(x-\Delta x)}\right)\\ +B_{i}\left(\delta_{j-1}e^{ik_{e}(x+\Delta x)} - \delta_{j-1}e^{ik_{e}(x-\Delta x)}\right)\end{array}\right]\delta_{n,j-1}^{\alpha} \tag{17.80}$$

$$\delta_{n+1} - \delta_0 = \sum_{j=0}^{n} \left[A_i \left(\delta_j e^{ik_e\Delta x} - 2\delta_j + \delta_j e^{-ik_e\Delta x} \right) + B_i \left(\delta_j e^{ik_e\Delta x} - \delta_j e^{-ik_e\Delta x} \right) \right] \delta_{n,j}^{\alpha}$$

$$- \sum_{j=0}^{n} \left[A_i \left(\delta_{j-1} e^{ik_e\Delta x} - 2\delta_{j-1} + \delta_{j-1} e^{-ik_e\Delta x} \right) + B_i \left(\delta_{j-1} e^{ik_e\Delta x} - \delta_{j-1} e^{-ik_e\Delta x} \right) \right] \delta_{n,j-1}^{\alpha} \quad (17.81)$$

We know that:

$$e^{ik_e\Delta x} - 2 + e^{-ik_e\Delta x} = -4\sin^2\left[\frac{k_e\Delta x}{2} \right] \quad (17.82)$$

Thus, we have:

$$\delta_{n+1} - \delta_0 = \sum_{j=0}^{n} A_i \left[-4\sin^2\left(\frac{k_e\Delta x}{2} \right) + B_i \left(2i\operatorname{Sin}^2\left(k_e\Delta x \right) \right) \right] \delta_{n,j}^{\alpha} \delta_j$$

$$- \sum_{j=0}^{n} \left[-4\sin^2\left(\frac{k_e\Delta x}{s} \right) A_i + B_i \sin\left(k_e\Delta x \right) \right] \delta_{n,j}^{\alpha} \delta_{j-1} \quad (17.83)$$

The above can be reformulated as:

$$\delta_{n+1} - \delta_0 = \sum_{j=0}^{n-1} \left[-4A_i \sin^2\left(\frac{k_e\Delta x}{2} \right) + 2iB_i \sin\left(k_e\Delta x \right) \right] \delta_{n,j}^{\alpha} \delta_j - \left[\begin{array}{c} -4A_i \sin^2\left(\frac{k_e\Delta x}{2} \right) \\ +2iB_i \sin\left(k_e\Delta x \right) \end{array} \right]$$

$$\delta_{n,j}^{\alpha} \delta_n + \sum_{j=0}^{n-1} \left[-4A_i \sin^2\left(\frac{k_e\Delta x}{2} \right) + 2iB_i \sin\left(k_e\Delta x \right) \right] \delta_{n,j-1}^{\alpha} \delta_j$$

$$+ \left[-4A_i \sin^2\left(\frac{k_e\Delta x}{2} \right) + 2i\operatorname{Sin}\left(k_e\Delta x \right) \right] \delta_n^{\alpha} \delta_{n-1} \quad (17.84)$$

We prove the stability by introducing "n", thus when h = 0 we have:

$$\delta_1 - \delta_0 = \left[-4A_i \sin^2\left(\frac{k_e\Delta x}{2} \right) + 2iB_i \sin\left(k_e\Delta x \right) \right] \delta_{0,0}^{\alpha} \quad (17.85)$$

Therefore, we have:

$$\frac{\delta_1}{\delta_0} = \left[1 - 4A_i \sin^2\left(\frac{k_e\Delta x}{2} \right) + 2iB_i \sin\left(k_e\Delta x \right) \right] \delta_{0,0}^{\alpha} \quad (17.86)$$

$$\left| \frac{\delta_1}{\delta_0} \right| = \delta_{0,0}^{\alpha} \sqrt{ \left(1 - 4A_i \sin^2\left(\frac{k_e\Delta x}{2} \right) \right)^2 + 4B_i^2 \sin^2\left(k_e\Delta x \right) } \cdot \quad (17.87)$$

17.11   CONCLUSION

In this chapter we have considered the introduction of a fractional differential operator, the Caputo derivative, to the equations proposing a new model for groundwater transport. The Caputo derivative was considered because it is deemed suitable for modeling real-world problems. The strength of the application of fractional differentiation has proved to be very efficient in describing the memory effects. Also, it has proved to have an ability to model the effect of fracture, a differential operator based on power law kernel, and was used to extend the classical advection dispersion equation with a variable coefficient. The Laplace transform technique was applied to the Caputo derivative and integral. Furthermore, numerical and Lagrange approximation, and Riemann–Liouville, have been applied to the Caputo derivative and integral. Finally, the results were introduced into the proposed new groundwater transport equations.

REFERENCES

Abdeljawad, T. (2011). On Riemann and Caputo fractional differences. *Computers and Mathematics with Applications*, 62(3), 1602–1611.

Atangana, A., & Gómez-Aguilar, J.F. (2018). Decolonisation of fractional calculus rules: Breaking commutativity and associativity to capture more natural phenomena. *The European Physical Journal Plus*, 133(4), 166. doi:10.1140/epjp/i2018-12021-3.

Benson, D.A., Wheatcraft, S.W., & Meerschaert, M.M. (2000). Application of a fractional advection-dispersion equation. *Water Resources Research*, 36(6), 1403–1412. doi:10.1029/2000WR900031.

Caputo, M. (1967). Linear models of dissipation whose Q is almost Frequency Independent – II. *Geophysical Journal International*, 13(5), 529–539. doi:10.1111/j.1365-246X.1967.tb02303.x.

Deyi, M., & Atangana, A. (2020). A new model for groundwater contamination transport in dual media. PhD Thesis, University of the Free State, Bloemfontein, South Africa.

Eckhoff, K.S., & Wasberg, C.E. (1995). On the numerical approximation of derivatives by a modified Fourier collocation method (Report No. 99). University of Bergen.

Gorenflo, R., Kilbas, A.A., Mainardi, F., & Rogosin, S.V. (2014). *Mittag-Leffler Functions, Related Topics and Applications*. Springer Berlin Heidelberg.

Haubold, H.J., Mathai, A.M., & Saxena, R.K. (2011). Mittag-Leffler functions and their applications [Research article]. *Journal of Applied Mathematics*. doi: 10.1155/2011/298628

Kilbas, A.A., Srivastava, H.M., & Trujillo, J.J. (2006). *Theory and Applications of Fractional Differential Equations* (Vol. 13). Elsevier B.V., Amsterdam.

Kumar, K., Pandey, R.K., & Sharma, S. (2019). Approximations of fractional integrals and Caputo derivatives with application in solving Abel's integral equations. *Journal of King Saud University-Science*, 31(4), 692–700. doi:10.1016/j.jksus.2017.12.017.

Odibat, Z. (2006). Approximations of fractional integrals and Caputo fractional derivatives. *Applied Mathematics and Computation*, 178(2), 527–533. doi:10.1016/j.amc.2005.11.072.

Owolabi, K.M., & Atangana, A. (2019). *Numerical Methods for Fractional Differentiation* (Vol. 54).

Quarteroni, A., & Valli, A. (2008). *Numerical Approximation of Partial Differential Equations*. Springer Science & Business Media.

Tateishi, A.A., Ribeiro, H.V., & Lenzi, E.K. (2017). The role of fractional time-derivative operators on anomalous diffusion. *Frontiers in Physics*, 5.

Toufik, M., & Atangana, A. 2017. New numerical approximation of fractional derivative with non-local and non-singular kernel: Application to chaotic models. *The European Physical Journal Plus* 132(10), 444.

Yeh, K.C., & Kwan, K.C. (1978). A comparison of numerical integrating algorithms by trapezoidal, Lagrange, and spline approximation. *Journal of Pharmacokinetics and Biopharmaceutics*, 6(1), 79–98. doi:10.1007/BF01066064.

18 New Groundwater Transport in Dual Media with the Atangana–Baleanu Differential Operators

Mpafane Deyi
University of the Free State, Bloemfontein, South Africa

Abdon Atangana
China Medical University Hospital, China Medical University, Taichung, Taiwan
University of the Free State, Bloemfontein, South Africa

CONTENTS

18.1 INTRODUCTION

Fractional calculus, a field within the domain of the study of mathematics, has grown out of the tradition definitions of calculus integral and derivative operators in much the same way as fractional exponents in an outgrowth of exponents with inter value (Rahimy, 2010). In recent years, fractional calculus has been applied in different disciplines such as engineering (electrical and mechanical) and the sciences (chemistry and biology) as well as in the field of economics, notably as control theory, and signal and image processing (Atangana & Secer, 2013; Gómez-Aguilar et al., 2016; Abro et al., 2018). As an example of its relevance and application to the groundwater studies, Atangana and Alkahtani (2015) developed a new model of groundwater flow within a confined aquifer in the Caputo–Fabrizio derivative sense, the advantage of this fractional derivative being that it is able to capture and replicate the memory effect. Furthermore, their derivative is able to take into account the effect of the porosity with small permeable media. This derivative was found to be able to describe the diffusion of material on different and variable scales.

Atangana and Baleanu introduced a new derivative with a nonlocal and non-singular kernel. The two fractional operators are based on the generalized Mittag–Leffler function (Atangana & Baleanu, 2016). The kernel considered is nonlocal, non-singular and has all the benefits of the fractional operators of Riemann–Liouville, Liouville–Caputo, and Caputo–Fabrizio (Gómez-Aguilar et al., 2016).

DOI: 10.1201/9781003266266-18

The Atangana–Baleanu derivative was considered for this study because in previous studies it was used to model the flow of heat in material with different scales and also those with heterogeneous media such as a matrix–fracture scenario. Tateishi et al. (2017), when applying the Atangana–Baleanu operator, found a crossover from stretched exponential to power law between the two diffusive regimes: normal for shorter times, and a sub-diffusive for longer periods, a feature observed in several empirical systems. They also found that the operator can be associated with a fractional diffusion equation with derivatives of distributed order.

18.2 INTRODUCING ATANGANA–BALEANU OPERATORS INTO THE MATRIX–FRACTURE EQUATIONS

Deyi and Atangana (2020) proposed matrix–fracture equations and showed that Equations (18.1) and (18.2) for matrix and fracture can be discretized, and proved that the equations are stable after including von Neumann's method of analysis in their study.

Equation for the matrix:

$$\left(1 + \beta k_D\right)\frac{\partial Cm}{\partial t} = D_h \frac{\partial^2 C_m}{\partial x^2} - v_m \frac{\partial C_m}{\partial x} \tag{18.1}$$

Equation for the fracture:

$$\left(1 + \beta k_D\right)\frac{\partial Cf}{\partial t} = D_h \frac{\partial^2 C_f}{\partial x^2} - v_m \frac{\partial C_f}{\partial x} \tag{18.2}$$

In this chapter we consider the introduction of the Atangana–Baleanu fractional derivative with nonlocal and non-singular kernel on the above equations. Because of similarities in the equations we shall use only one equation – Equation (18.1).

18.3 ATANGANA–BALEANU DERIVATIVE AND INTEGRAL

The Mittag–Leffler kernel was used to construct the Atangana–Baleanu fractional derivative. Atangana and Baleanu introduced new fractional derivatives in the Riemann–Liouville and Caputo sense.

The kernel Atangana–Baleanu fractional derivative (ABFD) shown in Equation (18.3) is based on the generalized Mittag–Leffler function without singularity and locality (Atangana & Baleanu, 2016). ABFD is suitable for describing physical and real-world phenomena. The non-singularity and nonlocality of the kernel gives a better description of the memory within the structure with different scales. Furthermore, the Atangana–Baleanu operators satisfy all the mathematical principles under the scope of fractional calculus (Saqib et al., 2018).

$$^{ABC}_{\ b}D^\alpha_t f\left(t\right) = \frac{AB(\alpha)}{(1-\alpha)} \int_0^t \frac{d}{d\tau} f\left(\tau\right) E_\alpha\left[-\alpha \frac{\left(t-\tau\right)^\alpha}{1-\alpha}\right] d\tau \tag{18.3}$$

Saqib et al. (2018) applied the Atangana–Baleanu derivative fractional in the Caputo sense to the convective flow of carboxymethyl cellulose (CMC)-based carbon nanotubes (CNTs') nanofluid in a vertical microchannel. The kernel was used in the mathematical formulation to obtain the time fractional governing equations subject to physical initial and boundary conditions.

An integral of Atangana–Baleanu derivative is provided:

$$^{ABC}_0 J^\alpha_t f(t) = f(0) + \frac{1-\alpha}{AB(\alpha)} f(t) + \frac{\alpha}{\Gamma(\alpha) AB(\alpha)} \int_0^t f(\tau)(t-\tau)^{\alpha-1} d\tau \quad (18.4)$$

18.4 LAPLACE TRANSFORM

The Laplace transform functions can be used to derive exact solutions of linear ordinary and partial differential equations. Laplace transforms are also widely used in solving groundwater flow problems (Mathias & Zimmerman, 2003). With the Laplace transform, it is often possible to avoid working with equations of differential order directly by translating the equation into a domain where the solution presents itself algebraically.

Now the Atangana–Baleanu derivative from Equation (18.3) is also brought forward.

$$^{ABC}_0 D^\alpha_t f(t) = \frac{AB(\alpha)}{(1-\alpha)} \int_0^t \frac{d}{d\tau} f(\tau) E_\alpha\left(-\frac{\alpha}{1-\alpha}(t-\tau)^\alpha\right) d\tau \quad (18.5)$$

Rearranging the equation, we get:

$$^{ABC}_0 D^\alpha_t f(t) = \int_0^t \frac{d}{d\tau} f(\tau) \frac{AB(\alpha)}{(1-\alpha)} F_\alpha\left(\frac{\alpha}{1-\alpha}(t-\tau)^\alpha\right) d\tau \quad (18.6)$$

Then we have:

$$^{ABC}_0 D^\alpha_t f(t) = \frac{d}{dt} f(t) * \frac{AB(\alpha)}{(1-\alpha)} E_\alpha\left(-\frac{\alpha}{1-\alpha} t^\alpha\right) \quad (18.7)$$

Now finding the Laplace transform for Equation (18.7):

$$\mathcal{L}\left(^{ABC}_0 D^\alpha_t f(t)\right) = \mathcal{L}\left(\frac{d}{dt} f(t) \frac{AB(\alpha)}{1-\alpha} E_\alpha\left[-\frac{\alpha}{1-\alpha} t^\alpha\right]\right) \quad (18.8)$$

$$\mathcal{L}\left(^{ABC}_0 D^\alpha_t f(t)\right) = \mathcal{L}\left(\frac{d}{dt} f(t) \frac{AB(\alpha)}{1-\alpha}\right) \mathcal{L}\left(E_\alpha\left[-\frac{\alpha}{1-\alpha} t^\alpha\right]\right) \quad (18.9)$$

$$\mathcal{L}\left(^{ABC}_0 D^\alpha_t f(t)\right) = \left[S\tilde{f}(s) - f(0)\right] \frac{AB(\alpha)}{1-\alpha} \frac{S^{\alpha-1}}{S^\alpha + \frac{\alpha}{1-\alpha}} \quad (18.10)$$

Then finally we have:

$$\mathcal{L}\left(^{ABC}_0 D^\alpha_t f(t)\right) = \frac{S^\alpha \tilde{f}(s) - S^{\alpha-1} f(0)}{S^\alpha + \frac{\alpha}{1-\alpha}} AB(\alpha) \quad (18.11)$$

18.4.1 Applying the Laplace Transform Technique to the Atangana–Baleanu Integral

An integral of the Atangana–Baleanu derivative is brought forward from Equation (18.4).

$$^{ABC}_{0}J^\alpha_t f(t) = f(0) + \frac{1-\alpha}{AB(\alpha)} f(t) + \frac{\alpha}{\Gamma(\alpha)AB(\alpha)} \int_0^t f(\tau)(t-\tau)^{\alpha-1} d\tau \tag{18.12}$$

Rewriting the integral we obtain:

$$^{ABC}_{0}J^\alpha_t f(t) = f(0) + \frac{1-\alpha}{AB(\alpha)} f(t) + \int_0^t f(\tau) \frac{\alpha(t-\tau)^{\alpha-1}}{\Gamma(\alpha)AB(\alpha)} d\tau \tag{18.13}$$

Now applying the Laplace transform technique to Equation (18.12).

$$\mathcal{L}\left(^{ABC}_{0}J^\alpha_t f(t)\right) = \mathcal{L}\left(f(0)\right) + \mathcal{L}\left(\frac{1-\alpha}{AB(\alpha)} f(t)\right) + \mathcal{L}\left(\int_0^t f(\tau) \frac{\alpha(t-\tau)^{\alpha-1}}{\Gamma(\alpha)AB(\alpha)} d\tau\right) \tag{18.14}$$

$$\mathcal{L}\left(^{ABC}_{0}J^\alpha_t f(t)\right) = \frac{f(0)}{S} + \frac{1-\alpha}{AB(\alpha)} \tilde{f}(s) + \mathcal{L}\left(f(s)\right)\mathcal{L}\left(\frac{\alpha(t-\tau)^{\alpha-1}}{\Gamma(\alpha)AB(\alpha)}\right) \tag{18.15}$$

Finally we have:

$$\mathcal{L}\left(^{ABC}_{0}J^\alpha_t f(t)\right) = \frac{f(0)}{S} + \frac{1-\alpha}{AB(\alpha)} \tilde{f}(s) + \frac{\tilde{f}(s)}{S^\alpha} \frac{\alpha}{AB(\alpha)}. \tag{18.16}$$

18.5 NUMERICAL APPROXIMATION

Numerical approximation and discretization of the Atangana–Baleanu derivate and integral is applied. Numerical approximation of partial differential equations is regarded as an important branch of numeral analysis (Quarteroni & Valli, 2008). Furthermore, the numerical approximation or schemes have been recognized as powerful mathematical tools to solve non-linear ordinary differential equations with local and nonlocal operators. In the main, they are normally used when all applied analytical methods have failed (Toufik and Atangana, 2017).

18.5.1 Numerical Approximation of the Atangana–Baleanu Derivative

The purpose of numerical approximation is to calculate accurate approximation to the derivates. Numerical approximation of derivates have been applied by many scholars when solving numerical schemes, including Eckhoff & Wasberg, 1995; Quarteroni & Valli, 2008; and Toufik and Atangana, 2017).

Now we do the numerical approximation and the discretization of the Atangana–Baleanu derivative.

$$^{ABC}_{0}D^\alpha_t f(t) = \frac{AB(\alpha)}{(1-\alpha)} \int_0^t \frac{d}{d\tau} f(\tau) E_\alpha\left(-\frac{\alpha}{1-\alpha}(t-\tau)^\alpha\right) d\tau \tag{18.17}$$

$$^{ABC}_0 D_t^\alpha f(x,t) = \frac{AB(\alpha)}{(1-\alpha)} \int_0^t \frac{\partial}{\partial \tau} f(x,\tau) E_\alpha \left(-\frac{\alpha(t-\tau)^\alpha}{1-\alpha} \right) d\tau \tag{18.18}$$

At (x_i, t_{n+1}) we present the numerical approximation, as follows:

$$^{ABC}_0 D_t^\alpha f(x,t) = \frac{AB(\alpha)}{(1-\alpha)} \int_0^{t_{n+1}} \frac{d}{d\tau} f(x_i,\tau) E_\alpha \left(-\frac{\alpha(t_{n+1}-\tau)^\alpha}{1-\alpha} \right) d\tau \tag{18.19}$$

$$^{ABC}_0 D_t^\alpha f(x,t) = \frac{AB(\alpha)}{(1-\alpha)} \sum_0^n \frac{f_i^{j+1}-f_i^j}{\Delta t} \int_{t_j}^{t_{j+1}} E_\alpha \left(-\frac{\alpha(t_{n+1}-\tau)^\alpha}{1-\alpha} \right) d\tau \tag{18.20}$$

$$^{ABC}_0 D_t^\alpha f(x,t) = \frac{AB(\alpha)}{(1-\alpha)} \sum_0^n \frac{f_i^{j+1}-f_i^j}{\Delta t} \int_{t_j}^{t_{j+1}} \sum_{k=0}^\infty \frac{\left[\dfrac{-\alpha(t_{n+1}-\tau)^\alpha}{1-\alpha} \right]^k}{\Gamma(\alpha k+1)} d\tau \tag{18.21}$$

$$^{ABC}_0 D_t^\alpha f(x,t) = \frac{AB(\alpha)}{(1-\alpha)} \sum_0^n \frac{f_i^{j+1}-f_i^j}{\Delta t} \sum_{k=0}^\infty \frac{\left(\dfrac{-\alpha}{1-\alpha} \right)^k}{\Gamma(\alpha k+1)} \int_{t_j}^{t_{j+1}} (t_{n+1}-\tau)^{\alpha k} d\tau \tag{18.22}$$

$$= \frac{AB(\alpha)}{(1-\alpha)} \sum_0^n \frac{f_i^{j+1}-f_i^j}{\Delta t} \sum_{k=0}^\infty \frac{\left(\dfrac{-\alpha}{1-\alpha} \right)^k}{\Gamma(\alpha k+1)} (\Delta t)^{\alpha k+1} \left[(n-j+1)^{\alpha k+1} - (n-j)^{\alpha k+1} \right] \tag{18.23}$$

$$= \frac{AB(\alpha)}{(1-\alpha)} \sum_0^n \frac{f_i^{j+1}-f_i^j}{\Delta t} \left[\sum_{k=0}^\infty \left(\frac{-\alpha}{1-\alpha} \right)^k \frac{\left((n-j+1)\Delta t \right)^{\alpha k+1}}{\Gamma(\alpha k+2)} - \sum_{k=0}^\infty \frac{\left(\dfrac{-\alpha}{1-\alpha} \right)^k \left((n-j+1)\Delta t \right)^{\alpha k+1}}{\Gamma(\alpha k+2)} \right] \tag{18.24}$$

$$= \frac{AB(\alpha)}{(1-\alpha)} \sum_0^n \frac{f_i^{j+1}-f_i^j}{\Delta t} \left[(n-j+1)\Delta t \sum_{k=0}^\infty \frac{\left(\dfrac{(-\alpha(n-j+1)\Delta t)^\alpha}{1-\alpha} \right)^k}{\Gamma(\alpha k+2)} - (n-j)\Delta t \sum_{k=0}^\infty \frac{\left(\dfrac{(-\alpha(n-j)\Delta t)^\alpha}{1-\alpha} \right)^k}{\Gamma(\alpha k+2)} \right] \tag{18.25}$$

$$= \frac{AB(\alpha)}{(1-\alpha)} \sum_0^n \frac{f_i^{j+1}-f_i^j}{\Delta t} \left[(n-j+1)\Delta t E_{\alpha,2} \left[\frac{(-\alpha(n-j+1)\Delta t)^\alpha}{1-\alpha} - (n-j)\Delta t E_{\alpha,2} \frac{\left((-\alpha(n-j)\Delta t)^\alpha \right)}{1-\alpha} \right] \right] \tag{18.26}$$

Therefore, the discretized derivative for the Atangana–Baleanu derivative is shown below:

$$
{}^{ABC}_0D^\alpha_t f(x,t) = \frac{AB(\alpha)}{(1-\alpha)} \sum_0^n \frac{f_i^{j+1}-f_i^j}{\Delta t} \left[(n-j+1)\Delta t E_{\alpha,2} \left[\frac{\dfrac{(-\alpha(n-j+1)\Delta t)^\alpha}{1-\alpha}}{-(n-j)\Delta t E_{\alpha,2} \dfrac{\left((-\alpha(n-j)\Delta t)^\alpha\right)}{1-\alpha}} \right] \right] \tag{18.27}
$$

18.5.2 Numerical Approximation of the Atangana–Baleanu Integral

Numerical approximation of integrals is considered when analytical integrations are difficult to evaluate analytically (Odibat, 2006). According to Kumar et al. (2019) numerical integrations for the fractional integrals also become important in developing the algorithms for solving applied problems defined using fractional derivatives.

Now we shall do numerical approximation for the Atangana–Baleanu integral:

$$
{}^{ABC}_0J^\alpha_t f(t) = f(0) + \frac{1-\alpha}{AB(\alpha)} f(t) + \frac{\alpha}{\Gamma(\alpha)AB(\alpha)} \int_0^t f(\tau)(t-\tau)^{\alpha-1} d\tau \tag{18.28}
$$

At (x_i, t_{n+1}) we present the numerical approximation as follows:

$$
{}^{ABC}_0J^\alpha_t f(x,t) = f(x_i,0) + \frac{1-\alpha}{AB(\alpha)} f(x_i,t_{n+1}) + \frac{\alpha}{AB(\alpha)\Gamma(\alpha)} \int_0^{t_{n+1}} f(x_i,\tau)(t_{n+1}-\tau)^{\alpha-1} d\tau \tag{18.29}
$$

$$
{}^{ABC}_oJ^\alpha_t f(x,t) = f(x_i,0) + \frac{1-\alpha}{AB(\alpha)} \left(\frac{f_i^{n+1}+f_i^n}{2} \right)
$$
$$
+ \frac{\alpha}{AB(\alpha)\Gamma(\alpha)} \sum_{j=0}^n \frac{f_i^{j+1}+f_i^j}{2} \int_0^{t_{j+1}} (t_{n+1}-\tau)^{\alpha-1} d\tau \tag{18.30}
$$

$$
{}^{ABC}_oJ^\alpha_t f(x,t) = f(x_i,0) + \frac{1-\alpha}{AB(\alpha)} \left(\frac{f_i^{n+1}+f_i^n}{2} \right) + \frac{\alpha}{AB(\alpha)\Gamma(\alpha)} \sum_{j=0}^n \frac{f_i^{j+1}+f_i^j}{2\alpha} \left[\begin{array}{l} (t_{n+1}-t_i)^\alpha \Delta t^\alpha \\ -(t_{n+1}-t_{j+1})^\alpha \Delta t^\alpha \end{array} \right] \tag{18.31}
$$

$$
{}^{ABC}_oJ^\alpha_t f(x,t) = f(x_i,0) + \frac{1-\alpha}{AB(\alpha)} \left(\frac{f_i^{n+1}+f_i^n}{2} \right) + \frac{\alpha}{AB(\alpha)\Gamma(\alpha)} \sum_{j=0}^n \frac{f_i^{j+1}+f_i^j}{2\alpha} (\Delta t)^\alpha \left[\begin{array}{l} (n-j+1)^\alpha \\ -(n-j)^\alpha \end{array} \right] \tag{18.32}
$$

Then, finally, we have:

$$
{}^{ABC}_oJ^\alpha_t f(x,t) = f(x_i,0) + \frac{1-\alpha}{AB(\alpha)} \left(\frac{f_i^{n+1}+f_i^n}{2} \right) + \frac{1}{AB(\alpha)\Gamma(\alpha)}
$$
$$
\times \sum_{j=0}^n \frac{f_i^{j+1}+f_i^j}{2} \Delta t^\alpha \left[(n-j+1)^\alpha - (n-j)^\alpha \right] \tag{18.33}
$$

18.6 MODEL WITH ATANGANA–BALEANU

The Atangana–Baleanu fractional operator has been applied in different research areas as it has been found to be suitable for modeling real-world problems when compared to other fractional operators. By and large, this fractional operator has been applied in different fields include electronic circuits exhibiting chaotic behavior, flow dynamics (convection and channel flow), and the spread of diseases (Ebola virus and tuberculosis) (Alkahtani, 2016; Abro et al., 2018; Altaf Khan et al., 2018; Koca, 2018; Saqib et al., 2018).

The Atangana–Baleanu operator may be associated with a fractional diffusion equation with derivates of distributed order. Furthermore, this operator has a crossover from stretched exponent to power law when considering waiting time distribution, and it captures the memory effects better (Tateishi et al., 2017).

The Atangana–Baleanu fractional operator is also considered in the equation below:

$$\left(1+\beta k_D\right)\frac{\partial Cm}{\partial t} = D_h \frac{\partial^2 C_m}{\partial x^2} - v_m \frac{\partial C_m}{\partial x} \tag{18.34}$$

The Atangana–Baleanu derivative is as follows:

$$^{ABC}_0 D_t^\alpha f(t) = \frac{AB(\alpha)}{(1-\alpha)} \int_0^t \frac{d}{d\tau} f(\tau) E_\alpha\left(-\frac{\alpha}{1-\alpha}(t-\tau)^\alpha\right)d\tau \tag{18.35}$$

Now introducing the derivative to the derived equation for the matrix:

$$\left(1+\beta k_D\right)\frac{AB(\alpha)}{(1-\alpha)} \int_0^t \frac{d}{d\tau} f(\tau) E_\alpha\left(-\frac{\alpha}{1-\alpha}(t-\tau)^\alpha\right)d\tau = D_h \frac{\partial^2 C_m}{\partial x^2} - v_m \frac{\partial C_m}{\partial x} \tag{18.36}$$

Dividing by the retardation factor $(1 + \beta k_D)$ on both sides we get:

$$\frac{AB(\alpha)}{(1-\alpha)} \int_0^t \frac{\partial}{\partial\tau} C_m(x,\tau) E_\alpha\left(-\frac{\alpha}{1-\alpha}(t-\tau)^\alpha\right)d\tau = \frac{1}{1+\beta k_D}\left[D_h \frac{\partial^2 C_m}{\partial x^2} - v_m \frac{\partial C_m}{\partial x}\right] \tag{18.37}$$

To simplify the equation we let $\dfrac{1}{1+\beta k_D}\left[D_h \dfrac{\partial^2 C_{m(x,t)}}{\partial x^2} - v_m \dfrac{\partial C_{m(x,t)}}{\partial x}\right] = F\left(x,t,C_m(x,t)\right)$

Therefore we can represent Equation (18.37) as:

$$\frac{AB(\alpha)}{(1-\alpha)} \int_0^t \frac{\partial}{\partial\tau} C_m(x,\tau) E_\alpha\left(-\frac{\alpha}{1-\alpha}(t-\tau)^\alpha\right)d\tau = F\left(x,t,C_m(x,t)\right) \tag{18.38}$$

We then apply an integral on both sides to obtain:

$$C_m(x,t) - C_m(x_i,0) = \frac{1-\alpha}{AB(\alpha)} F\left(x,t,C_m(x,t)\right) + \frac{\alpha}{AB(\alpha)\Gamma(\alpha)} \int_0^t (t-\tau)^{\alpha-1} F\left(x,\tau,C_m(x,\tau)\right)d\tau$$

$$\tag{18.39}$$

At (x_i, t_{n+1}) we have the following:

$$C_m\left(x_i, t_{n+1}\right) - C_m\left(x_i, 0\right) = \frac{1-\alpha}{AB(\alpha)} F\left(x_i, t_n, C_m\left(x_i, t_n\right)\right) + \frac{\alpha}{AB(\alpha)\Gamma(\alpha)} \int_0^{t_{n+1}} \left(t_{n+1} - \tau\right)^{\alpha-1} \quad (18.40)$$
$$\times F\left(x_i, \tau, C_m\left(x_i, \tau\right)\right) d\tau$$

Then we represent the equation as:

$$C_m\left(x_i, t_{n+1}\right) - C_m\left(x_i, 0\right) = \frac{1-\alpha}{AB(\alpha)} F\left(x_i, t_n, C_m\left(x_i, t_n\right)\right) + \frac{\alpha}{AB(\alpha)\Gamma(\alpha)} \sum_{j=0}^{n} \int_{t_j}^{t_{j+1}} \left(t_{n+1} - \tau\right)^{\alpha-1}$$
$$\times F\left(x_i, \tau, C_m\left(x_i, \tau\right)\right) d\tau \qquad (18.41)$$

Within the $[t_j, t_{j+1}]$ interval, we approximate the function $F(x_i, \tau, C_m(x_i, \tau))$ to a polynomial $P_j(\tau)$.

$$F(x_i, \tau, C_m\left(x_i, \tau\right)) = P_j\left(\tau\right) \qquad (18.42)$$

We then use the well-known Lagrange polynomial.

$$P_j\left(\tau\right) = \frac{\tau - t_{j-1}}{t_j - t_{j-1}} F\left(x_i, t_j, C_m\left(x_i, t_j\right)\right) + \frac{\tau - t_j}{t_{j-1} - t_j} F\left(x_i, t_{j-1}, C_m\left(x_i, t_{j-1}\right)\right) \quad . \qquad (18.43)$$

Replacing $P_j(\tau)$ in the general equation we get:

$$C_m\left(x_i, t_{n+1}\right) - C_m\left(x_i, 0\right) = \frac{1-\alpha}{AB(\alpha)} F\left(x_i, t_n, C_m\left(x_i, t_n\right)\right) + \frac{\alpha}{AB(\alpha)\Gamma(\alpha)}$$
$$\sum_{j=0}^{n} \int_{t_j}^{t_{j+1}} \left(t_{n+1} - \tau\right)^{\alpha-1} \left[\begin{array}{l} \dfrac{\tau - t_{j-1}}{t_j - t_{j-1}} F\left(x_i, t_j, C_m\left(x_i, t_j\right)\right) \\[2mm] + \dfrac{\tau - t_j}{t_{j-1} - t_j} F\left(x_i, t_{j-1}, C_m\left(x_i, t_{j-1}\right)\right) \end{array} \right] d\tau \qquad (18.44)$$

Now after integration, we obtain the following approximate solution:

$$C_m\left(x_i, t_{n+1}\right) - C_m\left(x_i, 0\right) = \frac{1-\alpha}{AB(\alpha)} F\left(x_i, t_n, C_m\left(x_i, t_n\right)\right) + \frac{\alpha}{AB(\alpha)} \sum_{j=0}^{n} \left(\frac{\left(\Delta t\right)^2 F\left(x_i, t_j, C_m\left(x_i, t_j\right)\right)}{\Gamma(\alpha+2)} \right)$$
$$\left\{ \left(n-j+1\right)^{\alpha} \left(n-j+2-\alpha\right) - \left(n-j\right)^{\alpha} \left(n-j+2+2\alpha\right) \right\}$$
$$- \frac{\left(\Delta t\right)^{\alpha}}{\Gamma(\alpha+2)} F\left(x_i, t_{j-1}, C_m\left(x_i, t_{j-1}\right)\right) \left\{ \left(n-j+1\right)^{\alpha+1} - \left(n-j\right)^{\alpha} \left(n-j+\alpha+1\right) \right\}$$

$$(18.45)$$

where:

$$F\left(x_i,t_j,C_m\left(x_i,t_j\right)\right)=\frac{1}{1+\beta k_d}\left[\begin{array}{l}D_h\left(x_i\right)\dfrac{C_{m,i+1}^j-2C_{m,i}^j+C_{m,i-1}^j}{\left(\Delta x\right)^2}\\[2ex]-v_m\left(x_i\right)\dfrac{C_{m,i+1}^j-C_{m,i-1}^j}{2\Delta x}\end{array}\right]F\left(x_i,t_{j-1},C_m\left(x_i,t_{j-1}\right)\right)$$

$$=\frac{1}{1+\beta k_D}\left[\begin{array}{l}D_h\left(x_i\right)\dfrac{C_{m,i+1}^{j-1}-2C_{m,i}^{j-1}+C_{m,i-1}^{j-1}}{\left(\Delta x\right)^2}\\[2ex]-v_m\left(x_i\right)\dfrac{C_{m,i+1}^{j-1}-C_{m,i-1}^{j-1}}{2\Delta x}\end{array}\right] \tag{18.46}$$

Now replacing $F(x_i,t_j,C_m(x_i,t_j))$ with Equation (18.45) above we get:

$$C_{m,i}^{n+1}-C_{m,i}^0=\frac{1-\alpha}{AB\left(\alpha\right)}\frac{1}{1+\beta k_D}\left[D_h\left(x_i\right)\frac{C_{m,i+1}^n-2C_{m,i}^n+C_{m,i-1}^n}{\left(\Delta x\right)^2}-v_m\left(x_i\right)\frac{C_{m,i+1}^{n-1}-C_{m,i-1}^{n-1}}{2\Delta x}\right]$$

$$+\frac{\alpha}{AB\left(\alpha\right)}\sum_{j=0}^n\left(\frac{\left(\Delta t\right)^2\dfrac{1}{1+\beta k_D}\left[D_h\left(x_i\right)\dfrac{C_{m,i+1}^j-2C_{m,i}^j+C_{m,i-1}^j}{\left(\Delta x\right)^2}-v_m\left(x_i\right)\dfrac{C_{m,i+1}^j-C_{m,i-1}^j}{2\Delta x}\right]}{\Gamma\left(\alpha+2\right)}\right.$$

$$\left\{\left(n-j+1\right)^\alpha\left(n-j+2+\alpha\right)-\left(n-j\right)^\alpha\left(n-j+2+2\alpha\right)\right\}$$

$$-\frac{\left(\Delta t\right)^\alpha}{\Gamma\left(\alpha+2\right)}\frac{1}{1+\beta k_D}\left[D_h\left(x_i\right)\frac{C_{m,i+1}^{j-1}-2C_{m,i}^{j-1}+C_{m,i-1}^{j-1}}{\left(\Delta x\right)^2}-v_m\left(x_i\right)\frac{C_{m,i+1}^{j-1}-C_{m,i-1}^{j-1}}{2\Delta x}\right]$$

$$\left\{\left(n-j+1\right)^{\alpha+1}-\left(n-j\right)^\alpha\left(n-j+\alpha+1\right)\right\} \tag{18.47}$$

$$C_{m,i}^{n+1}-C_{m,i}^0=\frac{1-\alpha}{AB\left(\alpha\right)}\frac{1}{1+\beta k_D}\left[D_h\left(x_i\right)\frac{C_{m,i+1}^n-2C_{m,i}^n+C_{m,i-1}^n}{\left(\Delta x\right)^2}-v_m\left(x_i\right)\frac{C_{m,i+1}^{n-1}-C_{m,i-1}^{n-1}}{2\Delta x}\right]$$

$$+\frac{\alpha}{AB\left(\alpha\right)}\sum_{j=0}^n\left(\frac{\left(\Delta t\right)^2\dfrac{1}{1+\beta k_D}\left[D_h\left(x_i\right)\dfrac{C_{m,i+1}^j-2C_{m,i}^j+C_{m,i-1}^j}{\left(\Delta x\right)^2}-v_m\left(x_i\right)\dfrac{C_{m,i+1}^j-C_{m,i-1}^j}{2\Delta x}\right]}{\Gamma\left(\alpha+2\right)}\right.$$

$$\left\{\left(n-j+1\right)^\alpha\left(n-j+2+\alpha\right)-\left(n-j\right)^\alpha\left(n-j+2+2\alpha\right)\right\}$$

$$-\frac{\left(\Delta t\right)^\alpha}{\Gamma\left(\alpha+2\right)}\frac{1}{1+\beta k_D}\left[D_h\left(x_i\right)\frac{C_{m,i+1}^{j-1}-2C_{m,i}^{j-1}+C_{m,i-1}^{j-1}}{\left(\Delta x\right)^2}-v_m\left(x_i\right)\frac{C_{m,i+1}^{j-1}-C_{m,i-1}^{j-1}}{2\Delta x}\right]$$

$$\left\{\left(n-j+1\right)^{\alpha+1}-\left(n-j\right)^\alpha\left(n-j+\alpha+1\right)\right\} \tag{18.48}$$

Now we let:

$$\frac{1-\alpha}{AB(\alpha)}\frac{1}{(1+\beta k_D)}\frac{1}{(\Delta x)^2} = a_1 \tag{18.49}$$

and

$$\frac{1-\alpha}{AB(\alpha)}\frac{1}{(1+\beta k_D)}\frac{1}{2\Delta x} = a_2 \tag{18.50}$$

and

$$\frac{\alpha}{AB(\alpha)}\frac{(\Delta t)^2}{(1+\beta k_D)}\frac{1}{\Gamma(\alpha+2)}\frac{1}{(\Delta x)^2}\left\{(n-j+1)^\alpha(n-j+2+\alpha)-(n-j)^\alpha(n-j+2+2\alpha)\right\} = \delta_{n,\alpha}^j \tag{18.51}$$

and

$$\frac{\alpha}{AB(\alpha)}\frac{(\Delta t)^2}{(1+\beta k_D)}\frac{1}{\Gamma(\alpha+2)}\frac{1}{2\Delta x}\left\{(n-j+1)^\alpha(n-j+2+\alpha)-(n-j)^\alpha(n-j+2+2\alpha)\right\} = \delta_{n,\alpha}^{j,1} \tag{18.52}$$

and

$$\frac{(\Delta t)^\alpha}{\Gamma(\alpha+2)}\frac{1}{(1+\beta k_D)}\frac{1}{(\Delta x)^2}\left\{(n-j+1)^{\alpha+1}-(n-j)^\alpha(n-j+\alpha+1)\right\} = \delta_{n,\alpha}^{j,2} \tag{18.53}$$

and

$$\frac{(\Delta t)^\alpha}{\Gamma(\alpha+2)}\frac{1}{(1+\beta k_D)}\frac{1}{2\Delta x}\left\{(n-j+1)^{\alpha+1}-(n-j)^\alpha(n-j+\alpha+1)\right\} = \delta_{n,\alpha}^{j,3} \tag{18.54}$$

Now, therefore, finally we get a simplified version of the equation:

$$C_{m,i}^{n+1} - C_{m,i}^0 = D_h(x_i)\left(C_{m,i+1}^n - 2C_{m,i}^n + C_{m,i-1}^n\right)a_1 - v_m(x_i)\left(C_{m,i+1}^{n-1} - C_{m,i-1}^{n-1}\right)a_2$$
$$+ \sum_{j=0}^n\left[D_h(x_i)\left(C_{m,i+1}^j - 2C_{m,i}^j + C_{m,i-1}^j\right)\delta_{n,\alpha}^j - v_m(x_i)\left(C_{m,i+1}^j - C_{m,i-1}^j\right)\delta_{n,\alpha}^{j,1}\right] \tag{18.55}$$
$$- D_h(x_i)\left(C_{m,i+1}^{j-1} - 2C_{m,i}^{j-1} + C_{m,i-1}^{j-1}\right)\delta_{n,\alpha}^{j,2} - v_m(x_i)\left(C_{m,i+1}^{j-1} - C_{m,i-1}^{j-1}\right)\delta_{n,\alpha}^{j,3}$$

Doing stability for the above equation:

$$C_{m,i}^{n+1} - C_{m,i}^0 = D_h(x_i)\left(C_{m,i+1}^n - 2C_{m,i}^n + C_{m,i-1}^n\right)a_1 - v_m(x_i)\left(C_{m,i+1}^{n-1} - C_{m,i-1}^{n-1}\right)a_2$$
$$+ \sum_{j=0}^n\left[D_h(x_i)\left(C_{m,i+1}^j - 2C_{m,i}^j + C_{m,i-1}^j\right)\delta_{n,\alpha}^j - v_m(x_i)\left(C_{m,i+1}^j - C_{m,i-1}^j\right)\delta_{n,\alpha}^{j,1}\right] \tag{18.56}$$
$$- D_h(x_i)\left(C_{m,i+1}^{j-1} - 2C_{m,i}^{j-1} + C_{m,i-1}^{j-1}\right)\delta_{n,\alpha}^{j,2} - v_m(x_i)\left(C_{m,i+1}^{j-1} - C_{m,i-1}^{j-1}\right)\delta_{n,\alpha}^{j,3}$$

Now $C_{m,i}^{n+1} - C_{m,i}^0$ replacing with and we get $\delta_{n+1}e^{ik_m(x)} - \delta_0 e^{ik_m(x)}$ then:

$$
\begin{aligned}
\delta_{n+1}e^{ik_m(x)} - \delta_0 e^{ik_m(x)} = {} & D_h\left(x_i\right)\left(\delta_n e^{ik_m(x+\Delta x)} - 2\delta_n e^{ik_m(x)} + \delta_n e^{ik_m(x-\Delta x)}\right)a_1 \\
& - v_m\left(x_i\right)\left(\delta_{n-1}e^{ik_m(x+\Delta x)} - \delta_{n-1}e^{ik_m(x-\Delta x)}\right)a_2 \\
& + \sum_{j=0}^{n}\left[\begin{array}{l} D_h\left(x_i\right)\left(\delta_j e^{ik_m(x+\Delta x)} - 2\delta_j e^{ik_m(x)} + \delta_j e^{ik_m(x-\Delta x)}\right)\delta_{n,\alpha}^{j} \\ -v_m\left(x_i\right)\left(\delta_j e^{ik_m(x+\Delta x)} - \delta_j e^{ik_m(x-\Delta x)}\right)\delta_{n,\alpha}^{j,1} \end{array}\right] \\
& - D_h\left(x_i\right)\left(\delta_{j-1}e^{ik_m(x+\Delta x)} - 2\delta_{j-1}e^{ik_m(x)} + \delta_{j-1}e^{ik_m(x-\Delta x)}\right)\delta_{n,\alpha}^{j,2} \\
& - v_m\left(x_i\right)\left(\delta_{j-1}e^{ik_m(x+\Delta x)} - \delta_{j-1}e^{ik_m(x-\Delta x)}\right)\delta_{n,\alpha}^{j,3}
\end{aligned}
\tag{18.57}
$$

Then we have:

$$
\begin{aligned}
e^{ik_m(x)}\left(\delta_{n+1} - \delta_0\right) = {} & D_h\left(x_i\right)\left(\delta_n\left(e^{ik_m(x)}e^{ik_m(\Delta x)}\right) - 2\delta_n e^{ik_m(x)} + \left(\delta_n\left(e^{ik_m(x)}e^{ik_m(-\Delta x)}\right)\right)\right)a_1 \\
& - v_m\left(x_i\right)\left(\delta_{n-1}\left(e^{ik_m(x)}e^{ik_m(\Delta x)}\right) - \left(\delta_{n-1}\left(e^{ik_m(x)}e^{ik_m(-\Delta x)}\right)\right)\right)a_2 \\
& + \sum_{j=0}^{n}\left[\begin{array}{l} D_h\left(x_i\right)\left(\delta_j\left(e^{ik_m(x)}e^{ik_m(\Delta x)}\right) - 2\delta_j e^{ik_m(x)} + \left(\delta_j\left(e^{ik_m(x)}e^{ik_m(-\Delta x)}\right)\right)\right)\delta_{n,\alpha}^{j} \\ -v_m\left(x_i\right)\left(\delta_j\left(e^{ik_m(x)}e^{ik_m(\Delta x)}\right) - \delta_j\left(e^{ik_m(x)}e^{ik_m(-\Delta x)}\right)\right)\delta_{n,\alpha}^{j,1} \end{array}\right] \\
& - D_h\left(x_i\right)\left(\delta_{j-1}\left(e^{ik_m(x)}e^{ik_m(\Delta x)}\right) - 2\delta_{j-1}e^{ik_m(x)} + \delta_{j-1}\left(e^{ik_m(x)}e^{ik_m(-\Delta x)}\right)\right)\delta_{n,\alpha}^{j,2} \\
& - v_m\left(x_i\right)\left(\delta_{j-1}\left(e^{ik_m(x)}e^{ik_m(\Delta x)}\right) - \delta_{j-1}\left(e^{ik_m(x)}e^{ik_m(-\Delta x)}\right)\right)\delta_{n,\alpha}^{j,3}
\end{aligned}
\tag{18.58}
$$

Then we have:

$$
\begin{aligned}
\left(\delta_{n+1} - \delta_0\right) = {} & D_h\left(x_i\right)\left(\delta_n e^{ik_m(\Delta x)} - 2\delta_n + \delta_n e^{ik_m(-\Delta x)}\right)a_1 - v_m\left(x_i\right)\left(\delta_{n-1}e^{ik_m(\Delta x)} - \delta_{n-1}\left(e^{ik_m(-\Delta x)}\right)\right)a_2 \\
& + \sum_{j=0}^{n}\left[D_h\left(x_i\right)\left(\delta_j e^{ik_m(\Delta x)} - 2\delta_j + \delta_j\left(e^{ik_m(-\Delta x)}\right)\right)\delta_{n,\alpha}^{j} - v_m\left(x_i\right)\left(\delta_j e^{ik_m(\Delta x)} - \delta_j e^{ik_m(-\Delta x)}\right)\delta_{n,\alpha}^{j,1}\right] \\
& - D_h\left(x_i\right)\left(\delta_{j-1}e^{ik_m(\Delta x)} - 2\delta_{j-1} + \delta_{j-1}e^{ik_m(-\Delta x)}\right)\delta_{n,\alpha}^{j,2} - v_m\left(x_i\right)\left(\delta_{j-1}\left(e^{ik_m(\Delta x)}\right) - \delta_{j-1}\left(e^{ik_m(-\Delta x)}\right)\right)\delta_{n,\alpha}^{j,3}
\end{aligned}
\tag{18.59}
$$

Now simplifying the equation further, we get:

$$
\begin{aligned}
\left(\delta_{n+1} - \delta_0\right) = {} & D_h\left(x_i\right)\left(\delta_n\left(e^{ik_m(\Delta x)} - 2 + e^{ik_m(-\Delta x)}\right)\right)a_1 - v_m\left(x_i\right)\left(\delta_{n-1}\left(e^{ik_m(\Delta x)} - e^{ik_m(-\Delta x)}\right)\right)a_2 \\
& + \sum_{j=0}^{n}\left[D_h\left(x_i\right)\left(\delta_j\left(e^{ik_m(\Delta x)} - 2 + e^{ik_m(-\Delta x)}\right)\right)\delta_{n,\alpha}^{j} - v_m\left(x_i\right)\left(\delta_j\left(e^{ik_m(\Delta x)} - e^{ik_m(-\Delta x)}\right)\right)\delta_{n,\alpha}^{j,1}\right. \\
& \left. - D_h\left(x_i\right)\left(\delta_{j-1}\left(e^{ik_m(\Delta x)} - 2 + e^{ik_m(-\Delta x)}\right)\right)\delta_{n,\alpha}^{j,2} - v_m\left(x_i\right)\left(\delta_{j-1}\left(e^{ik_m(\Delta x)} - e^{ik_m(-\Delta x)}\right)\right)\delta_{n,\alpha}^{j,3}\right]
\end{aligned}
\tag{18.60}
$$

Now we let:

$$k_m(\Delta x) = \theta \tag{18.61}$$

And we let:

$$-k_m(\Delta x) = -\theta \tag{18.62}$$

Therefore when replacing with θ and $-\theta$ we get the following equation:

$$
\begin{aligned}
(\delta_{n+1} - \delta_0) &= D_h(x_i)\left(\delta_n\left(e^{i\theta} - 2 + e^{-i\theta}\right)\right)a_1 - v_m(x_i)\left(\delta_{n-1}\left(e^{i\theta} - e^{-i\theta}\right)\right)a_2 \\
&+ \sum_{j=0}^{n}\left[D_h(x_i)\left(\delta_j\left(e^{i\theta} - 2 + e^{-i\theta}\right)\right)\delta_{n,\alpha}^{j} - v_m(x_i)\left(\delta_j\left(e^{i\theta} - e^{-i\theta}\right)\right)\delta_{n,\alpha}^{j,1}\right. \\
&\left. - D_h(x_i)\left(\delta_{j-1}\left(e^{i\theta} - 2 + e^{-i\theta}\right)\right)\delta_{n,\alpha}^{j,2} - v_m(x_i)\left(\delta_{j-1}\left(e^{i\theta} - e^{-i\theta}\right)\right)\delta_{n,\alpha}^{j,3}\right]
\end{aligned} \tag{18.63}
$$

Then we know that:

$$e^{i\theta} + e^{-i\theta} = 2\mathrm{Cos}\,\theta \tag{18.64}$$

and, that:

$$e^{i\theta} - e^{-i\theta} = 2i\mathrm{Sin}\,\theta \tag{18.65}$$

Then Equation (4.205) becomes:

$$
\begin{aligned}
(\delta_{n+1} - \delta_0) &= D_h(x_i)\left(\delta_n\left(2\mathrm{Cos}\,\theta - 2\right)\right)a_1 - v_m(x_i)\left(\delta_{n-1}\left(2i\mathrm{Sin}\,\theta\right)\right)a_2 \\
&+ \sum_{j=0}^{n}\left[D_h(x_i)\left(\delta_j\left(2\mathrm{Cos}\,\theta - 2\right)\right)\delta_{n,\alpha}^{j} - v_m(x_i)\left(\delta_j\left(2i\mathrm{Sin}\,\theta\right)\right)\delta_{n,\alpha}^{j,1}\right. \\
&\left. - D_h(x_i)\left(\delta_{j-1}\left(2\mathrm{Cos}\,\theta - 2\right)\right)\delta_{n,\alpha}^{j,2} - v_m(x_i)\left(\delta_{j-1}\left(2i\mathrm{Sin}\,\theta\right)\right)\delta_{n,\alpha}^{j,3}\right]
\end{aligned} \tag{18.66}
$$

$$
\begin{aligned}
\delta_{n+1} &= \delta_0 + 4D_h(x_i)\mathrm{Sin}^2\left(\frac{\theta}{2}\right)a_1\delta_n - 2i\mathrm{Sin}\,\theta\int_{n-1} v_m(x_i)a_2 \\
&+ \sum_{j=0}^{n}\left[\begin{array}{l} -D_h(x_i)4\mathrm{Sin}^2\left(\dfrac{\theta}{2}\right)\delta_{n,\alpha}^{j} + 4D_h(x_i)\delta_{j-1}\mathrm{Sin}^2\left(\dfrac{\theta}{2}\right)\delta_{n,\alpha}^{j,2} \\ -v_m(x_i)\delta_{j-1}2i\mathrm{Sin}\,\theta\,\delta_{n,\alpha}^{j,3} \end{array}\right]
\end{aligned} \tag{18.67}
$$

When h = 0, we have:

$$\delta_1 = \delta_0 - 4D_h(x_i)\mathrm{Sin}^2\left(\frac{\theta}{2}\right)a_1\delta_0 \tag{18.68}$$

We want to show that:

$$\left|\frac{\delta_1}{\delta_0}\right| < 1 \tag{18.69}$$

Then:

$$\left|\delta_1\right| = \left|\delta_0\right|\left|1 - 4D_h\left(x_i\right)Sin^2\left(\frac{\theta}{2}\right)a_1\right| \tag{18.70}$$

$$\left|\frac{\delta_1}{\delta_0}\right| < 1 \, imply \, that \left|1 - 4D_h\left(x_i\right)Sin^2\left(\frac{\theta}{2}\right)a_1\right| < 1 \tag{18.71}$$

$$\left|1 - 4D_h\left(x_i\right)Sin^2\left(\frac{\theta}{2}\right)a_1\right| < 1 \, for \, all \, values \, of \, m \, are \, equal \, to \, k_m\Delta x \tag{18.72}$$

$$\left|1 - 4D_h\left(x_i\right)a_i\right| < 1 \tag{18.73}$$

$$\left|1 - 4D_h\left(x_i\right)a_i\right| = \begin{cases} -1 + 4D_h\left(x_i\right)a_1 - 2 < 1 - 4D_h\left(x\right)a_1 < 1 \\ -1 + 4D_h\left(x_i\right)a_1 - 2 < -4D_h\left(x_i\right)a_1 < 0 \end{cases} \tag{18.74}$$

This means that:

$$4D_h\left(x_i\right)a_i < 2 \tag{18.75}$$

$$D_h\left(x_i\right)a_i < \frac{1}{2} \tag{18.76}$$

$$D_h\left(x_i\right) < \frac{1}{2a_1} \tag{18.77}$$

This we consider to satisfy the 1st condition.
Therefore we have:

$$\left|\frac{\delta_1}{\delta_0}\right| < 1 => D_h\left(x_i\right) < \frac{1}{2}a_1 \tag{18.78}$$

We assume that for all values of h > 1, and therefore:

$$\left|\frac{\delta_h}{\delta_0}\right| < 1 \, we \, want \, to \, show \, that \left|\frac{\delta_{n+1}}{\delta_0}\right| < 1 \tag{18.79}$$

However,

$$\left|\delta_{n+1}\right| = \left|\begin{array}{l} \delta_0 - 4D_h\left(x_i\right)Sin^2\left(\frac{\theta}{2}\right)a_1\delta_n - 2iSin\theta\,\delta_{n-1}\,v_m\left(x_1\right)a_2 \\ + \sum_{j=0}^{n}\left[\begin{array}{l} -D_h\left(x_i\right)4Sin^2\left(\frac{\theta}{2}\right)\delta_{n,\alpha}^j + 4D_h\left(x_i\right)\delta_{j-1}Sin^2\left(\frac{\theta}{2}\right)\delta_{n,\alpha}^{j,2} \\ -4v_m\left(x_i\right)\delta_{j-1}Sin\theta\,\delta_{n,\alpha}^{j,3} \end{array}\right] \end{array}\right| \tag{18.80}$$

$$\left|\delta_{n+1}\right| = \left\|\begin{matrix} \delta_0 - 4D_h\left(x_i\right)\operatorname{Sin}^2\left(\dfrac{\theta}{2}\right)a_1\delta_n + \sum_{j=0}^{n}\left(-4\operatorname{Sin}^2\left(\dfrac{\theta}{2}\right)\delta_{n,\alpha}^{j}D_h\left(x_i\right)\delta_n\right) \\ +i\left(-2\operatorname{Sin}\theta\delta_{n-1}v_m\left(x_i\right)a_2 + \sum_{j=0}^{n}-2v_m\left(x_i\right)\delta_{j-1}\operatorname{Sin}\theta\,\delta_{n,\alpha}^{j,3}\right) \end{matrix}\right\| \quad (18.81)$$

$$\left|\delta_{n+1}\right| < \left|\begin{matrix} \left|\delta_0\right| - 4D_h\left(x_i\right)\operatorname{Sin}^2\left(\dfrac{\theta}{2}\right)a_1\left|\delta_n\right| + \sum_{j=0}^{n}\left(-4\operatorname{Sin}^2\left(\dfrac{\theta}{2}\right)\delta_{n,\alpha}^{j}D_h\left(x_i\right)\right)\left|\delta_j\right| \\ +i\left(2\operatorname{Sin}\theta\,v_m\left(x_i\right)a_2\left|\delta_{n-1}\right| + \sum_{j=0}^{n}-2v_m\left(x_i\right)\operatorname{Sin}\theta\,\delta_{n,2}^{j,3}\left|\delta_{j-1}\right|\right) \end{matrix}\right| \quad (18.82)$$

By introduction of hypothesis:

$$\left|\delta_{n+1}\right| < \left|\delta_0\right|\left\|1 - 4D_h\left(x_i\right) + i\left(2\operatorname{Sin}\theta\,v_m\left(x_i\right)a_2 + \sum_{j=0}^{n}\left(-2v_m\left(x_i\right)\operatorname{Sin}\theta\,\delta_{n,2}^{j,3}\right)\right)\right\| \quad (18.83)$$

see the final equations
Then we have

$$\left|\dfrac{\delta_{n+1}}{\delta_0}\right| < 1 \Rightarrow \sqrt{\begin{matrix}\left(1 - 4D_h\left(x_i\right)\operatorname{Sin}^2\left(\dfrac{\theta}{2}\right)a_2 + \sum_{j=0}^{n}\left(-4\operatorname{Sin}^2\left(\dfrac{\theta}{2}\right)\delta_{n,j}^{j}D_h\left(x_i\right)\right)\right)^2 \\ +\left(2\operatorname{Sin}\theta v_m\left(x_i\right)a_2 + \sum_{j=0}^{n}\left(-2v_m\left(x_i\right)\operatorname{Sin}\theta\,\delta_{n,2}^{j,3}\right)\right)^2\end{matrix}} \quad (18.84)$$

The numerical scheme will be stable if:

$$Min\left\{D_h\left(x_i\right) < \dfrac{1}{2a_2}, \sqrt{\begin{matrix}\left(1 - 4D_h\left(x_i\right)\operatorname{Sin}^2\left(\dfrac{\theta}{2}\right)a_2 + \sum_{j=0}^{n}\left(-4\operatorname{Sin}^2\left(\dfrac{\theta}{2}\right)\delta_{n,j}^{j}D_h\left(x_i\right)\right)\right)^2 \\ +\left(2\operatorname{Sin}\theta v_m\left(x_i\right)a_2 + \sum_{j=0}^{n}\left(-2v_m\left(x_i\right)\operatorname{Sin}\theta * \delta_{n,2}^{j,3}\right)\right)^2\end{matrix}} < 1\right\} \quad (18.85)$$

18.7 CONCLUSION

In this chapter we introduced the Atangana–Baleanu fractional derivative and integral operators to the dispersion–advection equation because of their benefits. The Atangana–Baleanu derivative was considered for this study because in previous studies it was used to model the flow of heat in material with different scales and those with heterogeneous media a situation similar to a matrix–fracture scenario. Applying the Laplace transform theorem on both the Atangana–Baleanu derivative and integral we constructed a new solution. Numerical approximation was applied both on the Atangana–Baleanu derivate and integral, and then at last introduced into the advection–dispersion equation. A numerical scheme was performed on the new model with the Antangana–Baleanu operator. Simulations applying the new model are recommended for future studies.

REFERENCES

Abro, K.A., Khan, I., & Tassaddiq, A. (2018). Application of Atangana–Baleanu fractional derivative to convection flow of MHD Maxwell fluid in a porous medium over a vertical plate. *Mathematical Modelling of Natural Phenomena*, 13(1), 1. doi:10.1051/mmnp/2018007.

Ahmed, S., Jayakumar, R., & Salih, A. (2008). *Groundwater Dynamics in Hard Rock Aquifer: Sustainable Management and Optimal Monitoring Network Design*. Springer, Dordrecht, The Netherlands with Capital Publishing Company, New Delhi, India.

Alkahtani, B.S.T. (2016). Chua's circuit model with Atangana–Baleanu derivative with fractional order. *Chaos, Solitons & Fractals*, 89, 547–551. doi:10.1016/j.chaos.2016.03.020.

Altaf Khan, M., Ullah, S., & Farooq, M. (2018). A new fractional model for tuberculosis with relapse via Atangana–Baleanu derivative. *Chaos, Solitons & Fractals*, 116, 227–238 doi:10.1016/j.chaos.2018.09.039.

Atangana, A., & Alkahtani, B.S.T. (2015). New model of groundwater flowing within a confined aquifer: Application of Caputo–Fabrizio derivative. *Arabian Journal of Geosciences*, 9(1), 8. doi:10.1007/s12517-015-2060-8.

Atangana, A., & Baleanu, D. (2016). New fractional derivatives with nonlocal and non-singular kernel: Theory and application to heat transfer model. *Journal of Thermal Science*, 20, 763–769.

Atangana, A., & Secer, A. (2013). A note on fractional order derivatives and table of fractional derivatives of some special functions [Research article]. *Abstract and Applied Analysis*, doi:10.1155/2013/279681.

Deyi, M., & Atangana, A. (2020). A new model for groundwater contamination transport in dual media. PhD Thesis, University of the Free State, Bloemfontein, South Africa.

Eckhoff, K.S., & Wasberg, C.E. (1995). On the numerical approximation of derivatives by a modified Fourier collocation method (Report No. 99), University of Bergen.

Gómez-Aguilar, J.F., Escobar-Jiménez, R.F., López-López, M.G., & Alvarado-Martínez, V.M. (2016). Atangana–Baleanu fractional derivative applied to electromagnetic waves in dielectric media. *Journal of Electromagnetic Waves and Applications*. doi:10.1080/09205071.2016.1225521.

Koca, I. (2018). Modelling the spread of Ebola virus with Atangana–Baleanu fractional operators. *The European Physical Journal Plus*, 133(3), 100. doi:10.1140/epjp/i2018-11949-4.

Mathias, S.A., & Zimmerman, R.W. (2003). Laplace transform inversion for late-time behavior of groundwater flow problems. *Water Resources Research*, 39(10). doi:10.1029/2003WR002246.

Odibat, Z. (2006). Approximations of fractional integrals and Caputo fractional derivatives. *Applied Mathematics and Computation*, 178(2), 527–533. doi:10.1016/j.amc.2005.11.072.

Quarteroni, A., & Valli, A. (2008). *Numerical Approximation of Partial Differential Equations*. Springer Science & Business Media.

Rahimy, M. (2010). Applications of fractional differential equations. *Applied Mathematical Sciences*, 4(50), 2453–2461.

Saqib, M., Khan, I., & Shafie, S. (2018). Application of Atangana–Baleanu fractional derivative to MHD channel flow of CMC-based-CNT's nanofluid through a porous medium. *Chaos, Solitons & Fractals*, 116, 79–85. doi:10.1016/j.chaos.2018.09.007.

Tateishi, A.A., Ribeiro, H.V., & Lenzi, E.K. (2017). The role of fractional time-derivative operators on anomalous diffusion. *Frontiers in Physics*, 5. doi:10.3389/fphy.2017.00052.

Toufik, M., & Atangana, A. (2017). New numerical approximation of fractional derivative with non-local and non-singular kernel: Application to chaotic models. *The European Physical Journal Plus*, 132(10), 444.

19 Modeling Soil Moisture Flow
New Proposed Models

Tshanduko Mutandanyi and Abdon Atangana
University of the Free State, Bloemfontein, South Africa

CONTENTS

19.1 INTRODUCTION

The sub-surface water flow is controlled by various factors in a range of flow media. Therefore each sub-surface water system is associated with a unique level of complexities. Understanding groundwater flow seems to be less complicated, since aquifers' hydraulic properties do not vary randomly in space. This is not the case when dealing with flow in the unsaturated zone; soil hydraulic properties are not fixed throughout the soil volume. The spatial evolution of soil hydraulic properties makes the modeling process complicated because they need to be measured throughout the soil volume. Regardless of complexities associated with the unsaturated zone, there are mathematical formulations that were developed to describe flow in this zone. The unsaturated flow in porous media was formulated mathematically by Henry Darcy in 1856 (Whitaker, 1986; Zhang, 1991). Darcy explained sub-surface flow in a porous media by means of Darcy's law (Soulaine, 2015), which expresses the relationship between specific discharge, water discharge per unit area, and the hydraulic gradient with Darcy's permeability as the proportionality constant. This law assumes a steady state flow in porous media (Gordon, 1989). Therefore it does not cater for transient flows in soils. There is another equation developed to model water flow in the unsaturated zone: the well-known Richards equation (Zhang, 1991; Ojha et al., 2017). Richards equation is a better mathematical way of expressing the unsaturated flow processes in the unsaturated zone. However, the effectiveness of Richards equation seems to decrease with an increase in complexities and heterogeneities of the system. Spatial heterogeneities of water content in soils result in nonlinear parametric functions to Richards' equation which makes it impossible to solve analytically. Despite the

DOI: 10.1201/9781003266266-19

weaknesses associated with Richards equation, it remains the governing equation to express soil water flow processes occurring in the unsaturated zone. Richards equation requires a knowledge of soil hydraulic parameters such as unsaturated hydraulic conductivity and the water retention curve. These parameters are, most of the time, impossible to obtain in the field. As a result, researchers proposed ways of reproducing the unsaturated system theoretically (e.g. Childs and Collis-George, 1950; Burdine, 1953; van Genuchten, 1980; Assouline et al., 1998) by using soil hydraulic parameters that are easy to obtain. These parametric models are highly nonlinear, and their combination with the Richards equation yields a highly nonlinear model that requires a computer code to solve it. This study attempts to solve Richards equation both numerically and analytically to obtain an exact solution and numerical solutions.

19.2 THE UNSATURATED FLOW MODEL

Modeling flow in the unsaturated zone is essential for understanding the sub-surface water system. There are mathematical formulations found in the literature that were suggested as describing unsaturated flow systems, and great success has been achieved by these formulations in describing the unsaturated flow processes. The most common formulation of the unsaturated system that researchers have been using for decades to describe flow in unsaturated zone is Richards equation (Vereecken et al., 2008; Barari et al., 2009), introduced by Richards (1931) after conducting studies with capillary tubes to mimic water movement in unsaturated porous media. To describe moisture flow in soils, Richards (1931) adopted the concept of water conductivity in porous media, suggested by Buckingham (1907) who also suggested that in unsaturated flows conductivity is highly affected by moisture content (Barari et al., 2009).

Richards equation is a generalized equation for describing flow in unsaturated porous media; it assumes a general nonsteady one-dimensional flow in the unsaturated zone. It is expected that water content will vary throughout the medium. As a result, the hydraulic conductivity as well as forces governing the water flux are also not fixed throughout the medium. This is the cause of the highly nonlinear nature of Richards equation. The spatial variation in conductivity also leads to complexities in the field of modlling of unsaturated flow, since it is quite difficult to measure non-uniform hydraulic conductivities throughout the medium. Richards (1931) formulated a nonlinear partial differential equation to describe one-dimensional vertical flow in unsaturated non-swelling porous soils by combining mass conservation equation

$$\frac{\partial \theta}{\partial t} = -\frac{\partial q}{\partial z} \tag{19.1}$$

with Darcy's law

$$q = -K(\theta)\frac{\partial h}{\partial z} \tag{19.2}$$

to obtain

$$\frac{\partial \theta}{\partial t} = \frac{\partial}{\partial z}\left(K(\theta)\frac{\partial H}{\partial z}\right) \tag{19.3}$$

where $H = h + z$. By substituting H into Equation (19.3) the following is obtained

$$\frac{\partial \theta}{\partial t} = \frac{\partial}{\partial z}\left[K(\theta)\left(\frac{\partial h}{\partial z} + \frac{\partial z}{\partial z}\right)\right] \tag{19.4}$$

which is simplified into

$$\frac{\partial \theta}{\partial t} = \frac{\partial}{\partial z}\left[K(\theta)\left(\frac{\partial h}{\partial z} + 1\right)\right] \tag{19.5}$$

where q is the water flux; θ is the dimensionless volumetric water content or moisture content; t is the time; z is the vertical distance; h is the pore water pressure head, also referred to as capillary pressure; and $K(\theta)$ is the unsaturated hydraulic conductivity.

Equation (19.5) above is the well-known mixed-form Richards equation used in modeling unsaturated flows; it is commonly used to describe flows in unsaturated soils. Given its mixed-form state, it is more expensive and time-consuming to obtain all the required data for this equation because it has two independent variables, namely soil volumetric moisture content θ, and soil pore water pressure head h. Moreover, the mixed-form Richards equation does not cater for all circumstances in unsaturated flow, such as conditions with wetting fronts and where the flow is not stable. This is because of the nonlinear nature of Richards equation caused by the strong dependence of the unsaturated hydraulic conductivity on capillary pressure head h and volumetric water content θ. Alternatively, there are two formulations of Richards equation that consist of only one independent variable each. These formulations are: soil moisture content based form, θ-based form; and pore water pressure head based form, h-based form. To obtain a head base form θ needs to be eliminated from Equation (19.5) above to yield an equation that has only capillary pressure as an independent variable. θ is eliminated by introducing the derivative of the soil water retention curve, the specific water capacity concept, into Equation (19.5) which yields

$$C(h)\frac{\partial h}{\partial t} = \frac{\partial}{\partial z}\left(K(\theta)\frac{\partial h}{\partial z}\right) + \frac{\partial K(\theta)}{\partial z} \tag{19.6}$$

where

$$C(h) = \frac{d\theta}{dh} \tag{19.7}$$

is the rate of change of water content or saturation in relation to the matric pressure head.

A θ-based form Richards equation is formulated by incorporating soil pore water diffusivity D term into the mixed-form Richards equation to produce a transient unsaturated flow equation with only one independent variable θ. This is achieved by assuming that the flow is governed by water content only, and neglecting the effect of soil matrix potential h. By incorporating a pore-water diffusivity term

$$D = \frac{K(\theta)}{C(h)} = K(\theta)\frac{dh}{d\theta} \tag{19.8}$$

into Equation (19.5), the following is obtained

$$\frac{\partial \theta}{\partial t} = \frac{\partial}{\partial z}\left(D\frac{\partial \theta}{\partial z}\right) + \frac{\partial K(\theta)}{\partial z} \tag{19.9}$$

Equation (19.9) above is a θ-based form of Richards equation. Both forms are effective in mimicking flows in unsaturated systems, and the choice of which one to use depends on the available data.

Both D and $K(\theta)$ depend highly on θ, but during data collection, D is much easier to measure compared to $K(\theta)$ and h. In Equation (19.9) a decrease in $K(\theta)$ because of a decrease in θ is compensated by a typical increase of $\dfrac{dh}{d\theta}$ with a decline in θ. Therefore D varies less than $K(\theta)$ in the field. It seems to be much easier to conduct θ measurements compared to h measurements in the field. Moreover, θ measurement are more reliable than h measurements, because they cover the water content of the entire range whereas h measurements does not cover the entire media but only that part of the water retention curve that is wet. In this study, only a θ-based form of Richards equation will be analyzed.

Despite the success of Richards equation, its application is problematic since it requires knowledge of many soil hydraulic parameters that are often hard to obtain (Ojha et al., 2017). Unlike groundwater flow, unsaturated zone flow is characterized by parameters that require spatial and temporal characterization since they differ randomly in space as a result of system heterogeneity. To add parameterizing functions required to solve Richards equation are highly nonlinear. This is because the soil hydraulic parameters depend strongly on volumetric water content, which evolves randomly throughout a soil volume. As a result, Richards equation is highly nonlinear and difficult to solve. Obtaining analytical solutions requires oversimplification of the natural unsaturated system, which may lead to the underestimation or overestimation of flow parameters. For example, unsaturated hydraulic conductivity is commonly mispredicted in models where the effect of capillarity is overemphasized. Numerical solutions, on the other hand, are the most reliable. However, they are sometimes associated with convergence problems because of the highly nonlinear nature of this equation. Moreover, computational procedures are complicated, time-consuming and expensive.

19.3 METHODS AND MATERIALS

19.3.1 Development of a Linear Unsaturated Hydraulic Conductivity Model

It has been proved that hydraulic conductivity depends strongly on the evolving water content (van Genuchten, 1980). As water migrates in a downward direction, pores are emptied in upper parts of the unsaturated zone and water is replaced by air. Water content tends to decrease with depth. This is because of water that is retained in disconnected pores as residual water content. Eventually, a number of pores conducting water flow declines. As a result, the flow becomes more tortuous, since there are only a few connected pores that are filled with water, and flow has to follow those pores. The decline in water content and number of connected pores with water content is the reason for spatial variation of the unsaturated hydraulic conductivity.

Unsaturated hydraulic conductivity is a function of effective saturation and pressure, it is highly variable in space and requires spatial characterization. The function of unsaturated hydraulic conductivity is given by the product of relative hydraulic conductivity and saturated hydraulic conductivity

$$K(\theta) = K_s k_r \tag{19.10}$$

Estimation of unsaturated hydraulic conductivity requires the substitution of a relative hydraulic conductivity estimation model into Equation (19.10). There are quite a number of models in the literature for predicting relative hydraulic conductivity (e.g. Childs and Collis-George, 1950; Burdine, 1953; Brooks and Corey, 1964; Mualem, 1976; van Genuchten, 1980; Assouline et al., 1998; Vereecken et al., 2008; Ghanbarian-Alavijeh and Hunt, 2012;). The models differ in the theory they are based on, and in quantification of governing soil properties. For example, there are models that conceptualize the soil pores as bundles of capillary tubes (Childs and Collis-George, 1950; Burdine, 1953; Brooks and Corey, 1964; Mualem, 1976; van Genuchten, 1980). Other models, such as Ghanbarian-Alavijeh and Hunt (2012), are based on percolation theory. Capillary based models are believed to over-emphasize the effect of capillarity on total water conduction while neglecting the effect of film flow. In addition, relative hydraulic conductivity models are combined with water retention functions (Brooks and Corey, 1964; van Genuchten, 1980) to obtain required flow parameters such as saturated hydraulic conductivity and some basic soil properties that are related to pore-size distribution.

In this chapter, the capillary theory proposed by Childs and Collis-George, Burdine, and Mualem is adopted to develop an unsaturated hydraulic conductivity model. The adopted capillary theory is expressed mathematically as:

$$\frac{K(\theta)}{K_s} = \left(Se\right)^b \left[\left. \int_0^{Se} \frac{dx}{\psi(x)} \middle/ \int_0^1 \frac{dx}{\psi^{2-r}(x)} \right. \right]^m \tag{19.11}$$

Childs and Collis-George, Burdine, and Mualem proposed different values for model parameter b, r and m. For example, Mualem suggested that $b = 0.5$; $r = 1$; and $m = 2$; and Burdine, on the other hand, suggested that $b = 2$; $r = 0$; and $m = 1$.

When Mualem's parameter values are adopted, Equation (19.11) becomes:

$$K_r(\theta) = Se^{0.5} \left[\left. \int_0^{\theta} \frac{d\theta}{\psi} \middle/ \int_0^{\theta_{sat}} \frac{d\theta}{\psi} \right. \right]^2 \tag{19.12}$$

where

$$Se = \frac{\theta - \theta_r}{\theta_s - \theta_r} \tag{19.13}$$

0.5 in Equation (19.12) is a value representing an empirical term for the pore-size interaction term. The values of this term have been carefully selected by researchers. The most common ones are 0.5, as indicated in the equation above; 2 which is associated with the Burdine model. Brooks and Corey suggested that a pore-size interaction term is not necessary when a pore-size distribution index is used, and considered it to be zero. A pore-size interaction term adds flexibility to relative hydraulic conductivity models. Leij et al. (1994) proved that the pore-size interaction term is not confined to these values only; however, it can have any value including negative values. Therefore the wide range for values of this term, referred to as n from now on, allows both linear and non-linear relationships between effective saturation and relative hydraulic conductivity. Considering that $n = -2$ and adopting other soil physical parameters proposed by Mualem, Equation (19.12) becomes:

$$K_r(\theta) = Se^{-2} \left[\left. \int_0^{\theta} \frac{d\theta}{\psi} \middle/ \int_0^{\theta_{sat}} \frac{d\theta}{\psi} \right. \right]^2 \tag{19.14}$$

Incorporating a water retention characteristic function described by Brooks and Corey:

$$Se = \left(\varphi / \omega_{cr}\right)^{-\lambda} \tag{19.15}$$

into Equation (19.14) yields:

$$K_r = Se^w = Se^{-2+2+\frac{2}{\lambda}} \tag{19.16}$$

where λ is an empirical parameter representing the pore-size distribution index. Considering a completely homogeneous soil volume with pore-size distribution index of 2, the exponent in ~Equation (19.16) becomes 1 when the value of λ is substituted in Equation (19.1). Therefore, Equation (19.16) becomes:

$$K_r = Se = \frac{\theta - \theta_r}{\theta_s - \theta_r} \tag{19.17}$$

The relative permeability of the soil given in the equation above is equivalent to effective saturation. Therefore, unsaturated hydraulic conductivity varies linearly across space. The soil profile is assumed to be completely homogeneous, and the resultant unsaturated hydraulic conductivity function is given by:

$$K(\theta) = K_r K_s = K_s \left(\frac{\theta - \theta_r}{\theta_s - \theta_r}\right) \tag{19.18}$$

Equation (19.18) above gives an expression of the unsaturated hydraulic conductivity model proposed in this chapter.

19.3.1.1 The Linear Unsaturated Flow Model

Proposed unsaturated hydraulic conductivity models are linear, therefore a linear model will be obtained when it is combined with Richards equation. As already mentioned, this chapter will cover analysis of the water-content-based form of Richards equation given by Equation (19.9). Since unsaturated hydraulic conductivity is difficult to obtain, a proposed linear model will be substituted into Equation (19.9) to remove unsaturated hydraulic conductivity. The resulting equation is given by:

$$\left(D\frac{\partial \theta}{\partial z}\right) + \frac{\partial}{\partial z}\left(K_s\left(\frac{\theta - \theta_r}{\theta_s - \theta_r}\right)\right)\frac{\partial \theta}{\partial t} = \frac{\partial}{\partial z} \tag{19.19}$$

Equation (19.19) is the linearized Richards equation and it can be simplified to obtain:

$$\frac{\partial \theta}{\partial t} = D\frac{\partial^2 \theta}{\partial z^2} + \frac{\partial}{\partial z}\left(K_s\frac{\theta}{\theta_s - \theta_r}\right) \tag{19.20}$$

$$\frac{\partial \theta}{\partial t} = D\frac{\partial^2 \theta}{\partial z^2} + \frac{\partial \theta}{\partial z}\left(K_s\frac{1}{\theta_s - \theta_r}\right) \tag{19.21}$$

If $K_s \dfrac{1}{\theta_s - \theta_r} = \beta$ then Equation (19.21) can be written as:

$$\frac{\partial \theta}{\partial t} = D\frac{\partial^2 \theta}{\partial z^2} + \beta\frac{\partial \theta}{\partial z} \tag{19.22}$$

The above equation is a one-dimension partial differential equation for unsaturated flow in a vertical direction which will be solved in the following section using Laplace transform and Green's function to obtain an exact solution.

19.3.2 THE EXACT SOLUTION TO RICHARDS EQUATION

In this section, the partial differential Equation (19.22) obtained in the previous section is solved using Laplace transform to obtain the exact solution. The transform is applied to Equation (19.22) with an assumption that variable t meets the following condition: $0 < t < \infty$, t is transformed from being a variable to a parameter so that Equation (19.22) becomes an algebraic equation that is easy to solve compared to differential equations. The Laplace transform of a function is given by:

$$\mathcal{L}\big(f(t)\big) = \tilde{f}(s) \tag{19.23}$$

where $\mathcal{L}$ is a Laplace transform operator, $\tilde{f}(s)$ is a function in Laplace space, $f(t)$ is a function in its original space where $0 < t < \infty$, and s is a parameter representing t in Laplace space. Application of Laplace transform on both sides of Equation (19.22):

$$\mathcal{L}\left(\frac{\partial \theta}{\partial t}\right) = D\mathcal{L}\left(\frac{\partial^2 \theta}{\partial z^2}\right) + \beta \mathcal{L}\left(\frac{\partial \theta}{\partial z}\right) \tag{19.24}$$

yields:

$$s\tilde{\theta} - \theta(0) = D\left(\frac{\partial^2 \tilde{\theta}}{\partial z^2}\right) + \beta\left(\frac{\partial \tilde{\theta}}{\partial z}\right) \tag{19.25}$$

By rearranging the equation above, we obtain:

$$D\frac{\partial^2 \tilde{\theta}}{\partial z^2} + \beta \frac{\partial \tilde{\theta}}{\partial z} - s\tilde{\theta} = \theta(0) \tag{19.26}$$

where $\theta(0)$ represents the initial soil water content; and $s\tilde{\theta}$ is an expression of Laplace transform of function $\frac{\partial \theta}{\partial z}$ involving parameter s and which corresponds to t in the original space. The above equation is non-homogeneous because $\theta(0)$ is not equal to zero. Just for the purpose of obtaining a particular solution of the above equation, soil water content $\theta(0)$ is considered to be 0. If $\theta(0)$ is zero, Equation (19.26) becomes a homogeneous equation and is written as:

$$D\frac{\partial^2 \tilde{\theta}}{\partial z^2} + \beta \frac{\partial \tilde{\theta}}{\partial z} - s\tilde{\theta} = 0 \tag{19.27}$$

The above equation is in the form of a quadratic equation, if $\tilde{\theta} = Ae^{rz}$, then equations becomes

$$\frac{\partial \tilde{\theta}}{\partial z} = rAe^{rz} \tag{19.28}$$

$$\frac{\partial^2 \tilde{\theta}}{\partial z} = r^2 Ae^{rz} \tag{19.29}$$

$$\frac{\partial^2 \tilde{\theta}}{\partial z} = r^2 Ae^{rz} \tag{19.30}$$

are obtained and substituted into the Laplacian quadratic equation in Equation (19.27) to obtain:

$$Dr^2 Ae^{rz} + \beta rAe^{rz} - sAe^{rz} = 0 \tag{19.31}$$

Equation (19.31) is further simplified into:

$$Dr^2 + r\beta - s = 0 \tag{19.32}$$

A quadratic formula is used to solve the above equation for r values and the following values are obtained:

$$r_- = \frac{-\beta - \sqrt{\beta^2 + 4Ds}}{2D}$$

and

$$r_+ = \frac{-\beta + \sqrt{\beta^2 + 4Ds}}{2D} \tag{19.33}$$

Since soil water content cannot increase to infinity, r_- is used and $\tilde{\theta}(z, s)$ can be expressed as:

$$\tilde{\theta}(z, s) = Ae^{z\left(\frac{-\beta - \sqrt{\beta^2 + 4Ds}}{2D}\right)} \tag{19.34}$$

The inverse Laplace transform of Equation (19.32) yields a particular solution to Equation (19.21) in its original space. The inverse Laplace transform of Equation (19.32) is given by:

$$\mathcal{L}^{-1}\left(\tilde{\theta}(z,s)\right) = A\mathcal{L}^{-1}\left(e^{z\left(\frac{-\beta - \sqrt{\beta^2 + 4Ds}}{2D}\right)}\right) = \tilde{\theta}(z,t) \tag{19.35}$$

$$\theta_1(z,t) = A.\left(\frac{ze^{\left(\frac{-(-\beta t + z) - t}{4Dt}\right)}}{2\sqrt{\frac{\pi}{D}}t^{\frac{3}{2}}}\right) \tag{19.36}$$

Since Equation (19.26) is non-homogeneous with $\theta(0) \neq 0$, finding the exact solution will require the use of Green's function. Green's function gives a solution to non-homogenous linear differential equations defined on a domain with boundary problems. In Green terms, Equation (19.26) can be expressed as:

$$D\frac{\partial^2 G}{\partial z^2} + \beta \frac{\partial G}{\partial z} - sG = \delta(z) \tag{19.37}$$

The above Green's function is solved by applying a Laplace transform to both sides to obtain:

$$\mathcal{L}\left(D\frac{\partial^2 G}{\partial z^2} + \beta \frac{\partial G}{\partial z} - sG\right) = \mathcal{L}\left(\delta(z!)\right) \tag{19.38}$$

$$D\left(p^2\tilde{G} - pG' - G(0)\right) + \beta\left(p\tilde{G} - G(0)\right) - s\tilde{G} = 1 \tag{19.39}$$

where p a parameter in the second Laplace is space, and $p\tilde{G}$ is a function sG in the second Laplace space. Grouping and rearranging of Equation (19.40) yields:

$$\left(Dp^2 + \beta p - s\right)\tilde{G} = 1 \tag{19.40}$$

$$\tilde{G} = \frac{1}{\left(Dp^2 + \beta p - s\right)} \tag{19.41}$$

Using the quadratic formula Equation (19.41) can be written as:

$$\tilde{G} = \frac{1}{\left(\dfrac{-\beta - \sqrt{\beta^2 + 4Ds}}{2D}\right)}$$

or

$$\tilde{G} = \frac{1}{\left(\dfrac{-\beta + \sqrt{\beta^2 + 4Ds}}{2D}\right)} \tag{19.42}$$

From the above equation, the values of Δ; p_+; and p_- are obtained and given as follows:

$$\Delta = \beta^2 + 4sD \tag{19.43}$$

$$p_- = \frac{-\beta - \sqrt{\Delta}}{2D} = a_1 \tag{19.44}$$

$$p_+ = \frac{-\beta + \sqrt{\Delta}}{2D} = a_2 \tag{19.45}$$

Using the above values, Equation (19.41) can be written as

$$\tilde{G} = \frac{1}{\left(p - a_1\right)\left(p - a_2\right)} \tag{19.46}$$

$\tilde{G}$ is a convolution of two functions and can be expressed in the form of:

$$\mathcal{L}\left(f(z) * h(z)\right) = \mathcal{L}\left(f(z)\right)\mathcal{L}\left(h(z)\right) = \tilde{G} \tag{19.47}$$

where $\tilde{f}(p)$ and $\tilde{h}(p)$ are chosen as:

$$\frac{1}{\left(p - a_1\right)} = \tilde{f}$$

and

$$\frac{1}{\left(p - a_2\right)} = \tilde{h}$$

Equation (19.46) can be written in the form of Equation (19.47) as:

$$\tilde{G} = \frac{1}{p-a_1} \cdot \frac{1}{p-a_2},$$ (19.48)

The inverse Laplace transform of a convolution is given by the product of the inverse of the individual functions. In this case, the inverse of $\tilde{G}$ will be a product inverse of $\tilde{f}(p)$ and $\tilde{h}(p)$ and is given by:

$$\mathcal{L}^{-1}(\tilde{f}).\mathcal{L}^{-1}(\tilde{h}) = e^{a_1 z}.e^{a_2 z}$$ (19.49)

The inverse Laplace transform of Equation (19.48) can be obtained using the convolution theorem given by the following integrals:

$$\mathcal{L}^{-1}(\tilde{G}) = \int_0^z f(\tau)h(z-\tau)d\tau$$ (19.50)

Using the above equation and Equation (19.49), G becomes:

$$G = \int_0^z e^{a_1 \tau} e^{a_2(z-\tau)}.d\tau$$ (19.51)

$$G = e^{a_2 z} \int_0^z e^{(a_1 \tau - a_2 \tau)}.d\tau$$ (19.52)

$$G = e^{a_2 z} \int_0^z e^{\tau(a_1 - a_2)}.d\tau$$ (19.53)

After the integration, we get:

$$G = e^{a_2 z} \frac{1}{a_1 - a_2} e^{\tau(a_1 - a_2)} \Big|_0^z$$ (19.54)

$$G = e^{a_2 z}\left(\frac{1}{a_1 - a_2} e^{z(a_1 - a_2)} - \frac{1}{a_1 - a_2} \right) = G(z,s)$$ (19.55)

Substituting Equations (19.44 and 19.45) into the above equation yields:

$$G(z,s) = \left[e^{z\left(\frac{-\beta-\sqrt{s}}{2D}\right)} \right] \cdot \left[\frac{1}{\left(\frac{-\beta+\sqrt{s}}{2D}\right)-\left(\frac{-\beta+\sqrt{s}}{2D}\right)} e^{z\left(\frac{-\beta+\sqrt{s}}{2D}\frac{-\beta-\sqrt{s}}{2D}\right)} - \frac{1}{\left(\frac{-\beta+\sqrt{s}}{2D}\right)-\left(\frac{-\beta+\sqrt{s}}{2D}\right)} \right]$$ (19.56)

The above equation is simplified into:

$$G(z,s) = \left[e^{z\left(\frac{-\beta-\sqrt{s}}{2D}\right)} \right] \left[\cfrac{1}{\left(\frac{-\beta+\sqrt{s}}{2D}\right) - \left(\frac{-\beta+\sqrt{s}}{2D}\right)} \right] \left[e^{z\left(\frac{-\beta+\sqrt{s}}{2D} - \frac{-\beta-\sqrt{s}}{2D}\right)} - 1 \right] \qquad (19.57)$$

Substituting Equation (19.43) into the above equation yields:

$$G(z,s) = \left[e^{z\left(\frac{-\beta-\sqrt{\beta^2+4Ds}}{2D}\right)} \right] \left[\cfrac{1}{\left(\frac{-\beta+\sqrt{\beta^2+4Ds}}{2D}\right) - \left(\frac{-\beta+\sqrt{\beta^2+4Ds}}{2D}\right)} \right]$$
$$\left[e^{z\left(\frac{-\beta+\sqrt{\beta^0+4Ds}}{2D} - \frac{-\beta-\sqrt{\beta^2+4Ds}}{2D}\right)} - 1 \right] \qquad (19.58)$$

The exact solution in Laplace space is given by:

$$\theta(z,s) = \theta_1(z,s) + \int_0^z \theta(0,s) G(z-\tau,s)\,d\tau \qquad (19.59)$$

where $\theta(z,s)$ is the term for exact solution; $\theta_1(z,s)$ is the particular solution given by Equation (19.35); $\theta(0,s)$ is the term for initial soil water content; and $\int_0^z G(z-\tau,s)\,d\tau$ is the integral of Equation (19.58) in τ direction.

To obtain the exact solution, each term must be substituted into the above equation, and substitution of the particular solution, Equation (19.35) yields:

$$\theta(z,s) = A.e^{z\left(\frac{-\beta-\sqrt{\beta^2+4Ds}}{2D}\right)} + \theta(0,s) \int_0^z G(z-\tau,s)\,d\tau \qquad (19.60)$$

To find $\int_0^z G(z-\tau,s)\,d\tau$, Equation (19.58) is again expressed as Equation (19.55). By simplifying it can be expressed as:

$$G(z,s) = e^{a_2 z} \left(\frac{1}{a_1 - a_2} \right) \left(e^{z(a_1-a_2)} - 1 \right) \qquad (19.61)$$

and its integral is expressed as:

$$\int_0^z G(z-\tau,s)d\tau = \frac{1}{a_1-a_2}\int_0^z e^{a_2 z}\cdot\left(e^{z(a_1-a_2)}-1\right).d\tau \tag{19.62}$$

$$\int_0^z G(z-\tau,s)d\tau = \frac{1}{a_1-a_2}\int_0^z e^{a_2(z-\tau)}\cdot e^{(z-\tau)(a_1-a_2)}-e^{a_2(z-\tau)}.d\tau = \frac{1}{a_1-a_2}\int_0^z\left(e^{a_1(z-\tau)}-e^{a_2(z-\tau)}\right).d\tau \tag{19.63}$$

Let $(z-\tau)$ be y; when $\tau=0$, y = z and when z = 0, y = 0; thus dy = $-$ dτ
Thus,

$$\int_0^z G(z-\tau,s)d\tau = \frac{1}{a_1-a_2}\int_z^0\left(e^{a_2 y}-e^{a_2 y}\right)(-dy) = \frac{1}{a_1-a_2}\int_z^0\left(e^{a_1 y}-e^{a_2 y}\right).dy \tag{19.64}$$

and after integration, the following is obtained:

$$\int_0^z G(z-\tau,s)d\tau = \frac{1}{a_1-a_2}\left(\frac{1}{a_1}e^{a_1 y}-\frac{1}{a_2}e^{a_2 y}\right)\Big|_0^z \tag{19.65}$$

By substituting Equations (19.43–19.45) into the above equation, the following equation is obtained:

$$\int_0^z G(z-\tau,s)d\tau = \left(\frac{1}{\left(\dfrac{-\beta-\sqrt{\beta^2+4Ds}}{2D}\right)-\left(\dfrac{-\beta+\sqrt{\beta^2+4Ds}}{2D}\right)}\right)$$
$$\left(\frac{1}{\left(\dfrac{-\beta-\sqrt{\beta^2+4Ds}}{2D}\right)}e^{z\left(\frac{-\beta-\sqrt{\beta^2+4Ds}}{2D}\right)}\right.$$
$$-\frac{1}{\left(\dfrac{-\beta+\sqrt{\beta^2+4Ds}}{2D}\right)}e^{z\left(\frac{-\beta+\sqrt{\beta^2+4Ds}}{2D}\right)}$$
$$\left.-\frac{1}{\left(\dfrac{-\beta-\sqrt{\beta^2+4Ds}}{2D}\right)}+\frac{1}{\left(\dfrac{-\beta+\sqrt{\beta^2+4Ds}}{2D}\right)}\right) \tag{19.66}$$

Now $\int_0^z G(z-\tau,s)d\tau$ in Equation (19.60) can be replaces by Equation (19.66) to give:

$$\theta(z,s) = A.e^{z\left(\frac{-\beta - \sqrt{\beta^2 + 4Ds}}{2D}\right)}$$

$$+ \left[\theta(0,s)\right]\left(-\frac{D}{\sqrt{\beta^2 + 4Ds}}\right)\left[\begin{array}{c}\left(\dfrac{1}{\left(\dfrac{-\beta - \sqrt{\beta^2 + 4Ds}}{2D}\right)}e^{z\left(\frac{-\beta - \sqrt{\beta^2 + 4Ds}}{2D}\right)}\right) \\[2em] -\dfrac{1}{\left(\dfrac{-\beta + \sqrt{\beta^2 + 4Ds}}{2D}\right)}e^{z\left(\frac{-\beta + \sqrt{\beta^2 + 4Ds}}{2D}\right)} \\[2em] -\dfrac{1}{\sqrt{\beta^2 + 4Ds}}\end{array}\right] \qquad (19.67)$$

The equation above is an expression of the exact solution of the linearized Richards equation. The equation above is still in Laplace space; to remove the solution from Laplace space to its original space, inverse Laplace transform is applied as follows:

$$\theta(z,t) = \mathcal{L}^{-1}\left(\theta(z,s)\right) = A\left(\frac{ze^{\left(\frac{-(-\beta t + z) - t}{4Dt}\right)}}{2\sqrt{\dfrac{\pi}{D}}t^{\frac{3}{2}}}\right)$$

$$+ \mathcal{L}^{-1}\left\{+\left[\theta(0,s)\right]\left[\left(-\frac{D}{\sqrt{\beta^2 + 4Ds}}\right)\left[\begin{array}{c}\left(\dfrac{1}{\left(\dfrac{-\beta - \sqrt{\beta^2 + 4Ds}}{2D}\right)}e^{z\left(\frac{-\beta - \sqrt{\beta^2 + 4Ds}}{2D}\right)}\right) - \\[2em] \dfrac{1}{\left(\dfrac{-\beta + \sqrt{\beta^2 + 4Ds}}{2D}\right)}e^{z\left(\frac{-\beta + \sqrt{\beta^2 + 4Ds}}{2D}\right)} - \dfrac{1}{\sqrt{\beta^2 + 4Ds}}\end{array}\right]\right]\right\} \qquad (19.68)$$

$$\theta(z,t) = \theta_0 \sum_{j=0}^{\infty} e^{-\lambda_n t}\left(Ae^{-z\left(\sqrt{\beta^2 - 4D\lambda^2}\right)} + Be^{z\left(\sqrt{\beta^2 - 4D\lambda^2}\right)}\right) \qquad (19.69)$$

$$\theta(z,0) = \theta_L \qquad (19.70)$$

$$\theta(z,0) = \theta_0 \sum_{j=0}^{\infty} e^{-\lambda_n t}\left(Ae^{-rz}\right) \qquad (19.71)$$

where

$$r = -\sqrt{\beta^2 - 4D\lambda^2} \qquad (19.72)$$

and

$$\beta = K_s \frac{1}{\theta_s - \theta_r} \qquad (19.73)$$

The term

$$Be^{z\left(\sqrt{\beta^2 - 4D\lambda^2}\right)}$$

in Equation (19.73) was dropped because water cannot increase to infinity.

19.3.3 Numerical Analysis

19.3.3.1 Numerical Analysis of Richards Equation Combined with Pre-Existing Nonlinear Models

For the considered soil volume, two pre-existing unsaturated hydraulic conductivity models were selected to model flow in the unsaturated zone. The selected models are the Brooks and Corey (1964) model

$$K(\theta) = K_s \left(\frac{\theta - \theta_r}{\theta_s - \theta_r}\right)^5 \qquad (19.74)$$

and the Mualem (1976) model

$$K(\theta) = K_s \left(\frac{\theta - \theta_r}{\theta_s - \theta_r}\right)^{3.5} \qquad (19.75)$$

The exponents given on both equations depend on the proposed values of soil parameters for each model. The value of the pore-size distribution index is considered to be 2 for this soil volume. Since both models are nonlinear, their exponents will be expressed by ω and the resulting equation is:

$$K(\theta) = K_s \left(\frac{\theta - \theta_r}{\theta_s - \theta_r}\right)^\omega \qquad (19.76)$$

Combined with Richards equation to obtain:

$$\frac{\partial \theta}{\partial t} = D \frac{\partial^2 \theta}{\partial z^2} + \frac{\partial}{\partial z} \left[K_s \left(\frac{\theta - \theta_r}{\theta_s - \theta_r}\right)^\omega \right] \qquad (19.77)$$

where ω represents the exponent of effective saturation for both equations, $\omega = 5$ for the Brooks and Corey model, and $\omega = 3.5$ for the Mualem model. The above nonlinear equation is discretized using the Crank–Nicolson method and Laplace Adams–Bashforth method. First, Equation (19.77) is simplified into:

$$\frac{\partial \theta}{\partial t} = D\frac{\partial^2 \theta}{\partial z^2} + \frac{K_s}{\left(\theta_s - \theta_r\right)^\omega}\frac{\partial}{\partial z}\left(\theta - \theta_r\right)^\omega \quad \frac{\partial \theta}{\partial t} = D\frac{\partial^2 \theta}{\partial z^2} + \frac{K_s}{\left(\theta_s - \theta_r\right)^\omega}\omega\frac{\partial \theta}{\partial z}\left(\theta - \theta_r\right)^{\omega-1} \quad (19.78)$$

19.3.3.1.1   Crank–Nicolson Scheme

The Crank–Nicolson finite-difference approximation to the above equation is given by:

$$\frac{\theta_i^{j+1} - \theta_i^j}{\Delta t} = \left[\frac{1}{2\left(\Delta z\right)^2}D\left(\left(\theta_{i+1}^{j+1} - 2\theta_i^{j+1} + \theta_{i-1}^{j+1}\right) + \left(\theta_{i+1}^j - 2\theta_i^j + \theta_{i-1}^j\right)\right)\right]$$
$$+\left[\frac{1}{2}\cdot\left(\frac{K_s}{\left(\theta_s - \theta_r\right)^\omega}\cdot\omega\right)\left(\left(\frac{\theta_{i+1}^{j+1} - \theta_{i-1}^{j+1}}{2\Delta z} + \frac{\theta_{i+1}^j - \theta_{i-1}^j}{2\Delta z}\right)\left(\frac{\theta_i^{j+1} + \theta_i^j}{2} - \theta_r\right)^{\omega-1}\right)\right] \quad (19.79)$$

19.3.3.1.2   Laplace Adams–Bashforth Scheme

The governing partial differential equation is solved by applying a Laplace transform on both sides of the equation in order to transform the equation from a partial differential equation into a differential equation

$$\mathcal{L}\left(\frac{\partial \theta}{\partial t}\right) = \mathcal{L}\left(D\frac{\partial^2 \theta}{\partial z^2} + \frac{K_s}{\left(\theta_s - \theta_r\right)^\omega}\omega\frac{\partial \theta}{\partial z}\left(\theta - \theta_r\right)^{\omega-1}\right) \quad (19.80)$$

where $\mathcal{L}$ is the Laplace transform operator. The resulting equation is given by

$$\frac{d\theta\left(s,t\right)}{dt} = \mathcal{L}\left(D\frac{\partial^2 \theta}{\partial z^2} + \frac{K_s}{\left(\theta_s - \theta_r\right)^\omega}\omega\frac{\partial \theta}{\partial z}\left(\theta - \theta_r\right)^{\omega-1}\right) \quad (19.81)$$

s can be silenced by:

$$\frac{d\theta\left(s,t\right)}{dt} = F\left(\theta,t\right) \quad (19.82)$$

$$\theta\left(t\right) = \theta\left(s,t\right) \quad (19.83)$$

and Equation (19.81) can be written as:

$$F\left(\theta,t\right) = \mathcal{L}\left(D\frac{\partial^2 \theta}{\partial z^2} + \frac{K_s}{\left(\theta_s - \theta_r\right)^\omega}\omega\frac{\partial \theta}{\partial z}\left(\theta - \theta_r\right)^{\omega-1}\right) \quad (19.84)$$

Applying the fundamental theorem of calculus on the equation above yields:

$$\theta\left(t\right) = \theta\left(t_0\right) + \int_0^t F\left(\theta,\tau\right)d\tau \quad (19.85)$$

which is also:

$$\theta(t) = \theta_0 + \int_0^t F(\theta, \tau) d\tau \tag{19.86}$$

When $t = t_{n+1}$, the equation becomes:

$$\theta_{n+1} = \theta(t_{n+1}) = \theta_0 \int_0^{t_{n+1}} F(\theta, \tau) d\tau \tag{19.87}$$

When $t = t_n$, the equation becomes:

$$\theta_n = \theta(t_n) = \theta_0 \int_0^{t_n} F(\theta, \tau) d\tau \tag{19.88}$$

and

$$\theta_{n+1} - \theta_n = \int_0^{t_{n+1}} F(\theta, \tau) d\tau - \int_0^{t_n} F(\theta, \tau) d\tau \tag{19.89}$$

$$\theta_{n+1} - \theta_n = \int_n^{t_{n+1}} F(\theta, \tau) d\tau \tag{19.90}$$

The Langrange polynomial is used for the approximation of $F(\theta, t)$ to obtain:

$$P(t)\big(\approx F(\theta, t)\big) = \frac{t - t_{n-1}}{t_n - t_{n-1}} F(\theta, t_n) + \frac{t - t_n}{t_{n-1} - t_n} F(\theta, t_{n-1}) \tag{19.91}$$

$$P(t) = \frac{t - t_{n-1}}{t_n - t_{n-1}} F_n + \frac{t - t_n}{t_{n-1} - t_n} F_{n-1} \tag{19.92}$$

Therefore,

$$\theta_{n+1} - \theta_n = \int_n^{t_{n+1}} F(\theta, \tau) d\tau \tag{19.93}$$

$$\theta_{n+1} - \theta_n = \int_{t_n}^{t_{n+1}} \left(\frac{t - t_{n-1}}{t_n - t_{n-1}} F_n + \frac{t - t_n}{t_{n-1} - t_n} F_{n-1} \right) dt \tag{19.94}$$

$$\theta_{n+1} - \theta_n = \frac{F_n}{t_n - t_{n-1}} \int_{t_n}^{t_{n+1}} (t - t_{n-1}) dt + \frac{F_{n-1}}{t_{n-1} - t_n} \int_n^{t_{n+1}} (t - t_n) dt \tag{19.95}$$

$$\theta_{n+1} - \theta_n = \frac{F_n}{t_n - t_{n-1}} \left[\frac{1}{2}t^2 - tt_{n-1} \right]_{t_n}^{t_{n+1}} + \frac{F_{n-1}}{t_{n-1} - t_n} \left[\frac{1}{2}t^2 - tt_n \right]_{t_n}^{t_{n+1}} \tag{19.96}$$

If $h = t_n - t_{n-1}$, then the following is obtained:

$$\theta_{n+1} - \theta_n = \frac{F_n}{h} \left(\frac{1}{2}t^2_{n+1} - t_{n+1}t_{n-1} - \frac{1}{2}t^2_n + t_n t_{n-1} \right) - \frac{F_{n-1}}{h} \left(\frac{1}{2}t^2_{n+1} - t_{n+1}t_n - \frac{1}{2}t^2_n + t^2_n \right) \tag{19.97}$$

Further simplification gives:

$$\theta_{n+1} - \theta_n = \frac{F_n}{h} \left(\frac{1}{2}(t_{n+1} - t_n)(t_{n+1} + t_n) - t_{n-1}(t_{n+1} - t_n) \right)$$
$$- \frac{F_{n-1}}{h} \left(\frac{1}{2}(t_{n+1} - t_n)(t_{n+1} + t_n) - t_n(t_{n+1} - t_n) \right) \tag{19.98}$$

$$\theta_{n+1} - \theta_n = \frac{F_n}{h} \left(\frac{1}{2}h(t_{n+1} + t_n) - ht_{n-1} \right) - \frac{F_{n-1}}{h} \left(\frac{1}{2}h(t_{n+1} + t_n) - ht_n \right) \tag{19.99}$$

$$\theta_{n+1} - \theta_n = F_n \left(\frac{1}{2}(t_{n+1} + t_n) - t_{n-1} \right) - F_{n-1} \left(\frac{1}{2}(t_{n+1} + t_n) - t_n \right) \tag{19.100}$$

$$\theta_{n+1} - \theta_n = F_n \left(\frac{1}{2}((n+1)h + nh) - (n-1)h \right) - F_{n-1} \left(\frac{1}{2}((n+1)h + nh) - nh \right) \tag{19.101}$$

$$\theta_{n+1} - \theta_n = F_n \left(nh + \frac{1}{2}h - nh + h \right) - F_{n-1} \left(nh + \frac{1}{2}h - nh \right) \tag{19.102}$$

$$\theta_{n+1} = \theta_n + h \left(\frac{3}{2}F_n - \frac{1}{2}F_{n-1} \right) \tag{19.103}$$

Applications of inverse Laplace transform to take back the equation to its real space are given by:

$$\mathcal{L}^{-1}[\theta_{n+1}] = \mathcal{L}^{-1} \left[\theta_n + h \left(\frac{3}{2}F_n - \frac{1}{2}F_{n-1} \right) \right] \tag{19.104}$$

$$\theta(z, t_{n+1}) = \theta(z, t_n) + h\frac{3}{2} \left(D\frac{\partial^2 \theta(z, t_n)}{\partial z^2} + \frac{K_s}{(\theta_s - \theta_r)^\omega} \omega \frac{\partial \theta(z, t_n)}{\partial z} (\theta(z, t_n) - \theta_r)^{\omega-1} \right)$$
$$- h\frac{1}{2} \left(D\frac{\partial^2 \theta(z, t_{n-1})}{\partial z^2} + \frac{K_s}{(\theta_s - \theta_r)^\omega} \omega \frac{\partial \theta(z, t_{n-1})}{\partial z} (\theta(z, t_{n-1}) - \theta_r)^{\omega-1} \right) \tag{19.105}$$

The equation above is discretized forward and backward in space to yield:

$$\theta\left(z_i, t_{n+1}\right) = \theta\left(z_i, t_n\right) + h\frac{3}{2}\left(D\frac{\partial^2\theta\left(z_i, t_n\right)}{\partial z^2} + \frac{K_s}{\left(\theta_s - \theta_r\right)^\omega}\omega\frac{\partial\theta\left(z_i, t_n\right)}{\partial z}\left(\theta\left(z_i, t_n\right) - \theta_r\right)^{\omega-1}\right)$$

$$-h\frac{1}{2}\left(D\frac{\partial^2\theta\left(z_i, t_{n-1}\right)}{\partial z^2} + \frac{K_s}{\left(\theta_s - \theta_r\right)^\omega}\omega\frac{\partial\theta\left(z_i, t_{n-1}\right)}{\partial z}\left(\theta\left(z_i, t_{n-1}\right) - \theta_r\right)^{\omega-1}\right) \tag{19.106}$$

Let $\theta\left(z_i, t_n\right) = \theta_i^n$ and $\Delta z = l$, then the above equation becomes:

$$\theta_i^{n+1} = \theta\left(z_i, t_n\right) + h\frac{3}{2}\left[D\left(\frac{\theta_{i+1}^n - 2\theta_i^n + \theta_{i-1}^n}{\left(\Delta l\right)^2}\right) + \frac{K_s}{\left(\theta_s - \theta_r\right)^\omega}\omega\left(\frac{\theta_i^n - \theta_{i-1}^n}{\Delta l}\right)\left(\theta_i^n - \theta_r\right)^{\omega-1}\right]$$

$$-h\frac{1}{2}\left[D\left(\frac{\theta_{i+1}^{n-1} - \theta_i^{n-1} + \theta_{i-1}^{n-1}}{\left(\Delta l\right)^2}\right) + \frac{K_s}{\left(\theta_s - \theta_r\right)^\omega}\omega\left(\frac{\theta_i^{n-1} - \theta_{i-1}^{n-1}}{\Delta l}\right)\left(\theta_i^{n-1} - \theta_r\right)^{\omega-1}\right] \tag{19.107}$$

The above is the numerical solution to Richards equation obtained using the Laplace Adams–Bashforth numerical approximation scheme.

19.3.3.2 Numerical Analysis of the Proposed Linear Model

19.3.3.2.1 Crank–Nicolson Finite-Difference Approximation Scheme

This sub-section will provide a numerical approximation of the linearized Richards equation using the Crank–Nicolson finite-difference approximation method. Numerical approximation of Equation (19.22) is given by:

$$\frac{\theta_i^{j+1} - \theta_i^j}{\Delta t} = 0.5\left(D\frac{\theta_{i+1}^{j+1} - 2\theta_i^{j+1} + \theta_{i-1}^{j+1}}{\left(\Delta z\right)^2} + \beta\frac{\theta_{i+1}^{j+1} - \theta_{i-1}^{j+1}}{2\Delta z}\right)$$

$$+ 0.5\left(D\frac{\theta_{i+1}^j - 2\theta_i^j + \theta_{i-1}^j}{\left(\Delta z\right)^2} + \beta\frac{\theta_{i+1}^j - \theta_{i-1}^j}{2\Delta z}\right) \tag{19.108}$$

Expanding and rearranging give:

$$2\left(\frac{D}{\left(\Delta z\right)^2} + \frac{1}{\Delta t}\right)\theta_i^{j+1} = 2\left(\frac{1}{\Delta t} - \frac{D}{\left(\Delta z\right)^2}\right)\theta_i^j + \left(\frac{D}{\left(\Delta z\right)^2} + \frac{\beta}{2\Delta y}\right)\theta_{i+1}^{j+1} + \left(\frac{D}{\left(\Delta z\right)^2} - \frac{\beta}{2\Delta y}\right)\theta_{i-1}^{j+1}$$

$$+ \left(\frac{D}{\left(\Delta z\right)^2} + \frac{\beta}{2\Delta y}\right)\theta_{i+1}^j + \left(\frac{D}{\left(\Delta z\right)^2} - \frac{\beta}{2\Delta y}\right)\theta_{i-1}^j \tag{19.109}$$

If the following constants are used

$$a = 2\left(\frac{D}{\left(\Delta z\right)^2} + \frac{1}{\Delta t}\right)$$

$$b = 2\left(\frac{1}{\Delta t} - \frac{D}{(\Delta z)^2}\right)$$

$$c = \left(\frac{D}{(\Delta z)^2} + \frac{\beta}{2\Delta y}\right)$$

$$d = \left(\frac{D}{(\Delta z)^2} - \frac{\beta}{2\Delta y}\right)$$

Then Equation (19.109) becomes:

$$a\theta_i^{j+1} = b\theta_i^j + c\theta_{i+1}^{j+1} + d\theta_{i-1}^{j+1} + c\theta_{i+1}^j + d\theta_{i-1}^j \tag{19.110}$$

19.3.3.2.2 Laplace Adams–Bashforth Scheme

This sub-section provides a numerical approximation method of the linearized Richards equation using the Laplace Adams–Bashforth method. The application of Laplace transform on both sides of Equation (19.22) transforms the equation from a partial differential equation to a differential equation

$$\mathcal{L}\left(\frac{\partial\theta}{\partial t}\right) = \mathcal{L}\left(D\frac{\partial^2\theta}{\partial z^2} + \beta\frac{\partial\theta}{\partial z}\right) \tag{19.111}$$

The resulting equation is given by:

$$\frac{d\theta}{dt} = \mathcal{L}\left(D\frac{\partial^2\theta}{\partial z^2} + \beta\frac{\partial\theta}{\partial z}\right) \tag{19.112}$$

s can be silenced by:

$$\frac{d\theta(s,t)}{dt} = F(\theta,t)$$

$$\theta(t) = \theta(s,t)$$

$$F(\theta,t) = \mathcal{L}\left(D\frac{\partial^2\theta}{\partial z^2} + \beta\frac{\partial\theta}{\partial z}\right) \tag{19.113}$$

The same procedure that was followed when discretizing the nonlinear equation is followed here to obtain:

$$\theta_{n+1} = \theta_n + h\left(\frac{3}{2}F_n - \frac{1}{2}F_{n-1}\right) \tag{19.114}$$

Applications of an inverse Laplace transform to take back the equation to its real space is given by:

$$\mathcal{L}^{-1}\left[\theta_{n+1}\right] = \mathcal{L}^{-1}\left[\theta_n + h\left(\frac{3}{2}F_n - \frac{1}{2}F_{n-1}\right)\right] \tag{19.115}$$

which results in:

$$\theta(z, t_{n+1}) = \theta(z, t_n) + h\frac{3}{2}\left(D\frac{\partial^2\theta(z, t_n)}{\partial z^2} + \beta\frac{\partial\theta(z, t_n)}{\partial z}\right)$$
$$-h\frac{1}{2}\left(D\frac{\partial^2\theta(z, t_{n-1})}{\partial z^2} + \beta\frac{\partial\theta(z, t_{n-1})}{\partial z}\right) \tag{19.1·16}$$

Forward and backward discretization in space variable yields:

$$\theta_i^{n+1} = \theta_i^n + h\frac{3}{2}\left[D\left(\frac{\theta_{i+1}^n - 2\theta_i^n + \theta_{i-1}^n}{(\Delta z)^2}\right) + \beta\left(\frac{\theta_{i+1}^n - \theta_i^n}{\Delta z}\right)\right]$$
$$-h\frac{1}{2}\left[D\left(\frac{\theta_{i+1}^{n-1} - 2\theta_i^{n-1} + \theta_{i-1}^{n-1}}{(\Delta z)^2}\right) + \beta\left(\frac{\theta_{i+1}^{n-1} - \theta_i^{n-1}}{\Delta z}\right)\right] \tag{19.117}$$

where $\theta(z_i, t_n) = \theta_i^n$. If $\Delta z = l$ then the equation above becomes:

$$\theta_i^{n+1} = \theta_i^n + h\frac{3}{2}\left[D\left(\frac{\theta_{i+1}^n - 2\theta_i^n + \theta_{i-1}^n}{(l)^2}\right) + \beta\left(\frac{\theta_{i+1}^n - \theta_i^n}{l}\right)\right]$$
$$-h\frac{1}{2}\left[D\left(\frac{\theta_{i+1}^{n-1} - 2\theta_i^{n-1} + \theta_{i-1}^{n-1}}{(l)^2}\right) + \beta\left(\frac{\theta_{i+1}^{n-1} - \theta_i^{n-1}}{l}\right)\right] \tag{19.118}$$

Expanding

$$\theta_i^{n+1} = \theta_i^n + \frac{3hD}{2(l)^2}\left(\theta_{i+1}^n - 2\theta_i^n + \theta_{i-1}^n\right) + \frac{3h\beta}{2l}\left(\theta_{i+1}^n - \theta_i^n\right)$$
$$-\frac{hD}{2(l)^2}\left(\theta_{i+1}^{n-1} - 2\theta_i^{n-1} + \theta_{i-1}^{n-1}\right) - \frac{h\beta}{2l}\left(\theta_{i+1}^{n-1} - \theta_i^{n-1}\right) \tag{19.119}$$

If the following constants are substituted into Equation (19.119) above

$$a = \frac{3hD}{2(l)^2}$$

$$b = \frac{3h\beta}{2l}$$

$$c = \frac{hD}{2(l)^2}$$

$$d = \frac{h\beta}{2l}$$

then the following equation is obtained:

$$\theta_i^{n+1} = \theta_i^n + a\left(\theta_{i+1}^n - 2\theta_i^n + \theta_{i-1}^n\right) + b\left(\theta_{i+1}^n - \theta_i^n\right) - c\left(\theta_{i+1}^{n-1} - 2\theta_i^{n-1} + \theta_{i-1}^{n-1}\right) - d\left(\theta_{i+1}^{n-1} - \theta_i^{n-1}\right) \tag{19.120}$$

which is simplified into:

$$\theta_i^{n+1} = (1 - 2a - b)\theta_i^j + (a+b)\theta_{i+1}^n + a\theta_{i-1}^n + (2c+d)\theta_i^{n-1} - (c+d)\theta_{i+1}^{n-1} - c\theta_{i-1}^{n-1} \tag{19.121}$$

19.3.4 NUMERICAL STABILITY ANALYSIS

Stability analysis is conducted to evaluate the performance of numerical approximation methods. It is essential for ensuring that discrete errors do not spread to the entire simulation (Allwright and Atangana, 2018). In this study, the Fourier expansion in space variables will be used.

$$\theta(z,t) = \sum_f \hat{\theta}(t)\exp(jfl) \tag{19.122}$$

19.3.4.1 Crank–Nicolson Finite-Difference Approximation Scheme

The stability analysis of the solution obtained using the Crank–Nicolson method is provided in this sub-section. Using the Fourier expansion, Equation (19.109) becomes:

$$a\hat{\theta}_{n+1}e^{jifl} = b\hat{\theta}_n e^{jifl} + c\hat{\theta}_n e^{j(i+1)fl} + d\hat{\theta}_n e^{j(i-1)fl} + c\hat{\theta}_{n+1}e^{j(i+1)fl} + d\hat{\theta}_{n+1}e^{j(i-1)fl} \tag{19.123}$$

where

$$\theta_i^{j+1} = \hat{\theta}_{n+1}e^{jifl} \tag{19.124}$$

$$\theta_i^j = \hat{\theta}_n e^{jifl} \tag{19.125}$$

$$\theta_{i+1}^{j+1} = \hat{\theta}_{n+1}e^{j(i+1)fl} \tag{19.126}$$

$$\theta_{i-1}^{j+1} = \hat{\theta}_{n+1}e^{j(i-1)fl} \tag{19.127}$$

Dividing Equation (19.123) by e^{jifl} yields:

$$a\hat{\theta}_{n+1} = b\hat{\theta}_n + c\hat{\theta}_n e^{jfl} + d\hat{\theta}_n e^{-jfl} + c\hat{\theta}_{n+1}e^{jfl} + d\hat{\theta}_{n+1}e^{-jfl} \tag{19.128}$$

At $n = 0$

The equation becomes:

$$a\hat{\theta}_1 = b\hat{\theta}_0 + c\hat{\theta}_0 e^{ifl} + d\hat{\theta}_0 e^{-ifl} + c\hat{\theta}_1 e^{ifl} + d\hat{\theta}_1 e^{-ifl} \tag{19.129}$$

Rearranging:

$$a\hat{\theta}_1 - c\hat{\theta}_1 e^{ifl} - d\hat{\theta}_1 e^{-ifl} = b\hat{\theta}_0 + c\hat{\theta}_0 e^{ifl} + d\hat{\theta}_0 e^{-ifl} \tag{19.130}$$

Simplifying:

$$\hat{\theta}_1\left(a - ce^{ifl} - de^{-ifl}\right) = \hat{\theta}_0\left(b + ce^{ifl} + de^{-ifl}\right) \tag{19.131}$$

Rearranging:

$$\frac{\hat{\theta}_1}{\hat{\theta}_0} = \frac{b + ce^{ifl} + de^{-ifl}}{a - ce^{ifl} - de^{-ifl}} \tag{19.132}$$

If,

$$e^{ifl} = \cos(fl) + j\sin(fl)$$

$$e^{-ifl} = \cos(fl) - j\sin(fl)$$

Then Equation (19.132) becomes:

$$\frac{\hat{\theta}_1}{\hat{\theta}_0} = \frac{b + c\left(\cos(fl) + j\sin(fl)\right) + d\left(\cos(fl) - j\sin(fl)\right)}{a - c\left(\cos(fl) + j\sin(fl)\right) - d\left(\cos(fl) - j\sin(fl)\right)} \tag{19.133}$$

Expanding and rearranging:

$$\frac{\hat{\theta}_1}{\hat{\theta}_0} = \frac{b + c\cos(fl) + d\cos(fl) + jc\sin(fl) - jd\sin(fl)}{a - c\cos(fl) - d\cos(fl) + jd\sin(fl) - jc\sin(fl)(d)} \tag{19.134}$$

Simplifying:

$$\frac{\hat{\theta}_1}{\hat{\theta}_0} = \frac{b + \cos(fl)(c+d) + jd\sin(fl)(c-d)}{a - \cos(fl)(c+d) + jd\sin(fl)(c-d)} \tag{19.135}$$

The solution will be obtained when

$$\frac{\left(b + (c+d)\cos(fl)\right)^2 + (c-d)^2\sin^2(fl)}{\left(a - (c+d)\cos(fl)\right)^2 + (c-d)^2\sin^2(fl)} < 1 \tag{19.136}$$

$$\left(b+(c+d)\cos(fl)\right)^2 < \left(a-(c+d)\cos(fl)\right)^2 \tag{19.137}$$

$$b^2 + 2\cos(fl)(c+d) + \cos^2(fl)(c+d)^2 < a^2 - 2\cos(fl)(c+d) + \cos^2(fl)(c+d)^2 \tag{19.138}$$

The assumed condition for stability is true when

$$b^2 - a^2 < -4\cos(fl)(c+d) \tag{19.139}$$

Substituting for the constants $a = 2\left(\dfrac{D}{(\Delta z)^2} + \dfrac{1}{\Delta t}\right)$; $b = 2\left(\dfrac{1}{\Delta t} - \dfrac{D}{(\Delta z)^2}\right)$; $c = \left(\dfrac{D}{(\Delta z)^2} + \dfrac{\beta}{2\Delta y}\right)$; and

$d = \left(\dfrac{D}{(\Delta z)^2} - \dfrac{\beta}{2\Delta y}\right)$ into Equation (19.140) yields:

$$\left(\frac{2}{\Delta t} - \frac{2D}{(\Delta z)^2}\right)^2 - \left(\frac{2}{\Delta t} + \frac{2D}{(\Delta z)^2}\right)^2 < -4\cos(fl)\left(\frac{D}{(\Delta z)^2} + \frac{\beta}{2\Delta y} + \frac{D}{(\Delta z)^2} - \frac{\beta}{2\Delta y}\right) \tag{19.140}$$

By simplifying, the following is obtained:

$$\left(\frac{2}{\Delta t} - \frac{2D}{(\Delta z)^2}\right)^2 - \left(\frac{2}{\Delta t} + \frac{2D}{(\Delta z)^2}\right)^2 < -8\cos(fl)\frac{D}{(\Delta z)^2} \tag{19.141}$$

It is concluded that the present solution is stable for $\forall n \leq 0$ when this condition is met, and can be used to obtain reliable numerical simulations.

19.3.4.1.1 The Laplace Adams–Bashforth Scheme

For the stability analysis of Equation (19.121), the term θ_i^n will be replaced by θ_i^j

The stability analysis of the solution obtained using the Laplace Adams–Bashforth method is provided in this sub-section. Using the Fourier expansion, Equation (19.121) becomes:

$$\begin{aligned}
\hat{\theta}_{n+1}e^{jifl} &= (1-2a-b)\hat{\theta}_n e^{jifl} + (a+b)\hat{\theta}_n e^{j(i+1)fl} + a\hat{\theta}_n e^{j(i-1)fl} \\
&\quad + (2c+d)\hat{\theta}_{n-1}e^{jifl} - (c+d)\hat{\theta}_{n-1}e^{j(i+1)fl} - c\hat{\theta}_{n-1}e^{j(i-1)fl}
\end{aligned} \tag{19.142}$$

By dividing Equation (19.142) above by e^{jifl} the following is obtained:

$$\hat{\theta}_{n+1} = (1-2a-b)\hat{\theta}_n + (a+b)\hat{\theta}_n e^{ifl} + a\hat{\theta}_n e^{-ifl} + (2c+d)\hat{\theta}_{n-1} - (c+d)\hat{\theta}_{n-1}e^{ifl} - c\hat{\theta}_{n-1}e^{-ifl} \tag{19.143}$$

and by grouping $\hat{\theta}_n$ and $\hat{\theta}_{n-1}$ terms together, Equation (19.143) becomes:

$$\hat{\theta}_{n+1} = \left(1-2a-b+(a+b)e^{ifl}+ae^{-ifl}\right)\hat{\theta}_n + \left(2c+d-(c+d)e^{ifl}-ce^{-ifl}\right)\hat{\theta}_{n-1} \tag{19.144}$$

and by expanding and rearranging becomes:

$$\hat{\theta}_{n+1} = (1-2a-b)\hat{\theta}_n + (a+b)\hat{\theta}_n e^{ifl} + a\hat{\theta}_n e^{-ifl} + 2c\hat{\theta}_{n-1} + d\hat{\theta}_{n-1} - (c+d)e^{ifl}\hat{\theta}_{n-1} - ce^{-ifl}\hat{\theta}_{n-1} \tag{19.145}$$

If,

$$e^{jfl} = \cos(fl) + j\sin(fl)$$

$$e^{-jfl} = \cos(fl) - j\sin(fl)$$

Then Equation (19.145) becomes:

$$\hat{\theta}_{n+1} = (1-2a-b)\hat{\theta}_n + (a+b)\big(\cos(fl)+j\sin(fl)\big)\hat{\theta}_n + a\big(\cos(fl)-j\sin(fl)\big)\hat{\theta}_n$$
$$+2c\hat{\theta}_{n-1} + d\hat{\theta}_{n-1} - (c+d)\big(\cos(fl)+j\sin(fl)\big)\hat{\theta}_{n-1} - c\big(\cos(fl)-j\sin(fl)\big)\hat{\theta}_{n-1} \qquad (19.146)$$

Expanding Equation (19.146) gives:

$$\hat{\theta}_{n+1} = (1-2a-b)\hat{\theta}_n + (a+b)\cos(fl)\hat{\theta}_n + (a+b)j\sin(fl)\hat{\theta}_n + a\cos(fl)\hat{\theta}_n$$
$$-aj\sin(fl)\hat{\theta}_n + 2c\hat{\theta}_{n-1} + d\hat{\theta}_{n-1} - (c+d)\cos(fl)\hat{\theta}_{n-1}$$
$$-(c+d)j\sin(fl)\hat{\theta}_{n-1} - c\cos(fl)\hat{\theta}_{n-1} + cj\sin(fl)\hat{\theta}_{n-1} \qquad (19.147)$$

By grouping and simplifying, the following is obtained:

$$\hat{\theta}_{n+1} = \big(1-2a-b+(2a+b)\cos(fl)\big)\hat{\theta}_n$$
$$+(2c+d)\big(1-\cos(fl)\big)\hat{\theta}_{n-1} + j\sin(fl)\big(b\hat{\theta}_n - d\hat{\theta}_{n-1}\big) \qquad (19.148)$$

$$\hat{\theta}_{n+1} = \big(1-2a-b+(2a+b)\cos(fl)+j\sin(fl)b\big)\hat{\theta}_n$$
$$+\big((2c+d)\big(1-\cos(fl)\big)-j\sin(fl)d\big)\hat{\theta}_{n-1} \qquad (19.149)$$

Equation (19.149) above can be written as:

$$\hat{\theta}_{n+1} = \hat{\theta}_n A + \hat{\theta}_{n-1} B \qquad (19.150)$$

where

$$A = 1-2a-b+(2a+b)\cos(fl)+j\sin(fl)b$$

$$B = (2c+d)\big(1-\cos(fl)\big)-j\sin(fl)d$$

At $n = 0$, Equation (19.150) becomes

$$\hat{\theta}_1 = \hat{\theta}_0 A + \hat{\theta}_{-1} B \qquad (19.151)$$

The condition required for the solution to be stable is $\left|\dfrac{\theta_1}{\theta_0}\right| < 1$

If Equation (19.151) is considered to be

$$\hat{\theta}_1 = A\hat{\theta}_0 \qquad (19.152)$$

then

$$\left| \frac{\hat{\theta}_1}{\hat{\theta}_0} \right| = A \tag{19.153}$$

The condition above is assumed to be true if

$$|A| = \sqrt{x^2 + y^2} < 1 \tag{19.154}$$

where x and y are obtained by splitting

$$A = 1 - 2a - b + (2a + b)\cos(fl) + j\sin(fl)b \tag{19.155}$$

into

$$x = 1 - 2a - b + (2a + b)\cos(fl) \tag{19.156}$$

and

$$y = j\sin(fl)b \tag{19.157}$$

Substituting the x and y into Equation (19.154) yields

$$|A| = \sqrt{\left(1 - 2a - b + (2a + b)\cos(fl)\right)^2 + \left(j\sin(fl)b\right)^2} \tag{19.158}$$

$$\frac{\left| \left(1 - \frac{3hD}{(l)^2} - \frac{3h\beta}{2l} + \left(\frac{3hD}{(l)^2} + \frac{3h\beta}{2l}\right)\cos(fl)\right)^2 + \left(j\sin(fl)\frac{3h\beta}{2l}\right)^2 \right|}{= \sqrt{\left(1 - \frac{3hD}{(l)^2} - \frac{3h\beta}{2l} + \left(\frac{3hD}{(l)^2} + \frac{3h\beta}{2l}\right)\cos(fl)\right)^2 + \left(j\sin(fl)\frac{3h\beta}{2l}\right)^2}} \tag{19.159}$$

$$\sqrt{\left(1 - \frac{3hD}{(l)^2} - \frac{3h\beta}{2l} + \left(\frac{3hD}{(l)^2} + \frac{3h\beta}{2l}\right)\cos(fl)\right)^2 + \left(j\sin(fl)\frac{3h\beta}{2l}\right)^2} < 1 \tag{19.160}$$

It is concluded that the solution is stable for $\forall n = 0$. To prove that the solution is stable for

$\forall n \leq 0$ the following condition is assumed $\left| \frac{\hat{\theta}_{n+1}}{\hat{\theta}_0} \right| < 1$, then:

$$\hat{\theta}_{n+1} = \hat{\theta}_n A + \hat{\theta}_{n-1} B \tag{19.161}$$

$$\left|\hat{\theta}_{n+1}\right| = \left|\hat{\theta}_n A + \hat{\theta}_{n-1} B\right| \tag{19.162}$$

$$\left|\hat{\theta}_{n+1}\right| = \left|\hat{\theta}_n A + \hat{\theta}_{n-1} B\right| \le \left|\hat{\theta}_n\right|\left|A\right| + \left|\hat{\theta}_{n-1}\right|\left|B\right| \tag{19.163}$$

According to induction theory, the solution is stable when:

$$\left|\hat{\theta}_n\right|\left|A\right| + \left|\hat{\theta}_{n-1}\right|\left|B\right| < \left|\hat{\theta}_0\right|\left|A\right| + \left|\hat{\theta}_0\right|\left|B\right| \tag{19.164}$$

$$\frac{\left|\hat{\theta}_{n+1}\right|}{\left|\hat{\theta}_n\right|} < |A| + |B| < 1 \tag{19.165}$$

Expanding by substituting for A and B

$$\frac{\left|\hat{\theta}_{n+1}\right|}{\left|\hat{\theta}_n\right|} < \left|1 - 2a - b + (2a + b)\cos(fl) + j\sin(fl)b\right| + \left|(2c + d)(1 - \cos(fl)) - j\sin(fl)d\right| < 1 \tag{19.166}$$

which is also

$$\frac{\left|\hat{\theta}_{n+1}\right|}{\left|\hat{\theta}_n\right|} < \left|1 - \frac{3hD}{(l)^2} - \frac{3h\beta}{2l} + \left(\frac{3hD}{(l)^2} + \frac{3h\beta}{2l}\right)\cos(fl) + j\sin(fl)\frac{3h\beta}{2l}\right|$$
$$+ \left|\left(\frac{hD}{(l)^2} + \frac{h\beta}{2l}\right)(1 - \cos(fl)) - j\sin(fl)\frac{h\beta}{2l}\right| < 1 \tag{19.167}$$

It can be concluded that the solution obtained using the Laplace Adams–Bashforth numerical method is stable.

19.4   NUMERICAL SIMULATIONS

A computer program called MATHEMATICA was used to produce numerical simulations of water movement in the unsaturated zone. The program was used to solve both the highly nonlinear Richards equation for vertical flow obtained from using the Brooks and Corey and Mualem models, and the linearized Richards equation based on system oversimplifying assumptions. The following initial and boundary conditions were considered:

$$\theta(z,0) > \theta_r$$

$$\theta(z,l) = \theta_L$$

$$\theta(0,t) > \theta_r$$

Flow in a homogeneous volume of soil was simulated for three different models: the Brooks and Corey model; the Mualem model; and the proposed linear model. The following soil hydraulic parameters were considered in this study for numerical simulations: $D = 0.05$, $\theta_r = 0.043$, $\theta_s = 0.44$, $\theta_L = 0.3$, $K_s = 6.935$, and $\theta_I = 0.43$.

19.4.1 Results and Discussion

To reproduce the unsaturated flow system, three unsaturated hydraulic conductivity models were combined with Richards equation. The first was the Brooks and Corey (1964) model; the second one was the Mualem (1976) model; and the third is the linear model proposed in this study. Simulation results obtained when the Richards equation was combined with the Brooks and Corey model showed that soil water content evolves across space and time. However, the evolution presented by this model is non-realistic, because for the simulated time water is less than initial water content during early and mid-simulation time. Moreover, water seems to be constant at a shallow depth and there is a sudden increase in water content. The numerical solution presented in Figure 19.1 shows a rapid increase in water content near the lower flow boundary; this is questionable, because water content rises above initial values. Figure 19.2 present a ContourPlot of water content. Similarly, there is one characteristic contour from the surface extending toward the lower boundary. Then, a sudden change in the contour's slope is present, and there are steep contours close to the boundary. This model implies that, for this particular soil volume, almost all soil moisture is located near the lower boundary.

The Mualem model also yielded results with a water content evolution trend that is similar to the one explained above. This is expected, provided that the models theories are not completely different.

It was found necessary to revise these models to see if it will be possible to obtain realistic results. In order to revise the Brooks and Corey and Mualem models, Equations (19.73) and (19.74), respectively, close attention has to be paid to how the model parameters are related. The relationship between relative hydraulic conductivity and effective saturation suggested by these models is highly nonlinear. Therefore, realistic results can be obtained if the relationship is made less nonlinear. In this case, it is suggested that the exponent in Equation (19.73) and Equation (19.74) is given to $\theta - \theta_r$ only. The exclusion of $\theta_s - \theta_r$ from the base of the exponent makes the resultant model less

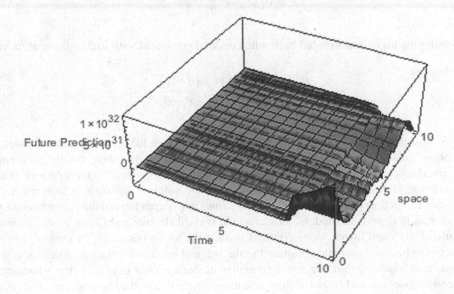

FIGURE 19.1 Numerical solution of Brooks and Corey (1964) model combined with Richards equation.

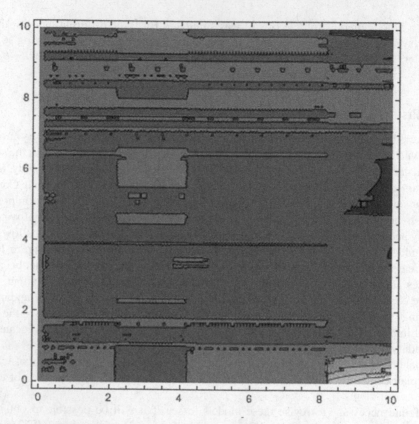

FIGURE 19.2 ContourPlot of Brooks and Corey (1964) model combined with Richards equation.

nonlinear. The proposed nonlinear model for estimating unsaturated hydraulic conductivity is given by:

$$K(\theta) = K_s \frac{(\theta - \theta_r)^\omega}{\theta_s - \theta_r}$$

(19.168)

Combining the above unsaturated hydraulic conductivity model with Richards equation yields:

$$\frac{\partial \theta}{\partial t} = D \frac{\partial^2 \theta}{\partial z^2} + \frac{\partial}{\partial z}\left[K_s \frac{(\theta - \theta_r)^\omega}{\theta_s - \theta_r} \right]$$

(19.169)

where ω is 5 for the revised Books and Corey model, and 3.5 for the revised Mualem model.

Simulation runs using these revised models yielded results that were totally different from those obtained using original models. The water content evolution trend is the opposite of the one explained above. Here, water is distributed across the soil volume; and there is high water content at the beginning. As time passes, water content declines with depth toward the lower boundary. This trend is seen in both revised models. However, the revised Brooks and Corey model shows some oscillations of high and lower water contents toward the lower boundary. In general, the revised models seem to be more realistic compared to the original models for the considered soil volume. To expand, it is expected that the water content will decline over time and depth because some water becomes disconnected from the flow, resulting in a decline in the residual flow. The results for revised models are presented in Figures 19.3–19.11 below.

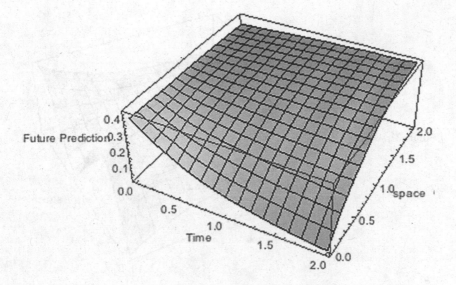

FIGURE 19.3 Numerical solution of the proposed linear model.

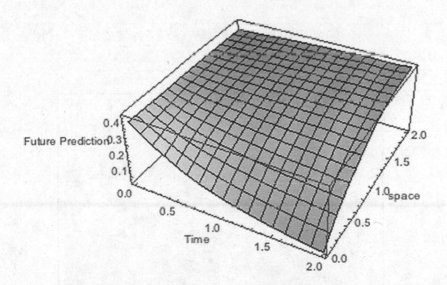

FIGURE 19.4 Numerical solution of the proposed nonlinear model obtained from revising the Mualem model.

Simulation runs using the proposed unsaturated hydraulic conductivity model combined with Richards equation yielded results that are presented in Figures 19.3, 19.6 and 19.9 above. The numerical solution shows an evolution of water content with depth into the soil volume. Water content is decreasing as expected, the ContourPlot and Density Plots show how the water content decreases with time and space. Water content is high close to the surface and low toward the lower boundary. Therefore the results are realistic for the considered soil volume.

The proposed linear and nonlinear models performed similarly. Though there is a slight difference in solutions, models were able to produce valid results. The performance of the revised Mualem model and the proposed model is very close. The revised Brooks and Corey model also showed a similar performance at early to mid-simulation times; however, at late simulation time, water was distributed in a wave-like form.

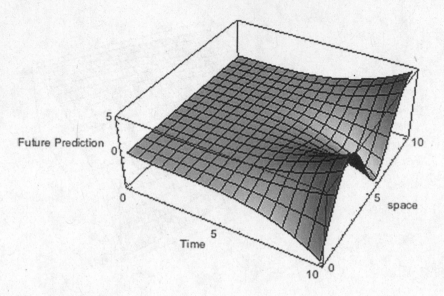

FIGURE 19.5   Numerical solution of the proposed nonlinear model obtained from revising Brooks and Corey model.

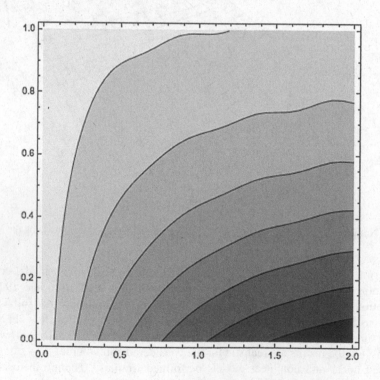

FIGURE 19.6   ContourPlot of the proposed linear model.

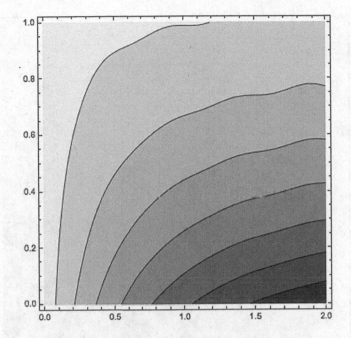

FIGURE 19.7 ContourPlot of the proposed nonlinear model obtained from revising the Mualem model.

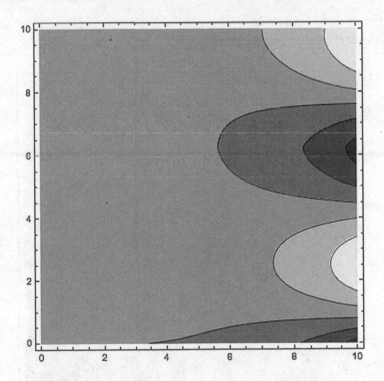

FIGURE 19.8 ContourPlot of the proposed nonlinear model obtained from revising the Brooks and Corey model.

FIGURE 19.9   Density Plot of the proposed linear model.

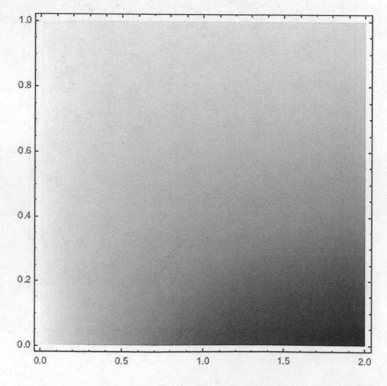

FIGURE 19.10   Density Plot of the proposed nonlinear model obtained from revising the Mualem model.

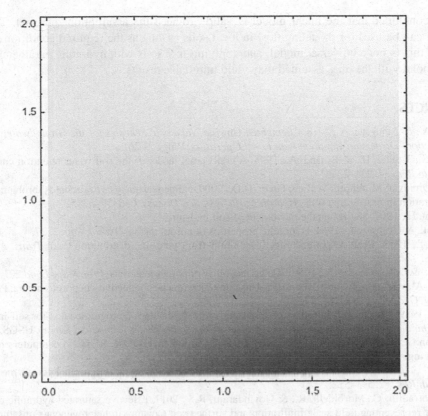

FIGURE 19.11   Density Plot of the proposed nonlinear model obtained from revising the Brooks and Corey model.

19.5   CONCLUSION

The purpose of this study was to model sub-surface water flow in the unsaturated zone using selected pre-existing nonlinear models and a proposed linear model. A volume of unsaturated soil with characteristic soil hydraulic properties was considered to address the aforementioned purpose. Incorporation of pre-existing unsaturated hydraulic conductivity models in Richards equation yielded highly nonlinear models that required numerical analysis. Application of the proposed linear unsaturated hydraulic conductivity model to Richards equation resulted in a linearized Richards equation which was easy to solve both numerically and analytically. The exact solution of the linearized Richards equation obtained using the Laplace transform and Green's function is valid. Therefore, if parametric soil hydraulic properties are available, a system can be solved without a computer program.

Numerical analysis was performed for all models, using two numerical approximation methods for more reliable results. The stability of resultant solutions of linearized Richards equation tested using the Fourier expansion stability analysis method, the equation was stable for both approximation methods provided the required conditions were met.

Numerical simulations were obtained for all models, and the results showed that for a vertical water flow soil water content evolves with depth. Nonlinear results showed poor performance and the resulting numerical solutions were not realistic. To expand, simulations showed that there is an increase in water content with depth which is not realistic. Water content is expected to decline with depth because of the concept of water retention. Revision of these models yielded results that were found to be more realistic. The results corresponded with results obtained using a linear model.

It can be concluded that the linear model is valid provided the assumptions are met. Therefore, this model can be used for modeling flow in local soils as long as the required conditions are met. Moreover, this is not a universal model, and applying it in soils with hydraulic parameters that do not correspond with the ones assumed may yield unreliable results.

REFERENCES

Allwright, A., & Atangana, A., 2018. *Augmented Upwind Numerical Schemes for the Groundwater Transport Advection–Dispersion Equation with Local Operators*. Wiley, 1–26.

Assouline, S., Tessier, D., & Bruand, A., 1998. A conceptual model of the soil water retention curve. *Water Resour. Res.* 34, 223–231.

Barari, A., Omidvar, M., Ghotbi, A.R., & Ganji, D.D., 2009. Numerical analysis of Richards' problem for water penetration in unsaturated soils. *Hydrol. Earth Syst. Sci. Discuss.* 6, 6359–6385.

Buckingham, E., 1907. Studies on the movement of soil moisture.

Brooks, R.H., & Corey, A.T., 1964. Hydraulic properties of porous media. *Hydrol. Pap.* 3, 27.

Burdine, N.T., 1953. Relative permeability calculation from pore-size distribution data. *Trans. AIME* 198, 71–78.

Childs, E.C., & Collis-George, N., 1950. The permeability of porous materials. *Proc R Soc Lond* 201, 392–405.

Ghanbarian-Alavijeh, B., & Hunt, A.G., 2012. Unsaturated hydraulic conductivity in porous media: Percolation theory. *Geoderma* 187–188, 77–84.

Gordon, D., 1989. Chapter B2: Introduction to ground-water hydraulics, a programed text for self-instruction, in *Techniques of Water-Resources Investigations of the United States Geological Survey*. USGS, pp. 1–29.

Leij, F.J., van Genuchten, M.T., Yates, S.R., Russel, W.B., & Kaveh, F., 1994. RETC: A computer program for analyzing soil water retention and hydraulic conductivity data.

Mualem, Y., 1976. A new model for predicting the hydraulic conductivity of unsaturated porous media. *Water Resour. Res.* 12, 513–522.

Ojha, R., Corradini, C., Morbidelli, R., & Govindaraju, R.S., 2017. Effective saturated hydraulic conductivity for representing field-scale infiltration and surface soil moisture in heterogeneous unsaturated soils subjected to rainfall events. *Water* 9.

Richards, L.A., 1931. Capillary conduction of liquids through porous mediums. *Physics*, 1(5), 318–333.

Soulaine, C., 2015. Chapter 1: On the origin of Darcy's law. Stanford.

van Genuchten, M.T., 1980. A closed-form equation for predicting the hydraulic conductivity of unsaturated soils. *Soil Sci. Soc. Am. J.* 44, 892–898.

Vereecken, H., Huisman, J.A., Bogena, H., Vanderborght, J., Vrugt, J.A., & Hopmans, J.W., 2008. On the value of soil moisture measurements in vadose zone hydrology: A review. *Water Resour. Res.* 44(4).

Whitaker, S., 1986. Flow in porous media I: A theoretical derivation of Darcy's Law. *Reidel Publ. Co.* 3–25.

Zhang, T., 1991. A statistical approach for water movement in the unsaturated zone (Report No. 1010). Department of Water Resources Engineering Lund Institute of Technology, Lund University, Lund, Sweden.

20 Deterministic and Stochastic Analysis of Groundwater in Unconfined Aquifer Model

Dineo Ramakatsa and Abdon Atangana

Institute for Groundwater Studies, University of the Free State, South Africa

CONTENTS

20.1 INTRODUCTION

Groundwater is the water preserved in permeable media such as ground pore spaces or rock formations under the surface of the earth. Groundwater flow is the motion of water in the underground porous media. An aquifer is regarded as a geological formation or stratum that comprises groundwater and enables large amounts of this groundwater to pass through it under normal ground conditions (Neuman, 1974; Dagan & Kroszynski, 1975). The groundwater behavior during the groundwater movement process through an aquifer system depends on the characteristics of the water itself and the medium in which it flows (Andersine, 1992; Berg & Illma, 2012). The purpose of groundwater flow modeling is to estimate flow speeds and head forecasts. However, speed estimates are generally based on hydraulic head differences and are therefore much more sensitive to numerical modeling errors than hydraulic head estimates alone. Meaningful transport predictions often involve the calculation of the speed range on a fine spatial grid. Analytical solutions therefore have a certain advantage over numerical processes (Neumann, 1974). Deterministic and stochastic

DOI: 10.1201/9781003266266-20

models in groundwater studies are developed to describe the flow mechanisms in groundwater systems to understand groundwater behavior (Gordon, 1989; Kock & Aralai, 2004; Pool et al., 2015; Maliva, 2016). Deterministic and stochastic groundwater mathematical models are usually used through partial differential equations to simulate flow and transport processes in aquifer systems. A stochastic model is a predicament in which there is uncertainty. In other words, this is a model for a process that is random. The word stochastic comes from the Greek word *stokhazesthai*, meaning aiming or speculating. In the real word, uncertainty is a part of the norm; therefore a stochastic model can depict literally anything. In contrast, the deterministic model is one that predicts the output with 100 per cent certainty. Deterministic models always have an equation set that accurately describes the system inputs and outputs. Stochastic models, on the other hand, will probably yield different results each time the model is run (Youngs & Poulovassilis, 1976; Yen & Gary, 1990; Zhang, 1991; Zhu & Yeh, 2005; Yang & Shi, 2008; Philippe et al., 2013).

The movement of water within a confined, unconfined or leaky aquifer cannot be captured using the same mathematical models. The Theis model was introduced to capture flow within a confined aquifer, while the Hantush model was suggested to predict the flow within a leaky aquifer, these two models are unable to account for the flow within an unconfined aquifer, and more precisely, they are of linear form. For example, (Theis, 1935) equation for groundwater flow is widely used to solve groundwater related problems in a confined aquifer system. Thus it is one of the primary solutions for groundwater flow in the deterministic approach to solve groundwater water problems. This equation was obtained based on the following assumptions: aquifer homogeneity, uniform thickness, isotropic, infinite aerial extend and constant pumping rate. However, in reality, the opposite is true, in that the aquifer is usually heterogeneous, anisotropic, pumped at different discharge rates, and has finite aerial extend resulting from impermeable boundaries. To capture flow within an unconfined aquifer, a new stochastic mathematical equation was suggested and happens to be an integro-differential type. The study of this model is not popular, perhaps because of the complexity of the mathematical setting. In this study, we considered the model of groundwater flowing within an unconfined aquifer. We derived the conditions under which the exact solution could be obtained. We suggested numerical solutions using different schemes, including forward Euler, Crank–Nicolson and Atangana–Batogna schemes. For each of them, we presented a detailed study underpinning the stability of the scheme in use. To conclude, we suggested a new numerical scheme that combines the fundamental theorem of calculus, Adams–Bashforth and the trapezoidal rule.

20.2   DETERMINISTIC APPROACH

Deterministic mathematical models usually simulate flow and transportation processes in groundwater systems using partial differential equations (Bear, 1972; Anderson and Woessner, 1992; Konikow, 2001). Presuming the groundwater flow is a time-reliant fundamental problem; full expressions of the deterministic mathematical models include statements of equations, initial and boundary conditions. Mathematical deterministic models can easily be resolved, either analytically or numerically. Analytical solutions, however, require generally highly idealized and detailed parameters and boundaries.

The governing equations are mathematical relations that describe groundwater movement approximations via aquifer mechanisms. They can be derived directly from different concepts of groundwater systems by incorporating water mass balance mathematically with Darcy's law. There are two different concepts of groundwater systems: Equation (20.1), the viewpoint of the aquifer; and Equation (20.2), the viewpoint of the flow system. The view of aquifers is based on concepts of confined and unconfined aquifers (Anderson & Woessner, 1992). A general form of the governing equation in a confined aquifer can be expressed as:

$$\frac{\partial}{\partial x}\left(T_x \frac{\partial h}{\partial x}\right) + \frac{\partial}{\partial y}\left(T_y \frac{\partial h}{\partial y}\right) = S\frac{\partial h}{\partial t} - R + L \qquad (20.1)$$

where h is the hydraulic head, T_x and T_y are horizontal transmissivity components, S represents the coefficient of storage, R depicts a sink/source phrase that is identified as intrinsically valuable for recharge, and L represents leakage in a confined bed (Anderson and Woessner, 1992).

In an unconfined aquifer, the components of transmissivity that are practically assumed, T_x and T_y in Equation (20.1) are substituted by $T_x = K_x h$ and $T_y = K_y h$, respectively, and the L component which represents leakage in Equation (20.1) is equivalent to zero. This result in a nonlinear governing equation, also referred to as the Boussinesq equation (Bear, 1972; Anderson and Woessner, 1992), representing an unconfined aquifer flow as:

$$\frac{\partial}{\partial x}\left(K_x h \frac{\partial h}{\partial x}\right) + \frac{\partial}{\partial y} K_y h \left(\frac{\partial h}{\partial y}\right) = S_y \frac{\partial h}{\partial t} - R \tag{20.2}$$

where h represents the saturated thickness in an unconfined aquifer, K_x and K_y are tensor conductivity horizontal components, and S_y is an unconfined aquifer's specific output. The flow mechanism viewpoint relates to three-dimensional head distribution, hydraulic conductivity and groundwater storage properties. It enables vertical and horizontal flow components to be analyzed and therefore enables evaluations of two-dimensional or three-dimensional groundwater flow. A standard form of the regulatory equation for the flow system is:

$$\frac{\partial}{\partial x}\left(K_x \frac{\partial h}{\partial x}\right) + \frac{\partial}{\partial y}\left(K_y \frac{\partial h}{\partial y}\right) + \frac{\partial}{\partial z}\left(K_z \frac{\partial h}{\partial z}\right) = S_s \frac{\partial h}{\partial t} - R \tag{20.3}$$

where K_x is the vertical conductivity tensor component, S_s represents the specific storage, R represents a sink/source phrase denoted as the system input volume per unit volume of aquifer per unit time (Anderson & Woessner, 1992). Since groundwater flow motions are described in general, the governing equations do not produce any information on the characteristics of unique groundwater flow situations. To obtain solutions for special cases, the limits and initial conditions must also be specified, together with the relevant equations.

. The boundary conditions represent mathematical statements that specify the dependent variable or derivative of the dependent variable at the boundary of problems relating to groundwater (Anderson & Woessner, 1992).

20.3 STOCHASTIC APPROACH

Historically, groundwater movement in aquifers was modeled deterministically, which assumes that the information required in the modeling, such as aquifer parameters and boundary conditions, should be known with certainty. Therefore a unique solution correlated with the set of deterministic conditions can be obtained. Hydrological events are better described in reality as random encounters. The boundary conditions, such as the river stage, can be uncertain or even unknown, more especially when future predictions are to be made. The aquifer parameters, such as hydraulic conductivity, transmissivity and storativity are also random variables because of the lack of information and/or the underlying complexity of the geologic process. Therefore, groundwater flow is more realistically modeled using the stochastic approach.

The hydraulic head distribution equation in a transient two-dimensional aquifer system with spatially randomly chosen transmissivity and spatio-temporally recharge is expressed as:

$$\frac{\partial}{\partial_{xi}}\left[T(x)\frac{\partial H}{\partial_{xi}}\right] + R(x, t) = S_y \frac{\partial H}{\partial t} \tag{20.4}$$

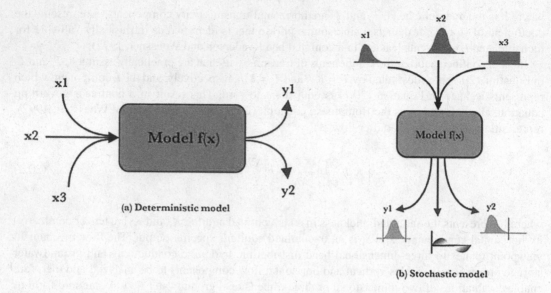

FIGURE 20.1   Representation of the two models: deterministic and stochastic (Gallage et al., 2013).

where: $x = (x_1, x_2)$ represents a vector point in the horizontal plane, $T(x)$ represents transmissivity at a specific location x, $R(x, t)$ denotes transient recharge at a particular time t and location x. The transient hydraulic head is represented by $H(x, t)$ at a specific time t and location x, S_y represents the specific yield, which is the volume of water produced per unit area per unit head decline, which serves as a known constant.

Pore water velocity vector for two-dimensional transient flow is:

$$Vi(x,t) = \frac{T(x)}{bn} \frac{\partial H(x,t)}{\partial_{xi}}, i = 1.2 \tag{20.5}$$

where in $Vi(x, t)$ at vector location x and time t is the pore velocity, $T(x)$ is the transmissivity, b is the thickness of the aquifer, and n represents the porosity. Here, both the thickness of the aquifer and porosity are generally known to be constants.

Figure 20.1 shows two different model types: the deterministic, by which the output depends entirely on the initial conditions and the parameter values; the stochastic model, on the contrary, has a certain randomness.

20.4   NUMERICAL APPROXIMATION

Boulton (1954b) extended the transient confined theory of Theis to include the effect of delayed yields in unconfined aquifers resulting from the movement of the water table. The solutions proposed by Boulton (1954b, 1963) reproduce the unconfined time-drawing curve in all three segments. In its formulation of delayed yields, it assumed that, as the water table falls, water is gradually released from storage (drainage) rather than instantly, as in the Boulton (1954a) and Dagan (1967) free-surface solutions. This approach resulted in an integrated differential flow equation in terms of average drawdown s^* as:

$$\frac{\partial^2 s^*}{\partial r^2} + \frac{1}{r} \frac{\partial s^*}{\partial r} = \left[\frac{S}{T} \frac{\partial s^*}{\partial t} \right] + \left\{ \alpha S_y \int_0^t \frac{\partial s^*}{\partial \tau} e^{-\alpha(t-\tau)} d\tau \right\} \tag{20.6}$$

Boulton linearized by denoting T a constant value. The phrase in square brackets is instantaneous confined storage, the phrase in braces represents a convolution integral which represents the storage that has been released steadily since the pumping started as a result of the decline of the water table. Boulton (1963) showed that the time when delayed yield effects are negligible is equivalent to α, leading to the term referred to as 'the delay index' (Prickett,1965). Prickett (1965) used this principle and developed an empirical relationship between the delay index and the physical aquifer properties by analyzing large amounts of field drawdown data using Boulton's (1963) solution. Prickett (1965) implemented an estimation methodology for S, S_y, K, and α of unconfined aquifers by analyzing pumping tests using the Boulton (1963) solution.

Though Boulton's model could replicate all three segments of the time-drawing curve in an unconfined aquifer, the physical mechanism of the delayed yield process was not explained by the non-physical nature of the 'delay index' in the Boulton (1963) solution. Streltsova (1972) invented an estimated solution for the water table decline and the complete penetration of pumping and observation wells. Like Boulton (1954b), Streltsova (1972) represented the water table as a sharp material boundary, and wrote the two-dimensional flow equation with average depth as ∂^2.

$$\frac{\partial^2 s^*}{\partial r^2} + \frac{1}{r}\frac{\partial s^*}{\partial r} = \frac{S}{T}\left(\frac{\partial s^*}{\partial t} - \frac{\partial \xi}{\partial t}\right) \tag{20.7}$$

The rate of decline in the water table was then presumed to be linearly proportional to the difference between and the vertically averaged head, $b - s^*$ and the elevation ξ of the water table.

$$\frac{\partial \xi}{\partial t} = \frac{K_z}{S_y b_z}\left(s^* - b + \xi\right) \tag{20.8}$$

where $bz = b/3$ represent the effective thickness of the aquifer over which the water table is recharged to the deep aquifer. Equation (20.8) can be seen as an estimate to Boulton's (1954a) and Dagan's (1967) zero-order linearized free-surface boundary condition. Streltsova considered the initial condition of $\xi(r,t = 0) = b$, making use of a similar boundary condition at the pumping well and the outer boundary $r \to \infty$ of Theis (1935) and Boulton (1963), respectively. The solution is Equation (1.7) (Streltsov, 1972b)

$$\frac{\partial \xi}{\partial t} = -\alpha_T \int_0^t e^{-\alpha_T(t-r)}\frac{\partial s^*}{\partial \tau}d\tau \tag{20.9}$$

where $\alpha_T = K_z/(S_y bz)$. The substitution of Equation (20.9) into Equation (20.8) generates solution Equation (20.6) (Boulton 1954b, 1963); both solutions are roughly equal. The delayed yield theory of Boulton (as that of Streltsova) does not account at all for water flow in unsaturated areas but treats the water table as a material boundary moving vertically downwards under gravitational influence. Streltsova (1973) used raw data gathered by Meyer (1962) to show the unsaturated flow of water, which had virtually no effect on the delayed process observed. While Streltsova's solution linked the delay index of Boulton to physical aquifer properties, it later became a function of r (Neuman, 1975; Herrera et al., 1978).

Boulton and Streltsova's delayed yield solutions do not account for vertical flow in an unconfined aquifer by simplifying assumptions. These solutions cannot be broadened to account for partially penetrating pumping and observation wells. The pumping test carried out by Prickett near Lawrenceville, Illinois, USA (Prickett, 1965) later depicted that specific storage in unconfined aquifers could be much higher than the observed values in confined aquifers – likely to be a result of trapped air bubbles or aggregated shallow sediments. The elastic characteristics of unconfined aquifers are obviously too crucial to be ignored.

The models of Boulton (1954b, 1963) experienced conceptual difficulties outlining the physical mechanism of full release of water from storage in unconfined aquifers. Neuman (1972) described a physically focused mathematical model that regarded the unconfined aquifer as compressible (Boulton (1954b, 1963) & Streltsova (1972)) as well as the water table as a moving boundary (Boulton (1954a) & Dagan (1967)). In Neuman's delayed response to the aquifer, triggered by the liberation of the physical water table, he implemented the replacement of the phrase 'delayed yield' with 'delayed water table response'.

The Laplace equation of Boulton (1954a) and Dagan (1967) was replaced by the diffusion equation. Neuman (1972); it is denoted as:

$$\frac{\partial^2 s_D}{\partial r^2_D} + \frac{1}{r_D}\frac{\partial s_D}{\partial r_D} + K_D \frac{\partial^2 s_D}{\partial z^2_D} = \frac{\partial s_D}{\partial t_D} \qquad (20.10)$$

Neuman regarded the water table as a moving boundary, like that of Boulton (1954a) and Dagan (1967). He later linearized it and viewed the anisotropic aquifer as a three-dimensional symmetrical axis. Neumann (1974) then partially accounted for the penetration. Like Dagan (1967), Neuman was able to reconstruct all three parts of the detected unconfined time-drawdown curve to produce estimates of parameters (along with the ability to approximate K_z) similar to the models of delay yield by the use of confined storage in the governing Equation (20.10).

In comparison with the models of delay index, Neuman's solution achieved similar data adjustments (Neuman 1975, 1979) interrogated the physical nature of the delay index of Boulton. He carried out a regression between the solutions of Boulton (1954b) and Neuman (1972), ultimately resulting in a correlation.

$$\alpha = \frac{K_z}{S_y b}\left[3.063 - 0.567\log\left(\frac{K_D r^2}{b^2}\right)\right] \qquad (20.11)$$

Expressing α significantly reduces linearly with log r and thus is not an aquifer constant. If the logarithmic phrase in Equation (20.11) is ignored, the relation $\alpha = 3K_z/S_y b$ implemented by Streltsova (1972) is approximately recovered. After a comprehensive analysis of various strategies for determining specific yields, Neuman (1987) deduced that the reaction of the water table to pumping is much quicker than drainage in the unsaturated area just above it. Malama (2011) has lately implemented a new linearization of roughly including the impacts of overlooked second-order terms, which ultimately leads to the boundary condition of an alternative water table of:

$$S_y \frac{\partial s}{\partial t} = -K_z\left(\frac{\partial s}{\partial z} + \beta \frac{\partial^2 s}{\partial z^2}\right) z = h_0 \qquad (20.12)$$

where β represents a coefficient of linearization[L]. The variable β offers additional alteration of the physical shape of the intermediate part of the time-drawdown curve, leading to enhanced approximations of S_y

20.5   ANALYSIS OF THE DETERMINISTIC MODEL

Using the forward Euler to discretize the differential equation, we find that:

$$\int_0^t \frac{\partial s^*}{\partial \tau} e^{-\alpha(t-\tau)} d\tau = \int_0^{t_n} \frac{\partial s^*}{\partial \tau}\exp\left(-\alpha\left(t_n - \tau\right)\right)d\tau \qquad (20.13)$$

$$= \sum_{j=0}^{n} \int_{t_j}^{t_{j+1}} \frac{f^{j+1} - f^j}{\Delta t} \exp\left[-\alpha\left(t_n - \tau\right)\right] d\tau \tag{20.14}$$

$$= \sum_{j=0}^{n} \frac{f^{j+1} - f^j}{\Delta t} \int_{t_j}^{t_{j+1}} \exp\left[-\alpha\left(t_n - \tau\right)\right] d\tau \tag{20.15}$$

$$= \sum_{j=0}^{n} \frac{f^{j+1} - f^j}{\Delta t} \int_{t_n - t_{j+1}}^{t_n - t_j} \exp\left[-\alpha y\right] dy \tag{20.16}$$

$$= \sum_{j=0}^{n} \frac{f^{j+1} - f^j}{\Delta t} \left(-\frac{1}{\alpha}\right) \int_{t_n - t_{j+1}}^{t_n - t_j} -\alpha \exp\left[-\alpha y\right] dy \tag{20.17}$$

$$= \sum_{j=0}^{n} \frac{f^{j+1} - f^j}{\Delta t} \left[-\frac{1}{\alpha} \exp(-\alpha y)\right] \tag{20.18}$$

$$= \sum_{j=0}^{n} \frac{f^{j+1} - f^j}{\Delta t} \left\{\left(-\frac{1}{\alpha}\right) - \exp\left(t_n - t_j\right) - \exp\left(t_n - t_{j+1}\right)\right\} \tag{20.19}$$

$$= \sum_{j=0}^{n} \frac{f^{j+1} - f^j}{\Delta t} \left(-\frac{1}{\alpha} \exp\left(t_n - t_j\right) + \exp\left(t_n - t_{j+1}\right)\right) \tag{20.20}$$

$$= \sum_{j=0}^{n} \frac{f^{j+1} - f^j}{\Delta t} \left\{\left[\frac{\exp\left[\Delta t(n - j)\right] - \exp\left[\Delta t(n - j - 1)\right]}{\alpha}\right]\right\} \tag{20.21}$$

Applying the forward Euler method to discretize the unconfined aquifer equation, we can write that:

$$\frac{\partial^2 s}{\partial r^2} + \frac{1}{r}\frac{\partial s}{\partial r} = \frac{S}{T}\frac{\partial s}{\partial t} + \left\{\alpha S_y \int_0^t \frac{\partial s^*}{\partial \tau} e^{-\alpha(t - \tau)} d\tau\right\} \tag{20.22}$$

$$\frac{S\left(r_{i+1},\, t_{n+1}\right) - 2S\left(r_i,\, t_{n+1}\right) + S\left(r_{i-1},\, t_{n+1}\right)}{\left(\Delta r\right)^2} + \frac{1}{r_i}\left(\frac{S\left(r_{i+1},\, t_{n+1}\right) - S\left(r_i,\, t_{n+1}\right)}{\Delta r}\right)$$

$$= \frac{S}{T}\left(\frac{S\left(r_i,\, t_{n+1}\right) - S\left(r_i,\, t_n\right)}{\Delta t}\right) + \alpha S_y \sum_{j=0}^{n} \frac{S\left(r_i,\, t_{j+1}\right) - S\left(r_i,\, t_j\right)}{\Delta t} \tag{20.23}$$

We let $\qquad\qquad\qquad\qquad\qquad\qquad\qquad S\left(r_i, t_n\right) = S_i^n$

$$\frac{S_{i+1}^{n+1} - 2S_i^{n+1} + S_{i-1}^{n+1}}{(\Delta r)^2} + \frac{1}{r_i}\left(\frac{S_{i+1}^{n+1} - S_i^{n+1}}{\Delta r}\right) = \frac{S}{T}\left(\frac{S_i^{n+1} - S_i^n}{\Delta t}\right)$$

$$+ \alpha S_y \sum_{j=0}^{n} \frac{S_i^{j+1} - S_i^j}{\Delta t}\left\{\left[\frac{\exp\left[\Delta t(n-j) - \exp\left[\Delta t(n-j-1)\right]\right]}{\alpha}\right]\right\} \qquad (20.24)$$

$$\frac{S_{i+1}^{n+1} - 2S_i^{n+1} + S_{i-1}^{n+1}}{(\Delta r)^2} + \frac{1}{r_i}\left(\frac{S_{i+1}^{n+1} - S_i^{n+1}}{\Delta r}\right) = \frac{S}{T}\left(\frac{S_i^{n+1} - S_i^n}{\Delta t}\right) + \alpha S_y \sum_{j=0}^{n} \frac{S_i^{j+1} - S_i^j}{\Delta t}\delta_{j,n}^{\infty} \qquad (20.25)$$

$$\frac{S_{i+1}^{n+1} - 2S_i^{n+1} + S_{i-1}^{n+1}}{(\Delta r)^2} + \frac{1}{r_i}\left(\frac{S_{i+1}^{n+1} - S_i^{n+1}}{\Delta r}\right) =$$

$$\frac{S}{T}\left(\frac{S_i^{n+1} - S_i^n}{\Delta t}\right) + \alpha S_y \sum_{j=0}^{n-1} \frac{S_i^{j+1} - S_i^j}{\Delta t} + \sum_{j=0}^{n+1} \frac{S_i^{n+1} - S_i^n}{\Delta t}\frac{1-e^{-\Delta t}}{\alpha} \qquad (20.26)$$

This discretized equation will be used to perform many other analyses, such as the stability analysis for both models. That is, stochastic and deterministic.

20.5.1 Von Neumann Stability Analysis

Von Neumann's stability method, also referred to as Fourier's stability method, has been and still is in operation to date, generally being used to perform stability analysis on finite differential equations. The estimate is based on the decay of numerical errors in Fourier. This scheme was implemented at the Los Alamos Regional Lab by the British researchers Crank and Nicolson in 1947. The scheme became popular and many scientists began to use it.

The stability of a finite difference scheme is reached if the error at that particular point does not result in an increase in errors as the mathematical calculations continue. A rationally stable system remains constant when calculations are carried out. Therefore, if the errors decline and finally dampen, the numerical system is stable. In contrast, if the errors increase over time, the numerical scheme becomes unstable. Thus the Von Neumann method becomes a crucial tool in examining the stability of any linear or differential equation.

In essence, the numerical errors (rounded off because of the final precision of computers) should not be allowed to grow unboundedly, and the numerical solution itself should remain uniformly bounded. For time-dependent problems, stability guarantees that the numerical method produces a bounded solution whenever the solution of the exact differential equation is bounded. Stability, in general, can be difficult to investigate, especially when the equation under consideration is nonlinear.

In some cases, Von Neumann stabilization is appropriate and feasible for stability analysis in the perception of Lax–Richtmyer (used in the Lax equivalence theory): the finite differential model as well as the partial differential equation (PDE) are linear. The PDE is fairly constant with continuous limit conditions but has only two independent parameters. Stability of Von Neumann is essential in a much wider range of situations. It is mainly used to replace a more complicated and detailed stability analysis. This is done so that one achieves a good understanding of the restrictions on the steps used in the scheme because it is relatively straightforward.

The Von Neumann mechanism is based solely on the error breakdown into the series of Fourier. Consider the following one-dimensional heat equation to emphasize the method:

$$\frac{\partial u}{\partial t} = \alpha \frac{\partial^2 u}{\partial x^2}$$

(20.27)

The discretization of the equation is as follows:

$$u_j^{n+1} = u_j^n + r\left(u_{j+1}^n - 2u_j^n + u_{j-1}^n\right)$$

(20.28)

where:

$$r = \frac{\alpha \Delta t}{\Delta x^2}$$

(20.29)

The solution u_j^n of the linear equation simulates the PDE analytical solution $u(x,t)$ on the national grid.

Define the round-off error ϵ_j^n as

$$\epsilon_j^n = N_j^n - u_j^n$$

(20.30)

The u_j^n is the ultimate solution of the discretized equation, which would be calculated in the lack of a round-off error, and N_j^n is the alternative of the numerical solution acquired in finite accuracy arithmetic. Because the exact solution should comply accurately with the discrete equation, the error ϵ_j^n must also comply with the discrete equation. Here, however, we presumed that the N_j^n complies with the equation. Thus:

$$\epsilon_j^{n+1} = \epsilon_j^n + r\left(\epsilon_{j+1}^n - 2\epsilon_j^n + \epsilon_{j-1}^n\right)$$

(20.31)

The above equation represents an error recurrence relationship. For linear partial differential equations with a sporadic boundary condition, the spatial variability of error in a Fourier series can be broadened in the L interval,

$$\epsilon(x) = \sum_{m=1}^{M} A_m e^{ik_m x}$$

(20.32)

Here, the wavenumber $k_m = \frac{\pi m}{L}$ with $m = 1, 2, 3, \ldots, M$ and $M = \frac{L}{\Delta x}$. The time dependency of the error is featured in the presumption that the error amplitude A_m is a time parameter. Seeing that the error tends to exponentially increase or decrease over time, it is logical to assume that the amplitude also fluctuates exponentially over time, where the wavenumber with $m = 1, 2, 3, \ldots, M$ and $M = \frac{L}{\Delta x}$. The time dependence of the error is included by assuming that the amplitude of error A_m is a function of time. Since the error tends to grow or decay exponentially with time, it is reasonable to assume that the amplitude also varies exponentially with time; hence

$$\epsilon(x,t) = \sum_{m=1}^{M} e^{at} e^{ik_m x}$$

(20.33)

where a is a constant.

Because the error difference equation is linear (the conduct of every phrase in the series is the same as the series itself), the error growth of a fairly typical phrase is sufficient:

$$\epsilon_m(x,t) = e^{at}e^{ik_m x} \tag{20.34}$$

The characteristics of stability can be examined by using this form for the error without loss in general.

$$\varepsilon_j^{\ n} = \exp\left[at\right]\exp\left[ik_m x\right] \tag{20.35}$$

$$\varepsilon_j^{\ n+1} = \exp\left[at\right]\exp\left[ik_m\left(x+\Delta x\right)\right] \tag{20.36}$$

$$\varepsilon_{j+1}^{\ n+1} = \exp\left[a\left(t+\Delta t\right)\right]\exp\left[ik_m\left(x+\Delta x\right)\right] \tag{20.37}$$

$$\varepsilon_{j-1}^{\ n-1} = \exp\left[a\left(t-\Delta t\right)\right]\exp\left[ik_m\left(x-\Delta x\right)\right] \tag{20.38}$$

to yield (after simplification)

$$e^{a\Delta t} = 1 + r\left(e^{ik_m \Delta x} + e^{-ik_m \Delta x} - 2\right) \tag{20.39}$$

Using the identities

$$\sin\left(\frac{k_m \Delta x}{2}\right) = \frac{e^{ik_m \Delta x/2} - e^{-ik_m \Delta x/2}}{2i} \rightarrow \sin^2\left(\frac{k_m \Delta x}{2}\right) = -\frac{e^{ik_m \Delta x} + e^{-ik_m \Delta x} - 2}{4} \tag{20.40}$$

Equation (20.39) may be written as

$$e^{a\Delta t} = 1 - 4r\sin^2\left(\frac{k_m \Delta x}{2}\right) \tag{20.41}$$

Define the amplification factor

$$G = \frac{\epsilon_j^{\ n+1}}{\epsilon_j^{\ n}} \tag{20.42}$$

The necessary and sufficient condition for the error to remain bounded is that $|G| \leq 1$.

Applying the Von Neumann stability method to the unconfined aquifer discretized equation, we find that:

$$\frac{S_{i+1}^{\ n+1} - 2S_i^{\ n+1} + S_{i-1}^{\ n+1}}{(\Delta r)^2} + \frac{1}{r_i}\left(\frac{S_{i+1}^{\ n+1} - S_i^{\ n+1}}{\Delta r}\right) = \frac{S}{T}\left(\frac{S_i^{\ n+1} - S_i^{\ n}}{\Delta t}\right) + \alpha S_y \sum_{j=0}^{n-1} \frac{S_i^{\ j+1} - S_i^{\ j}}{\Delta t}\delta^\alpha_{\ i,j} \tag{20.43}$$

Expanding

$$\frac{S_{i+1}^{\ n+1} - 2S_i^{\ n+1} + S_{i-1}^{\ n+1}}{(\Delta r)^2} + \frac{1}{r_i}\left(\frac{S_{i+1}^{\ n+1} - S_i^{\ n+1}}{\Delta r}\right) =$$

$$\frac{S}{T}\left(\frac{S_i^{\ n+1} - S_i^{\ n}}{\Delta t}\right) + S_y \sum_{j=0}^{n-1} \frac{S_i^{\ j+1} - S_i^{\ j}}{\Delta t}\delta^\alpha_{\ i,j} + \frac{S_i^{\ n+1} - S_i^{\ n}}{\Delta t}\delta^\alpha_{\ i,n}S_y\alpha \tag{20.44}$$

Rearranging

$$\frac{S_{i+1}^{n+1} - 2S_i^{n+1} + S_{i-1}^{n+1}}{(\Delta r)^2} + \frac{1}{r_i}\left(\frac{S_{i+1}^{n+1} - S_i^{n+1}}{\Delta r}\right)$$

$$= \frac{S_i^{n+1} - S_i^n}{\Delta t}\left(\frac{S}{T} + S_y\alpha, \delta^\alpha_{i,j}\right) + S_y\sum_{j=0}^{n-1}\frac{S_i^{j+1} - S_i^j}{\Delta t}\delta^\alpha_{i,j} \qquad (20.45)$$

Factorizing

$$S_i^{n+1}\left[\frac{-2}{\Delta r^2} + \frac{1}{r_i}\left(\frac{1}{\Delta r}\right) - \frac{S}{T\Delta t} - S_y\alpha\delta^\alpha_{i,n}\right] = -S_i^n\left(\frac{1}{\Delta t}\frac{S}{T} + S_y\alpha\delta^\alpha_{i,n}\right)$$

$$-S_{i+1}^{n+1}\left[\frac{1}{\Delta r^2} - \frac{1}{r_i}\left(\frac{1}{\Delta r}\right)\right] - S_{i-1}^{n+1}\left(\frac{1}{\Delta r^2}\right) + S_y\sum_{j=0}^{n-1}\frac{S_i^{j+1} - S_i^j}{\Delta t}\delta^\alpha_{i,j} \qquad (20.46)$$

Simplifying

$$\alpha_1\left(S_l^{n+1}\right) = -\alpha_2\left(S_i^n\right) - \alpha_3\left(S_{i+1}^{n+1}\right) - \alpha_4\left(S_{i-1}^{n+1}\right) + S_y\sum_{j=0}^{n-1}\frac{S_i^{j+1} - S_i^j}{\Delta t}\delta^\alpha_{i,j} \qquad (20.47)$$

Note that:

$$\alpha_1 = \frac{1}{r_i}\left(\frac{1}{\Delta r}\right) - \frac{S}{T\Delta t} - \propto S_y\delta_{i,n}^\infty \qquad (20.48)$$

$$\alpha_2 = \frac{1}{\Delta t}\frac{S}{T} + \propto S_y\delta_{i,n}^\infty \qquad (20.49)$$

$$\alpha_3 = \left(\frac{1}{r^2} - \frac{1}{r_i}\left(\frac{1}{\Delta r}\right)\right) \qquad (20.50)$$

$$\alpha_4 = \left(\frac{1}{\Delta r^2}\right) \qquad (20.51)$$

The Von Neumann stability method can be expressed in the form:

$$\varepsilon_{m(x,t)} = \exp\left[at\right]\exp\left[ik_m\right] \qquad (20.52)$$

Such that:

$$\varepsilon_j^n = \exp\left[at\right]\exp\left[ik_m x\right] \qquad (20.53)$$

$$\varepsilon_j^{n+1} = \exp\left[at\right]\exp\left[ik_m\left(x + \Delta x\right)\right] \qquad (20.54)$$

$$\varepsilon_{j+1}{}^{n+1} = \exp\left[a\left(t+\Delta t\right)\right]\exp\left[ik_m\left(x+\Delta x\right)\right] \tag{20.55}$$

$$\varepsilon_{j-1}{}^{n-1} = \exp\left[a\left(t-\Delta t\right)\right]\exp\left[ik_m\left(x-\Delta x\right)\right] \tag{20.56}$$

From the above discretized Equation (20.43), we apply the Von Neumann stability method to be expressed as:

$$\begin{aligned}
\propto_1 \exp\left[a\left(t+\Delta t\right)\right]\exp\left[ik_m x\right] &= -\propto_2 \exp\left[at\right]\exp\left[ik_m x\right] - \propto_3 \exp\left[a\left(t+\Delta t\right)\right] \\
&\quad \exp\left[ik_m\left(x+\Delta x\right)\right] - \propto_4 \exp\left[a\left(t+\Delta t\right)\right]\exp\left[ik_m\left(x-\Delta x\right)\right] \\
&\quad + S_y \sum_{j=0}^{n-1} \left\{ \frac{\exp\left[a\left(t+\Delta t\right)\right]\exp\left[ik_m x\right] - \exp\left[at\right]\exp\left[ik_m x\right]}{\Delta t} \delta^{\alpha}{}_{i,j} \right\}
\end{aligned} \tag{20.57}$$

Taking out $\exp\left[at\right]\exp\left[ik_m x\right]$ as a common factor, we have:

$$\begin{aligned}
\propto_1 \exp\left[a\Delta t\right] &= -\propto_2 - \propto_3 \exp\left[a\Delta t\right]\exp\left[ik_m \Delta x\right] \\
&\quad + \propto_4 \exp\left[a\Delta t\right]\exp\left[-ik_m \Delta x\right] + S_y \sum_{j=0}^{n-1} \left\{ \frac{\exp\left[\left(a\Delta t\right)\right]-1}{\Delta t} \delta^{\alpha}{}_{i,j} \right\}
\end{aligned} \tag{20.58}$$

Taking out $\exp\left[\left(a\Delta t\right)\right]$ as a common factor, we have:

$$\begin{aligned}
&\exp\left[\left(a\Delta t\right)\right]\left\{ \propto_1 + \propto_3 \exp\left[ik_m \Delta x\right] - \propto_4 \exp\left[-ik_m \Delta x\right] - \frac{\propto S_y}{\Delta t}\left[\frac{\exp\left[\left(\Delta tn\right)\right]-1}{\Delta t}\right] \right\} = \\
&-\propto_2 - \frac{\propto S_y}{\Delta t}\frac{\exp\left[\left(\Delta tn\right)\right]-1}{\Delta t} = -\propto_2 - \frac{\propto S_y}{\Delta t}\frac{\exp\left[\left(\Delta tn\right)\right]-1}{\Delta t}
\end{aligned} \tag{20.59}$$

From Equations (20.53) and (20.54), we find that:

$$\frac{\varepsilon_i{}^{n+1}}{\varepsilon_i{}^{n}} = \exp\left[\left(a\Delta t\right)\right] < 1 \tag{20.60}$$

Expressing in terms of Equation (20.60), we have:

$$\exp\left[\left(a\Delta t\right)\right] = \frac{-\propto_2 - \dfrac{\propto S_y}{\Delta t}\dfrac{\exp\left[\left(\Delta tn\right)\right]-1}{\Delta t}}{\left\{ \propto_1 + \propto_3 \exp\left[ik_m \Delta x\right] - \propto_4 \exp\left[-ik_m \Delta x\right] - \dfrac{\propto S_y}{\Delta t}\left[\dfrac{\exp\left[\left(\Delta tn\right)\right]-1}{\Delta t}\right] \right\}} \tag{20.61}$$

If

$$\exp\left[ik_m \Delta x\right] = Cos\left[k_m \Delta x\right] + iSin\left[k_m \Delta x\right] \tag{20.62}$$

$$\exp\left[-ik_m\Delta x\right] = Cos\left[k_m\Delta x\right] - iSin\left[k_m\Delta x\right] \tag{20.63}$$

Then,

$$\exp\left[(a\Delta t)\right] = \cfrac{-\propto_2 - \cfrac{\propto S_y \exp\left[(\Delta tn)\right] - 1}{\Delta t}}{\left[\begin{aligned} &\propto_1 + \propto_3 \left(Cos\left[k_m\Delta x\right] + iSin\left[k_m\Delta x\right]\right) \\ &-\propto_4 \left(Cos\left[k_m\Delta x\right] - iSin\left[k_m\Delta x\right]\right) - \left[\cfrac{\propto S_y \exp\left[(\Delta tn)\right] - 1}{\Delta t}\right] \end{aligned}\right]} \tag{20.64}$$

If we express the equation in terms of $\exp[(a\Delta t)]$, we find that:

$$\exp\left[(a\Delta t)\right] = \cfrac{-\propto_2 - \cfrac{\propto S_y \exp\left[(\Delta tn)\right] - 1}{\Delta t}}{\left\{\begin{aligned} &\propto_1 + \left\{Cos\left[k_m\Delta x\right]\left(\propto_3 - \propto_4\right)\right\} \\ &-\left[\cfrac{\propto S_y \exp\left[(\Delta tn)\right] - 1}{\Delta t}\right] + iSin\left[k_m\Delta x\right]\left(\propto_3 - \propto_4\right) \end{aligned}\right\}} \tag{20.65}$$

Since $exp[(a\Delta t)] < 1$, then

$$\left|\exp\left[(a\Delta t)\right]\right| < 1, \cfrac{-\propto_2 - \cfrac{\propto S_y \exp\left[(\Delta tn)\right] - 1}{\Delta t}}{\propto_1 + \left\{Cos\left[k_m\Delta x\right]\left(\propto_3 - \propto_4\right)\right\} - \left[\cfrac{\propto S_y \exp\left[(\Delta tn)\right] - 1}{\Delta t}\right] + iSin\left[k_m\Delta x\right]\left(\propto_3 - \propto_4\right)} < 1 \tag{20.66}$$

Similarly,

$$\cfrac{\left|\begin{aligned} &\propto_1 + \left\{Cos\left[k_m\Delta x\right]\left(\propto_3 - \propto_4\right)\right\} \\ &-\left[\cfrac{\propto S_y \exp\left[(\Delta tn)\right] - 1}{\Delta t}\right] + iSin\left[k_m\Delta x\right]\left(\propto_3 - \propto_4\right) \end{aligned}\right|}{\sqrt{\left(\propto_1 + \left\{Cos\left[k_m\Delta x\right]\left(\propto_3 - \propto_4\right)\right\} - \left[\cfrac{\propto S_y \exp\left[(\Delta tn)\right] - 1}{\Delta t}\right]\right)^2 + \left(iSin\left[k_m\Delta x\right]\left(\propto_3 - \propto_4\right)\right)^2}} = \tag{20.67}$$

Rearranging

$$\frac{\left| -\alpha_2 - \dfrac{\alpha S_y \exp\left[(\Delta tn)\right] - 1}{\Delta t} \right|}{\sqrt{\left(\alpha_1 + \left\{ Cos\left[k_m \Delta x\right]\left(\alpha_3 - \alpha_4\right)\right\} - \left[\dfrac{\alpha S_y \exp\left[(\Delta tn)\right] - 1}{\Delta t}\right] \right)^2 + \left(i Sin\left[k_m \Delta x\right]\left(\alpha_3 - \alpha_4\right)\right)^2}} < 1 \qquad (20.68)$$

Simplifying

$$\left| -\alpha_2 - \frac{\alpha S_y \exp\left[(\Delta tn)\right] - 1}{\Delta t} \right| < \sqrt{\left(\alpha_1 + \left\{ Cos\left[k_m \Delta x\right]\left(\alpha_3 - \alpha_4\right)\right\} - \left[\dfrac{\alpha S_y \exp\left[(\Delta tn)\right] - 1}{\Delta t}\right] \right)^2 + \left(i Sin\left[k_m \Delta x\right]\left(\alpha_3 - \alpha_4\right)\right)^2} \qquad (20.69)$$

From the stability analysis performed using the Von Neumann method, it can be concluded that the delayed yield unconfined response equation is stable, provided that the conditions are met.

20.6   ANALYSIS OF THE STOCHASTIC MODEL

Despite comprehensive research in stochastic sub-surface hydrology since around 1990, the capacity for analyzing and modeling heterogeneous groundwater structures continues to be extremely limited. Though the 'stochastic uprising' has generated a vast number of mathematical publications and influenced our thinking significantly about variability, it has had very little impact on realistic modeling. In practice today, the prevailing groundwater simulation paradigm is still predominantly deterministic based on the classical conjectures established generations ago.

A crucial question emerges: if heterogeneity is so essential, why is it that no one in reality uses theories of stochastic modeling? A number of recent reviews carried out research on this issue in detail and came up with the following reasons.

Stochastic coding is likely to be completely inconsistent with today's conventional new technologies. This is because the standard raw data are sometimes too restricted to supply the necessary geostatistic variables for stochastic modeling. Relatively new measuring technologies, new data sources with a much higher resolution and absolutely usable data conversion approaches are required to describe aquifer heterogeneity. However, it is also important to note that it is very difficult to apply stochastic analytical ideas to so many problems of practical complexity. These arguments are based on several restrictive requirements that are useful in practice. The presumptions of ergodicity, stationarity, Gaussian distribution, average uniform flow, and minor disturbance, should be relaxed significantly.

Stochastic conjectures are mathematical and complicated, and even for experts who have developed them are difficult to implement. A general, integrated computer platform is urgently needed before stochastic modeling can be popularized. Because of these crucial evaluations, we address a number of important conceptual, computational and implementation problems in groundwater modeling in this project. This study represents our attempt to reduce the gap between them.

Stochastic ideas and programs are, fundamentally, the practical modeling of groundwater. Stochastic theories of underground flow and transport have also had a major impact on our thinking about uncertainty and heterogeneity. However, the way predictions are obtained and disclosed in practical groundwater modeling studies did not have much impact.

It becomes abundantly clear that stochastic modeling must be made much more general and flexible if it is to become a feasible critical tool. In particular, a stochastic model should be able to integrate immediately site-specific aquifer structures before they can be applied regularly in practice. It must enable the modeling of flexible zones, layers and general trends, as most real-world aquifers not only exhibit 'fairly random' heterogeneity but also purposeful 'structural' heterogeneity, and the statistics portraying aquifer heterogeneity can vary from area to area in reaction to systematic adjustments in the dispersion of aquifer materials.

A vast amount of numerical tactics could also be used to analyze statistically uniform flows and transport, especially inhomogeneous statistical aquifers. These include, for example, Monte Carlo strategies and disturbance methodologies, such as moment equation techniques and second-moment first-order techniques based primarily on Taylor's growth.

All these techniques are, however, computer-intensive when implemented to flow, and transport related problems of realistic size, primarily because they still involve the resolution of huge numbers of partial differential equations on some very fine statistical discretizations in order to predict the effect of heterogeneous small dynamics.

20.6.1 LOG-NORMAL DISTRIBUTION

In the theory of probability, a log-normal distribution is a continuous statistical probability distribution of a randomly chosen parameter normally distributed by the logarithm. Therefore, if the seemingly random parameter X is dispersed log-normally, $Y = ln(X)$ has a normal distribution. The exponential function of Y, $X = \exp(Y)$ also has a log-normal distribution if Y has a normal distribution. A randomly chosen parameter that is dispersed log-normally only takes positive real values.

A log-normal process refers to a probabilistic notion of the multiplier product of many entirely independent, positive random variables. This is enforced by taking into account the central mathematics of the log domain. In addition, a log-normal distribution represents the maximum probability of entropy for a randomly chosen variant X specifying the mean and variance of $ln(X)$.

20.6.2 NOTATION

Given a log-normally distributed random variable X and two parameters, μ and σ, that are, respectively, the mean and standard deviation of the variable's natural logarithm, then the logarithm of X is normally distributed, and we can write X as

$$X = e^{\mu + \sigma Z} \tag{20.70}$$

with Z a standard normal variable.

This relationship is true regardless of the base of the logarithmic or exponential function. If $\log_a(Y)$ is normally distributed, then so is $\log_b(Y)$, for any two positive numbers a, $b \neq 1$. Likewise, if e^x is log-normally distributed, then so is a^x, where a is a positive number $\neq 1$. The two parameters μ and σ are not location and scale parameters for a log-normal distributed random variable X, but they are respectively location and scale parameters for the normally distributed logarithm $ln(X)$. The quantity e^μ is a scale parameter for the family of log-normal distributions.

In contrast, the mean and variance of the non-logarithmized sample values are respectively denoted m and v in this section. The two sets of parameters can be related as

$$\mu = \ln\left(\frac{m}{\sqrt{1 + \dfrac{v}{m^2}}} \right) \tag{20.71}$$

$$\sigma = \ln\left(1 + \frac{v}{m^2}\right) \tag{20.72}$$

20.6.3 Probability Density Function

A positive random variable X is distributed log-normally if the X logarithm is dispersed normally,

$$\ln(X) \sim \mathcal{N}\left(\mu, \sigma^2\right) \tag{20.73}$$

Let the two functions, Φ and φ, be the cumulative probability distribution function and the density function of the $N(0,1)$ distribution, respectively.

$$f_X(x) = \frac{d}{dx} \Pr(X \le x) = \frac{d}{dx} \Pr(\ln X \le \ln x) \tag{20.74}$$

$$= \frac{d}{dx} \Phi\left(\frac{\ln x - \mu}{\sigma}\right) \tag{20.75}$$

$$= \varphi\left(\frac{\ln x - \mu}{\sigma}\right) \frac{d}{dx}\left(\frac{\ln x - \mu}{\sigma}\right) \tag{20.76}$$

$$= \varphi\left(\frac{\ln x - \mu}{\sigma}\right) \frac{d}{dx}\left(\frac{1}{\sigma x}\right) \tag{20.77}$$

$$= \frac{1}{x} \cdot \frac{1}{\sigma\sqrt{2\pi}} \exp\left(-\frac{(\ln x - \mu)^2}{2\sigma^2}\right). \tag{20.78}$$

20.6.4 Cumulative Distributive Function

The cumulative distribution function is:

$$f_X(x) = \Phi\left(\frac{\ln x - \mu}{\sigma}\right) \tag{20.79}$$

where Φ depicts the cumulative distribution function of the standard normal distribution (i.e. $N(0,1)$). This can also be denoted as follows:

$$\frac{1}{2}\left[1 + erf\left(\frac{\ln x - \mu}{\sigma\sqrt{2}}\right)\right] = \frac{1}{2} erfc\left(\frac{\ln x - \mu}{\sigma\sqrt{2}}\right) \tag{20.80}$$

The following figures illustrate semi log-normal density functions and a cumulative distribution function of the log-normal distribution.

20.6.5 The Stochastic Model

By varying aquifer parameters, we develop the stochastic model. From our defined unconfined aquifer equation from Chapter 1, i.e., Equation (20.6), all the parameters including storativity, transmissivity, specific yield and the delay index are varied, through which a mean will be calculated with the aim of developing a stochastic model, through which analysis will be made from.

$$\frac{\partial^2 s^*}{\partial r^2}+\frac{1}{r}\frac{\partial s^*}{\partial r}=\left[\frac{S}{T}\frac{\partial s^*}{\partial t}\right]+\left\{\alpha S_y\int_0^t\frac{\partial s^*}{\partial \tau}e^{-\alpha(t-\tau)}d\tau\right\} \tag{20.81}$$

Introducing the stochastic component

$$\frac{\partial^2 s^*}{\partial r^2}+\frac{1}{r}\frac{\partial s^*}{\partial r}-\frac{\bar{S}+\lambda\dfrac{1}{x}.\dfrac{1}{\sigma\sqrt{2\pi}}\exp\left(\dfrac{-(lnx-\mu)^2}{2\sigma^2}\right)}{\bar{T}+\lambda\dfrac{1}{x}.\dfrac{1}{\sigma\sqrt{2\pi}}\exp\left(\dfrac{-(lnx-\mu)^2}{2\sigma^2}\right)}\frac{\partial s}{\partial t}$$

$$+\alpha\bar{S}_y\lambda\frac{1}{x}.\frac{1}{\sigma\sqrt{2\pi}}\exp\left(\frac{-(lnx-\mu)^2}{2\sigma^2}\right)\int_0^t\frac{\partial s}{\partial \tau}\exp\left[-\alpha(t-\tau)\right]d_\tau \tag{20.82}$$

After discretization using the forward Euler method, we have:

$$\frac{S_{i+1}^{n+1}-2S_i^{n+1}+S_{i-n}^{n+1}}{(\Delta r)^2}+\frac{1}{r_i}\left(\frac{S_{i+1}^{n+1}-S_i^{n+1}}{\Delta r}\right)=\frac{\bar{S}+\lambda\dfrac{1}{r_i}.\dfrac{1}{\sigma\sqrt{2\pi}}\exp\left(\dfrac{-(lnr_i-\mu)^2}{2\sigma^2}\right)}{\bar{T}+\lambda\dfrac{1}{x}.\dfrac{1}{\sigma\sqrt{2\pi}}\exp\left(\dfrac{-(lnx-\mu)^2}{2\sigma^2}\right)}\left(\frac{S_i^{n+1}-S_i^n}{\Delta t}\right)$$

$$+\alpha\bar{S}_y+\lambda\frac{1}{r_i}.\frac{1}{\sigma\sqrt{2\pi}}\exp\left(\frac{-(lnr_i-\mu)^2}{2\sigma^2}\right)\sum_{j=0}^n\frac{S_i^{j+1}-S_i^j}{\Delta t}\frac{1+e^{-\Delta t}}{\alpha}$$

$$\left\{\left[\frac{\exp\left[\Delta t(n-j)-\exp\left[\Delta t(n-j-1)\right]\right]}{\alpha}\right]\right\} \tag{20.83}$$

$$\frac{S_{i+1}^{n+1}-2S_i^{n+1}+S_{i-n}^{n+1}}{(\Delta r)^2}+\frac{1}{r_i}\left(\frac{S_{i+1}^{n+1}-S_i^{n+1}}{\Delta r}\right)=\frac{\bar{S}+\lambda\dfrac{1}{r_i}.\dfrac{1}{\sigma\sqrt{2\pi}}\exp\left(\dfrac{-(lnr_i-\mu)^2}{2\sigma^2}\right)}{\bar{T}+\lambda\dfrac{1}{x}.\dfrac{1}{\sigma\sqrt{2\pi}}\exp\left(\dfrac{-(lnx-\mu)^2}{2\sigma^2}\right)}\left(\frac{S_i^{n+1}-S_i^n}{\Delta t}\right)$$

$$+\alpha\bar{S}_y+\lambda\frac{1}{r_i}.\frac{1}{\sigma\sqrt{2\pi}}\exp\left(\frac{-(lnr_i-\mu)^2}{2\sigma^2}\right)\sum_{j=0}^n\frac{S_i^{j+1}-S_i^j}{\Delta t}\frac{1+e^{-\Delta t}}{\alpha}\delta_{j,n}^\alpha \tag{20.84}$$

where

$$\delta_{j,n}^\alpha=\frac{1+e^{-\Delta t}}{\alpha}\left\{\left[\frac{\exp\left[\Delta t(n-j)-\exp\left[\Delta t(n-j-1)\right]\right]}{\alpha}\right]\right\} \tag{20.85}$$

This completes the stochastic discretization.

20.6.6 VON NEUMANN STABILITY ANALYSIS

$$\frac{S_{i+1}^{n+1} - 2S_i^{n+1} + S_{i-n}^{n+1}}{(\Delta r)^2} + \frac{1}{r_i}\left(\frac{S_{i+1}^{n+1} - S_i^{n+1}}{\Delta r}\right) = \frac{\bar{S} + \lambda\frac{1}{r_i}.\frac{1}{\sigma\sqrt{2\pi}}\exp\left(\frac{-(lnr_i - \mu)^2}{2\sigma^2}\right)}{\bar{T} + \lambda\frac{1}{x}.\frac{1}{\sigma\sqrt{2\pi}}\exp\left(\frac{-(lnx - \mu)^2}{2\sigma^2}\right)}$$

$$\left(\frac{S_i^{n+1} - S_i^n}{\Delta t}\right) + \alpha\bar{S}_y + \lambda\frac{1}{r_i}.\frac{1}{\sigma\sqrt{2\pi}}\exp\left(\frac{-(lnr_i - \mu)^2}{2\sigma^2}\right)\sum_{j=0}^{n}\frac{S_i^{j+1} - S_i^j}{\Delta t}\frac{1 + e^{-\Delta t}}{\alpha}$$

$$\left\{\left[\frac{\exp\left[\Delta t(n-j) - \exp\left[\Delta t(n-j-1)\right]\right]}{\alpha}\right]\right\} \tag{20.86}$$

We let:

$$\bar{S} + \lambda\frac{1}{r_i}.\frac{1}{\sigma\sqrt{2\pi}}\exp\left(\frac{-(lnr_i - \mu)^2}{2\sigma^2}\right) = S_i \tag{20.87}$$

$$\bar{T} + \lambda\frac{1}{r_i}.\frac{1}{\sigma\sqrt{2\pi}}\exp\left(\frac{-(lnr_i - \mu)^2}{2\sigma^2}\right) = T_i \tag{20.88}$$

$$\bar{S}_y + \lambda\frac{1}{r_i}.\frac{1}{\sigma\sqrt{2\pi}}\exp\left(\frac{-(lnr_i - \mu)^2}{2\sigma^2}\right) = S_{yi} \tag{20.89}$$

Simplifying

$$\frac{S_{i+1}^{n+1} - 2S_i^{n+1} + S_{i-n}^{n+1}}{(\Delta r)^2} + \frac{1}{r_i}\left(\frac{S_{i+1}^{n+1} - S_i^{n+1}}{\Delta r}\right) = \frac{S_i}{T_i}\left(\frac{S_i^{n+1} - S_i^n}{\Delta t}\right) + \alpha S_y\sum_{j=0}^{n=1}\frac{S_i^{j+1} - S_i^j}{\Delta t}\delta_{i,j}^\alpha$$

$$+ \frac{S_i^{n+1} - S_i^n}{\Delta t}\delta_{n,j}^\alpha S_y\alpha \tag{20.90}$$

Grouping

$$\frac{S_{i+1}^{n+1} - 2S_i^{n+1} + S_{i-n}^{n+1}}{(\Delta r)^2} + \frac{1}{r_i}\left(\frac{S_{i+1}^{n+1} - S_i^{n+1}}{\Delta r}\right) = \left(\frac{S_i^{n+1} - S_i^n}{\Delta t}\right)\left(\frac{S_i}{T_i} + S_{yi}\alpha\delta_{n,j}^\alpha\right)$$

$$+ \alpha S_y\sum_{j=0}^{n-1}\frac{S_i^{j+1} - S_i^j}{\Delta t}\delta_{i,j}^\alpha \tag{20.91}$$

Factorizing

$$S_i^{n+1}\left[\frac{-2}{\Delta r^2}+\frac{1}{r_i}\left(\frac{1}{\Delta r}\right)-\frac{S}{T\Delta t}-S_y\alpha\delta^\alpha{}_{i,n}\right]=-S_i^n\left(\frac{1}{\Delta t}\frac{S}{T}+S_y\alpha\delta^\alpha{}_{i,n}\right)$$
$$-S_{i+1}{}^{n+1}\left[\frac{1}{\Delta r^2}-\frac{1}{r_i}\left(\frac{1}{\Delta r}\right)\right]-S_{i-1}{}^{n+1}\left(\frac{1}{\Delta r^2}\right)+\alpha S_y\sum_{j=0}^{n-1}\frac{S_i^{j+1}-S_i^{j}}{\Delta t}\delta^\alpha{}_{i,j}\qquad(20.92)$$

Applying the Von Neumann stability notation

$$\delta_{n+1}e^{ikmx}\left(\alpha_1\right)=\delta_n e^{ikmx}\left(\alpha_2\right)-\delta_{n+1}e^{ikmx}e^{ikm\Delta x}\left(\alpha_3\right)-\delta_{n+1}e^{ikmx}e^{-ikm\Delta x}\left(\alpha_4\right)$$
$$+\alpha S_y\sum_{j=0}^{n-1}\delta_{n-j}\frac{e^{ikmx}e^{ikm\Delta x}-e^{ikmx}}{\Delta t}\delta_{n,j}\qquad(20.93)$$

Note that:

$$\alpha_1=\left[\frac{-2}{\Delta r^2}+\frac{1}{r_i}\left(\frac{1}{\Delta r}\right)-\frac{S}{T\Delta t}-S_y\alpha\delta^\alpha{}_{i,n}\right]\qquad(20.94)$$

$$\alpha_2=\left(\frac{1}{\Delta t}\frac{S}{T}+S_y\alpha\delta^\alpha{}_{i,n}\right)\qquad(20.95)$$

$$\alpha_3=\left[\frac{1}{\Delta r^2}-\frac{1}{r_i}\left(\frac{1}{\Delta r}\right)\right]\qquad(20.96)$$

$$\alpha_4=\left(\frac{1}{\Delta r^2}\right)\qquad(20.97)$$

$$\delta_{n+1}e^{ikmx}\left(\alpha_1\right)=\delta_n e^{ikmx}\left(\alpha_2\right)-\delta_{n+1}e^{ikmx}e^{ikm\Delta x}\left(\alpha_3\right)-\delta_{n+1}e^{ikmx}e^{-ikm\Delta x}\left(\alpha_4\right)$$
$$+\alpha S_y\sum_{j=0}^{n-1}\delta_{n-j}\frac{e^{ikmx}e^{ikm\Delta x}-e^{ikmx}}{\Delta t}\delta_{n,j}\qquad(20.98)$$

Dividing by e^{ikmx}, we get:

$$\delta_{n+1}\left(\alpha_1\right)=\delta_n\left(\alpha_2\right)-\delta_{n+1}e^{ikm\Delta x}\left(\alpha_3\right)-\delta_{n+1}e^{-ikm\Delta x}\left(\alpha_4\right)+\alpha S_y\sum_{j=0}^{n-1}\delta_{n-j}\left(\frac{e^{ikm\Delta x}}{\Delta t}\right)\qquad(20.99)$$

Grouping

$$\delta_{n+1}\left(\alpha_1-\alpha_3 e^{ikm\Delta x}-e^{-ikm\Delta x}\alpha_4\right)=\delta_n\left(\alpha_2\right)+\alpha S_y\sum_{j=0}^{n-1}\delta_{n-j}\delta_{n,j}\qquad(20.100)$$

If

$$\exp\left[ik_m\Delta x\right] = Cos\left[k_m\Delta x\right] + iSin\left[k_m\Delta x\right] \tag{20.101}$$

$$\exp\left[-ik_m\Delta x\right] = Cos\left[k_m\Delta x\right] - iSin\left[k_m\Delta x\right] \tag{20.102}$$

Then

$$\delta_{n+1}\left(\alpha_1 - \alpha_3\left(Cos\left(ikm\Delta x\right) + iSin\left(ikm\Delta x\right)\right) - \alpha_4\left(Cos\left(ikm\Delta x\right) + iSin\left(ikm\Delta x\right)\right)\right) =$$
$$\delta_n\left(\alpha_2\right) + \alpha S_y \sum_{j=0}^{n-1}\delta_{n-j}\delta_{n,j} \tag{20.103}$$

We prove the stability via induction technique when $n = 0$

$$\delta_1\left\{\alpha_1 - Cos\left(ikm\Delta x\right)\left(\alpha_4 + \alpha_3\right) + iSin\left(ikm\Delta x\right)\left(\alpha_4 - \alpha_3\right)\right\} = \delta_0\alpha_2 \tag{20.104}$$

Resulting

$$\left|\frac{\delta_1}{\delta_0}\right| < 1 \rightarrow \frac{\alpha_2}{\alpha_1 - Cos\left(ikm\Delta x\right)\left(\alpha_4 + \alpha_3\right) + iSin\left(ikm\Delta x\right)\left(\alpha_4 - \alpha_3\right)} < 1 \tag{20.105}$$

Which is time in general, so we assume that for all $n \geq 0$

$$\left|\frac{\delta_n}{\delta_0}\right| < 1 \tag{20.106}$$

We want to prove that

$$\left|\frac{\delta_{n+1}}{\delta_0}\right| < 1 \tag{20.107}$$

But

$$\delta_{n+1} = \frac{\alpha_2}{\varphi}\delta_n + \sum_{j=0}^{n-1}\frac{\alpha S_{y\delta_{n,j}}}{\varphi} \tag{20.108}$$

$$\left|\delta_{n+1}\right| = \left|\frac{\alpha_2}{\varphi}\delta_n + \sum_{j=0}^{n-1}\frac{\alpha S_{y\delta_{n,j}\delta_{n-j}}}{\varphi}\right| \leq \left|\frac{\alpha_2}{\varphi}\right|\left|\delta_n\right| + \sum_{j=0}^{n-1}\left|\frac{\alpha S_{y\delta_{n,j}}}{\varphi}\right|\left|\delta_{n-j}\right| \tag{20.109}$$

From inductive principles, we have

$$\left|\delta_{n+1}\right| < \left|\frac{\alpha_2}{\varphi}\right|\left|\delta_0\right| + \sum_{j=0}^{n-1}\left|\frac{\alpha S_{y\delta_{n,j}}}{\varphi}\right|\left|\delta_0\right| \tag{20.110}$$

$$\left|\delta_{n+1}\right| < \left|\delta_0\right| \left|\left(\frac{\alpha_2}{\varphi} + \sum_{j=0}^{n-1} \frac{\alpha S_{y\delta_{n,j}}}{\varphi}\right)\right| \tag{20.111}$$

$$\left|\frac{\delta_{n+1}}{\delta_0}\right| < \left|\frac{\alpha_2}{\varphi} + \sum_{j=0}^{n-1} \frac{\alpha S_{y\delta_{n,j}}}{\varphi}\right| \tag{20.112}$$

$$\left|\frac{\delta_{n+1}}{\delta_0}\right| < 1 \rightarrow \alpha_2 + \sum_{j=0}^{n-1} \alpha S_{y\delta_{n,j}} < \varphi \tag{20.113}$$

From the stability analysis performed above using the inductive principle, we can conclude that the equation is stable, provided that the conditions are met.

20.7   NEW NUMERICAL SCHEME: LAGRANGE POLYNOMIAL INTERPOLATION AND THE TRAPEZOIDAL RULE

In past decades, multiple numerical methods have been established to resolve partial differential equations that entail integrals as well as derivatives. To mention a few, Crank and Nicolson merged the forward and backward method to produce a numerical scheme that is used to discretize partial differential equations and seems to be very stable. While these numerical schemes are still in use to date, new numerical schemes are constantly being established to solve differential equations based on the complexity of the problem. For example, Adams and Bashforth established an advanced numerical scheme employing the Lagrange polynomial interpolation. Additionally, Batogna & Atangana (2018) established a new numerical scheme employing the Laplace transform as well as the Adams–Bashforth procedure. All the new numerical schemes that are constantly being established are still in use to date. These numerical schemes, however, have limitations and as a result more advanced numerical schemes are still appreciated. In this unit, we employ a new numerical scheme that integrates the Lagrange polynomial interpolation as well as the trapezoidal rule. Using this numerical scheme on the following partial differential equation, we can employ that:

$$\frac{\partial^2 s^*}{\partial r^2} + \frac{1}{r}\frac{\partial s^*}{\partial r} = \left[\frac{S}{T}\frac{\partial s^*}{\partial t}\right] + \left\{\alpha S_y \int_0^t \frac{\partial s^*}{\partial \tau} e^{-\alpha(t-\tau)} d\tau\right\} \tag{20.114}$$

$$\frac{\partial s}{\partial t} = \frac{T}{S}\left(\frac{\partial^2 s}{\partial r^2} + \frac{1}{r}\frac{\partial s}{\partial r} - \alpha S_y \int_0^t \frac{\partial s}{\partial \tau} e^{-\alpha(t-\tau)} d\tau\right) \tag{20.115}$$

$$\frac{\partial s}{\partial t} = f(S,r,t) + \int_0^t F(S,r,\tau) d\tau \tag{20.116}$$

where

$$f(S,r,t) = \frac{T}{S}\left(\frac{\partial^2 s}{\partial r^2} + \frac{1}{r}\frac{\partial s}{\partial r}\right) \tag{20.117}$$

$$F(S, r, t) = -\frac{T}{S} \alpha S_y \left(\int_0^t \frac{\partial s}{\partial \tau} e^{-\alpha(t-\tau)} d\tau \right) \tag{20.118}$$

This equation is an integro-differential equation.

$$\int_0^t \frac{\partial s}{\partial \tau} d\tau = \int_0^t f(S, r, t) d\tau + \int_0^t \int_0^\tau F(S, r, l) dl \, d\tau \tag{20.119}$$

$$S(r, t) - S(r, 0) = \int_0^t f(S, r, \tau) d\tau + \int_0^t \int_0^\tau F(S, r, l) dl \, d\tau \tag{20.120}$$

At the point $t = t_{n+1}$, we have

$$S(r, t_{n+1}) - S(r, 0) = \int_0^{t_{n+1}} f(S, r, \tau) d\tau + \int_0^{t_{n+1}} \int_0^t F(S, r, l) dl \, d\tau \tag{20.121}$$

Also at

$$t = t_n \tag{20.122}$$

$$S(r, t_n) - S(r, 0) = \int_0^{t_n} f(S, r, t) dt + \int_0^{t_n} \int_0^t F(S, r, l) dl \, dt \tag{20.123}$$

$$S(r, t_{n+1}) - S(r, t_n) = \int_0^{t_{n+1}} f(S, r, t) dt + \int_0^{t_{n+1}} \int_0^t F(S, r, l) dl \, dt \tag{20.124}$$

We know that using the Lagrange polynomial

$$\int_a^b f(t) d_t = \frac{3}{2}(b-a) f(b) - \frac{(b-a)}{2} f(a) \tag{20.125}$$

$$\int_{t_n}^{t_{n+1}} f(S, r, t) d_t = \frac{3(t_{n+1} - t_n)}{2} f(S, r, t_n) - \frac{(t_{n+1} - t_n)}{2} f(S, r, t_{n-1}) \tag{20.126}$$

$$\int_{t_n}^{t_{n+1}} f(S, r, t) d_t = \frac{3(\Delta t)}{2} f(S^n, r, t_n) - \frac{\Delta t}{2} f(S^{n-1}, r, t_{n-1}) \tag{20.127}$$

$$\int_{t_n}^{t_{n+1}} \int_0^t F(S, r, l) dl \, dt = \int_{t_n}^{t_{n+1}} g(t) dt \tag{20.128}$$

$$\int_a^b g(t) dt = \frac{(b-a)}{2} \left[f(b) + f(a) \right] \tag{20.129}$$

$$\int\limits_{t_n}^{t_{n+1}} g(t)\,dt = \frac{(t_{n+1}-t_n)}{2}\Big[g(t_{n+1})+g(t_n)\Big] = \frac{\Delta t}{2}\left[\int\limits_{0}^{t_{n+1}} F(S,r,\tau)\,d\tau + \int\limits_{0}^{t_n} F(S,r,\tau)\,d\tau\right] \quad (20.130)$$

$$= \frac{\Delta t}{2}\left[\int\limits_{0}^{t_n} F(S,r,\tau)\,d\tau + \int\limits_{t_n}^{t_{n+1}} F(S,r,\tau)\,d\tau + \int\limits_{0}^{t_n} F(S,r,\tau)\,d\tau\right] \quad (20.131)$$

$$= \frac{\Delta t}{2}\left[2\int\limits_{0}^{t_n} F(S,r,\tau)\,d\tau + \int\limits_{t_n}^{t_{n+1}} F(S,r,\tau)\,d\tau\right] \quad (20.132)$$

$$= \frac{\Delta t}{2}\left[2\int\limits_{0}^{t_1} F(S,r,\tau)\,d\tau + 2\int\limits_{t_1}^{t_n} F(S,r,\tau)\,d\tau + \frac{3}{2}\Delta t\,F(S^n,r,t_n) - \frac{\Delta t}{2}F(S^{n-1},r,t_{n-1})\right] \quad (20.133)$$

$$= \frac{\Delta t}{2}\left[\frac{2\Delta t}{2}\begin{pmatrix}\left(F(S^1,r,t_1)+F(S^0,r,t_0)\right)+\frac{3}{2}\Delta t\,F(S^n,r,t_n)\\[6pt]-\frac{\Delta t}{2}F(S^{n-1},r,t_{n-1})+2\sum\limits_{j=1}^{n-1}\int\limits_{t_j}^{t_{j+1}} F(S,r,\tau)\,d\tau\end{pmatrix}\right] \quad (20.134)$$

$$= \frac{\Delta t}{2}\left\{\Delta t\left[\begin{matrix}F(S^1,r,t_1)+F(S^0,r,t_0)+\frac{3}{2}\Delta t\,F(S^n,r,t_n)-\frac{\Delta t}{2}F(S^{n-1},r,t_{n-1})\\[6pt]+2\sum\limits_{j=1}^{n-1}\left[\frac{3}{2}\Delta t\,F(S^j,r,t_j)-\frac{\Delta t}{2}F(S^{j-1},r,t_{j-1})\right]\end{matrix}\right]\right\} \quad (20.135)$$

$$S(r,t_{n+1}) - S(r,t_n) \quad (20.136)$$

$$= \frac{3}{2}\Delta t\,f(S^n,r,t_n) - \frac{\Delta t}{2}f(S^{n-1},r,t_{n-1}) + \frac{(\Delta t)^2}{2}\left(F(S^1,r,t_1)-F(S^0,r,t_0)\right)$$
$$+\frac{3}{4}\Delta t^2 F(S^n,r,t_n) - \frac{(\Delta t)^2}{4}F(S^{n-1},r,t_{n-1}) + \sum\limits_{j=1}^{n-1}\left[\begin{matrix}\frac{3}{2}(\Delta t)^2 F(S^j,r,t_j)-\\[4pt]\frac{(\Delta t)^2}{2}F(S^{j-1},r,t_{j-1})\end{matrix}\right] \quad (20.137)$$

Now we discretize in space to have:
At $r = r_i$, we have

$$S_i^{n+1} - S_i^n = \frac{3}{2}\Delta t\,f(S_i^n,r_i,t_n) - \frac{\Delta t}{2}f(S_i^{n-1},r_i,t_{n-1}) + \frac{(\Delta t)^2}{2}\Big[F(S_i^1,r_i,t_1)-F(S_i^0,r_i,t_0)\Big]$$
$$+\frac{3}{4}(\Delta t)^2 F(S_i^n,r_i,t_n) - \frac{(\Delta t)^2}{4}F(S_i^{n-1},r_i,t_{n-1}) + \sum\limits_{j=1}^{n-1}\left[\begin{matrix}\frac{3}{2}(\Delta t)^2 F(S_i^j,r_i,t_j)-\\[4pt]\frac{(\Delta t)^2}{2}F(S_i^{j-1},r_i,t_{j-1})\end{matrix}\right] \quad (20.138)$$

Now we discretize in time to have

$$f(S, r, t) = \frac{T}{S}\left(\frac{\partial^2 s}{\partial r^2} + \frac{1}{r}\frac{\partial s}{\partial r}\right)$$

(20.139)

$$f\left(S_i^1, r_i, t_i\right) = \frac{T}{S}\left[\frac{S_{i+1}^1 - 2S_i^1 + S_{i-1}^1}{\left(\Delta x\right)^2} + \frac{1}{r_i}\frac{S_{i+1}^1 - S_{i-1}^1}{\Delta x}\right]$$

(20.140)

$$F(S, r, t) = -\frac{T}{S}\alpha S_y\left(\frac{\partial s}{\partial t}e^{-\alpha(t-\tau)}\right)$$

(20.141)

$$F\left(S_i^1, r_i, t_i\right) = -\frac{T}{S}\alpha S_y\left(\frac{S_i^1 - S_i^0}{\Delta t}e^{-\alpha(t-\tau)}\right)$$

(20.142)

$$f\left(S_i^n, r_i, t_i\right) = \frac{T}{S}\left[\frac{S_{i+1}^n - 2S_i^n + S_{i-1}^n}{\left(\Delta x\right)^2} + \frac{1}{r_i}\frac{S_{i+1}^n - S_{i-1}^n}{\Delta x}\right]$$

(20.143)

$$F\left(S_i^n, r_i, t_i\right) = -\frac{T}{S}\alpha S_y\left(\frac{S_i^n - S_i^0}{\Delta t}e^{-\alpha(t-\tau)}\right)$$

(20.144)

$$f\left(S_i^{n-1}, r_i, t_i\right) = \frac{T}{S}\left[\frac{S_{i+1}^{n-1} - 2S_i^{n-1} + S_{i-1}^{n-1}}{\left(\Delta x\right)^2} + \frac{1}{r_i}\frac{S_{i+1}^{n-1} - S_{i-1}^{n-1}}{\Delta x}\right]$$

(20.145)

$$F\left(S_i^{n-1}, r_i, t_i\right) = -\frac{T}{S}\alpha S_y\left(\frac{S_i^{n-1} - S_i^0}{\Delta t}e^{-\alpha(t-\tau)}\right)$$

(20.146)

$$f\left(S_i^0, r_i, t_i\right) = \frac{T}{S}\left[\frac{S_{i+1}^0 - 2S_i^0 + S_{i-1}^0}{\left(\Delta x\right)^2} + \frac{1}{r_i}\frac{S_{i+1}^0 - S_{i-1}^0}{\Delta x}\right]$$

(20.147)

$$F\left(S_i^0, r_i, t_i\right) = -\frac{T}{S}\alpha S_y\left(\frac{S_i^{0-1} - S_i^0}{\Delta t}e^{-\alpha(t-\tau)}\right)$$

(20.148)

$$F\left(S_i^j, r_i, t_i\right) = \frac{T}{S}\left[\frac{S_{i+1}^j - 2S_i^j + S_{i-1}^j}{\left(\Delta x\right)^2} + \frac{1}{r_i}\frac{S_{i+1}^j - S_{i-1}^j}{\Delta x}\right]$$

(20.149)

$$F\left(S_i^j, r_i, t_i\right) = -\frac{T}{S}\alpha S_y\left(\frac{S_i^j - S_i^0}{\Delta t}e^{-\alpha(t-\tau)}\right)$$

(20.150)

$$F\left(S_i^{j-1}, r_i, t_i\right) = \frac{T}{S}\left[\frac{S_{i+1}^{j-1} - 2S_i^{j-1} + S_{i-1}^{j-1}}{\left(\Delta x\right)^2} + \frac{1}{r_i}\frac{S_{i+1}^{j-1} - S_{i-1}^{j-1}}{\Delta x}\right]$$

(20.151)

$$F\left(S_i^{j-1}, r_i, t_i\right) = -\frac{T}{S}\alpha S_y\left(\frac{S_i^{j-1}-S_i^0}{\Delta t}e^{-\alpha(t-\tau)}\right) \tag{20.152}$$

$$S_i^{n+1}-S_i^n = \frac{3}{2}\Delta t\left[\frac{T}{S}\left[\frac{S_{i+1}^n-2S_i^n+S_{i-1}^n}{\left(\Delta x\right)^2}+\frac{1}{r_i}\frac{S_{i+1}^n-S_{i-1}^n}{\Delta x}\right]\right]$$

$$-\frac{\Delta t}{2}\left[\frac{T}{S}\left[\frac{S_{i+1}^{n-1}-2S_i^{n-1}+S_{i-1}^{n-1}}{\left(\Delta x\right)^2}+\frac{1}{r_i}\frac{S_{i+1}^{n-1}-S_{i-1}^{n-1}}{\Delta x}\right]\right]$$

$$+\frac{\left(\Delta t\right)^2}{2}\left[-\frac{T}{S}\alpha S_y\left(\frac{S_i^1-S_i^0}{\Delta t}e^{-\alpha(t-\tau)}\right)-\left[-\frac{T}{S}\alpha S_y\left(\frac{S_i^{0-1}-S_i^0}{\Delta t}e^{-\alpha(t-\tau)}\right)\right]\right]$$

$$+\frac{3}{4}\left(\Delta t\right)^2\left[-\frac{T}{S}\alpha S_y\left(\frac{S_i^n-S_i^0}{\Delta t}e^{-\alpha(t-\tau)}\right)\right]-\frac{\left(\Delta t\right)^2}{4}\left[-\frac{T}{S}\alpha S_y\left(\frac{S_i^{n-1}-S_i^0}{\Delta t}e^{-\alpha(t-\tau)}\right)\right]$$

$$+\sum_{j=1}^{n-1}\left[\begin{array}{l}\frac{3}{2}\left(\Delta t\right)^2\left[-\frac{T}{S}\alpha S_y\left(\frac{S_i^j-S_i^0}{\Delta t}e^{-\alpha(t-\tau)}\right)\right]\\-\frac{\left(\Delta t\right)^2}{2}\left[-\frac{T}{S}\alpha S_y\left(\frac{S_i^{j-1}-S_i^0}{\Delta t}e^{-\alpha(t-\tau)}\right)\right]\end{array}\right] \tag{20.153}$$

This newly derived numerical scheme uses Lagrange polynomial interpolation and the trapezoidal rule to discretize in space and time.

20.8 NUMERICAL SIMULATIONS

We present below the numerical simulations of the mathematical model for different values of stochastic parameters. The numerical results are depicted in Figures 20.2 to 20.15. The figures

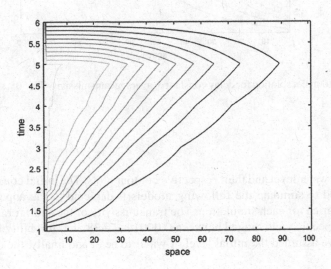

FIGURE 20.2 Contour plot solution for delay constant 3, average transmissivity $T = 405$, specific yield $= 0.002$

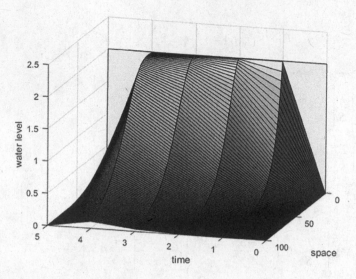

FIGURE 20.3   Reduction of water level for delay constant 3, average transmissivity T = 405, specific yield = 0.002.

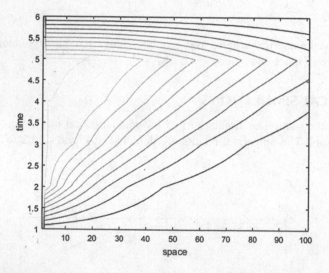

FIGURE 20.4   Contour plot solution for delay constant 3, average transmissivity T = 505, specific yield = 0.025.

simulate change in water level and their respective contour plots. The initial conditions and boundary conditions used to simulate the following models is depicted in the appendix. The aquifer parameters were varied for each simulation. The transmissivity of the aquifer ranged between 400 and 800m²/d, the specific yield ranged between 0.0001and 0.38, the storability of course between 0.001 and 0.009, we assumed the initial level of water to be 2, and finally, the delay index varied from 0.001 to 6.

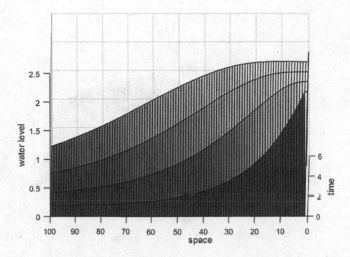

FIGURE 20.5   Reduction of water level for delay constant 3, average transmissivity T = 505, specific yield = 0.025

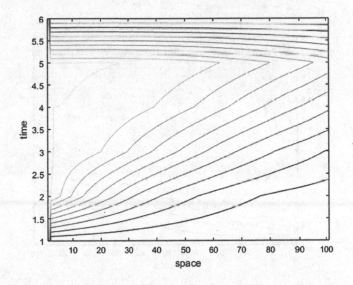

FIGURE 20.6   Contour plot solution for delay constant 3, average transmissivity T = 555, specific yield = 0.3.

20.9   RESULTS AND DISCUSSIONS

The above simulations depict that aquifer parameters such as transmissivity, specific yield and stor-ativity vary with a reduction of water level resulting from heterogeneities of a geologic nature. The deterministic model which assumes homogeneity falls short in that natural occurring processes can never be known with certainty. Thus, even if the physics of the system is fairly easy and compre-hensible by deterministic equations, it is difficult to fully support the solutions of deterministic models mainly because the input variables, model geometry, initial and boundary conditions and so on are not very well known or, in the ideal case, never known extensively. Stochastic techniques can be considered as a mechanism for combining physics, statistics and uncertainty in a meaningful

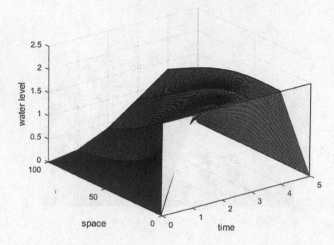

FIGURE 20.7   Reduction of water level for delay constant 3, average transmissivity T = 555, specific yield = 0.3.

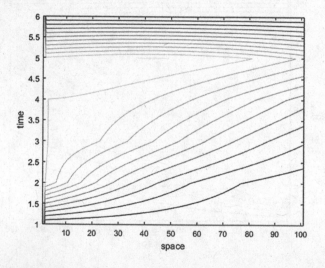

FIGURE 20.8   Contour plot solution for delay constant 3, average transmissivity T = 600, specific yield = 0.303.

theoretical context. On the one hand, statistic distributions describe the unknown parameters, while on the other hand, the various parameters that describe the problem are related to each other and the (uncertain) model parameters through physical laws (deterministic). The resulting models are partial differential stochastic equations. Returning to the basic groundwater example, recharge (r) and inflow (q) are not very well known. These two variables can therefore be seen as random functions that vary widely in space and/or time, and have statistical properties (such as a variance, covariance and mean) that can be inferred from samples. It therefore becomes essential to take into account the heterogeneity that comes with naturally occurring processes when modeling.

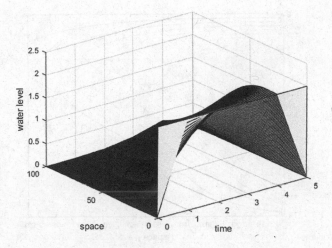

FIGURE 20.9   Reduction of water level for delay constant 3, average transmissivity T = 600, specific yield = 0.303.

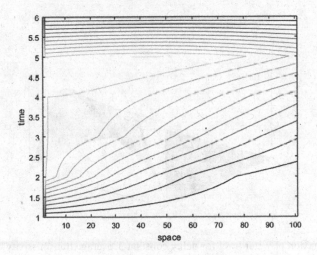

FIGURE 20.10   Contour plot solution for delay constant 3, average transmissivity T = 650, specific yield = 0.304.

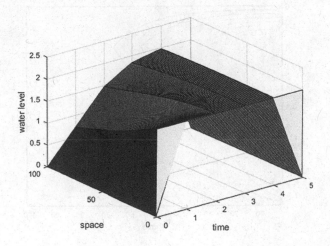

FIGURE 20.11   Reduction of water level for delay constant 3, average transmissivity T = 650, specific yield = 0.304

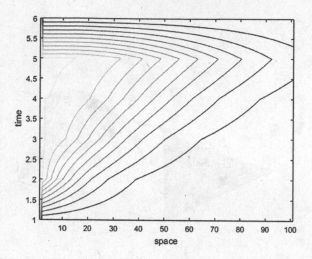

FIGURE 20.12   Contour plot solution for delay constant 3, average transmissivity T = 700, specific yield = 0.32.

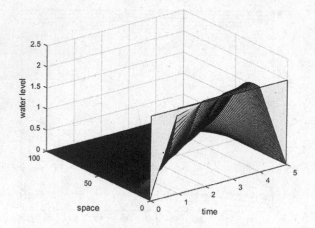

FIGURE 20.13   Reduction of water level for delay constant 3, average transmissivity T = 700, specific yield = 0.32

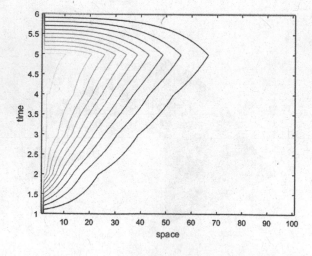

FIGURE 20.14   Contour plot solution for delay constant 3, average transmissivity T = 800, specific yield = 0.35.

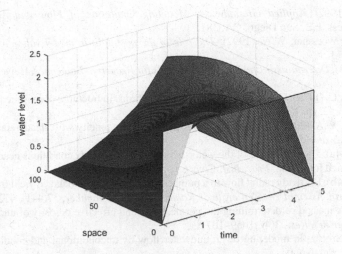

FIGURE 20.15 Reduction of water level for delay constant 3, average transmissivity T = 800, specific yield = 0.35

20.10   CONCLUSION

Traditionally, the movement of groundwater was modeled in a deterministic fashion, which assumes that aquifer parameters and boundary conditions are known with certainty. In addition to the already mentioned uncertainty, it is also assumed that the aquifer's parameters are constant everywhere within the geological formation, which in practice cannot be validated as the sub-surfaces are subjected to heterogeneities. In reality, hydrological events are better described as random phenomena. The aquifer properties, such as hydraulic conductivity, transmissivity and storativity will depend on the properties of the soil, which is naturally different from one point to another because of the variability of the geology. From already published research works, it was well established that modeling groundwater flow with constant parameters could only be applicable if the geological formation is homogeneous. Otherwise, it is preferable to capture some heterogeneities using the concept of stochastic modeling, where a given aquifer parameter is considered to be a distribution. Groundwater flow is therefore modeled more realistically via the stochastic approach. The aim of this thesis is to distinguish between the deterministic and stochastic models, thus saying which is realistically fit to model the flow of groundwater in the unconfined aquifer system. The main aim of this research was achieved through the analysis performed from the two numerical schemes. While some researchers still use the deterministic approach to model the flow of groundwater, results have proved that the stochastic approach is realistically fit to model the flow of groundwater, since it relies wholly on uncertainty, which is expected to happen in nature as a result of the heterogeneity of natural occurring geological processes. This research enabled us to see how the assumption of homogeneity on natural processes can lead to incorrect interpretation. The numerical simulations showed graphical representations of the flow of groundwater, and how varying aquifer parameters as well as keeping them constant affect the simulations that are appropriate for the modeling of regional groundwater models.

REFERENCES

Aguirre, C.G., & Haghighi, K. (2003). Stochastic modelling of transient contaminant transport, *J. Hydrol.*, 276, 1–4 (224).

Allwright, A., & Atangana, A. (2018). Augmented upwind numerical schemes for the groundwater transport advection–dispersion equation with local operators. *Int. J. Numer. Meth. Fluids*, 33(2), 1–20.

Andersine, M.P. (1992). *Applied Groundwater Modeling-Simulation of Flow and Advective Transport.* Academic Press, Inc., San Diego, CA, 381.

Anderson, M.P., & Woessner, W.W. (1992). *The role of the postaudit in model validation. Adv. Water Res.,* 15(3), 167–173.

Bartlett, M.S. (2008). *Deterministic and Stochastic Models for Recurrent Epidemics,* University of Manchester, England.

Bear, J., & Braester, C. (1972). On the flow of two immscible fluids in fractured porous media. In *Developments in Soil Science,* 2, pp. 177–202. Elsevier.

Berg, S.J., & Illma, W.A. (2012). Field study of subsurface heterogeneity with steady-state hydraulic tomography. *Groundwater,* 51, 1.

Boulton, N. S. (1954a). The drawdown of the water-table under non-steady conditions near a pumped 437 well in an unconfined formation. *Proc. Inst. Civil Eng.,* 3(4), 564–579, 438.

Boulton, N. S. (1954b). Unsteady radial flow to a pumped well allowing for delayed yield from 440 storage. *Int. Ass. Sci. Hydrol, Assembl´ee G´en´erale de Rome, Publication number,* 37(441), 472–477.

Dagan, G. (1967). A method of determining the permeability and effective porosity of unconfined anisotropie aquifers. *Water Res. Res.,* 3(4), 1059–1071.

Dagan, G. (1982). Stochastic modeling of groundwater flow by unconditional and conditional probabilities. *Water Resour. Res.* No. 04.

Dryden, I.L., Markus, L., & Taylor, C.C. (2005). Non-stationary spatiotemporal analysis of Karst water levels. *J. Royal Statistical Soc.: Series C Appl. Statis.,* 54, 3, 673.

Gallag, C., Kodikara, J., & Uchimara, T. (2013). Laboratory measurement of hydraulic conductivity functions of two unsaturated sandy soils during drying and wetting processes. *Soils Found,* 53, 417–430.

Gnitchogna, R., & Atangana A. (2018). New two step Laplace Adams–Bashforth method for integer a noninteger order partial differential equations. *Number Methods Partial Differ. Eq* 34, 1738–1739.

Gordon, D. (1989). Chapter B2: Introduction to ground-water hydraulics: A programmed text for self-instruction, in: Techniques of water-resources investigations of the United States Geological Survey. USGS, pp. 1–29.

Kock, M., & Aralai, P. (2004). Deterministic and Stochastic Modeling of Groundwater Flow and Solute Transport in the Heavily Stressed Bangkok Aquifer, Thailand, An Investigation of Optimal Management Strategies for Possible Restoration, Department of Water Resources, Thailand.

Konikow, L.F., August, L.L., & Voss, C.I. (2001). Effects of clay dispersion on aquifer storage and recovery in coastal aquifers. *Transport Porous Media,* 43(1), 45–64.

Kroszynski, U.I. & Dagan, G. 1975. Well pumping in unconfined aquifers: The influence of the unsaturated zone. *Water Res. Res.,* 11(3), 479–490.

Maliva, R.G. (2016). Geostatistical methods and applications, *Aquifer Characterization Tech.,* doi:10.1007/978-3-319-32137-0_20, (595–671).

Feddes, R.A., Bresler, E., & Neuman, S.P. (1974). Field test of a modified numerical model for water uptake by root systems. *Water Res. Res.,* 10(6), 1199–1206.

Pool, M., Carrera, J., Alcolea, A., & Bocanegra, E.M. (2015). A comparison of deterministic and stochastic approaches for regional scale inverse modelling on the Mar del Plata aquifer, *J. Hydrol.*

Philippe, R., Alcolea, A., & Ginsbourger, D. (2013). Stochastic versus Deterministic Environmental modelling finding simplicity in complexity.

Prickett, T.A. (1965). Type-curve solution to aquifer tests under water-table conditions. *Groundwater,* 3(3), 5–14.

Streltsova, T.D. (1972). Unsteady radial flow in an unconfined aquifer. *Water Res. Res.,* 8(4), 1059–1066.

Theis, C.V. (1935). The relation between the lowering of the piezometric surface and the rate and duration of discharge of a well using ground-water storage. *Eos, Transactions American Geophysical Union,* 16(2), 519–524.

Yang, J., & Shi, L. (2008). Stochastic analysis of groundwater flow subject to random boundary conditions, *J. Hydrodynamics, Ser,* 20, 5 (553–560). doi: 10.1016/5100)-6058 (08) 60094-3

Yen, C.C., & Gary, G. (1990). An efficient deterministic-probabilistic approach to modeling regional groundwater flow. doi:10.1029/WR026i007p01559

Youngs, E.A., & Poulovassilis, A. (1976). The distribution of moisture profile developed during the distribution of soil water infiltration. *Water Resour. Res.* 12.

Zhang, T. (1991). A statistical approach for water movement in the unsaturated zone (Report No. 1010). Lund Institute of Technology, Lund University, Lund, Sweden.

Zhu, J., & Yeh, T.J. (2005). Characterization of aquifer heterogeneity using transient hydraulic tomography, *Water Res.,* 41, 7.

21 A New Method for Modeling Groundwater Flow Problems
Fractional–Stochastic Modeling

Mohau Mahantane and Abdon Atangana
University of the Free State, Bloemfontein, South Africa

CONTENTS

21.1 INTRODUCTION

One of the real-world's most complicated and challenging issues to be represented by means of simple mathematical models and/or equations is that facing groundwater investigations, as it requires the modeler's detailed understanding of the geologic structure and the response of the aquifer through which groundwater moves. Thus modeling such a problem remains a challenging task, because the response and geological structure of the aquifer through which groundwater travels is invisible and changes with space and time (Atangana and Bildik, 2013). For example, when dealing with groundwater contamination, scientists and other researchers are generally faced with the challenge of predicting the concentration of a migrating contaminant in the aquifer with respect to time and distance (Bobba & Singh, 1995). This is one reason that many researchers from various fields of science have focused much of their attention on formulating new mathematical equations and models that can be used to capture and understand the behavior of groundwater flow. Numerous models have been suggested in the literature, with classical derivatives using different concepts, including

DOI: 10.1201/9781003266266-21

those with fractional derivatives with variable order derivatives with stochastic setting (Atangana & Bonyah, 2019).

In this study, modeling of groundwater flow problems will be based on the modification of the classical advection–dispersion equation (ADE) for predicting transport behavior by coupling suitable fractional differential operators and a stochastic approach in order to account for random movement of groundwater and solute transport with respect to heterogeneity of aquifer systems. Some literature, however, approaches the solution using ADE along with a set of initial and boundary conditions by assuming dispersion and velocity as constants (Jaiswal & Kumar, 2011). However, in reality, parameters influencing flow and solute transport are not constants, but depend on the temporal and spatial scales of the aquifer heterogeneities through which they flow and transport takes place (Kumar et al., 2012). It is therefore very preferable to capture such uncertainties and heterogeneities using the concept of stochastic modeling in which all input parameters of interest are converted into distributions.

None the less, some models fail to provide reliable groundwater flow estimates because of their inability to account for the heterogeneity, viscoelasticity and memory effect. Therefore, to solve such a problem we employed the concept of fractional differentiation known to be a better mathematical tool to describe real-world problems with accuracy (Cushman & Ginn, 2000; Podlubny, 2002; Atangana & Alkahtani, 2015). Both the concept of fractional differential operators and stochastic approach play an important role in the field of science, though they have always been used separately for the solution of problems arising from different scenarios. For example, one of the capabilities of stochastic technique is that it is able to address some complexity of situations in heterogeneous environments such as the Markovian process, though the nonlocal fractional operators can be applied to solve problems of non-Markovian processes (Atangana & Bonyah, 2019). Therefore, the main question of importance in this study is: could we perhaps couple these two techniques and formulate a new mathematical equation to solve problems associated with pollution transport in complex real-world situations? To answer this question, we used a classical 1-d ADE and suggested a new numerical scheme to solve such a problem.

21.2   FRACTIONAL–STOCHASTIC MODELING

In this section, we present the application of the concept of fractional differentiation and stochastic approach to groundwater transport equation. This approach of fractional–stochastic modeling in this study is presented using different fractional differential operators where we substitute the time derivative with a time-fractional operator for each given differential operator. In addition, the ADE input parameters are also transformed into their probability distributions. Thus we consider the following groundwater transport equation in this chapter.

$$\frac{\partial C(x,t)}{\partial t} = D\frac{\partial^2 C(x,t)}{\partial x^2} - v\frac{\partial C(x,t)}{\partial x} - \lambda R C(x,t) \qquad (21.1)$$

Our input parameters of interest in the equation above are D, which is the dispersion coefficient, v is the average linear groundwater velocity, and R is the retardation factor. Therefore, to convert these input parameters into their distributions we follow the approach suggested by Atangana and Bonyah (2019) where we suppose a range of D in $[a_1 + a_2 \cdots + a_n]$, v in $[b_1 + b_2 \cdots + b_n]$, and R in $[c_1 + c_2 \cdots + c_n]$.

We then present the distributrion for each sample as follows:

$$\hat{D} = \bar{D} + \gamma N(0,1), \hat{v} = \bar{v} + \gamma N(0,1), \hat{R} = \bar{R} + \gamma N(0,1)$$

Thus, Equation (2.1) becomes:

$$\frac{\partial C(x,t)}{\partial t} = \left(\bar{D} + \gamma N\left(\bar{D}, \sigma^2\right)\right)\frac{\partial^2 C(x,t)}{\partial x^2} - \left(\bar{v} + \gamma N\left(\bar{v}, \sigma^2\right)\right)\frac{\partial C(x,t)}{\partial x}$$
$$- \left(\bar{R} + \gamma N\left(\bar{R}, \sigma^2\right)\right)\lambda C(x,t) \tag{21.2}$$

or,

$$\hat{D}\frac{\partial^2 C(x,t)}{\partial x^2} - \hat{v}\frac{\partial C(x,t)}{\partial x} - \lambda \hat{R} C(x,t) = \frac{\partial C(x,t)}{\partial t} \tag{21.3}$$

From the above Equation (21.3), we then provide an illustration based on the theory of fractional differentiation to generate Equation (21.4) below. We substitute the time derivative $\left(\frac{\partial C}{\partial t}\right)$ with a time-fractional operator $\left({}_{0}^{F}D_{t}^{\alpha}\right)$ and obtain the following form of equation:

$$_{0}^{F}D_{t}^{\alpha}C(x,t) = \hat{D}\frac{\partial^2 C(x,t)}{\partial x^2} - \hat{v}\frac{\partial C(x,t)}{\partial x} - \lambda \hat{R} C(x,t) \tag{21.4}$$

21.3 NUMERICAL SOLUTIONS

The numerical approximation to the solutions of ordinary differential equations are normally achieved through the application of numerical methods for ordinary differential equations (Süli, 2014). The most widely applied methods when presenting the approximated solutions to the initial value problem for ordinary differential equations are the Adams methods, based on approximating the integral by means of a polynomial integration within the intervals, say (t_n, t_{n+1}) (Peinado et al, 2010). In addition, there exist two types of Adams methods, being the explicit and the implicit Adams–Bashforth (AB) methods. These methods are derived from the fundamental theorem of calucus by means of polynomial interpolation in the Lagrange form. In this section, we apply the AB approach based on the Caputo, Caputo–Fabrizio and the Atangana–Baleanu fractional derivative to solve the advection–dispersion transport.

21.3.1 NUMERICAL SOLUTION OF THE NEW MODEL WITH CAPUTO FRACTIONAL DERIVATIVE

In this section, we apply the approach using the Caputo differential operator which is based on the power law. We recall our Equation (21.3), then, by replacing the time derivative with a Caputo time-fractional derivative we obtain the following:

$$_{0}^{C}D_{t}^{\alpha}C(x,t) = \hat{D}\frac{\partial^2 C(x,t)}{\partial x^2} - \hat{v}\frac{\partial C(x,t)}{\partial x} - \lambda \hat{R} C(x,t)$$

or,

$$\frac{1}{\Gamma(1-\alpha)}\int_{0}^{t}(t-\tau)^{-\alpha} f'(\tau)d\tau = \hat{D}\frac{\partial^2 C(x,t)}{\partial x^2} - \hat{v}\frac{\partial C(x,t)}{\partial x} - \lambda \hat{R} C(x,t) \tag{21.5}$$

where the terms on the right-hand side can be replaced by the function, $f(x,t,C(x,t))$ such that:

$$\hat{D}\frac{\partial^2 C(x,t)}{\partial x^2} - \hat{v}\frac{\partial C(x,t)}{\partial x} - \lambda \hat{R} C(x,t) = f\left(x,t,C(x,t)\right) \tag{21.6}$$

Therefore, we consider the following nonlinear fractional ordinary equation:

$$\,_0^C D_t^\alpha C(x,t) = f(x,t,C(x,t)) \tag{21.7}$$

or,

$$f(x,t,C(x,t)) = \frac{1}{\Gamma(\alpha)} \int_0^t (t-\tau)^{\alpha-1} f'(\tau) d\tau \tag{21.8}$$

When applying the fundamental theorem of calculus to Equation (21.8), we obtain:

$$C(x,t) - C(x,0) = \frac{1}{\Gamma(\alpha)} \int_0^t (t-\tau)^{\alpha-1} f(x,\tau,C(x,\tau)) d\tau \tag{21.9}$$

We then consider at point (t_{n+1}), where $n = 0, 1, 2, 3, \ldots$, the equation above is reformulated as follows:

$$C_i^{n+1} - C_i^0 = \frac{1}{\Gamma(\alpha)} \int_0^{t_{n+1}} (t_{n+1}-\tau)^{\alpha-1} f(x_i,\tau,C(x_i,\tau)) d\tau \tag{21.10}$$

$$= \frac{1}{\Gamma(\alpha)} \sum_{j=0}^n \int_{t_j}^{t_{j+1}} (t_{n+1}-\tau)^{\alpha-1} f(x_i,\tau,C(x_i,\tau)) d\tau \tag{21.11}$$

When we approximate the function $f(x_i,\tau,C(x_i,\tau))$ within the interval $[t_j, t_{j+1}]$ using the Lagrange polynomial method, the following equation is obtained:

$$C_i^{n+1} = C_i^0 + \frac{1}{\Gamma(\alpha)} \sum_{j=0}^n \int_{t_j}^{t_{j+1}} (t_{n+1}-\tau)^{\alpha-1} P_j(\tau) d\tau \tag{21.12}$$

Therefore,

$$C_i^{n+1} = C_i^0 + \frac{(\Delta t)^{-\alpha}}{\Gamma(2-\alpha)} \sum_{j=0}^n \left[\begin{array}{l} f(x_i,t_j,C_i^j)\left\{(n-j+1)^\alpha(n-j+2+\alpha)-(n-j)^\alpha(n-j+2+2\alpha)\right\} \\ -f(x_i,t_{j-1},C_i^{j-1})\left\{(n-j+1)^{\alpha+1}-(n-j)^\alpha(n-j+1+\alpha)\right\} \end{array} \right] \tag{21.13}$$

Thus,

$$f(x_i,t_j,C_i^j) = \hat{D}\frac{C_{i+1}^j - 2C_i^j + C_{i-1}^j}{(\Delta x)^2} - \hat{v}\frac{C_{i+1}^j - C_{i-1}^j}{\Delta x} - \lambda\hat{R}C_i^j \tag{21.14}$$

$$f(x_i,t_{j-1},C_i^{j-1}) = \hat{D}\frac{C_{i+1}^{j-1} - 2C_i^{j-1} + C_{i-1}^{j-1}}{(\Delta x)^2} - \hat{v}\frac{C_{i+1}^{j-1} - C_{i-1}^{j-1}}{\Delta x} - \lambda\hat{R}C_i^{j-1} \tag{21.15}$$

Rewriting Equation (21.13) by substituting with the results of the functions obtained in Equations (21.14) and (21.15), we then have the following form of equation:

$$
C_i^{n+1} = C_i^0 + \frac{(\Delta t)^{-\alpha}}{\Gamma(2-\alpha)} \sum_{j=0}^{n} \left[
\begin{array}{l}
\left(\hat{D} \dfrac{C_{i+1}^j - 2C_i^j + C_{i-1}^j}{(\Delta x)^2} - \hat{v} \dfrac{C_{i+1}^j - C_{i-1}^j}{\Delta x} \right) \left\{ \begin{array}{l} (n-j+1)^\alpha (n-j+2+\alpha) \\ -(n-j)^\alpha (n-j+2+2\alpha) \end{array} \right\} \\[4pt]
- \lambda \hat{R} C_i^j \\[8pt]
- \left(\hat{D} \dfrac{C_{i+1}^{j-1} - 2C_i^{j-1} + C_{i-1}^{j-1}}{(\Delta x)^2} - \hat{v} \dfrac{C_{i+1}^{j-1} - C_{i-1}^{j-1}}{\Delta x} - \lambda \hat{R} C_i^{j-1} \right) \\[8pt]
\left\{ (n-j+1)^{\alpha+1} - (n-j)^\alpha (n-j+1+\alpha) \right\}
\end{array}
\right]
$$

$$(21.16)$$

21.3.2 Numerical Solution of the New Model with Caputo–Fabrizio Fractional Derivative

In this section, we apply the approach using the Caputo–Fabrizio differential operator, which is based on the exponential decay law. We then recall our Equation (21.3) and replace the time derivative with the Caputo–Fabrizio time-fractional operator to generate the following:

$$
{}^{CF}_0 D_t^\alpha C(x,t) = \hat{D} \frac{\partial^2 C(x,t)}{\partial x^2} - \vartheta \frac{\partial C(x,t)}{\partial x} - \lambda \hat{R} C(x,t)
$$

or,

$$
\frac{M(\alpha)}{1-\alpha} \int_a^t \exp\left[-\frac{\alpha}{1-\alpha}(t-\tau) \right] f'(\tau) d\tau = \hat{D} \frac{\partial^2 C(x,t)}{\partial x^2} - \hat{v} \frac{\partial C(x,t)}{\partial x} - \lambda \hat{R} C(x,t) \quad (21.17)
$$

where the terms on the right-hand side can be replaced by the function, $f(x, t, C(x,t))$ such that:

$$
\hat{D} \frac{\partial^2 C(x,t)}{\partial x^2} - \hat{v} \frac{\partial C(x,t)}{\partial x} - \lambda \hat{R} C(x,t) = f(x, t, C(x,t)) \quad (21.18)
$$

We then consider the following nonlinear fractional equation expressed in terms of the Caputo–Fabrizio fractional operator $\left({}^{CF}_0 D_t^\alpha \right)$ as:

$$
{}^{CF}_0 D_t^\alpha C(x,t) = f(x, t, C(x,t)) \quad (21.19)
$$

By using the fundamental theorem of calculus, the above equation becomes:

$$
C(x,t) - C(x,0) = \frac{1-\alpha}{M(\alpha)} f(x, t, C(x,t)) + \frac{\alpha}{M(\alpha)} \int_0^t f(x, \tau, C(x,\tau)) d\tau \quad (21.20)
$$

We consider at a given point (x_i, t_{n+1}), where $n = 0, 1, 2, 3, \ldots$, then the above equation is reformulated as follows:

$$
C(x_i, t_{n+1}) = C(x_i, 0) + \frac{1-\alpha}{M(\alpha)} f(x_i, t_n, C(x_i, t_n)) + \frac{\alpha}{M(\alpha)} \int_0^{t_{n+1}} f(x_i, \tau C(x_i, \tau)) d\tau \quad (21.21)
$$

At a point (x_i, t_n),

$$C(x_i, t_n) = C(x_i, 0) + \frac{1-\alpha}{M(\alpha)} f(x_i, t_{n-1}, C(x_i, t_{n-1})) + \frac{\alpha}{M(\alpha)} \int_0^{t_n} f(x_i, \tau\, C(x_i, \tau)) d\tau \qquad (21.22)$$

By subtracting Equation (21.22) from Equation (21.21), the following is obtained:

$$C_i^{n+1} - C_i^n = \frac{1-\alpha}{M(\alpha)} \Big[f(x_i, t_n, C_i^n) - f(x_i, t_{n-1}, C_i^{n-1}) \Big] + \frac{\alpha}{M(\alpha)} \int_{t_n}^{t_{n+1}} f(x_i, \tau, C(x_i, \tau)) d\tau \qquad (21.23)$$

$$C_i^{n+1} = C_i^n + \frac{1-\alpha}{M(\alpha)} \Big[f(x_i, t_n, C_i^n) - f(x_i, t_{n-1}, C_i^{n-1}) \Big]$$
$$+ \frac{\alpha}{M(\alpha)} \Big[\frac{3}{2} \Delta t f(x_i, t_n, C_i^n) - \frac{\Delta t}{2} f(x_i, t_{n-1}, C_i^{n-1}) \Big] \qquad (21.24)$$

At this point, the functions $f(x_i, t_n, C_i^n)$ and $f(x_i, t_{n-1}, C_i^{n-1})$ from Equation (21.24) can be presented as:

$$f(x_i, t_n, C_i^n) = \hat{D} \frac{C_{i+1}^n - 2C_i^n + C_{i-1}^n}{(\Delta x)^2} - \hat{v} \frac{C_{i+1}^n - C_{i-1}^n}{\Delta x} - \lambda \hat{R} C_i^n \qquad (21.25)$$

$$f(x_i, t_{n-1}, C_i^{n-1}) = \frac{C_{i+1}^{n-1} - 2C_i^{n-1} + C_{i-1}^{n-1}}{(\Delta x)^2} - \hat{v} \frac{C_{i+1}^{n-1} - C_{i-1}^{n-1}}{\Delta x} - \lambda \hat{R} C_i^{n-1} \qquad (21.26)$$

Substituting with the results of the functions, the new model is presented as follows, based on the discretized Caputo–Fabrizio integral operator:

$$C_i^{n+1} = C_i^n + \frac{1-\alpha}{M(\alpha)} \left\{ \begin{array}{l} \left(\hat{D} \dfrac{C_{i+1}^n - 2C_i^n + C_{i-1}^n}{(\Delta x)^2} - \hat{v} \dfrac{C_{i+1}^n - C_{i-1}^n}{\Delta x} - \lambda \hat{R} C_i^n \right) \\[2ex] - \left(\hat{D} \dfrac{C_{i+1}^{n-1} - 2C_i^{n-1} + C_{i-1}^{n-1}}{(\Delta x)^2} - \hat{v} \dfrac{C_{i+1}^{n-1} - C_{i-1}^{n-1}}{\Delta x} - \lambda \hat{R} C_i^{n-1} \right) \end{array} \right\}$$
$$+ \frac{\alpha}{M(\alpha)} \left\{ \begin{array}{l} \dfrac{3\Delta t}{2} \left(\hat{D} \dfrac{C_{i+1}^n - 2C_i^n + C_{i-1}^n}{(\Delta x)^2} - \hat{v} \dfrac{C_{i+1}^n - C_{i-1}^n}{\Delta x} - \lambda \hat{R} C_i^n \right) \\[2ex] - \dfrac{\Delta t}{2} \left(\hat{D} \dfrac{C_{i+1}^{n-1} - 2C_i^{n-1} + C_{i-1}^{n-1}}{(\Delta x)^2} - \hat{v} \dfrac{C_{i+1}^{n-1} - C_{i-1}^{n-1}}{\Delta x} - \lambda \hat{R} C_i^{n-1} \right) \end{array} \right\} \qquad (21.27)$$

21.3.3　Numerical Solution of the New Model with Atangana–Baleanu Fractional Derivative Caputo Sense

In this section, we apply the approach using the Atangana–Baleanu differential operator, which is based on the Mittag-Leffler law. Likewise, we recall Equation (21.3) and substitute the time derivative with Atangana–Baleanu nonlocal fractional operator in the Caputo sense to obtain the following:

$$^{ABC}_0 D^\alpha_t C(x,t) = \hat{D}\frac{\partial^2 C(x,t)}{\partial x^2} - \hat{v}\frac{\partial C(x,t)}{\partial x} - \lambda\hat{R}C(x,t)$$

or,

$$\frac{B(\alpha)}{1-\alpha}\int_a^t E_\alpha\left[-\frac{\alpha}{1-\alpha}(t-\tau)^\alpha\right]f'(\tau)d\tau = \hat{D}\frac{\partial^2 C(x,t)}{\partial x^2} - \hat{v}\frac{\partial C(x,t)}{\partial x} - \lambda\hat{R}C(x,t) \qquad (21.28)$$

where the terms on the right-hand side can be replaced by the function, $f(x,t,C(x,t))$ such that:

$$\hat{D}\frac{\partial^2 C(x,t)}{\partial x^2} - \hat{v}\frac{\partial C(x,t)}{\partial x} - \lambda\hat{R}C(x,t) = f\left(x,t,C(x,t)\right)$$

or,

$$f\left(x,t,C(x,t)\right) = \frac{AB(\alpha)}{1-\alpha}\int_0^t f'(\tau)E_\alpha\left[-\alpha\frac{(t-\tau)^\alpha}{1-\alpha}\right]d\tau \qquad (21.29)$$

Applying the fundamental theorem of calculus to the above Equation (21.29) yields:

$$C(x,t) - C(x,0) = \frac{1-\alpha}{AB(\alpha)}f\left(x,t,C(x,t)\right) + \frac{\alpha}{AB(\alpha)\Gamma(\alpha)}\int_0^t (t-\tau)^{\alpha-1}f\left(x,\tau,C(x,\tau)\right)d\tau$$

$$(21.30)$$

We then consider at point (t_{n+1}), where $n = 0, 1, 2, 3,\ldots$, that the above equation is reformulated as:

$$C_i^{n+1} - C_i^0 = \frac{1-\alpha}{AB(\alpha)}f\left(x_i,t_n,C_i^n\right) + \frac{\alpha}{AB(\alpha)\Gamma(\alpha)}\int_0^{t_{n+1}} (t_{n+1}-\tau)^{\alpha-1}f\left(x_i,\tau,C(x_i,\tau)\right)d\tau \qquad (21.31)$$

$$= \frac{1-\alpha}{AB(\alpha)}f\left(x_i,t_n,C_i^n\right) + \frac{\alpha}{AB(\alpha)\Gamma(\alpha)}\sum_{j=0}^n\int_{t_j}^{t_{j+1}} (t_{n+1}-\tau)^{\alpha-1}f\left(x_i,\tau,C(x_i,\tau)\right)d\tau \qquad (21.32)$$

We approximate of the function $f(x_i,\tau,C(x_i,\tau))$ within the interval $[t_j,t_{j+1}]$ using the Lagrange polynomial method, the following equation is obtained:

$$C_i^{n+1} - C_i^0 = \frac{1-\alpha}{AB(\alpha)}f\left(x_i,t_n,C_i^n\right) + \frac{\alpha}{AB(\alpha)\Gamma(\alpha)}\sum_{j=0}^n\int_{t_j}^{t_{j+1}} P_j(\tau)(t-\tau)^{\alpha-1}d\tau \qquad (21.33)$$

We further generate the following form of equation:

$$C_i^{n+1} = C_i^0 + \frac{1-\alpha}{AB(\alpha)} f\left(x_i, t_n, C_i^n\right)$$

$$+ \frac{(\Delta t)^\alpha}{AB(\alpha)\Gamma(\alpha+2)} \sum_{j=0}^{n} \left[\begin{array}{l} f\left(x_i, t_j, C_i^j\right)\left\{\begin{array}{l}(n-j+1)^\alpha (n-j+2+\alpha) \\ -(n-j)^\alpha (n-j+2+2\alpha)\end{array}\right\} \\ -f\left(x_i, t_{j-1}, C_i^{j-1}\right)\left\{(n-j+1)^{\alpha+1} - (n-j)^\alpha (n-j+1+\alpha)\right\} \end{array} \right]$$

$$\tag{21.34}$$

At this point, the functions $f\left(x_i, t_n, C_i^n\right), f\left(x_i, t_j, C_i^j\right)$ and $f\left(x_i, t_{j-1}, C_i^{j-1}\right)$ from Equation (21.34) can be presented as:

$$f\left(x_i, t_j, C_i^j\right) = \hat{D}\frac{C_{i+1}^j - 2C_i^j + C_{i-1}^j}{(\Delta x)^2} - \hat{v}\frac{C_{i+1}^j - C_{i-1}^j}{\Delta x} - \lambda \hat{R} C_i^j \tag{21.35}$$

$$f\left(x_i, t_{j-1}, C_i^{j-1}\right) = \hat{D}\frac{C_{i+1}^{j-1} - 2C_i^{j-1} + C_{i-1}^{j-1}}{(\Delta x)^2} - \hat{v}\frac{C_{i+1}^{j-1} - C_{i-1}^{j-1}}{\Delta x} - \lambda \hat{R} C_i^{j-1} \tag{21.36}$$

$$f\left(x_i, t_n, C_i^n\right) = \hat{D}\frac{C_{i+1}^n - 2C_i^n + C_{i-1}^n}{(\Delta x)^2} - \hat{v}\frac{C_{i+1}^n - C_{i-1}^n}{\Delta x} - \lambda \hat{R} C_i^n \tag{21.37}$$

Equation (21.24) can now be written by substituting with the results of the functions in Equations (21.35)–(21.37), we then have the following:

$$C_i^{n+1} = C_i^0 + \frac{1-\alpha}{AB(\alpha)}\left[\hat{D}\frac{C_{i+1}^n - 2C_i^n + C_{i-1}^n}{(\Delta x)^2} - \hat{v}\frac{C_{i+1}^n - C_{i-1}^n}{\Delta x} - \lambda \hat{R} C_i^n \right]$$

$$+ \frac{(\Delta t)^\alpha}{AB(\alpha)\Gamma(\alpha+2)} \sum_{j=0}^{n} \left[\begin{array}{l} \left(\hat{D}\frac{C_{i+1}^j - 2C_i^j + C_{i-1}^j}{(\Delta x)^2} - \hat{v}\frac{C_{i+1}^j - C_{i-1}^j}{\Delta x} - \lambda \hat{R} C_i^j\right) \\ \left\{(n-j+1)^\alpha (n-j+2+\alpha) - (n-j)^\alpha (n-j+2+2\alpha)\right\} \\ -\left(\hat{D}\frac{C_{i+1}^{j-1} - 2C_i^{j-1} + C_{i-1}^{j-1}}{(\Delta x)^2} - \hat{v}\frac{C_{i+1}^{j-1} - C_{i-1}^{j-1}}{\Delta x} - \lambda \hat{R} C_i^{j-1}\right) \\ \left\{(n-j+1)^{\alpha+1} - (n-j)^\alpha (n-j+1+\alpha)\right\} \end{array} \right]$$

$$\tag{21.38}$$

21.3.4 NUMERICAL STABILITY ANALYSIS OF THE NEW MODEL USING THE VON NEUMANN METHOD

In this section, the conditions for stability of numerical schemes for the newly generated groundwater transport model is analyzed using von Neumann stability analysis. Analysis for stability is essential, because during discretization of partial differential equations (PDEs), numerical errors are likely to have been generated, hence the need for stability analysis (Delahaies, 2012). A finite difference scheme is said to be stable if the associated error remains constant or decreases with time throughout the entire process of computation. On the other hand, when the generated error increases with time, then the scheme becomes unstable. Note that the numerical scheme is said to be stable if $|\xi| \leq 1$, and unstable if $|\xi| > 1$. The von Neumann method is based on the decay of errors into Fourier series.

Consequently, the Fourier expansion can be presented in terms of space as follows:

$$\rho(x,t) = \sum_f \hat{\rho}(t) \exp(ik_m x) \tag{21.39}$$

For von Neumann stability analysis, we assume the following:

$$C_i^{n+1} = \hat{\rho}_{n+1} e^{ik_m x} \tag{21.40}$$

$$C_i^n = \hat{\rho}_n e^{ik_m x} \tag{21.41}$$

$$C_{i+1}^n = \hat{\rho}_n e^{ik_m(x+\Delta x)} \tag{21.42}$$

$$C_{i-1}^n = \hat{\rho}_n e^{ik_m(x-\Delta x)} \tag{21.43}$$

$$C_i^{n-1} = \hat{\rho}_{n-1} e^{ik_m x} \tag{21.44}$$

$$C_{i+1}^{n-1} = \hat{\rho}_{n-1} e^{ik_m(x+\Delta x)} \tag{21.45}$$

$$C_{i-1}^{n-1} = \hat{\rho}_{n-1} e^{ik_m(x-\Delta x)} \tag{21.46}$$

21.3.4.1 Stability Analysis of the New Numerical Scheme for Solution of PDEs Derived in Terms of the Caputo–Fabrizio Fractional Derivative

In this section, we present the condition of stability to our discretized advection–dispersion transport Equation (21.27) in the case of the Caputo–Fabrizio fractional order derivative. The approach employed is the von Neumann stability analysis.

We recall our discretized Equation (21.38) and simplify using the constants a, b, f, g, h, k and z, where:

$$C_i^{n+1} = C_i^n + f \left\{ \begin{bmatrix} a\left(C_{i+1}^n - 2C_i^n + C_{i-1}^n\right) - b\left(C_{i+1}^n - C_{i-1}^n\right) - zC_i^n \end{bmatrix} \\ -\begin{bmatrix} a\left(C_{i+1}^{n-1} - 2C_i^{n-1} + C_{i-1}^{n-1}\right) - b\left(C_{i+1}^{n-1} - C_{i-1}^{n-1}\right) - zC_i^{n-1} \end{bmatrix} \right\} \\ + g \left\{ \begin{aligned} h\left[a\left(C_{i+1}^n - 2C_i^n + C_{i-1}^n\right) - b\left(C_{i+1}^n - C_{i-1}^n\right) - zC_i^n \right] \\ -k\left[a\left(C_{i+1}^{n-1} - 2C_i^{n-1} + C_{i-1}^{n-1}\right) - b\left(C_{i+1}^{n-1} - C_{i-1}^{n-1}\right) - zC_i^{n-1} \right] \end{aligned} \right\} \tag{21.47}$$

We further simplify the above Equation (21.47) by grouping the like terms and plug in Equations (21.39) to (21.46) as follows:

$$\begin{aligned}
\hat{\rho}_{n+1} e^{ik_m x} &= \left(1 - 2af - zf\right)\hat{\rho}_n e^{ik_m x} + af\left(e^{ik_m \Delta x} + e^{-ik_m \Delta x}\right)\hat{\rho}_n e^{ik_m x} \\
&\quad - bf\left(e^{ik_m \Delta x} - e^{-ik_m \Delta x}\right)\hat{\rho}_n e^{ik_m x} + \left(2af + zf\right)\hat{\rho}_{n-1} e^{ik_m x} - af\left(e^{ik_m \Delta x} + e^{-ik_m \Delta x}\right)\hat{\rho}_{n-1} e^{ik_m x} \\
&\quad + bf\left(e^{ik_m \Delta x} - e^{-ik_m \Delta x}\right)\hat{\rho}_{n-1} e^{ik_m x} - \left(2agh + zgh\right)\hat{\rho}_n e^{ik_m x} + agh\left(e^{ik_m \Delta x} + e^{-ik_m \Delta x}\right)\hat{\rho}_n e^{ik_m x} \\
&\quad - bgh\left(e^{ik_m \Delta x} - e^{-ik_m \Delta x}\right)\hat{\rho}_n e^{ik_m x} + \left(2agk + zgk\right)\hat{\rho}_{n-1} e^{ik_m x} - agk\left(e^{ik_m \Delta x} + e^{-ik_m \Delta x}\right)\hat{\rho}_{n-1} e^{ik_m x} \\
&\quad + bgk\left(e^{ik_m \Delta x} - e^{-ik_m \Delta x}\right)\hat{\rho}_{n-1} e^{ik_m x}
\end{aligned} \tag{21.48}$$

Dividing both sides of Equation (21.48) by $e^{ik_m x}$ yields the following:

$$
\begin{aligned}
\hat{\rho}_{n+1} ={}& \left(1-2af-zf\right)\hat{\rho}_n + af\left(e^{ik_m\Delta x}+e^{-ik_m\Delta x}\right)\hat{\rho}_n - bf\left(e^{ik_m\Delta x}-e^{-ik_m\Delta x}\right)\hat{\rho}_n \\
&+ \left(2af+zf\right)\hat{\rho}_{n-1} - af\left(e^{ik_m\Delta x}+e^{-ik_m\Delta x}\right)\hat{\rho}_{n-1} + bf\left(e^{ik_m\Delta x}-e^{-ik_m\Delta x}\right)\hat{\rho}_{n-1} \\
&- \left(2agh+zgh\right)\hat{\rho}_n + agh\left(e^{ik_m\Delta x}+e^{-ik_m\Delta x}\right)\hat{\rho}_n - bgh\left(e^{ik_m\Delta x}-e^{-ik_m\Delta x}\right)\hat{\rho}_n \\
&+ \left(2agk+zgk\right)\hat{\rho}_{n-1} - agk\left(e^{ik_m\Delta x}+e^{-ik_m\Delta x}\right)\hat{\rho}_{n-1} + bgk\left(e^{ik_m\Delta x}-e^{-ik_m\Delta x}\right)\hat{\rho}_{n-1}
\end{aligned}
\tag{21.49}
$$

If,

$$
e^{ik_m\Delta x} = \cos\left(k_m\Delta x\right) + i\sin\left(k_m\Delta x\right)
\tag{21.50}
$$

and,

$$
e^{-ik_m\Delta x} = \cos\left(k_m\Delta x\right) - i\sin\left(k_m\Delta x\right)
\tag{21.51}
$$

Then, by using the double angle for *cos* and factorization we can simplify Equation (21.49) into:

$$
\hat{\rho}_{n+1} =
\begin{bmatrix}
1-zf-4af\sin^2\left(\dfrac{k_m\Delta x}{2}\right)-bf\left(2i\sin\left(k_m\Delta x\right)\right)-zgh \\[2mm]
-4agh\sin^2\left(\dfrac{k_m\Delta x}{2}\right)-bgh\left(2i\sin\left(k_m\Delta x\right)\right)
\end{bmatrix}\hat{\rho}_n
$$
$$
+
\begin{bmatrix}
zf+4af\sin^2\left(\dfrac{k_m\Delta x}{2}\right)+bf\left(2i\sin\left(k_m\Delta x\right)\right)+zgk \\[2mm]
+4agk\sin^2\left(\dfrac{k_m\Delta x}{2}\right)+bgk\left(2i\sin\left(k_m\Delta x\right)\right)
\end{bmatrix}\hat{\rho}_{n-1}
\tag{21.52}
$$

Let us suppose from the above equation that:

$$
A =
\begin{bmatrix}
1-zf-4af\sin^2\left(\dfrac{k_m\Delta x}{2}\right)-bf\left(2i\sin\left(k_m\Delta x\right)\right)-zgh \\[2mm]
-4agh\sin^2\left(\dfrac{k_m\Delta x}{2}\right)-bgh\left(2i\sin\left(k_m\Delta x\right)\right)
\end{bmatrix}
$$

and,

$$
B =
\begin{bmatrix}
zf+4af\sin^2\left(\dfrac{k_m\Delta x}{2}\right)+bf\left(2i\sin\left(k_m\Delta x\right)\right)+zgk \\[2mm]
+4agk\sin^2\left(\dfrac{k_m\Delta x}{2}\right)+bgk\left(2i\sin\left(k_m\Delta x\right)\right)
\end{bmatrix}
$$

Then Equation (21.52) can be written as:

$$
\hat{\rho}_{n+1} = A\hat{\rho}_n + B\hat{\rho}_{n-1}
\tag{21.53}
$$

From Equation (21.53), when $n = 0$ we have:

$$
\hat{\rho}_1 = A\hat{\rho}_0
\tag{21.54}
$$

By rearranging, we get:

$$\frac{\hat{\rho}_1}{\hat{\rho}_0} = A \qquad (21.55)$$

Thus, we find the stability condition for which:

$$\left| \frac{\hat{\rho}_1}{\hat{\rho}_0} \right| < 1 \qquad (21.56)$$

Which implies that:

$$\left| \frac{\hat{\rho}_1}{\hat{\rho}_0} \right| = |A| < 1$$

$$|A| < 1 \qquad (21.57)$$

Therefore, by applying the absolute value of a complex number, where we assume that $s = a + ib$ and $|s| - \sqrt{a^2 + b^2}$. We now have:

$$|A| = \sqrt{\left(1 - zf - 4af \sin^2\left(\frac{k_m \Delta x}{2} \right) - zgh - 4agh \sin^2\left(\frac{k_m \Delta x}{2} \right) \right)^2 - \left(2bf \sin\left(k_m \Delta x \right) + 2bgh \sin\left(k_m \Delta x \right) \right)^2} \qquad (21.58)$$

Thus,

$$|A| = \sqrt{\left(1 - zf - 4af \sin^2\left(\frac{k_m \Delta x}{2} \right) - zgh - 4agh \sin^2\left(\frac{k_m \Delta x}{2} \right) \right)^2 - \left(2bf \sin\left(k_m \Delta x \right) + 2bgh \sin\left(k_m \Delta x \right) \right)^2} < 1 \qquad (21.59)$$

Therefore, we can infer that $\left| \hat{\rho}_1 \right| < \left| \hat{\rho}_0 \right|$ when:

$$\sqrt{\left(1 - zf - 4af \sin^2\left(\frac{k_m \Delta x}{2} \right) - zgh - 4agh \sin^2\left(\frac{k_m \Delta x}{2} \right) \right)^2 - \left(2bf \sin\left(k_m \Delta x \right) + 2bgh \sin\left(k_m \Delta x \right) \right)^2} < 1 \qquad (21.60)$$

We again recall from Equation (21.52) and suppose that:

$$|B| = \sqrt{\left(zf + 4af \sin^2\left(\frac{k_m \Delta x}{2} \right) + zgk + 4agk \sin^2\left(\frac{k_m \Delta x}{2} \right) \right)^2 + \left(2bf \sin\left(k_m \Delta x \right) + 2bgk \sin\left(k_m \Delta x \right) \right)^2} \qquad (21.61)$$

$\forall n > 0$, we assume that:

$$\left| \hat{\rho}_n \right| < \left| \hat{\rho}_0 \right| \Rightarrow \left| \frac{\hat{\rho}_n}{\hat{\rho}_0} \right| < 1$$

Using the above assumption, we can prove that:

$$\left|\hat{\rho}_{n+1}\right| < \left|\hat{\rho}_0\right| \Rightarrow \left|\frac{\hat{\rho}_{n+1}}{\hat{\rho}_0}\right| < 1$$

Thus,

$$\left|\hat{\rho}_{n+1}\right| = \left|A\hat{\rho}_n + B\hat{\rho}_{n-1}\right| \tag{21.62}$$

Which means that,

$$\left|\hat{\rho}_{n+1}\right| \le \left|A\right|\left|\hat{\rho}_n\right| + \left|B\right|\left|\hat{\rho}_{n-1}\right| \tag{21.63}$$

By applying the inductive hypothesis, we suppose that:

$$\left|\hat{\rho}_n\right| < 0 \quad \text{and} \quad \left|\hat{\rho}_{n-1}\right| < \left|\hat{\rho}_0\right|$$

$$\left|\hat{\rho}_{n+1}\right| < \left|A\right|\left|\hat{\rho}_0\right| + \left|B\right|\left|\hat{\rho}_0\right| \tag{21.64}$$

By factorization, we can infer that:

$$\left|\hat{\rho}_{n+1}\right| < \left|\hat{\rho}_0\right|\left(\left|A\right| + \left|B\right|\right) \tag{21.65}$$

We further have:

$$\left|\frac{\hat{\rho}_{n+1}}{\hat{\rho}_0}\right| < \left|A\right| + \left|B\right| \tag{21.66}$$

Remember that:

$$\left|\frac{\hat{\rho}_{n+1}}{\hat{\rho}_0}\right| < 1$$

This also implies that:

$$\left|A\right| + \left|B\right| < 1 \tag{21.67}$$

Similarly,

$$\sqrt{\begin{gathered}\left(1 - zf - 4af\sin^2\left(\frac{k_m\Delta x}{2}\right) - zgh - 4agh\sin^2\left(\frac{k_m\Delta x}{2}\right)\right)^2 \\ -\left(2bf\sin\left(k_m\Delta x\right) + 2bgh\sin\left(k_m\Delta x\right)\right)^2\end{gathered}}$$

$$+\sqrt{\begin{gathered}\left(zf + 4af\sin^2\left(\frac{k_m\Delta x}{2}\right) + zgk + 4agk\sin^2\left(\frac{k_m\Delta x}{2}\right)\right)^2 \\ +\left(2bf\sin\left(k_m\Delta x\right) + 2bgk\sin\left(k_m\Delta x\right)\right)^2\end{gathered}} < 1 \tag{21.68}$$

Therefore, we conclude that under this condition our numerical method is conditionally stable.

21.3.4.2 Stability Analysis of the New Numerical Scheme for Solution of PDEs Derived in Terms of the Atangana–Baleanu Fractional Derivative in the Caputo Sense

This section presents the analysis of stability condition to our discretized advection–dispersion transport equation (Equation 21.38) in the case of Atangana–Baleanu fractional order derivative in the Caputo sense. The method of approach employed to test for stability is the von Neumann stability method.

Therefore we recall our discretized Equation (21.38) and simplify using the constants a, b, f, g, z, $\delta_n^{\alpha,1}$ and $\delta_n^{\alpha,2}$, where:

$$
\begin{aligned}
C_i^{n+1} = C_i^n &+ f\left[a\left(C_{i+1}^{n+1} - 2C_i^{n+1} + C_{i-1}^{n+1}\right) - b\left(C_{i+1}^{n+1} - C_{i-1}^{n+1}\right) - zC_i^{n+1}\right] \\
&- f\left[a\left(C_{i+1}^n - 2C_i^n + C_{i-1}^n\right) - b\left(C_{i+1}^n - C_{i-1}^n\right) - zC_i^n\right] \\
&+ g\left[a\left(C_{i+1}^n - 2C_i^n + C_{i-1}^n\right) - b\left(C_{i+1}^n - C_{i-1}^n\right) - zC_i^n\right]\delta_n^{\alpha,1} \\
&- g\left[a\left(C_{i+1}^{n-1} - 2C_i^{n-1} + C_{i-1}^{n-1}\right) - b\left(C_{i+1}^{n-1} - C_{i-1}^{n-1}\right) - zC_i^{n-1}\right]\delta_n^{\alpha,2}
\end{aligned}
\tag{21.69}
$$

We further simplify the above Equation (21.69) by grouping like terms and plug in Equations (21.40) to (21.46) as follows:

$$
\begin{aligned}
\hat{\rho}_{n+1}e^{ik_m x} = &\left(1 + 2af + zf\right)\hat{\rho}_n e^{ik_m x} - af\left(e^{ik_m\Delta x} + e^{-ik_m\Delta x}\right)\hat{\rho}_n e^{ik_m x} \\
&+ bf\left(e^{ik_m\Delta x} - e^{-ik_m\Delta x}\right)\hat{\rho}_n e^{ik_m x} - \left(2ag\delta_n^{\alpha,1} + zg\delta_n^{\alpha,1}\right)\hat{\rho}_n e^{ik_m x} + ag\delta_n^{\alpha,1}\left(e^{ik_m\Delta x} + e^{-ik_m\Delta x}\right)\hat{\rho}_n e^{ik_m x} \\
&- bg\delta_n^{\alpha,1}\left(e^{ik_m\Delta x} - e^{-ik_m\Delta x}\right)\hat{\rho}_n e^{ik_m x} - \left(2af + zf\right)\hat{\rho}_{n+1}e^{ik_m x} + af\left(e^{ik_m\Delta x} + e^{-ik_m\Delta x}\right)\hat{\rho}_{n+1}e^{ik_m x} \\
&- bf\left(e^{ik_m\Delta x} - e^{-ik_m\Delta x}\right)\hat{\rho}_{n+1}e^{ik_m x} + \left(2ag\delta_n^{\alpha,2} + zg\delta_n^{\alpha,2}\right)\hat{\rho}_{n-1}e^{ik_m x} \\
&- ag\delta_n^{\alpha,2}\left(e^{ik_m\Delta x} + e^{-ik_m\Delta x}\right)\hat{\rho}_{n-1}e^{ik_m x} + bg\delta_n^{\alpha,2}\left(e^{ik_m\Delta x} - e^{-ik_m\Delta x}\right)\hat{\rho}_{n-1}e^{ik_m x}
\end{aligned}
\tag{21.70}
$$

Dividing both sides of Equation (21.70) by $e^{ik_m x}$ yields the following:

$$
\begin{aligned}
\hat{\rho}_{n+1} = &\left(1 + 2af + zf\right)\hat{\rho}_n - af\left(e^{ik_m\Delta x} + e^{-ik_m\Delta x}\right)\hat{\rho}_n + bf\left(e^{ik_m\Delta x} - e^{-ik_m\Delta x}\right)\hat{\rho}_n \\
&- \left(2ag\delta_n^{\alpha,1} + zg\delta_n^{\alpha,1}\right)\hat{\rho}_n + ag\delta_n^{\alpha,1}\left(e^{ik_m\Delta x} + e^{-ik_m\Delta x}\right)\hat{\rho}_n - bg\delta_n^{\alpha,1}\left(e^{ik_m\Delta x} - e^{-ik_m\Delta x}\right)\hat{\rho}_n \\
&- \left(2af + zf\right)\hat{\rho}_{n+1} + af\left(e^{ik_m\Delta x} + e^{-ik_m\Delta x}\right)\hat{\rho}_{n+1} - bf\left(e^{ik_m\Delta x} - e^{-ik_m\Delta x}\right)\hat{\rho}_{n+1} \\
&+ \left(2ag\delta_n^{\alpha,2} + zg\delta_n^{\alpha,2}\right)\hat{\rho}_{n-1} - ag\delta_n^{\alpha,2}\left(e^{ik_m\Delta x} + e^{-ik_m\Delta x}\right)\hat{\rho}_{n-1} + bg\delta_n^{\alpha,2}\left(e^{ik_m\Delta x} - e^{-ik_m\Delta x}\right)\hat{\rho}_{n-1}
\end{aligned}
\tag{21.71}
$$

If,

$$
e^{ik_m\Delta x} = \cos\left(k_m\Delta x\right) + i\sin\left(k_m\Delta x\right)
$$

and,

$$
e^{-ik_m\Delta x} = \cos\left(k_m\Delta x\right) - i\sin\left(k_m\Delta x\right)
$$

Then, by using the double angle for cos and factorization we can simplify Equation (21.71) into:

$$\hat{\rho}_{n+1} = \begin{bmatrix} 1 + zf - 2af\left(1 - \cos\left(k_m\Delta x\right)\right) + bf\left(2i\sin\left(k_m\Delta x\right)\right) + zg\delta_n^{\alpha,1} \\ -2ag\delta_n^{\alpha,1}\left(1 - \cos\left(k_m\Delta x\right)\right) - bg\delta_n^{\alpha,1}\left(2i\sin\left(k_m\Delta x\right)\right) \end{bmatrix}\hat{\rho}_n$$
$$+ \left[zg\delta_n^{\alpha,2} + 2ag\delta_n^{\alpha,2}\left(1 - \cos\left(k_m\Delta x\right)\right) + bg\delta_n^{\alpha,2}\left(2i\sin\left(k_m\Delta x\right)\right) \right]\hat{\rho}_{n-1}$$
$$- \left[zf + 2af\left(1 - \cos\left(k_m\Delta x\right)\right) + bf\left(2i\sin\left(k_m\Delta x\right)\right) \right]\hat{\rho}_{n+1} \tag{21.72}$$

We further have:

$$\begin{bmatrix} 1 + zf + 4af\sin^2\left(\dfrac{k_m\Delta x}{2}\right) \\ + bf\left(2i\sin\left(k_m\Delta x\right)\right) \end{bmatrix}\hat{\rho}_{n+1} = \begin{bmatrix} 1 + zf + 4af\sin^2\left(\dfrac{k_m\Delta x}{2}\right) + bf\left(2i\sin\left(k_m\Delta x\right)\right) \\ -zg\delta_n^{\alpha,1} - 4ag\sin^2\left(\dfrac{k_m\Delta x}{2}\right)\delta_n^{\alpha,1} - bg\delta_n^{\alpha,1}\left(2i\sin\left(k_m\Delta x\right)\right) \end{bmatrix}\hat{\rho}_n$$
$$+ \left[zg\delta_n^{\alpha,2} + 4ag\sin^2\left(\dfrac{k_m\Delta x}{2}\right)\delta_n^{\alpha,2} + bg\delta_n^{\alpha,2}\left(2i\sin\left(k_m\Delta x\right)\right) \right]\hat{\rho}_{n-1} \tag{21.73}$$

From the above equation when $n = 0$, we have:

$$\begin{bmatrix} 1 + zf + 4af\sin^2\left(\dfrac{k_m\Delta x}{2}\right) \\ + bf\left(2i\sin\left(k_m\Delta x\right)\right) \end{bmatrix}\rho_1 = \begin{bmatrix} 1 + zf + 4af\sin^2\left(\dfrac{k_m\Delta x}{2}\right) + bf\left(2i\sin\left(k_m\Delta x\right)\right) \\ -zg\delta_n^{\alpha,1} - 4ag\sin^2\left(\dfrac{k_m\Delta x}{2}\right)\delta_n^{\alpha,1} - bg\delta_n^{\alpha,1}\left(2i\sin\left(k_m\Delta x\right)\right) \end{bmatrix}\hat{\rho}_0 \tag{21.74}$$

Let us suppose that:

$$A_m = \left[1 + zf + 4af\sin^2\left(\dfrac{k_m\Delta x}{2}\right) + bf\left(2i\sin\left(k_m\Delta x\right)\right) \right]$$

and,

$$B_m = \begin{bmatrix} 1 + zf + 4af\sin^2\left(\dfrac{k_m\Delta x}{2}\right) + bf\left(2i\sin\left(k_m\Delta x\right)\right) - zg\delta_n^{\alpha,1} \\ -4ag\sin^2\left(\dfrac{k_m\Delta x}{2}\right)\delta_n^{\alpha,1} - bg\delta_n^{\alpha,1}\left(2i\sin\left(k_m\Delta x\right)\right) \end{bmatrix}$$

Therefore, we can present Equation (21.74) as:

$$\hat{\rho}_1 A_m = \hat{\rho}_0 B_m \tag{21.75}$$

Thus we can find the condition for which:

$$\left|\frac{\hat{\rho}_1}{\hat{\rho}_0}\right| < 1 \tag{21.76}$$

Which implies that:

$$\left|\frac{\hat{\rho}_1}{\hat{\rho}_0}\right| = \left|\frac{B_m}{A_m}\right| < 1 \tag{21.77}$$

By applying the absolute value of a complex number, where we assume that $s = a + ib$ and $|s| = \sqrt{a^2 + b^2}$. Thus, we now have:

$$|A_m| = \sqrt{\left(1 + zf + 4af\sin^2\left(\frac{k_m\Delta x}{2}\right)\right)^2 + \left(2bf\sin\left(k_m\Delta x\right)\right)^2} \qquad (21.78)$$

and,

$$|B_m| = \sqrt{\begin{array}{l}\left(1 + zf + 4af\sin^2\left(\frac{k_m\Delta x}{2}\right) - zg\delta_n^{\alpha,1} - 4ag\sin^2\left(\frac{k_m\Delta x}{2}\right)\delta_n^{\alpha,1}\right)^2 \\ + \left(2bf\sin\left(k_m\Delta x\right) - 2bg\delta_n^{\alpha,1}\sin\left(k_m\Delta x\right)\right)^2\end{array}} \qquad (21.79)$$

Therefore, $\left|\dfrac{B_m}{A_m}\right| < 1$ is presented as:

$$\left|\frac{B_m}{A_m}\right| = \frac{\sqrt{\begin{array}{l}\left(1 + zf + 4af\sin^2\left(\frac{k_m\Delta x}{2}\right) - zg\delta_n^{\alpha,1} - 4ag\sin^2\left(\frac{k_m\Delta x}{2}\right)\delta_n^{\alpha,1}\right)^2 \\ + \left(2bf\sin\left(k_m\Delta x\right) - 2bg\delta_n^{\alpha,1}\sin\left(k_m\Delta x\right)\right)^2\end{array}}}{\sqrt{\left(1 + zf + 4af\sin^2\left(\frac{k_m\Delta x}{2}\right)\right)^2 + \left(2bf\sin\left(k_m\Delta x\right)\right)^2}} < 1 \qquad (21.80)$$

Simplifying,

$$\sqrt{\begin{array}{l}\left(1 + zf + 4af\sin^2\left(\frac{k_m\Delta x}{2}\right) - zg\delta_n^{\alpha,1} - 4ag\sin^2\left(\frac{k_m\Delta x}{2}\right)\delta_n^{\alpha,1}\right)^2 \\ + \left(2bf\sin\left(k_m\Delta x\right) - 2bg\delta_n^{\alpha,1}\sin\left(k_m\Delta x\right)\right)^2\end{array}}$$
$$< \sqrt{\left(1 + zf + 4af\sin^2\left(\frac{k_m\Delta x}{2}\right)\right)^2 + \left(2bf\sin\left(k_m\Delta x\right)\right)^2} \qquad (21.81)$$

We recall Equation (21.74) and suppose that:

$$C_m = \left[zg\delta_n^{\alpha,2} + 4ag\sin^2\left(\frac{k_m\Delta x}{2}\right)\delta_n^{\alpha,2} + bg\delta_n^{\alpha,2}\left(2i\sin\left(k_m\Delta x\right)\right)\right]$$

Then,

$$|C_m| = \sqrt{\left(zg\delta_n^{\alpha,2} + 4ag\sin^2\left(\frac{k_m\Delta x}{2}\right)\delta_n^{\alpha,2}\right)^2 + \left(bg\delta_n^{\alpha,2}\left(2\sin\left(k_m\Delta x\right)\right)\right)^2} \qquad (21.82)$$

Therefore, Equation (21.74) can be presented as:

$$\hat{\rho}_{n+1}A_m = \hat{\rho}_n B_m + \hat{\rho}_{n-1}C_m \qquad (21.83)$$

$\forall n > 0$, we assume that:

$$\left|\hat{\rho}_n\right| < \left|\hat{\rho}_0\right| \Delta \left|\frac{\hat{\rho}_n}{\Delta_0}\right| < 1$$

Using the above assumption, we can prove that:

$$\left|\frac{\hat{\rho}_{n+1}}{\hat{\rho}_0}\right| < 1$$

Thus,

$$\left|\hat{\rho}_{n+1}\right|\left|A_m\right| = \left|\hat{\rho}_n B_m + \hat{\rho}_{n-1} C_m\right| \tag{21.84}$$

Which means that:

$$\left|\hat{\rho}_{n+1}\right|\left|A_m\right| \le \left|\hat{\rho}_n\right|\left|B_m\right| + \left|\hat{\rho}_{n-1}\right|\left|C_m\right| \tag{21.85}$$

By applying the inductive hypothesis, we suppose that:

$$\left|\hat{\rho}_n\right| < 0 \text{ and } \left|\hat{\rho}_{n-1}\right| < \left|\hat{\rho}_0\right|$$

$$\left|\hat{\rho}_{n+1}\right|\left|A_m\right| < \left|\hat{\rho}_0\right|\left|B_m\right| + \left|\hat{\rho}_0\right|\left|C_m\right| \tag{21.86}$$

By factorization and simplification, we have:

$$\left|\hat{\rho}_{n+1}\right|\left|A_m\right| < \left|\hat{\rho}_0\right|\left(\left|B_m\right| + \left|C_m\right|\right)$$

$$\left|\frac{\hat{\rho}_{n+1}}{\hat{\rho}_0}\right| < \frac{\left|B_m\right| + \left|C_m\right|}{\left|A_m\right|} \tag{21.87}$$

Remember that:

$$\left|\frac{\hat{\rho}_{n+1}}{\hat{\rho}_0}\right| < 1$$

Which also implies that:

$$\frac{\left|B_m\right| + \left|C_m\right|}{\left|A_m\right|} < 1 \tag{21.88}$$

$$\frac{\sqrt{\begin{array}{l}\left(1 + zf + 4af\sin^2\left(\frac{k_m\Delta x}{2}\right) - zg\delta_n^{\alpha,1} - 4ag\sin^2\left(\frac{k_m\Delta x}{2}\right)\delta_n^{\alpha,1}\right)^2 \\ + \left(2bf\sin\left(k_m\Delta x\right) - 2bg\delta_n^{\alpha,1}\sin\left(k_m\Delta x\right)\right)^2\end{array}} + \sqrt{\left(zg\delta_n^{\alpha,2} + 4ag\sin^2\left(\frac{k_m\Delta x}{2}\right)\delta_n^{\alpha,2}\right)^2 + \left(bg\delta_n^{\alpha,2}\left(2\sin\left(k_m\Delta x\right)\right)\right)^2}}{\sqrt{\left(1 + zf + 4af\sin^2\left(\frac{k_m\Delta x}{2}\right)\right)^2 + \left(2bf\sin\left(k_m\Delta x\right)\right)^2}} < 1$$

We can therefore conclude that, under this condition, our numerical method is conditionally stable.

21.3.5 Numerical Simulations

Modeling real-world problems requires three major steps: observation, analysis and prediction. The last two steps are performed within the framework of mathematical models, where observed facts are converted into a mathematical equation or set of mathematical equations. To perform the analysis, one needs to solve such mathematical problems by using either analytical methods, which normally provide exact solutions. However, when the model is highly nonlinear, analytical methods are replaced by numerical methods. Numerical methods are now able to provide an approximate solution of the model. In the previous section, we presented some applications of new numerical methods to solve the new model of groundwater flow with stochastic coefficients. The stability analysis was performed to guarantee the accuracy of the method. In this section, we present the numerical simulations of the new suggested method for modeling advection and dispersion transport problems via ABC derivatives with different alpha values used. The figures are plotted using MATLAB software and the numerical simulations via this software are portrayed from Figures 21.1 to 21.19.

To perform the simulation, we consider the following theoretical parameters $0.2 < D < 2$, $0.4 < V < 2$, the following initial condition is considered $c(0,0) = 1000$, we consider the boundary condition to follow the exponential decay law with respect to time, and consider a fixed decay rate of 0.9. For each set of dispersion we consider the normal distribution as its statistical representation. Since the sub-surface is concerned and heterogeneity, one will expect a crossover behavior in transport distribution. To account for this crossover, we chose to simulate the model with the Atangana–Baleanu fractional derivative as this derivative was found to be a powerful mathematical operator able to capture crossover from waiting time distribution to probability distribution. In general, we are aware that within a geological formation with heterogeneity, the pollution path always follow a non-Gaussian distribution; this also allows us not to simulate with Caputo–Fabrizio to avoid a steady state situation.

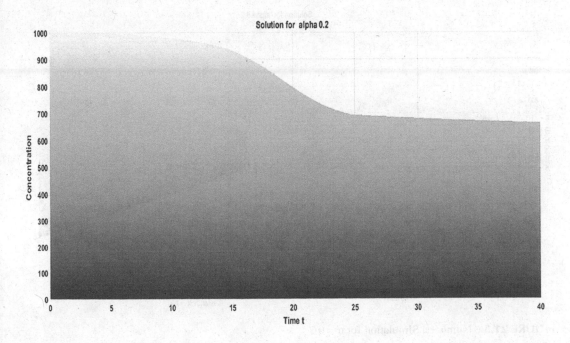

FIGURE 21.1 Numerical Simulation for $\alpha = 0.2$

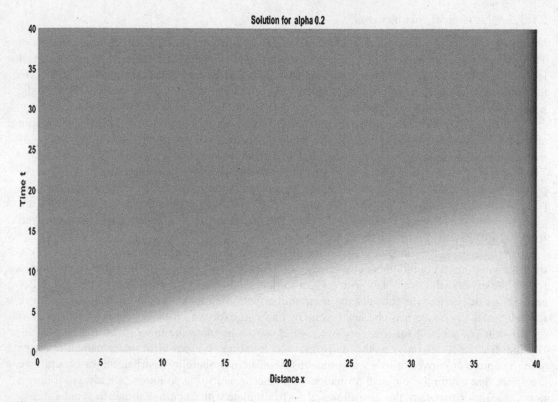

FIGURE 21.2   Numerical Simulation for $\alpha = 0.2$

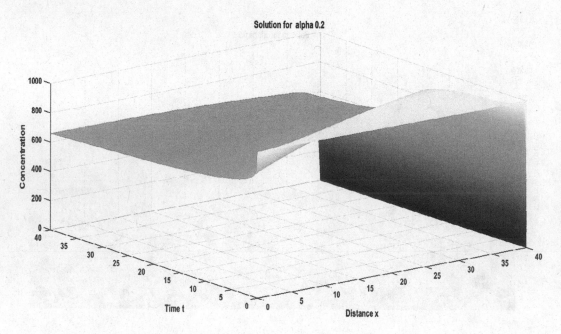

FIGURE 21.3   Numerical Simulation for $\alpha = 0.2$

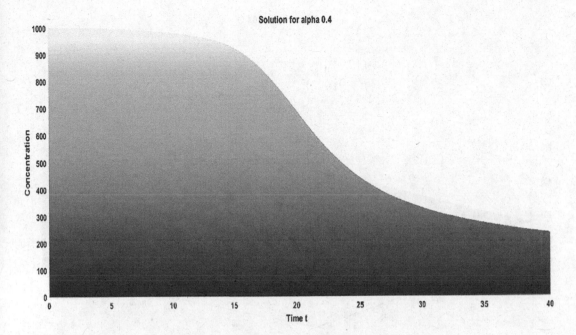

FIGURE 21.4 Numerical Simulation for $\alpha = 0.4$

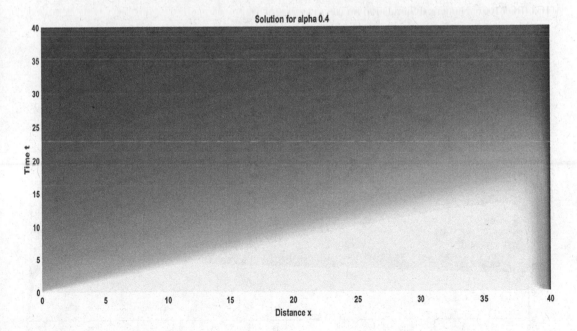

FIGURE 21.5 Numerical Simulation for $\alpha = 0.4$

21.3.6 RESULTS AND DISCUSSIONS

From the numerical simulations presented in the previous section, it can be seen that the results depict the behavior of certain real-world situation in which the contaminant concentration changes

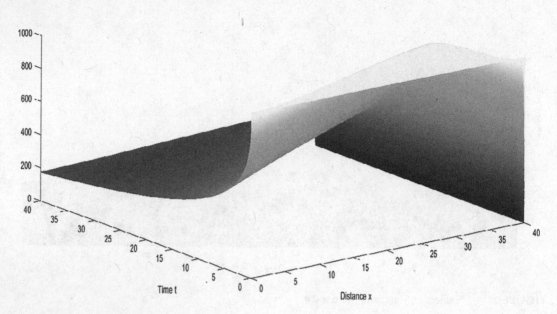

FIGURE 21.6   Numerical Simulation for $\alpha = 0.4$

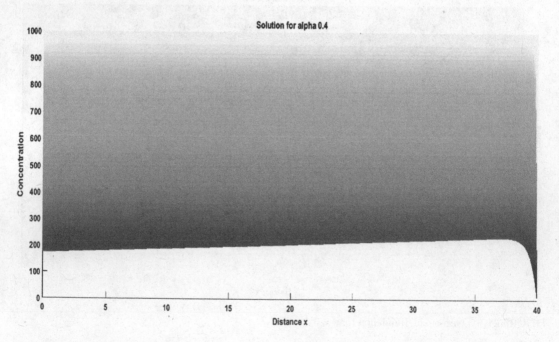

FIGURE 21.7   Numerical Simulation for $\alpha = 0.4$

with respect to time and space. This also indicates that the presence of heterogeneity with the aquifer systems has an impact on the groundwater velocity and dispersion of pollution. From the above
figures, we observe the gradual decrease of the contaminant concentration with time. That is, in
high values of alpha the pollution concentration decreases within a short time as compared to low

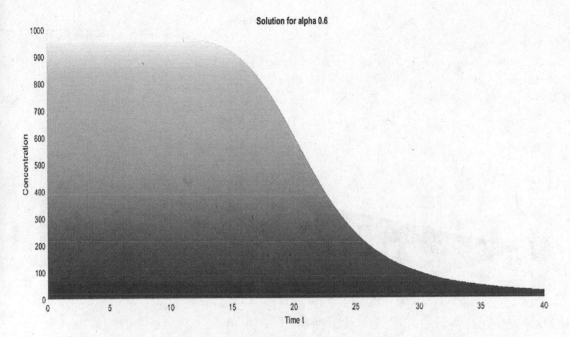

FIGURE 21.8 Numerical Simulation for $\alpha = 0.6$

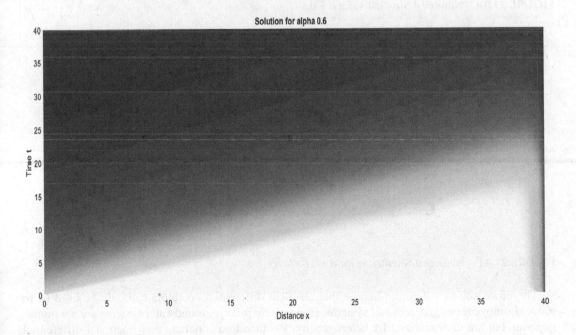

FIGURE 21.9 Numerical Simulation for $\alpha = 0.6$

alpha values. This means that increasing or decreasing the scale factor has a direct influence on the concentration of pollution. The numerical simulations also show that there is a crossover from Gauss to non-Gaussian probability distribution. Thus the above figures depict a normal distribution at some point but as the pollution travels with respect to time and distance, then we observe a change to non-Gaussian distribution. This is the main reason why the Atangana–Baleanu fractional derivative was used as it can capture these crossovers.

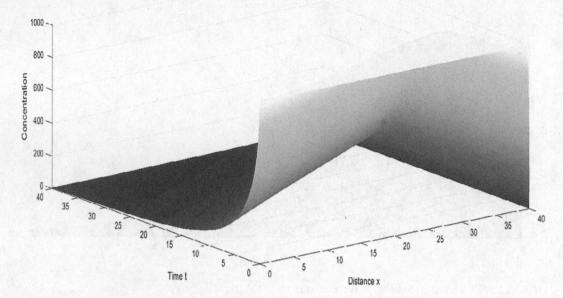

FIGURE 21.10   Numerical Simulation for $\alpha = 0.6$

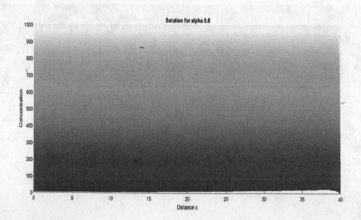

FIGURE 21.11   Numerical Simulation for $\alpha = 0.6$

Therefore, the new theorem here is that classical differential operators can only be used be to model homogeneous groundwater systems. However, in reality, groundwater systems are not homogeneous but are characterized by heterogeneity. We therefore conclude that fractional differential operators must be used to model natural groundwater systems because they can capture heterogeneity. The numerical obviously show that the fractional order differential operator can replicate very accurately the fast, slow and normal flow depending on the value of the used alpha. Another

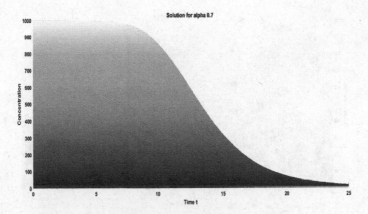

FIGURE 21.12 Numerical Simulation for α = 0.7

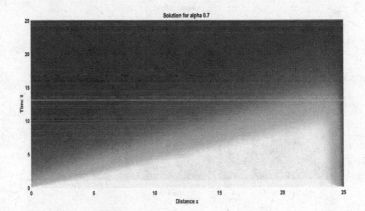

FIGURE 21.13 Numerical Simulation for α = 0.7

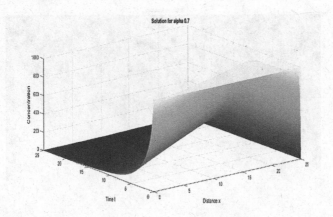

FIGURE 21.14 Numerical Simulation for α = 0.7

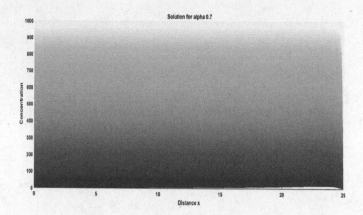

FIGURE 21.15 Numerical Simulation for α = 0.7

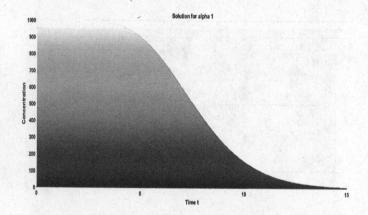

FIGURE 21.16 Numerical Simulation for α = 0.1

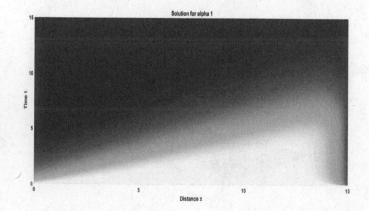

FIGURE 21.17 Numerical Simulation for α = 0.1

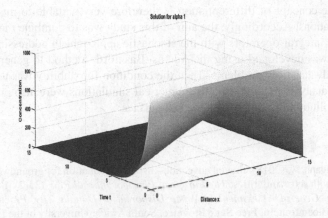

FIGURE 21.18 Numerical Simulation for $\alpha = 0.1$

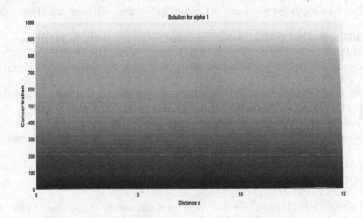

FIGURE 21.19 Numerical Simulation for $\alpha = 0.1$

important fact is that the concept of arithmetic average used in groundwater problems is not suitable as it gives exaggerated results. We suggest geometric means if all the same are different from zero.

21.4 CONCLUSION

Modeling groundwater flow behavior using mathematical equations has always been a challenge as it requires a detailed understanding about the geological formation through which groundwater moves. Some literature however, assumes that the aquifer parameters are constant at every point within the geological formation. Nevertheless, such assumptions become practically invalid because the sub-surface is characterized by heterogeneity, and aquifer parameters are not known with certainty. It is therefore very preferable to capture such uncertainties and heterogeneities using the concept of stochastic modeling. Thus, a stochastic approach has the ability to capture some physical processes with statistical setting. None the less, some models fail to provide reliable groundwater flow estimates because of their inability to account for the heterogeneity, viscoelasticity and memory effect. The concept of differentiation is therefore very suitable to model different types of geological formations. Accordingly, the aim of this study was to combine the concept of nonlocal differential and integral operators with the stochastic approach. In addressing the main, a new numerical scheme was developed using the Adams–Bashforth method to generate a new solution of modeling groundwater flow problems. Then the condition for stability was tested using the von Neumann stability analysis method. Finally, numerical simulations were also presented to observe the behavior of a pollution moving within the sub-surface.

REFERENCES

Allwright, A., & Atangana, A. (2018). Fractal advection–dispersion equation for groundwater transport in fractured aquifers with self similarities. *The European Physical Juornal Plus* 133(2), 1–20.

Atangana, A. (2013). *General Assessment of Waste Disporsal at Douala City. Practices, Principles and Uncertainties*. Bloemfontein, Free State Province, South Africa: University of the Free State, January.

Atangana, A. (2020). Prof. (S. Manundu, Interviewer), December 3.

Atangana, A., & Alkahtani, B.S. (2015). Analysis of the Keller–Segel model with a fractional derivative without a singular kernel. *Entropy*, 2015(17), 44394453. doi:10.3390/e17064439.

Atangana, A., & Araz, S.I. (2020). New concept in Calculus: Piecewise differential and integral. *Chaos Solution and Fractal*.

Atangana, A., & Bildik, N. (2013). The use of fractional order derivative to predict the groundwater flow. *Mathematical Problems in Engineering*, 2013(543026), 1–9. doi: 10.1155/2013/543026.

Atangana, A., & Bonyah, E. (2019). Fractional stochastic modeling: New approach to capture more heterogeneity. *Chaos*, 29(013118), 1–13.

Atangana, A., & Baleanu, D. (2016). New fractional derivatives with nonlocal and non-singular kernel: Theory and application to heat transfer model. *Thermal Science*. 20(2), 763–769.

Atangana, A., & Alqahtani, R. (2016). Numerical approximation of the space–time Caputo–Fabrizio fractional derivative and application to the groundwater pollution equation. *Advances in Difference Equations*. (1), 1–13

Batogna, G.R., & Atangana, A. (2019). Generalised class of time fractional Black–Scholes equation and numerical analysis. *Discrete & Continuous Dynamical Systems*.

Bear, J. (1972). Dynamics of Fluids in Porous Media. *Elsevier*, 579–582.

Bobba, A.G., & Singh, V.P. (1995). Groundwater contamination modelling. In V. P. Singh (Ed.), *Environmental Hydrology* (Vol. 15, p. 226). Water Science and Technology Library Springer, Dordrecht. doi:10.1007/978-94-017-1439-6_8.

Caputo, M., & Fabrizio, M. (2015). A new definition of fractional derivative without singular kernel. *Progress in Fractional Differentiation and Applications*, 73–85.

Chen, C.X. (1996). Theory and model of groundwater solute transport. *China University of Geosciene Press*, 4–7.

Chilakapati, A. (1995). *RAFT: A simulator for Reactive Flow and Transport of groundwater contaminants*. Pacific Northwest Laboratory, Richland, WA.

Cho, C. (1971). Convective transport of ammonium with nitrification in soil. *Canadian Journal of Soil Science*, 51(3), 339–350.

Chua, T.D. (1984). The design of a variable step integrator for the simulation of gas transmission network. *Journal of Numerical Methods Engineering*, 20(10), 1797–1813.

Clement, T.P., Sun, Y., Hooker, B.S., & Petersen, J.N. (1998). Modeling multispecies reactive transport in ground water. *Groundwater Monitoring & Remediation*, 18(2), 79–92.

Company, R.P. (2009). A second order numerical method for solving advection–diffusion models. *Math Computing Model*, 806–811.

Cordero, J.R., Eleazar, M.S., & Deane, R. (2019). Integrated discrete fracture and dual porosity: Dual permeability models for fluid flow in deformable fractured media. *Journal of Petroleum Science and Engineering*,175, 644–653.

Cortis, A. (2004). Analomous transport in classical soil and sand column. *Soil Science Society of America*, 1539–1548.

Cushman, J., & Ginn, T. (2000). Fractional advection–dispersion equation: A classical mass balance. *Water Resource Research*, 2000(36), 3763–3766.

Delahaies, S. (2012). *Numerical Solutions for Partial Differential Equations*. Retrieved from https://www. maths.surrey.ac.uk/st/S.B/ MAT3015 notes 2 2012.pdf.

Domenico, P.A. (1987). An analytical model for multidimensional transport of a decaying contaminant species. *Journal of Hydrology*, 49–58.

Jaiswal, D. K., & Kumar, A. (2011). Analytical solutions of advection-dispersion equation for varying pulse type input point source in one-dimension. *International Journal of Engineering, Science and Technology*, 3(1), 22–29.

McCartin, B. L. (2003). Accurate and efficient pricing of vanilla stock options via the Crandall Douglas Scheme. *Applied Math and Computing*, 143(1), 39–60.

Oishi, C.M. (2015). Stability analysis of Crank–Nicolson and Euler schemes. *BIT Numerical Mathematics*, 55(2), 487–513.

Partha, P.A. (2017). Comparison of Deterministic and Stochastic methods to predict spatial variation of groundwater depth. *Applied Water Science*, 7(1), 339–348.

Peinado, J., Ibáñez, J., Arias, E., & Hernández, V. (2010). Adams–Bashforth and Adams–Moulton methods for solving differential Riccati equations. *Computers and Mathematics with Applications*, 2010(60), 3032–3045.

Pool, M., Carrera, J., Alcolea, A., & Bocanegra, E.M. (2015). A comparison of deterministic and stochastic approaches for regional scale inverse modeling on the Mar del Plata aquifer. *Journal of Hydrology*, 531. 214–229.

Podlubny, I. (2002). Geometric and physical interpretation of fractional integration and fractional differentiation. *Fractional Calculus and Applied Analysis*, 2002(5), 367–396.

Qingdong, Z., & Jun, Y. (2015). Numerical simulation of fluid–solid coupling in fractured porous media with discrete fracture model and extended finite element method. *Computation*.

Sharma, P.K. (2018). A review on groundwater contaminant transport and remediation. *ISH Journal of Hydraulic Engineering*, 26(1), 112–121.

Simpson, J.M. (2015). Analytical model of reactive transport processes with spatially variable coefficients. *Royal Society of Open Science* 2(5), 140348.

Skinner, B.J., Stephen, P.C., Jeffrey, J.P., & Harold, L.L. (2004). *Dynamic Earth: An Introduction to Physical Geology*.

Stagnitti, F., Parlange, J.Y., Steenhuis, T.S., Barry, D.A., Li, L., Lockington, D.A., & Sander, G.C. (1995). Mathematical equations of the spread of pollution in soils. *Hydrological Systems Modeling*, II.

Süli, E. (2014). *Numerical Solution of Ordinary Differential Equations*. Mathematical Institute, Oxford of Exford.

Sun, Y.P. (1999). Analytical solutions for multiple species reactive transport in multiple dimensions. *Journal of Contaminant Hydrology*, 35(4), 429–440.

Tateishi, A.A., Ribeiro, H.V., & Lenzi, E.K. (2017). The role of fractional time-derivative operators on anomalous diffusion. *Frontiers in Physics*, 5, 52.

Taylor, H.M., & Karlin, S. (1998). *An Introduction to Stochastic Modeling*. California, USA: Academic Press Limited.

Toufik, M., & Atangana, A. (2017). New numerical approximation of fractional derivative with nonlocal and non-singular kernel: Application to chaotic models. *The European Physical Journal Plus*, 132(10), 1–16.

Warren, J.E., & Root, P.J. (1963). The behavior of naturally fractured reservoirs. *Society of Petroleum Engineers Journal*, 3(03), 245–255.

Wilson, J.L. (1978). Two-dimensional plume in uniform groundwater flow. *Journal of the Hydraulics Division*, 104(4), 503–514.

Witek, M.L. (2008). On stable and explicit numerical ethods for the advection–diffusion equation. *Maths, Computing and Simulation*, 79(3), 561–570.

Yunwei, S. (2003). Anlytical solutions for reactive transport of N-member radionuclide chains in a single fracture. *Journal of Contaminant Hydrology*, 62, 695–712.

22 Modelling a Conversion of a Confined to an Unconfined Aquifer Flow with Classical and Fractional Derivatives

Awodwa Magingi and Abdon Atangana
University of the Free State, Bloemfontein, South Africa

CONTENTS

22.1 INTRODUCTION

Freshwater demand is becoming an issue in most parts of the world. This demand includes groundwater resources to meet requirements for domestic, industrial, and agricultural use and even for recreational and environmental activities. Due to this increase in freshwater demand, driven by amongst other things an increasing population, groundwater is likely to be over-abstracted as demand cannot be met by surface water resources alone. Many confined aquifers have been reported as being pumped intensively all around the world (Springer and Bair, 1992; Wang and Zhan, 2009). With minimal knowledge and understanding of the different aquifers, over-abstraction can easily change the natural state of an aquifer, most likely from a confined to an unconfined one. It is therefore crucial to understand this conversion as it helps in the management of groundwater resources. Wang and Zhan (2009), Hu and Chen (2008), and Wang et al. (2009) report that a confined aquifer can be changed to an unconfined aquifer flow when the pumping rate is extremely high or the pumping period is too long.

A number of research studies have been conducted (e.g., Moench and Prickett, 1972; Chen et al., 2006; Hu and Chen, 2008). However, we should point out that several aspects have not been touched on in recent decades. An understanding of any natural problem, especially groundwater flow, since it is out of sight, starts with the good construction of a mathematical model. So far, the studies done

DOI: 10.1201/9781003266266-22

were based on local differential operators. Classical differential operators have been recognized to only predict physical problems following processes with no memory, which means they cannot capture the heterogeneity effect that is involved in groundwater flow systems. Very recently, some new concepts of differentiation have been suggested, called non-local operators; they are able to capture the flow within heterogeneous media, the flow that can follow Brownian motion as well as the random walk. The newly introduced mathematical operator has the ability to describe a statistical setting like Gaussian distribution. More precisely, the operator can capture normal and sub-diffusion, with a crossover in waiting time distribution that ranges from the exponential decay law to the power law. The aim of this chapter is to analyse the existing model adopted from Wang and Zhan (2009) using some newly introduced numerical schemes that have been recognized to be very efficient and powerful mathematical tools. Secondly, the aim is to extend the existing model using the newly introduced differential operator known as the Atangana–Baleanu derivative to provide a numerical scheme that can be used to solve such a model, present the condition under which the scheme is stable, and finally present some numerical simulations.

22.2   MODEL OUTLINE

The model is based on the schema shown in Figure 22.1. Consider pumping a confined aquifer at a constant rate (Q). The aquifer has a piezometric head (H) greater than the thickness of the aquifer (b). As the pumping period increases, the hydraulic head drops to below the upper confining unit, making H less than the thickness of the aquifer. As the piezometric surface drops, the water table also drops, converting the conditions of the immediate area around the pumping well from confined to unconfined aquifer conditions. The piezometric head decreases and the converted zone becomes larger with continuous pumping. The two equations below were adopted from Wang and Zhan (2009); they are models for a transient confined–unconfined flow. The confined zone is represented in Equation (22.1) and the unconfined zone in (22.2).

$$S_c \frac{\partial h}{\partial t} = \frac{KB}{r} \frac{\partial}{\partial r} \left(r \frac{\partial h}{\partial r} \right) h \geq B \tag{22.1}$$

$$S_y \frac{\partial h}{\partial t} = \frac{K}{r} \frac{\partial}{\partial r} \left(rh \frac{\partial h}{\partial r} \right) 0 < h \leq B \tag{22.2}$$

Equation (22.1) can be rewritten as:

$$\frac{S_c}{T} \frac{\partial h}{\partial t} = \frac{1}{r} \frac{\partial h}{\partial r} + \frac{\partial^2 h}{\partial r^2} \tag{22.3}$$

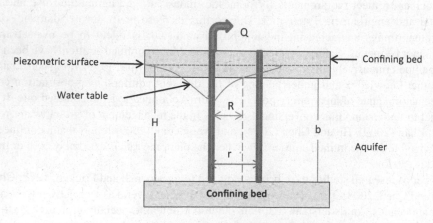

FIGURE 22.1   Schematic illustration of the Moench and Prickett model.

Equation (22.2) can be rewritten as:

$$S_y \frac{\partial h}{\partial t} = \frac{Kh}{r} \frac{\partial h}{\partial r} + K \left(\frac{\partial h}{\partial r} \right)^2 + Kh \frac{\partial^2 h}{\partial r^2}$$ (22.4)

B represents the thickness of the aquifer, S_c is the storage coefficient in the confined aquifer, r is the radial distance from the centre of the pumping well, t is the time since the commencement of pumping, h is the hydraulic head, K is the hydraulic conductivity, T is transmissivity, and S_y is the specific yield of the unconfined zone.

22.3   NUMERICAL SOLUTIONS

Often, systems described by differential equations are complex, or large, so much so that a pure analytical solution to the equations is not tractable. It is in these complex systems where computer simulations and numerical methods are useful. In this section we will apply existing numerical schemes to obtain solutions for Equations (22.1) and (22.2).

22.3.1   ADAMS–BASHFORTH METHOD (AB)

The AB method is one of the two Adams multi-methods; it is an explicit type of method. It is based on approximating the integral with a polynomial within the interval $(t_n, (t_{n+1}))$. This method uses transient and steady state information to develop the numerical solution of a complex function (Zeltkevic, 1998). Due to the scope of this chapter, we will not focus on the AB method as it cannot solve Partial Differential Equations with both local and non-local operators. However, Batogna and Atangana, 2017 modified this efficient and common AB method and developed a numerical scheme that can solve PDEs with both local and non-local operators.

22.3.2   ATANGANA–GNITCHOGNA NUMERICAL METHOD (NEW TWO-STEP LAPLACE ADAM-BASHFORTH METHOD)

The AB method has been proven to be very useful and good for differential equations with classical derivatives and those with non-integer order derivatives (Batogna and Atangana, 2017; Alkahtani, 2018). Atangana and Gnitchogna modified the AB method and combined it with the Laplace transform into a PDE to transform it into an Ordinary Differential Equations that can be analysed in Laplace space. The resulting ODE equation was further solved in a Laplace space using the AB method; a solution in inverse is given, which takes back the Laplace space into the real space and can be used to approximate PDEs with both local and non-local operators. Their modification commences with a general partial differential Equation (22.5) which is converted into an ordinary differential equation (22.6) using the Laplace transform; see Equation (22.7):

$$\frac{\partial u(x,t)}{\partial t} = Lu(x,t) + Nu(x,t)$$ (22.5)

The Laplace transform is given as:

$$\mathcal{L}\{f(t)\} = F(s)$$ (22.6)

where $F(s)$ is the Laplace transform and $f(t)$ together with $F(s)$ are called a Laplace transform pair. Applying the Laplace integral to Equation (22.6):

$$\mathcal{L}\left(\frac{\partial u(x,t)}{\partial t} \right) = \mathcal{L}\left(Lu(x,t) + Nu(x,t) \right)$$ (22.7)

The Laplace integral is used in the above equation so that there is only one variable remaining to be solved:

$$\frac{d}{dt}\big(u(p,t)\big) = \mathcal{L}\big(Lu(x,t) + Nu(x,t)\big) \tag{22.8}$$

Equation (22.8) can be changed to simple and similar terms:

$$\frac{d}{dt}\big(u(t)\big) = F(u,t) \tag{22.9}$$

where $(u(t))$ is $(u(p,t))$ and $F(u,t)$ is $\mathcal{L}\big(Lu(x,t) + Nu(x,t)\big)$.

Introducing the fundamental theorem of calculus results in the following integrated equations (integrated on both sides), where the integration boundaries are t and 0:

$$\int_0^t \frac{d}{dt}\big(u(t)\big) = \int_0^t F(u,\tau)\,d\tau \tag{22.10}$$

$$u(t) - u(0) = \int_0^t F(u,\tau)\,d\tau \tag{22.11}$$

$$u(t) = u(0) + \int_0^t F(u,\tau)\,d\tau \tag{22.12}$$

The boundary conditions can also be given as $t = (t_n)$ and $t = (t_{n+1})$, and the equation will be given by the following two equations:

When $t = (t_n)$:

$$u(t_n) = u(0) + \int_0^{t_n} F(u,\tau)\,d\tau \tag{22.13}$$

When $t = (t_{n+1})$:

$$u(t_{n+1}) = u(0) + \int_0^{t_{n+1}} F(u,\tau)\,d\tau \tag{22.14}$$

$$u(t_{n+1}) - u(t_n) = \left[u(0) + \int_0^{t_{n+1}} F(u,\tau)\,d\tau \right] - \left[u(0) + \int_0^{t_n} F(u,\tau)\,d\tau \right] \tag{22.15}$$

Conventions: applying the reversing limit integral:

$$u(t_{n+1}) - u(t_n) = \left[u(0) + \int_0^{t_{n+1}} F(u,\tau)\,d\tau \right] - \left[u(0) + \int_{t_n}^{0} F(u,\tau)\,d\tau \right] \tag{22.16}$$

Addition of integration intervals and subtraction of like terms into Equation (22.10):

$$u(t_{n+1}) - u(t_n) = \int_{t_n}^{t_{n+1}} F(u,\tau)\,d\tau \tag{22.17}$$

As stated in the introduction, the AB method uses the Lagrange polynomial method (LPM) to augment the new proposal; LPM is used for polynomial interpolation, meaning that this method is used for verification in theoretical arguments to support what is studied or in question. It is therefore used to prove that the solution can actually be used for non-Euclidean equations. $F(u,t)$ is approximated using the Lagrange Polynomial Difference method:

$$P(t) \approx F(u,t) = \frac{t - t_{n-1}}{t_n - t_{n-1}} F(u,t_n) + \frac{t - t_n}{t_{n-1} - t_n} F(u,t_{n-1}) \tag{22.18}$$

$$P(t) = \frac{t - t_{n-1}}{t_n - t_{n-1}} F_n + \frac{t - t_n}{t_{n-1} - t_n} F(u,t_{n-1}) \tag{22.19}$$

Equation (22.17) can be given as:

$$u(t_{n+1}) - u(t_n) = \int_{t_n}^{t_{n+1}} \left[\frac{t - t_{n-1}}{t_n - t_{n-1}} F_n + \frac{t - t_n}{t_{n-1} - t_n} F(u,t_{n-1}) \right] dt \tag{22.20}$$

$$u(t_{n+1}) - u(t_n) = \frac{F_n}{t_n - t_{n-1}} \int_{t_n}^{t_{n+1}} (t - t_{n-1}) dt + \frac{F_{n-1}}{t_{n-1} - t_n} \int_{t_n}^{t_{n+1}} (t - t_n) dt \tag{22.21}$$

Integrate Equation (22.21) and let $h = t_n - t_{n-1}$

$$u(t_{n+1}) - u(t_n)$$

$$= \frac{F_n}{t_n - t_{n-1}} \left[\frac{t^2}{2} - t t_{n-1} \right]_{t_n}^{t_{n+1}} + \frac{F_{n-1}}{t_{n-1} - t_n} \left[\frac{t^2}{2} - t t_n \right]_{t_n}^{t_{n+1}} \tag{22.22}$$

$$u(t_{n+1}) - u(t_n) = \frac{F_n}{h} \left[\frac{t^2}{2} - t t_{n-1} \right]_{t_n}^{t_{n+1}} + \frac{F_{n-1}}{(-h)} \left[\frac{t^2}{2} - t t_n \right]_{t_n}^{t_{n+1}} \tag{22.23}$$

Substituting the boundaries into Equation (22.22) where there is t:

$$u(t_{n+1}) - u(t_n) = \frac{F_n}{h} \left[\left(\frac{t_{n+1}^2}{2} - \frac{t_n^2}{2} \right) - (t_{n+1} t_{n-1} - t_n t_{n-1}) \right] - \frac{F_{n-1}}{h} \left[\left(\frac{t_{n+1}^2}{2} - \frac{t_n^2}{2} \right) - (t_{n+1} t_n - t_n t_n) \right] \tag{22.24}$$

$$u(t_{n+1}) - u(t_n) = \frac{F_n}{h} \left[\frac{1}{2} \left[(t_{n+1} - t_n)(t_{n+1} + t_n) \right] - t_{n-1} (t_{n+1} - t_n) \right]$$
$$- \frac{F_{n-1}}{h} \left[\frac{1}{2} \left[(t_{n+1} - t_n)(t_{n+1} + t_n) \right] - t_n (t_{n+1} - t_n) \right] \tag{22.25}$$

It should be noted that $h = t_{n+1} - t_n$

$$u(t_{n+1}) - u(t_n) = \frac{F_n}{h} \left(\frac{h}{2} (t_{n+1} + t_n) - h t_{n-1} \right) - \frac{F_{n-1}}{h} \left(\frac{h}{2} (t_{n+1} t_n) - h t_n \right) \tag{22.26}$$

Then h cancels out to obtain:

$$u(t_{n+1}) - u(t_n) = F_n - F_{n-1} \left(\frac{1}{2} (t_{n+1} t_n) - t_n \right) \tag{22.27}$$

$$u(t_{n+1}) - u(t_n) = F_n \left[\frac{1}{2} \big((n+1)h + nh \big) - (n-1)h \right] - F_{n-1} \left[\frac{1}{2} \big((n+1)h + nh \big) - nh \right] \quad (22.28)$$

$$u(t_{n+1}) - u(t_n) = F_n \left(\frac{nh}{2} + \frac{h}{2} + \frac{nh}{2} - nh + h \right) - F_{n-1} \left(\frac{nh}{2} + \frac{h}{2} + \frac{nh}{2} - nh \right) \quad (22.29)$$

$$u(t_{n+1}) - u(t_{n+1}) = F_n \left(h + \frac{h}{2} \right) - F_{n-1} \left(nh - nh + \frac{h}{2} \right) \quad (22.30)$$

$$u(t_{n+1}) - u(t_n) = F_n \left(\frac{3h}{2} \right) - F_{n-1} \left(\frac{h}{2} \right) \quad (22.31)$$

22.3.3 NUMERICAL SOLUTION FOR THE UNCONFINED AQUIFER ZONE

Using Equation (22.31) on the equation for the unconfined zone, we rearrange Equation (22.2) by dividing both sides by S_y.

Given in time:

$$\frac{\partial h}{\partial t} = \frac{K}{S_y} \frac{h}{r} \frac{\partial h}{\partial r} + \frac{K}{S_y} \left(\frac{\partial h}{\partial r} \right)^2 + \frac{K}{S_y} h \frac{\partial^2 h}{\partial r^2} \quad (22.32)$$

Given in space:

$$h(x_i, t_{n+1}) - h(x_i, t_n) = \frac{3}{2} \Delta t F_i^n - \frac{\Delta t}{2} F_i^{n-1} \quad (22.33)$$

Now substitute Equation (22.32) into Equation (22.31) to obtain:

$$
\begin{aligned}
h_i^{n+1} - h_i^n = \frac{3}{2} \Delta t & \left[\frac{K}{S_y} \left(\frac{h_{i+1}^n - h_{i-1}^n}{4\Delta r} + \frac{h_{i+1}^{n-1} - h_{i-1}^{n-1}}{4\Delta r} \right) \frac{h_i^n}{r_i} + \frac{K}{S_y} \left(\frac{h_{i+1}^n - h_{i-1}^n}{4\Delta r} + \frac{h_{i+1}^{n-1} - h_{i-1}^{n-1}}{4\Delta r} \right)^2 \right. \\
& \left. + \frac{K}{S_y} h_i^n \left(\frac{h_{i+1}^n - 2h_i^n + h_{i-1}^n}{(\Delta r)^2} + \frac{h_{i+1}^{n-1} - 2h_i^{n-1} + h_{i-1}^{n-1}}{(\Delta r)^2} \right) \right] \\
- \frac{\Delta t}{2} & \left[\frac{K}{S_y} \left(\frac{h_{i+1}^{n-1} - h_{i-1}^{n-1}}{4\Delta r} + \frac{h_{i+1}^{n-2} - h_{i-1}^{n-2}}{4\Delta r} \right) \frac{h_i^{n-1}}{r_i} + \frac{K}{S_y} \left(\frac{h_{i+1}^{n-1} - h_{i-1}^{n-1}}{4\Delta r} + \frac{h_{i+1}^{n-2} - h_{i-1}^{n-2}}{4\Delta r} \right)^2 \right. \\
& \left. + \frac{K}{S_y} h_i^{n-1} \left(\frac{h_{i+1}^{n-1} - 2h_i^{n-1} + h_{i-1}^{n-1}}{(\Delta r)^2} + \frac{h_{i+1}^{n-2} - 2h_i^{n-2} + h_{i-1}^{n-2}}{(\Delta r)^2} \right) \right]
\end{aligned}
$$

$$(22.34)$$

From Equation (22.34), it is noticeable that like terms cannot be grouped and therefore no stability analysis can be performed.

22.4 APPLICATION OF THE NON-CLASSIC ATANGANA–BATOGNA NUMERICAL SCHEME

As opposed to the application of the classical AB numerical scheme explained and used in Section 22.3.3 above, in this section we will apply the *non-classic* Atangana–Batogna numerical scheme which can be taken back to the AB method if $\alpha = 1$ in Equation (22.35) (Batogna and Atangana, 2017):

$$u_{n+1} - u_n = \frac{h^\alpha}{\Gamma(\alpha)} \left[\left(\frac{2(n+1)^\alpha - n^\alpha}{\alpha} + \frac{n^{\alpha+1} - (n+1)^{\alpha+1}}{\alpha+1} \right) f_n - \left(\frac{(n+1)^\alpha}{\alpha} + \frac{n^{\alpha+1} - (n+1)^{\alpha+1}}{\alpha+1} \right) f_{n-1} \right]$$

$$(22.35)$$

The derivation of this method is as follows. Consider a general PDE

$$\frac{\partial^\alpha u(x,t)}{\partial t^\alpha} = Lu(x,t) + Nu(x,t)$$ (22.36)

Let L be a linear operator and N a nonlinear operator, then apply the Laplace transform to both sides of the equation to obtain:

$$\mathcal{L}\left(\frac{\partial^\alpha u(x,t)}{\partial t^\alpha}\right) = \mathcal{L}\big(Lu(x,t) + Nu(x,t)\big)$$ (22.37)

Taken into a Caputo-type fractional partial derivative, Equation (22.37) will then be:

$$_a^C D_t^\alpha u(p,t) = \mathcal{L}\big(Lu(x,t) + Nu(x,t)\big)$$ (22.38)

$$_a^C D_t^\alpha u(t) = F(u,t)$$ (22.39)

where $u(t) = u(p,t)$ and $F(u,t) = \mathcal{L}\big(Lu(x,t) + Nu(x,t)\big)$

With the application of the Caputo fractional integral operator to Equation (22.39) we obtain:

$$u(t) - u(t_0) = \frac{1}{\Gamma(\alpha)} \int_0^t (t-\tau)^{\alpha-1} F(u,\tau)d\tau$$

Then when $t = t_{n+1}$

$$u(t_{n+1}) = u_0 + \frac{1}{\Gamma(\alpha)} \int_0^{t_{n+1}} (t_{n+1}-\tau)^{\alpha-1} F(u,\tau)d\tau$$ (22.40)

When $t = t_n$

$$u(t_n) = u_0 + \frac{1}{\Gamma(\alpha)} \int_0^{t_n} (t_n-\tau)^{\alpha-1} F(u,\tau)d\tau$$ (22.41)

Then Equations (22.40) and (22.41) result to:

$$u_{n+1} - u_n = \frac{1}{\Gamma(\alpha)} \left[\int_0^{t_{n+1}} (t_{n+1}-\tau)^{\alpha-1} F(u,\tau)d\tau - \int_0^{t_n} (t_n-\tau)^{\alpha-1} F(u,\tau)d\tau \right]$$ (22.42)

$$\int_0^{t_{n+1}} (t_{n+1}-\tau)^{\alpha-1} F(u,\tau)d\tau = \sum_{j=0}^n \int_{t_j}^{t_{j+1}} (t_{n+1}-\tau)^{\alpha-1} F(u,\tau)d\tau$$

Then approximate $F(u,t)$ with a Lagrange polynomial:

$$P(t) \approx F(u,t) = \frac{t-t_{n-1}}{t_n-t_{n-1}} F(u,t_n) + \frac{t-t_n}{t_{n-1}-t_n} F(u,t_{n-1})$$

$$P(t) = \frac{t-t_{n-1}}{t_n-t_{n-1}} F_n + \frac{t-t_n}{t_{n-1}-t_n} F_{n-1}$$ (22.43)

The expression of the first fractional integral in Equation (22.42) is given by:

$$\int_0^{t_{n+1}} (t_{n+1}-\tau)^{\alpha-1} F(u,\tau)d\tau = \sum_{j=0}^n \int_{t_j}^{t_{j+1}} (t_{n+1}-t)^{\alpha-1}\left(\frac{t-t_{n-1}}{t_n-t_{n-1}}F_n + \frac{t-t_n}{t_{n-1}-t_n}F_{n-1}\right)dt$$

$$\int_0^{t_{n+1}} (t_{n+1}-\tau)^{\alpha-1} F(u,\tau)d\tau = \sum_{j=0}^n \left[\begin{array}{c} \dfrac{F_n}{t_n-t_{n-1}}\displaystyle\int_{t_j}^{t_{j+1}} (t_{n+1}-t)^{\alpha-1}(t-t_{n-1})dt \\[4mm] + \dfrac{F_{n-1}}{t_{n-1}-t_n}\displaystyle\int_{t_j}^{t_{j+1}} (t_{n+1}-t)^{\alpha-1}(t-t_n)dt \end{array} \right]$$

$$= \sum_{j=0}^n \left[\begin{array}{c} \dfrac{F_n}{h}\displaystyle\int_{t_j}^{t_{j+1}} (t_{n+1}-t)^{\alpha-1}(t-t_{n-1})dt \\[4mm] -\dfrac{F_{n-1}}{h}\displaystyle\int_{t_j}^{t_{j+1}} (t_{n+1}-t)^{\alpha-1}(t-t_n)dt \end{array} \right] \tag{22.44}$$

The implementation of the change of variable is in the following manner: let $y = t_{n+1}-t$, $dt = -dy$, and $t = t_{n+1}-y$

$$\int_{t_j}^{t_{j+1}} (t_{n+1}-t)^{\alpha-1}(t-t_{n-1})dt = \int_{t_{n+1}-t_j}^{t_{n+1}-t_{j+1}} y^{\alpha-1}(t_{n+1}-y-t_{n-1})dy$$

$$= \int_{t_{n+1}-t_j}^{t_{n+1}-t_{j+1}} \left(y^\alpha - 2hy^{\alpha-1}\right)dy$$

$$= \frac{1}{\alpha+1}\left[y^{\alpha+1}\right]_{t_{n+1}-t_j}^{t_{n+1}-t_{j+1}} - \frac{2h}{\alpha}\left[y^\alpha\right]_{t_{n+1}-t_j}^{t_{n+1}-t_{j+1}}$$

$$= \frac{1}{\alpha+1}\left((t_{n+1}-t_{j+1})^{\alpha+1} - (t_{n+1}-t_j)^{\alpha+1}\right)$$

$$- \frac{2h}{\alpha}\left((t_{n+1}-t_{j+1})^\alpha - (t_{n+1}-t_j)^\alpha\right) \tag{22.45}$$

And on the other side

$$\int_{t_j}^{t_{j+1}} (t_{n+1}-t)^{\alpha-1}(t-t_{n-1})dt = -\int_{t_{n+1}-t_j}^{t_{n+1}-t_{j+1}} y^{\alpha-1}(t_{n+1}-y-t_n)dy$$

$$= \int_{t_{n+1}-t_j}^{t_{n+1}-t_{j+1}} \left(y^\alpha - hy^{\alpha-1}\right)dy = \frac{1}{\alpha+1}\left[y^{\alpha+1}\right]_{t_{n+1}-t_j}^{t_{n+1}-t_{j+1}} - \frac{h}{\alpha}\left[y^\alpha\right]_{t_{n+1}-t_j}^{t_{n+1}-t_{j+1}}$$

$$= \frac{1}{\alpha+1}\left((t_{n+1}-t_j)^{\alpha+1} - (t_{n+1}-t_{j+1})^{\alpha+1}\right) - \frac{h}{\alpha}\left((t_{n+1}-t_{j+1})^\alpha - (t_{n+1}-t_j)^\alpha\right) \tag{22.46}$$

Following this will be:

$$\int_0^{t_{n+1}} (t_{n+1}-\tau)^{\alpha-1} F(u,\tau)d\tau = \frac{F_n}{h} \left\{ \begin{array}{l} \frac{1}{\alpha+1} \sum_{j=0}^{n} \left[\left((t_{n+1}-t_{j+1})^{\alpha+1} - (t_{n+1}-t_j)^{\alpha+1} \right) \right] \\ -\frac{2h}{\alpha} \sum_{j=0}^{n} \left[\left((t_{n+1}-t_{j+1})^{\alpha} - (t_{n+1}-t_j)^{\alpha} \right) \right] \end{array} \right\}$$

$$-\frac{F_{n-1}}{h} \left\{ \frac{1}{\alpha+1} \sum_{j=0}^{n} \left[\left((t_{n+1}-t_{j+1})^{\alpha+1} - (t_{n+1}-t_j)^{\alpha+1} \right) \right] - \frac{h}{\alpha} \sum_{j=0}^{n} \left[\left((t_{n+1}-t_{j+1})^{\alpha} - (t_{n+1}-t_j)^{\alpha} \right) \right] \right\}$$

(22.47)

The evaluation of the second fractional integral in Equation (22.47) is as follows:

$$\int_0^{t_n} (t_n-\tau)^{\alpha-1} F(u,\tau)d\tau = \sum_{j=0}^{n-1} \int_{t_j}^{t_{j+1}} (t_n-t)^{\alpha-1} \left(\frac{t-t_{n-1}}{t_n-t_{n-1}} F_n + \frac{t-t_n}{t_{n-1}-t_n} F_{n-1} \right) dt$$

$$= \frac{F_n}{h} \sum_{j=0}^{n-1} \int_{t_j}^{t_{j+1}} (t_n-t)^{\alpha-1} (t-t_{n-1}) dt - \frac{F_{n-1}}{h} \sum_{j=0}^{n-1} \int_{t_j}^{t_{j+1}} (t_n-t)^{\alpha-1} (t-t_n) dt \quad (22.48)$$

Then the change of variable is implemented as follows: let $y = t_n - t$, $dt = -dy$, and $t = t_n - y$

$$\int_0^{t_n} (t_n-\tau)^{\alpha-1} F(u,\tau)d\tau = \frac{F_n}{h} \sum_{j=0}^{n-1} \int_{t_n-t_j}^{t_n-t_{j+1}} -(y)^{\alpha-1} (t-t_{n-1}-y) dy - \frac{F_{n-1}}{h} \sum_{j=0}^{n-1} \int_{t_n-t_j}^{t_n-t_{j+1}} y^{\alpha} dy$$

$$\int_0^{t_n} (t_n-\tau)^{\alpha-1} F(u,\tau)d\tau = \frac{F_n}{h} \sum_{j=0}^{n-1} \int_{t_n-t_j}^{t_n-t_{j+1}} (y^{\alpha} - hy^{\alpha-1}) dy - \frac{F_{n-1}}{h} \sum_{j=0}^{n-1} \int_{t_n-t_j}^{t_n-t_{j+1}} y^{\alpha} dy$$

$$\int_0^{t_n} (t_n-\tau)^{\alpha-1} F(u,\tau)d\tau = \frac{F_n}{h} \sum_{j=0}^{n-1} \left[\frac{y^{\alpha+1}}{\alpha+1} - \frac{h}{\alpha} y^{\alpha} \right]_{t_n-t_j}^{t_n-t_{j+1}} - \frac{F_{n-1}}{h} \sum_{j=0}^{n-1} \left[\frac{y^{\alpha+1}}{\alpha+1} \right]_{t_n-t_j}^{t_n-t_{j+1}}$$

$$\int_0^{t_n} (t_n-\tau)^{\alpha-1} F(u,\tau)d\tau = \frac{F_n}{h} \sum_{j=0}^{n-1} \left(\frac{(t_n-t_{j+1})^{\alpha+1}}{\alpha+1} - \frac{h}{\alpha} (t_n-t_{j+1})^{\alpha} - \frac{(t_n-t_j)^{\alpha+1}}{\alpha+1} - \frac{h}{\alpha} (t_n-t_j)^{\alpha} \right)$$

$$-\frac{F_{n-1}}{h} \sum_{j=0}^{n-1} \left(\frac{(t_n-t_{j+1})^{\alpha+1}}{\alpha+1} - \frac{(t_n-t_j)^{\alpha+1}}{\alpha+1} \right)$$

$$= \frac{F_n}{h} \left\{ \sum_{j=0}^{n-1} \left(\frac{(t_n-t_{j+1})^{\alpha+1}}{\alpha+1} - \frac{(t_n-t_j)^{\alpha+1}}{\alpha+1} \right) - \frac{h}{\alpha} \sum_{j=0}^{n-1} \left((t_n-t_{j+1})^{\alpha} - (t_n-t_j)^{\alpha} \right) \right\}$$

$$-\frac{F_{n-1}}{h} \sum_{j=0}^{n-1} \left(\frac{(t_n-t_{j+1})^{\alpha+1}}{\alpha+1} - \frac{(t_n-t_j)^{\alpha+1}}{\alpha+1} \right)$$

$$= \frac{F_n}{h}\left\{-\frac{(t_n-t_0)^{\alpha+1}}{\alpha+1} - \frac{h}{\alpha}\left(-(t_n-t_0)^{\alpha}\right)\right\} - \frac{F_{n-1}}{h(\alpha+1)}\left(-(t_n-t_0)^{\alpha+1}\right)$$

$$= \frac{F_n}{h}\left(\frac{-n^{\alpha+1}h^{\alpha+1}}{\alpha+1} + \frac{n^{\alpha}h^{\alpha+1}}{\alpha}\right) + \frac{n^{\alpha+1}h^{\alpha+1}}{h(\alpha+1)}F_{n-1} \tag{22.49}$$

This therefore means that:

$$\int_0^{t_n}(t_n-\tau)^{\alpha-1}F(u,\tau)d\tau = h^{\alpha}\left(\left(\frac{n^{\alpha}}{\alpha} - \frac{n^{\alpha+1}}{\alpha+1}\right)F_n + \frac{n^{\alpha+1}}{(\alpha+1)}F_{n-1}\right)$$

Now Equation (22.42) can be rewritten, substituting in the results to get:

$$u_{n+1}-u_n = \frac{h^{\alpha}}{\Gamma\alpha}\left[\begin{array}{l}\left(\frac{2(n+1)^{\alpha}}{\alpha} - \frac{(n+1)^{\alpha+1}}{\alpha+1}\right)F_n - \left(\frac{(n+1)^{\alpha}}{\alpha} - \frac{(n+1)^{\alpha+1}}{\alpha+1}\right)F_{n-1} \\ -\left(\left(\frac{n^{\alpha}}{\alpha} - \frac{n^{\alpha+1}}{\alpha+1}\right)F_n + \frac{n^{\alpha+1}}{(\alpha+1)}F_{n-1}\right)\end{array}\right] \tag{22.50}$$

Equation (22.50) can be simplified into Equation (22.35) and be used for both confined and unconfined zones.

Equation (22.3) is substituted into Equation (22.35) with its derivation as shown above:

$$\frac{S_c}{T}\frac{\partial h}{\partial t} = \frac{1}{r}\frac{\partial h}{\partial r} + \frac{\partial^2 h}{\partial r^2}$$

$$\frac{S_c}{T}\left(h_i^{n+1}-h_i^n\right) = \frac{h^{\alpha}}{\Gamma(\alpha)}\left[\begin{array}{l}\left(\frac{2(n+1)^{\alpha}-n^{\alpha}}{\alpha} + \frac{n^{\alpha+1}-(n+1)^{\alpha+1}}{\alpha+1}\right)f_i^n \\ -\left(\frac{(n+1)^{\alpha}}{\alpha} + \frac{n^{\alpha+1}-(n+1)^{\alpha+1}}{\alpha+1}\right)f_i^{n-1}\end{array}\right]$$

Using Equation (22.3) the definition of f_i^n and the definition of f_i^{n-1} are given as:

$$F_i^n = \frac{T}{S_c}\left[\frac{1}{r_i}\frac{h_{i+1}^n-h_{i-1}^n}{4\Delta r} + \frac{h_{i+1}^{n-1}-h_{i-1}^{n-1}}{4\Delta r} + \frac{h_{i+1}^n-2h_i^n+h_{i-1}^n}{(\Delta r)^2} + \frac{h_{i+1}^{n-1}-2h_i^{n-1}+h_{i-1}^{n-1}}{(\Delta r)^2}\right] \tag{22.51}$$

$$F_i^{n-1} = \frac{T}{S_c}\left[\frac{1}{r_i}\frac{h_{i+1}^{n-1}-h_{i-1}^{n-1}}{4\Delta r} + \frac{h_{i+1}^{n-2}-h_{i-1}^{n-2}}{4\Delta r} + \frac{h_{i+1}^{n-1}-2h_i^{n-1}+h_{i-1}^{n-1}}{(\Delta r)^2} + \frac{h_{i+1}^{n-2}-2h_i^{n-2}+h_{i-1}^{n-2}}{(\Delta r)^2}\right] \tag{22.52}$$

Now we substitute Equation (22.3) into Equation (22.35) using the defined equations (22.51) and (22.52) to get Equation (22.53):

$$h_i^{n+1}-h_i^n$$

$$
= \frac{\Delta t^{\alpha}}{\Gamma(\alpha)} \left[\left(\frac{2(n+1)^{\alpha} - n^{\alpha}}{\alpha} + \frac{n^{\alpha+1} - (n+1)^{\alpha+1}}{\alpha+1} \right) \frac{T}{S_c} \left[\frac{1}{r_i} \frac{h_{i+1}^n - h_{i-1}^n}{4\Delta r} + \frac{h_{i+1}^{n-1} - h_{i-1}^{n-1}}{4\Delta r} \right. \right.
$$

$$
\left. + \frac{h_{i+1}^n - 2h_i^n + h_{i-1}^n}{(\Delta r)^2} \right.
$$

$$
\left. + \frac{h_{i+1}^{n-1} - 2h_i^{n-1} + h_{i-1}^{n-1}}{(\Delta r)^2} \right]
$$

$$
- \left(\frac{(n+1)^{\alpha}}{\alpha} + \frac{n^{\alpha+1} - (n+1)^{\alpha+1}}{\alpha+1} \right) \frac{T}{S_c} \left[\frac{1}{r_i} \frac{h_{i+1}^{n-1} - h_{i-1}^{n-1}}{4\Delta r} + \frac{h_{i+1}^{n-2} - h_{i-1}^{n-2}}{4\Delta r} \right.
$$

$$
\left. + \frac{h_{i+1}^{n-1} - 2h_i^{n-1} + h_{i-1}^{n-1}}{(\Delta r)^2} \right.
$$

$$
\left. \left. + \frac{h_{i+1}^{n-2} - 2h_i^{n-2} + h_{i-1}^{n-2}}{(\Delta r)^2} \right) \right]
$$

$$ \text{(22.53)} $$

$$
h_i^{n+1} = h_i^n \left\{ 1 - \frac{\Delta t^{\alpha}}{\Gamma(\alpha)} \left(\frac{4T(n+1)^{\alpha} - 2Tn^{\alpha}}{\alpha S_c (\Delta r)^2} \right) - \frac{\Delta t^{\alpha}}{\Gamma(\alpha)} \left(\frac{2Tn^{\alpha+1} - 2T(n+1)^{\alpha+1}}{(\alpha+1) S_c (\Delta r)^2} \right) \right\}
$$

$$
+ h_{i+1}^n \left\{ \frac{\Delta t^{\alpha}}{\Gamma(\alpha)} \left(\frac{2T(n+1)^{\alpha} - Tn^{\alpha}}{2\Delta r S_c r_i \alpha} \right) + \frac{\Delta t^{\alpha}}{\Gamma(\alpha)} \left(\frac{Tn^{\alpha+1} - T(n+1)^{\alpha+1}}{(\alpha+1)2\Delta r S_c r_i} \right) \right.
$$

$$
\left. + \frac{\Delta t^{\alpha}}{\Gamma(\alpha)} \left(\frac{2T(n+1)^{\alpha} - Tn^{\alpha}}{\alpha S_c (\Delta r)^2} \right) + \frac{\Delta t^{\alpha}}{\Gamma(\alpha)} \left(\frac{2Tn^{\alpha+1} - 2T(n+1)^{\alpha+1}}{(\alpha+1) S_c (\Delta r)^2} \right) \right\}
$$

$$
+ h_{i-1}^n \left\{ \frac{\Delta t^{\alpha}}{\Gamma(\alpha)} \left(\frac{-2T(n+1)^{\alpha} + Tn^{\alpha}}{2\Delta r S_c r_i \alpha} \right) + \frac{\Delta t^{\alpha}}{\Gamma(\alpha)} \left(\frac{-Tn^{\alpha+1} + T(n+1)^{\alpha+1}}{(\alpha+1)2\Delta r S_c r_i} \right) \right.
$$

$$
\left. + \frac{\Delta t^{\alpha}}{\Gamma(\alpha)} \left(\frac{2T(n+1)^{\alpha} - Tn^{\alpha}}{\alpha S_c (\Delta r)^2} \right) + \frac{\Delta t^{\alpha}}{\Gamma(\alpha)} \left(\frac{2Tn^{\alpha+1} - 2T(n+1)^{\alpha+1}}{(\alpha+1) S_c (\Delta r)^2} \right) \right\}
$$

$$
+ h_{i+1}^{n-1} \left\{ \frac{\Delta t^{\alpha}}{\Gamma(\alpha)} \left(\frac{2T(n+1)^{\alpha} - Tn^{\alpha}}{2\Delta r S_c r_i \alpha} \right) + \frac{\Delta t^{\alpha}}{\Gamma(\alpha)} \left(\frac{Tn^{\alpha+1} - T(n+1)^{\alpha+1}}{(\alpha+1)2\Delta r S_c r_i} \right) \right.
$$

$$
+ \frac{\Delta t^{\alpha}}{\Gamma(\alpha)} \left(\frac{2T(n+1)^{\alpha} - Tn^{\alpha}}{\alpha S_c (\Delta r)^2} \right) + \frac{\Delta t^{\alpha}}{\Gamma(\alpha)} \left(\frac{2Tn^{\alpha+1} - 2T(n+1)^{\alpha+1}}{(\alpha+1) S_c (\Delta r)^2} \right)
$$

$$
- \frac{\Delta t^{\alpha}}{\Gamma(\alpha)} \left(\frac{T(n+1)^{\alpha}}{2\Delta r S_c r_i \alpha} + \frac{Tn^{\alpha+1} - T(n+1)^{\alpha+1}}{(\alpha+1)2\Delta r S_c r_i} \right)
$$

$$
\left. - \frac{\Delta t^{\alpha}}{\Gamma(\alpha)} \left(\frac{T(n+1)^{\alpha}}{\alpha S_c (\Delta r)^2} + \frac{Tn^{\alpha+1} - T(n+1)^{\alpha+1}}{(\alpha+1) S_c (\Delta r)^2} \right) \right\}
$$

$$
\begin{aligned}
+ h_{i-1}^{n-1} & \left\{ \frac{\Delta t^{\alpha}}{\Gamma(\alpha)} \left(\frac{2T(n+1)^{\alpha} - Tn^{\alpha}}{2\Delta r S_c r_i \alpha} \right) + \frac{\Delta t^{\alpha}}{\Gamma(\alpha)} \left(\frac{Tn^{\alpha+1} - T(n+1)^{\alpha+1}}{(\alpha+1)2\Delta r S_c r_i} \right) \right. \\
& + \frac{\Delta t^{\alpha}}{\Gamma(\alpha)} \left(\frac{2T(n+1)^{\alpha} - Tn^{\alpha}}{\alpha S_c (\Delta r)^2} \right) + \frac{\Delta t^{\alpha}}{\Gamma(\alpha)} \left(\frac{Tn^{\alpha+1} - T(n+1)^{\alpha+1}}{(\alpha+1)S_c (\Delta r)^2} \right) \\
& - \frac{\Delta t^{\alpha}}{\Gamma(\alpha)} \left(\frac{T(n+1)^{\alpha}}{2\Delta r S_c r_i \alpha} + \frac{Tn^{\alpha+1} - T(n+1)^{\alpha+1}}{(\alpha+1)2\Delta r S_c r_i} \right) \\
& \left. - \frac{\Delta t^{\alpha}}{\Gamma(\alpha)} \left(\frac{T(n+1)^{\alpha}}{\alpha S_c (\Delta r)^2} + \frac{Tn^{\alpha+1} - T(n+1)^{\alpha+1}}{(\alpha+1)S_c (\Delta r)^2} \right) \right\}
\end{aligned}
$$

$$
\begin{aligned}
+ h_i^{n-1} & \left\{ \frac{\Delta t^{\alpha}}{\Gamma(\alpha)} \left(\frac{-4T(n+1)^{\alpha} + 2Tn^{\alpha}}{\alpha S_c (\Delta r)^2} \right) + \frac{\Delta t^{\alpha}}{\Gamma(\alpha)} \left(\frac{-2Tn^{\alpha+1} + 2T(n+1)^{\alpha+1}}{(\alpha+1)S_c (\Delta r)^2} \right) \right. \\
& \left. - \frac{\Delta t^{\alpha}}{\Gamma(\alpha)} \left(\frac{-2T(n+1)^{\alpha}}{\alpha S_c (\Delta r)^2} - \frac{2Tn^{\alpha+1} + 2T(n+1)^{\alpha+1}}{(\alpha+1)S_c (\Delta r)^2} \right) \right\}
\end{aligned}
$$

$$
+ h_{i+1}^{n-2} \left\{ - \frac{\Delta t^{\alpha}}{\Gamma(\alpha)} \left(\frac{T(n+1)^{\alpha}}{2\Delta r S_c r_i \alpha} + \frac{Tn^{\alpha+1} - T(n+1)^{\alpha+1}}{(\alpha+1)2\Delta r S_c r_i} \right) - \frac{\Delta t^{\alpha}}{\Gamma(\alpha)} \left(\frac{T(n+1)^{\alpha}}{\alpha S_c (\Delta r)^2} + \frac{Tn^{\alpha+1} - T(n+1)^{\alpha+1}}{(\alpha+1)S_c (\Delta r)^2} \right) \right\}
$$

$$
+ h_{i-1}^{n-2} \left\{ - \frac{\Delta t^{\alpha}}{\Gamma(\alpha)} \left(\frac{T(n+1)^{\alpha}}{2\Delta r S_c r_i \alpha} + \frac{Tn^{\alpha+1} - T(n+1)^{\alpha+1}}{(\alpha+1)2\Delta r S_c r_i} \right) - \frac{\Delta t^{\alpha}}{\Gamma(\alpha)} \left(\frac{T(n+1)^{\alpha}}{\alpha S_c (\Delta r)^2} + \frac{Tn^{\alpha+1} - T(n+1)^{\alpha+1}}{(\alpha+1)S_c (\Delta r)^2} \right) \right\}
$$

$$
+ h_i^{n-2} \left\{ \frac{\Delta t^{\alpha}}{\Gamma(\alpha)} \left(\frac{2T(n+1)^{\alpha}}{\alpha S_c (\Delta r)^2} - \frac{2Tn^{\alpha+1} + 2T(n+1)^{\alpha+1}}{(\alpha+1)S_c (\Delta r)^2} \right) \right\}
$$

$$
h_i^{n+1} = Ah_i^n + Bh_{i+1}^n + Ch_{i-1}^n + Dh_{i+1}^{n-1} + Eh_{i-1}^{n-1} + Fh_i^{n-1} + Gh_{i+1}^{n-2} + Hh_{i-1}^{n-2} + Ih_i^{n-2} \tag{22.54}
$$

$$
\begin{aligned}
\hat{\rho}_{n+1} e^{ik_m x} = {} & A\hat{\rho}_n e^{ik_m x} + B\hat{\rho}_n e^{ik_m(x+\Delta x)} + C\hat{\rho}_n e^{ik_m(x-\Delta x)} \\
& + D\hat{\rho}_{n-1} e^{ik_m(x+\Delta x)} + E\hat{\rho}_{n-1} e^{ik_m(x-\Delta x)} + F\hat{\rho}_{n-1} e^{ik_m x} \\
& + G\hat{\rho}_{n-2} e^{ik_m(x+\Delta x)} + H\hat{\rho}_{n-2} e^{ik_m(x-\Delta x)} + I\hat{\rho}_{n-2} e^{ik_m x}
\end{aligned} \tag{22.55}
$$

Then we take the multiple out to obtain:

$$
\begin{aligned}
\hat{\rho}_{n+1} e^{ik_m x} = {} & A\hat{\rho}_n e^{ik_m x} + B\hat{\rho}_n e^{ik_m x} e^{ik_m \Delta x} + C\hat{\rho}_n e^{ik_m x} e^{-ik_m \Delta x} \\
& + D\hat{\rho}_{n-1} e^{ik_m x} e^{ik_m \Delta x} + E\hat{\rho}_{n-1} e^{ik_m x} e^{-ik_m \Delta x} + F\hat{\rho}_{n-1} e^{ik_m x} \\
& + G\hat{\rho}_{n-2} e^{ik_m x} e^{ik_m \Delta x} + H\hat{\rho}_{n-2} e^{ik_m x} e^{-ik_m \Delta x} + I\hat{\rho}_{n-2} e^{ik_m x}
\end{aligned} \tag{22.56}
$$

Divide both sides by $e^{ik_m x}$ to get:

$$
\begin{aligned}
\hat{\rho}_{n+1} = {} & A\hat{\rho}_n + B\hat{\rho}_n e^{ik_m \Delta x} + C\hat{\rho}_n e^{-ik_m \Delta x} + D\hat{\rho}_{n-1} e^{ik_m \Delta x} \\
& + E\hat{\rho}_{n-1} e^{-ik_m \Delta x} + F\hat{\rho}_{n-1} + G\hat{\rho}_{n-2} e^{ik_m \Delta x} + H\hat{\rho}_{n-2} e^{-ik_m \Delta x} + I\hat{\rho}_{n-2}
\end{aligned}
$$

$$\hat{\rho}_{n+1} = \left(A + Be^{ik_m\Delta x} + Ce^{-ik_m\Delta x}\right)\hat{\rho}_n + \left(De^{ik_m\Delta x} + Ee^{-ik_m\Delta x} + F\right)\hat{\rho}_{n-1} + \left(Ge^{ik_m\Delta x} + He^{-ik_m\Delta x} + I\right)\hat{\rho}_{n-2}$$

(22.57)

Let $k_m\Delta x = \theta$

$$e^{i\theta} = \cos\theta + i\sin\theta \text{ and } e^{-i\theta} = \cos\theta - i\sin\theta$$

$$e^{i\theta} + e^{-i\theta} = \cos\theta + i\sin\theta + \cos\theta - i\sin\theta = 2\cos\theta$$

$$\hat{\rho}_{n+1} = \left[A + B(\cos\theta + i\sin\theta) + C(\cos\theta - i\sin\theta)\right]\hat{\rho}_n$$
$$+ \left[D(\cos\theta + i\sin\theta) + E(\cos\theta - i\sin\theta) + F\right]\hat{\rho}_{n-1}$$
$$+ \left[G(\cos\theta + i\sin\theta) + H(\cos\theta - i\sin\theta) + I\right]\hat{\rho}_{n-2}$$

$$\hat{\rho}_{n+1} = \left[A + (B+C)\cos\theta + i(B-C)\sin\theta\right]\hat{\rho}_n$$
$$+ \left[(D+E)\cos\theta + i(D-E)\sin\theta + F\right]\hat{\rho}_{n-1} + \left[(G+H)\cos\theta + i(G-H)\sin\theta + I\right]\hat{\rho}_{n-2}$$

(22.58)

Now we shall prove the stability using the inductive approach on the natural numbers. For $n = 0$, $\hat{\rho}_1$ and $\hat{\rho}_2$ will not be applicable and therefore we have:

$$\hat{\rho}_1 = \left[A + (B+C)\cos0 + (B-C)i\sin0\right]\hat{\rho}_0$$

$$\frac{\hat{\rho}_1}{\hat{\rho}_0} = \left[A + (B+C)\cos\theta + i(B-C)\sin\theta\right]$$

$$\left|\frac{\hat{\rho}_1}{\hat{\rho}_0}\right| < 1 \Rightarrow \left[A + (B+C)\cos\theta + i(B-C)\sin\theta\right] < 1$$

(22.59)

$$\left(A + (B+C)\cos\theta\right)^2 + i(B-C)^2\sin^2\theta < 1$$

(22.60)

$$A^2 + 2A(B+C)\cos\theta + (B+C)^2\cos^2\theta + (B-C)^2\sin^2\theta < 1$$

$$A^2 + 2A(B+C)\cos\theta + (B^2 + 2BC + C^2)\cos^2\theta + (B^2 - 2BC + C^2)\sin^2\theta < 1$$

$$A^2 + 2A(B+C)\cos\theta + B^2(\cos^2\theta + \sin^2\theta) + 2BC(\cos^2\theta - \sin^2\theta) + C^2(\cos^2\theta + \sin^2\theta) < 1$$

$$A^2 + 2A(B+C)\cos\theta + B^2 + C^2 + 2BC(1 - 2\sin^2\theta) < 1$$

(22.61)

Therefore the first condition is given in Equation (22.62) as

$$A^2 + B^2 + C^2 + 2A(B+C)\cos\theta < 1 + 2BC(2\sin^2\theta - 1)$$

(22.62)

We make the assumption that the formula is true for $\forall n > 0$, then we verify at $n + 1$

$$\hat{\rho}_{n+1} = \left\{\begin{array}{l}\left[A + (B+C)\cos\theta + (B-C)i\sin\theta\right]\hat{\rho}_n \\ + \left[(D+E)\cos\theta + (D-E)i\sin\theta + F\right]\hat{\rho}_{n-1} + \left[(G+H)\cos\theta + (G-H)i\sin\theta + I\right]\hat{\rho}_{n-2}\end{array}\right\}$$

(22.63)

Let $A_1 = A + (B + C)cos\theta$; $B_1 = (B - C)sin\theta$; $A_2 = F + (D + E)cos\theta$; $B_2 = (D - E)sin\theta$; $A_3 = (G + H)cos\theta + I$; $B_3 = (G - H)sin\theta$, then we will have:

$$\left|\hat{\rho}_{n+1}\right| < \left|A_1 + iB_1\right|\left|\hat{\rho}_n\right| + \left|A_2 + iB_2\right|\left|\hat{\rho}_{n-1}\right| + \left|A_3 + iB_3\right|\left|\hat{\rho}_{n-2}\right| \tag{22.64}$$

Given the stability condition in Equation (22.62),

$$\left|\hat{\rho}_{n+1}\right| < \left|A_1 + iB_1\right|\left|\hat{\rho}_0\right| + \left|A_2 + iB_2\right|\left|\hat{\rho}_0\right| + \left|A_3 + iB_3\right|\left|\hat{\rho}_0\right| \tag{22.65}$$

$$\frac{\left|\hat{\rho}_{n+1}\right|}{\left|\hat{\rho}_0\right|} < \left|\left(A_1 + iB_1\right) + \left(A_2 + iB_2\right) + \left(A_3 + iB_3\right)\right| < 1 \tag{22.66}$$

$$\frac{\left|\hat{\rho}_{n+1}\right|}{\left|\hat{\rho}_0\right|} < \left|\left(A_1 + A_2 + A_3\right) + i\left(B_1 + B_2 + B_3\right)\right| < 1 \tag{22.67}$$

$$\sqrt{\left(A_1 + A_2 + A_3\right)^2 + \left(B_1 + B_2 + B_3\right)^2} < 1 \tag{22.68}$$

$$\left(A_1 + A_2 + A_3\right)^2 + \left(B_1 + B_2 + B_3\right)^2 < 1 \tag{22.69}$$

Now it can be concluded that this numerical scheme, the non-classical Atangana–Batogna, is stable if and only if:

$$A^2 + B^2 + C^2 + 2A\left(B + C\right)\cos\theta < 1 + 2BC\left(2\sin^2\theta - 1\right) \text{ and } \left(A_1 + A_2 + A_3\right)^2 + \left(B_1 + B_2 + B_3\right)^2 < 1. \tag{22.70}$$

22.5 FRACTIONAL DIFFERENTIATION

Fractional calculus is known as a framework for dealing with complex systems (Tateishi et al., 2017) and the literature shows that fractional differentiation has been and can be successfully used to model complex real-world problems in the field of science including Earth Sciences. With the theory gathered from the literature, one can therefore state that fractional differentiation can be used to understand the conversion of a confined to an unconfined flow. However, it is of utmost importance to understand which fractional order derivatives should be applied to model the complexity and the reality of the system such that it provides the memory effect which has a significant effect on the field data used. It is imperative for one to note that the permeable geological formations that can transmit large quantities of water, called aquifers, are mostly very complex systems and can be a combination of both heterogeneity and the anisotropic.

In fractional differentiation, it has been believed for a fairly long time that its nature and its complexities can be modeled using the power law: $x - \alpha$. The power law owes its importance to the Riemann–Liouville power-law differential operator which has been the only commonly used operator to model physical problems. In cases like complex heterogeneous aquifers with transforming flow from the confined to the unconfined, the power law and the decay law cannot yield correct results. Recently, researchers conducted studies and proposed new fractional-time operators, namely the Caputo–Fabrizio and the Atangana–Baleanu, which are then defined by a non-singular memory kernel (Tateishi et al., 2017; Atangana and Gomez-Aguilar, 2018b). Due to the advantages of the Atangana–Baleanu derivative over the Caputo–Fabrizio derivative, in the subsection below, we will use the former. As stipulated by Tateishi et al. (2017) and Atangana and Gomez-Aguilar (2018a), amongst other advantages of the former operator, the Atangana–Baleanu fractional derivative is deterministic and stochastic (as it is comparable to the Riemann–Liouville one and Brownian motion, respectively), the probability distribution is a Gaussian to non-Gaussian crossover, and its waiting time distribution is a crossover from the stretched exponential to the power law.

22.5.1 APPLICATION OF THE ATANGANA–BALEANU DERIVATIVE

The Atangana–Baleanu fractional derivative, which is based on the Mittag–Leffler function, is associated with a non-singular and non-local function, allowing the behaviour of groundwater flow in lithological material showing viscoelastic effects to be addressed. The definition of the Atangana–Baleanu derivative is:

$$^{ABC}_{0}D_t^\alpha \{f(t)\} = \frac{AB(\alpha)}{1-\alpha} \int_0^t \frac{d}{dx} f(x) E_\alpha \left(-\alpha \frac{(t-x)^\alpha}{1-\alpha} \right) dx \tag{22.71}$$

Then we substitute Equation (22.1) with Equation (22.71):

$$Sc \left[\frac{AB(\alpha)}{1-\alpha} \int_0^t \frac{\partial h}{\partial x}(r,x) E_\alpha \left[-\frac{\alpha}{1-\alpha}(t_n-x)^\alpha \right] dx \right] = \frac{KB}{r} \frac{\partial}{\partial r} \left(r \frac{\partial h}{\partial r} \right) \tag{22.72}$$

Now we discretize Equation (22.71) using the Atangana–Baleanu in the Caputo sense to obtain:

$$^{ABC}_{0}D_t^\alpha h(r,t) = \frac{AB(\alpha)}{1-\alpha} \int_0^t \frac{d}{dx} f(x) E_\alpha \left[\frac{\alpha}{1-\alpha}(t_n \quad x)^\alpha \right] dx \tag{22.73}$$

$$^{ADC}_{0}D_t^\alpha h(r,t) = \frac{AB(\alpha)}{1-\alpha} \int_0^t \frac{\partial h}{\partial x}(r_i,t) E_\alpha \left[-\frac{\alpha}{1-\alpha}(t_n-x)^\alpha \right] dx \tag{22.74}$$

$$^{ABC}_{0}D_t^\alpha h(r,t) = \frac{AB(\alpha)}{1-\alpha} \sum_{j=0}^n \frac{h_i^{j+1} - h_i^j}{\Delta t} \int_{t_j}^{t_{j+1}} E_\alpha \left[-\frac{\alpha}{1-\alpha}(t_n-x)^\alpha \right] dx \tag{22.75}$$

$$= \frac{AB(\alpha)}{1-\alpha} \sum_{j=0}^n \frac{h_i^{j+1} - h_i^j}{\Delta t} \delta_{n,j}^\alpha \tag{22.76}$$

Now substitute into the main equation of the confined zone:

$$\frac{S_c}{T} \left[\frac{AB(\alpha)}{1-\alpha} \int_0^t \frac{\partial h}{\partial x}(r,x) E_\alpha \left[-\frac{\alpha}{1-\alpha}(t_n-x)^\alpha \right] dx \right] = \frac{1}{r}\frac{\partial h}{\partial r} + \frac{\partial^2 h}{\partial r^2} \tag{22.77}$$

$$\frac{S_c}{T} \left[\frac{AB(\alpha)}{1-\alpha} \sum_{j=0}^n \frac{h_i^{j+1} - h_i^j}{\Delta t} \int_{t_j}^{t_{j+1}} E_\alpha \left(-\frac{\alpha}{1-\alpha}(t_n-x)^\alpha \right) dx \right] = \frac{1}{r_i}\frac{1}{2}\left[\frac{h_{i+1}^{n+1} - h_{i-1}^{n+1}}{\Delta r} + \frac{h_{i+1}^n - h_{i-1}^n}{\Delta r} \right]$$

$$+ \frac{1}{2}\left[\frac{h_{i+1}^{n+1} - 2h_i^{n+1} + h_{i-1}^{n+1}}{(\Delta r)^2} + \frac{h_{i+1}^n - 2h_i^n + h_{i-1}^n}{(\Delta r)^2} \right]. \tag{22.78}$$

22.5.2 STABILITY ANALYSIS

Von Neumann stability analysis, which is also known as Fourier stability analysis, is a way of assessing the stability of a finite difference scheme such as a PDE. It is therefore used to assess the stability of numerical solutions. The analysis of stability incorporates Fourier's series and Euler formulas the finite difference scheme is then classified as stable if the errors incurred at a discrete time step are not propagated throughout the simulation (Seta and Takahashi, 2002).

Fourier's series in space is given as:

$$\rho(x,t) = \sum_n \hat{\rho}(t) e^{ik_m x} \tag{22.79}$$

Now Equation (22.42) can be written as:

$$\frac{S_c}{T}\left(\frac{h_i^{n+1} - h_i^n}{\Delta t}\delta_{n,n}^{\alpha}\right) + \frac{S_c}{T}\left[\sum_{j=0}^{n-1}\frac{h_i^{j+1} - h_i^j}{\Delta t}\delta_{n,j}^{\alpha}\right] = \frac{1}{2r_i}\left[\frac{h_{i+1}^{n+1} - h_{i-1}^{n+1}}{\Delta r} + \frac{h_{i+1}^n - h_{i-1}^n}{\Delta r}\right]$$

$$+ \frac{1}{2}\left[\frac{h_{i+1}^{n+1} - 2h_i^{n+1} + h_{i-1}^{n+1}}{(\Delta r)^2} + \frac{h_{i+1}^n - 2h_i^n + h_{i-1}^n}{(\Delta r)^2}\right] \tag{22.80}$$

$$\frac{S_c h_i^{n+1}}{T\Delta t}\delta_{n,n}^{\alpha} = \frac{S_c h_i^n}{T\Delta t}\delta_{n,n}^{\alpha} - \frac{S_c}{T}\left[\sum_{j=0}^{n-1}\frac{h_i^{j+1} - h_i^j}{\Delta t}\delta_{n,j}^{\alpha}\right] + \frac{1}{2r_i}\left[\frac{h_{i+1}^{n+1} - h_{i-1}^{n+1}}{\Delta r} + \frac{h_{i+1}^n - h_{i-1}^n}{\Delta r}\right]$$

$$+ \frac{1}{2}\left[\frac{h_{i+1}^{n+1} - 2h_i^{n+1} + h_{i-1}^{n+1}}{(\Delta r)^2} + \frac{h_{i+1}^n - 2h_i^n + h_{i-1}^n}{(\Delta r)^2}\right] \tag{22.81}$$

$$+ h_i^{n+1}\left\{\frac{S_c}{T\Delta t}\delta_{n,n}^{\alpha} + \frac{1}{(\Delta r)^2}\right\} + h_i^n\left\{\frac{S_c}{T\Delta t}\delta_{n,n}^{\alpha} - \frac{1}{(\Delta r)^2}\right\} + h_{i+1}^{n+1}\left\{\frac{1}{2r_i\Delta r} + \frac{1}{2(\Delta r)^2}\right\}$$

$$+ h_{i-1}^{n+1}\left\{\frac{1}{2r_i\Delta r} + \frac{1}{2(\Delta r)^2}\right\} + h_{i+1}^n\left\{\frac{1}{2r_i\Delta r} + \frac{1}{2(\Delta r)^2}\right\} + h_{i-1}^n\left\{\frac{1}{2r_i\Delta r} + \frac{1}{2(\Delta r)^2}\right\}$$

$$- \frac{S_c}{T}\left[\sum_{j=0}^{n-1}\frac{h_i^{j+1} - h_i^j}{\Delta t}\delta_{n,j}^{\alpha}\right]$$

$$ah_i^{n+1} = bh_i^n + ch_{i+1}^{n+1} + dh_{i-1}^{n+1} + eh_{i+1}^n + fh_{i-1}^n - \frac{S_c}{T}\left[\sum_{j=0}^{n-1}\frac{h_i^{j+1} - h_i^j}{\Delta t}\delta_{n,j}^{\alpha}\right]$$

$$a\hat{\rho}_{n+1}e^{ik_m x} = b\hat{\rho}_n e^{ik_m x} + c\hat{\rho}_{n+1}e^{ik_m(x+\Delta x)}$$
$$+ d\hat{\rho}_{n+1}e^{ik_m(x-\Delta x)} + e\hat{\rho}_n e^{ik_m(x+\Delta x)} + f\hat{\rho}_n e^{ik_m(x-\Delta x)}$$
$$- \left[\sum_{j=0}^{n-1}\frac{\hat{\rho}_{j+1}e^{ik_m x} - \hat{\rho}_j e^{ik_m x}}{\Delta t}\delta_{n,j}^{\alpha}\right] \tag{22.82}$$

Then factorise, divide both sides by e^{ixk_m} and rearrange to obtain:

$$\left(a - ce^{ik_m\Delta x} - de^{-ik_m\Delta x}\right)\hat{\rho}_{n+1} = \left(b + ee^{ik_m\Delta x} + fe^{ik_m\Delta x}\right)\hat{\rho}_n - \sum_{j=0}^{n-1}\left[\frac{\hat{\rho}_{j+1} - \hat{\rho}_j}{\Delta t}\delta_{n,j}^{\alpha}\right] \quad (22.83)$$

If $n = 0$ then:

$$\left(a - ce^{ik_m\Delta x} - de^{-ik_m\Delta x}\right)\hat{\rho}_1 = \left(b + ee^{ik_m\Delta x} + fe^{ik_m\Delta x}\right)\hat{\rho}_0 \quad (22.84)$$

And the condition required translates to:

$$\left|\frac{\hat{\rho}_1}{\hat{\rho}_0}\right| < 1$$

$$\frac{\hat{\rho}_1}{\hat{\rho}_0} = \frac{\left(b + ee^{-ik_m\Delta x} + fe^{ik_m\Delta x}\right)}{\left(a - ce^{ik_m\Delta x} - de^{-ik_m\Delta x}\right)}$$

$$\left|\frac{\left(b + ee^{-ik_m\Delta x} + fe^{ik_m\Delta x}\right)}{\left(a - ce^{ik_m\Delta x} - de^{-ik_m\Delta x}\right)}\right| < 1 \quad (22.85)$$

$$e^{i\theta} = \cos\theta + i\sin\theta \text{ and } e^{-i\theta} = \cos\theta - i\sin\theta$$

$$\left|\frac{b + 2c\cos\theta}{a - 2c\cos\theta}\right| < 1$$

If $a - 2c\cos\theta > 0$

$$\left|\frac{b + 2c\cos\theta}{a - 2c\cos\theta}\right| < 1$$

$$b + 2c\cos\theta < a - 2c\cos\theta$$

$$b + 4\cos\theta < a$$

Remembering that $a = \dfrac{S_c}{T\Delta t}\delta_{n,n}^{\alpha} + \dfrac{1}{(\Delta r)^2}$ and $b = \dfrac{S_c}{T\Delta t}\delta_{n,n}^{\alpha} - \dfrac{1}{(\Delta r)^2}$ then we get:

$$\frac{S_c}{T\Delta t}\delta_{n,n}^{\alpha} - \frac{1}{(\Delta r)^2} + 4\cos\theta < \frac{S_c}{T\Delta t}\delta_{n,n}^{\alpha} + \frac{1}{(\Delta r)^2} \quad (22.86)$$

$$4\cos\theta < \frac{2}{(\Delta r)^2}$$

This therefore means that $\theta < \cos^{-1}\left(\dfrac{1}{2(\Delta r)^2}\right)$

Since the formula is true for $i = 0$, we assume true for $\forall i > 0$, we verify the case for $n + 1$

$$\hat{\rho}_{n+1}\left(a - ce^{ik_m\Delta x} - de^{-ik_m\Delta x}\right) = \hat{\rho}_n\left(b + ee^{ik_m\Delta x} + fe^{ik_m\Delta x}\right) - \sum_{j=0}^{n-1}\left(\frac{\hat{\rho}_{j+1} - \hat{\rho}_j}{\Delta t}\right)\delta_{n,j}^{\alpha}$$

$$\sum_{j=0}^{n-1}\left(\frac{\hat{\rho}_{j+1}-\hat{\rho}_{j}}{\Delta t}\right)\delta_{n,j}^{\alpha}=\frac{\hat{\rho}_{n-1}+\hat{\rho}_{1}}{\Delta t}\delta_{n,n-1}^{\alpha}$$

$$\hat{\rho}_{n+1}\left(a-ce^{ik_{m}\Delta x}-de^{-ik_{m}\Delta x}\right)=\hat{\rho}_{n}\left(b+ee^{ik_{m}\Delta x}+fe^{ik_{m}\Delta x}\right)-\frac{\hat{\rho}_{n-1}+\hat{\rho}_{1}}{\Delta t}\delta_{n,n-1}^{\alpha} \qquad (22.87)$$

$$\left|\hat{\rho}_{n+1}\right|\left|\left(a-ce^{ik_{m}\Delta x}-de^{-ik_{m}\Delta x}\right)\right|=\left|\hat{\rho}_{n}\left(b+ee^{ik_{m}\Delta x}+fe^{ik_{m}\Delta x}\right)-\frac{\hat{\rho}_{n-1}}{\Delta t}\delta_{n,n-1}^{\alpha}+\frac{\hat{\rho}_{1}}{\Delta t}\delta_{n,1}^{\alpha}\right| \quad (a+b)\leq|a|+|b|$$

$$\left|\hat{\rho}_{n+1}\right|\left|\left(a-ce^{ik_{m}\Delta x}-de^{-ik_{m}\Delta x}\right)\right|\leq\left|\hat{\rho}_{0}\right|\left|\left(b+ee^{ik_{m}\Delta x}+fe^{ik_{m}\Delta x}\right)\right|+\left|\frac{\hat{\rho}_{n-1}}{\Delta t}\delta_{n,n-1}^{\alpha}\right|+\left|\frac{\hat{\rho}_{1}}{\Delta t}\delta_{n,1}^{\alpha}\right| \quad (22.88)$$

Now using the recursive hypothesis we get:

$$\left|\hat{\rho}_{n+1}\right|\left|a-2c\cos\theta\right|<\left|\hat{\rho}_{0}\right|\left|b+2c\cos\theta\right|+\left|\hat{\rho}_{0}\right|\frac{\delta_{n,n-1}^{\alpha}}{\Delta t}+\left|\hat{\rho}_{0}\right|\frac{\delta_{n,1}^{\alpha}}{\Delta t} \qquad (22.89)$$

$$\left|\hat{\rho}_{n+1}\right|\left|a-2c\cos\theta\right|<\left|\hat{\rho}_{0}\right|\left(b+2c\cos\theta+\frac{\delta_{n,n-1}^{\alpha}}{\Delta t}+\frac{\delta_{n,1}^{\alpha}}{\Delta t}\right) \qquad (22.90)$$

$$\frac{\left|\hat{\rho}_{n+1}\right|}{\left|\hat{\rho}_{0}\right|}<\frac{b+2c\cos\theta+\dfrac{\delta_{n,n-1}^{\alpha}}{\Delta t}+\dfrac{\delta_{n,1}^{\alpha}}{\Delta t}}{a-2c\cos\theta}<1 \qquad (22.91)$$

$$\frac{S_{c}}{T\Delta t}\delta_{n,n}^{\alpha}-\frac{1}{\left(\Delta r\right)^{2}}+4\cos\theta<\frac{S_{c}}{T\Delta t}\delta_{n,n}^{\alpha}+\frac{1}{\left(\Delta r\right)^{2}}+\frac{\delta_{n,n-1}^{\alpha}}{\Delta t}+\frac{\delta_{n,1}^{\alpha}}{\Delta t} \qquad (22.92)$$

$$\cos\theta<\frac{1}{2\left(\Delta r\right)^{2}}+\frac{\delta_{n,n-1}^{\alpha}}{4\Delta t}+\frac{\delta_{n,1}^{\alpha}}{4\Delta t}$$

$$\theta<\cos^{-1}\left(\frac{1}{2\left(\Delta r\right)^{2}}+\frac{\delta_{n,n-1}^{\alpha}}{4\Delta t}+\frac{\delta_{n,1}^{\alpha}}{4\Delta t}\right) \qquad (22.93)$$

Therefore it can be concluded that the numerical scheme is stable when

$$\theta<\cos^{-1}\left(\frac{1}{2\left(\Delta r\right)^{2}}+\frac{\delta_{n,n-1}^{\alpha}}{4\Delta t}+\frac{\delta_{n,1}^{\alpha}}{4\Delta t}\right).$$

22.6   NUMERICAL SIMULATIONS

This section presents and discusses the numerical simulations used for the solutions obtained in the previous sections. There are many numerical methods used for the simulation of complex problems, among which the finite difference method (FD) and the finite element method (FE) are the most commonly used. There is quite a number of existing software packages that are used in numerical modelling; however, in this study, we have used Matrix Laboratory, commonly known as MATLAB.

MATLAB is a software package developed by MathWorks, which allows computation, data analysis, development of algorithms, simulation and modelling, and produces graphical displays and graphical user interfaces (Knight, 2000). In order to successfully present numerical simulations, we develop mathematical code using the numerical schemes presented in Sections 22.3 and 22.4 to simulate numerical solutions for different values of alpha. Theoretical parameter values are used to obtain simulations.

The hydraulic parameter values used to obtain the numerical simulations and contour plots of the solutions (see Figures 22.2–22.11), include a storage coefficient of 0.0001, a specific yield of 0.001, and a hydraulic conductivity value of 40 m/d with a transmissivity value equal to 400 m²/d and an aquifer thickness of 2 m. Figures 22.2 and 22.4 show numerical simulations for $\alpha = 1$ and 0.8, respectively; however, we may notice how the hydraulic head (h), as a function of space and time, changes when the value of alpha decreases; this gradual change of h is depicted in Figures 22.6, 22.8, and 22.10.

Observing the contour plots as illustrated for different values of alpha, we notice the change in water within the aquifer as shown by the hydraulic head. The water level decreases as the value of alpha increases from 0.2 to 1, a decrease in water level means the water table becomes deeper and

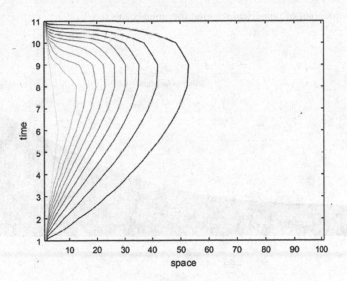

FIGURE 22.2 Contour plot of numerical solution for alpha equals 1

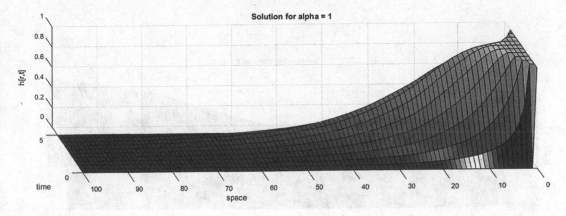

FIGURE 22.3 Numerical simulation for alpha equals 1

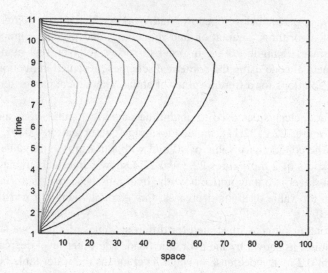

FIGURE 22.4   Contour plot of numerical solution for alpha equals 0.8

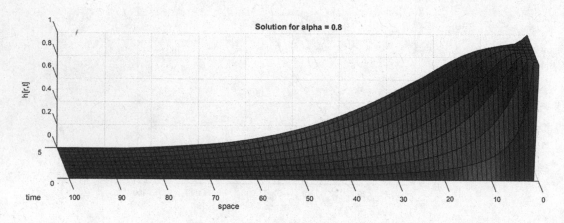

FIGURE 22.5   Numerical simulation for alpha equals 0.8

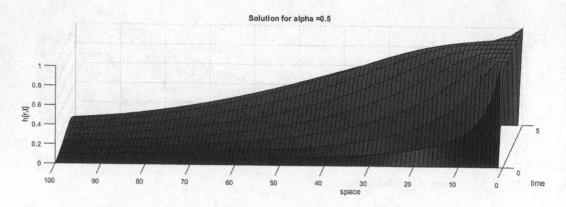

FIGURE 22.6   Numerical simulation for alpha equals 0.5

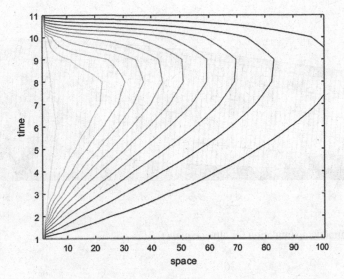

FIGURE 22.7   Contour plot of numerical solution for alpha equals 0.5

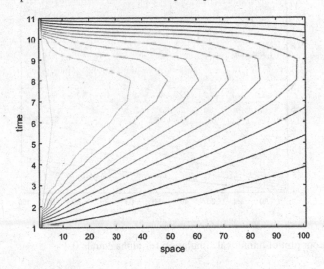

FIGURE 22.8   Contour plot of numerical solution for alpha equals 0.3

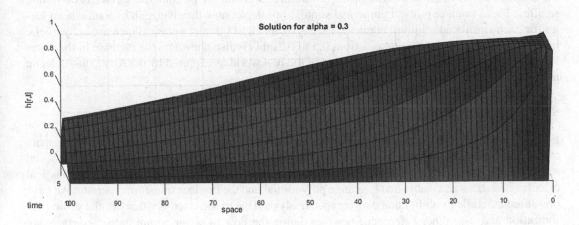

FIGURE 22.9   Numerical simulation for alpha equals 0.3

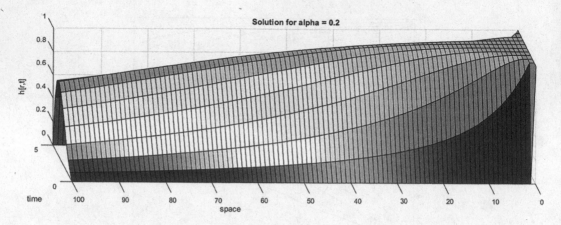

FIGURE 22.10   Numerical simulation for alpha equals 0.2

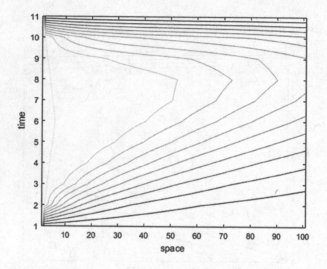

FIGURE 22.11   Contour plot of numerical simulation for alpha equals 0.2

the distance between the bottom of the top confining layer and the piezometric surface is becoming smaller. These contour plots of numerical simulations depict how the change in hydraulic head varies for each simulated solution. Figures 22.8, 22.10, and 22.11 depict a gradual increase in hydraulic head as the value of alpha decreases from 0.5 to 0.2; this is also shown by the increase in the water level. This points out what is to be expected in the real world as captured by the Atangana–Baleanu non-local operator.

22.7   CONCLUSION

In this chapter, we have reviewed the existing models and pointed out their limitations with respect to capturing formation heterogeneity. We have presented a study of the numerical analysis of the nonlinear case using a novel numerical scheme that employs the fundamental theorem of classical calculus and the well-established Lagrange polynomial and the Laplace transform operator. We have presented a detailed stability and convergence analysis. Due to the visco-elasticity of the geological formation and also other heterogeneities that define the flow of water within the sub-surface, we argue that the existing model cannot capture such a flow and thus we replace the local differential operator by a non-local one with the Mittag–Leffler function. The new model has the advantage that it can capture flow by following the power law and the decay law. The new model is also able to capture flow with Gaussian and non-Gaussian distribution and finally it can replicate very efficiently normal and sub-diffusion. From the numerical simulations, it can be seen that the results show that the groundwater system's hydraulic head changes over time. We may then conclude that the conversion of a confined to an unconfined aquifer flow in a heterogeneous medium is well defined and described by the concept of fractional differentiation with the Mittag–Leffler law using the Atangana–Baleanu fractional derivative.

REFERENCES

Alkahtani, B.S.T., 2018. Atangana-Batogna numerical scheme applied on a linear and non-linear fractional differential equation. *European Physical Journal Plus* 133: 111.

Atangana, A., and Gomez-Aguilar, J.F., 2018a. Decolonisation of fractional calculus rules: Breaking commutativity and associativity to capture more natural phenomena. *European Physical Journal Plus* 133: 1–23 a.

Atangana, A., and Gomez-Aguilar, J.F., 2018b. Fractional derivatives with no-index law property: Chaos. *Solitons and Fractals* 114 (2018): 516–535.

Batogna, R.G., and Atangana, A., 2017. New Two Step Laplace Adams-Bashforth Method for Integer an Non integer Order Partial Differential Equations. arXiv-1708.

Chen, C.X., Hu, L.T., and Wang, X.S., 2006. Analysis of steady ground water flow toward wells in a confined–unconfined aquifer. *Ground Water* 44 (4): 609–612.

Hu, L., and Chen, C., 2008. Analytical methods for transient flow to a well in a confined-unconfined aquifer. *Groundwater* 46(4): 642–646.

Knight, A., 2000. *Basics of MATLAB and Beyond*. CRC Press.

LLC, 2000 *N.W. Corporate Blvd*, Boca Raton, Florida.

Moench, A.F., and Prickett, T.A., 1972. Radial flow in an infinite aquifer undergoing conversion from artesian to water table conditions. *Water Resources Research* (2): 494–499.

Seta, T., and Takahashi, R. 2002. *Journal of Statistical Physics* 107: 557.

Springer, A.E., and Bair, E.S. 1992. Comparison of methods used to delineate capture zones of wells: 2. Stratified-drift buried-valley aquifer. *Groundwater* 30(6): 908–917.

Tateishi, A.A., Ribeiro, H.V., and Lenzi, E.K. 2017. The role of fractional time-derivative operators on anomalous diffusion. *Front Physical* 5:1–9.

Wang, X.S., Wan, L., and Hu, B., 2009. New approximate solutions of horizontal confined–unconfined flow. *Journal of Hydrology* 376: 417–427.

Wang, X.S., and Zhan, H.B., 2009. A new solution of transient confined–unconfined flow driven by a pumping well. *Advances in Water Resources* 32 (8): 1213–1222.

Zeltkevic, M., 1998. Adams methods. Retrieved from web.mit.edu/100/001/web/course_notes/Differential_Equations_notes/notes6.html

23 New Model to Capture the Conversion of Flow from Confined to Unconfined Aquifers

Makosha Ishmaeline Charlotte Morakaladi and Abdon Atangana
University of the Free State, Bloemfontein, South Africa

CONTENTS

23.1 INTRODUCTION

The increase in the human population over recent decades has caused a rapid increase in the demand for fresh groundwater all over the world. Fresh groundwater is required for human consumption, agricultural use, industrial use, and environmental activities. Groundwater is over-abstracted to meet requirements not satisfied by surface water alone. Less knowledge and understanding of different aquifers results in confined aquifers being pumped heavily over a long period of time and in the conversion of conditions from confined to unconfined, which changes the natural state of an aquifer. An extensive pumping rate and period results in the piezometric surface close to the abstraction dropping to beneath the confining zone. The confined aquifer consequently becomes unconfined close to the pumping well (Wang et al., 2009; Wang & Zhan, 2009; Xiao et al., 2018). Heavy pumping may occur during the process of groundwater over-exploration and mine dewatering (Springer and Bair 1992; Chen et al., 2006; Xiao et al., 2018). This conversion occurs in many large aquifers around the world (Wang & Zhan, 2009; Xiao, 2014). The conversion from confined to unconfined can result in variations in hydraulic properties such as storativity, transmissivity, and diffusion between confined and unconfined zones (Xiao et al., 2018). The understanding of the conversion is an important factor as it helps with the management of groundwater resources.

Study on confined to unconfined conversion has been done in the past five decades. Studies were based on numerical and analytical solutions and were carried out by several investigators. Rushton

DOI: 10.1201/9781003266266-23

and Wedderburn (1971) for numerical solutions used a resistance-capacitance electrical analogue to investigate the behaviour of aquifers during confined to unconfined conversion. In the numerical solution, the specific yield for unconfined aquifers replaced the storativity of the confined region. Elango and Swaminathan (1980) proposed a finite-element numerical solution for confined to unconfined conversion. Using Dupuit's assumptions, they simulated the conversion of flow from confined to unconfined by applying finite element methods that consisted of four-sided mixed-curved isoperimetric elements that were restricted to only analysing a steady-state flow (Xiao, 2014). A seminumerical solution for the confined to unconfined flow was proposed by Wang and Zhan (2009), which took into consideration both changes in transmissivity and storativity throughout the conversion. This solved the nonlinearity of the unconfined flow using the Runge–Kutta method.

Analytical solutions were introduced by Moench and Prickett (1972), Chen et al., (2006), and Hu and Chen (2008) which have improved the understanding of the conversion of flow from confined to unconfined over the past years. The models were later known as the MP (Moench and Prickett) and Chen models, respectively. Moench and Prickett (1972) suggested a mathematical solution for the conversion of flow by making use of a constant transmissivity for the unconfined layer. The MP model was acquired based on the similar case of heat flow in a cylindrical symmetry in which both melting or freezing can take place (Xiao, 2014). Hu and Chen (2008) explained their model by deriving it from Girinskii's potential function, a potential of the steady-state flow of groundwater in a porous medium that is horizontally layered and used to outline a variation of transmissivity in an unconfined zone of the Chen model (Xiao, 2014). The Chen model was obtained by looking at the assumption that the Theis equation, that is used for transient flow, was correct in the calculation of Girinskii's potential (Wang and Zhan, 2009). However, the error that is made by the approximation is still unclear. To understand a natural problem such as the flow of groundwater, since it is out of sight, starts with the good construction of a mathematical model.

23.2   AN EXISTING MODEL: THE MOENCH AND PRICKETT MODEL (MP MODEL)

The existing model is an approximate solution where the governing equation of the unconfined flow is linearized, depending on the assumption of a constant transmissivity rather than making use of a varying transmissivity (Wang and Zhan, 2009). If the variation in the water table is less than the saturated thickness of the unconfined layer, it is appropriate to make use of the idea of a constant transmissivity (Bear, 1972). This is not the same if the water table differences are similar to the saturated thickness of the unconfined layer, where the nonlinearity of unconfined flow is considered (Kompani-Zare and Zhan, 2006).

The existing model is based on the schema shown in Figure 23.1. Consider a confined aquifer that is pumped and discharges with a constant rate (Q). The piezometric head (H) of the aquifer is higher than the aquifer thickness (b). When the extensive pumping rate and pumping period is long, the hydraulic head near the pumping well drops below the confining layer, making H less than the aquifer thickness. The drop in the piezometric surface will result in a drop in the water table and cause the conversion from confined to unconfined flow close to the abstraction wells. With continuous pumping the piezometric head drops and the converted zone from confined to unconfined becomes larger. The conversion is described in Equation (23.1), which shows the equation of flow for the confined region, and Equation (23.7), which is the flow equation in the unconfined region. Both equations are symmetrical forms of the Boussinesq equation and are given as follows.

Equation of the flow for the confined region:

$$Sc\frac{dh}{dt} = \frac{KB}{r}\frac{\partial}{\partial r}\left(r\frac{\partial h}{\partial r}\right), h \geq B \tag{23.1}$$

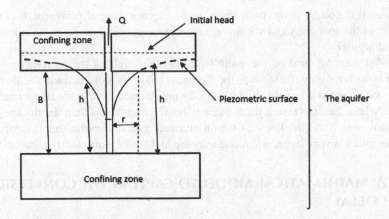

FIGURE 23.1   Sketch illustrating the confined to unconfined conversion of flow for the MP model (Wang & Zhan, 2009).

Equation (23.1) can be expanded to:

$$Sc\frac{\partial h}{\partial t} = \frac{K}{r}\frac{\partial}{\partial r}\left(rh\frac{\partial h}{\partial r}\right) \tag{23.2}$$

$$Sc = \frac{\partial h}{\partial t}(x,t) = \frac{K}{r}\frac{\partial}{\partial r}\left[rh(r,t)\frac{\partial h}{\partial r}(r,t)\right] \tag{23.3}$$

$$Sc\frac{\partial h(r,t)}{\partial t} = \frac{K}{r}\left[\frac{\partial}{\partial r}\left[rh(r,t)\right]\frac{\partial h(r,t)}{\partial r} + \frac{\partial^2 h(r,t)}{\partial r^2}r\frac{\partial h}{\partial r}(r,t)\right] \tag{23.4}$$

$$Sc\frac{\partial h(r,t)}{\partial t} = \frac{K}{r}\left[\left(h(r,t) + \frac{\partial h(r,t)}{\partial r}r\right)\frac{\partial h(r,t)}{\partial r} + r\frac{\partial^2 h(r,t)}{\partial r^2}\frac{\partial h(r,t)}{\partial r}\right] \tag{23.5}$$

$$Sc\frac{\partial h(r,t)}{\partial t} = \frac{K}{r}\left[h(r,t)\frac{\partial h(r,t)}{\partial r} + r\left(\frac{\partial h(r,t)}{\partial r}\right)^2 + r\frac{\partial^2 h(r,t)}{r^2}\frac{\partial h(r,t)}{r^2}\right] \tag{23.6}$$

Equation of the flow in the unconfined zone:

$$Sy\frac{dh}{dt} = \frac{K}{r}\frac{\partial}{\partial r}\left(rh\frac{\partial h}{\partial r}\right), 0 \le h \le B \tag{23.7}$$

Equation (23.1) can be rewritten as:

$$\frac{Sc}{T}\frac{\partial h}{\partial t} = \frac{1}{r}\frac{\partial h}{\partial r} + \frac{\partial^2 h}{\partial r^2} \tag{23.8}$$

Equation (23.7) can be rewritten as:

$$Sy\frac{dh}{dt} = \frac{Kh}{r}\frac{\partial h}{\partial r} + K\left(\frac{\partial h}{\partial r}\right)^2 + Kh\frac{\partial^2 h}{\partial r^2} \tag{23.9}$$

where b is the aquifer thickness, h represents the hydraulic head, r shows the radial distance from the pumping well centre, rs is the radial separation of the interface that occurs between the confined and the unconfined regions, t is the time since the commencement of pumping, K is the hydraulic conductivity, Sy is the specific yield within an unconfined aquifer, and Sc is the storage coefficient for the confined aquifer.

The MP model was proposed on the assumption that the height of the piezometric surface (h) of the unconfined zone is roughly the same as the thickness (B) of the confined aquifer, therefore the MP model can only be accepted when the thickness in the unsaturated zone of the unconfined layer which is near the area where the conversion takes place is notably much less than the thickness of the confined aquifer. However, if the thickness of the unsaturated zone in the area that is far from where the conversion takes place is very large, when one uses the MP model, sufficiently great errors can occur.

23.3 A NEW MATHEMATICAL MODEL TO CAPTURE THE CONVERSION WITH DELAY

The Theis transient principles on confined theory was further analysed and improved by Boulton (1954b) to take into account the delayed yield effect. This extension was because of the changes in the rising and lowering of the water table of unconfined aquifers. Boulton (1954b, 1963) suggested solutions where three segments of the unconfined aquifer time-drawdown curve were able to be reproduced. Using the effect of the delayed yield Boulton assumed that, at the same time the water table drops, water moves freely from storage through drainage, slowly instead of instantly, as depicted in the free-surface solutions proposed by him (1954b) and Dagan (1967) (Mishra & Kuhlman, 2013).

Boulton's (1954b) proposed solutions produced an integro-differential flow equation with regard to an averaged drawdown s^* as:

$$\frac{\partial^2 s^*}{\partial r^2} + \frac{1}{r}\frac{\partial s^*}{\partial r} = \left[\frac{S}{T}\frac{\partial s^*}{\partial t}\right] + \left\{\alpha S_y \int_0^t \frac{\partial s^*}{\partial \tau} e^{-\alpha(t-\tau)}d\tau\right\} \tag{23.10}$$

The model proposed by Boulton (1954b, 1963) was made on an assumption that the quantity of water allowed to move from drainage per unit horizontal area of the aquifer due to a unit drawdown that occurs at τ (time) is due to the combination of two components. The components include S where an amount of water instantaneously escapes for storage at τ S_y that represents the amount of water in which the discharge of the water is delayed with time due to $\alpha S_y \ exp \ [-\alpha(t-\tau)]$ which is the empirical formula, where α is an empirical constant. The solution is linearized by taking T as a constant.

For this chapter we propose a new mathematical model that depicts the conversion from confined to unconfined by taking into account the delay process. Equations (23.11) and (23.12) describe the conversion of flow in the confined and unconfined zones respectively, where h, the hydraulic head (from Equations 23.1 and 23.7), is replaced by s, due to the drawdown in Equation (23.11).

The following equation is for the flow in the confined zone:

$$Sc\frac{dh}{dt} = \frac{KB}{r}\frac{\partial}{\partial r}\left(r\frac{\partial s}{\partial r}\right), h \geq B \tag{23.11}$$

The equation for the flow in the unconfined zone is:

$$Sy\frac{\partial s}{\partial t} = \frac{K}{r}\frac{\partial}{\partial r}\left(rs\frac{\partial s}{\partial r}\right), 0 \leq s \geq B \tag{23.12}$$

The equation below is used for the purposes of this chapter:

$$\frac{S}{T}\frac{\partial h(r,t)}{\partial t} + \left\{\alpha S_y \int_0^t \frac{\partial h(r,t)}{\partial \tau} e^{-\alpha(t-\tau)}d\tau\right\} = \frac{1}{r}\frac{\partial h(r,t)}{\partial r} + \frac{\partial^2 h(r,t)}{\partial r^2} \tag{23.13}$$

Equation (23.13) is highly nonlinear and provides information that is somewhat exaggerated in which the results are non-realistic, giving a higher flow than expected. For this work we include a delay process. Although this has been used before by many other investigators, we realized that the second part of the equation is highly nonlinear and does not take into account the delay process in terms of the water being introduced from confined to unconfined regions. From previous studies it is assumed that the soil is homogeneous in nature, but in reality the opposite is true because it can happen that the soil does not have the same properties, meaning that the heterogeneous nature of the geological formation is not considered in the existing model. We assume a kind of heterogeneity that will delay the water arriving at the precise time. By suggesting the delay process we will be able to reduce the nonlinearity of the equation and provide something that is approximately what is seen in the field.

Thus, in this work the confined to unconfined conversion will be presented by the following mathematical formula:

$$
\begin{cases}
Sc\dfrac{\partial h(r,t)}{\partial t} = \dfrac{KB}{r}\dfrac{\partial}{\partial r}\left(r\dfrac{hs}{\partial r}\right), h > B \\
\dfrac{S}{T}\dfrac{\partial h(r,t)}{\partial t} = \dfrac{1}{r}\dfrac{\partial h(r,t)}{\partial r} + \dfrac{\partial^2 h(r,t)}{\partial r^2} - \alpha Sy\displaystyle\int_0^t \dfrac{\partial h(r,t)}{\partial \tau}e^{-\alpha(t-\tau)}d\tau, h \le B
\end{cases}
\tag{23.14}
$$

From this equation we can say that the system is now well defined, where $e^{-\alpha(t-\tau)}$ is the delay process.

23.4 DERIVATION OF AN EXACT AND NUMERICAL SOLUTION OF THE NEW MODEL

When a confined aquifer is over-pumped at a constant rate by a fully penetrating well, the influence of the discharge rate extends outwards as time changes. The storativity multiplied by the rate of decline of the head, totalled over the area of influence, is equal to the discharge rate Q. Using the existing models, when the extensive pumping rate and pumping period is long, the hydraulic head near the pumping well drops below the confining zone, making the piezometric head less than the aquifer's thickness (Figure 23.1). The drop in the piezometric surface will result in a drop in the water table and cause the conversion from confined to unconfined flow close to the abstraction wells. As mentioned before in this work, the derivation for the flow in the confined flow is given by:

$$
Sc\frac{\partial h(r,t)}{\partial t} = \frac{KB}{r}\frac{\partial}{\partial r}\left(r\frac{h(r,t)}{\partial r}\right), h \ge B
\tag{23.15}
$$

where h represents the hydraulic head that is measured from the bottom part of the confining layer of the aquifer (Figure 23.1), t is the time, r is the radial distance from the centre of the well, B represents the aquifer's thickness, Sc is the storage coefficient in the confined aquifer, and K is the hydraulic conductivity. We should mention that the derivation presented here can be found in other sources.

The equation solution is:

$$
\frac{\partial h(r,t)}{\partial t} = \frac{T}{S}\left\{\frac{\partial h(r,t)}{r\partial r} + \frac{\partial^2 h(r,t)}{\partial r^2}\right\}
\tag{23.16}
$$

where $T = Kb$ is the constant transmissivity of the confined zone.

The solution can be derived using the Boltzmann transform $h(\lambda) = h(r,t)$ with the following transformation:

$$\lambda = \frac{r^2}{t}$$

(23.17)

using the derivative:

$$\frac{\partial h}{\partial t} = \frac{\partial h}{\partial \lambda} \cdot \frac{\partial \lambda}{\partial t}$$

(23.18)

This is because:

$$\frac{\partial \lambda}{\partial t} = -\frac{r^2}{t^2}$$

(23.19)

$$\frac{\partial h}{\partial t} = \frac{\partial h}{\partial \lambda}\left(-\frac{r^2}{t^2}\right)$$

(23.20)

Deriving $\lambda = \frac{r^2}{t}$ with respect to r:

$$\frac{\partial \lambda}{\partial r} = \frac{2r}{t}$$

(23.21)

$$\frac{\partial h}{\partial r} = \frac{\partial h}{\partial \lambda} \cdot \frac{\partial \lambda}{\partial r} = \frac{\partial h}{\partial \lambda} \cdot \frac{2r}{t}$$

(23.22)

$$\frac{\partial^2 h}{\partial r^2} = \frac{\partial h}{\partial \lambda}\left(\frac{\partial h}{\partial \lambda} \cdot \frac{2r}{t}\right)\frac{\partial \lambda}{\partial r} = \frac{\partial^2 h}{\partial \lambda^2} \cdot \frac{4r^2}{t^2}$$

(23.23)

Substituting back into Equation (23.16):

$$\frac{\partial h}{\partial \lambda} \cdot \frac{-r^2}{t^2} = \frac{T}{S}\left(\frac{\partial^2 h}{\partial \lambda^2} \cdot \frac{4r^2}{t^2} + \frac{1}{r} \cdot \frac{\partial h}{\partial \lambda} \cdot \frac{2r}{t}\right)$$

(23.24)

Dividing by $-\frac{r^2}{t^2}$:

$$\frac{\partial h}{\partial \lambda} = \frac{T}{S}\left(-\frac{4\partial^2 h}{\partial \lambda^2} - \frac{2t}{r^2} \cdot \frac{\partial h}{\partial \lambda}\right)$$

(23.25)

Take the inverse of $\lambda = \frac{r^2}{t}$:

$$\frac{\partial h}{\partial \lambda} = \frac{T}{S}\left(\frac{-4\partial^2 h}{\partial \lambda^2} - \frac{2}{\lambda} \frac{\partial h}{\partial \lambda}\right)$$

(23.26)

Cross-multiply and group like-terms together:

$$\frac{S}{T} \frac{\partial h}{\partial \lambda} + \frac{2\partial h}{\lambda \partial \lambda} = \left(\frac{-4\partial^2 h}{\partial \lambda^2}\right)$$

(23.27)

Taking out the common factor and division:

$$\frac{\partial^2 h}{\partial \lambda^2} = \frac{\partial h}{\partial \lambda}\left(\frac{-S}{4T} - \frac{1}{2\lambda}\right)$$

(23.28)

If $A = \dfrac{\partial h}{\partial \lambda}$ then $A' = \dfrac{\partial^2 h}{\partial \lambda^2}$

$$A' = \left(\frac{-S}{4T} - \frac{1}{2\lambda}\right) A \tag{23.29}$$

$$\frac{A'}{A} = \left(\frac{-S}{4T} - \frac{1}{2\lambda}\right) \tag{23.30}$$

Integrating:

$$\ln A = -\left(\frac{S\lambda}{4T} + \frac{1}{2}\ln\lambda\right) \tag{23.31}$$

$$\int \frac{A'}{A} = \int \frac{S}{4T} + \int \frac{1}{2\lambda} \tag{23.32}$$

$$\ln A = -\left(\frac{S}{4T}\int 1 + \frac{1}{2}\int \frac{1}{\lambda}\right) \tag{23.33}$$

$$\ln A = -\left(\frac{S}{4T} + \frac{1}{2}\ln\lambda\right) \tag{23.34}$$

$$A - \exp\left\{\left[-\left(\frac{S\lambda}{4T} + \frac{1}{2}\ln\lambda\right)\right] + C\right\} \tag{23.35}$$

where C is a constant

$$\frac{\partial h}{\partial \lambda} = \exp\left\{\left[-\left(\frac{S\lambda}{4T} + \frac{1}{2}\ln\lambda\right)\right] \cdot e^C\right\} \tag{23.36}$$

From the equation above, $e^C = W$:

$$\frac{\partial h}{\partial \lambda} = W \cdot \exp\left\{\left[-\left(\frac{S\lambda}{4T} + \frac{1}{2}\ln\lambda\right)\right]\right\} \tag{23.37}$$

$$\lim_{n \to \infty} \frac{\partial h}{\partial r} = \frac{-Q}{2\pi T} = W \tag{23.38}$$

Using boundary addition, if $r \longrightarrow 0$ then $\lambda \longrightarrow 0$

$$\frac{\partial h}{\partial \lambda} = \frac{-Q}{2\pi T} \cdot \exp\left[-\left(\frac{S\lambda}{4T} + \frac{1}{2}\ln\lambda\right)\right] \tag{23.39}$$

But $n \ln x = \ln x^n$

$$\frac{\partial h}{\partial \lambda} = \frac{-Q}{4\pi T} \cdot \exp\left[-\left(\frac{S\lambda}{4T} + \ln\lambda^{\frac{1}{2}}\right)\right] \tag{23.40}$$

Separating exponential terms:

$$\frac{\partial h}{\partial \lambda} = \frac{-Q}{4\pi T} \cdot \exp\frac{-S\lambda}{4T} + \exp\ln\lambda^{-\frac{1}{2}} \tag{23.41}$$

Let $\lambda = u$

$$\frac{\partial h}{\partial u} = \frac{-Q}{4\pi T} \int_0^\lambda \exp\left(\frac{-Su}{4T}\right) \cdot \frac{1}{\sqrt{u}} \, du + C \tag{23.42}$$

$$0 = \frac{-Q}{4\pi T} \int_0^\infty \exp\left(\frac{-Su}{4T}\right) \cdot \frac{1}{\sqrt{u}} \, du + C \tag{23.43}$$

$$C = \frac{Q}{4\pi T} \int_0^\infty \exp\left(\frac{-Su}{4T}\right) \cdot \frac{1}{\sqrt{u}} \, du \tag{23.44}$$

$$\int \frac{\partial h}{\partial u} = \frac{Q}{4\pi T} \int_0^\alpha \exp\left(\frac{-Su}{4T}\right) \cdot \frac{1}{\sqrt{u}} \, du + \frac{Q}{4\pi} \int_0^\infty \exp\left(\frac{-Su}{4T}\right) \frac{1}{\sqrt{u}} \tag{23.45}$$

$$h(\lambda) = \frac{Q}{4\pi T} \int_0^\alpha \exp\left(\frac{-Su}{4T}\right) \cdot \frac{1}{\sqrt{u}} \, du + \frac{Q}{4\pi} \int_0^\infty \exp\left(\frac{-Su}{4T}\right) \frac{1}{\sqrt{u}} \tag{23.46}$$

$$h(\lambda) = \frac{Q}{4\pi T} \int_u^\infty \exp\left(\frac{-Su}{4T}\right) \cdot \frac{1}{\sqrt{u}} \, du \tag{23.47}$$

$$h(\lambda, u) = \frac{Q}{4\pi T} \int_\alpha^\infty \exp\left(\frac{-Su}{4T}\right) \cdot \frac{1}{\sqrt{u}} \, du \tag{23.48}$$

Let $r = \sqrt{u}$

$$r^2 = u \tag{23.49}$$

$$\lambda = u$$

where $\lambda = \dfrac{r^2}{t}, \dfrac{r^2 S}{4Tt} = u$

If we substitute u into $\lambda = \dfrac{r^2}{t}$

$$\frac{u}{t} = \frac{r^2 S}{4Tt} \tag{23.50}$$

For the Theis solution the drawdown will be a function of Q, r, t, T, and S. Both S and Q are not included in the definition of the dimensionless variable because S is a dimensionless variable and Q is in the boundary condition only. There is no characteristic length or time for the Theis problem, so a dimensionless combination of r and T should be found. The chosen dimensionless variable is:

$$U = \frac{r^2 S}{4T} = y \tag{23.51}$$

Let $y = \dfrac{Sr^2}{4T}$

The Theis (1935) equation derived from the comparison between groundwater flow and heat conduction is given by the solution:

$$h(r, t) = \frac{Q}{4\pi T} \int_u^\infty \exp \frac{(-y)}{y} \cdot dy \tag{23.52}$$

23.5 APPLYING THE LAPLACE TRANSFORM TO OUR EQUATION

In this section we apply the Laplace transform to reduce our differential equation to a much simpler form that can be expressed as an algebraic expression that will be able to be solved using algebraic rules.

When we apply the Laplace transform, we get:

$$\frac{S}{T}\frac{\partial h(r,t)}{\partial t} + \left\{\alpha S_y \int_0^t \frac{\partial h(r,t)}{\partial \tau} e^{-\alpha(t-\tau)} d\tau\right\} = \frac{1}{r}\frac{\partial h(r,t)}{\partial r} + \frac{\partial^2 h(r,t)}{\partial r^2} \tag{23.53}$$

$$\mathcal{L}_t\left(\frac{S}{T}\frac{\partial h(r,t)}{\partial t} + \left\{\alpha S_y \int_0^t \frac{\partial h(r,t)}{\partial \tau} e^{-\alpha(t-\tau)} d\tau\right\}\right) = L\left(\frac{1}{r}\frac{\partial h(r,t)}{\partial r} + \frac{\partial^2 h(r,t)}{\partial r^2}\right) \tag{23.54}$$

$$\frac{S}{T}\mathcal{L}_t\left(\frac{\partial h(r,t)}{\partial r}\right) + \alpha S_y \mathcal{L}_t\left(\int_0^t \frac{\partial h(r,\tau)}{\partial \tau} e^{-\alpha(t-\tau)} d\tau\right) = \frac{1}{r}\frac{\partial \tilde{h}(r,p)}{\partial r} + \frac{\partial^2 \tilde{h}(r,p)}{\partial r^2} \tag{23.55}$$

$$\frac{S}{T}\left[p\tilde{h}(r,p) - h(r,0)\right] + \alpha S_y \mathcal{L}\left(\frac{\partial h(r,t)}{\partial t}\right)\mathcal{L}\left(e^{(-\alpha t)}\right) = \frac{1}{r}\frac{\partial \tilde{h}(r,p)}{\partial r} + \frac{\partial^2 \tilde{h}(r,p)}{\partial r^2} \tag{23.56}$$

$$\frac{S}{T}\left[p\tilde{h}(r,p) - h(r,0)\right] + \alpha S_y \left[p\tilde{h}(r,p) - h(r,0)\right]\frac{1}{p+\alpha} = \frac{1}{r}\frac{\partial \tilde{h}(r,p)}{\partial r} + \frac{\partial^2 \tilde{h}(r,p)}{\partial r^2} \tag{23.57}$$

By grouping, put $\tilde{h}(r,p)$ together to get:

$$\tilde{h}(r,p)\left[\frac{Sp}{T} + \alpha\frac{Syp}{p+\alpha}\right] - h(r,0)\left[\frac{S}{T} + \frac{\alpha Sy}{p+\alpha}\right] = \frac{1}{r}\frac{\partial \tilde{h}(r,p)}{\partial r} + \frac{\partial^2 \tilde{h}(r,p)}{\partial r^2} \tag{23.58}$$

By factorization, let $\beta(p) = \dfrac{S}{T}p + \dfrac{\alpha Sy}{p+\alpha}p$

$$\tilde{h}(r,p)\beta(p) - \frac{1}{r}\frac{\partial \tilde{h}(r,p)}{\partial r} - \frac{\partial^2 \tilde{h}(r,p)}{\partial r^2} = h(r,0)\beta(p) \tag{23.59}$$

For simplicity, we put $\tilde{h}(r,p) = B(r)$ to get:

$$B(r)\beta(p) - \frac{1}{r}\frac{d}{dr}B(r) - \frac{d^2 B(r)}{dr^2} = B_1(r)\beta(p) \tag{23.60}$$

Multiplying Equation (23.60) by r we get:

$$rB(r)\beta(p) - \frac{d}{dr}B(r) - \frac{rd^2 B(r)}{dr^2} = rB_1(r)\beta(p) \tag{23.61}$$

We know that $\mathcal{L}\{t.f(t)\} = \dfrac{d}{ds}F(s)$, so:

$$\mathcal{L}\left(\frac{d}{dr}B(r)\right) = -\frac{d}{ds}\left[\mathcal{L}\left(\frac{d}{dr}B(r)\right)\right] \tag{23.62}$$

$$= -\frac{d}{ds}\left[s\tilde{B}(s) - B(0)\right] \tag{23.63}$$

$$= -s\frac{d}{ds}\tilde{B}(s) + \tilde{B}(s) \tag{23.64}$$

Applying the Laplace transform to the second order derivative, we get:

$$\mathcal{L}\left(r\frac{d^2}{dr^2}B(r)\right) = -\frac{d}{ds}\left[\mathcal{L}\left(\frac{d^2}{dr^2}B(r)\right)\right] \tag{23.65}$$

$$= -\frac{d}{ds}\left[s^2\tilde{B}(s) - sB(0) + B'(0)\right] \tag{23.66}$$

$$= -\left[2s\tilde{B}(s) + s^2\frac{d}{ds}\tilde{B}(s) - B(0)\right] \tag{23.67}$$

$$= -2s\tilde{B}(s) + s^2\frac{d}{ds}\tilde{B}(s) - B(0) \tag{23.68}$$

Applying the Laplace transform to $B(r)\beta(p)$, we get:

$$\mathcal{L}\left(rB(r)\beta(p)\right) = -\beta(p)\frac{d}{ds}\tilde{B}(s) \tag{23.69}$$

Applying the Laplace transform to $B_1(r)\beta(p)$, we get:

$$\mathcal{L}\left(B_1(r)\beta(p)\right) = -\beta(p)\mathcal{L}\left(B_1(r)\right) \tag{23.70}$$

$$= -\beta(p)\tilde{B}_1(s) \tag{23.71}$$

Now the equation becomes:

$$-\beta(p)\frac{d}{ds}\tilde{B}(s) - s\frac{d}{ds}\tilde{B}(s) + \tilde{B}(s) - 2s\tilde{B}(s) + s^2\frac{d}{ds}\tilde{B}(s) - B(0) = -\beta(p)\tilde{B}_1(s) \tag{23.72}$$

By factorizing $\dfrac{d}{ds}\tilde{B}(s)$ we get:

$$\frac{d}{ds}\tilde{B}(s)\left[s^2 - s - \beta(p)\right] + \left[1 - 2s\right]\tilde{B}(s) = B(0) - \beta(p)\tilde{B}_1(s) \tag{23.73}$$

Divide everything by $s^2 - s - \beta(p)$:

$$\frac{d}{ds}\tilde{B}(s) + \frac{1 - 2s}{s^2 - s - \beta(p)}\tilde{B}(s) = \frac{B(0)}{s^2 - s - \beta(p)} - \frac{\beta(p)\tilde{B}_1(s)}{s^2 - s - \beta(p)} \tag{23.74}$$

$$\frac{d}{ds}\tilde{B}(s) + q(s)\tilde{B}(s) = g(s) \tag{23.75}$$

where $q(s) = \dfrac{1 - 2s}{s^2 - s - \beta(p)}$ and $g(s) = \dfrac{B(0)}{s^2 - s - \beta(p)} - \dfrac{\beta(p)\tilde{B}_1(s)}{s^2 - s - \beta(p)}$

Now we can solve the above equation using linear differential equations.

23.6   LINEAR DIFFERENTIAL EQUATIONS

The type of first order differential equation focused on is the linear one, where a formula is derived for the general solution, in this case making it different from the majority of the first order cases. It is important not to memorize the formula, but instead to memorize and have a clear understanding of the process that is used to derive the formula. In this case, this makes it easier to solve problems by using the process instead of the formula.

After using initial and boundaries conditions in Laplace space, we apply the inverse Laplace transform with respect to r to obtain:

$$\bar{h}(r,s) = C_0 \sum_{n=0}^{\infty} \frac{(-1)^n s^n r^{2n}}{2^{2n}(n!)^2 (s+\alpha)^{\frac{n}{2}}} + \frac{a}{s} \tag{23.76}$$

Using the inverse Laplace transform with respect to s we get the following general solution:

$$h(r,t) = a\delta(t) + \frac{Q}{2\pi T} \sum_{n=0}^{\infty} \frac{(-1)^n r^{2n}}{2^{2n}(n!)^2} t^{-1-\frac{n}{2}} Hypergeometric1F1Regularized\left[\frac{n}{2}, -\frac{n}{2}, -at\right] \tag{23.77}$$

The above is the exact solution of the second equation, first derived here in this chapter. Therefore, the exact solution of our system is given by:

$$\left\{ \begin{array}{l} h(r,t) = \dfrac{Q}{4\pi T} \displaystyle\int_u^{\omega} \exp\frac{(-y)}{y} \cdot dy, h > B \\[4mm] h(r,t) = a\delta(t) + \dfrac{Q}{2\pi T} \displaystyle\sum_{n=0}^{\infty} \dfrac{(-1)^n r^{2n}}{2^{2n}(n!)^2} t^{-1-\frac{n}{2}} Hypergeometric1F1Regularized\left[\dfrac{n}{2}, -\dfrac{n}{2}, -at\right], h \le B \end{array} \right. \tag{23.78}$$

While we have completed with great success the derivation of the exact solution, which is novel, we must point out that it is in the form of a series, which could be problematic when dealing with real world data. In particular, one will not be able to consider all the terms of the series. For this case, we shall now deviate our focus to derive the numerical solution using some accurate numerical scheme.

23.7   NEW NUMERICAL SCHEME USING THE ADAMS–BASHFORTH METHOD

The Laplace transform was successfully used to derive the exact solution. Due to the complicated nature of the equation, we used an appropriate numerical scheme, the Adams–Bashforth method, to derive the numerical solution.

Using our equation, we have:

$$\frac{\partial h(r,t)}{\partial t} = \frac{1}{r}\frac{\partial h(r,t)}{\partial r} + \frac{\partial^2 h(r,t)}{\partial r^2} - \alpha S_y \int_0^t \frac{\partial h(r,t)}{\partial \tau} e^{-\alpha(t-\tau)} d\tau \tag{23.79}$$

Now

$$\frac{\partial h(r,t)}{\partial t} = f\left(r,t,h(r,t)\right) \tag{23.80}$$

At the point t_{n+1}
we can let $y(t_{n+1}) = y^{n+1}$, $y^n = y(t_n)$
So:

$$y^p\left(t_{n+1}\right) = y\left(t_n\right) + \Delta t f\left(y\left(t_n\right), t_n\right) \tag{23.81}$$

$$y\left(t_{n+1}\right) = y\left(t_n\right) + \frac{h}{2}\left[f\left(y^p\left(t_{n+1}\right), t_{n+1}\right) + f t_n, y\left(t_n\right)\right] \tag{23.82}$$

At r_i, t_{n+1} and (r_i, t_n):

$$h^p\left(r_i, t_{n+1}\right) = h\left(r_i, t_n\right) + \Delta t f\left(r_i, t_n, h\left(r_i, t_n\right)\right) \tag{23.83}$$

$$h\left(r_i, t_{n+1}\right) = h\left(r_i, t_n\right) + \frac{\Delta t}{2}\left[f(r_i, t_{n+1}, h^p\left(r_i, t_{n+1}\right) + f(r_i, t_n, h\left(r_i, t_n\right)\right)\right] \tag{23.84}$$

For the central difference approximation, we have:

$$\frac{\partial h\left(r_i, t_n\right)}{\partial r} = \frac{h\left(r_{i+1}, t_n\right) - h\left(r_{i-1}, t_n\right)}{2\Delta r} \tag{23.85}$$

To obtain the finite difference approximation for a second order partial differential equation $\left(\dfrac{\partial^2 h\left(r_i, t_n\right)}{\partial r^2}\right)$ we have:

$$\frac{\partial^2 h\left(r_i, t_n\right)}{\partial r^2} = \frac{h\left(r_{i+1}, t_n\right) - 2h\left(r_i, t_n\right) + h\left(r_{i-1}, t_n\right)}{\Delta r^2} \tag{23.86}$$

Now going back to our equation, the time derivative is approximated using the forward difference approximation to give

$$\int_0^{t_n} \frac{\partial h\left(r_i, t_n\right)}{\partial \tau} \exp\left[-\alpha\left(t_n - \tau\right)\right] d\tau = \sum_{j=0}^{n-1} \int_{t_j}^{t_{j+1}} \frac{h\left(r_i, t_{j+1}\right) - h\left(r_i, t_j\right)}{\Delta t} \exp\left[-\alpha\left(t_n - \tau\right)\right] d\tau \tag{23.87}$$

$$= \sum_{j=0}^{n-1} \frac{h_i^{j+1} - h_i^{j}}{\Delta t} \int_{t_j}^{t_{j+1}} \exp\left[-\alpha\left(t_n - \tau\right)\right] d\tau \tag{23.88}$$

Now going back to (r_i, t_{n+1}) and (r_i, t_n), we know that

$$h^p\left(r_i, t_{n+1}\right) = h\left(r_i, t_n\right) + \Delta t f(r_i, t_n, h\left(r_i, t_n, h\left(r_i, t_n\right)\right)) \tag{23.89}$$

and

$$h\left(r_i, t_{n+1}\right) = h\left(r_i, t_n\right) + \frac{\Delta t}{2}\left[f\left(r_i, t_n, h^p\left(r_i, t_{n+1}\right)\right) + f\left(r_i, t_n, h(r_i, t_n)\right)\right] \tag{23.90}$$

So now what is $f(r_i, t_n, h(r_i, t_n))$ equal to?

$$f\left(r_i, t_n, h\left(r_i, t_n\right)\right) = \frac{1}{r_i}\frac{\partial h\left(r_i, t_n\right)}{\partial r} + \frac{\partial^2 h\left(r_i, t_n\right)}{\partial r^2} - \alpha S_y \int_0^{t_n} \frac{\partial h\left(r_i, t_n\right)}{\partial \tau} \exp\left[-\alpha\left(t_n - \tau\right)\right] d\tau \tag{23.91}$$

We let $y = t_n - \tau$ when $e = t_j$, $y = t_n - t_j$, $\tau = t_{j+1}$, and $y = t_n - t_{j+1}$

$$\int\limits_{t_j}^{t_{j+1}} \exp\left[-\alpha\left(t_n - \tau\right)\right] d\tau = \int\limits_{t_n - t_j}^{t_n - t_{j+1}} \exp\left[-\alpha y\right](-dy) \tag{23.92}$$

The equation becomes:

$$-\int\limits_{t_n - t_j}^{t_n - t_{j+1}} \exp\left[-\alpha y\right] dy \tag{23.93}$$

We know that an integral changes signs when the boundaries are switched.
Now the equation becomes:

$$\int\limits_{t_j}^{t_{j+1}} \exp\left[-\alpha\left(t_n - \tau\right)\right] d\tau = \int\limits_{t_n - t_{j+1}}^{t_n - t_j} \exp\left[-\alpha y\right] dy \tag{23.94}$$

where the sign of the integral has changed and the bounds of integration swapped.
Therefore:

$$\int\limits_{a}^{b} \exp\left(ay\right) dy - \frac{1}{a}\exp\left(ay\right) \tag{23.95}$$

$$\int\limits_{t_n - t_{j+1}}^{t_n - t_j} \exp\left[-\alpha y\right] dy = -\frac{1}{\alpha}\exp\left(-\alpha y\right)\bigg|_{t_n - t_{j+1}}^{t_n - t_j} = \frac{1}{\alpha}\exp\left[-\alpha\left(t_{n-j}\right)\right] + \frac{1}{\alpha}\exp\left[-\alpha\left(t_n - t_{j+1}\right)\right]$$

$$\tag{23.96}$$

which can be simplified to:

$$= \frac{1}{\alpha}\left[\exp\left[-\alpha\left(t_n - t_{j+1}\right)\right] - \exp\left[-\alpha\left(t_n - t_j\right)\right]\right] \tag{23.97}$$

$$\int\limits_{0}^{t_n} \frac{\partial h\left(r_i, t_n\right)}{\partial \tau}\exp\left[-\alpha\left(t_n - \tau\right)\right] d\tau = \sum_{j=0}^{n-1} \frac{h_i^{j+1} - h_i^j}{\alpha\Delta r}\left[\exp\left[-\alpha\left(t_n - t_{j+1}\right)\right] - \exp\left[-\alpha\left(t_n - t_j\right)\right]\right] \tag{23.98}$$

We know that at $f(r_i, t_n, h(r_i, t_n))$ we have:

$$f\left(r_i, t_n, h\left(r_i, t_n\right)\right) = \frac{1}{r_i}\frac{\partial h\left(r_i, t_n\right)}{\partial r} + \frac{\partial^2 h\left(r_i, t_n\right)}{\partial r^2} - \alpha Sy\int\limits_{0}^{t_n} \frac{\partial h\left(r_i, t_n\right)}{\partial r}\exp\left[-\alpha\left(t_n - \tau\right)\right] d\tau \tag{23.99}$$

$$= \frac{1}{r_i}\frac{h_{i+1}^n - h_{i-1}^n}{2\Delta r} + \frac{h_{i+1}^n - 2h_i^n + h_{i-1}^n}{\Delta r^2} - \alpha Sy\sum_{j=0}^{n-1} \frac{h_i^{j+1} - h_i^j}{\alpha\Delta r}\left\{\exp\left[-\alpha\left(t_n - t_{j+1}\right)\right] - \exp\left[-\alpha\left(t_n - t_j\right)\right]\right\}$$

$$\tag{23.100}$$

Also:

$$f\left(r_i, t_n, h^p\left(r_i, t_{n+1}\right)\right) = \frac{1}{r_i}\frac{\partial h^p\left(r_i, t_{n+1}\right)}{\partial r} + \frac{\partial^2 h^p\left(r_i, t_{n+1}\right)}{\partial r^2} - \alpha Sy\int\limits_{0}^{t_n} \frac{\partial h^p\left(r_i, t_{n+1}\right)}{\partial r}\exp\left[-\alpha\left(t_{n+1} - \tau\right)\right] d\tau$$

$$\tag{23.101}$$

23.8 VON NEUMANN STABILITY ANALYSIS

The stability analysis is generally used to stabilize finite differential equations. The method of analysis is also called the Fourier stability analysis method.

$$
\frac{h_i^{n+1} - h_i^n}{\Delta t} = \frac{1}{r_i} \frac{h_{i+1}^{p,n+1} - h_{i-1}^{p,n+1}}{2\Delta r} + \frac{h_{i+1}^{p,n+1} - 2h_i^{p,n+1} + h_{i-1}^{p,n+1}}{\Delta r^2}
$$
$$
- \alpha Sy \sum_{j=0}^{n-1} \frac{h_i^{p,j+1} - h_i^{p,j}}{\alpha \Delta r} \left\{ \exp\left[-\alpha\left(t_{n+1} - t_{j+1}\right)\right] - \exp\left[-\alpha\left(t_{n+1} - t_j\right)\right] \right\} \tag{23.102}
$$

Now let

$$
\left\{ \exp\left[-\alpha\left(t_{n+1} - t_{j+1}\right)\right] - \exp\left[-\alpha\left(t_{n+1} - t_j\right)\right] \right\} = \delta_n^j \tag{23.103}
$$

The equation now becomes:

$$
= \frac{1}{r_i} \frac{h_{i+1}^{p,n+1} - h_{i-1}^{p,n+1}}{2\Delta r} + \frac{h_{i+1}^{p,n+1} - 2h_i^{p,n+1} + h_{i-1}^{p,n+1}}{\Delta r^2} - \alpha Sy \sum_{j=0}^{n-1} \frac{h_i^{p,j+1} - h_i^{p,j}}{\alpha \Delta r} \delta_n^j \tag{23.104}
$$

The above equation was simplified to make its solving much easier.
Now going to the predictor-corrector, at (r_i, t_{n+1}) and (r_i, t_n) we have:

$$
\left\{
\begin{aligned}
h^p\left(r_i, t_{n+1}\right) &= h\left(r_i, t_n\right) + \Delta t f\left(r_i, t_n, h\left(r_i, t_n\right)\right) \\
h\left(r_i, t_{n+1}\right) &= h\left(r_i, t_n\right) + \frac{\Delta t}{2}\left[f\left(r_i, t_{n+1}, h^p\left(r_i, t_{n+1}\right)\right) + f\left(r_i, t_n, h\left(r_i, t_n\right)\right)\right]
\end{aligned}
\right) \tag{23.105}
$$

By linking the equations at (r_i, t_{n+1}) and (r_i, t_n) we now have:

$$
\left\{
\begin{aligned}
h^{p,n+1} &= h_i^n + \Delta t f\left(r_i, t_n, h_i^n\right) \\
h_i^{n+1} &= h_i^n + \frac{\Delta t}{2}\left[f\left(r_i, t_{n+1}, h_i^n + \Delta t f\left(r_i, t_n, h_i^n\right)\right) + f\left(r_i, t_n, h_i^n\right)\right]
\end{aligned}
\right) \tag{23.106}
$$

This results in the following system:

$$
h_i^{p,n+1} = h_i^n + \Delta t\left[\frac{1}{r_i} \frac{h_{i+1}^n - h_{i-1}^n}{2\Delta r} + \frac{h_{i+1}^n - 2h_i^n + h_{i-1}^n}{\Delta r^2} - \alpha Sy \sum_{j=0}^{n-1} \frac{h_i^{j+1} - h_i^j}{\alpha \Delta r} \delta_n^\alpha\right]
$$
$$
\left\{
h_i^{n+1} = h_i^n + \frac{\Delta t}{2}
\left\{
f\left[h_i^n + \Delta t\left[\frac{1}{r_i} \frac{h_{i+1}^n - h_{i-1}^n}{2\Delta r} + \frac{h_{i+1}^n - 2h_i^n + h_{i-1}^n}{\Delta r^2} \atop - \alpha Sy \sum_{j=0}^{n-1} \frac{h_i^{j+1} - h_i^j}{\alpha \Delta r} \delta_n^\alpha\right]\right] +
\atop
\frac{1}{r_i} \frac{h_{i+1}^{n+1} - h_{i-1}^{n+1}}{2\Delta r} + \frac{h_{i+1}^{n+1} - 2h_i^{n+1} + h_{i-1}^{n+1}}{\Delta r^2} \atop - \alpha Sy \sum_{j=0}^{n-1} \frac{h_i^{j+1} - h_i^j}{\alpha \Delta r} \delta_n^j
\right\}
\right) \tag{23.107}
$$

From the von Neumann stability analysis that was performed, we can say that the equation is now stable and the conditions of the system are met.

23.9   NUMERICAL SIMULATIONS

In order to access the efficiency of the suggested mathematical model depicting the conversion from confined to unconfined aquifers, together with the used numerical scheme, we present a numerical simulation for different theoretical values of aquifer parameters. The numerical simulations are depicted in Figures 23.2 to 23.11 and in Table 23.1.

23.10   RESULTS AND DISCUSSION

The main aim of a mathematical model is to replicate real world observed facts. Therefore, the translation from observed facts into mathematical equations should be able to capture reality. In fact, such a model should be in good agreement with purely obtained experimental data if any are available. The numerical simulation, when it agrees with experimental data, can now be used for prediction, thus if such simulations are inaccurate, the prediction will be misleading and this can cause damage, or even be deadly in some cases.

The mathematical models that were suggested to depict the conversion of flow from confined to unconfined aquifers in the literature have included high nonlinearity that leads to the exaggeration of determined aquifer parameters, as sometimes the numerical solution predicts high flow while the actual flow may be less. Using such a model will with no doubt lead to wrong prediction. In this work, the obtained numerical simulation predicts the flow within the unconfined aquifer with no high nonlinearity; however, the model is able to account for delay and fading memory processes.

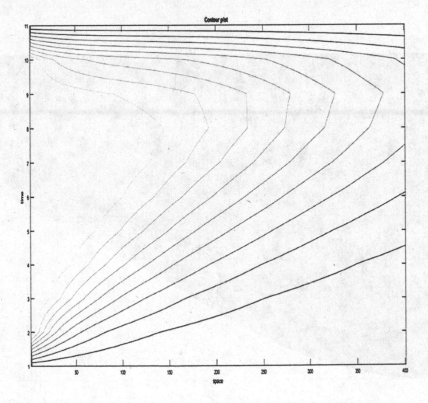

FIGURE 23.2   Contour plots showing flow in unconfined aquifer with high permeability.

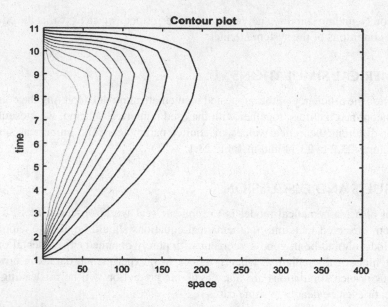

FIGURE 23.3   Contour plot showing the flow with low permeability.

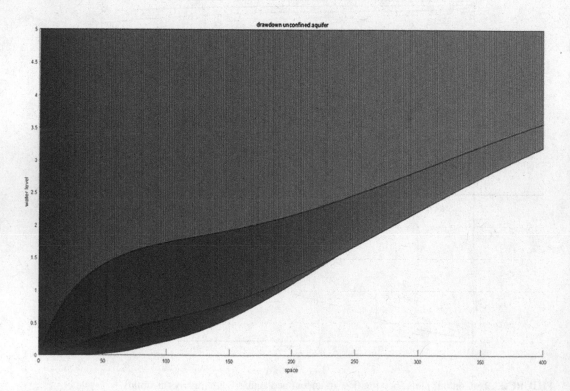

FIGURE 23.4   Drawdown with high transmissivity.

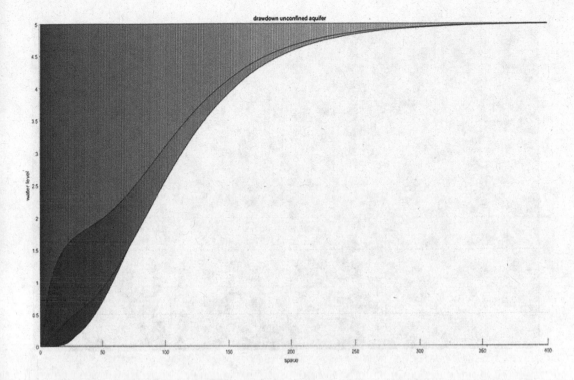

FIGURE 23.5   Drawdown with low transmissivity.

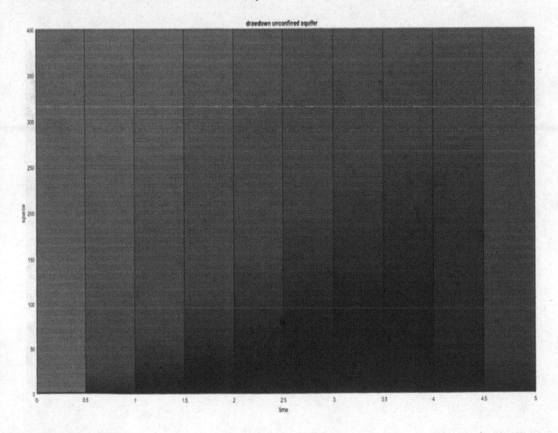

FIGURE 23.6   Histogram showing flow with high transmissivity.

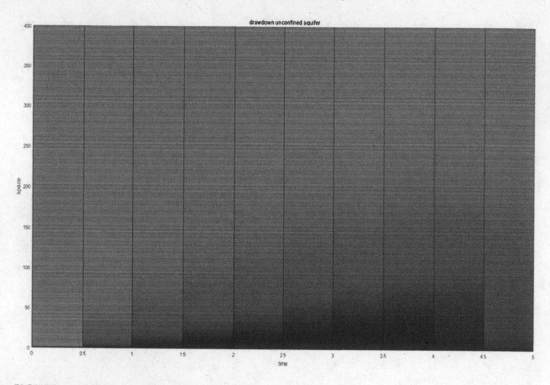

FIGURE 23.7   Histogram showing flow with low transmissivity.

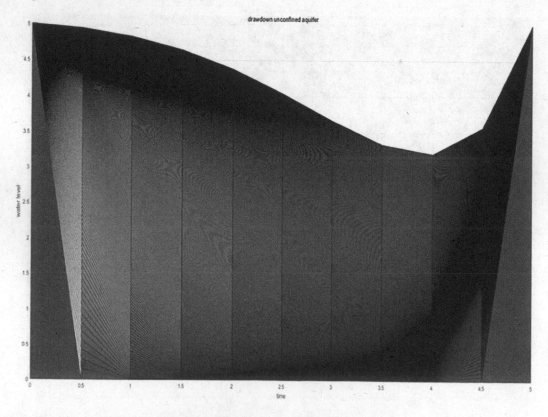

FIGURE 23.8   Flow in time with high transmissivity.

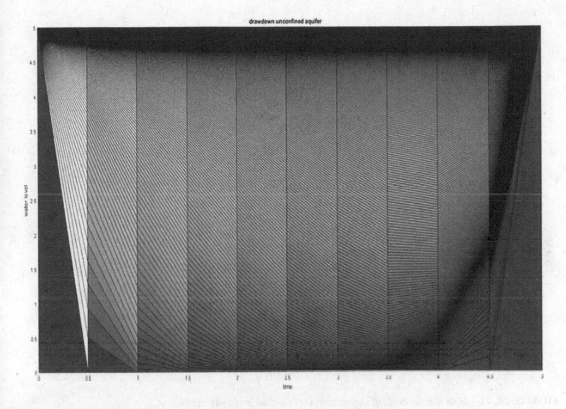

FIGURE 23.9   Flow in time with low transmissivity.

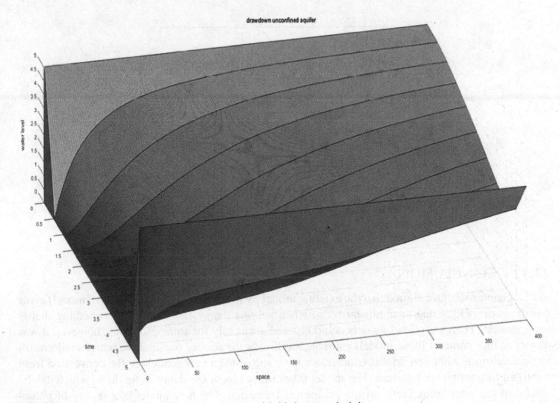

FIGURE 23.10   Space–time cone of depression with high transmissivity.

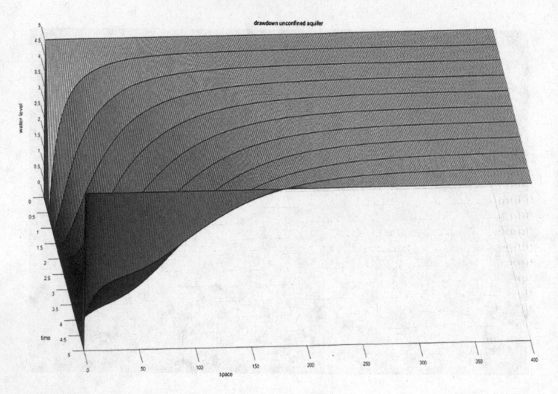

FIGURE 23.11   Space–time cone of depression for low transmissivity.

TABLE 23.1
**Theoretical used Aquifer Parameters to
Perform Simulations**

Parameters	Values
S_c	0.001 to 0.009
S_y	0.0001 to 0.00091
B	4
T	800
α	(0.04 to 0.1) × 100

23.11   CONCLUSION

In this chapter we have argued that the existing model by Moench and Prickett (the MP model) gives non-realistic results due to nonlinearity, so when solving the system this gives something that is exaggerated. Their suggested models could be used accurately for some problems; however, it was observed that some of those models estimate too highly the aquifer parameters, which could result in misleading predictions. In this chapter, we have suggested a new model for the conversion from confined to unconfined aquifers. The model takes into account the delay in the flow, which can be linked to the retardation factor of the geological formation. The new model is a system of partial differential equations, where the first equation was suggested by Theis and has been used with some success in recent decades; the second equation has a fading memory element that could be used to capture memory. The solution of the first equation is well-known as is the derivation of its exact solution. The second equation was solved analytically and numerically. The Laplace transform operator was used to obtain the solution in Laplace space, while a newly introduced numerical scheme was used to solve such an equation. The conditions under which the used method is stable and converges were presented. Some numerical simulations were performed. We can conclude that our method is suitable in predicting the conversion from confined to unconfined aquifers with a delay.

REFERENCES

Bear, J. 1972. *Dynamics of Fluids in Porous Media*. New York: Elsevier.

Boulton, N.S. 1954a. The drawdown of the water-table under non-steady conditions near a pumped well in an unconfined formation. *Proceedings Institution of Civil Engineers*, 3, 564–579.

Boulton, N.S. 1954b. Unsteady radial flow to a pumped well allowing for delayed yield from storage. *International Association Science Hydrology*, 37, 472–477.

Boulton, N.S. 1963. Analysis of data from non-equilibrium pumping tests allowing for delayed yield from storage. *Proceedings Institution of Civil Engineers*, 26, 469–482.

Chen, C.X., Hu, L.T. and Wang, X.S. 2006. Analysis of steady ground water flow toward wells in a confined–unconfined aquifer. *Groundwater*, 44, 609–612.

Dagan, G. 1967. A method of determining the permeability and effective porosity of unconfined anisotropic aquifers. *Water Resources Research*, 3, 1059–1071.

Elango, K. and Swaminathan, K. 1980. Finite-element model for concurrent confined–unconfined zones in an aquifer. *Journal of Hydrology*, 46, 289–299.

Hu, L. and Chen, C. 2008. Analytical methods for transient flow to a well in a confined-unconfined aquifer. *Groundwater*, 46, 642–646.

Kompani-Zare, M. and Zhan, H. 2006. Steady flow to a horizontal drain in an unconfined aquifer with variable thickness. *Journal of Hydrology*, 327, 74–85.

Mishra, P.K. and Kuhlman, K.L. 2013. Unconfined aquifer flow theory: From Dupuit to present. *Advances in Hydrogeology*, 185–202.

Moench, A. F. and Prickett, T.A. 1972. Radial flow in an infinite aquifer undergoing conversion from artesian to water table conditions. *Water Resources Reasearch*, 8, 494–499.

Rushton, K. R. and Wedderburn, L.A. 1971. Aquifers changing between the confined and unconfined state. *Groundwater*, 9, 30–38.

Springer, A.E. and Bair, E.S. 1992. Comparison of methods used to delineate capture zones of wells: 2. Stratified-drift buried-valley aquifer. *Groundwater*, 30, 908–917.

Theis, C.V. 1935. The relation between the lowering of the piezometric surface and the rate and duration of discharge of a well using ground-water storage. *American Geophysical Union*, 16, 519–524.

Wang, X., Li, W and Hu, B. 2009. New approximate solutions of horizontal confined–unconfined flow. *Journal of Hydrology*, 376(3–4), 417–427.

Wang, X. and Zhan, H. 2009. A new solution of transient confined–unconfined flow driven by a pumping well. *Advances in Water Resources*, 32, 1213–1222.

Xiao, L. 2014. *Evaluation of Groundwater Flow Theories and Aquifer Parameters Estimation*. Doctor of Philosophy, University of the Western Cape.

Xiao, L., Ye, M. and Xu, Y. 2018. A new solution for confined-unconfined flow toward a fully penetrating well in a confined aquifer. *Ground Water*, 56, 959–968.

24 Modeling the Diffusion of Chemical Contamination in Soil with Non-Conventional Differential Operators

Palesa Myeko and Abdon Atangana
University of the Free State, Bloemfontein, South Africa

CONTENTS

24.1 INTRODUCTION

The heterogeneity of fractured rock aquifers has made it challenging for the advection-dispersion transport equation to account for the movement of groundwater in fractures, faults, and dykes in geological formations. Allwright & Atangana (2018) used the fractal advection-dispersion equation to model the impact that fractures have on groundwater flow; the model was run such that it functions independently of location. One of the fatal effects of groundwater pollution recorded in the history of humankind is perhaps that of the Love Canal.

The diffusion of chemical contamination in soil is a result of the movement of water through the soil. A solute can undergo some chemical reactions in the subsurface that will influence the plume regarding composition, location, and size. Stagnitti et al. (1995) explain the movement of groundwater in the subsurface in terms of average water velocity for one dimension using the convective-diffusive equation:

DOI: 10.1201/9781003266266-24

$$\frac{\partial c}{\partial t} + v\frac{\partial c}{\partial x} = \frac{\partial}{\partial x}\left[D\frac{\partial c}{\partial x}\right] - f(c) - \frac{\partial c_a}{\partial x} \tag{24.1}$$

where D is dispersion, v is the mean velocity, c_a is the concentration of the adsorbed chemical, and f is the irreversible reaction decay rate. D, v, c_a, and f can be complex functions of c, x, and t. In certain instances, the concentration of the adsorbed chemical can be considered as $c_a = (R-1)c$ when D and v are constant if f is a linear function of c. The model is subjected to the following boundary conditions, defined in terms of the following equation:

$$c_f = c - (D/v)\frac{\partial c}{\partial x} = C_0 \text{ at } x = 0 \tag{24.2}$$

where C_f is the flux concentration and c is the resident concentration. Utilizing a cylinder with limited length L in a laboratory experiment, the boundary conditions are to be defined for $x = 0$ and $x = L$. Based on the process of insertion, c will be forced at $x = 0$. The chemical contaminants found in the subsurface are transported through the geological formation to the saturated zone. As discussed earlier, the above equation was constructed using the concept of local differential operators, thus the equation cannot really replicate the long range behavior of the pollution within a fracture, nor can it trace the trajectory of the pollution. This implies the model is Markovian, as it only depends on the initial state that is considered here, as the initial condition, and on the mathematical formula which is the exact solution. It is important to recall that the movement of solids within a geological formation is conditioned by the structure of the soil, which is not purely homogeneous. With the properties of the used differential operator, one cannot expect heterogeneity to be considered. Thus, the above equation cannot accurately replicate the observed facts and so needs to be modified. This has been demonstrated in several already published studies in which the concept of non-local differential operators was suitable to replicate such behavior, those non-local differential operators being built using the power law, the exponential decay law, and the generalized Mittag–Leffler function. In the following section, a detailed presentation of the properties and applications of these three kernels is given.

24.2　NUMERICAL SOLUTIONS FOR THE CLASSICAL CASE

We use classical differential operators to model the solute flow of chemicals in groundwater through a porous geological formation. The numerical analysis for fractional differential operators is derived using numerical methods. The type of numerical approximation method utilized is dependent on whether the equation is a Partial Differential Equation (PDE) or an Ordinary Differential Equation (ODE). The three approximation methods that we apply to the convective-diffusive equation are the forward Euler, the backward Euler, and the Crank–Nicolson.

24.2.1　Forward Euler Numerical Scheme

Apply the forward Euler approximation method to Equation (24.1):

$$\frac{c(x_i,t_{n+1}) - c(x_i,t_n)}{\Delta t} + v\frac{c(x_{i+1},t_{n+1}) - c(x_i,t_{n+1})}{\Delta x} =$$
$$D\frac{c(x_{i+1},t_{n+1}) - 2c(x_i,t_{n+1}) + c(x_{i-1},t_{n+1})}{\Delta x^2} -$$
$$f(c(x_i,t_{n+1})) - \frac{c_a(x_{i+1},t_{n+1}) - c_a(x_i,t_{n+1})}{\Delta x} \tag{24.3}$$

which can be rewritten as:

$$\frac{c_i^{n+1} - c_i^n}{\Delta t} + v\frac{c_{i+1}^{n+1} - c_i^{n+1}}{\Delta x} = D\frac{c_{i+1}^{n+1} - 2c_i^{n+1} + c_{i-1}^{n+1}}{\Delta x^2} - f\left(c_i^{n+1}\right) - \frac{c_{a(i+1)}^{n+1} - c_{ai}^{n+1}}{\Delta x} \qquad (24.4)$$

To simplify the equation let

$$\frac{1}{\Delta t} = r_1, \frac{v}{\Delta x} = r_2, \frac{D}{\Delta x^2} = r_3, \frac{1}{\Delta x} = r_4 \text{ and } \left(c_i^{n+1}\right) = \lambda\left(c_i^{n+1}\right)$$

Substitute into Equation (24.4):

$$r_1\left(c_i^{n+1} - c_i^n\right) + r_2\left(c_{i+1}^{n+1} - c_i^{n+1}\right) = r_3\left(c_{i+1}^{n+1} - 2c_i^{n+1} + c_{i-1}^{n+1}\right) - \lambda\left(c_i^{n+1}\right) - r_4\left(c_{a(i+1)}^{n+1} - c_{ai}^{n+1}\right) \qquad (24.5)$$

Factorize:

$$r_1 c_i^{n+1} - r_1 c_i^n + r_2 c_{i+1}^{n+1} - r_2 c_i^{n+1} = r_3 c_{i+1}^{n+1} - 2r_3 c_i^{n+1} + r_3 c_{i-1}^{n+1} + \lambda\left(c_i^{n+1}\right) - r_4 c_{a(i+1)}^{n+1} - r_4 c_{ai}^{n+1} \qquad (24.6)$$

$$r_1 c_i^{n+1} - r_1 c_i^n - r_2 c_i^{n+1} + 2r_3 c_i^{n+1} = r_3 c_{i+1}^{n+1} + r_2 c_{i+1}^{n+1} - \lambda\left(c_i^{n+1}\right) - r_4\left(c_{a(i+1)}^{n+1} - c_{ai}^{n+1}\right) \qquad (24.7)$$

$$c_i^{n+1}\left(r_1 + r_2 + 2r_3\right) - r_1 c_i^n = c_{i+1}^{n+1}\left(r_3 - r_2\right) - \lambda\left(c_i^{n+1}\right) - r_4\left(c_{a(i+1)}^{n+1} - c_{ai}^{n+1}\right) \qquad (24.8)$$

$$c_i^{n+1}\left(r_1 + r_2 + 2r_3\right) = c_{i+1}^{n+1}\left(r_3 - r_2\right) - \lambda\left(c_i^{n+1}\right) - r_4\left(c_{a(i+1)}^{n+1} - c_{ai}^{n+1}\right) + r_1 c_i^n \qquad (24.9)$$

Substitute the Fourier series into the numerical solution derived with the forward Euler approximation method to determine the stability of the solution in a conventional medium:

$$e^{a(t+\Delta t)}e^{ik_m x}\left(r_1 - r_2 + 2r_3\right) = e^{a(t+\Delta t)}e^{ik_m(x+\Delta x)}\left(r_3 - r_2\right) - \lambda e^{a(t+\Delta t)}e^{ik_m x} - r_4\left(e^{a(t+\Delta t)}e^{ik_m(x+\Delta x)} - e^{a(t+\Delta t)}e^{ik_m x}\right) + r_1 e^{at}e^{ik_m x} \qquad (24.10)$$

Divide by $e^{at}e^{ik_m x}$:

$$e^{a\Delta t}\left(r_1 - r_2 + 2r_3 + \lambda + 1\right) = \left(r_3 - r_2\right)e^{a\Delta t}e^{i\Delta x k_m} - r_4 e^{a\Delta t}e^{i\Delta x k_m} + r_1 \qquad (24.11)$$

Simplify:

$$e^{a\Delta t}\left[r_1 - r_2 + 2r_3 + \lambda + 1 + r_4 e^{i\Delta x k_m} + \left(r_2 - r_3\right)e^{i\Delta x k_m}\right] = r_1 \qquad (24.12)$$

Make $e^{a\Delta t}$ the subject:

$$e^{a\Delta t} = \frac{r_1}{r_1 - r_2 + 2r_3 + \lambda + 1 + r_4 e^{i\Delta x k_m} + \left(r_2 - r_3\right)e^{i\Delta x k_m}} \qquad (24.13)$$

where

$$e^{i\Delta xkm} = \cos\left(\Delta xk_m\right) + i\sin\left(\Delta xk_m\right) \text{ and } e^{-i\Delta xkm} = \cos\left(\Delta xk_m\right) - i\sin\left(\Delta xk_m\right)$$

$$e^{a\Delta t} = \frac{r_1}{r_1 - r_2 + 2r_3 + \lambda + 1 + r_4\left(\cos\left(\Delta xk_m\right) + i\sin\left(\Delta xk_m\right)\right) + \left(r_2 - r_3\right)\left(\cos\left(\Delta xk_m\right) + i\sin\left(\Delta xk_m\right)\right)}$$

(24.14)

Make $\Delta xk_m = \theta$, $r_1 - r_2 + 2r_3 + \lambda + 1 = A$:

$$e^{a\Delta t} = \frac{r_1}{A + r_4\cos\theta + r_4 i\sin\theta + r_2\cos\theta + r_2 i\sin\theta - r_3\cos\theta - r_3 i\sin\theta}$$

(24.15)

Rewrite:

$$e^{a\Delta t} = \frac{r_1}{A + r_4\cos\theta + r_2\cos\theta - r_3\cos\theta + i\left(r_4\sin\theta + r_2\sin\theta - r_3\sin\theta\right)}$$

(24.16)

Let $A + r_4\cos\theta + r_2\cos\theta - r_3\cos\theta = M$, $r_4\sin\theta + r_2\sin\theta - r_3\sin\theta = M_1$

$$\left|e^{a\Delta t}\right| = \frac{r_1}{M + iM_1} < 1$$

(24.17)

$$\left|e^{a\Delta t}\right| = \left|\frac{r_1}{M + iM_1}\right| < 1$$

(24.18)

$$\left|e^{a\Delta t}\right| = \frac{r_1}{\sqrt{M^2 + M_1^2}} < 1$$

(24.19)

The numerical solution for the forward Euler is stable unconditionally.

24.2.2 BACKWARD EULER NUMERICAL SCHEME

Apply the backward Euler method to Equation (24.1):

$$\frac{c\left(x_i, t_n\right) - c\left(x_i, t_{n-1}\right)}{\Delta t} + v\frac{c\left(x_{i+1}, t_n\right) - c\left(x_i, t_n\right)}{\Delta x} = D\frac{c\left(x_{i+1}, t_n\right) - 2c\left(x_i, t_n\right) + c\left(x_{i-1}, t_n\right)}{\Delta x^2} -$$
$$f\left(c\left(x_i, t_n\right)\right) - \frac{c_a\left(x_{i+1}, t_n\right) - c_a\left(x_i, t_n\right)}{\Delta x}$$

(24.20)

Rewrite as:

$$\frac{c_i^n - c_i^{n-1}}{\Delta t} + v\frac{c_{i+1}^n - c_i^n}{\Delta x} = D\frac{c_{i+1}^n - 2c_i^n + c_{i-1}^n}{\Delta x^2} - f\left(c_i^n\right) - \frac{c_{a(i+1)}^n - c_{ai}^n}{\Delta x}$$

(24.21)

Simplify by making

$$\frac{1}{\Delta t} = r_1, \frac{v}{\Delta x} = r_2, \frac{D}{\Delta x^2} = r_3 \ , \ \frac{1}{\Delta x} = r_4 \text{ and } f\left(c_i^{n+1}\right) = \lambda\left(c_i^{n+1}\right)$$

Substitute into Equation (24.21):

$$r_1\left(c_i^n - c_i^{n-1}\right) + r_2\left(c_{i+1}^n - c_i^n\right) = r_3\left(c_{i+1}^n - 2c_i^n + c_{i-1}^n\right) - \lambda\left(c_i^n\right) - r_4\left(c_{ai+1}^n - c_{ai}^n\right) \tag{24.22}$$

Factorize:

$$r_1c_i^n - r_1c_i^{n-1} + r_2c_{i+1}^n - r_2c_i^n = r_3c_{i+1}^n - 2r_3c_i^n + r_3c_{i-1}^n - \lambda\left(c_i^n\right) - r_4c_{ai+1}^n - r_4c_{ai}^n \tag{24.23}$$

Simplify:

$$c_i^n\left(r_1 - r_2 + 2r_3\right) - r_1c_i^{n-1} = c_{i+1}^n\left(r_3 - r_2\right) + r_3c_{i-1}^n - \lambda c_i^n - r_4\left(c_{ai+1}^n - c_{ai}^n\right) \tag{24.24}$$

$$c_i^n\left(r_1 - r_2 + 2r_3\right) = c_{i+1}^n\left(r_3 - r_2\right) + r_3c_{i-1}^n - \lambda c_i^n - r_4\left(c_{ai+1}^n - c_{ai}^n\right) + r_1c_i^{n-1} \tag{24.25}$$

The Fourier von Neumann stability analysis is conducted for the convective-diffusive partial equation discretized using the backward Euler approximation method:

$$e^{at}e^{ik_mx}\left(r_1 - r_2 + 2r_3\right) = e^{at}e^{ik_m(x+\Delta x)}\left(r_3 - r_2\right) + r_3e^{at}e^{ik_m(x+\Delta x)} - \\ \lambda e^{at}e^{ik_mx} - r_4\left(e^{at}e^{ik_mx}\right) + r_1e^{a(t+\Delta t)}e^{ik_mx} \tag{24.26}$$

Divide by $e^{at}e^{ik_mx}$:

$$r_1 - r_2 + 2r_3 + \lambda + r_4 = e^{i\Delta xk_m}\left(r_3 - r_2\right) + r_3e^{i\Delta k_m} + r_1e^{-a\Delta t} \tag{24.27}$$

Simplify:

$$r_1 - r_2 + 2r_3 + \lambda + r_4 + e^{i\Delta xk_m}\left(r_2 - r_3\right) - r_3e^{i\Delta k_m} = r_1e^{-a\Delta t} \tag{24.28}$$

Make $e^{a\Delta t}$ the subject:

$$e^{a\Delta t} = \frac{r_1}{r_1 - r_2 + 2r_3 + \lambda + r_4 + e^{i\Delta xk_m}\left(r_2 - r_3\right) - r_3e^{i\Delta xk_m}} \tag{24.29}$$

where

$$e^{i\Delta xkm} = \cos\left(\Delta xk_m\right) + i\sin\left(\Delta xk_m\right) \text{ and } e^{-i\Delta xkm} = \cos\left(\Delta xk_m\right) - i\sin\left(\Delta xk_m\right)$$

$$e^{a\Delta t} = \frac{r_1}{r_1 - r_2 + 2r_3 + \lambda + r_4 + \left(r_2 - r_3\right)\left(\cos\left(\Delta xk_m\right) + i\sin\left(\Delta xk_m\right)\right) - r_3\left(\cos\left(\Delta xk_m\right) + i\sin\left(\Delta xk_m\right)\right)} \tag{24.30}$$

Let $r_1 - r_2 + 2r_3 + \lambda + r_4 = \mathrm{A}$ and $\Delta x k_m = \theta$:

$$e^{a\Delta t} = \frac{r_1}{A + (r_2 - r_3)(\cos\theta + i\sin\theta) - r_3(\cos\theta + i\sin\theta)} \tag{24.31}$$

$$e^{a\Delta t} = \frac{r_1}{A + r_2\cos\theta - r_3\cos\theta + r_2 i\sin\theta - r_3 i\sin\theta - r_3\cos\theta - r_3 i\sin\theta} \tag{24.32}$$

$$e^{a\Delta t} = \frac{r_1}{A + r_2\cos\theta - 2r_3\cos\theta + r_2 i\sin\theta - 2r_3 i\sin\theta} \tag{24.33}$$

$$e^{a\Delta t} = \frac{r_1}{A + \cos\theta(r_2 - 2r_3) + i\sin\theta(r_2 - 2r_3)} \tag{24.34}$$

Let $M = A + \cos\theta(r_2 - 2r_3)$:

$$\left| e^{a\Delta t} \right| = \left| \frac{r_1}{M + i\sin\theta(r_2 - 2r_3)} \right| < 1 \tag{24.35}$$

$$\left| e^{a\Delta t} \right| = \frac{r_1}{\sqrt{M^2 + \sin^2\theta(r_2 - 2r_3)^2}} < 1 \tag{24.36}$$

The numerical solution for the backward Euler is stable on condition that:

$$\frac{r_1^2}{M^2 + \sin^2\theta(r_2 - 2r_3)^2} < 1 \tag{24.37}$$

24.2.3   CRANK–NICOLSON NUMERICAL SCHEME

Apply the Crank–Nicolson method to Equation (24.1):

$$\frac{c(x_i, t_{n+1}) - c(x_i, t_n)}{\Delta t} + v\frac{c(x_{i+1}, t_{n+1}) - c(x_i, t_{n+1})}{\Delta x} =$$
$$\frac{1}{2}\left[\begin{pmatrix} D\dfrac{c(x_{i+1}, t_{n+1}) - 2c(x_i, t_{n+1}) + c(x_{i-1}, t_{n+1})}{\Delta x^2} - f(c(x_i, t_{n+1})) - \\ \dfrac{c_a(x_{i+1}, t_{n+1}) - c_a(x_i, t_{n+1})}{\Delta x} \end{pmatrix} + \\ \left(D\dfrac{c(x_{i+1}, t_n) - 2c(x_i, t_n) + c(x_{i-1}, t_n)}{\Delta x^2} - f(c(x_i, t_n)) - \dfrac{c_a(x_{i+1}, t_n) - c_a(x_i, t_n)}{\Delta x} \right) \right] \tag{24.38}$$

$$\frac{c_i^{n+1} - c_i^n}{\Delta t} + v\frac{c_{i+1}^{n+1} - c_i^{n+1}}{\Delta x} =$$
$$\frac{1}{2}\left[\left(D\frac{c_{i+1}^{n+1} - 2c_i^{n+1} + c_{i-1}^{n+1}}{\Delta x^2} - \frac{c_{ai+1}^{n+1} - c_{ai}^{n+1}}{\Delta x} \right) + \left(D\frac{c_{i+1}^n - 2c_i^n + c_{i-1}^n}{\Delta x^2} - \frac{c_{ai+1}^n - c_{ai}^n}{\Delta x} \right) \right]$$
$$- f(c_i^{n+1}) \tag{24.39}$$

Let $\dfrac{1}{\Delta t} = r_1, \dfrac{v}{\Delta x} = r_2, \dfrac{D}{\Delta x^2} = r_3 , \dfrac{1}{\Delta x} = r_4$ and $f\left(c_i^{n+1}\right) = \lambda\left(c_i^{n+1}\right)$

Substitute into Equation (24.39):

$$r_1\left(c_i^{n+1} - c_i^n\right) + r_2\left(c_{i+1}^{n+1} - c_i^{n+1}\right) = \frac{1}{2}\left[\begin{array}{l} r_3\left(c_{i+1}^{n+1} - 2c_i^{n+1} + c_{i-1}^{n+1}\right) - r_4\left(c_{ai+1}^{n+1} - c_{ai}^{n+1}\right) + \\ r_3\left(c_{i+1}^n + 2c_i^n + c_{i-1}^n\right) - r_4\left(c_{ai+1}^n - c_{ai}^n\right) \end{array}\right] - \lambda c_i^{n+1} \quad (24.40)$$

Factorize:

$$r_1 c_i^{n+1} - r_1 c_i^n + r_2 c_{i+1}^{n+1} - r_2 c_i^{n+1} = \frac{1}{2}\left[\begin{array}{l} r_3 c_{i+1}^{n+1} - 2r_3 c_i^{n+1} + r_3 c_{i-1}^{n+1} - r_4 c_{ai+1}^{n+1} - r_4 c_{ai}^{n+1} + \\ r_3 c_{i+1}^n - 2r_3 c_i^n - r_3 c_{i-1}^n - r_4 c_{ai+1}^n + r_4 c_{ai+1}^n \end{array}\right] - \lambda c_i^{n+1} \quad (24.41)$$

Simplify:

$$c_i^{n+1}\left(r_1 - r_2 + 2r_3\right) - r_1 c_i^n = \frac{1}{2}\left[\begin{array}{l} c_i^n\left(r_3 - r_2\right) + r_3 c_{i-1}^{n+1} - r_4 c_{ai+1}^{n+1} + r_4 c_{ai}^{n+1} + r_3 c_{i+1}^n - \\ 2r_3 c_i^n - r_3 c_{i+1}^n - r_4 c_{ai+1}^n + r_4 c_{ai}^n \end{array}\right] - \lambda c_i^{n+1} \quad (24.42)$$

$$c_i^{n+1}\left(r_1 - r_2 + 2r_3\right) = \frac{1}{2}c_{i+1}^{n+1}\left(r_3 - r_2\right) + r_3 c_{i-1}^{n+1} - r_4 c_{ai+1}^{n+1} - r_4 c_{ai}^{n+1} - c_i^n\left(2r_3 + r_1\right) -$$
$$r_3 c_{i-1}^n - r_4 c_{ai+1}^n - r_4 c_{ai}^n - \lambda c_i^{n+1} \quad (24.43)$$

We further investigate the conditions under which the stability is reached when using the Crank–Nicolson scheme for solving the problem under investigation:

$$e^{a(t+\Delta t)}e^{ik_m x}\left(r_1 - r_2 + 2r_3\right) = \frac{1}{2}\left[\begin{array}{l} e^{a(t+\Delta t)}e^{ik_m(x+\Delta x)}\left(r_3 - r_2\right) + \\ r_3 e^{a(t+\Delta t)}e^{ik_m(x-\Delta x)} - \\ r_4 e^{a(t+\Delta t)}e^{ik_m(x+\Delta x)} - r_4 e^{a(t+\Delta t)}e^{ik_m x} - \\ e^{at}e^{ik_m x} - r_3 e^{at}e^{ik_m(x-\Delta x)} - \\ r_4 e^{at}e^{ik_m(x+\Delta x)} - r_4 e^{at}e^{ik_m x} \end{array}\right] - \lambda e^{a(t+\Delta t)}e^{ik_m x} \quad (24.44)$$

$$e^{a\Delta t}\left(r_1 - r_2 + 2r_3\right) = \frac{1}{2}\left[\begin{array}{l} e^{a\Delta t}\left(r_3 - r_2\right) + r_3 e^{a\Delta t}e^{-i\Delta x k_m} - r_4 e^{a\Delta t}e^{i\Delta x k_m} \\ -r_4 e^{a\Delta t} - 1 - r_3 e^{-i\Delta x k_m} - r_4 e^{i\Delta x k_m} - r_4 \end{array}\right] - \lambda e^{a\Delta t} \quad (24.45)$$

$$e^{a(t+\Delta t)}e^{ik_m x}\left(r_1 - r_2 + 2r_3\right) = \frac{1}{2}\left[\begin{array}{l} e^{a(t+\Delta t)}e^{ik_m(x+\Delta x)}\left(r_3 - r_2\right) + \\ r_3 e^{a(t+\Delta t)}e^{ik_m(x-\Delta x)} - \\ r_4 e^{a(t+\Delta t)}e^{ik_m(x+\Delta x)} - r_4 e^{a(t+\Delta t)}e^{ik_m x} - \\ -e^{at}e^{ik_m x} - r_3 e^{at}e^{ik_m(x-\Delta x)} - \\ -r_4 e^{at}e^{ik_m(x+\Delta x)} - r_4 e^{at}e^{ik_m x} \end{array}\right] - \lambda e^{a(t+\Delta t)}e^{ik_m x} \quad (24.46)$$

$$e^{a\Delta t}\left(r_1 - r_2 + 2r_3\right) = \frac{1}{2}\left[\begin{array}{l} e^{a\Delta t}\left(r_3 - r_2\right) + r_3 e^{a\Delta t} e^{-i\Delta x k_m} - r_4 e^{a\Delta t} e^{i\Delta x k_m} \\ -r_4 e^{a\Delta t} - 1 - r_3 e^{-i\Delta x k_m} - r_4 e^{i\Delta x k_m} - r_4 \end{array}\right] - \lambda e^{a\Delta t} \tag{24.47}$$

$$e^{a\Delta t}\left(r_1 - r_2 + 2r_3\right) + \left(r_2 - r_3\right) - r_3 e^{-i\Delta x k_m} + r_4 e^{i\Delta x k_m} + r_4 + 1 + r_4 + \lambda = \frac{1}{2}\left(-r_3 e^{-i\Delta x k_m}\right) \tag{24.48}$$

$$e^{a\Delta t} = \frac{-\dfrac{1}{2} r_3 e^{-i\Delta x k_m}}{r_1 - r_2 + 2r_3 + \left(r_2 - r_3\right) - r_3 e^{-i\Delta x k_m} + r_4 e^{i\Delta x k_m} + 2r_4 + 1 + \lambda} \tag{24.49}$$

$$e^{a\Delta t} = \frac{-\dfrac{r_3 e^{-i\Delta x k_m}}{2}}{r_1 - r_2 + 2r_3 + \left(r_2 - r_3\right) - r_3 e^{-i\Delta x k_m} + r_4 e^{i\Delta x k_m} + 2r_4 + 1 + \lambda} \tag{24.50}$$

$$e^{a\Delta t} = \frac{-\dfrac{r_3 \cos\theta - r_3 i \sin\theta}{2}}{r_1 - r_2 + 2r_3 + \left(r_2 - r_3\right) - r_3\left(\cos\theta - i\sin\theta\right) + r_4\left(\cos\theta + i\sin\theta\right) + 2r_4 + 1 + \lambda} \tag{24.51}$$

$$e^{a\Delta t} = \frac{-\dfrac{r_3 \cos\theta - r_3 i \sin\theta}{2}}{r_1 - r_2 + 2r_3 + \left(r_2 - r_3\right) + r_4\cos\theta - r_3\cos\theta + r_4 i\sin\theta + r_3 i\sin\theta + 2r_4 + 1 + \lambda} \tag{24.52}$$

$$e^{a\Delta t} = \frac{\left|\dfrac{r_3 \cos\theta - r_3 i \sin\theta}{2}\right|}{r_1 - r_2 + 2r_3 + \left(r_2 - r_3\right) + r_4\cos\theta - r_3\cos\theta + r_4 i\sin\theta + r_3 i\sin\theta + 2r_4 + 1 + \lambda} \tag{24.53}$$

For simplicity we let:

$$M = r_1 - r_2 + 2r_3 + \left(r_2 - r_3\right) + r_4 \cos\theta - r_3 \cos\theta 2r_4 + 1 + \lambda$$

$$\left|e^{a\Delta t}\right| = \left|\frac{\dfrac{r_3}{2}}{M + i\sin\theta\left(r_4 + r_3\right)}\right| < 1 \tag{24.54}$$

$$\left|e^{a\Delta t}\right| = \frac{r_3}{2\sqrt{M^2 + \sin\theta^2\left(r_4 + r_3\right)^2}} < 1 \tag{24.55}$$

$$\frac{r_3^2}{4\left[M^2 + \sin\theta^2\left(r_4 + r_3\right)^2\right]} < 1 \tag{24.56}$$

$$\frac{r_3^2}{M^2 + \sin\theta^2 \left(r_4 + r_3\right)^2} < 4 \tag{24.57}$$

The numerical solution is conditionally stable for Crank–Nicolson numerical approximation.

24.2.4 DISCRETIZE THE CONVECTIVE-DIFFUSIVE EQUATION BASED ON TIME

Discrete time convolution is a process in which the input interacts in linear time-invariant systems to produce an output.

Replace $\dfrac{\partial c_a}{\partial x}$ with $\dfrac{\partial \left(R-1\right)c}{\partial x}$ in Equation (24.1) to be:

$$\frac{\partial c}{\partial t} = -v\frac{\partial c}{\partial x} + D\frac{\partial^2 c}{\partial x^2} - f\left(c\right) - \frac{\partial \left(R-1\right)c}{\partial x} \tag{24.58}$$

$$\frac{c^{n+1} - c^n}{\Delta t} = -v\frac{\partial c^n}{\partial x} + D\frac{\partial^2 c^n}{\partial x^2} - f\left(c^n\right) - \left(R-1\right)\frac{\partial c^n}{\partial x} \tag{24.59}$$

Simplify equation by dividing by Δt:

$$c^{n+1} = c^n - v\Delta t\frac{\partial c^n}{\partial x} + D\Delta t\frac{\partial^2 c^n}{\partial x^2} - \Delta t f\left(c^n\right) - \Delta t\left(R-1\right)\frac{\partial c^n}{\partial x} \tag{24.60}$$

Convolute with the function g:

$$\left(c^{n+1}, g\right) = \left(c^n, g\right) - v\Delta t\left(\frac{\partial c^n}{\partial x}, \frac{\partial g}{\partial x}\right) + \Delta t D\left(\frac{\partial^2 c^{n+1}}{\partial x^2}, \frac{\partial^2 g}{\partial x^2}\right) - \Delta t\left(R-1\right)\left(\frac{\partial c^n}{\partial x}, \frac{\partial g}{\partial x}\right) \tag{24.61}$$

where

$$\left(f, g\right) = \int fg \, dx_\Omega \, \|g\|_0 = \sqrt{g, g}, \|g\|_1 = \sqrt{\|g\|_0^2 + \|\frac{d^2 g}{dx}\|} \tag{24.62}$$

Prove that:

$$\|c^1\|_0 < \|c^0\|_0 \left(c^1, c^1\right) = \left(c^0, c^1\right) - v\Delta t\left(\frac{\partial c^0}{\partial x}, \frac{\partial c^0}{\partial x}\right) + \Delta t D\left(\frac{\partial^2 c^1}{\partial x^2}, \frac{\partial^2 c^1}{\partial x^2}\right)$$

$$-\Delta t\left(R-1\right)\left(\frac{\partial c^0}{\partial x}, \frac{\partial c^0}{\partial x}\right) + \Delta t\left(f\left(c^0\right), f\left(c^0\right)\right) \tag{24.63}$$

$$\|c^1\|^2 = \|c^0\|_0 \|c^1\|_0 - \Delta t\|f\left(c^0\right)\|\|\|c^1\|_1^2 \le \|c^0\|_0\|c^1\|_1 - \Delta t\|c^0\|_0 \tag{24.64}$$

The function f must satisfy:

$$\|f\left(c\right)^0\|_0 < \theta_0\|c^0\|_0 \tag{24.65}$$

$$\|c^1\|_1^2 \leq \|c^0\|_0 \|c^1\|_1 - \Delta t \theta_0 \|c^0\|_0 \tag{24.66}$$

$$\|c^1\|_1^2 \leq \|c^0\|_0 \|c^1\|_1 - \Delta t \theta_0 \frac{\|c^0\|_0 \|c^1\|_1}{\|c^1\|_1} \tag{24.67}$$

$$\|c^1\|_1 \leq \|c^0\|_0 \left(1 - \frac{\Delta t \theta_0}{\|c^1\|_1}\right) \tag{24.68}$$

$$\frac{\|c^1\|_1}{\|c^0\|_0} < 1 \tag{24.69}$$

To have $\|c^1\|_1 \leq \|c^0\|_0 \ 1 - \dfrac{\Delta t \theta_0}{\|c^1\|_1} \left\langle 1 \dfrac{\Delta t \theta_0}{\|c^1\|_1} \right\rangle 0$

The convolution of Equation (24.58) is expressed as:

$$\frac{c^{n+1} - c^n}{\Delta t} = -v \frac{\partial c^n}{\partial x} + D \frac{\partial^2 c^n}{\partial x^2} - f(c^n) - (R-1) \frac{\partial c^n}{\partial x} \tag{24.70}$$

$$c^{n+1} = c^n - v\Delta t \frac{\partial c^n}{\partial x} + D\Delta t \frac{\partial^2 c^n}{\partial x^2} - \Delta t f(c^n) \Delta t - (R-1) \frac{\partial c^n}{\partial x} \tag{24.71}$$

$$\left(c^{n+1}, g\right) = \left(c^n, g\right) - v\Delta t \left(\frac{\partial c^n}{\partial x}, \frac{\partial g}{\partial x}\right) + \Delta t D \left(\frac{\partial^2 c^{n+1}}{\partial x^2}, \frac{\partial^2 g}{\partial x^2}\right) - \Delta t (R-1) \left(\frac{\partial c^n}{\partial x}, \frac{\partial g}{\partial x}\right) \tag{24.72}$$

We assume by the inductive principle that $\forall n \geq 0 \ \|c^0\|_1 < \|c^0\|_0$

$$\left(c^{n+1}, c^{n+1}\right) = \left(c^n, c^{n+1}\right) - v\Delta t \left(\frac{\partial c^n}{\partial x}, \frac{\partial c^{n+1}}{\partial x}\right) + D\Delta t \left(\frac{\partial^2 c^{n+1}}{\partial x^2}, \frac{\partial^2 c^{n+1}}{\partial x^2}\right)$$
$$- \Delta t (R-1) \left(\frac{\partial c^n}{\partial x}, \frac{\partial c^{n+1}}{\partial x}\right) - \Delta t (f(c^n), f(c^{n+1})) \tag{24.73}$$

$$\|c^{n+1}\|_1^2 = \|c^n\|_0 \|c^{n+1}\|_0 - v\Delta t + \Delta t (R-1) \| \frac{\partial c^n}{\partial x} \|_0 \| \frac{\partial c^{n+1}}{\partial x} \|_0^1 - \Delta t \|f(c^n)\|_0 \|f(c^{n+1})\|_0 \tag{24.74}$$

$$\leq \|c^n\|_0 \|c^{n+1}\|_0 - v\Delta t + \Delta t (R-1) \theta_1 \|c^n\|_1 \theta_2 \|c^{n+1}\|_0 - \theta_4 \theta_3 \|c^n\|_0 \|c^{n+1}\|_0 \tag{24.75}$$

$$< \|c^0\|_0 \|c^{n+1}\|_0 - \lambda \|c^0\|_0 \|c^{n+1}\|_0 - \lambda_2 \|c^0\|_0 \|c^{n+1}\|_0 \tag{24.76}$$

$$\|c^{n+1}\|_1 < \|c^0\|_0 \tag{24.77}$$

$$\|c^{n+1}\|_1^2 < \|c^0\|_0 \|c^{n+1}\|_1 - \lambda \|c^0\|_0 \|c^{n+1}\|_1 - \lambda_2 \|c^0\|_0 \|c^{n+1}\|_1 \tag{24.78}$$

$$\|c^{n+1}\|_1 < (1 - \lambda - \lambda_2) \|c^0\|_0 \tag{24.79}$$

$$1 - \lambda - \lambda_2 < 1 \tag{24.80}$$

$$\lambda + \lambda_2 > 0 \tag{24.81}$$

24.2.5 NUMERICAL ANALYSIS WITH THE TWO-STEP LAPLACE ADAM–BASHFORTH METHOD

Atangana and Gnitchogna developed a method to generalize the use of Adam–Bashforth to PDEs. The Lagrange polynomial interpolation is applied to the convective-diffusive equation:

$$u_i^{n+1} = u_i^n + \frac{3\Delta t}{2}\left[-v\frac{c_{i+1}^n - c_{i-1}^n}{x\Delta} + D\frac{c_{i+1}^n - 2c_i^n + c_{i-1}^n}{(\Delta x)^2} + f\left(c_i^n\right) - \frac{c_{ai+1}^n - c_{ai-1}^n}{\Delta x}\right] -$$

$$\frac{\Delta t}{2}\left[-v\frac{c_{i+1}^{n-1} - c_i^n}{x\Delta} + D\frac{c_{i+1}^{n-1} - 2c_i^{n-1} + c_{i-1}^{n-1}}{(\Delta x)^2} + f\left(c_i^{n-1}\right) - \frac{c_{ai+1}^{n-1} - c_{ai-1}^{n-1}}{\Delta x}\right]. \tag{24.82}$$

24.3 FRACTAL FORMULATION

A fractal is a continuous pattern that repeats itself at various scales. Fractals are used to investigate anomalous systems occurring in nature. They are one of the latest developments in statistics and inhabit scaling and self-similar properties. Scaling is the ability of objects to increase or decrease in size by a scale factor that is the same in all dimensions. Self-similarity is the capacity of an object to duplicate itself. Fractal derivatives are of a self-similar nature and repeat themselves at various scales in space and time. Fractals are believed to have links to fractional derivatives, Levy statistics, Brownian motion, and empirical power-law scaling (Chen, 2005). The fractal derivative can be used describe motion in turbulence, fractal flow, and viscos-elastic behavior (Chen et al., 2017). Chen (2017) replaces the inter-order time derivative with a fractal derivative using the power law to describe the anomalous diffusion in heterogeneous media. The following equation gives the fractal derivative:

$$\frac{\partial u}{\partial t^\alpha} = \lim_{t_1 \to t}\frac{u(t_1) - u(t)}{t_1^\alpha - t^\alpha} \quad \alpha > 0 \tag{24.83}$$

Allwright and Atangana developed the fractal integral by considering the fractal time derivative of a function $u(t)$.

24.3.1 FRACTAL FORMULATION OF THE CONVECTIVE-DIFFUSIVE EQUATION

Consider Equation (24.1) where adsorbed chemical $c_a = (R - 1)c$. We apply the formulation to the convective-diffusive equation:

$$\frac{\partial}{\partial t}c(x,t) = -v\frac{\partial}{\partial x^\alpha}c(x,t) + D\frac{\partial}{\partial x^\alpha}\left[\frac{\partial}{\partial x^\alpha}c(x,t)\right] - f(c) - (R-1)\frac{\partial}{\partial x^\alpha}c(x,t) \tag{24.84}$$

where α is the fractal dimension.

To express the fractal dimension in terms of integer order, the determined fractal property is considered to be:

$$\frac{d}{dt^\alpha}f(t) = f'(t)\frac{1}{\alpha t^{\alpha-1}}$$

$$\frac{\partial}{\partial x^\alpha}c(x, t) = \frac{\partial}{\partial x}c(x, t) \cdot \left(\frac{x^{1-\alpha}}{\alpha}\right) \tag{24.85}$$

The mean velocity can be described as:

$$-v\frac{\partial}{\partial x^{\alpha}}c(x,t)=-v\frac{\partial}{\partial x}c(x,t).\left(\frac{x^{1-\alpha}}{\alpha}\right) \tag{24.86}$$

The following equation describes the dispersion:

$$D\frac{\partial}{\partial x^{\alpha}}\left[\frac{\partial}{\partial x^{\alpha}}c(x,t)\right]=D\frac{\partial^{2}}{\partial x^{2}}c(x,t).\left(\frac{x^{1-\alpha}}{\alpha}\right)^{2}+D\frac{\partial}{\partial x}c(x,t).\frac{1-\alpha}{\alpha^{2}}.x^{1-2\alpha} \tag{24.87}$$

$$=D\frac{\partial^{2}}{\partial x^{2}}c(x,t).\frac{x^{2-2\alpha}}{\alpha^{2}}+D\frac{\partial}{\partial x}c(x,t).\frac{1-\alpha}{\alpha^{2}}.x^{1-2\alpha} \tag{24.88}$$

The concentration adsorbed chemical is determined as:

$$(R-1)\frac{\partial}{\partial x^{\alpha}}c(x,t)=(R-1)\frac{\partial}{\partial x}c(x,t).\left(\frac{x^{1-\alpha}}{\alpha}\right) \tag{24.89}$$

The function $f(c)$ remains unchanged.
Substitute back into the convective-diffusive equation:

$$\frac{\partial}{\partial t}c(x,t)=-v\frac{\partial}{\partial x}c(x,t).\left(\frac{x^{1-\alpha}}{\alpha}\right)+D\left[\frac{\partial^{2}}{\partial x^{2}}c(x,t).\frac{x^{2-2\alpha}}{\alpha^{2}}+\frac{\partial}{\partial x}c(x,t).\frac{1-\alpha}{\alpha^{2}}.x^{1-2\alpha}\right]-$$
$$f(c)-(R-1)\frac{\partial}{\partial x}c(x,t).\left(\frac{x^{1-\alpha}}{\alpha}\right) \tag{24.90}$$

Let the function $V_F^{\alpha}(x)$ be equal to:

$$V_F^{\alpha}(x)=-v\left(\frac{x^{1-\alpha}}{\alpha}\right)+D\left(\frac{1-\alpha}{\alpha}\right).x^{1-2\alpha} \tag{24.91}$$

Let the function $D_F^{\alpha}(x)$ be equal to:

$$D_F^{\alpha}(x)=D\frac{x^{2-2\alpha}}{\alpha^{2}} \tag{24.92}$$

Let the function R_F^{α} be equal to:

$$R_F^{\alpha}(x)=(R-1)\left(\frac{x^{1-\alpha}}{\alpha}\right) \tag{24.93}$$

Simplify as:

$$\frac{\partial}{\partial t}c(x,t)=V_F^{\alpha}(x)c(x,t)+D_F^{\alpha}(x)\frac{\partial^{2}}{\partial x^{2}}c(x,t)-f(c)-R_F^{\alpha}(x)c(x,t) \tag{24.94}$$

where $V_F^\alpha(x)$ is the mean velocity with a fractal dimension with respect to x, $D_F^\alpha(x)$ is the dispersion with a fractal dimension with respect to x, and $R_F^\alpha(x)$ is the adsorbed chemical with a fractal dimension with respect to x. The function $f(c)$ does not have a fractal dimension. The fractal convective-diffusive equation should be changed back to the classical formulation when $\alpha = 1$.

Consider a fractal dimension mean velocity term $\alpha = 1$:

$$V_F^\alpha(x) = D\left(\frac{1-\alpha}{\alpha}\right) \cdot x^{1-2\alpha} - v\left(\frac{x^{1-\alpha}}{\alpha}\right) \tag{24.95}$$

$$= -v \tag{24.96}$$

Consider a fractal dimension dispersion term $\alpha = 1$:

$$D_F^\alpha(x) = D\frac{x^{2-2\alpha}}{\alpha^2} \tag{24.97}$$

$$= D \tag{24.98}$$

Considering a fractal dimension adsorbed chemical term $\alpha = 1$:

$$R_F^\alpha(x) = (R-1)\left(\frac{x^{1-\alpha}}{\alpha}\right) \tag{24.99}$$

$$= R-1 \tag{24.100}$$

24.3.1.1 Numerical Analysis with the Forward Euler Method

In this section we use the forward Euler classical method to find the numerical solution of the fractal formulation of a convective-diffusive equation, where the fractal concerns the following equation:

$$\frac{\partial}{\partial t} c(x,t) = V_F^\alpha(x) c(x,t) + D_F^\alpha(x) c(x,t) - f\big(c(x,t)\big) - R_F^\alpha(x) c(x,t) \tag{24.101}$$

where $V_F^\alpha = V_F^\alpha(x_i)$ $D_F^\alpha = D_F^\alpha(x_i)$ $R_F^\alpha = R_F^\alpha(x_i)$ $f(c(x,t)) = \lambda(c(x,t))$, where λ is the decay factor

$$\frac{c(x_i, t_{n+1}) - c(x_i, t)}{\Delta t} = V_{F,i}^\alpha \frac{c(x_{i+1}, t_{n+1}) - c(x_i, t_{n+1})}{\Delta x} +$$
$$D_{F,i}^\alpha \frac{c(x_{i+1}, t_{n+1}) - 2c(x_i, t_{n+1}) + c(x_{i-1}, t_{n+1})}{\Delta x^2} -$$
$$\lambda\big(c(x, t_{n+1})\big) - R_{F,i}^\alpha \frac{c(x_{i+1}, t_{n+1}) - c(x_i, t_{n+1})}{\Delta x} \tag{24.102}$$

$$\frac{c_i^{n+1} - c_i^n}{\Delta t} = V_{F,i}^\alpha \frac{c_{i+1}^{n+1} - c_i^{n+1}}{\Delta x} + D_{F,i}^\alpha \frac{c_{i+1}^{n+1} - 2c_i^{n+1} + c_{i-1}^{n+1}}{\Delta x^2} - \lambda c_i^{n+1} - R_{F,i}^\alpha \frac{c_{i+1}^{n+1} - c_i^{n+1}}{\Delta x} \tag{24.103}$$

To simplify let:

$$\frac{1}{\Delta t} = V_1 \quad \frac{V_{F,i}^{\alpha}}{\Delta x} = V_2 \quad \frac{D_{F,i}^{\alpha}}{\Delta x^2} = V_3 \quad \frac{R_{F,i}^{\alpha}}{\Delta x} = V_4$$

$$V_1\left(c_i^{n+1} - c_i^n\right) = V_2\left(c_{i+1}^{n+1} - c_i^{n+1}\right) + V_3\left(c_{i+1}^{n+1} - 2c_i^{n+1} + c_{i-1}^{n+1}\right) - \lambda c_i^{n+1} - V_4\left(c_{i+1}^{n+1} - c_i^{n+1}\right) \quad (24.104)$$

$$V_1 c_i^{n+1} - V_1 c_i^n = V_2 c_{i+1}^{n+1} - V_2 c_i^{n+1} + V_3 c_{i+1}^{n+1} - 2V_3 c_i^{n+1} + V_3 c_{i-1}^{n+1} - \lambda c_i^{n+1} - V_4 c_{i+1}^{n+1} + V_4 c_i^{n+1} \quad (24.105)$$

$$c_i^{n+1}\left(V_1 + V_2 + 2V_3 + \lambda - V_4\right) - V_1 c_i^n = c_{i+1}^{n+1}\left(V_2 + V_3 - V_4\right) + V_3 c_{i-1}^{n+1} \quad (24.106)$$

$$c_i^{n+1}\left(V_1 + V_2 + 2V_3 + \lambda - V_4\right) = c_{i+1}^{n+1}\left(V_2 + V_3 - V_4\right) + V_3 c_{i-1}^{n+1} - V_1 c_i^n \quad (24.107)$$

We apply the Fourier series to analyze for the stability of the derived (forward Euler) numerical solution for the fractal formualtion of the convective-diffusive equation:

$$e^{a(t+\Delta t)}e^{ikmx}\left(V_1 + V_2 + 2V_3 + \lambda - V_4\right) = e^{a(t+\Delta t)}e^{ikm(x+\Delta x)}\left(V_2 + V_3 - V_4\right) + $$
$$e^{a(t+\Delta t)}e^{ikm(x-\Delta x)}V_3 + e^{a(t+\Delta t)}e^{ikmx}V_1 \quad (24.108)$$

Divide by $e^{at}e^{ikmx}$:

$$e^{a\Delta t}\left(V_1 + V_2 + 2V_3 + \lambda - V_4\right) = e^{a\Delta t}e^{ikm\Delta x}\left(V_2 + V_3 - V_4\right) + e^{a\Delta t}e^{-ikm\Delta x}V_3 + V_1 \quad (24.109)$$

$$e^{a\Delta t}\left(V_1 + V_2 + 2V_3 + \lambda - V_4\right) - e^{a\Delta t}e^{ikm\Delta x}\left(V_2 + V_3 - V_4\right) - e^{a\Delta t}e^{-ikm\Delta x}V_3 = V_1 \quad (24.110)$$

Simplify:

$$e^{a\Delta t}\left[\left(V_1 + V_2 + 2V_3 + \lambda - V_4\right) - e^{ikm\Delta x}\left(V_2 + V_3 - V_4\right) - e^{-ikm\Delta x}V_3\right] = V_1 \quad (24.111)$$

$$e^{a\Delta t} = \frac{V_1}{V_1 + V_2 + 2V_3 + \lambda - V_4 - e^{ikm\Delta x}\left(V_2 + V_3 - V_4\right) - e^{-ikm\Delta x}V_3} \quad (24.112)$$

where

$$e^{ikm\Delta x} = \cos(\Delta xkm) + i sin(\Delta xkm) \text{ and } e^{-ikm\Delta x} = \cos(\Delta xkm) - i sin(\Delta xkm)$$

$$e^{a\Delta t} = \frac{V_1}{V_1 + V_2 + 2V_3 + \lambda - V_4 - \left[\cos(\Delta xkm) + i\sin(\Delta xkm)\right]\left(V_2 + V_3 - V_4\right) - \left[\cos(\Delta xkm) - i\sin(\Delta xkm)\right]V_3} \quad (24.113)$$

Let $\Delta xkm = \theta$ and $V_1 + V_2 + 2V_3 + \lambda - V_4 = A$

$$e^{a\Delta t} = \frac{V_1}{A - V_2 \cos\theta - V_2 i \sin\theta - V_3 \cos\theta - V_3 i \sin\theta - V_4 \cos\theta - V_4 i \sin\theta - V_3 \cos\theta + V_3 i \sin\theta} \quad (24.114)$$

$$e^{a\Delta t} = \frac{V_1}{A - V_2 \cos\theta - 2V_3 \cos\theta - V_4 \cos\theta - V_4 i \sin\theta - i\left(V_2 \sin\theta + V_4 \sin\theta\right)} \quad (24.115)$$

Let $A - V_2\cos\theta - 2V_3\cos\theta - V_4\cos\theta = M$ and $V_2\sin\theta + V_4\sin\theta = M_1$

$$\left|e^{a\Delta t}\right| = \left|\frac{V_1}{M - iM_1}\right| < 1 \quad (24.116)$$

$$\left|e^{a\Delta t}\right| = \frac{1}{\Delta t\sqrt{M^2 - M_1^2}} < 1 \quad (24.117)$$

$$\frac{1}{\Delta t\left(M^2 - M_1^2\right)} < 1 \quad (24.118)$$

We thus conclude that the numerical solution is unconditionally stable.

24.3.1.2 Numerical Analysis with Backward Euler

We apply the backward Euler classical method to derive the numerical solution for the fractal formulation of a convective-diffusive equation, where the fractal concerns Equation (24.101):

$$\frac{c(x_i, t_n) - c(x_i, t_{n-1})}{\Delta t} = V_{F,i}^{\alpha} \frac{c(x_{i+1}, t_n) - c(x_i, t_n)}{\Delta x} +$$
$$D_{F,i}^{\alpha} \frac{c(x_{i+1}, t_n) - 2c(x_i, t_n) + c(x_{i-1}, t_n)}{\Delta x^2} -$$
$$\lambda\left(c(x, t_n)\right) - R_{F,i}^{\alpha} \frac{c(x_{i+1}, t_n) - c(x_i, t_n)}{\Delta x} \quad (24.119)$$

$$\frac{c_i^n - c_i^{n-1}}{\Delta t} = V_{F,i}^{\alpha} \frac{c_{i+1}^n - c_i^n}{\Delta x} + D_{F,i}^{\alpha} \frac{c_{i+1}^n - 2c_i^n + c_{i-1}^n}{\Delta x^2} - \lambda c_i^n - R_{F,i}^{\alpha} \frac{c_{i+1}^n - c_i^n}{\Delta x} \quad (24.120)$$

For simplicity let:

$$\frac{1}{\Delta t} = V_1, \quad \frac{V_{F,i}^{\alpha}}{\Delta x} = V_2, \quad \frac{D_{F,i}^{\alpha}}{\Delta x^2} = V_3, \quad \frac{R_{F,i}^{\alpha}}{\Delta x} = V_4$$

$$V_1\left(c_i^n - c_i^{n-1}\right) = V_2\left(c_{i+1}^n - c_i^n\right) + V_3\left(c_{i+1}^n - 2c_i^n + c_{i-1}^n\right) - \lambda c_i^n - V_4\left(c_{i+1}^n - c_i^n\right) \quad (24.121)$$

$$V_1 c_i^n - V_1 c_i^{n-1} = V_2 c_{i+1}^n - V_2 c_i^n + V_3 c_{i+1}^n - 2V_3 c_i^n + V_3 c_{i-1}^n - \lambda c_i^n - V_4 c_{i+1}^n + V_4 c_i^n \quad (24.122)$$

$$c_i^n\left(V_1 + V_2 - V_3 + \lambda - V_4\right) = c_{i+1}^n\left(V_2 + V_3 - V_4\right) + V_3 c_{i-1}^n + V_1 c_i^{n-1} \quad (24.123)$$

We test for stability using von Neumann analysis for the numerical solution derived by backward Euler for the fractal formulation:

$$e^{at}e^{ikmx}\left(V_1 + V_2 + 2V_3 + \lambda - V_4\right) = e^{at}e^{ikm(x+\Delta x)}\left(V_2 + V_3 - V_4\right) + e^{at}e^{ikm(x-\Delta x)}V_3 + e^{a(t-\Delta t)}e^{ikmx}V_1 \quad (24.124)$$

Divide by $e^{at}e^{ikmx}$:

$$V_1 + V_2 + 2V_3 + \lambda - V_4 = e^{ikm\Delta x}\left(V_2 + V_3 - V_4\right) - e^{-ikm\Delta x}V_3 + e^{-a\Delta t}V_1 \quad (24.125)$$

$$V_1 + V_2 + 2V_3 + \lambda - V_4 + e^{ikm\Delta x}\left(V_2 - V_3 - V_4\right) - e^{-ikm\Delta x}V_3 = e^{-a\Delta t}V_1 \quad (24.126)$$

where

$$e^{ikm\Delta x} = \cos\left(\Delta xkm\right) + i\sin\left(\Delta xkm\right) \quad \text{and} \quad e^{-ikm\Delta x} = \cos\left(\Delta xkm\right) - i\sin\left(\Delta xkm\right)$$

For simplicity let:

$$V_1 + V_2 + 2V_3 + \lambda - V_4 = A$$

$$e^{a\Delta t} = \frac{V_1}{A + \left(V_2 - V_3 - V_4\right)\left(\cos\left(\Delta xkm\right) + i\sin\left(\Delta xkm\right)\right) + V_3\cos\left(\Delta xkm\right) - i\sin\left(\Delta xkm\right)} \quad (24.127)$$

Let $\Delta xkm = \theta$:

$$= \frac{V_1}{A + V_4\cos\theta + V_4 i\sin\theta - V_2\cos\theta - V_2 i\sin\theta - V_3\cos\theta - V_3 i\sin\theta + V_3\cos\theta - V_3 i\sin\theta} \quad (24.128)$$

$$e^{a\Delta t} = \frac{V_1}{A + V_4\cos\theta - V_2\cos\theta + V_4 i\sin\theta - V_2 i\sin\theta - 2V_3 i\sin\theta} \quad (24.129)$$

Let:

$$A + V_4\cos\theta - V_2\cos\theta = M \quad \text{and} \quad V_4\sin\theta - V_2\sin\theta - 2V_3\sin\theta = M_1$$

$$\left|e^{a\Delta t}\right| = \left|\frac{V_1}{M + M_1}\right| < 1 \quad (24.130)$$

$$\left|e^{a\Delta t}\right| = \frac{1}{\Delta t\sqrt{M^2 + M_1^2}} < 1 \quad (24.131)$$

$$\frac{1}{\Delta t\left(M^2 + M_1^2\right)} < 1 \quad (24.132)$$

The backward Euler numerical approximation is unconditionally stable for the fractal formulation of the convective-diffusive equation.

24.3.1.3 Numerical Analysis with a Crank–Nicolson Numerical Scheme

We conduct a numerical analysis for the fractal formulation of the convective-diffusive equation using the Crank–Nicolson approximation method:

$$
\frac{c\left(x_i, t_{n+1}\right) - c\left(x_i, t_n\right)}{\Delta t} = V_{F,i}^\alpha \frac{c\left(x_{i+1}, t_{n+1}\right) - c\left(x_i, t_{n+1}\right)}{\Delta x}
$$

$$
+ \frac{1}{2}\left[\left(\begin{array}{l} D_{F,i}^\alpha \dfrac{c\left(x_{i+1}, t_{n+1}\right) - 2c\left(x_i, t_{n+1}\right) + c\left(x_{i-1}, t_{n+1}\right)}{\Delta x^2} - \\ \lambda\left(c\left(x, t_{n+1}\right)\right) - R_{F,i}^\alpha \dfrac{c\left(x_{i+1}, t_{n+1}\right) - c\left(x_i, t_{n+1}\right)}{\Delta x} \end{array} \right) \right.
$$

$$
+ \left. \left(\begin{array}{l} D_{F,i}^\alpha \dfrac{c\left(x_{i+1}, t_n\right) - 2c\left(x_i, t_n\right) + c\left(x_{i-1}, t_n\right)}{\Delta x^2} - \\ \lambda\left(c\left(x, t_n\right)\right) - R_{F,i}^\alpha \dfrac{c\left(x_{i+1}, t_n\right) - c\left(x_i, t_n\right)}{\Delta x} \end{array} \right) \right] \tag{24.133}
$$

$$
\frac{c_i^{n+1} - c_i^n}{\Delta t} = V_{F,i}^\alpha \frac{c_{i+1}^{n+1} - c_i^{n+1}}{\Delta x} + \frac{1}{2}\left[\left(D_{F,i}^\alpha \frac{c_{i+1}^{n+1} - 2c_i^{n+1} + c_{i-1}^{n+1}}{\Delta x^2} - \lambda\left(c_i^{n+1}\right) - R_{F,i}^\alpha \frac{c_{i+1}^{n+1} - c_i^{n+1}}{\Delta x} \right) \right.
$$

$$
+ \left. \left(D_{F,i}^\alpha \frac{c_{i+1}^n - 2c_i^n + c_{i-1}^n}{\Delta x^2} - \lambda\left(c_i^n\right) - R_{F,i}^\alpha \frac{c_{i+1}^n - c_i^n}{\Delta x} \right) \right] \tag{24.134}
$$

$$
\frac{1}{\Delta t} = V_1 \qquad \frac{V_{F,i}^\alpha}{\Delta x} = V_2 \qquad \frac{D_{F,i}^\alpha}{\Delta x^2} = V_3 \qquad \frac{R_{F,i}^\alpha}{\Delta x} = V_4
$$

$$
V_1\left(c_i^{n+1} - c_i^n\right) = V_2\left(c_{i+1}^{n+1} - c_i^{n+1}\right) + \frac{1}{2}\left[V_3\left(c_{i+1}^{n+1} - 2c_i^{n+1} + c_{i-1}^{n+1}\right) - V_4\left(c_{i+1}^{n+1} - c_i^{n+1}\right) \right.
$$

$$
\left. + V_3\left(c_{i+1}^n - 2c_i^n + c_{i-1}^n\right) - V_4\left(c_{i+1}^n - c_i^n\right) \right] - \lambda c_i^{n+1} \tag{24.135}
$$

$$
V_1 c_i^{n+1} - V_1 c_i^n = V_2 c_{i+1}^{n+1} - V_2 c_i^{n+1} + \frac{1}{2}\left(V_3 c_{i+1}^{n+1} - 2V_3 c_i^{n+1} + V_3 c_{i-1}^{n+1} - V_4 c_{i+1}^{n+1} - V_4 c_i^{n+1} \right.
$$

$$
\left. + V_3 c_{i+1}^n - 2V_3 c_i^n + V_3 c_{i-1}^n - V_4 c_i^n - V_4 c_{i-1}^n \right) - \lambda c_i^{n+1} \tag{24.136}
$$

$$
c_i^{n+1}\left(V_1 + V_2 + 2V_3 + V_4 + \lambda\right) = \frac{1}{2} c_{i+1}^{n+1}\left(V_2 + V_3 - V_4\right) + c_{i-1}^n\left(V_3 - V_4\right)
$$

$$
- c_i^n\left(2V_3 - V_4 + V_1\right) + V_3\left(c_{i-1}^{n+1} + c_{i-1}^n\right) \tag{24.137}
$$

Stability analysis using von Neumann analysis for the numerical solution derived by the Crank–Nicolson approximation method for the fractal formulation of the convective-diffusive equation:

$$
e^{a(t+\Delta t)} e^{ikmx}\left(V_1 + V_2 + 2V_3 + V_4 + \lambda\right) = \frac{1}{2}\left[e^{a(t+\Delta t)} e^{ikm(x+\Delta x)}\left(V_2 + V_3 - V_4\right) \right.
$$

$$
+ e^{at} e^{ikm(x-\Delta x)}\left(V_3 - V_4\right) - e^{at} e^{ikmx}\left(2V_3 - V_4 + V_1\right)
$$

$$
\left. + V_3\left(e^{a(t+\Delta t)} e^{ikm(x-\Delta x)} + e^{at} e^{ikm(x+\Delta x)}\right) \right] \tag{24.138}
$$

Divide by $e^{at}e^{ikmx}$:

$$e^{a\Delta t}\left(V_1 + V_2 + 2V_3 + V_4 + \lambda\right) = \frac{1}{2}\left[e^{a\Delta t}e^{ikm\Delta x}\left(V_2 + V_3 - V_4\right) + e^{-ikm\Delta}\left(V_3 + V_4\right) - 2V_3 - V_4 - V_1 + V_3 e^{a\Delta t}e^{-ikm\Delta x} + V_3 e^{ikmx}\right] \tag{24.139}$$

$$e^{a\Delta t}\left(V_1 + V_2 + 2V_3 - V_4 + \lambda\right) + e^{ikm\Delta x}\left(V_2 + V_3 - V_4\right) - V_3 e^{-ikm\Delta x} = \frac{1}{2}e^{ikm\Delta x}\left(V_3 + V_4\right) - 2V_3 - V_4 - V_1 + V_3 e^{ikm\Delta x} \tag{24.140}$$

$$e^{a\Delta t} = \frac{\dfrac{1}{2}e^{-ikm\Delta x}\left(V_3 + V_4\right) - 2V_3 - V_4 - V_1 + V_3 e^{ikm\Delta x}}{V_1 + V_2 + 2V_3 + V_4 + \lambda + e^{ikm\Delta x}\left(V_2 + V_3 - V_4\right) - V_3 e^{ikm\Delta x}} \tag{24.141}$$

$$e^{a\Delta t} = \frac{\dfrac{e^{-ikm\Delta x}\left(V_3 + V_4\right) - 2V_3 - V_4 - V_1 + V_3 e^{ikm\Delta x}}{2}}{V_1 + V_2 + 2V_3 + V_4 + \lambda + e^{ikm\Delta x}\left(V_2 + V_3 - V_4\right) - V_3 e^{ikm\Delta x}} \tag{24.142}$$

where $e^{ikm\Delta x} = \cos(\Delta xkm) + i\sin(\Delta xkm)$ and $e^{-ikm\Delta x} = \cos(\Delta xkm) - i\sin(\Delta xkm)$.
Let $\Delta xkm = \theta$ and $V_1 + V_2 + 2V_3 + V_4 + \lambda = A$

$$e^{a\Delta t} = \frac{V_1}{A - V_2\cos\theta - V_2 i\sin\theta - V_3\cos\theta - V_3 i\sin\theta - V_4\cos\theta - V_4 i\sin\theta - V_3\cos\theta + V_3 i\sin\theta} \tag{24.143}$$

$$e^{a\Delta t} = \frac{V_1}{A - V_2\cos\theta - 2V_3\cos\theta - V_4\cos\theta - i\left(V_2\sin\theta + V_4\sin\theta\right)} \tag{24.144}$$

Let $A - V_2\cos\theta - 2V_3\cos\theta - V_4\cos\theta = M$ and $V_2\sin\theta + V_4\sin\theta = M_1$

$$\left|e^{a\Delta t}\right| = \left|\frac{V_1}{M - iM_1}\right| < 1 \tag{24.145}$$

$$\left|e^{a\Delta t}\right| = \frac{1}{\Delta t\sqrt{M^2 + M_1^2}} < 1 \tag{24.146}$$

The Crank–Nicolson numerical scheme for the convective-diffusive equation is conditionally stable.

24.4 CAPUTO–FABRIZIO FRACTIONAL DIFFERENTIAL OPERATOR

The operator can without doubt well describe the fatigue effect, which is also observed in the process of groundwater pollution. It is practically true that if there is no preferential path in the geological formation, the initial concentration will be higher than the next layer. And of course at the last boundary one will not expect any concentration. This is the fading memory process and can be explained by a Caputo–Fabrizio derivative. For those readers that are not familiar with this operator, the Caputo–Fabrizio of a function f that is differentiable in the classical sense is given as:

$$
{}^{CF}_0 D^\alpha_t f(t) = \frac{M(\alpha)}{(1-\alpha)} \int_0^t \frac{df(\tau)}{d\tau} \exp\left[-\frac{\alpha(t-\tau)}{1-\alpha}\right] d\tau \tag{24.147}
$$

where $M(\alpha)$ is a normalization function; thus $M(0) = M(1) = 1$. This new derivative can be used for functions that do not belong to $H^1(a,b)$ (Caputo & Fabrizio, 2015). In the case of a PDE, the following definition is used:

$$
{}^{CF}_0 D^\alpha_t c(x,t) = \frac{M(\alpha)}{(1-\alpha)} \int_0^t \frac{\partial}{\partial \tau} c(x,\tau) \exp\left[-\frac{\alpha(t-\tau)}{1-\alpha}\right] d\tau \tag{24.148}
$$

To include in the mathematical model the exponential decay waiting time distribution we replace the classical differentiation with the Caputo–Fabrizio derivative to obtain:

$$
\frac{M(\alpha)}{(1-\alpha)} \int_0^t \frac{\partial}{\partial \tau} c(x,\tau) \exp\left[-\frac{\alpha(t-\tau)}{1-\alpha}\right] d\tau =
$$
$$
-v \frac{\partial}{\partial x} c(x,t) + D \frac{\partial^2}{\partial x^2} c(x,t) - f(c(x,t)) - \frac{\partial c_a}{\partial x} \tag{24.149}
$$

where $f(c(x,t)) = \lambda(c(x,t))$ and $c_a = (R-1)c$.

This can be rewritten as:

$$
\frac{M(\alpha)}{(1-\alpha)} \int_0^t \frac{\partial}{\partial \tau} c(x,\tau) \exp\left[-\frac{\alpha}{1-\alpha}(t-\tau)\right] d\tau =
$$
$$
-v \frac{\partial}{\partial x} c(x,t) + D \frac{\partial^2}{\partial x^2} c(x,t) - \lambda(c(x,t)) - \frac{\partial}{\partial x}(R-1)c \tag{24.150}
$$

For

$$
\frac{M(\alpha)}{(1-\alpha)} \int_0^t \frac{\partial}{\partial \tau} c(x,\tau) \exp\left[-\frac{\alpha}{1-\alpha}(t-\tau)\right] d\tau \tag{24.151}
$$

At t_n we have:

$$
\frac{M(\alpha)}{(1-\alpha)} \int_0^{t_n} \frac{\partial}{\partial \tau} c(x,\tau) \exp\left[-\frac{\alpha}{1-\alpha}(t_n-\tau)\right] d\tau \tag{24.152}
$$

Capture the memory in terms of t_0 to t_{n-1}:

$$
{}^{CF}_0 D^\alpha_{t_n} c(x_i,t) = \frac{M(\alpha)}{(1-\alpha)} \left[\int_{t_0}^{t_1} \frac{c(x_i,t_1) - c(x_i,t_0)}{t_1 - t_0} \exp\left[-\frac{\alpha}{1-\alpha}(t_n-\tau)\right] d\tau + \ldots \right] \tag{24.153}
$$

Rewrite in terms of t_j to t_{j+1}:

$$\prescript{CF}{0}{D}_{t_n}^{\alpha}c(x_i,t) = \frac{M(\alpha)}{(1-\alpha)}\sum_{j=0}^{n-1}\int_{t_j}^{t_{j+1}}\frac{c_i^{j+1}-c_i^{j}}{\Delta t}\exp\left[-\frac{\alpha}{1-\alpha}(t_n-\tau)\right]d\tau \qquad (24.154)$$

Rearrange the equation:

$$\prescript{CF}{0}{D}_{t_n}^{\alpha}c(x_i,t) = \frac{M(\alpha)}{(1-\alpha)}\sum_{j=0}^{n-1}\frac{c_i^{j+1}-c_i^{j}}{\Delta t}\int_{t_j}^{t_{j+1}}\exp\left[-\frac{\alpha}{1-\alpha}(t_n-\tau)\right]d\tau \qquad (24.155)$$

Let $a = \dfrac{\alpha}{1-\alpha}$ and $y = t_n - \tau$.

When $\tau = t_j$ then $y = t_n - t_j$.

And when $\tau = t_{j+1}$ then $y = t_n - t_{j+1}$ and $dy = -d\tau$.

Substitute the y values back into Equation (24.155):

$$\prescript{CF}{0}{D}_{t_n}^{\alpha}c(x_i,t_n) = \frac{M(\alpha)}{(1-\alpha)}\sum_{j=0}^{n-1}\frac{c_i^{j+1}-c_i^{j}}{\Delta t}\int_{t_n-t_j}^{t_n-t_{j+1}}\exp(-ay)(-dy) \qquad (24.156)$$

When $-\int_{t_n-t_j}^{t_n-t_{j+1}}dy = \int_{t_n-t_{j+1}}^{t_n-t_j}dy$ then Equation (24.156) can be rewritten as:

$$\prescript{CF}{0}{D}_{t_n}^{\alpha}c(x_i,t_n) = \frac{M(\alpha)}{(1-\alpha)}\sum_{j=0}^{n-1}\frac{c_i^{j+1}-c_i^{j}}{\Delta t}\int_{t_n-t_{j+1}}^{t_n-t_j}\exp(-ay)dy \qquad (24.157)$$

$$\prescript{CF}{0}{D}_{t_n}^{\alpha}c(x_i,t_n) = \frac{M(\alpha)}{(1-\alpha)}\sum_{j=0}^{n-1}\frac{c_i^{j+1}-c_i^{j}}{\Delta t}-\frac{1}{a}\exp(-ay)\Big|_{t_n-t_{j+1}}^{t_n-t_j} \qquad (24.158)$$

$$\prescript{CF}{0}{D}_{t_n}^{\alpha}c(x_i,t_n) = \frac{M(\alpha)}{(1-\alpha)}\sum_{j=0}^{n-1}\frac{c_i^{j+1}-c_i^{j}}{\Delta t}$$
$$\left(-\frac{1-\alpha}{\alpha}\exp\left[-a(t_n-t_j)\right]+\frac{1-\alpha}{\alpha}\exp\left[-\frac{\alpha}{1-\alpha}(t_n-t_{j+1})\right]\right) \qquad (24.159)$$

Simplify:

$$\frac{M(\alpha)}{\alpha}\sum_{j=0}^{n-1}\frac{c_i^{j+1}-c_i^{j}}{\Delta t}\left(\exp\left[-\frac{\alpha}{1-\alpha}(t_n-t_j)\right]-\exp\left[\frac{\alpha}{1-\alpha}(t_n-t_{j+1})\right]\right) \qquad (24.160)$$

Let $\delta_{n,j}^{\alpha} = \exp\left[-\dfrac{\alpha}{1-\alpha}(t_n-t_j)\right]-\exp\left[\dfrac{\alpha}{1-\alpha}(t_n-t_{j+1})\right]$ and substitute back into the time-based numerical solution:

$$\prescript{CF}{0}{D}_{t_n}^{\alpha}c(x_i,t_n) = \frac{M(\alpha)}{\alpha}\sum_{j=0}^{n-1}\frac{c_i^{j+1}-c_i^{j}}{\Delta t}\delta_{n,j}^{\alpha} \qquad (24.161)$$

The spatial and temporal numerical solution of the convective-diffusive equation is given by:

$$\frac{M(\alpha)}{\alpha}\sum_{j=0}^{n-1}\frac{c_i^{j+1}-c_i^j}{\Delta t}\delta_{n,j}^{\alpha}=-v\frac{c_{i+1}^n-c_{i-1}^n}{\Delta x}+D\frac{c_{i+1}^n-2c_i^n+c_{i-1}^n}{\Delta x^2}-\lambda c_i^n-(R-1)\frac{c_{i+1}^n-c_{i-1}^n}{\Delta x} \quad (24.162)$$

For simplicity let $\Delta_{n,j}^{\alpha}=\delta_{n,j}^{\alpha}\dfrac{M(\alpha)}{\alpha}$

$$\sum_{j=0}^{n-1}\frac{c_i^{j+1}-c_i^j}{\Delta t}\Delta_{n,j}^{\alpha}=-v\frac{c_{i+1}^n-c_{i-1}^n}{\Delta x}+D\frac{c_{i+1}^n-2c_i^n+c_{i-1}^n}{\Delta x^2}-\lambda c_i^n-(R-1)\frac{c_{i+1}^n-c_{i-1}^n}{\Delta x} \quad (24.163)$$

$$\sum_{j=0}^{n-2}\frac{c_i^{j+1}-c_i^j}{\Delta t}\Delta_{n,j}^{\alpha}+\frac{c_i^n-c_i^{n-1}}{\Delta t}\Delta_{n,n-1}^{\alpha}=-v\frac{c_{i+1}^n-c_{i-1}^n}{\Delta x}+$$
$$D\frac{c_{i+1}^n-2c_i^n+c_{i-1}^n}{\Delta x^2}-\lambda c_i^n-(R-1)\frac{c_{i+1}^n-c_{i-1}^n}{\Delta x} \quad (24.164)$$

$$\frac{c_i^n}{\Delta t}\Delta_{n,n-1}^{\alpha}+\frac{2Dc_i^n}{\Delta x^2}+\lambda c_i^n=-\sum_{j=0}^{n-2}\frac{c_i^{j+1}-c_i^j}{\Delta t}\Delta_{n,j}^{\alpha}+\frac{c_i^{n-1}}{\Delta t}\Delta_{n,n-1}^{\alpha}$$
$$+c_{i+1}^n\left(\frac{v}{\Delta x}+\frac{D}{\Delta x^2}+\frac{R-1}{\Delta x}\right)+c_{i-1}^n\left(\frac{v}{\Delta x}+\frac{D}{\Delta x^2}+\frac{R-1}{\Delta x}\right) \quad (24.165)$$

$$c_i^n\left(\frac{\Delta_{n,n-4}^{\alpha}}{\Delta t}+\frac{2D}{\Delta x^2}-\frac{R-1}{\Delta x}\right)=c_i^{n-1}\frac{\Delta_{n,n-1}^{\alpha}}{\Delta t}+c_{i+1}^n$$
$$\left(-\frac{v}{\Delta x}+\frac{D}{\Delta x^2}-\frac{R-1}{\Delta x}\right)+C_{i-1}^n\left(\frac{v}{\Delta x}+\frac{D}{\Delta x^2}-\frac{R-1}{\Delta x}\right)$$
$$\sum_{j=0}^{n-2}\frac{c_i^{j+1}-c_i^j}{\Delta t}\Delta_{n,n-1}^{\alpha} \quad (24.166)$$

For simplicity let

$$\frac{\Delta_{n,n-1}^{\alpha}}{\Delta t}+\frac{2D}{\Delta x^2}-\frac{R-1}{\Delta x}=A\quad\frac{\Delta_{n,n-1}^{\alpha}}{\Delta t}=B-\frac{v}{\Delta x}+\frac{D}{\Delta x^2}-\frac{R-1}{\Delta x}=C\text{ and }\frac{v}{\Delta x}+\frac{D}{\Delta x^2}-\frac{R-1}{\Delta x}=D$$

$$\delta_n e^{ikmx}A=\delta_{n-1}e^{ikmx}B+\delta_n e^{ikm(x+\Delta x)}C+\delta_n e^{ikm(x-\Delta x)}D-\sum_{j=0}^{n-2}\delta_{n-j}e^{ikmx}\Delta_{n,j}^{\alpha} \quad (24.167)$$

$$\delta_n A=\delta_{n-1}B+\delta_n e^{ikm\Delta x}C+\delta_n e^{-ikm\Delta x}+\sum_{j=0}^{n-2}\delta_{n-j}\Delta_{n,j}^{\alpha} \quad (24.168)$$

$$\delta_1 A=\delta_0 B+\delta_1 e^{ikm\Delta x}C+\delta_1 e^{-ikm\Delta x}D \quad (24.169)$$

$$\delta_1\left(A-Ce^{ikm\Delta x}-De^{-ikm\Delta x}\right)=\delta_0 B \quad (24.170)$$

$$\frac{\delta_1}{\delta_0} = \frac{B}{A - Ce^{ikm\Delta x} - De^{-ikm\Delta x}} \tag{24.171}$$

$$\left|\frac{\delta_1}{\delta_0}\right| = \left|\frac{B}{A - Ce^{ikm\Delta x} - De^{-ikm\Delta x}}\right| < 1 \tag{24.172}$$

Let

$$e^{ikm\Delta x} = \cos(km\Delta x) + i\sin(ikm\Delta x) \quad \text{and} \quad e^{-ikm\Delta x} = \cos(km\Delta x) - i\sin(ikm\Delta x)$$

We rewrite the equation in terms of cos and sin to be

$$\left|\frac{\delta_1}{\delta_0}\right| = \left|\frac{B}{A - C\left(\cos(km\Delta x) + i\sin(ikm\Delta x)\right) - D\left(\cos(km\Delta x) - i\sin(ikm\Delta x)\right)}\right| < 1 \tag{24.173}$$

Rearrange the equation:

$$\left|\frac{\delta_1}{\delta_0}\right| = \left|\frac{B}{A - (C+D)\cos(km\Delta x) - i(D-C)\sin(ikm\Delta x)}\right| < 1 \tag{24.174}$$

$$\left|\frac{\delta_1}{\delta_0}\right| = \frac{|B|}{\sqrt{\left(A - (C+D)\cos(km\Delta x)\right)^2 + \left((D-C)\sin(ikm\Delta x)\right)^2}} < 1 \tag{24.175}$$

$$\left|\frac{\delta_1}{\delta_0}\right| = \frac{\Delta_{n,n-1}^{\alpha}}{\Delta t\sqrt{\left(A - (C+D)\cos(km\Delta x)\right)^2 + \left((D-C)\sin(ikm\Delta x)\right)^2}} < 1 \tag{24.176}$$

We expand $\Delta_{n,n-1}^{\alpha}$ in terms of Caputo–Fabrizio:

$$\left|\frac{\delta_1}{\delta_0}\right| = \frac{\dfrac{M(\alpha)}{\alpha}\left[\exp\left(-\dfrac{\alpha}{1-\alpha}\Delta t\right) - \exp\left(-\dfrac{\alpha}{1-\alpha}2\Delta t\right)\right]}{\Delta t\sqrt{\left(A - (C+D)\cos(km\Delta x)\right)^2 + \left((D-C)\sin(ikm\Delta x)\right)^2}} < 1 \tag{24.177}$$

We assume that for $n \geq 1$
$|\delta_{n-1}| < |\delta_0|$ is true and that $|\delta_n| < |\delta_0|$

$$\delta_n\left(A - Ce^{ikm\Delta x} - De^{-ikm\Delta x}\right) = \delta_{n-1} + \sum_{j=0}^{n-2}\delta_{n-j}\Delta_{n,j}^{\alpha} \tag{24.178}$$

$$\delta_n\left(A - (C+D)\cos(km\Delta x) - i(D-C)\sin(km\Delta x)\right) = \delta_{n-1} + \sum_{j=0}^{n-2}\delta_{n-j}\Delta_{n,j}^{\alpha} \tag{24.179}$$

$$\delta_n = \frac{\delta_{n-1}B}{A - (C+D)\cos(km\Delta x) - i(D-C)\sin(km\Delta x)} - \sum_{j=0}^{n-2} \frac{\delta_{n-j}\Delta_{n,j}^{\alpha}}{A - (C+D)\cos(km\Delta x) - i(D-C)\sin(km\Delta x)} \quad (24.180)$$

Let $A - (C+D)\cos(km\Delta x) - i(D-C)\sin(km\Delta x) = K$

$$\delta_n = \delta_{n-1}\frac{B}{K} - \sum_{j=0}^{n-2} \delta_{n-j}\frac{\Delta_{n,j}^{\alpha}}{K} \quad (24.181)$$

$$|\delta_n| = \left| \delta_{n-1}\frac{B}{K} - \sum_{j=0}^{n-2} \delta_{n-j}\frac{\Delta_{n,j}^{\alpha}}{K} \right| \quad (24.182)$$

$$\leq |\delta_{n-1}|\left|\frac{B}{K}\right| - \left|\sum_{j=0}^{n-2} \frac{\delta_{n-j}}{K}\Delta_{n,j}^{\alpha}\right| \quad (24.183)$$

$$\leq |\delta_{n-1}|\left|\frac{B}{K}\right| - \sum_{j=0}^{n-2} \frac{|\delta_{n-j}|}{|K|}\left|\Delta_{n,j}^{\alpha}\right| \quad (24.184)$$

By inductive hypothesis:

$$|\delta_n| < |\delta_0|\left|\frac{B}{K}\right| + |\delta_0|\sum_{j=0}^{n-2} \frac{\Delta_{n,j}^{\alpha}}{K} \quad (24.185)$$

$$|\delta_n| < \delta_0\left(\frac{B}{K} + \sum_{j=0}^{n-2} \frac{\Delta_{n,j}^{\alpha}}{K}\right) \quad (24.186)$$

$$\frac{|\delta_n|}{|\delta_0|} < \frac{B + \sum_{j=0}^{n-2} \Delta_{n,j}^{\alpha}}{K} < 1 \quad (24.187)$$

The numerical solution is stable provided that $\frac{\delta_1}{\delta_0} < 1$ is equal to $B + \sum_{j=0}^{n-2}\Delta_{n,j}^{\alpha} < K$.

24.4.1 NEW NUMERICAL SCHEME THAT COMBINES THE TRAPEZOIDAL RULE AND THE LAGRANGE POLYNOMIAL

In recent years many numerical methods have been introduced to solve ordinary and partial differential equations with classical derivatives and classical integrals. For instance Crank and Nicolson combined the forward and backward Euler methods to form a numerical scheme that seems to be stable. Adam and Bashforth introduced a new numerical scheme using the Lagrange polynomial interpolation. Atangana and Gnitchogna (2018) introduced a new numerical scheme using the Laplace transform and the Adam–Bashforth procedure. While this introduced method has been used to solve linear and nonlinear, partial and ordinary differential equations, they all have advantages and limitations and therefore new numerical schemes are still welcome. In this section we introduce a new numerical scheme that combines the trapezoidal rule and the Lagrange polynomial interpolation. We consider the following general PDE:

$$\frac{\partial}{\partial t} c(x, t) = f(c, x, t) \tag{24.188}$$

We integrate equation (24.188):

$$c(x, t) - c(x, 0) = \int_0^t f(c, x, \tau) d\tau \tag{24.189}$$

Consider the integration at t_{n+1}:

$$c(x, t_{n+1}) - c(x, 0) = \int_0^{t_{n+1}} f(c, x, \tau) d\tau \tag{24.190}$$

And at t_n:

$$c(x, t_n) - c(x, 0) = \int_0^{t_n} f(c, x, \tau) d\tau \tag{24.191}$$

where

$$c(x, t_{n+1}) - c(x, t_n) = \int_0^{t_{n+1}} f(c, x, \tau) d\tau - \int_0^{t_n} f(c, x, \tau) d\tau \tag{24.192}$$

We expand the equation to be:

$$(x, t_{n+1}) - c(x, t_n) = \int_{t_0}^{t_1} f(c, x, \tau) + \int_{t_1}^{t_{n+1}} f(c, x, \tau) - \int_{t_0}^{t_1} f(c, x, \tau) - \int_{t_1}^{t_n} f(c, x. \tau) \tag{24.193}$$

We apply the trapezoidal rule to $c(x, t_{n+1})$ and the Lagrange polynomial interpolation to $c(x, t_n)$ of the equation:

$$= \frac{(t_1 - t_0)}{2} \left[f(c_1, x, t_1) + f(c_0, x, t_0) \right] + \sum_{j=1}^{n} \int_{t_j}^{t_{j+1}} f(c, x, \tau) d\tau$$

$$- \left[\frac{3}{2} h f(c_1, x, t_1) - \frac{h}{2} f(c_0, x, t_0) \right] - \sum_{j=1}^{n-1} \int_{t_j}^{t_{j+1}} f(c, x, \tau) d\tau \tag{24.194}$$

$$\frac{h}{2}\big[f(c_1, x, t_1) + f(c_0, x, t_o)\big] + \sum_{j=1}^{n}\left(\frac{t_{j+1}-t_j}{2}\right)\big[f(c_{j+1}, x, t_{j+1}) + f(c_j, x, t_j)\big]$$

$$-\frac{3}{2}hf(c_1, x, t_1) + \frac{h}{2}f(c_0, x, t_0)$$

$$-\sum_{j=0}^{n-1}\left[\frac{3}{2}hf(t_j, c_j) - \frac{h}{2}f(t_{j-1}, x, c_{j-1})\right] \tag{24.195}$$

$$c(x, t_{n+1}) - c(x, t_n) = f(c_0, x, t_0) - \frac{h}{2}f(c_1, x, t_1) +$$

$$\sum_{j=1}^{n}\frac{h}{2}\left[f(c_{j+1}, x, c_{j+1}) = f(c_j, x, t_j) - \sum_{j=1}^{n-1}\frac{3}{2}h\big[f(c_{j-1}, x, t_{j-1})\big]\right] \tag{24.196}$$

At x_i we have:

$$c(x_i, t_{n+1}) - c(x_i, t_n) = f(c_0^i, x_i, t_0) - \frac{h}{2}f(c_1^i, x_i, t_1) +$$

$$\sum_{j=1}^{n}\frac{h}{2}\big[f(c_{j+1}^i, x_i, t_{j+1}) + f(c_j^i, x_i, t_j)\big] - \sum_{j=1}^{n-1}\left[\frac{3}{2}hf(c_j^i, x_i, t_j) - \frac{h}{2}f(c_{j-1}^i, x_i, t_{j-1})\right] \tag{24.197}$$

The new numerical scheme is denoted as:

$$c_{n+1}^i - c_n^i = f(c_0^i, x_i, t_0) - \frac{h}{2}f(c_1^i, x_i, t_1) + \sum_{j=1}^{n}\frac{h}{2}\big[f(c_{j+1}^i, x_i, t_{j+1}) + f(c_j^i, x_i, t_j)\big]$$

$$-\sum_{j=1}^{n-1}\left[\frac{3}{2}hf(c_j^i, x_i, t_j) - \frac{h}{2}f(c_{j-1}^i, x_i, t_{j-1})\right] \tag{24.198}$$

We now apply integration to the convective-diffusive equation:

$$c_{n+1}^i - c_n^i = \left[-v\frac{c_0^{i+1}-c_0^{i-1}}{\Delta x} + D\frac{c_0^{i+1}-2c_0^i+c_0^{i-1}}{\Delta x^2} - \lambda c_0^i - (R-1)\frac{c_0^{i+1}-c_0^{i-1}}{\Delta x}\right] -$$

$$\frac{h}{2}\left[-v\frac{c_1^{i+1}-c_1^{i-1}}{\Delta x} + D\frac{c_1^{i+1}-2c_1^i+c_1^{i-1}}{\Delta x^2} - \lambda c_1^i - (R-1)\frac{c_1^{i+1}-c_1^{i-1}}{\Delta x}\right] +$$

$$\sum_{j=1}^{n}\frac{h}{2}\left\{\begin{array}{l}\left[-v\frac{c_{j+1}^{i+1}-c_{j+1}^{i-1}}{\Delta x} + D\frac{c_{j+1}^{i+1}-2c_{j+1}^i+c_{j+1}^{i-1}}{\Delta x^2} - \lambda c_{j+1}^i - (R-1)\frac{c_{j+1}^{i+1}-c_{j+1}^{i-1}}{\Delta x}\right] \\ +\left[-v\frac{c_j^{i+1}-c_j^{i-1}}{\Delta x} + D\frac{c_j^{i+1}-2c_j^i+c_j^{i-1}}{\Delta x^2} - \lambda c_j^i - (R-1)\frac{c_j^{i+1}-c_j^{i-1}}{\Delta x}\right]\end{array}\right\}$$

$$-\sum_{j=1}^{n-1}\left\{\begin{array}{l}\frac{3}{2}h\left[-v\frac{c_j^{i+1}-c_j^{i-1}}{\Delta x} + D\frac{c_j^{i+1}-2c_j^i+c_j^{i-1}}{\Delta x^2} - \lambda c_j^i - (R-1)\frac{c_j^{i+1}-c_j^{i-1}}{\Delta x}\right] - \\ \frac{h}{2}\left[-v\frac{c_{j-1}^{i+1}-c_{j-1}^{i-1}}{\Delta x} + D\frac{c_{j-1}^{i+1}-2c_{j-1}^i+c_{j-1}^{i-1}}{\Delta x^2} - \lambda c_{j-1}^i - (R-1)\frac{c_{j-1}^{i+1}-c_{j-1}^{i-1}}{\Delta x}\right]\end{array}\right\} \tag{24.199}$$

24.5 NUMERICAL SIMULATIONS

The real aim of modeling real world problems is to understand the observed facts using mathematical equations. Mathematical models can be solved using analytical approaches which provide an exact solution of the problem, which can then be used for analysis and possible prediction according to the input parameters. Nevertheless, there exist some mathematical models that cannot be solved analytically; modelers therefore rely on numerical approximation as we have done in this chapter. To obtain numerical simulation, one first approximates the given model using a given numerical method; the numerical method will later be transformed into computational code. There exist in the literature nowadays several software packages that can be used for this purpose. The numerical methods presented in this section will be checked and their efficiency will also be evaluated according to their rate of convergence. We present the numerical simulation using the software Maple, and we start with the classical case where the differential operator is based on the concept of the rate of change. The numerical simulations are represented in Figures 24.1–24.5.

The numerical simulations of the model with Caputo–Fabrizio are represented in Figures 24.5–24.8.

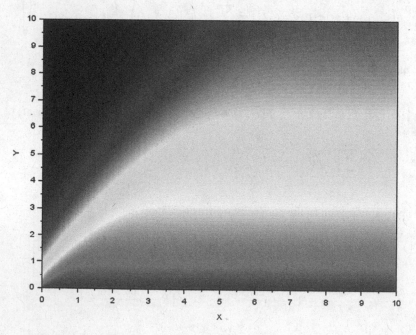

FIGURE 24.1   Contour plot of concentration for $\alpha = 1$.

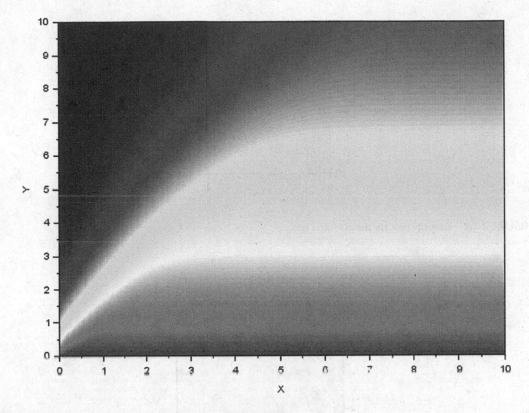

FIGURE 24.2   Contour plot of concentration for $\alpha = 0.96$.

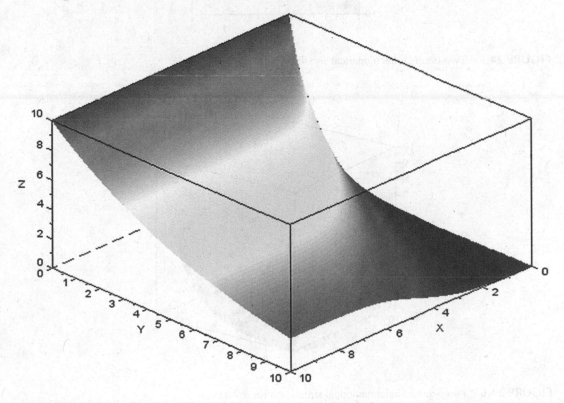

FIGURE 24.3   Three-dimensional numerical simulation for the classical case.

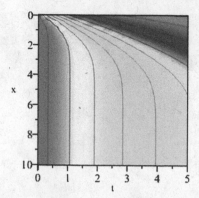

FIGURE 24.4   Density plot for the classical case.

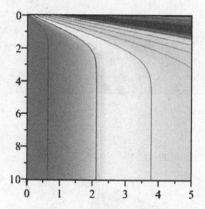

FIGURE 24.5   Two-dimensional numerical simulations for $\alpha = 0.96$.

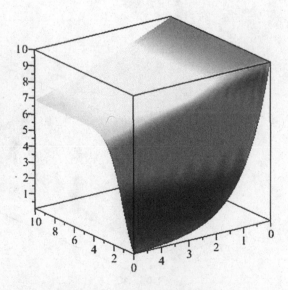

FIGURE 24.6   Three-dimensional numerical simulation for $\alpha = 0.96$.

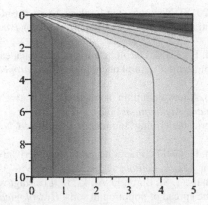

FIGURE 24.7   Two-dimensional numerical simulation for $\alpha = 0.98$.

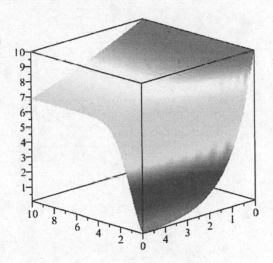

FIGURE 24.8   Three-dimensional numerical simulation for $\alpha = 0.98$.

24.6   CONCLUSION

In this chapter, we have used the fractal integral to capture the transport within a geological formation with self-similar properties. Classical numerical schemes including forward and backward Euler and Crank–Nicolson have been used to solve numerically the equation under investigation. For each case we derived the conditions under which the scheme is stable. Additionally after converting the equation under investigation to a Volterra type, we suggested a new numerical scheme that combines the trapezoidal rule and the Lagrange interpolation in time and space (forward/backward). Numerical simulations were used to underpin the seriousness of the suggested numerical scheme.

REFERENCES

Ahokpossi, D. P., Atangana, A., & Vermeulen, D. P. (2018). Modelling of fractal flow in dual media with fractional differentiation with power and generalized Mattig-Leffler laws kernels. *The European Physical Journal*, 226(16), 3705–3727.

Alkahtani, B. S., Atangana, A., & Koca, I. (2016). A new nonlinear triadic model of predator-derivative with non-singular kernel. *Advances in Mechanical Engineering*, 8(11), 1–9.

Allwright, A., & Atangana, A. (2018). Fractal advection-dispersion equation for groundwater transport in fractured aquifers with self-similarities. *The European Physical Journal Plus*, 133(2), 1–20.

Atangana, A., & Gnitchogna, R. (2018). New two-step Laplace Adam-Bashforth method for integer and non-integer order Partial Differential Equations. *Numerical Methods for Partial Differential Equations*, 34(5), 1739–1758.

Atangana, A., & Gomez-Aguilar, J. F. (2018). Decolonisation of fractional calculus rules: Breaking commutativity and associativity to capture more natural phenomena. *The European Physical Journal Plus*, 133(4), 1–22.

Caputo, M., & Fabrizio, M. (2015). A new definition of fractional derivative without singular kernel. *Progress in Fractional Differentiation and Applications*, 1(2), 1–13.

Chen, W. (2005). Time-space fabric underlying anomalous diffusion. *Chaos, Soliton and Fractals*, 28(4), 923–929.

Chen, W., Zhang, Y., Li, Z., & Sun, H. (2017). Fractional and fractal derivative models for trasient anomalous diffusion: Model Comparison. *Chaos, Solitons & Fractals*, 102, 346–353.

Stagnitti, F., Parlange, J. Y., Steenhuis, T. S., Barry, D. A., Li, L., Lockington, D. A., & Sander, G. C. (1995). Mathematical equations of the spread of pollution in soils. *Hydrological Systems Modeling, II*, 2, 106.

Tateishi, A. A., Ribeiro, H. V., & Lenzi, E. K. (2017). The role of fractional time-derivative operators on anomalous diffusion. *Frontiers in Physics*, 5, 52.

25 Modelling Groundwater Flow in a Confined Aquifer with Dual Layers

Disebo Venoliah Chaka and Abdon Atangana
University of the Free State, Bloemfontein, South Africa

CONTENTS

25.1 INTRODUCTION

Aquifers have been divided into different types, based on the geology and the existence or non-existence of the water table, as confined, unconfined, and leaky (Fetter, 1994). The aquifer is typically confined and has two types of layers in the saturated layer. These layers are classified as separate quantities because they have different lithologies and therefore cannot be simplified or reduced to one homogeneous layer. There are different types of groundwater flow conditions, which depend on the behaviour of water in certain situations, usually how it enters a system and how it exits the system. There are steady state, unsteady state, and pseudo-state flows. The modelling of groundwater was developed to conceptualize hydro-geological processes as well as to analyse information from the field by providing a quantitative framework. According to Wang and Anderson (1982), models are described as a "representation of the complex nature of the world". The most common purpose of a groundwater model is to predict the behaviour of some hydrological actions and can be used to model conditions that occurred in the past (as hindcasting and as interpretative tools).

Groundwater flow in a confined aquifer has been modelled in the past using deterministic mathematical models such as the Theis (1935) equation, which provides fundamental solutions. For the Theis equation to be applicable, certain assumptions have to be considered. These include the homogeneity of the aquifer, a uniform thickness, an infinite aerial extent, the fact that the aquifer is isotropic, and the discharge rate at which it is pumped is constant. These assumptions ignore the fact that in the field, the aquifers are of a heterogeneous nature, anisotropic, have finite aerial extent due to hydraulic boundaries, and are pumped at different discharge rates. The Theis equation is therefore simplified and does not account for the high order terms in the modelling of the flow of groundwater in a confined aquifer with dual layers. This research will account for these problems by developing a model for a groundwater flow equation for a confined aquifer with dual layers. The complexity and imperfect flow in the aquifers and special aquifers (confined with a dual layer) will be taken into consideration so as to not limit the equation by assumptions and simplifications.

DOI: 10.1201/9781003266266-25

25.2   FRACTAL CALCULUS

Fractal calculus is a somewhat new branch of mathematics dealing with kinetics (He, 2018). Fractal derivation is a non-standard derivation used in mathematics and applied mathematics and is scaled according to t^a. It is used to model the physical problems encountered in Fick's law, Darcy's law, and Fourier's law, to name a few. These problems cannot be applied to the media that consist of a non-integral fractal dimension because they are based on Euclidian geometry (Atangana, 2017).

25.3   CONNECTING FRACTIONAL AND FRACTAL DERIVATIONS

Fractal-fractional differentiation is a new concept combining both fractional differentiation and the fractal derivative, and is used to replicate physical problems exhibiting self-similarity and non-locality (Atangana, 2017). For example diffusion with anomalous behaviours have posed a challenge to modelers that use local and fractional differential operators, as these behaviours cannot be replicated accurately using these operators. In some cases, real world problems exhibit behaviours presenting similarities and non-locality at the same time. It is important to note that either fractional derivative nor fractal operators can be used to replicate these behaviours. Chen et al. (2010) introduced fractal differential operators to handle processes with similar properties, also fractional differential operators were introduced to capture nonlocalities. Nevertheless, by combining these two concepts, fractal and fractional, one will be able to capture both non-locality and self-similarities. To accommodate readers that are not acquainted with these notions, we present here some useful definitions.

The fractional derivative is defined by:

$$
{}^{c}_{o}D^{\alpha}_{t} f(t) = \frac{1}{\Gamma(1-\alpha)} \int_{0}^{t} \frac{df(\tau)}{d\tau} (t-\tau)^{-\alpha} \, dt
\tag{25.1}
$$

The fractal derivative is defined by:

$$
{}^{F}_{o}D^{\beta}_{t} f(t) = \lim_{t \to t_1} \frac{f(t) - f(t_i)}{t^{\beta} - t_1^{\beta}}
\tag{25.2}
$$

Therefore, the combination of the two derivatives is defined by:

$$
{}^{FF}_{o}D^{\alpha,\beta}_{t} f(t) = \frac{1}{\Gamma(1-\alpha)} \frac{d}{dt^{\beta}} \int_{o}^{t} f(\tau)(t-\tau)^{-\alpha} \, d\tau
\tag{25.3}
$$

The connection between the fractional and the fractional derivatives can be defined as:

$$
{}^{FF}_{o}D^{\alpha,\beta}_{t} f(t) = u(t)
\tag{25.4}
$$

The fractal derivation can be written as:

$$
\lim_{t \to t_1} \frac{f(t) - f(t_1)}{t - t_1} \frac{t - t_1}{t^{\beta} - t_1^{\beta}}
\tag{25.5}
$$

Applying the first principle:

$$
f'(t) = \frac{\dfrac{1}{t^{\beta} - t_1^{\beta}}}{t - t_1}
\tag{25.6}
$$

If $f(t) = t^\beta$, and

$$\frac{t - t_1}{t^\beta - t_1^\beta} \frac{f(t) - f(t_1)}{t - t_1} = f(t) \qquad (25.7)$$

then:

$$f'(t) = \left(t^\beta\right)' = \beta t^{\beta-1} \qquad (25.8)$$

Then the fractional-fractal derivation is:

$$^{FF}_{o}D_t^{\alpha,\beta} f(t) = \frac{1}{\Gamma(1-\alpha)} \frac{d}{dt} f(t) \qquad (25.9)$$

where

$$f(t) = \int_0^t t(\tau)(t-\tau)^{-\alpha} \, d\tau \qquad (25.10)$$

Therefore:

$$^{FF}_{o}D_t^{\alpha,\beta} f(t) = \frac{1}{\Gamma(1-\alpha)} \frac{d}{dt^\beta} f(t) * \frac{1}{\beta t^{\beta-1}} \qquad (25.11)$$

Replace $f(t)$ by a value:

$$^{FF}_{o}D_t^{\alpha,\beta} f(t) = \frac{1}{\Gamma(1-\alpha)} \frac{d}{dt} \int_0^t f(\tau)(t-\tau)^{-\alpha} \, d\tau \, \frac{1}{\beta t^{\beta-1}} = u(t) \qquad (25.12)$$

Simplifying the equation:

$$^{FF}_{o}D_t^{\alpha,\beta} f(t) = \frac{1}{\Gamma(1-\alpha)} \frac{d}{dt} \int_0^t f(\tau)(t-\tau)^{-\alpha} \, d\tau = \beta t^{\beta-1} u(t) \qquad (25.13)$$

Apply the Riemann–Liouville definition of a fractional derivative on both sides to obtain:

$$f(t) = \frac{\beta}{\Gamma(\alpha)} \int_0^t \tau^{\beta-1}(t-\tau)^{\alpha-1} u(\tau) \, d\tau \qquad (25.14)$$

Therefore, the fractal-fractional derivative is defined by:

$$^{FF}_{o}J_t^{\alpha,\beta} f(t) = \frac{\beta}{\Gamma(\alpha)} \int_0^t \tau^{\beta-1}(t-\tau)^{\alpha-1} f(\tau) \, d\tau. \qquad (25.15)$$

25.4 NUMERICAL SOLUTIONS

The purpose of the numerical approximation in this study is to create acceptable operators that can be subsequently used to represent the groundwater flow by making use of computer simulations. The numerical approximation application allows the numerical analysis of the partial differential equations to achieve some simplicity of the algorithms.

Fractal-fractional derivative with power law kernel is defined as follow:

Suppose that $y(t)$ be continuous, a fractal- fractional derivative of $y(t)$ with fractal dimension β and fractional order α is given as:

$$
{}^{FFP}_{0}J^{\alpha,\beta}_{t}\left(y(t)\right) = \frac{\beta}{\Gamma(1-\alpha)}\frac{d}{dt}\int_{0}^{t}(t-\tau)^{-\alpha}y(\tau)d\tau\frac{1}{\beta t^{\beta-1}} \tag{25.16}
$$

Representing the terms by $f(t)$, then:

$$
{}^{FFP}_{0}J^{\alpha,\beta}_{t}\left(y(t)\right) = \frac{d}{dt}f(t) \tag{25.17}
$$

Moving from 0 to t_{n+1}

$$
\frac{df(t)}{dt} = \frac{f(t_{n+1})-f(t_n)}{\Delta t} \tag{25.18}
$$

Replacing t with t_{n+1} in Equation (25.18):

$$
\frac{df(t_{n+1})}{dt} = \int_{0}^{t_n}\frac{(t_{n+1}-\tau)^{-\alpha}}{\Gamma(1-\alpha)}y(t)d\tau\frac{1}{\beta t^{\beta-1}_{n+1}} \tag{25.19}
$$

To represent the sum factor, the equation is modified as follows:

$$
\frac{df(t_{n+1})}{dt} = \frac{1}{\Gamma(1-\alpha)}\frac{1}{\beta t^{\beta-1}_{n+1}}\sum_{j=0}^{n}\int_{t_j}^{t_{j+1}}(t_{n+1}-\tau)^{-\alpha}y(t_j)d\tau \tag{25.20}
$$

Let $z = t_{n+1} - \tau$, therefore $z = t_{n+1} - t_j$ and $z = t_{n+1} - t_{j+1}$. Then $dz = -d\tau$, therefore:

$$
\frac{df(t_{n+1})}{dt} = \frac{1}{\Gamma(1-\alpha)}\frac{1}{\beta t^{\beta-1}_{n+1}}\sum_{j=0}^{n}y^{j}\int_{t_{n+1}-t_j}^{t_{n+1}-t_{j+1}}z^{-\alpha}(-dz) \tag{25.21}
$$

Apply the rule $-\int_{a}^{b}x = \int_{b}^{a}x$, integrate, and then simplify to get:

$$
\frac{df(t_{n+1})}{dt} = \frac{1}{\Gamma(1-\alpha)}\frac{1}{\beta t^{\beta-1}_{n+1}}\sum_{j=0}^{n}y^{j}\left[\frac{(t_{n+1}-t_j)^{1-\alpha}}{1-\alpha} - \frac{(t_{n+1}-t_{j+1})^{1-\alpha}}{1-\alpha}\right] \tag{25.22}
$$

Let the following be:

$$t_n = \Delta t_n, t_{n+1} = \Delta t (n+1), t_{j+1} = \Delta t (j+1), t_j = \Delta t_j$$

Therefore:

$$\frac{df(t_{n+1})}{dt} = \frac{1}{\Gamma(1-\alpha)} \frac{1}{\beta t_{n+1}^{\beta-1}} \sum_{j=0}^{n} y^j \left[\frac{((n+1)\Delta t - (j+1)\Delta t)^{1-\alpha}}{1-\alpha} - \frac{((n+1)\Delta t - (j+1)\Delta t)^{1-\alpha}}{1-\alpha} \right]$$

(25.23)

Taking out the common factor:

$$\frac{df(t_{n+1})}{dt} = \frac{(\Delta t)^{1-\alpha}}{\Gamma(2-\alpha)} \frac{1}{\beta t_{n+1}^{\beta-1}} \sum_{j=0}^{n} y^j \left[(n+1-j)^{1-\alpha} - (n-j)^{1-\alpha} \right]$$

(25.24)

The numerical approximation in terms of moving from 0 to t_n is:

$$\frac{df(t_n)}{dt} = \int_0^{t_n} \frac{(t_n - \tau)^{-\alpha}}{\Gamma(1-\alpha)} y(t) d\tau \frac{1}{\beta t_n^{\beta-1}}$$

(25.25)

$$\frac{df(t_n)}{dt} = \frac{1}{\Gamma(1-\alpha)} \frac{1}{\beta t_n^{\beta-1}} \int_0^{t_n} (t_n - \tau)^{-\alpha} y(\tau) d\tau$$

(25.26)

$$\frac{df(t_n)}{dt} = \frac{1}{\Gamma(1-\alpha)} \frac{1}{\beta t_n^{\beta-1}} \sum_{j=0}^{n} \int_{t_j}^{t_{j+1}} (t_n - \tau)^{-\alpha} y(t_j) d\tau$$

(25.27)

$$\frac{df(t_n)}{dt} = \frac{1}{\Gamma(1-\alpha)} \frac{1}{\beta t_n^{\beta-1}} \sum_{j=0}^{n} y^j \int_{t_j}^{t_{j+1}} (t_n - \tau)^{-\alpha} d\tau$$

(25.28)

Represent the following terms by $z = t_n - \tau$, therefore $z = t_n - t_j$ and $z = t_n - t_{j+1}$. Let $dz = -d\tau$, therefore:

$$\frac{df(t_n)}{dt} = \frac{1}{\Gamma(1-\alpha)} \frac{1}{\beta t_n^{\beta-1}} \sum_{j=0}^{n} y^j \int_{t_n - t_j}^{t_n - t_{j+1}} z^{-\alpha} (-dz)$$

(25.29)

Apply the rule $-\int_a^b x = \int_b^a x$, integrate, and then simplify to get:

$$\frac{df(t_n)}{dt} = \frac{1}{\Gamma(1-\alpha)} \frac{1}{\beta t_n^{\beta-1}} \sum_{j=0}^{n} y^j \left[\frac{(t_n - t_j)^{1-\alpha}}{1-\alpha} - \frac{(t_n - t_{j+1})^{1-\alpha}}{1-\alpha} \right]$$

(25.30)

Represent the following terms by $t_n = \Delta t_n$, $t_n = \Delta t(n)$, $t_{j+1} = \Delta t(j+1)$, $t_j = \Delta t_j$
Therefore:

$$\frac{df(t_n)}{dt} = \frac{1}{\Gamma(1-\alpha)} \frac{1}{\beta t_n^{\beta-1}} \sum_{j=0}^{n} y^j \left[\frac{(n\Delta t - j\Delta t)^{1-\alpha}}{1-\alpha} - \frac{(n\Delta t - (j+1)\Delta t)^{1-\alpha}}{1-\alpha} \right] \quad (25.31)$$

Taking out the common factor:

$$\frac{df(t_n)}{dt} = \frac{(\Delta t)^{1-\alpha}}{\Gamma(2-\alpha)} \frac{1}{\beta t_n^{\beta-1}} \sum_{j=0}^{n} y^j \left[(n-j)^{1-\alpha} - (n-j-1)^{1-\alpha} \right] \quad (25.32)$$

Substituting Equations (25.32 and 25.24) into (25.18):

$$\frac{df(t)}{dt} = \frac{1}{\Delta t} \left[\begin{array}{c} \left(\dfrac{(\Delta t)^{1-\alpha}}{\Gamma(2-\alpha)} \dfrac{1}{\beta t_{n+1}^{\beta-1}} \displaystyle\sum_{j=0}^{n} y^j \left[(n+1-j)^{1-\alpha} - (n-j)^{1-\alpha} \right] \right) \\ - \left(\dfrac{(\Delta t)^{1-\alpha}}{\Gamma(2-\alpha)} \dfrac{1}{\beta t_n^{\beta-1}} \displaystyle\sum_{j=0}^{n-1} y^j \left[(n-j)^{1-\alpha} - (n-j-1)^{1-\alpha} \right] \right) \end{array} \right] \quad (25.33)$$

When there is more than one variable in a function:

$$_{0}^{FFP}D_t^{\alpha,\beta} y(r_i, t_n) = \frac{\Delta t^{-\alpha}}{\Gamma(2-\alpha)} \left[\begin{array}{c} \displaystyle\sum_{j=0}^{n} y_i^j \left(\dfrac{1}{\beta t_{n+1}^{\beta-1}} \left[(n+1-j)^{1-\alpha} - (n-j)^{1-\alpha} \right] \right) \\ - \displaystyle\sum_{j=0}^{n-1} y_i^j \left(\dfrac{1}{\beta t_n^{\beta-1}} \left[(n-j)^{1-\alpha} - (n-j-1)^{1-\alpha} \right] \right) \end{array} \right] \quad (25.34)$$

Further simplification yields:

$$\frac{\Delta t^{-\alpha}}{\Gamma(2-\alpha)} \left[\sum_{j=0}^{n} \frac{1}{\beta t_{n+1}^{\beta-1}} h_i^j \delta_{n,j}^{\alpha,\beta} - \sum_{j=0}^{n-1} \frac{1}{\beta t_n^{\beta-1}} h_i^j \tau_{n,j}^{\alpha,\beta} \right] = \frac{T}{S} \left[\frac{1}{r_i} \frac{h_{i+1}^n - h_{i-1}^n}{\Delta r} + \frac{h_{i+1}^n - 2h_i^n + h_{i-1}^n}{(\Delta r)^2} \right] \quad (25.35)$$

From the first sum when $j = n$ and the second sum when $j = n - 1$:

$$\frac{\Delta t^{-\alpha}}{\Gamma(2-\alpha)} \left[\begin{array}{c} \dfrac{1}{\beta t_{n+1}^{\beta-1}} h_i^n \delta_{n,n}^{\alpha,\beta} + \dfrac{1}{\beta t_{n+1}^{\beta-1}} h_i^{n-1} \delta_{n,n-1}^{\alpha,\beta} + \displaystyle\sum_{j=0}^{n} \dfrac{1}{\beta t_{n+1}^{\beta-1}} h_i^j \delta_{n,j}^{\alpha,\beta} \\ - \dfrac{1}{\beta t_n^{\beta-1}} h_i^{n-1} \tau_{n,n-1}^{\alpha,\beta} - \displaystyle\sum_{j=0}^{n-2} \dfrac{h_i^j}{\beta t_n^{\beta-1}} \tau_{n,j}^{\alpha,\beta} \end{array} \right] = \frac{T}{S} \left[\frac{1}{r_i} \frac{h_{i+1}^n - h_i^n}{\Delta r} + \frac{h_{i+1}^n - 2h_i^n + h_{i-1}^n}{(\Delta r)^2} \right]$$

$$h_i^n \left(\frac{\Delta t^{-\alpha}}{\Gamma(2-\alpha)} \cdot \frac{\delta_{n,n}^{\alpha,\beta}}{\beta t_{n+1}^{\beta-1}} + \frac{T}{S} \frac{2}{(\Delta r)^2} \right)$$

$$+ h_i^{n-1} \left(\frac{\Delta t^{-\alpha}}{\Gamma(2-\alpha)} \frac{\delta_{n,n-1}^{\alpha,\beta}}{\beta t_{n+1}^{\beta-1}} - \frac{\Delta t^{-\alpha}}{\Gamma(2-\alpha)} \frac{\delta_{n,n-1}^{\alpha,\beta}}{\beta t_n^{\beta-1}} \right) - h_{i+1}^n \left(\frac{T}{S} \frac{1}{\Delta r(r_i)} - \frac{T}{S} \frac{1}{(\Delta r)^2} \right)$$

$$- h_{i-1}^n \left(\frac{T}{S} \frac{1}{\Delta r(r_i)} - \frac{T}{S} \frac{1}{(\Delta r)^2} \right) = \sum_{j=0}^{n} \frac{1}{\beta t_{n+1}^{\beta-1}} h_i^j \delta_{n,j}^{\alpha,\beta} + \sum_{j=0}^{n-1} \frac{1}{\beta t_n^{\beta-1}} h_i^j \tau_{n,j}^{\alpha,\beta} \qquad (25.36)$$

Fractal-fractional derivative with exponential decay kernel is defined as follow:

Suppose that $y(t)$ be continuous, a fractal- fractional derivative of $y(t)$ with fractal dimension β and fractional order α is given as:

$$^{FFE}_0 D_t^{\alpha,\beta} \left(y(t) \right) = \frac{M(\alpha)}{1-\alpha} \int_0^t f(\tau) \exp\left[-\frac{\alpha}{1-\alpha}(t-\tau) \right] d\tau \frac{1}{\beta t^{\beta-1}} \qquad (25.37)$$

$$^{FFE}_0 D_t^{\alpha,\beta} \left(y(t) \right) = \frac{M(\alpha)}{\beta t^{\beta-1}(1-\alpha)} \left[\exp\left[-\frac{\alpha}{1-\alpha}t \right] f(0) + \int_0^t \frac{d}{d\tau} f(\tau) \exp\left(-\frac{\alpha}{1-\alpha} \right) d\tau \right] \qquad (25.38)$$

At t_{n+1}:

$$^{FFE}_0 D_t^{\alpha,\beta} \left(y(t_{n+1}) \right) = \frac{M(\alpha)}{\beta t_{n+1}^{\beta-1}(1-\alpha)} \left[\exp\left[-\frac{\alpha}{1-\alpha}t_{n+1} \right] f(0) + \int_0^{t_{n+1}} \frac{d}{d\tau} f(\tau) \exp\left[-\frac{\alpha}{1-\alpha}(t_{n+1}-\tau) \right] d\tau \right]$$

$$(25.39)$$

$$^{FFE}_0 D_t^{\alpha,\beta} \left(y(t_{n+1}) \right) = \frac{1}{\beta t_{n+1}^{\beta-1}} \sum_{j=0}^{n} \int_{t_j}^{t_{j+1}} \exp\left[-\frac{\alpha}{1-\alpha}(t_{n+1}-\tau) \right] \frac{f^{j+1}-f^j}{\Delta t} d\tau \qquad (25.40)$$

Further simplify:

$$^{FFE}_0 D_t^{\alpha,\beta} \left(y(t_{n+1}) \right) = \frac{1}{\beta t_{n+1}^{\beta-1}} \sum_{j=0}^{n} \frac{f^{j+1}-f^j}{\Delta t} \int_{t_j}^{t_{j+1}} \exp\left[-\frac{\alpha}{1-\alpha}(t_{n+1}-\tau) \right] d\tau \qquad (25.41)$$

Representing the following terms by $= t_{n+1} - \tau, y = t_{n+1} - t_j, y = t_{n+1} - t_{j+1}, d\tau = -dy$

$$^{FFE}_0 D_t^{\alpha,\beta} \left(y(t_{n+1}) \right) = \frac{1}{\beta t_{n+1}^{\beta-1}} \sum_{j=0}^{n} \frac{f^{j+1}-f^j}{\Delta t} \int_{t_{n+1}-t_{j+1}}^{t_{n+1-t_j}} \exp\left(-\frac{\alpha}{1-\alpha}y \right) dy \qquad (25.42)$$

$$^{FFE}_0 D_t^{\alpha,\beta} \left(y(t_{n+1}) \right) = \frac{1}{\beta t_{n+1}^{\beta-1}} \sum_{j=0}^{n} \frac{f^{j+1}-f^j}{\Delta t} \frac{\alpha}{1-\alpha} \exp\left[-\frac{\alpha}{1-\alpha}y \right] \Big|_{t_{n+1}-t_{j+1}}^{t_{n+1}-t_j} \qquad (25.43)$$

Further simplify:

$$
{}^{FFE}_{0}D_t^{\alpha,\beta}\left(y(t_{n+1})\right) = \frac{M(\alpha)}{\beta t_{n+1}^{\beta-1}(1-\alpha)}\left[\begin{array}{l}\exp\left[-\dfrac{\alpha}{1-\alpha}(n-j)\Delta t\right]f(0)\\[2mm] +\displaystyle\sum_{j=0}^{n}\dfrac{f^{j+1}-f^j}{\Delta t}\dfrac{\alpha}{1-\alpha}\left[\begin{array}{l}\exp\left[-\dfrac{\alpha}{1-\alpha}(n-j+1)\Delta t\right]\\[2mm]-\exp\left[-\dfrac{\alpha}{1-\alpha}(n-j)\Delta t\right]\end{array}\right]\end{array}\right]
$$

$$(25.44)$$

When we have more than one variable forward, the solution then becomes:

$$
{}^{FFE}_{0}D_t^{\alpha,\beta}\left(y(t_{n+1})\right) = \frac{M(\alpha)}{\beta t_{n+1}^{\beta-1}(1-\alpha)}\left[\begin{array}{l}\exp\left[-\dfrac{\alpha}{1-\alpha}(n-j)\Delta t\right]h(r,0)\\[2mm] +\displaystyle\sum_{j=0}^{n}\dfrac{h_i^{j+1}-h_i^j}{\Delta t}\dfrac{\alpha}{1-\alpha}\left[\begin{array}{l}\exp\left[-\dfrac{\alpha}{1-\alpha}(n-j+1)\Delta t\right]\\[2mm]-\exp\left[-\dfrac{\alpha}{1-\alpha}(n-j)\Delta t\right]\end{array}\right]\end{array}\right]
$$

$$
= \frac{T}{S}\left[\frac{h_{i+1}^{n+1}-2h_i^{n+1}+h_{i-1}^{n+1}}{(\Delta r)^2}+\frac{1}{r_i}\frac{h_{i+1}^{n+1}-h_{i-1}^{n+1}}{\Delta r}\right]
$$

$$(25.45)$$

Fractal-fractional derivative with Mittag–Leffler function is defined as follow:

Suppose that $y(t)$ be continuous, a fractal- fractional derivative of $y(t)$ with fractal dimension β and fractional order α is given as:

$$
{}^{FFM}_{0}D_t^{\alpha,\beta}\left(y(t)\right) = \frac{AB(\alpha)}{1-\alpha}\frac{d}{dt^\beta}\int_0^t E_\alpha\left(-\frac{\alpha}{1-\alpha}(t-\tau)^\alpha\right)y(\tau)d\tau \tag{25.46}
$$

$$
{}^{FFM}_{0}D_t^{\alpha,\beta}\left(y(t)\right) = \frac{AB(\alpha)}{(1-\alpha)}\frac{d}{dt}\int_0^t E_\alpha\left(-\frac{\alpha}{1-\alpha}(t-\tau)^\alpha\right)y(\tau)d\tau\frac{1}{\beta t^{\beta-1}} \tag{25.47}
$$

At t_{n+1}:

$$
{}^{FFM}_{0}D_t^{\alpha,\beta}\left(y(t_{n+1})\right) = \frac{AB(\alpha)}{1-\alpha}\left[E_\alpha\left[-\frac{\alpha}{1-\alpha}t_{n+1}^\alpha\right]f(0)+\int_0^{t_{n+1}}\frac{d}{d\tau}f(\tau)E_\alpha\left[-\frac{\alpha}{1-\alpha}(t_{n+1}-\tau)\right]d\tau\right]\frac{1}{\beta t_{n+1}^{\beta-1}}
$$

$$(25.48)$$

$$
{}^{FFM}_{0}D_t^{\alpha,\beta}\left(y(t_{n+1})\right) = \int_0^{t_{n+1}}\frac{df(\tau)}{d\tau}E_\alpha\left[-\frac{\alpha}{1-\alpha}(t_{n+1}-\tau)^\alpha\right]d\tau\frac{1}{\beta t_{n+1}^{\beta-1}} \tag{25.49}
$$

$$
{}^{FFM}_{0}D_t^{\alpha,\beta}\left(y(t_{n+1})\right) = \sum_{j=0}^{n}\int_{t_j}^{t_{j+1}}E_\alpha\left[-\frac{\alpha}{1-\alpha}(t_{n+1}-\tau)^\alpha\right]\frac{f^{j+1}-f^j}{\Delta t}d\tau\frac{1}{\beta t_{n+1}^{\beta-1}} \tag{25.50}
$$

$$
{}^{FFM}_{0}D_t^{\alpha,\beta}\left(y\left(t_{n+1}\right)\right) = \sum_{j=0}^{n} \frac{f^{j+1}-f^j}{\Delta t} \int_{t_j}^{t_{j+1}} E_\alpha\left[-\frac{\alpha}{1-\alpha}\left(t_{n+1}-\tau\right)^\alpha\right] \frac{1}{\beta t_{n+1}^{\beta-1}}
\tag{25.51}
$$

Represent the following by $= t_{n+1}-\tau, y = t_{n+1}-t_j, y = t_{n+1}-t_{j+1}, d\tau = -dy$

$$
{}^{FFM}_{0}D_t^{\alpha,\beta}\left(y\left(t_{n+1}\right)\right) = \sum_{j=0}^{n} \frac{f^{j+1}-f^j}{\Delta t} \int_{t_{n+1}-t_{j+1}}^{t_{n+1}-t_j} E_\alpha\left(-\frac{\alpha}{1-\alpha}y^\alpha\right) dy \frac{1}{\beta t_{n+1}^{\beta-1}}
\tag{25.52}
$$

$$
{}^{FFM}_{0}D_t^{\alpha,\beta}\left(y\left(t_{n+1}\right)\right) = \sum_{j=0}^{n} \frac{f^{j+1}-f^j}{\Delta t} \int_{t_{n+1}-t_{j+1}}^{t_{n+1}-t_j} E_\alpha\left[-\frac{\alpha}{1-\alpha}y^\alpha\right] dy \frac{1}{\beta t_{n+1}^{\beta-1}}
\tag{25.53}
$$

$$
{}^{FFM}_{0}D_t^{\alpha,\beta}\left(y\left(t_{n+1}\right)\right) = \sum_{j=0}^{n} \frac{f^{j+1}-f^j}{\Delta t} E_{\alpha,2}\left[-\frac{\alpha}{1-\alpha}y^\alpha\right] - E_{\alpha,2}\left[-\frac{\alpha}{1-\alpha}y^\alpha\right]\Bigg|_{t_{n+1}-t_{j+1}}^{t_{n+1}-t_j} \frac{1}{\beta t_{n+1}^{\beta-1}}
\tag{25.54}
$$

$$
{}^{FFM}_{0}D_t^{\alpha,\beta}\left(y\left(t_{n+1}\right)\right) = \sum_{j=0}^{n} \frac{f^{j+1}-f^j}{\Delta t} E_{\alpha,2}\left[-\frac{\alpha}{1-\alpha}\left(n-j\right)^\alpha \Delta t\right] - E_{\alpha,2}\left[-\frac{\alpha}{1-\alpha}\left(n-j+1\right)^\alpha \Delta t\right] \frac{1}{\beta t_{n+1}^{\beta-1}}
\tag{25.55}
$$

Therefore:

$$
{}^{FFM}_{0}D_t^{\alpha,\beta}\left(y\left(t_{n+1}\right)\right) = \frac{AB(\alpha)}{(1-\alpha)}\left[E_\alpha\left[-\frac{\alpha}{1-\alpha}\left(n-j\right)^\alpha \Delta t\right] f(0) + \sum_{j=0}^{n} \frac{f^{j+1}-f^j}{\Delta t}\left[\begin{array}{c} E_{\alpha,2}\left[-\frac{\alpha}{1-\alpha}\left(n-j\right)^\alpha \Delta t\right] \\ -E_{\alpha,2}\left[-\frac{\alpha}{1-\alpha}\left(n-j+1\right)^\alpha \Delta t\right] \end{array} \right] \right] \frac{1}{\beta t_{n+1}^{\beta-1}}
\tag{25.56}
$$

When there is more than one variable:

$$
{}^{FFM}_{0}D_t^{\alpha,\beta}\left(y\left(t_{n+1}\right)\right) = \frac{AB(\alpha)}{(1-\alpha)}\left[E_\alpha\left[-\frac{\alpha}{1-\alpha}\left(n-j\right)^\alpha \Delta t\right] y(0) + \sum_{j=0}^{n}\left(h_i^{j+1}-h_i^j\right)\left[\begin{array}{c} \left(n-j\right)E_{\alpha,2}\left(-\frac{\alpha}{1-\alpha}\left(n-j\right)^\alpha \Delta t^\alpha\right) \\ -\left(n-j+1\right)E_{\alpha,2}\left(-\frac{\alpha}{1-\alpha}\left(n-j+1\right)^\alpha \Delta t^\alpha\right) \end{array} \right] \right] \frac{1}{\beta t_{n+1}^{\beta-1}}
\tag{25.57}
$$

Due to the appearance of the generalized Mittag–Leffler function on the discretized equation, we could revert the equation to the Volterra type then suggest a comparable approximation integral:

$$
{}_0^{AB}D_t^{\alpha,\beta}h(r,t) = \frac{T}{S}\left[\frac{\partial^2 h(r,t)}{\partial r^2} + \frac{1}{r}\frac{\partial h(r,t)}{\partial r}\right] \tag{25.58}
$$

Represent $\dfrac{T}{S}\left[\dfrac{\partial^2 h(r,t)}{\partial r^2} + \dfrac{1}{r}\dfrac{\partial h(r,t)}{\partial r}\right] = F(r,t,h(r,t))$

We then apply the fractal-fractional integral to both sides to obtain:

$$
h(r,t) - h(r,0) = \frac{(1-\alpha)\beta t^{\beta-1}}{AB(\alpha)}F(r,t,h(r,t)) + \frac{\alpha\beta}{AB\Gamma(\alpha)}\int_0^t \tau^{\beta-1}(t-\tau)^{\alpha-1}F(r,\tau,h(r,\tau))d\tau \tag{25.59}
$$

At the end we get:

$$
\begin{aligned}
h(r_i,t_{n+1}) - h(r_i,0) = {}&\frac{\alpha\beta}{AB(\alpha)}\sum_{j=0}^{n}\left[\frac{\Delta t^{\alpha}}{\Gamma(\alpha+2)}t_j^{\beta-1}\frac{T}{S}\left(\begin{array}{l}\dfrac{h(r_{i+1},t_j)-2h(r_i,t_j)+h(r_{i-1},t_j)}{\Delta r^2}\\[2mm]+\dfrac{1}{r_i}\dfrac{h(r_{i+1},t_j)-h(r_{i-1},t_j)}{2\Delta r}\end{array}\right)\right]\\[2mm]
&\left\{(n-j+1)^{\alpha}(n-j+2+2\alpha)-(n-j)^{\alpha}(n-j+2+2\alpha)\right\}\\[2mm]
&-\frac{\Delta t^{\alpha}}{\Gamma(\alpha+2)}\frac{T}{S}\left[\frac{h(r_{i+1},t_{j-1})-2h(r_i,t_{j-1})+h(r_{i-1},t_{j-1})}{\Delta r^2}+\frac{1}{r_i}\frac{h(r_{i+1},t_{j-1})-h(r_{i+1},t_{j-1})}{2\Delta r}\right]\\[2mm]
&\left\{(n-j+1)^{\alpha+1}-(n-j)^{\alpha}(n-j+1+\alpha)\right\}\\[2mm]
&+\frac{(1-\alpha)}{AB(\alpha)}\beta t_{n+1}^{\beta-1}\left[\frac{h(r_{i+1},t_n)-2h(r_i,t_n)+h(r_{i-1},t_n)}{\Delta r^2}+\frac{1}{r_i}\frac{h(r_{i+1},t_n)-h(r_{i-1},t_n)}{2\Delta r}\right].
\end{aligned} \tag{25.60}
$$

25.5 STABILITY ANALYSIS

Von Neumann stability analysis was developed by von Neumann at the Los Alamos National Laboratory but was only introduced to the public when it was described briefly in an article by Crank and Nicolson in 1947. It is known as the classical method to determine stability conditions including the problems with periodic boundary conditions. This method is mostly used to determine necessary and sufficient stability conditions (Sousa, 2009). The stability condition of each operator is given below.

First operator

$$
\max_{n\geq 0}\left[\left[\left(\left|\frac{\left(\dfrac{\Delta t^{-\alpha}}{\Gamma(2-\alpha)}\dfrac{\delta_{1,0}^{\alpha,\beta}}{\beta t_2^{\beta-1}}-\dfrac{\Delta t^{-\alpha}}{\Gamma(2-\alpha)}\dfrac{\delta_{1,0}^{\alpha,\beta}}{\beta t_1^{\beta-1}}\right)}{2i.\dfrac{T}{S}\dfrac{1}{\Delta r(r_i)}+\dfrac{T}{S}\dfrac{1}{(\Delta r)^2}\cos(k_m\Delta r)}\,-\frac{\dfrac{\delta_{1,0}^{\alpha,\beta}}{\beta t_2^{\beta-1}}+\dfrac{1}{\beta t_1^{\beta-1}}\tau_{1,0}^{\alpha,\beta}}{}\right.\right.\right.\right.
$$

$$
\left.\left.\left.\left.-\left(\frac{\Delta t^{-\alpha}}{\Gamma(2-\alpha)}\cdot\frac{\delta_{1,1}^{\alpha,\beta}}{\beta t_2^{\beta-1}}+\frac{T}{S}\frac{2}{(\Delta r)^2}-\frac{1}{\beta t_2^{\beta-1}}\delta_{1,1}^{\alpha,\beta}\right)\right|\right),\ \frac{|A|}{|B|}+\sum_{j=0}^{n-2}|\phi_{n,j}^{\alpha,\beta}|-\sum_{j=0}^{n-2}|\Phi_{n,j}^{\alpha,\beta}|\right)<1\right] \tag{25.61}
$$

Second operator

$$\frac{\Delta r}{r_i} < \frac{1}{2} \frac{\Delta r}{r_i} < \frac{1}{2}$$ (25.62)

Third operator

$$\max\left\{\frac{\sqrt{\left(1-4\sin^2\left(\frac{k_m\Delta r}{2}\right)\left(a_5^{j,0}-a_1^{j,0}\right)\right)^2 + \left(2\left(\sin\left(k_m\Delta r\right)\right)\left(a_6^{j,0}+a_2^{j,0}\right)\right)^2}}{\sqrt{\left(4\sin^2\left(\frac{k_m\Delta r}{2}\right)\left(a_1^{j,1}+a_3^{j,1}a_5^{j,1}\right)\right)^2 + \left(2\sin\left(k_m\Delta r\right)\left(a_2^{j,1}+a_4^{j,1}a_6^{j,1}\right)\right)^2}}\right\}, < 1.$$ (25.63)

25.6 NUMERICAL SIMULATIONS

Mathematical software was used to produce and represent numerical simulations of groundwater flow in a confined aquifer with dual layers using the modified mathematical approaches of fractal-fractional operators. Numerical simulations are depicted in Figures 25.1 to 25.15 for different values of fractional orders and fractal dimensions. The three concepts were used in this simulation including the model with fractal-fractionals with the power law, the model with fractal-fractionals with the exponential decay law, and finally the model with a crossover effect. Three majors flow types are observed from the numerical simulations including fast flow, normal flow, and slow flow. These three flows are mostly observed when the fractal dimension is reduced to zero, though this only helps to capture natural flow with no self-similar properties.

These are of course in agreement with the properties of the new concept as the fractal-fractional differential and integral operators can be reverted to classical differential and integral operators with fractional order. On the other hand, when the fractal dimensions are considered,

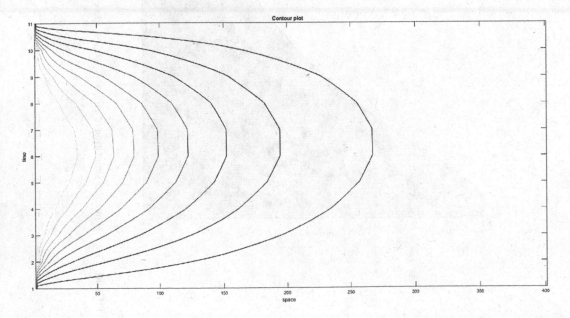

FIGURE 25.1 Contour plot numerical simulation for scale factor 1.

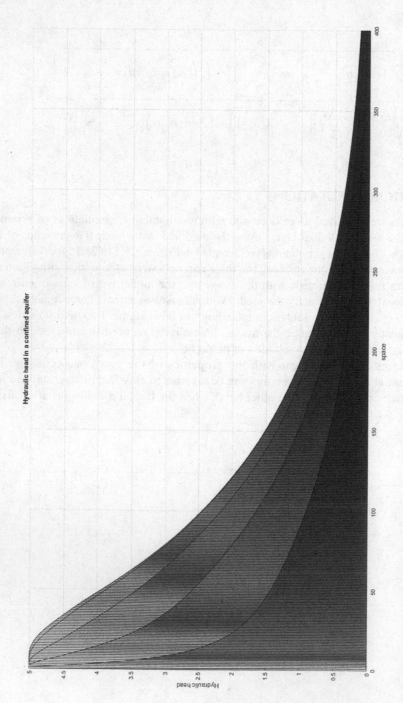

FIGURE 25.2   Numerical simulation of groundwater flow in a confined aquifer with respect to a hydraulic head and space, using a scale factor of 1 with $\alpha = 1$ and $\beta = 1$.

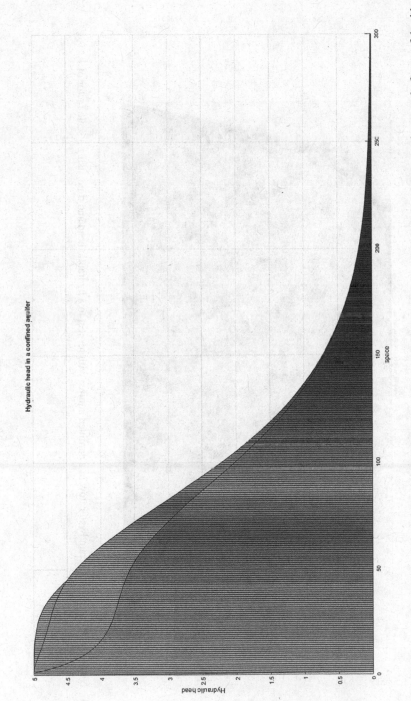

FIGURE 25.3 Numerical simulation of groundwater flow in a confined aquifer with respect to a hydraulic head and space, using a scale factor of 1 with $\alpha = 1$ and $\beta = 0.9$.

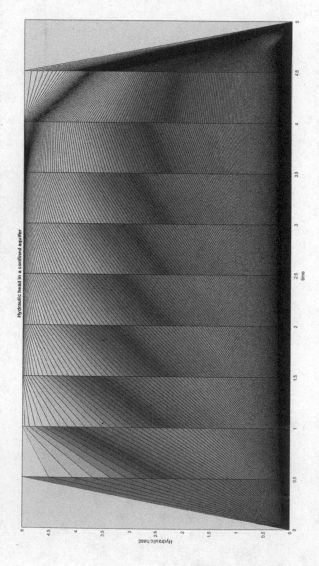

FIGURE 25.4   Numerical simulation of groundwater flow in a confined aquifer with respect to a hydraulic head and time, using a scale factor of 1.

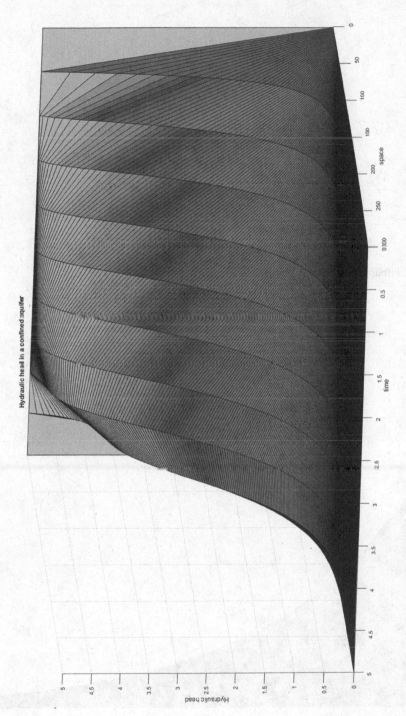

FIGURE 25.5 Numerical simulation of groundwater flow in a confined aquifer with respect to time and space showing also the hydraulic head, using a scale factor of 1.

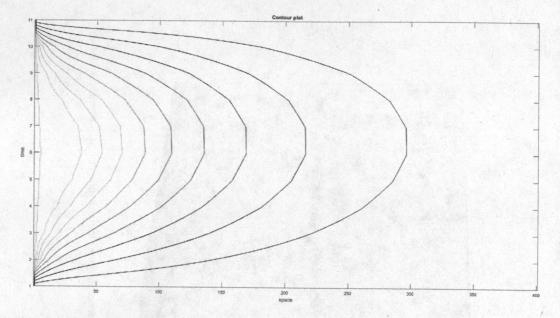

FIGURE 25.6   Contour plot numerical simulation for scale factor 0.9.

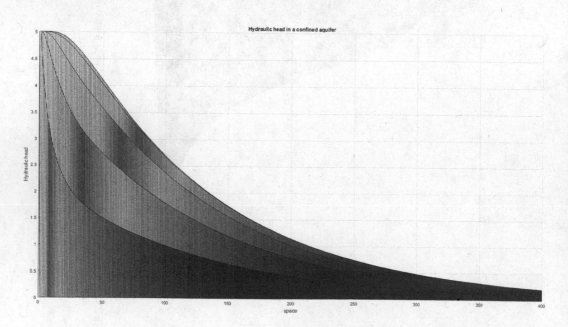

FIGURE 25.7   Numerical simulation of groundwater flow in a confined aquifer with respect to a hydraulic head and space, showing the hydraulic head, using a scale factor of 0.9.

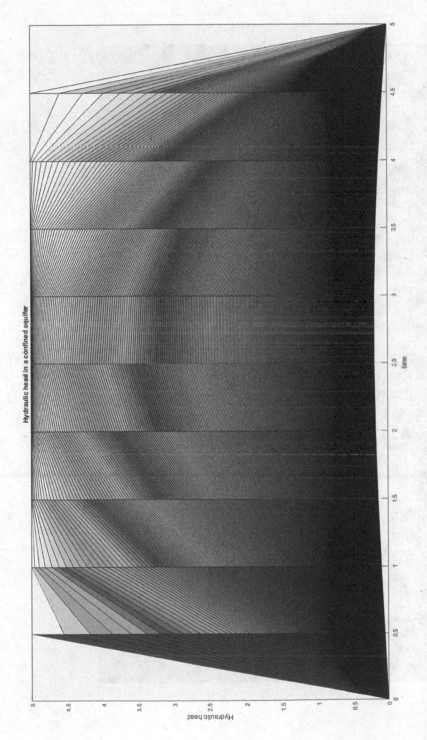

FIGURE 25.8 Numerical simulation of groundwater flow in a confined aquifer with respect to a hydraulic head and time, using a scale factor of 0.9.

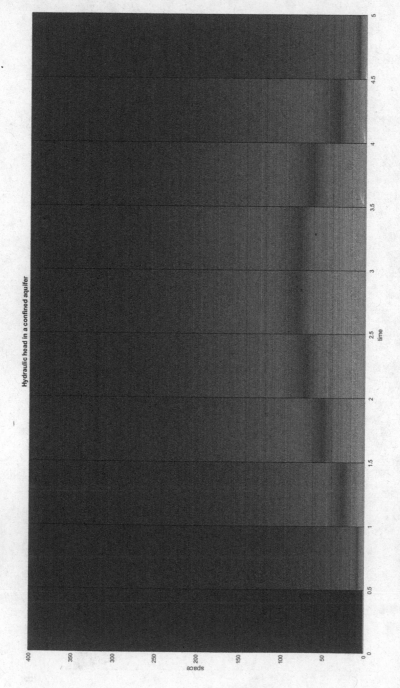

FIGURE 25.9   Numerical simulation of groundwater flow in a confined aquifer with respect to a hydraulic head and time, using a scale factor of 0.9.

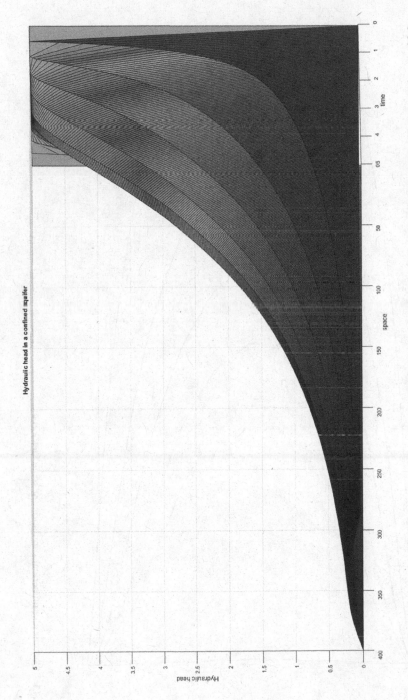

FIGURE 25.10 Numerical simulation of groundwater flow in a confined aquifer with respect to a hydraulic head and space, using a scale factor of 0.9.

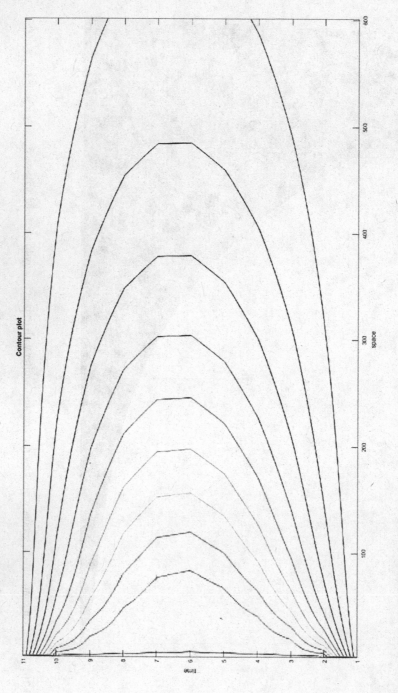

FIGURE 25.11 Contour plot numerical simulation for scale factor 0.4.

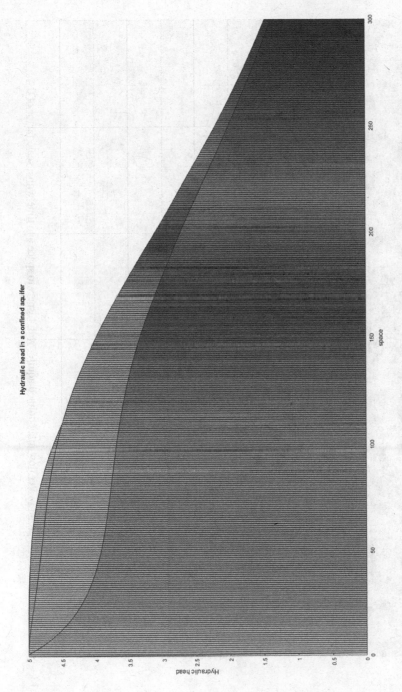

FIGURE 25.12 Numerical presentation groundwater flow in a confined aquifer with respect to a hydraulic head and space with a scale factor of 0.4.

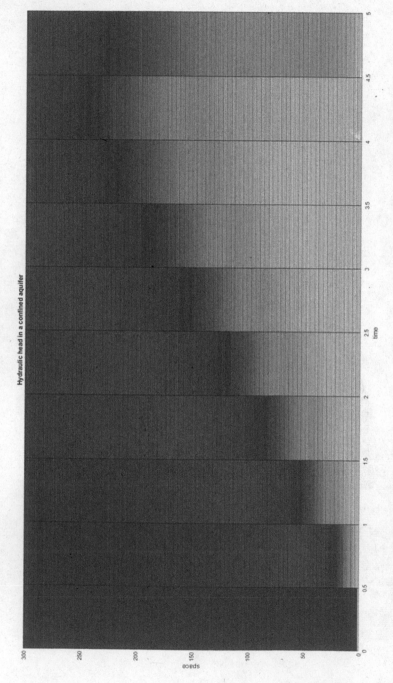

FIGURE 25.13 Numerical presentation of groundwater flow in a confined aquifer with respect to space and time with a scale factor of 0.4.

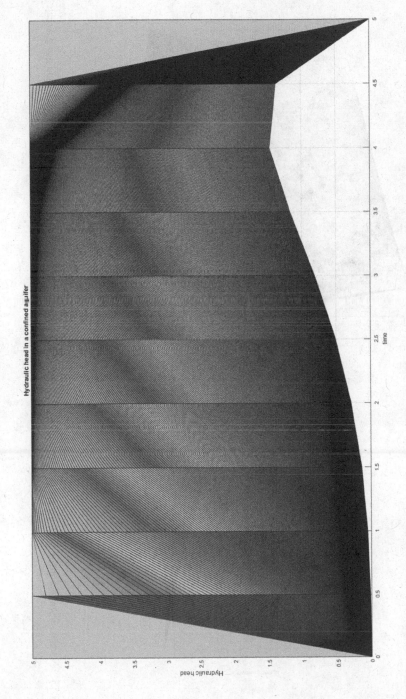

FIGURE 25.14 Numerical simulation of groundwater flow in a confined aquifer with respect to a hydraulic head and time, using a scale factor of 0.4.

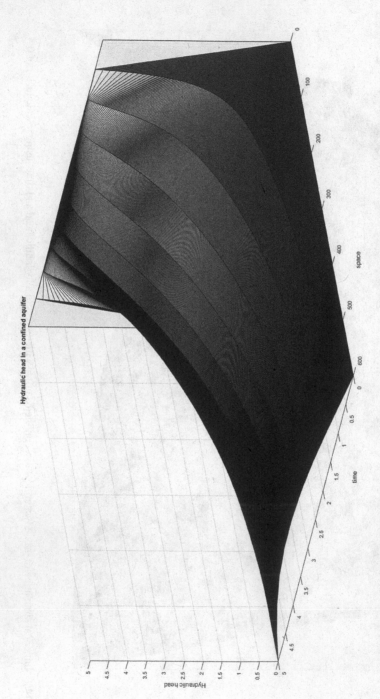

FIGURE 25.15　Numerical simulation of groundwater flow in a confined aquifer with respect to space and time, showing the hydraulic head and using a scale factor of 0.4

then the model exhibits fast flow with self-similarities, slow flow with self-similarities, and finally flow with crossover behaviour with self-similarities. In the real field observation, these results can be connected to a geological formation with dual media. The numerical solution obtained here suggest that non-conventional differential and integral operators are powerful mathematical tools able to replicate very accurately heterogeneity of the geological formation within which the sub-surface water flows.

25.7   CONCLUSION

Even though the Theis model was used successfully in many cases, the deviation of mathematical models from the observed facts leave no doubt that the suggested model has some limitations. These limitations are associated with the fact that the Theis model, in addition to the simplification made to obtain it, uses the concept of a differential operator based on the rate of change. This operator cannot account for non-Markovian flow processes. It cannot depict flow following the crossover process where the water flows from the soil matrix to a fracture or the reverse. This model cannot depict flow within the fracture, nor flow following fading velocity. However, to solve this problem, fractional differential operators based on the power law have been used intensively to model the flow within the geological formation as the power law has some important properties that help to capture long-range behaviour that can be associated with flow within a fracture, an important feature that classical differentiation cannot account for. Nevertheless, while such a model can also be used to replicate flow with fading velocity, the limitation of it is that fading velocity does not have a beginning and an end due to the very long tail of the power law. Additionally, the flow within a self-similar feature cannot be replicated here. In this thesis new operators called fractal-fractional derivatives have been used to generalize the flow within a confined-fractured aquifer with dual media, and three different physical laws were used. The new models are solved numerically using the newly suggested numerical scheme. Numerical simulation shows a connection between the fractional order, fractal dimension, and the heterogeneity of the geological formation.

REFERENCES

Anderson M., Woessner W., and Hunt R. (2015). *Applied Groundwater Modelling: Simulation of Flow and Advective Transport*. Academic Press. The United States of America.

Atangana A. (2017). Fractal-fractional differentiation and integration: Connecting fractal calculus and fractional calculus to predict complex system. *Chaos, Solitons and Fractals* 102, 396–406.

Atangana A. and Gomez-Aguilar J.F. (2018). Fractional derivatives with no-index law property: Application to chaos and statistics. *Chaos, Solitons and Fractals* 2018(114), 516–555.

Chen W., Sun H., Zhang X., and Korosvak D. (2010). Anomalous diffusion modelling by fractal and fractional derivatives. *Computers and Mathematics with Applications* 2010(59), 1754–1758.

Fetter C.W. (1994). *Applied Hydrogeology* 4th edition. Prentice-Hall, Inc, Upper Saddle River, New Jersey.

He J.H. (2018). Fractal calculus and its geometrical explanation. *Results in Physics* 2018(10), 272–276.

Liouville J. (1832). Mémoire sur le calcul des différentielles à indices quelconques. *Journal de l'École Polytechnique*, Paris, 13, 71–162.

Sousa E. (2009). On the edge of the stability analysis. *Applied Numerical Mathematics* 59(3), 1322–1336.

Theis C.V. (1935). The relation between the lowering of the piezometric surface and the rate and duration of discharge of well using groundwater storage. *American Geophysical union Trans* 16, 519–524.

Wang H.F. and Anderson M.P. (1982). *Introduction to Groundwater Modelling: Finite Difference and Finite Element Methods*, Academic Press, California.

26 The Dual Porosity Model

Siphokazi Simnikiwe Manundu and Abdon Atangana

University of the Free State, Bloemfontein, South Africa

CONTENTS

26.1 INTRODUCTION

26.1.1 DIFFERENT TYPES OF AQUIFERS

Groundwater exists beneath the earth's surface in faults, fractures, and pores of soils (Freeze & Cherry, 1979). Areas where groundwater exists are referred to as aquifers (saturated fractured rock or sand, from which usable volumes of groundwater can be pumped) (Freeze & Cherry, 1979). Aquifers are categorized into two types, namely porous media and fractured rock aquifers (Skinner et al., 2004). The porous media aquifer is composed of discrete soil; some examples are gravel, silt, and sand. These aquifers can be further divided into ones of unconsolidated porous media (where the grains do not adhere to each other, e.g., alluvial aquifers) and consolidated porous media (where the grains are cemented to each other, e.g., sandstone). However, fractured aquifers are defined as geological formations in which joints, cracks, and/or fractures act as conducive conduits for transporting groundwater. Examples of a fractured rock aquifer are basalt and granite.

For both categories, aquifers can be further grouped into three other types, namely unconfined, confined, and leaky aquifers (Skinner et al., 2004). The unconfined aquifer may be defined as a geological layer with no confining layer above the aquifer, whereas with the confined aquifer, the geological layer is bounded by confining beds above and beneath it, resulting in the groundwater being subjected to pressure greater than that of the atmosphere. Lastly, the leaky aquifer geological layer is bounded between aquitards that allow for minimal water to travel through.

Over the years, mathematical models have been continuously developed to assist in the sustainable management and safeguarding of the world's groundwater resources (Cordero, Eleazar, & Deane, 2009). These include the models which assist in understanding aquifers with double porosity.

DOI: 10.1201/9781003266266-26

26.1.2 Dual Media System

The dual-porosity media system theory is applied to studying the response of fractured and double porosity aquifers. The assumption that the dual-porosity model makes is that the porous medium consists of two continua: one related to the matrix block and fracture network, and the other to a less permeable pore system of the matrix block (Knaust et al., 2020).

The fracture network in this system is characterized by high hydraulic conductivity (K) and low storage (Ss) capability. These fractures are the main conduits of the system; however, their volume for storage capacity remains limited, whereas the matrix block section of the system is characterized by low hydraulic conductivity (K) and high storage (Ss) capability. These contribute a limited role in the transportation of fluids, though they play a significant role in fluid storage. Often, the fluid in the matrix is drained through fractures before it emerges out of the aquifer.

26.1.3 Existing Mathematical Models of the Dual Media System

The dual-porosity model is governed by the assumption that the medium can be separated into two distinct pore systems, microscopically composed of soil particles surrounded by fractures, both of which are treated as homogeneous media with separate hydraulic and solute transport properties. The dual-porosity medium is known to be a superposition of these two systems through the same volume (Warren & Root, 1963). This study seeks to modify a system of partial differential equations depicting the flow of subsurface water within a dual-porosity system.

26.2 PIECEWISE MODELLING

A piecewise derivative is given as follows:

$$
{}_{o}^{P}D_{g(t)} = \begin{cases} \dfrac{du(t)}{dt}, if\ 0 \le t \le T_1 \\[2ex] \dfrac{u'(t)}{g'(t)}, if\ T_1 \le t \le T_2 \end{cases}
\tag{26.1}
$$

Differential operators can be used to modify the existing equations. If we consider that the confined aquifer has two different parallel media then the mathematical model of groundwater flow in this system can be represented as:

$$
{}_{o}^{P}D_{g(t)}\big(h_1,h_2,t\big) = \begin{cases} T_1\left(\dfrac{\partial^2 h_1}{\partial x^2} + \dfrac{\partial^2 h_1}{\partial y^2}\right) = S_1 \dfrac{\partial h_1}{\partial t} + \pi a^2 \big(h_1 - h_2\big)\ if\ 0 \le t \le t_1 \\[3ex] T_2\left(\dfrac{\partial^2 h_2}{\partial x^2} + \dfrac{\partial^2 h_2}{\partial y^2}\right) = S_2 \dfrac{\partial h_2}{\partial t} g'(t) - \pi a^2 \big(h_1 - h_2\big)\ if\ t_1 \le t < T \end{cases}
\tag{26.2}
$$

Here the function g'(t) would be $\dfrac{t^{2-\alpha}}{2-\alpha}$ if $0 < \alpha \le 1$. It is worth noting that fractal derivatives are sub-classes of the global derivative.

Atangana and Goufo (2020) suggested that a fractal differential operator of a function f with fractal dimension $0 < \alpha \le 1$ is given as:

$$
{}_{0}^{F}D_{t}^{\alpha} f(t) = \lim_{t \to t_1} \frac{f(t) - f(t_1)}{t^{2-\alpha} - t_1^{2-\alpha}}(2 - \alpha).
\tag{26.3}
$$

26.2.1   NUMERICAL SOLUTION USING THE NEWTON POLYNOMIAL SCHEME

Mathematical models are largely used to depict real-world problems; however, to depict this problem, one needs to obtain a solution. The idea is to have exact solutions by using analytical methods, though the use of these methods is limited as some equations are so complex that they cannot be solved analytically. In these cases, researchers relied on numerical methods. The literature contains several numerical schemes, one of the most used being multi-step methods, for example, the Adams–Bashforth numerical scheme, the predictor-corrector method, or the Adams–Moulton method. These methods have been developed to solve mostly differential equations with classical derivatives. An extension has been made to solving fractional differential equations. For example, there is a version of Adams–Bashforth that has been developed in the case of the Caputo fractional derivative and its associate predictor-corrector. More versions have also been developed. In the case of non-singular kernels many new numerical schemes have been developed, for example by Batogna and Atangana (2019) using the Laplace transform for fractional differential operators. Toufik and Atangana (2017) suggested an alternative scheme for solving nonlinear fractional differential equations with Atangana–Baleanu fractional derivatives. Using the Newton polynomial interpolation, Atangana and Seda (2020) developed a numerical scheme that is able to handle fractional and fractal-fractional differential equations. This section will therefore present the numerical solution using the Newton polynomial scheme.

First, we will define the general Euler method of discretization in the different derivatives.

The general first derivative with change in time is defined as follows:

$$\frac{df(t)}{dt} = \lim_{t_1 \to t} \frac{f(t_1) - f(t)}{t_1 - t} \tag{26.4}$$

$$= \lim_{\Delta t \to 0} \frac{f(t + \Delta t) - f(t)}{\Delta t} \tag{26.5}$$

Therefore:

$$\left. \frac{df(t)}{dt} \right|_{t=t_j} \cong \frac{f(t_{j+1}) - f(t_j)}{\Delta t} \tag{26.6}$$

We will now apply discretization to the dual-porosity equation using Euler's method:

If $t_0 \leq t \leq t_1$:

$$S_1 \frac{\partial h_1(x,y,t)}{\partial t} = T_1 \left(\frac{\partial^2 h_1(x,y,t)}{\partial x^2} + \frac{\partial h_1(x,y,t)}{\partial y^2} \right) - \left(h_1(x,y,z) - h_2(x,y,z) \right) \pi a^2 \tag{26.7}$$

$$S_1 \frac{h_1(x_i,y_k,t_{j+1}) - h_1(x_i,y_k,t_j)}{\Delta t} = T_1 \left[\begin{array}{l} \left(\dfrac{h_1(x_{i+1},y_k,t_j) - 2h(x_i,y_k,t_j) + h_1(x_{i-1},y_k,t_j)}{\Delta x^2} \right) + \\[2mm] \left(\dfrac{h_1(x_i,y_{k+1},t_j) - 2h(x_i,y_k,t_j) + h_1(x_i,y_{k-1},t_j)}{\partial y^2} \right) \end{array} \right] -$$
$$\left(h_1(x_i,y_k,z_j) - h_2(x_i,y_k,z_j) \right) \pi a^2 \tag{26.8}$$

If $t_1 \leq t \leq T$:

$$S_2 \frac{\partial h_1(x,y,t)}{\partial t} \cdot \frac{dg(t)}{g(t)} = T_2 \left(\frac{\partial^2 h_1(x,y,t)}{\partial x^2} + \frac{\partial h_1(x,y,t)}{\partial y^2} \right) + \left(h_1(x,y,t) - h_2(x,y,t) \right) \pi a^2 \quad (26.9)$$

$$S_2 \frac{h_2(x_i,y_k,t_{j+1}) - h_2(x_i,y_k,t_j)}{\Delta t} \cdot \frac{g(t_{j+1}) - g(t_j)}{\Delta t}$$

$$= T_2 \left[\begin{array}{l} \left(\dfrac{h_2(x_{i+1},y_k,t_j) - 2h(x_i,y_k,t_j) + h_2(x_{i-1},y_k,t_j)}{\Delta x^2} \right) \\[3mm] + \left(\dfrac{h_2(x_i,y_{k+1},t_j) - 2h(x_i,y_k,t_j) + h_2(x_i,y_{k-1},t_j)}{\partial y^2} \right) \end{array} \right]$$

$$+ \left(h_1(x_i,y_k,z_j) - h_2(x_i,y_k,z_j) \right) \pi a^2 \quad (26.10)$$

In this section, we make use of the Newton polynomial-based method to provide an alternative numerical solution to the model. To achieve this, we first present the derivation of the method by considering a general nonlinear Cauchy problem as indicated in the following equation. Thus later we will convert this to a partial differential equation and apply it to our system of equations.

Within the interval $[t_n, t_{n+1}]$ we want to approximate the function $f(\tau, y(\tau))$

$$f(\tau, y(\tau)) \cong P_n(\tau) = f(t_{n-2}, y(t_{n-2})) +$$

$$\frac{f(t_{n-1}, y(t_{n-1})) - f(t_{n-2}, y(t_{n-2}))(\tau - t_{n-2})}{\Delta t} +$$

$$\frac{f(t_n, y(t_n)) - 2f(t_{n-1}, y(t_{n-1})) + f(t_{n-2}, y(t_{n-2}))(\tau - t_{n-2})(\tau - t_{n-1})}{2(\Delta t)^2} \quad (26.11)$$

$$y(t_{n+1}) - y(t_n) = \int_{t_n}^{t_{n+1}} P_n(\tau) d\tau \quad (26.12)$$

After integrating we obtain:

$$y(t_{n+1}) - y(t_n) = \frac{5}{12} f(t_{n-2}, y(t_{n-2})) \Delta t - \frac{4}{3} f(t_{n-1}, y(t_{n-1})) \Delta t + \frac{23}{12} f(t_n, y(t_n)) \Delta t \quad (26.13)$$

In our case

$$y(t_{n+1}) = h(x_i, y_j, t_{n+1})$$

$$f(t_n, y(t_n)) = f(x_i, y_j, h(x_i, y_j, t_n))$$

$$h\left(x_i,y_j,h\left(x_i,y_j,t_{n+1}\right)\right)-h\left(x_i,y_j,h\left(x_i,y_j,t_n\right)\right)$$

$$=\frac{5}{12}f\left(x_i,y_j,h\left(x_i,y_j,t_n\right)\right)\Delta t-\frac{4}{3}f\left(x_i,y_j,h\left(x_i,y_j,t_{n-1}\right)\right)$$

$$+\frac{23}{12}f\left(x_i,y_j,h\left(x_i,y_j,t_{n-2}\right)\right) \tag{26.14}$$

From our equation, if $t_0 \leq t \leq t_1$:

$$f_1\left(x_i,y_j,h\left(x_i,y_j,t_n\right)\right)=\frac{T_1}{S_1}\left(\frac{\partial^2 h_1\left(x_i,y_j,t_n\right)}{\partial x^2}+\frac{\partial^2 h_1\left(x_i,y_j,t_n\right)}{\partial y^2}\right)$$

$$-\pi a^2\left(h_1\left(x_i,y_j,t_n\right)-h_2\left(x_i,y_j,t_n\right)\right) \tag{26.15}$$

$$f_1\left(x_i,y_j,h\left(x_i,y_j,t_n\right)\right)=\frac{T_1}{S_1}\left(\begin{array}{c}\dfrac{h_1\left(x_{i+1},y_j,t_n\right)-2h_1\left(x_i,y_j,t_n\right)+h_1\left(x_{i-1},y_j,t_n\right)}{\left(\Delta x\right)^2}+\\[2mm]\dfrac{h_1\left(x_i,y_{j+1},t_n\right)-2h_1\left(x_i,y_j,t_n\right)+h_1\left(x_i,y_{j-1},t_n\right)}{\left(\Delta y\right)^2}\end{array}\right)$$

$$+\pi a^2\left(h_1\left(x_i,y_j,t_n\right)-h_2\left(x_i,y_j,t_n\right)\right) \tag{26.16}$$

$$h_1\left(x_i,y_j,h\left(x_i,y_j,t_{n+1}\right)\right)-h_1\left(x_i,y_j,h\left(x_i,y_j,t_n\right)\right)$$

$$=\frac{5}{12}\left\{\begin{array}{c}\dfrac{T_1}{S_1}\left(\begin{array}{c}\dfrac{h_1\left(x_{i+1},y_j,t_n\right)-2h_1\left(x_i,y_j,t_n\right)+h_1\left(x_{i-1},y_j,t_n\right)}{\left(\Delta x\right)^2}\\[2mm]+\dfrac{h_1\left(x_i,y_{j+1},t_n\right)-2h_1\left(x_i,y_j,t_n\right)+h_1\left(x_i,y_{j-1},t_n\right)}{\left(\Delta y\right)^2}\end{array}\right)\\[4mm]-\pi a^2\left(h_1\left(x_i,y_j,t_n\right)-h_2\left(x_i,y_j,t_n\right)\right)\end{array}\right\}\Delta t$$

$$-\frac{4}{3}\left\{\begin{array}{c}\dfrac{T_1}{S_1}\left(\begin{array}{c}\dfrac{h_1\left(x_{i+1},y_j,t_{n-1}\right)-2h_1\left(x_i,y_j,t_{n-1}\right)+h_1\left(x_{i-1},y_j,t_{n-1}\right)}{\left(\Delta x\right)^2}\\[2mm]+\dfrac{h_1\left(x_i,y_{j+1},t_{n-1}\right)-2h_1\left(x_i,y_j,t_{n-1}\right)+h_1\left(x_i,y_{j-1},t_{n-1}\right)}{\left(\Delta y\right)^2}\end{array}\right)\\[4mm]-\pi a^2\left(h_1\left(x_i,y_j,t_{n-1}\right)-h_2\left(x_i,y_j,t_{n-1}\right)\right)\end{array}\right\}$$

$$+\frac{23}{12}\left\{\begin{array}{c}\dfrac{T_1}{S_1}\left(\begin{array}{c}\dfrac{h_1\left(x_{i+1},y_j,t_{n-2}\right)-2h_1\left(x_i,y_j,t_{n-2}\right)+h_1\left(x_{i-1},y_j,t_{n-2}\right)}{\left(\Delta x\right)^2}\\[2mm]+\dfrac{h_1\left(x_i,y_{j+1},t_{n-2}\right)-2h_1\left(x_i,y_j,t_{n-2}\right)+h_1\left(x_i,y_{j-1},t_{n-2}\right)}{\left(\Delta y\right)^2}\end{array}\right)\\[4mm]-\pi a^2\left(h_1\left(x_i,y_j,t_{n-2}\right)-h_2\left(x_i,y_j,t_{n-2}\right)\right)\end{array}\right\} \tag{26.17}$$

And if $t_1 \leq t \leq T$

$$
\begin{aligned}
f_2\left(x_i, y_j, h\left(x_i, y_j, t_n\right)\right) = \frac{T_2}{S_2}\left(\frac{\partial^2 h_2\left(x_i, y_j, t_n\right)}{\partial x^2} + \frac{\partial^2 h_2\left(x_i, y_j, t_n\right)}{\partial y^2}\right) \\
+ \pi a^2\left(h_1\left(x_i, y_j, t_n\right) - h_2\left(x_i, y_j, t_n\right)\right)
\end{aligned}
\tag{26.18}
$$

$$
\begin{aligned}
f_2\left(x_i, y_j, h\left(x_i, y_j, t_n\right)\right) = \frac{T_2}{S_2}\left(\begin{array}{c}
\dfrac{h_2\left(x_{i+1}, y_j, t_n\right) - 2h_2\left(x_i, y_j, t_n\right) + h_2\left(x_{i-1}, y_j, t_n\right)}{\left(\Delta x\right)^2} \\
+ \dfrac{h_2\left(x_i, y_{j+1}, t_n\right) - 2h_2\left(x_i, y_j, t_n\right) + h_2\left(x_i, y_{j-1}, t_n\right)}{\left(\Delta y\right)^2}
\end{array}\right) \\
+ \pi a^2\left(h_1\left(x_i, y_j, t_n\right) - h_2\left(x_i, y_j, t_n\right)\right)
\end{aligned}
\tag{26.19}
$$

$$
\begin{aligned}
&h_2\left(x_i, y_j, h\left(x_i, y_j, t_{n+1}\right)\right) - h_2\left(x_i, y_j, h\left(x_i, y_j, t_n\right)\right) = \\
&= \frac{5}{12}\left\{\frac{T_2}{S_2}\left(\begin{array}{c}
\dfrac{h_2\left(x_{i+1}, y_j, t_n\right) - 2h_2\left(x_i, y_j, t_n\right) + h_2\left(x_{i-1}, y_j, t_n\right)}{\left(\Delta x\right)^2} \\
+ \dfrac{h_2\left(x_i, y_{j+1}, t_n\right) - 2h_2\left(x_i, y_j, t_n\right) + h_2\left(x_i, y_{j-1}, t_n\right)}{\left(\Delta y\right)^2}
\end{array}\right) \\
&\quad\quad + \pi a^2\left(h_1\left(x_i, y_j, t_n\right) - h_2\left(x_i, y_j, t_n\right)\right)\right\} \\
&\Delta t - \frac{4}{3}\left\{\frac{T_2}{S_2}\left(\begin{array}{c}
\dfrac{h_2\left(x_{i+1}, y_j, t_{n-1}\right) - 2h_2\left(x_i, y_j, t_{n-1}\right) + h_2\left(x_{i-1}, y_j, t_{n-1}\right)}{\left(\Delta x\right)^2} \\
+ \dfrac{h_2\left(x_i, y_{j+1}, t_{n-1}\right) - 2h_2\left(x_i, y_j, t_{n-1}\right) + h_2\left(x_i, y_{j-1}, t_{n-1}\right)}{\left(\Delta y\right)^2}
\end{array}\right) \\
&\quad\quad + \pi a^2\left(h_1\left(x_i, y_j, t_{n-1}\right) - h_2\left(x_i, y_j, t_{n-1}\right)\right)\right\} \\
&+ \frac{23}{12}\left\{\frac{T_2}{S_2}\left(\begin{array}{c}
\dfrac{h_2\left(x_{i+1}, y_j, t_{n-2}\right) - 2h_2\left(x_i, y_j, t_{n-2}\right) + h_2\left(x_{i-1}, y_j, t_{n-2}\right)}{\left(\Delta x\right)^2} \\
+ \dfrac{h_2\left(x_i, y_{j+1}, t_{n-2}\right) - 2h_2\left(x_i, y_j, t_{n-2}\right) + h_2\left(x_i, y_{j-1}, t_{n-2}\right)}{\left(\Delta y\right)^2}
\end{array}\right) \\
&\quad\quad + \pi a^2\left(h_1\left(x_i, y_j, t_{n-2}\right) - h_2\left(x_i, y_j, t_{n-2}\right)\right)\right\}.
\end{aligned}
\tag{26.20}
$$

26.3 STOCHASTIC MODEL

Models are applied to provide qualitative descriptions of natural phenomena using mathematical formulas. These models greatly assist in solving complex real-world problems. Models can be classified into two different versions: stochastic models and deterministic models. A deterministic model does not consist of random outputs, and the complete input and output characteristics of the model are conclusively determined. An example of a deterministic model would be a car being driven on a cruise control system. The predetermined input values of the system such as speed and distance travelled over time provide a definite arrival time. This arrival time represents the output from the predetermined set of circumstances (inputs) (Atangana, 2020), whereas with stochastic models, the model consists of random outputs and the complete output characteristics of the model are randomly determined. One example of a stochastic model is a pumping test where head change is observed over time. In this case the head change is dependent on the aquifer characteristics at the given time which is also influenced by the geological conditions.

This section discusses the stochastic model. This will be done by first introducing the stochastic processes; secondly, a comparison between stochastic and deterministic processes will be conducted; thirdly, the modified model with the stochastic approach will be presented; lastly, this modified model will be analysed.

26.3.1 MODIFIED MODEL WITH THE STOCHASTIC APPROACH

In this sub-section, the function of a random walk will be added to our equation. The general equation for a random walk is given as follows:

$$\frac{dy(t)}{dt} = f\left(t, y(t)\right) + 1_1 G\left(t, y(t)\right) \tag{26.21}$$

where 1_1 represents the density of randomness and G represents the random walk, which is then added to the dual-porosity model for the matrix section $(t_0 \leq t \leq t_1)$ and is represented as follows:

$$
\frac{\partial h_1\left(x_i, y_j, t_n\right)}{\partial t} = \frac{T_1}{S_1}\left(\frac{\partial^2 h_1\left(x_i, y_j, t_n\right)}{\partial x^2} + \frac{\partial^2 h_1\left(x_i, y_j, t_n\right)}{\partial y^2}\right)
$$
$$
- \pi a^2 \left(h_1\left(x_i, y_j, t_n\right) - h_2\left(x_i, y_j, t_n\right)\right) + 1_1 h_1 B_1\left(x_i, y_j, t_n\right) \tag{26.22}
$$

$$
h_1\left(x_i, y_j, t_n\right) - h_1\left(x_i, y_j, t_n\right)
$$
$$
= \int_0^t \frac{T_1}{S_1}\left(\frac{\partial^2 h_1\left(x_i, y_j, t_n\right)}{\partial x^2} + \frac{\partial^2 h_1\left(x_i, y_j, t_n\right)}{\partial y^2}\right)
$$
$$
- \pi a^2 \left(h_1\left(x_i, y_j, t_n\right) - h_2\left(x_i, y_j, t_n\right) + 1_1 h_1 dB_1\right)\left(x_i, y_j, t_n\right) \tag{26.23}
$$

At the point (x_i, y_j, t_{n+1}) and (x_i, y_j, t_n)

$$
h_1\left(x_i, y_j, t_{n+1}\right) - h_1\left(x_i, y_j, t_n\right)
$$
$$
= \int_{t_n}^{t_{n+1}} \frac{T_1}{S_1}\left(\frac{\partial^2 h_1\left(x_i, y_j, t_n\right)}{\partial x^2} + \frac{\partial^2 h_1\left(x_i, y_j, t_n\right)}{\partial y^2}\right)
$$
$$
- \pi a^2 \left(h_1\left(x_i, y_j, t_n\right) - h_2\left(x_i, y_j, t_n\right)\right) d\tau + \int_{t_n}^{t_{n+1}} \sigma_1 h_1 dB_1\left(x_i, y_j, \tau\right) \tag{26.24}
$$

We have shown the discretization of the first integral.

$$\int_{t_n}^{t_{n+1}} 1_1 h_1(x_i, y_j, \tau), dB_1(\tau) \tag{26.25}$$

We assume that $B_1(\tau)$ is differentiable:

$$dB(\tau) = B'(\tau) d\tau$$

Then Equation (26.25) becomes:

$$\int_{t_n}^{t_{n+1}} 1_1 h_1(x_i, y_j, \tau) B_1'(\tau) d\tau \tag{26.26}$$

Again, we approximate:

$$h_1(x_i, y_j, \tau) B_1'(\tau) = p_j'(\tau) \tag{26.27}$$

where $p_j'(\tau)$ is the Newton two-step polynomial.

$$1_1 \int_{t_n}^{t_{n+1}} p_j'(\tau) d\tau = \frac{5}{12} 1_1 \Delta t \, h_1(x_i, y_j, t_{n-2}) B_1'(t_{n-2})$$
$$- \frac{4}{3} 1_1 \Delta t h_1(x_i, y_j, t_{n-1}) B_1'(t_{n-1}) - \frac{23}{12} 1_1 \Delta t h_1(x_i, y_j, t_n) B_1'(t_n) \tag{26.28}$$

where:

$$B'_1(t_{n-2}) = \frac{B_1(t_{n-1}) - B_1(t_{n-2})}{\Delta t} \tag{26.29}$$

$$B'_1(t_{n-1}) = \frac{B_1(t_n) - B_1(t_{n-1})}{\Delta t} \tag{26.30}$$

$$B'_1(t_n) = \frac{B_1(t_{n+1}) - B_1(t_n)}{\Delta t} \tag{26.31}$$

$$1_1 \int_{t_n}^{t_{n+1}} h_1(x_i, y_j, \tau) B_1'(\tau) d\tau = \frac{5}{12} 1_1 \Delta t h_1(x_i, y_j, t_{n-2}) \frac{B_1(t_{n-1}) - B_1(t_{n-2})}{\Delta t}$$
$$- \frac{4}{3} 1_1 \Delta t h_1(x_i, y_j, t_{n-1}) \frac{B_1(t_n) - B_1(t_{n-1})}{\Delta t} - \frac{23}{12} 1_1 \Delta t h_1(x_i, y_j, t_n) \frac{B_1(t_{n+1}) - B_1(t_n)}{\Delta t} \tag{26.32}$$

Therefore:

$$
h_1\left(x_i, y_j, h\left(x_i, y_j, t_{n+1}\right)\right) - h_1\left(x_i, y_j, h\left(x_i, y_j, t_n\right)\right)
$$

$$
= \frac{5}{12}\left\{ \frac{T_1}{S_1} \left[\begin{array}{l} \dfrac{h_1\left(x_{i+1}, y_j, t_n\right) - 2h_1\left(x_i, y_j, t_n\right) + h_1\left(x_{i-1}, y_j, t_n\right)}{(\Delta x)^2} \\ + \dfrac{h_1\left(x_i, y_{j+1}, t_n\right) - 2h_1\left(x_i, y_j, t_n\right) + h_1\left(x_i, y_{j-1}, t_n\right)}{(\Delta y)^2} \end{array} \right] \right.
$$
$$
\left. -\pi a^2\left(h_1\left(x_i, y_j, t_n\right) - h_2\left(x_i, y_j, t_n\right)\right) + 1_1 h_1 B_1\left(x_i, y_j, t_n\right) \right\}
$$

$$
\Delta t - \frac{4}{3}\left\{ \frac{T_1}{S_1} \left[\begin{array}{l} \dfrac{h_1\left(x_{i+1}, y_j, t_{n-1}\right) - 2h_1\left(x_i, y_j, t_{n-1}\right) + h_1\left(x_{i-1}, y_j, t_{n-1}\right)}{(\Delta x)^2} \\ + \dfrac{h_1\left(x_i, y_{j+1}, t_{n-1}\right) - 2h_1\left(x_i, y_j, t_{n-1}\right) + h_1\left(x_i, y_{j-1}, t_{n-1}\right)}{(\Delta y)^2} \end{array} \right] \right.
$$
$$
\left. -\pi a^2\left(h_1\left(x_i, y_j, t_{n-1}\right) - h_2\left(x_i, y_j, t_{n-1}\right)\right) + 1_1 h_1 B_1\left(x_i, y_j, t_{n-1}\right) \right\}
$$

$$
+ \frac{23}{12}\left\{ \frac{T_1}{S_1} \left[\begin{array}{l} \dfrac{h_1\left(x_{i+1}, y_j, t_{n-2}\right) - 2h_1\left(x_i, y_j, t_{n-2}\right) + h_1\left(x_{i-1}, y_j, t_{n-2}\right)}{(\Delta x)^2} + \\ \dfrac{h_1\left(x_i, y_{j+1}, t_{n-2}\right) - 2h_1\left(x_i, y_j, t_{n-2}\right) + h_1\left(x_i, y_{j-1}, t_{n-2}\right)}{(\Delta y)^2} \end{array} \right] \right.
$$
$$
\left. -\pi a^2\left(h_1\left(x_i, y_j, t_{n-2}\right) - h_2\left(x_i, y_j, t_{n-2}\right)\right) + 1_1 h_1 B_1\left(x_i, y_j, t_{n-2}\right) \right\}
$$

$$
+ \left(\begin{array}{l} \dfrac{5}{12} 1_1 \Delta t\, h_1\left(x_i, y_j, t_{n-2}\right) \dfrac{B_1\left(t_{n-1}\right) - B_1\left(t_{n-2}\right)}{\Delta t} - \dfrac{4}{3} 1_1 \Delta t h_1\left(x_i, y_j, t_{n-1}\right) \dfrac{B_1\left(t_n\right) - B_1\left(t_{n-1}\right)}{\Delta t} \\ - \dfrac{23}{12} 1_1 \Delta t h_1\left(x_i, y_j, t_n\right) \dfrac{B_1\left(t_{n+1}\right) - B_1\left(t_n\right)}{\Delta t} \end{array} \right) \tag{26.33}
$$

In the following, the random walk is then introduced to the dual-porosity model for the fractured section ($t_1 < t < T$) and represented as follows:

$$
\frac{\partial h_2\left(x_i, y_j, t_n\right)}{\partial t} = \frac{T_2}{S_2}\left(\frac{\partial^2 h_2\left(x_i, y_j, t_n\right)}{\partial x^2} + \frac{\partial^2 h_2\left(x_i, y_j, t_n\right)}{\partial y^2} \right)
$$
$$
-\pi a^2\left(h_1\left(x_i, y_j, t_n\right) - h_2\left(x_i, y_j, t_n\right)\right) + 1_2 h_2 B_2\left(x_i, y_j, t_n\right) \tag{26.34}
$$

$$
h_2\left(x_i, y_j, t_n\right) - h_2\left(x_i, y_j, t_n\right) = \int_0^t \frac{T_2}{S_2}\left(\frac{\partial^2 h_2\left(x_i, y_j, t_n\right)}{\partial x^2} + \frac{\partial^2 h_2\left(x_i, y_j, t_n\right)}{\partial y^2} \right)
$$
$$
-\pi a^2\left(h_1\left(x_i, y_j, t_n\right) - h_2\left(x_i, y_j, t_n\right)\right) + 1_2 h_2\, dB_2\left(x_i, y_j, t_n\right) \tag{26.35}
$$

At the point (x_i, y_j, t_{n+1}) and (x_i, y_j, t_n)

$$
h_2\left(x_i, y_j, t_{n+1}\right) - h_2\left(x_i, y_j, t_n\right) = \int_{t_n}^{t_{n+1}} \frac{T_2}{S_2}\left(\frac{\partial^2 h_2\left(x_i, y_j, t_n\right)}{\partial x^2} + \frac{\partial^2 h_2\left(x_i, y_j, t_n\right)}{\partial y^2} \right)
$$
$$
-\pi a^2\left(h_1\left(x_i, y_j, t_n\right) - h_2\left(x_i, y_j, t_n\right)\right) + \int_{t_n}^{t_{n+1}} 1_2 h_2\, dB_2\left(x_i, y_j, \tau\right) \tag{26.36}
$$

We have shown the discretization of the first integral.

$$\int_{t_n}^{t_{n+1}} 1_2, h_2\left(x_i, y_j, \tau\right) B_2\left(\tau\right) \tag{26.37}$$

We assume that $B_1(\tau)$ is:

$$dB\left(\tau\right) = B'\left(\tau\right) d\tau$$

Then Equation (26.37)) becomes:

$$\int_{t_n}^{t_{n+1}} 1_2 h_2\left(x_i, y_j, \tau\right) B'_2\left(\tau\right) d\tau \tag{26.38}$$

Again, we approximate

$$h_2\left(x_i, y_j, \tau\right) B'_2\left(\tau\right) = p'_j\left(\tau\right) \tag{26.39}$$

where $p'_j\left(\tau\right)$ represents the Newton two-step polynomial.

$$1_2 \int_{t_n}^{t_{n+1}} p'_j\left(\tau\right) d\tau = \frac{5}{12} 1_2 \Delta t \, h_2\left(x_i, y_j, t_{n-2}\right) B'_2\left(t_{n-2}\right)$$

$$-\frac{4}{3} 1_2 \Delta t h_2\left(x_i, y_j, t_{n-1}\right) B'_2\left(t_{n-1}\right) - \frac{23}{12} 1_2 \Delta t h_2\left(x_i, y_j, t_n\right) B'_2\left(t_n\right) \tag{26.40}$$

where:

$$B'_2\left(t_{n-2}\right) = \frac{B_2\left(t_{n-1}\right) - B_2\left(t_{n-2}\right)}{\Delta t} \tag{26.41}$$

$$B'_2\left(t_{n-1}\right) = \frac{B_2\left(t_n\right) - B_2\left(t_{n-1}\right)}{\Delta t} \tag{26.42}$$

$$B'_2\left(t_n\right) = \frac{B_2\left(t_{n+1}\right) - B_2\left(t_n\right)}{\Delta t} \tag{26.43}$$

This is further represented as:

$$1_2 \int_{t_n}^{t_{n+1}} h_2\left(x_i, y_j, \tau\right) B'_2\left(\tau\right) d\tau = \frac{5}{12} 1_2 \Delta t \, h_2\left(x_i, y_j, t_{n-2}\right) \frac{B_2\left(t_{n-1}\right) - B_2\left(t_{n-2}\right)}{\Delta t}$$

$$-\frac{4}{3} 1_2 \Delta t h_2\left(x_i, y_j, t_{n-1}\right) \frac{B_2\left(t_n\right) - B_2\left(t_{n-1}\right)}{\Delta t} - \frac{23}{12} 1_2 \Delta t h_2\left(x_i, y_j, t_n\right) \frac{B_2\left(t_{n+1}\right) - B_2\left(t_n\right)}{\Delta t} \tag{26.44}$$

Therefore:

$$
\begin{aligned}
&h_2\big(x_i, y_j, h(x_i, y_j, t_{n+1})\big) - h_2\big(x_i, y_j, h(x_i, y_j, t_n)\big) \\
&= \frac{5}{12}\left\{
\begin{array}{l}
\dfrac{T_2}{S_2}\left(
\begin{array}{l}
\dfrac{h_2(x_{i+1}, y_j, t_n) - 2h_2(x_i, y_j, t_n) + h_2(x_{i-1}, y_j, t_n)}{(\Delta x)^2} \\
+ \dfrac{h_2(x_i, y_{j+1}, t_n) - 2h_2(x_i, y_j, t_n) + h_2(x_i, y_{j-1}, t_n)}{(\Delta y)^2}
\end{array}
\right) \\
-\pi a^2 \big(h_1(x_i, y_j, t_n) - h_2(x_i, y_j, t_n)\big) + 1_2, h_2, B_2(x_i, y_j, t_n)
\end{array}
\right\} \\[4pt]
&\Delta t - \frac{4}{3}\left\{
\begin{array}{l}
\dfrac{T_2}{S_2}\left(
\begin{array}{l}
\dfrac{h_2(x_{i+1}, y_j, t_{n-1}) - 2h_2(x_i, y_j, t_{n-1}) + h_2(x_{i-1}, y_j, t_{n-1})}{(\Delta x)^2} \\
+ \dfrac{h_2(x_i, y_{j+1}, t_{n-1}) - 2h_2(x_i, y_j, t_{n-1}) + h_2(x_i, y_{j-1}, t_{n-1})}{(\Delta y)^2}
\end{array}
\right) \\
-\pi a^2 \big(h_1(x_i, y_j, t_{n-1}) - h_2(x_i, y_j, t_{n-1})\big) + 1_2, h_2, B_2(x_i, y_j, t_{n-1})
\end{array}
\right\} \\[4pt]
&+ \frac{23}{12}\left\{
\begin{array}{l}
\dfrac{T_2}{S_2}\left(
\begin{array}{l}
\dfrac{h_2(x_{i+1}, y_j, t_{n-2}) - 2h_2(x_i, y_j, t_{n-2}) + h_2(x_{i-1}, y_j, t_{n-2})}{(\Delta x)^2} \\
+ \dfrac{h_2(x_i, y_{j+1}, t_{n-2}) - 2h_2(x_i, y_j, t_{n-2}) + h_2(x_i, y_{j-1}, t_{n-2})}{(\Lambda y)^2}
\end{array}
\right) \\
-\pi a^2 \big(h_1(x_i, y_j, t_{n-2}) - h_2(x_i, y_j, t_{n-2})\big) + 1_2, h_2, B_2(x_i, y_j, t_{n-2})
\end{array}
\right\} \\[4pt]
&+ \left(
\begin{array}{l}
\dfrac{5}{12} 1_2 \Delta t h_2(x_i, y_j, t_{n-2}) \dfrac{B_2(t_{n-1}) - B_2(t_{n-2})}{\Delta t} \\
-\dfrac{4}{3} 1_2 \Delta t h_2(x_i, y_j, t_{n-1}) \dfrac{B_2(t_n) - B_2(t_{n-1})}{\Delta t} \\
-\dfrac{23}{12} 1_2 \Delta t h_2(x_i, y_j, t_n) \dfrac{B_2(t_{n+1}) - B_2(t_n)}{\Delta t}
\end{array}
\right).
\end{aligned}
$$

$$\tag{26.45}$$

26.4 APPLICATION OF CAPUTO–FABRIZIO AND CAPUTO FRACTIONAL DERIVATIVES TO THE PIECEWISE MODEL

We now consider the effect of crossover, where the first section of the aquifer is considered as having fading memory flow while the second followed the power law process. In order to include this into a mathematical formulation, in the first equation the classical time differentiation is replaced by the Caputo–Fabrizio fractional derivative, while in the second equation the time differential operator is replaced by the Caputo fractional derivative. This section will present some of these definitions.

26.4.1 APPLICATION OF CAPUTO–FABRIZIO AND CAPUTO DERIVATIVE

The Caputo derivative is given as follows:

$$
{}_o^c D_t^\alpha f(t) = \frac{1}{\Gamma(1-\alpha)} \int_0^t \frac{d}{d\tau} f(\tau)(t-\tau)^{-\alpha}\, d\tau
$$

$$\tag{26.46}$$

where $0 < \alpha \leq 1$

The Caputo–Fabrizio fraction of a derivative is given as

$$_o^{cF}D_t^\alpha f(t) = \frac{M(\alpha)}{(1-\alpha)} \int_0^t \frac{d}{d\tau} f(\tau) \exp\left[-\frac{\alpha}{1-\alpha}(t-\tau)\right] d\tau \tag{26.47}$$

The associated integrals are given below as follows:
Riemann–Liouville integral:

$$_o^{RL}D_t^\alpha f(t) = \frac{1}{\Gamma(\alpha)} \int_0^t f(t)(t-\tau)^{\alpha-1} d\tau \tag{26.48}$$

Caputo–Fabrizio integral:

$$_o^{CF}D_t^\alpha f(t) = \frac{1-\alpha}{M(\alpha)} f(t) + \frac{\alpha}{M(\alpha)} \int_0^t f(\tau) d\tau \tag{26.49}$$

To include in the mathematical equation the effect of fading memory and the power law, we modify our system as follows with the application of Caputo–Fabrizio:

$$\frac{M(\alpha)}{1-\alpha} \int_0^t \frac{\partial h_1(x,y,\tau)}{\partial \tau} \exp\left[-\frac{\alpha}{1-\alpha}(t-\tau)\right] d\tau = \frac{T_1}{S_1}\left[\frac{\partial^2 h_1(x,y,t)}{\partial x^2} + \frac{\partial^2 h_1(x,y,t)}{\partial y^2}\right]$$
$$- \pi a^2\left(h_1(x,y,t) - h_2(x,y,t)\right) \tag{26.50}$$

With the application of Caputo:

$$\frac{1}{\Gamma(1-\alpha)} \int_0^t \frac{\partial h_2(x,y,\tau)}{\partial \tau}(t-\tau)^{\alpha-1} d\tau = \frac{T_2}{S_2}\left[\frac{\partial^2 h_2(x,y,t)}{\partial x^2} + \frac{\partial^2 h_2(x,y,t)}{\partial y^2}\right]$$
$$+ \pi a^2\left(h_1(x,y,t) - h_2(x,y,t)\right) \tag{26.51}$$

$$_o^{CF}D_t^\alpha h(x,y,t) = \frac{M(\alpha)}{1-\alpha} \int_0^t \frac{\partial h}{\partial \tau}(x,y,\tau)\exp\left[-\frac{\alpha}{1-\alpha}(t-\tau)\right] d\tau \tag{26.52}$$

Consider $t = t_{n+1} = \Delta t(n+1)$

$$_o^{CF}D_t^\alpha h(x_i, y_i, t_{n+1}) = \frac{M(\alpha)}{1-\alpha} \int_0^{n+1} \frac{\partial h}{\partial \tau}(x_i, y_i, \tau)\exp\left[-\frac{\alpha}{1-\alpha}(t_{n+1}-\tau)\right] d\tau$$
$$= \frac{M(\alpha)}{1-\alpha} \sum_{j=0}^n \int_{t_j}^{t_{j+1}} \frac{h(x_i, y_j, t_{j+1}) - h(x_i, y_j, t_j)}{\Delta t}\exp\left[-\frac{\alpha}{1-\alpha}(t_{n+1}-\tau)\right] d\tau \tag{26.53}$$

$$
{}^{CF}_{o}D^{\alpha}_t h\left(x_i, y_i, t_{n+1}\right) = \frac{M(\alpha)}{1-\alpha} \int_0^{n+1} \frac{\partial h}{\partial \tau}\left(x_i, y_i, \tau\right) \exp\left[-\frac{\alpha}{1-\alpha}\left(t_{n+1}-\tau\right)\right] d\tau
$$

$$
= \frac{M(\alpha)}{1-\alpha} \sum_{j=0}^{n} \frac{h\left(x_i, y_j, t_{j+1}\right) - h\left(x_i, y_j, t_j\right)}{\Delta t} \int_{t_j}^{t_{j+1}} \exp\left[-\frac{\alpha}{1-\alpha}\left(t_{n+1}-\tau\right)\right] d\tau \tag{26.54}
$$

$$
\int_{t_j}^{t_{j+1}} \exp\left[-\frac{\alpha}{1-\alpha}\left(t_{n+1}-\tau\right)\right] d\tau \tag{26.55}
$$

Let $\gamma = -\dfrac{\alpha}{1-\alpha}, y = t_{n+1} - d\tau, dy = d\left(t_{n+1}-\tau\right) = dt_{n+1} - d\tau = -d\tau$

Let $\tau = t_j, y = t_{n+1} - t_j$

And $\tau = t_{j+1}, y = t_{n+1} - t_{j+1}$

$$
\int_{t_{n+1}-t_j}^{t_{n+1}-t_{j+1}} -\exp\left[-\gamma y\right] dy \tag{26.56}
$$

Therefore:

$$
= \int_{t_{n+1}-t_j}^{t_{n+1}-t_{j+1}} \frac{-x}{-\gamma} \exp\left[-\gamma y\right] dy \tag{26.57}
$$

$$
\frac{1}{\gamma} \int_{t_{n+1}-t_j}^{t_{n+1}-t_{j+1}} -\gamma \exp\left[-\gamma y\right] dy \tag{26.58}
$$

$$
\left(\exp(-\gamma y)\right)' = -\gamma \exp(-\gamma y) \tag{26.59}
$$

$$
-\frac{1}{\gamma} \exp\left[-\gamma y\right]_{t_{n+1}-t_j}^{t_{j+1}-t_{j+1}} = \frac{1}{\gamma}\left[\exp\left(-\gamma\left(t_{n+1}-t_j\right)\right) - \exp\left(-\gamma\left(t_{n+1}-t_{j+1}\right)\right)\right] \tag{26.60}
$$

$$
= \frac{1}{\gamma}\left[\exp\left(-\gamma\left(\Delta t(n+1) - \Delta t j\right)\right)\right] - \exp\left[-\gamma\left(\Delta t(n+1) - \Delta t(j+1)\right)\right] \tag{26.61}
$$

$$
= \frac{1}{\gamma}\left[\exp\left(-\gamma\Delta t(n+1-j)\right)\right] - \exp\left[-\gamma\Delta t(n+j)\right] \tag{26.62}
$$

$$
= \frac{1-\alpha}{\alpha}\left[\exp\left[-\frac{\alpha}{1-\alpha}\Delta t(n-j)\right] + \exp\left[-\frac{\alpha}{1-\alpha}\Delta t(n+1-j)\right]\right] \tag{26.63}
$$

With the Caputo derivative, we have:

$$
{}^{c}_{o}D^{\alpha}_t h_2\left(x, y, t\right) = \frac{1}{\Gamma(1-\alpha)} \int_0^t \frac{\partial}{\partial \tau} h_2\left(x, y, t\right)\left(t-\tau\right)^{-\alpha} d\tau \tag{26.64}
$$

at the point (x_i, y_j, t_{n+1}). The above yields:

$$
{}_o^c D_t^\alpha h_2\left(x_i, y_j, t_{n+1}\right) = \frac{1}{\Gamma(1-\alpha)} \int_0^{t_{n+1}} \frac{\partial}{\partial \tau} h_2\left(x_i, y_j, \tau\right)\left(t_{n+1}-\tau\right)^{-\alpha} d\tau
$$

$$
= \frac{1}{\Gamma(1-\alpha)} \sum_{k=0}^n \int_{t_k}^{t_{k+1}} \frac{h_2\left(x_i, y_j, t_{k+1}\right)-h_2\left(x_i, y_j, t_k\right)}{\Delta t}\left(t_{n+1}-\tau\right)^{-\alpha} d\tau
$$

$$
= \frac{1}{\Gamma(1-\alpha)} \sum_{k=0}^n \frac{h_2\left(x_i, y_j, t_{k+1}\right)-h_2\left(x_i, y_j, t_k\right)}{\Delta t} \int_{t_k}^{t_{k+1}}\left(t_{n+1}-\tau\right)^{-\alpha} d\tau \quad (26.65)
$$

Here $\int_{t_k}^{t_{k+1}} \left(t_{n+1}-\tau\right)$, and we put $y = y = t_{n+1} - \tau$, $dy = -d\tau$ or $d\tau = -dy$

$$
\text{When } \tau \to t_k, y \to t_{n+1} - t_k
$$

When $\tau \to t_{k+1}, y \to t_{n+1} - t_{k+1}$.
Thus, the above integral is therefore presented as:

$$
\int_{t_{n+1}-t_k}^{t_{n+1}-t_{k+1}} y^{-\alpha} d\tau = \frac{\left(t_{n+1}-t_{k+1}\right)^{1-\alpha}-\left(t_{n+1}-t_k\right)^{1-\alpha}}{1-\alpha} \quad (26.66)
$$

Replacing this back into the original equation, we get:

$$
{}_o^c D_t^\alpha h_2\left(x_i, y_j, t_{n+1}\right) = \frac{1}{\Gamma(1-\alpha)} \sum_{k=0}^n \frac{\left(h_2\left(x_i, y_j, t_{k+1}\right)-h_2\left(x_i, y_j, t_k\right)\right)}{\Delta t}
$$

$$
\left[\frac{\left(t_{n+1}-t_k\right)^{1-\alpha}-\left(t_{n+1}-t_k\right)^{1-\alpha}}{1-\alpha}\right] = \frac{1}{\Gamma(1-\alpha)} \sum_{k=0}^n \frac{\left(h_2\left(x_i, y_j, t_{k+1}\right)-h_2\left(x_i, y_j, t_k\right)\right)}{\Delta t}
$$

$$
\left[\left(\Delta t(n+1)-k\Delta t\right)^{1-\alpha}-\left(\Delta t(n+1)-\Delta t(k+1)\right)^{1-\alpha}\right]
$$

$$
= \frac{(\Delta t)^{-\alpha}}{\Gamma(2-\alpha)} \sum_{k=0}^n \left(h_2\left(x_i, y_j, t_{k+1}\right)-h_2\left(x_i, y_j, t_k\right)\right)\left[(n+1-k)^{1-\alpha}-(n-k)^{1-\alpha}\right] \quad (26.67)
$$

With the above discretization in hand, we can now represent of the numerical of our system as:

$$
\frac{M(\alpha)}{\alpha \Delta t} \sum_{k=0}^n \left(h_1\left(x_i, y_j, t_{k+1}\right)-h_1\left(x_i, y_j, t_k\right)\right)\left\{\begin{array}{c} \exp\left[-\dfrac{\alpha}{1-\alpha}\Delta t(n+k)\right] \\[2mm] -\exp\left[-\dfrac{\alpha}{1-\alpha}\Delta t(n+1-k)\right] \end{array}\right\}
$$

$$
= \frac{T_1}{S_1}\left\{\begin{array}{c} \dfrac{h_1\left(x_{i+1}, y_j, t_{n+1}\right)-2h_1\left(x_i, y_j, t_{n+1}\right)+h_1\left(x_{i-1}, y_j, t_{n+1}\right)}{(\Delta x)^2} \\[4mm] +\dfrac{h_1\left(x_i, y_{j+1}, t_{n+1}\right)-2h_1\left(x_i, y_j, t_{n+1}\right)+h_1\left(x_i, y_{j+1}, t_{n+1}\right)}{(\Delta y)^2} \end{array}\right\}
$$

$$
-\pi a^2\left(h_1\left(x_i, y_j, t_{n+1}\right)-h_2\left(x_i, y_j, t_{n+1}\right)\right) \quad (26.68)
$$

$$\frac{(\Delta t)^{-\alpha}}{\Gamma(2-\alpha)} \sum_{k=0}^{n} \left(h_2(x_i, y_j, t_{k+1}) - h_2(x_i, y_j, t_k) \right) \left[(n+1-k)^{1-\alpha} - (n-k)^{1-\alpha} \right]$$

$$= \frac{T_2}{S_2} \left\{ \begin{array}{l} \dfrac{h_2(x_{i+1}, y_j, t_{n+1}) - 2h_2(x_i, y_j, t_{n+1}) + h_2(x_{i-1}, y_j, t_{n+1})}{(\Delta x)^2} \\ + \dfrac{h_2(x_i, y_{j+1}, t_{n+1}) - 2h_2(x_i, y_j, t_{n+1}) + h_2(x_i, y_{j+1}, t_{n+1})}{(\Delta y)^2} \end{array} \right\}$$

$$+ \pi a^2 \left(h_1(x_i, y_j, t_{n+1}) - h_2(x_i, y_j, t_{n+1}) \right) \tag{26.69}$$

$${}^{CF}_{o}D_t^\alpha h_1(x, y, t) = \frac{T_1}{S_1} \left[\frac{\partial^2 h_1}{\partial x^2} + \frac{\partial^2 h_1}{\partial y^2} \right] - \pi a^2 (h_1 - h_2) \tag{26.70}$$

We convert the above to an integral equation to obtain:

$${}^{CF}_{o}J_t^\alpha \left({}^{CF}_{o}D_t^\alpha h_1(x, y, t) \right) = {}^{CF}_{o}J_t^\alpha \left[\frac{T_1}{S_1} \left(\frac{\partial^2 h_1}{\partial x^2} + \frac{\partial^2 h_1}{\partial y^2} \right) - \pi a^2 (h_1 - h_2) \right] \tag{26.71}$$

$$h_1(x,y,t) - h_1(x,y,0) = \frac{1-\alpha}{M(\alpha)} \left[\frac{T_1}{S_1} \left(\frac{\partial^2 h_1}{\partial x^2} + \frac{\partial^2 h_1}{\partial y^2} \right) - \pi a^2 (h_1 - h_2) \right]$$

$$+ \frac{\alpha}{M(\alpha)} \int_0^t \left[\frac{T_1}{S_1} \left(\frac{\partial^2 h_1}{\partial x^2} + \frac{\partial^2 h_1}{\partial y^2} \right) - \pi a^2 (h_1 - h_2) \right] d\tau \tag{26.72}$$

For simplification we let:

$$F(t, h_1, h_2) = \frac{T_1}{S_1} \left(\frac{\partial^2 h_1}{\partial x^2} + \frac{\partial^2 h_1}{\partial y^2} \right) - \pi a^2 (h_1 - h_2) \tag{26.73}$$

$$h_1(x, y, t) - h_1(x, y, 0) = \frac{1-\alpha}{M(\alpha)} F(t, h_1, h_2) + \frac{\alpha}{M(\alpha)} \int_0^t F(t, h_1, h_2) d\tau \tag{26.74}$$

At (x_i, y_j, t_{n+1}) and (x_i, y_j, t_n) we have,

$$h(x_i, y_j, t_{n+1}) - h(x_i, y_j, 0) = \frac{1-\alpha}{M(\alpha)} F(x_i, y_j, t_n) + \frac{\alpha}{M(\alpha)} \int_0^{t_{n+1}} F(x_i, y_j, \tau) d\tau \tag{26.75}$$

$$h_1\left(x_i, y_j, t_n\right) - h_1\left(x_i, y_j, 0\right) = \frac{1-\alpha}{M(\alpha)} F\left(x_i, y_j, t_{n-1}\right) + \frac{\alpha}{M(\alpha)} \int_0^{t_n} F\left(x_i, y_j, \tau\right) d\tau \qquad (26.76)$$

Then we get,

$$h_1\left(x_i, y_j, t_{n+1}\right) - h_1\left(x_i, y_j, t_n\right) = \frac{1-\alpha}{M(\alpha)}\left[F\left(x_i, y_j, t_n\right) - F\left(x_i, y_j, t_{n-1}\right)\right] + \frac{\alpha}{M(\alpha)} \int_{t_n}^{t_{n+1}} F\left(x_i, y_j, \tau\right) d\tau \quad (26.77)$$

We apply the Newton interpolation to $F(x_i, y_j, \tau)$ with the interval $[t_n, t_{n+1}]$ and get the following expression:

$$h_1\left(x_i, y_j, t_{n+1}\right) - h_1\left(x_i, y_j, t_n\right) = \frac{1-\alpha}{M(\alpha)}\left[F\left(x_i, y_j, t_n\right) - F\left(x_i, y_j, t_{n-1}\right)\right]$$
$$+ \frac{\alpha}{M(\alpha)}\left[\frac{5}{12} \Delta t F\left(x_i, y_j, t_{n-2}\right) - \frac{4}{3} \Delta t F\left(x_i, y_j, t_{n-1}\right) + \frac{23}{12} \Delta t F\left(x_i, y_j, t_n\right)\right] \qquad (26.78)$$

where:

$$F\left(x_i, y_j, t_n\right) = \frac{T_1}{S_1} \left(\begin{array}{c} \dfrac{h_1\left(x_{i+1}, y_j, t_n\right) - 2h_1\left(x_i, y_j, t_n\right) + h_1\left(x_{i-1}, y_j, t_n\right)}{(\Delta x)^2} \\[2mm] + \dfrac{h_1\left(x_i, y_{j+1}, t_n\right) - 2h_1\left(x_i, y_j, t_n\right) + h_1\left(x_i, y_{j-1}, t_n\right)}{(\Delta y)^2} \end{array} \right)$$
$$- \pi a^2 \left(h_1\left(x_i, y_j, t_n\right) - h_2\left(x_i, y_j, t_n\right)\right) \qquad (26.79)$$

$$F\left(x_i, y_j, t_{n-1}\right) = \frac{T_1}{S_1} \left(\begin{array}{c} \dfrac{h_1\left(x_{i+1}, y_j, t_{n-1}\right) - 2h_1\left(x_i, y_j, t_{n-1}\right) + h_1\left(x_{i-1}, y_j, t_{n-1}\right)}{(\Delta x)^2} \\[2mm] + \dfrac{h_1\left(x_i, y_{j+1}, t_{n-1}\right) - 2h_1\left(x_i, y_j, t_{n-1}\right) + h_1\left(x_i, y_{j-1}, t_{n-1}\right)}{(\Delta y)^2} \end{array} \right)$$
$$- \pi a^2 \left(h_1\left(x_i, y_j, t_{n-1}\right) - h_2\left(x_i, y_j, t_{n-1}\right)\right) \qquad (26.80)$$

$$F\left(x_i, y_j, t_{n-2}\right) = \frac{T_1}{S_1} \left(\begin{array}{c} \dfrac{h_1\left(x_{i+1}, y_j, t_{n-2}\right) - 2h_1\left(x_i, y_j, t_{n-2}\right) + h_1\left(x_{i-1}, y_j, t_{n-2}\right)}{(\Delta x)^2} \\[2mm] + \dfrac{h_1\left(x_i, y_{j+1}, t_{n-2}\right) - 2h_1\left(x_i, y_j, t_{n-2}\right) + h_1\left(x_i, y_{j-1}, t_{n-2}\right)}{(\Delta y)^2} \end{array} \right)$$
$$- \pi a^2 \left(h_1\left(x_i, y_j, t_{n-2}\right) - h_2\left(x_i, y_j, t_{n-2}\right)\right) \qquad (26.81)$$

Therefore

$$
h_1\left(x_i, y_j, t_{n+1}\right) - h_1\left(x_i, y_j, t_n\right)
$$

$$
= \frac{1-\alpha}{M(\alpha)}\left\{\begin{bmatrix}\dfrac{T_1}{S_1}\begin{bmatrix}\dfrac{h_1\left(x_{i+1}, y_j, t_n\right) - 2h_1\left(x_i, y_j, t_n\right) + h_1\left(x_{i-1}, y_j, t_n\right)}{(\Delta x)^2} \\ +\dfrac{h_1\left(x_i, y_{j+1}, t_n\right) - 2h_1\left(x_i, y_j, t_n\right) + h_1\left(x_i, y_{j-1}, t_n\right)}{(\Delta y)^2}\end{bmatrix} \\ -\pi a^2\left(h_1\left(x_i, y_j, t_n\right) - h_2\left(x_i, y_j, t_n\right)\right)\end{bmatrix}\right.
$$

$$
\left. -\begin{bmatrix}\dfrac{T_1}{S_1}\begin{bmatrix}\dfrac{h_1\left(x_{i+1}, y_j, t_{n-1}\right) - 2h_1\left(x_i, y_j, t_{n-1}\right) + h_1\left(x_{i-1}, y_j, t_{n-1}\right)}{(\Delta x)^2} \\ +\dfrac{h_1\left(x_i, y_{j+1}, t_{n-1}\right) - 2h_1\left(x_i, y_j, t_{n-1}\right) + h_1\left(x_i, y_{j-1}, t_{n-1}\right)}{(\Delta y)^2}\end{bmatrix} \\ -\pi a^2\left(h_1\left(x_i, y_j, t_{n-1}\right) - h_2\left(x_i, y_j, t_{n-1}\right)\right)\end{bmatrix}\right\}
$$

$$
+\frac{\alpha}{M(\alpha)}\left\{\begin{bmatrix}\dfrac{5}{12}\Delta t\begin{bmatrix}\dfrac{T_1}{S_1}\begin{bmatrix}\dfrac{h_1\left(x_{i+1}, y_j, t_{n-2}\right) - 2h_1\left(x_i, y_j, t_{n-2}\right) + h_1\left(x_{i-1}, y_j, t_{n-2}\right)}{(\Delta x)^2} \\ +\dfrac{h_1\left(x_i, y_{j+1}, t_{n-2}\right) - 2h_1\left(x_i, y_j, t_{n-2}\right) + h_1\left(x_i, y_{j-1}, t_{n-2}\right)}{(\Delta y)^2}\end{bmatrix} \\ -\pi a^2\left(h_1\left(x_i, y_j, t_{n-2}\right) - h_2\left(x_i, y_j, t_{n-2}\right)\right)\end{bmatrix}\end{bmatrix}\right.
$$

$$
-\frac{4}{3}\Delta t\begin{bmatrix}\dfrac{T_1}{S_1}\begin{bmatrix}\dfrac{h_1\left(x_{i+1}, y_j, t_{n-1}\right) - 2h_1\left(x_i, y_j, t_{n-1}\right) + h_1\left(x_{i-1}, y_j, t_{n-1}\right)}{(\Delta x)^2} \\ +\dfrac{h_1\left(x_i, y_{j+1}, t_{n-1}\right) - 2h_1\left(x_i, y_j, t_{n-1}\right) + h_1\left(x_i, y_{j-1}, t_{n-1}\right)}{(\Delta y)^2}\end{bmatrix} \\ -\pi a^2\left(h_1\left(x_i, y_j, t_{n-1}\right) - h_2\left(x_i, y_j, t_{n-1}\right)\right)\end{bmatrix}
$$

$$
\left. +\frac{23}{12}\Delta t\begin{bmatrix}\dfrac{T_1}{S_1}\begin{bmatrix}\dfrac{h_1\left(x_{i+1}, y_j, t_n\right) - 2h_1\left(x_i, y_j, t_n\right) + h_1\left(x_{i-1}, y_j, t_n\right)}{(\Delta x)^2} \\ +\dfrac{h_1\left(x_i, y_{j+1}, t_n\right) - 2h_1\left(x_i, y_j, t_n\right) + h_1\left(x_i, y_{j-1}, t_n\right)}{(\Delta y)^2}\end{bmatrix} \\ -\pi a^2\left(h_1\left(x_i, y_j, t_n\right) - h_2\left(x_i, y_j, t_n\right)\right)\end{bmatrix}\right\} \tag{26.82}
$$

$$
{}_0^c D_t^\alpha h_2\left(x, y, t\right) = \frac{T_2}{S_2}\left[\frac{\partial^2 h_2}{\partial x^2} + \frac{\partial^2 h_2}{\partial y^2}\right] - \pi a^2\left(h_1 - h_2\right) \tag{26.83}
$$

We apply here the Riemann–Liouville integral on both sides to obtain:

$$
h_2\left(x, y, t\right) - h_2\left(x, y, 0\right) = \frac{1}{\Gamma(\alpha)}\int_0^t \frac{T_2}{S_2}\left[\frac{\partial^2 h_2}{\partial x^2} + \frac{\partial^2 h_2}{\partial y^2}\right] - \pi a^2\left(h_1 - h_2\right)\left(t - \tau\right)^{\alpha-1} d\tau \tag{26.84}
$$

$$
F_1\left(x, y, t\right) = \int_0^t \frac{T_2}{S_2}\left[\frac{\partial^2 h_2}{\partial x^2} + \frac{\partial^2 h_2}{\partial y^2}\right] - \pi a^2\left(h_1 - h_2\right)\left(t - \tau\right)^{\alpha-1} d\tau \tag{26.85}
$$

$$h_2(x, y, t) - h_2(x, y, 0) = \frac{1}{\Gamma(\alpha)} \int_0^t F_1(x, y, \tau)(t - \tau)^{-\alpha} \, d\tau \tag{26.86}$$

At (x_i, y_j, t_{n+1}) we have:

$$h_2(x_i, y_j, t_{n+1}) - h_2(x_i, y_j, 0) = \frac{1}{\Gamma(\alpha)} \sum_{k=0}^n \int_{t_k}^{t_{k+1}} (t_{n+1} - \tau)^{\alpha-1} F(x_i, y_j, \tau) \, d\tau \tag{26.87}$$

We approximate the function $F(x_i, y_j, \tau) \simeq P_k(\tau)$ where $P_k(\tau)$ is the Newton polynomial. Using the development done by Atangana and Seda, we obtain:

$$
\begin{aligned}
h_2(x_i, y_j, t_{n+1}) - h_2(x_i, y_j, 0) &= \frac{(\Delta t)^\alpha}{\Gamma(\alpha+1)} \sum_{k=0}^n F(t_{k-2}, x_i, y_j) \left\{ (n-k+1)^\alpha - (n-k)^\alpha \right\} \\
&+ \frac{(\Delta t)^\alpha}{\Gamma(2+\alpha)} \sum_{k=2}^n \left[F(t_{k-1}, x_i, y_j) - F(t_{k-2}, x_i, y_j) \right] \left\{ \begin{aligned} &(n-k+1)^\alpha (n-k+3+2\alpha) \\ &-(n-k)^\alpha (n-k+3+3\alpha) \end{aligned} \right\} \\
&+ \frac{(\Delta t)^\alpha}{2\Gamma(3+\alpha)} \left[F(t_k, x_i, y_j) - 2F(t_{k-1}, x_i, y_j) + F(t_{k-2}, x_i, y_j) \right] \\
&\left\{ \begin{aligned} &(n-k+1)^2 \left(2(n-k)^2 + (3\alpha+10)(n-j) + 2\alpha^2 + 9\alpha + 12 \right) \\ &-(n-k)^2 \left(2(n-k)^2 + (5\alpha+10)(n-k) + 6\alpha^2 + 18\alpha + 12 \right) \end{aligned} \right\}
\end{aligned}
\tag{26.88}
$$

where:

$$F(t_k, x_i, y_j) = \frac{T_2}{S_2} \left(\begin{aligned} &\frac{h_2(x_{i+1}, y_j, t_k) - 2h_2(x_i, y_j, t_k) + h_2(x_{i-1}, y_j, t_k)}{(\Delta x)^2} \\ &+ \frac{h_2(x_i, y_{j+1}, t_k) - 2h_2(x_i, y_j, t_k) + h_2(x_i, y_{j-1}, t_k)}{(\Delta y)^2} \end{aligned} \right) - \pi a^2 \left(h_1(x_i, y_j, t_k) - h_2(x_i, y_j, t_k) \right) \tag{26.89}$$

$$F(t_{k-1}, x_i, y_j) = \frac{T_2}{S_2} \left(\begin{aligned} &\frac{h_2(x_{i+1}, y_j, t_{k-1}) - 2h_2(x_i, y_j, t_{k-1}) + h_2(x_{i-1}, y_j, t_{k-1})}{(\Delta x)^2} \\ &+ \frac{h_2(x_i, y_{j+1}, t_{k-1}) - 2h_2(x_i, y_j, t_{k-1}) + h_2(x_i, y_{j-1}, t_{k-1})}{(\Delta y)^2} \end{aligned} \right) - \pi a^2 \left(h_1(x_i, y_j, t_{k-1}) - h_2(x_i, y_j, t_{k-1}) \right) \tag{26.90}$$

$$F(t_{k-2}, x_i, y_j) = \frac{T_2}{S_2} \left(\begin{aligned} &\frac{h_2(x_{i+1}, y_j, t_{k-2}) - 2h_2(x_i, y_j, t_{k-2}) + h_2(x_{i-1}, y_j, t_{k-2})}{(\Delta x)^2} \\ &+ \frac{h_2(x_i, y_{j+1}, t_{k-2}) - 2h_2(x_i, y_j, t_{k-2}) + h_2(x_i, y_{j-1}, t_{k-2})}{(\Delta y)^2} \end{aligned} \right) - \pi a^2 \left(h_1(x_i, y_j, t_{k-2}) - h_2(x_i, y_j, t_{k-2}) \right) \tag{26.91}$$

Therefore:

$$h_2(x_i, y_j, t_{n+1}) - h_2(x_i, y_j, 0)$$

$$= \frac{(\Delta t)^\alpha}{\Gamma(\alpha+1)} \sum_{k=0}^{n} \left[\frac{T_2}{S_2} \left(\begin{array}{c} \dfrac{h_2(x_{i+1}, y_j, t_{k-2}) - 2h_2(x_i, y_j, t_{k-2}) + h_2(x_{i-1}, y_j, t_{k-2})}{(\Delta x)^2} \\ + \dfrac{h_2(x_i, y_{j+1}, t_{k-2}) - 2h_2(x_i, y_j, t_{k-2}) + h_2(x_i, y_{j-1}, t_{k-2})}{(\Delta y)^2} \\ -\pi a^2 \left(h_1(x_i, y_j, t_{k-2}) - h_2(x_i, y_j, t_{k-2}) \right) \end{array} \right) \right]$$

$$\left\{ (n-k+1)^\alpha - (n-k)^\alpha \right\} + \frac{(\Delta t)^\alpha}{\Gamma(2+\alpha)}$$

$$\sum_{k=2}^{n} \left[\begin{array}{c} \left[\frac{T_2}{S_2} \left(\begin{array}{c} \dfrac{h_2(x_{i+1}, y_j, t_{k-1}) - 2h_2(x_i, y_j, t_{k-1}) + h_2(x_{i-1}, y_j, t_{k-1})}{(\Delta x)^2} \\ + \dfrac{h_2(x_i, y_{j+1}, t_{k-1}) - 2h_2(x_i, y_j, t_{k-1}) + h_2(x_i, y_{j-1}, t_{k-1})}{(\Delta y)^2} \\ -\pi a^2 \left(h_1(x_i, y_j, t_{k-1}) - h_2(x_i, y_j, t_{k-1}) \right) \end{array} \right) \right] \\ - \left[\frac{T_2}{S_2} \left(\begin{array}{c} \dfrac{h_2(x_{i+1}, y_j, t_{k-2}) - 2h_2(x_i, y_j, t_{k-2}) + h_2(x_{i-1}, y_j, t_{k-2})}{(\Delta x)^2} \\ + \dfrac{h_2(x_i, y_{j+1}, t_{k-2}) - 2h_2(x_i, y_j, t_{k-2}) + h_2(x_i, y_{j-1}, t_{k-2})}{(\Delta y)^2} \\ -\pi a^2 \left(h_1(x_i, y_j, t_{k-2}) - h_2(x_i, y_j, t_{k-2}) \right) \end{array} \right) \right] \end{array} \right]$$

$$\left\{ (n-k+1)^\alpha (h-k+3+2\alpha) - (n-k)^\alpha (n-k+3+3\alpha) \right\}$$

$$+ \frac{(\Delta t)^\alpha}{2\Gamma(3+\alpha)} \left[\begin{array}{c} \left[\frac{T_2}{S_2} \left(\begin{array}{c} \dfrac{h_2(x_{i+1}, y_j, t_k) - 2h_2(x_i, y_j, t_k) + h_2(x_{i-1}, y_j, t_k)}{(\Delta x)^2} \\ + \dfrac{h_2(x_i, y_{j+1}, t_k) - 2h_2(x_i, y_j, t_k) + h_2(x_i, y_{j-1}, t_k)}{(\Delta y)^2} \\ -\pi a^2 \left(h_1(x_i, y_j, t_k) - h_2(x_i, y_j, t_k) \right) \end{array} \right) \right] \\ -2 \left[\frac{T_2}{S_2} \left(\begin{array}{c} \dfrac{h_2(x_{i+1}, y_j, t_{k-1}) - 2h_2(x_i, y_j, t_{k-1}) + h_2(x_{i-1}, y_j, t_{k-1})}{(\Delta x)^2} \\ + \dfrac{h_2(x_i, y_{j+1}, t_{k-1}) - 2h_2(x_i, y_j, t_{k-1}) + h_2(x_i, y_{j-1}, t_{k-1})}{(\Delta y)^2} \\ -\pi a^2 \left(h_1(x_i, y_j, t_{k-1}) - h_2(x_i, y_j, t_{k-1}) \right) \end{array} \right) \right] \\ + \left[\frac{T_2}{S_2} \left(\begin{array}{c} \dfrac{h_2(x_{i+1}, y_j, t_{k-2}) - 2h_2(x_i, y_j, t_{k-2}) + h_2(x_{i-1}, y_j, t_{k-2})}{(\Delta x)^2} \\ + \dfrac{h_2(x_i, y_{j+1}, t_{k-2}) - 2h_2(x_i, y_j, t_{k-2}) + h_2(x_i, y_{j-1}, t_{k-2})}{(\Delta y)^2} \\ -\pi a^2 \left(h_1(x_i, y_j, t_{k-2}) - h_2(x_i, y_j, t_{k-2}) \right) \end{array} \right) \right] \end{array} \right]$$

$$\left\{ \begin{array}{c} (n-k+1)^2 \left(2(n-k)^2 + (3\alpha+10)(n-j) + 2\alpha^2 + 9\alpha + 12 \right) \\ -(n-k)^2 \left(2(n-k)^2 + (5\alpha+10)(n-k) + 6\alpha^2 + 18\alpha + 12 \right) \end{array} \right\}.$$

(26.92)

26.5 NUMERICAL SIMULATIONS

In previous sections, we have devoted our attention to the numerical analysis of the suggested models with different types of differential and integral operators. Our aim was to include in a mathematical formula the effect of non-localities that could be found within the geological formation. For example, we use the derivative with the generalized Mittag–Leffler function to include in the mathematical formula or the mathematical model depicting the flow within a dual-media geological formation the effect of crossover. Of course such a crossover has some limitations. For example, it was argued by Atangana and Seda that such a crossover cannot depict the passage from the power law to fading memory. Also one cannot really use this to identify in the field the time at which such a crossover will take place. We use the fractional derivative with the exponential decay law to express the fading memory effect of the flow with the geological formation. This differential is also able to account for the passage from a Gaussian process to a non-Gaussian one with a steady state. The Caputo derivative is also used, which is known to depict power-law processes due to its power-law kernel. But this power-law kernel cannot exhibit processes like fading memory or even a crossover. Atangana and Seda suggested then a new concept called piecewise derivative, a differential operator that is defined within a well-bounded interval. To perform this simulation, we used the numerical scheme based on the Newton polynomial to solve numerically the suggested models for different differential operators. The obtained numerical schemes were coded using MATLAB. We used some theoretical parameters to observe the behaviour of the model, for example, we used T1 = 200 and T2 = 400 where T1 and T2 are the transmissivity of the matrix rock and the fracture respectively. We used S1 = 0.001 and S2 = 0.0001, where S1 and S2 are the storativity of the rock matrix and fractures respectively. The numerical simulation was performed for parameter a = 0.5 for different fractional orders; here we chose the initial condition to be a constant 0.1 for both initial heads.

26.6 RESULTS AND DISCUSSION

Using the MATLAB code, we obtained the following simulations, which are depicted in the figures below. We used the model with the power-law kernel to see the effect of non-locality. The figures are obtained for different values of fractional orders. In these figures, x represents distance in kilometres and t represents time in hours. For a fractional order starting from 0.95 to 0.8, the obtained figures show high flow within fractures; this behaviour is known as long range dependency. Also within the matrix soil, we observed high flow, meaning in this case the fractional derivative is expressing the flow within a matrix rock with high transmissivity, such as sand/sandstone. This model therefore depicts flow behaviour in a sandy/sandstone aquifer. So from the mathematical equations, the function $h_1(x,t)$ and $h_2(x,t)$ represent the hydraulic heads in dolomite, in particular; the first function expresses the flow in the rock matrix and the second function expresses the flow in fractures. This behaviour has been recognized in the literature as fast flow or even super flow. Also it is shown in these figures that more water will be found within fractures than in the matrix rock. Figures 26.1 to 26.36 present contour plots and three-dimensional x–t phases for fluid flow in dual media for different fractional orders. For fractional orders ranging from 0.7 to 0.5, the obtained figures show normal flow behaviours, water flows with normal velocity within the matrix rock, and moderately high velocity within the fractures. Finally, from 0.5 to 0.2, we observed slow flow behaviour: the mathematical equations exhibit flow with low velocities, expressing the flow within a shale with limited or no transmissivity.

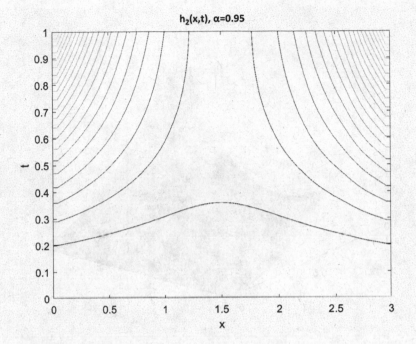

FIGURE 26.1 Contour plot representing h_2 in fractional order 0.95.

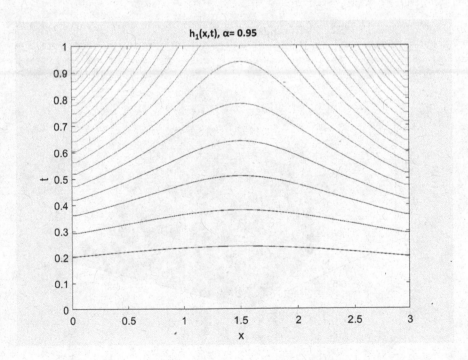

FIGURE 26.2 Contour plot representing h_1 in fractional order 0.95.

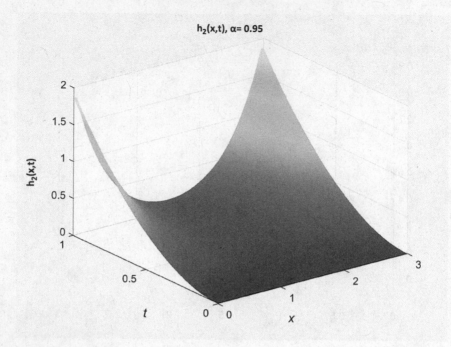

FIGURE 26.3   3D plot representing h_2 in fractional order 0.95.

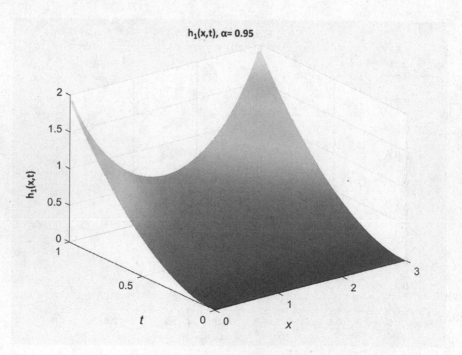

FIGURE 26.4   3D plot representing h_1 in fractional order 0.95.

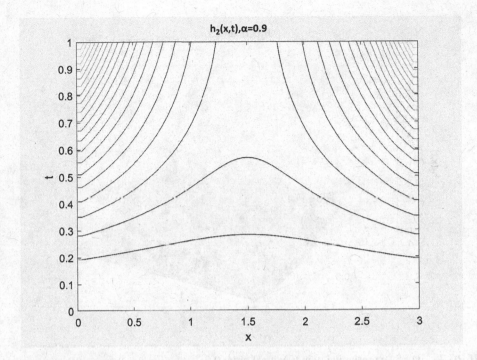

FIGURE 26.5   Contour plot representing h_2 in fractional order 0.9.

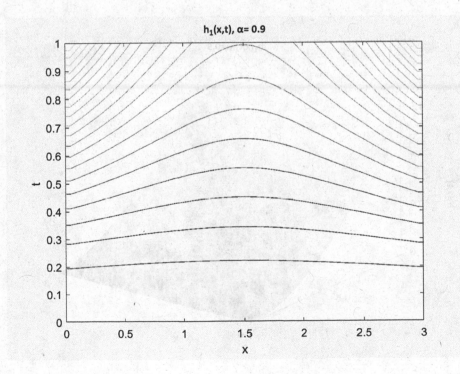

FIGURE 26.6   Contour plot representing h_1 in fractional order 0.9.

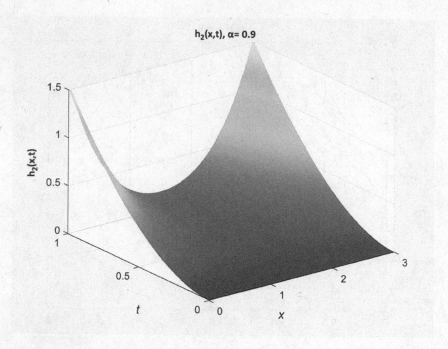

FIGURE 26.7   3D plot representing h_2 in fractional order 0.9.

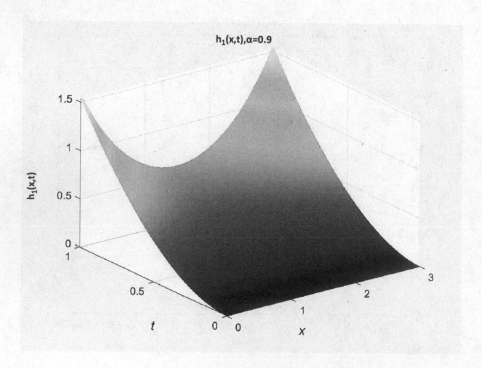

FIGURE 26.8   3D plot representing h_1 in fractional order 0.9.

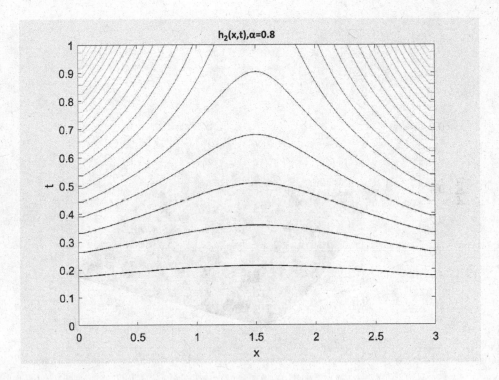

FIGURE 26.9 Contour plot representing h_2 in fractional order 0.8.

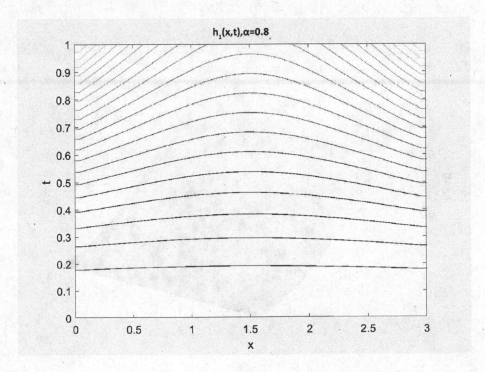

FIGURE 26.10 Contour plot representing h_1 in fractional order 0.8.

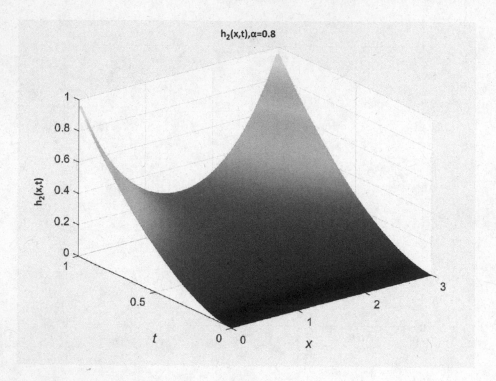

FIGURE 26.11   3D plot representing h_2 in fractional order 0.8.

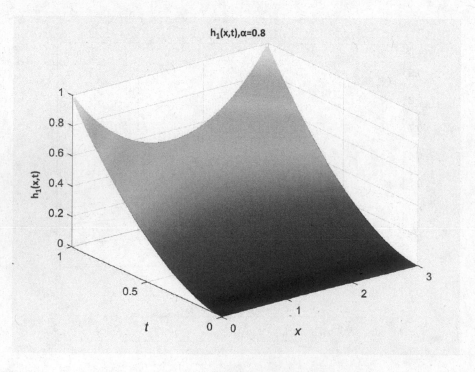

FIGURE 26.12   3D plot representing h_1 in fractional order 0.8.

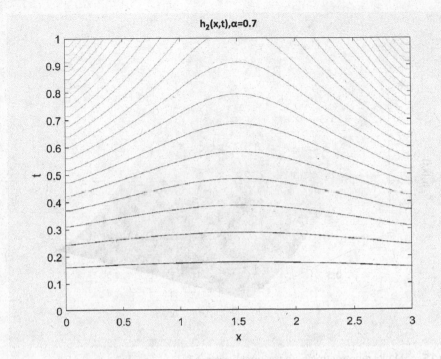

FIGURE 26.13 Contour plot representing h_2 in fractional order 0.7.

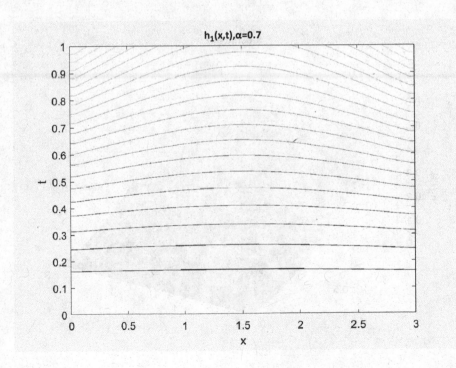

FIGURE 26.14 Contour plot representing h_1 in fractional order 0.7.

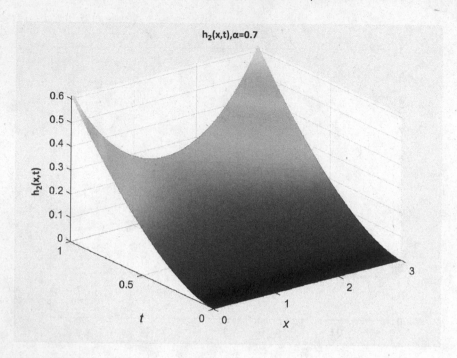

FIGURE 26.15　3D plot representing h_2 in fractional order 0.7.

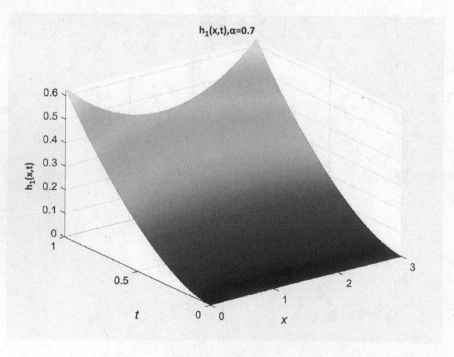

FIGURE 26.16　3D plot representing h_1 in fractional order 0.7.

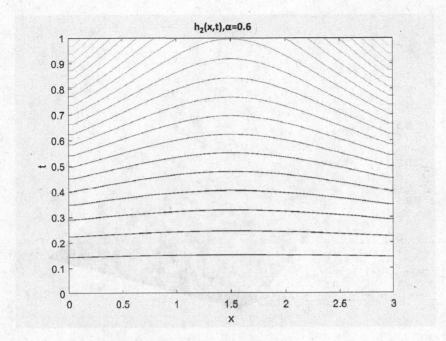

FIGURE 26.17 Contour plot representing h_2 in fractional order 0.6.

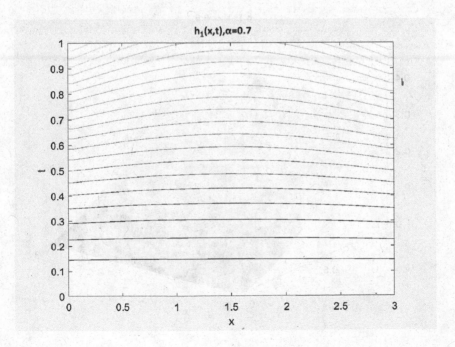

FIGURE 26.18 Contour plot representing h_1 in fractional order 0.6.

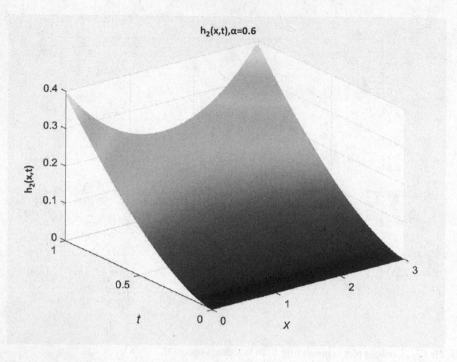

FIGURE 26.19   3D plot representing h_2 in fractional order 0.6.

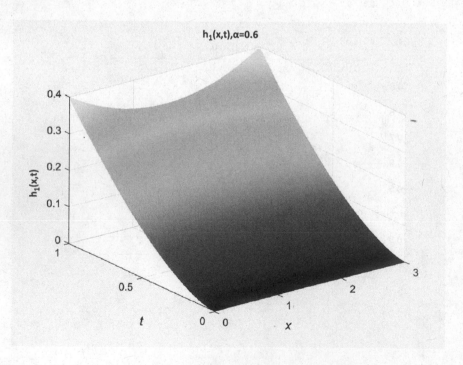

FIGURE 26.20   3D plot representing h_1 in fractional order 0.6.

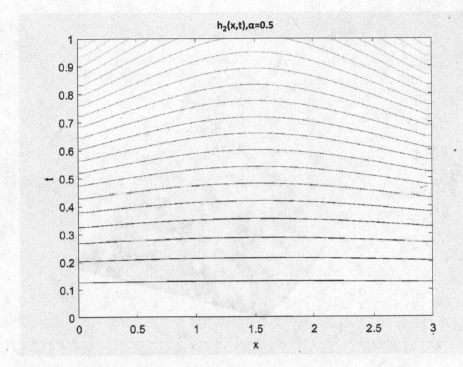

FIGURE 26.21 Contour plot representing h_2 in fractional order 0.5.

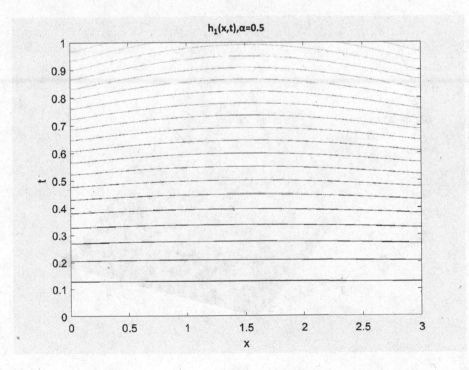

FIGURE 26.22 Contour plot representing h_1 in fractional order 0.5.

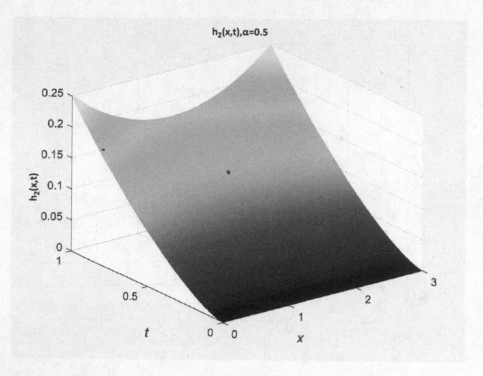

FIGURE 26.23   3D plot representing h_2 in fractional order 0.5.

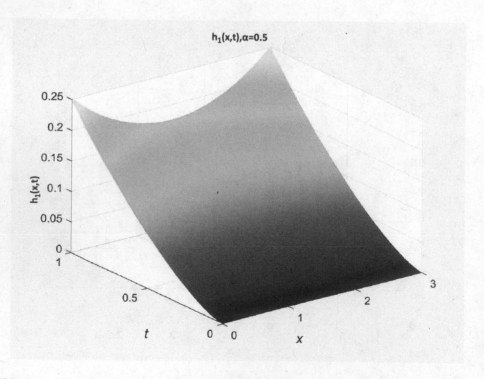

FIGURE 26.24   3D plot representing h_1 in fractional order 0.5.

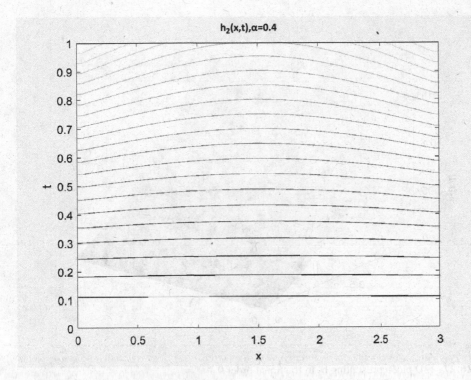

FIGURE 26.25   Contour plot representing h_2 in fractional order 0.4.

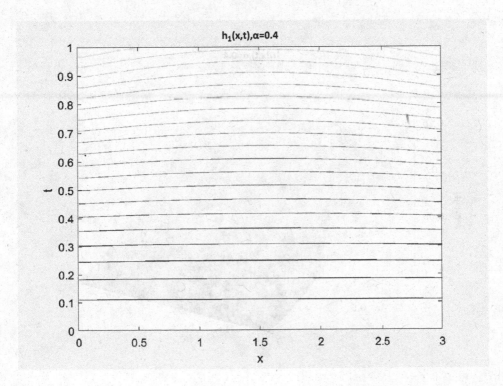

FIGURE 26.26   Contour plot representing h_1 in fractional order 0.4.

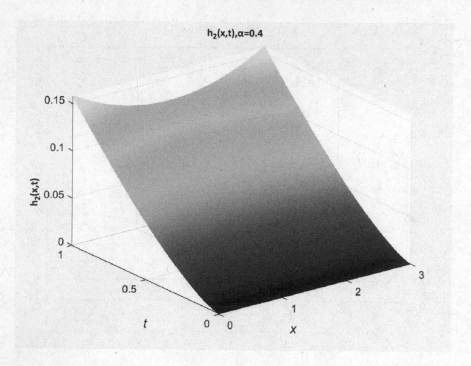

FIGURE 26.27   3D plot representing h_2 in fractional order 0.4.

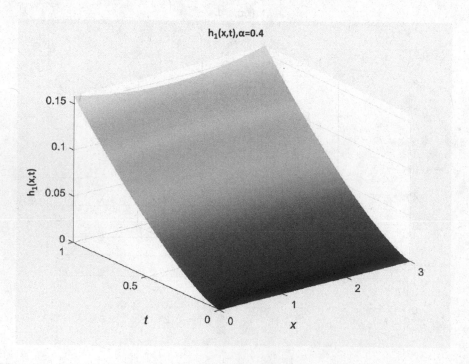

FIGURE 26.28   3D plot representing h_1 in fractional order 0.4.

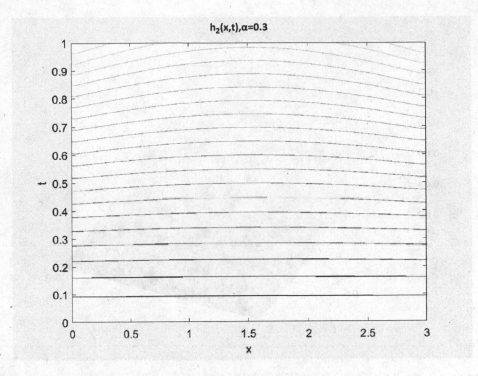

FIGURE 26.29   Contour plot representing h_2 in fractional order 0.3.

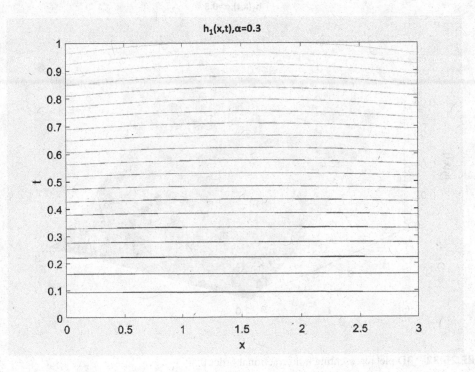

FIGURE 26.30   Contour plot representing h_1 in fractional order 0.3.

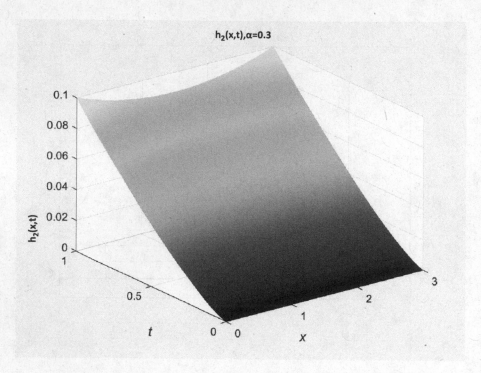

FIGURE 26.31   3D plot representing h_2 in fractional order 0.3.

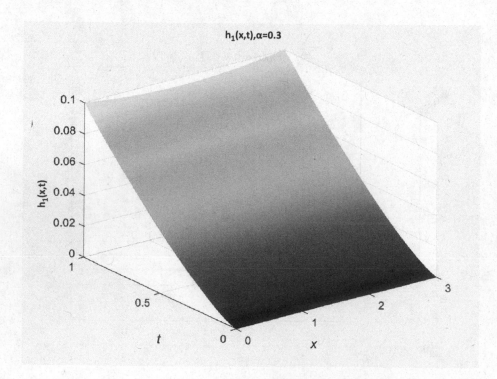

FIGURE 26.32   3D plot representing h_1 in fractional order 0.3.

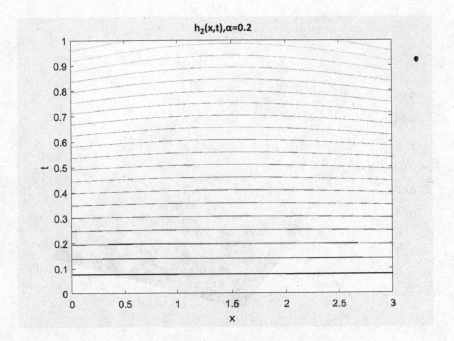

FIGURE 26.33   Contour plot representing h_2 in fractional order 0.2.

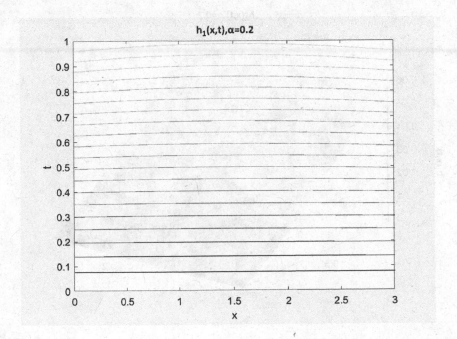

FIGURE 26.34   Contour plot representing h_1 in fractional order 0.2.

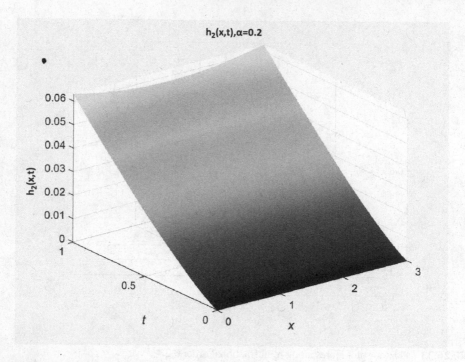

FIGURE 26.35   3D plot representing h_2 in fractional order 0.2.

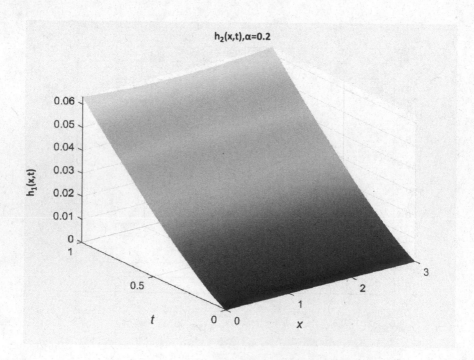

FIGURE 26.36   3D plot representing h_1 in fractional order 0.2.

26.7 CONCLUSION

Dual-porosity media have been considered in this chapter. In recent decades, many researchers have devoted their attention to modelling the flow of water within these types of aquifers. Some authors suggested the use of two hydraulic heads, where the first one accounts for the flow within the matrix rock and the second one the flow within the fractures. A system of partial differential equations has been suggested and applied in many scenarios with some limitations. The differential operators used in the past to model this situation is the classical derivative and also fractional derivatives. While the results obtained by these models were useful, one can still argue that they cannot depict with great precision the crossover effect and also the randomness of the geological formations. These properties cannot be accounted for when using the classical derivative, or even fractional derivatives. Very recently, piecewise differential and integral operators were suggested; these operators are defined in a given interval. The first part accounts for a given process and the second one for another process. The passage from one process to another is called a crossover. Additionally the concept of stochastic differential equations had been suggested to be included in the mathematical formulation of the effect of randomness. In this thesis, we modified a system of partial differential equations depicting the flow of subsurface water within a dual-porosity system. To achieve this, we considered fractional differential and integral operators with different kernels including the power law, exponential decay, and the generalized Mittag–Leffler functions. Additionally, we have included in the modified model randomness so as to obtain a piecewise fractional stochastic system of partial differential equations. We have presented an analysis to determine the conditions under which a unique system of solutions obtains. We have made use of the numerical method based on the Newton polynomial interpolation to solve numerically the obtained systems of partial differential equations. We used the software MATLAB to generate the figures in the case of the power law.

REFERENCES

Atangana, A., & Seda, İ.A. (2020). Extension of Atangana-Seda numerical method to partial differential equations with integer and non-integer order. *Alexandria Engineering Journal*, *59(4)*, 2355–2370.

Atangana, A., & Goufo, E.F.D. (2020). Cauchy problems with fractal–fractional operators and applications to groundwater dynamics. *Fractals*, 28(08), 2040043.

Batogna, G. R., & Atangana, A. (2019). Generalised class of Time Fractional Black Scholes equation and numerical analysis. *Discrete & Continuous Dynamical Systems*, 12(3), 435.

Freeze, A., & Cherry, J. (1979). *Groundwater*. Prentice-Hall, Englewood, Cliffs, New Jersey, 604.

Knaust, D., Dorador, J., & Rodrigues-Tovar, F. (2020). Burrowed matrix powering dual porosity systems– A case study from the Maastrichtian chalk of the Gullfaks Field, Norwegian North Sea. *Marine and Petroleum Geology*.

Skinner, B. J., Stephen, P. C., Jeffrey, J. P., & Harold, L. L. (2004). *Dynamic Earth: An introduction to physical geology*.

Tateishi, A. A., Ribeiro, H. V., & Lenzi, E. K. (2017). The role of fractional time-derivative operators on anomalous diffusion. *Frontiers in Physics*, 5, 52.

Toufik, M., & Atangana, A. (2017). New numerical approximation of fractional derivative with non-local and non-singular kernel: Application to chaotic models. *The European Physical Journal Plus*, 1–16.

Warren, J. E., & Root, P. J. (1963). The behavior of naturally fractured reservoirs. *Society of Petroleum Engineers Journal*, 245–255.

27 One-Dimensional Modelling of Reactive Pollutant Transport in Groundwater
The Case of Two Species

Hans Tah Mbah and Abdon Atangana
University of the Free State, Bloemfontein, South Africa

CONTENTS

27.1 INTRODUCTION

Groundwater pollution due to uncontrolled, toxic chemical effluent discharge is a leading cause of loss in water quality. The transport and fate of these materials cause their spread, which can potentially contaminate a borehole intersecting water-bearing structures. Solute spread in groundwater can be grouped as reactive or conservative (non-reactive) transport. For conservative transport, it is possible to predict the behaviour of a solute component without a detailed understanding of the species interaction with the subsurface material. However, as new chemicals are constantly being introduced in the environment, the chances of reactive transport tend to increase with the mixing rate. In this regard, it becomes vital to incorporate both sound and comprehensive scientific tools capable of predicting not just transport and redistribution, but also the chemical changes taking place and their impact on the quality of groundwater.

The modelling of chemical transport and transformation in groundwater has been largely achieved using the classical analytical advection-dispersion equation, which is based on the continuum approach. While this method has provided valuable insight into the transport behaviour of solutes, it is only valid for flow and transport in homogeneous media and thus often restricted to the interpretation of laboratory results (Genuchten & Wagenet, 1989). The use of analytical models to unpack the history of pollution transport in groundwater has been standard practice for over half a century and is widely used to this day (Bear, 1972; Chen, 1996), irrespective of the noted non-Fickian properties reported by other important contributions (Gelhar, 1993; Levy, 2003; Cortis and Berkowitz, 2004; Gouze et al., 2008). To account for the reactive component, the advection-dispersion equation has been extended in attempts to capture one or multiple chemical processes taking place in the geochemical system (Domenico, 1987; Helgeson, 1989; Chilakapati, 1995; Holstad, 2000; Yunwei, 2003; Simpson, 2015). These models have been reviewed multiple times (Ginn, 2017; Sharma, et al., 2018) and have increasingly been used as a reference for measuring the accurancy of newer and more complex modelling approaches (Gelhar, 1992; Pool, et al., 2015). Most recently, the model has been extended to better capture real-world problems by introducuing a the new Caputo–Fabrizio fractional derivative (Atangana & Alkhatani 2016; Atangana & Baleanu 2016). Analytical solutions for most reactive transport problems usually require detailed and complex mathematical manipulation; however, some standard solutions have been established (Cho, 1971) and are usually available for incorporation in new models. The challenge with obtaining analytical solutions comes with the intricacies involved in obtaining the Laplace transforms when using the inverse method. The Fourier transform has also been succesfully applied to arrive at an analytical solution for a one-dimensional reactive transport problem involving three species (Lunn, 1996).

As in analytical modelling, many numerical codes for multi-species reactive transport exist (Clement et al., 1998). These numerical codes are easy to use but often need validation by assesing them with respect to the analytical model. Simple first-order and sequential first-order reactions have been used to calibrate models and simulate reactive transport problems such as the dentrification process that produces amonia from the reaction of nitrate and nitrite (Sun et al., 1999), by using predetermined specific reaction-rate constants (Germon, 1985).

Never has it been the ambition of a single model to capture the entirety of chemical processes within a geochemical system given nature's inherent heterogeneities. The success of a model will thus depend on how reflective of the geochemical system is the mathematical model, where the rates of chemical reactions determine conversion ratios among the different solute species present. In kinetic modelling, the local equilibrium approach assumes reactive processes to be in local equilibrium with the rate of injection and conversion of reactants to products respectively. A second approach, the local chemical equilibrium, represents all reactions as rate equations, that is ordinary differential equations (ODEs). The third approach is the partial local chemical equilibrium method that describes a part of the reactions as ODEs while the others are expressed in the form of non-linear algebraic equations. Significant progress has been achieved in simulating single species solute transport in groundwater contamination problems (Bear, 1972; Wilson, 1978). Some of these solutions have been applied in different studies to assess the impact of natural plume attenuation.

In this chapter, the aim will not be to capture the effect of any chemical process within a geochemical system. Rather, a method has been developed to derive one-dimensional analytical and numerical solutions of the contributing species taking part in reactive transport involving two species, both before and after they react. The solution is based on existing single-species analytic solutions with first-order reactions (Bear, 1972; Atangana, 2013). Using this proposed method, any two single-species solutions can be extended (coupled) to describe the contribution of reactive transport to contaminant spread.

27.2 CONCEPTUAL MODEL AND MATHEMATICAL FORMULATION

For this purpose, we consider an initially pure and fractured aquifer, receiving pollution from two sources in reactive interaction with each other along zones of intense shear. We also assume that the aquifer is confined with no existing contaminant species and free from any geogenic sources. The general geometry of the hypothetical scenario is shown in Figure 27.1. The origin of the coordinate system is at the upper boundary and the positive Z-axis is downward. The bottom of the aquifer is lined by a no-flow boundary $Z = d$. The water table which separates the vadose and saturated zones is also considered a no-flow boundary. We assume that the slope of the water table is negligible and parallel with the lower boundary. The steady-state groundwater flow is along the x-axis.

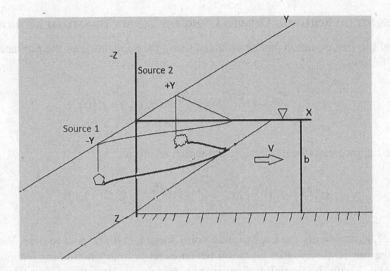

FIGURE 27.1 Section showing pollutant migration from source (U-1) along the x-direction with respect to time.

Solute transport at equilibrium can be represented for the one-dimensional system as follows:

$$
\begin{cases}
\dfrac{\partial C_A}{\partial t} - D_x \dfrac{\partial C^2 C_A}{\partial x^2} + V \dfrac{\partial C_A}{\partial x} = K_1 C_A \\[2ex]
\dfrac{\partial C_B}{\partial t} - D_x \dfrac{\partial C^2 C_B}{\partial x^2} + V \dfrac{\partial C_B}{\partial x} = K_2 C_B - K_1 C_A
\end{cases}
\qquad \text{(27.1a and b)}
$$

Initial pollution concentration:

$$
C(x,0) = c_0 = 0 \qquad (27.2)
$$

Boundary conditions:

$$
C(0,t) = c_1 \exp(-\gamma t) \, as \, t \to \infty \quad and \quad \frac{\partial C}{\partial x}(x,t) = 0 \qquad (27.3)
$$

Assume an initial pollution concentration of $c_0 = 0$ with boundary conditions as prescribed by Dirichlet.

where,

C is the solute concentration (kg/m³)

time, t, is measured in days

D_x I accounts for the coefficient of hydrodynamic dispersion in $x - direction$ (m²/day)

the groundwater flow velocity is measured as v in m/day

λ_1 is a retardation constant $\left(\dfrac{1}{\text{day}} \right)$; and

b is the aquifer's thickness (m).

Figure 27.1 A schematic diagram of two, one-dimensional source bodies within parallel non-penetrable boundaries. Both upper and lower boundaries are no-flow zones.

27.2.1 Case Study: Solution Derived Using the Laplace Transform Method

The solution of the first equation can be obtained using the application of the Laplace transform to both sides, as follows.

$$\mathcal{L}\left(\frac{\partial C_A}{\partial t} - D\frac{\partial^2 C_A}{\partial x^2} + V\frac{\partial C_A}{\partial x} + K_1 C_A\right) = \mathcal{L}(0) \tag{27.4}$$

where $\mathcal{L}$ is the Laplace transform operator, and which results as:

$$sC(x) - C(0) - D\frac{\partial^2 C_A}{\partial x^2} + V\frac{\partial C_A}{\partial x} + K_1 C(x) = 0 \tag{27.5}$$

Equation (27.4) represents the Laplace transform variable with respect to time. We can directly observe that the partial differential equation becomes an ordinary non-homogeneous differential equation of order 2 with the partial derivatives becoming total derivatives. Equation (27.5) breaks down to:

$$DC'' + VC' + (s + K_1)C = c_0 \tag{27.6}$$

The homogeneous solution associated with the above equation is given as:

$$C_h(x) = A_1 \exp(r_1 x) + A_2 \exp(r_2 x) \tag{27.7}$$

where A_1 and A_2 are arbitrary constants, and

$$r_2 = \frac{V_r \pm \sqrt{V^2 - 4D_r(s + K_1)}}{2D_r} \tag{27.8}$$

The particular solution associated with the given equation is given by:

$$C_q(x) = A_1 e^{(r_1 x)} + A_2 e^{(r_2 x)} \tag{27.9}$$

where $A_1(x)$ and $A_2(x)$ are functions satisfying the following system of equations:

$$\begin{cases} A_1'(x)\exp(r_1, x) + A_2'(x)\exp(r_2, x) = 0 \\ r_1 A'(x)\exp(r_1, x) + r_2 A_2'(x)\exp(r_2, x) = g(x) \end{cases} \tag{27.10, 27.11}$$

Hence the particular solution is given by:

$$C_p(x) = \frac{c_0}{s + K_1} \tag{27.12}$$

The general solution for the ordinary non-homogeneous differential equation is given by:

$$C(x) = A_1 \exp(r_1, x) + A_2 \exp(r_2, x) + \frac{c_0}{s + K_1} \tag{27.13}$$

We can obtain the coefficients A_1 and A_2 using our initial and boundary conditions in the Laplace space as follows:

$$\mathcal{L}\big(c(0,t)\big) = \frac{c_1}{\gamma + K_1} \tag{27.14}$$

Applying the following boundary condition:

$$\lim_{x \to \infty} c(x) \to 0$$

$$A_2 = 0 \tag{27.15}$$

Going forward, using the initial condition (27.14) gives:

$$A_1 = -\frac{c_0}{K_1 + s} + \frac{c_1}{\gamma + s} \tag{27.16}$$

But

$$c_0 = 0$$

So:

$$A_1 = \frac{c_1}{\gamma + s}$$

The solution of the ordinary differential equation follows, thus:

$$\frac{c_0}{K_1 + s} + \exp\left(x \frac{V_r - \sqrt{V_r^2 - 4D_r(s + K_1)}}{2D_r} \right)\left[\frac{c_1}{K_1 + s} - \frac{c_0}{\gamma + s} \right]$$

$$= \exp\left(x \frac{V_r - \sqrt{V_r^2 - 4D_r(s + K_1)}}{2D_r} \right) \quad \text{when} \quad c_0 = 0. \tag{27.17}$$

We know that the solution of the hydrodynamic dispersion equation is given as:

$$c(x,t) = \mathcal{L}^{-1}\big(c(x)\big) \tag{27.18}$$

where $\mathcal{L}^{-1}$ is the inverse Laplace transform operator.

The first term of the sum is invertible and gives:

$$\mathcal{L}^{-1}\left(\frac{c_0}{K_1 + s} \right) = c_0 \exp(-K_1 t) \tag{27.19}$$

The inverse Laplace transform of the second term can be obtained by applying the convolution theorem since we have a product of two functions, namely:

$$\exp\left(x \frac{V_r - \sqrt{V_r^2 - 4D_r(s + K_1)}}{2D_r} \right) \tag{27.20}$$

For the second expression, the inverse Laplace transform reads as:

$$\mathcal{L}^{-1}\left(\exp\left(x\frac{V_r-\sqrt{V_r^2-4D_r(s+K_1)}}{2D_r}\right)\right) = \frac{x\exp\left(-\frac{(-Vt+x)^2-tK_1}{4D_r}\right)}{\sqrt[2]{(\pi|D_r)}t^{3/2}} = f(t) \qquad (27.21)$$

$$\mathcal{L}^{-1}\left\{\frac{c_1}{K_1+s}-\frac{c_0}{\gamma+s}\right\} = c_1\exp(-\gamma t)-c_0\exp(-\gamma t)$$

The Laplace of the second term of the sum can be expressed as:

$$\int_0^t f(x)(t-x)dx \qquad (27.22)$$

Using the table of integrals we find that the above integral yields to:

$$\frac{c_1\exp(-\gamma t)}{2}\left\{\exp\left(x\frac{V_r-u_r}{2D_r}\right)erfc\left(\frac{x-u_rt}{2\sqrt{D_rt}}\right)+\exp\left(x\frac{V_r+u_r}{2D_r}\right)erfc\left(\frac{x+u_rt}{2\sqrt{D_rt}}\right)\right\}$$
$$-\frac{c_0\exp(-\gamma t)}{2}\left\{erfc\left(\frac{x-u_rt}{2\sqrt{D_rt}}\right)+\exp\left(x\frac{q_r}{2D_r}\right)erfc\left(\frac{x+u_rt}{2\sqrt{D_rt}}\right)\right\} \qquad (27.23)$$

$$\frac{c_1\exp(-\gamma t)}{2}\left\{\exp\left(x\frac{V_r-u_r}{2D_r}\right)erfc\left(\frac{x-u_rt}{2\sqrt{D_rt}}\right)+\exp\left(x\frac{V_r+u_r}{2D_r}\right)erfc\left(\frac{x+u_rt}{2\sqrt{D_rt}}\right)\right\}$$
$$-\frac{c_0\exp(-\gamma t)}{2}\left\{erfc\left(\frac{x-u_rt}{2\sqrt{D_rt}}\right)+\exp\left(x\frac{q_r}{2D_r}\right)erfc\left(\frac{x+u_rt}{2\sqrt{D_rt}}\right)\right\} \qquad (27.24)$$

where

$$u_r = \sqrt{V_r2+4D_r(K-\gamma)} \qquad (27.25)$$

Also $erfc(x) = \frac{2}{\sqrt{\pi}}\int_x^\infty \exp(-v^2)dv$ is an error function. Therefore, the analytical solution of the hydrodynamic dispersion equation given the prescribed initial and boundary conditions is given as:

$$c(x,t) = \frac{c_1\exp(-\gamma t)}{2}\left\{\exp\left(x\frac{V_r-u_r}{2D_r}\right)erfc\left(\frac{x-u_rt}{2\sqrt{D_rt}}\right)+\exp\left(x\frac{V_r+u_r}{2D_r}\right)erfc\left(\frac{x-u_rt}{2\sqrt{D_rt}}\right)+c_0\exp(-\gamma t)\right\}$$

$$(27.26)$$

This same analytical solution for the hydrodynamic dispersion equation for $c_0 = 0$ was presented by Atangana (2013) stated in Equation (27.29) as:

$$C(x,t) = \frac{c_1\exp(-\gamma t)}{2}\left\{\exp\left(x\frac{V_r-u_r}{2D_r}\right)erfc\left(\frac{x-u_rt}{2\sqrt{D_rt}}\right)+\exp\left(x\frac{V_r+u_r}{2D_r}\right)erfc\left(\frac{x-u_rt}{2\sqrt{D_rt}}\right)\right\} \qquad (27.27)$$

27.2.2 Solutions Obtained Using Green's Function Method

Equation (27.1b) can be solved using Green's function method. This is used in solving inhomogeneous linear differential equations by providing a split time representation of a systems response at a particular point (x', t), due to a field source point given at (x', τ) (Gringarten and Ramey, 1973). Even though this method is advantageous in providing easy handling of various types of no flow boundary and initial conditions, it has remained scarely utilized for solving groundwater pollution-related problems.

$$\frac{\partial C_B}{\partial t} - D_x \frac{\partial C^2 C_B}{\partial x^2} + V \frac{\partial C_B}{\partial x} = K_2 C_B - K_1 C_A$$

The above mathematical model has a heterogeneous and a homogeneous part. The heterogeneous part can be solved using Green's function method, which in this problem is defined as the concentration at (x, t) due to an instantaneous point source of strength unity generated at the point (x', τ), the aquifer being initially kept at zero concentration and the boundary surface also being kept at zero concentration (Gringarten and Ramey, 1973). Green's function of this problem can be obtained by solving the following differential equation (Equation 27.28) with initial and boundary conditions.

27.2.3 Solution of the Homogeneous System

The solution for the homogeneous part provides the particular solution in which the solution for the first equation is found by applying the Laplace transform, giving:

$$sC(x) - C(0) - D\frac{\partial^2 C_A}{\partial x^2} + V\frac{\partial C_A}{\partial x} + K_1 C(x) = 0 \tag{27.28}$$

where $\mathcal{L}$ is the Laplace transform operator, which results as:

$$sC(x) - C(0) - D\frac{\partial^2 C_B}{\partial x^2} + V\frac{\partial C_B}{\partial x} + K_1 C(x) = 0 \tag{27.29}$$

$$C_B(x,t) = \frac{c_1 \exp(-\gamma t)}{2} \left\{ \begin{array}{l} \exp\left(x\frac{V_r - u_r}{2D_r}\right) erfc\left(\frac{x - u_r t}{2\sqrt{D_r t}}\right) + \exp\left(x\frac{V_r + u_r}{2D_r}\right) erfc\left(\frac{x - u_r t}{2\sqrt{D_r t}}\right) \\ + c_0 \exp(-\gamma t) \end{array} \right\} \tag{27.30}$$

This same analytical solution for the hydrodynamic dispersion equation (Equation 29.28) which is also the solution for $c_0 = 0$ was presented by Atangana (2013).

$$C_B(x,t) = \frac{c_1 \exp(-\gamma t)}{2} \left\{ \exp\left(x\frac{V_r - u_r}{2D_r}\right) erfc\left(\frac{x - u_r t}{2\sqrt{D_r t}}\right) + \exp\left(x\frac{V_r + u_r}{2D_r}\right) erfc\left(\frac{x - u_r t}{2\sqrt{D_r t}}\right) \right\} \tag{27.31}$$

27.2.4 Solution of the Heterogeneous Part Using Green's Function

$$\frac{\partial C_B}{\partial t} - D_x \frac{\partial C^2 C_B}{\partial x^2} + V\frac{\partial C_B}{\partial x} = K_2 C_B - K_1 C_A \tag{27.32}$$

The solution for the homogeneous part can be stated as:

$$C_B(x,t) = \frac{c_1 \exp(-\gamma t)}{2} \left\{ \exp\left(x \frac{V_r - u_r}{2D_r} \right) erfc\left(\frac{x - u_r t}{2\sqrt{D_r t}} \right) + \exp\left(x \frac{V_r + u_r}{2D_r} \right) erfc\left(\frac{x - u_r t}{2\sqrt{D_r t}} \right) \right\} \quad (27.33)$$

Equation (27.1b) is inhomogeneous thus requiring the application of Green's function method.

$$\frac{\partial C_B}{\partial t} - D \frac{\partial^2 C_B}{\partial x^2} + V \frac{\partial C_B}{\partial x} = -K_1 C_A \quad (27.34)$$

Apply Laplace:

$$s\tilde{C} - C(0) - D \frac{\partial^2 \tilde{C}_B}{\partial x^2} + V \frac{\partial \tilde{C}_B}{\partial x} = -\tilde{C}K_1 \quad (27.35)$$

$$D \frac{\partial^2 \tilde{C}_B}{\partial x^2} - V \frac{\partial C_B}{\partial x} + s\tilde{C} - \tilde{C}K_1 = 0 \quad (27.36)$$

$$D \frac{\partial^2 \tilde{C}_B}{\partial x^2} + V \frac{\partial \tilde{C}_B}{\partial x} - \tilde{C}(s + K_1) = 0$$

Green's function method gives a solution to inhomogeneous linear differential equations defined on a domain with boundary problems. In Green's terms, this can be written as:

$$D \frac{\partial^2 G}{\partial x^2} + V \frac{\partial G}{\partial x} - G(s + K_1) = \sigma(0) \quad (27.37)$$

Applying the Laplace transform to Green's function:

$$\mathcal{L}\left(D \frac{\partial^2 G}{\partial x^2} + V \frac{\partial G}{\partial x} - G(s + K_1) \right) = \mathcal{L}(\sigma(0)) \quad (27.38)$$

$$D(p^2 \tilde{G} - pG' - G(0) + V(p\tilde{G} - G(0)) - \tilde{G}(s + K_1) = 1$$

$$Dp^2 \tilde{G} + Vp\tilde{G} + \tilde{G}(s + K_1) = 1$$

$$\left(Dp^2 + Vp + (s + K_1) \right) \tilde{G} = 1$$

$$\tilde{G} = \frac{1}{\left(Dp^2 + Vp - (s + K_1) \right)} \quad (27.39)$$

where:

$$p = \frac{-V \pm \sqrt{V^2 + 4D(s + K_1)}}{2D}; \quad p = \frac{-V - \sqrt{V^2 + 4D(s + K_1)}}{2D} \quad \text{or} \quad p = \frac{-V + \sqrt{V^2 + 4D(s + K_1)}}{2D}$$

Consider:

$$\frac{-V - \sqrt{V^2 + 4D(s + K_1)}}{2D} = \frac{-V - \sqrt{\Delta}}{2D} = \alpha_- \tag{27.40}$$

and:

$$\frac{-V + \sqrt{V^2 + 4D(s + K_1)}}{2D} = \frac{-V + \sqrt{\Delta}}{2D} = \alpha_+ \tag{27.41}$$

The roots of the above equation can be written as:

$$\tilde{G}(x,p) = \frac{1}{(p - \alpha_-)(p - \alpha_+)} \tag{27.42}$$

Equation (27.34) can be written in the form:

$$\frac{1}{(p - \alpha_-)} \cdot \frac{1}{(p - \alpha_+)} \tag{27.43}$$

The inverse of $\tilde{G}(p,s) - L\tilde{G}(x,p) - L\left(\frac{1}{(p - \alpha_-)} \cdot \frac{1}{(p - \alpha_+)}\right)$. Given the product of two functions, we use the convolution theory, which states that:

$$L^{-1}\{F(s)G(s)\} = \int_0^t f(y)g(t - y)dy \tag{27.44}$$

The inverse functions of $\tilde{G}$ is given as:

$$L^{-1}\left(\tilde{G}(x,s)\right) = \int_0^\tau e^{\alpha_- \tau} e^{\alpha_+(t - \tau)} d\tau \tag{27.45}$$

Simplifying gives:

$$L^{-1}\left(\tilde{G}(x,\tau)\right) = \int_0^x e^{\alpha_- \tau} e^{\alpha_+(x - \tau)} d\tau$$

$$G = e^x \int_0^x e^{(\alpha_- - \alpha_+)x} d\tau \tag{27.46}$$

Change of variables.

Let $e^{(\alpha_- - \alpha_+)x} = y$, $dy = (\alpha_- - \alpha_+)dx$

$$G = e^{x\alpha_+}\left[\frac{1}{\alpha_- - \alpha_+}.e^{x(\alpha_- - \alpha_+)}\right] \tag{27.47}$$

$$G = e^{x\alpha_+}\left[\frac{1}{\alpha_- - \alpha_+}.e^{x(\alpha_- - \alpha_+)}\right]_0^t$$

Substituting equations:

$$G(t) = e^{x\alpha_+}\left[\frac{1}{\alpha_- - \alpha_+}.e^{x(\alpha_- - \alpha_+)} - \frac{1}{\alpha_- - \alpha_+}\right] \tag{27.48}$$

$$G(t) = e^{x\alpha_+}\left[\frac{1}{\alpha_- - \alpha_+}.e^{x(\alpha_- - \alpha_+)} - \frac{1}{\alpha_- - \alpha_+}\right]$$

Substitute:

$$G(t,s) = e^{x\left(\frac{-V-\sqrt{\Delta}}{2V}\right)}\left[\frac{1}{\frac{-V-\sqrt{\Delta}}{2D} - \frac{-V+\sqrt{\Delta}}{2D}}.e^{x\left(\frac{-V-\sqrt{\Delta}}{2V} - \frac{-V+\sqrt{\Delta}}{2V}\right)} - \frac{1}{\frac{-V-\sqrt{\Delta}}{2D} - \frac{-V+\sqrt{\Delta}}{2D}}\right] \tag{27.49}$$

The above equation can be further simplified as:

$$G(t,s) = e^{x\left(\frac{-V-\sqrt{\Delta}}{2D}\right)}\left[\frac{1}{\frac{-V-\sqrt{\Delta}}{2D} - \frac{-V+\sqrt{\Delta}}{2D}}\right]e^{x\left(\frac{-V-\sqrt{\Delta}}{2D} - \frac{-V+\sqrt{\Delta}}{2D}\right)} - 1 \tag{27.50}$$

Substituting Equations (3.3.2) and (3.3.3):

$$G(t,s) = e^{x\left(\frac{-V-\sqrt{V^2+4D(s+K_1)}}{2D}\right)}\left[\frac{1}{\frac{-V-\sqrt{V^2+4D(s+K_1)}}{2D} - \frac{-V+\sqrt{V^2+4D(s+K_1)}}{2D}}\right]$$
$$\times e^{x\left(\frac{-V-\sqrt{V^2+4D(s+K_1)}}{2D} - \frac{-V+\sqrt{V^2+4D(s+K_1)}}{2D}\right)} - 1 \tag{27.51}$$

The exact solution in Laplace space can be written as:

$$C(x,s) = C_1(x,s) + \int_0^x C(0,s).G(x-\tau,s)d\tau \tag{27.52}$$

where $C(t,s)$ is the term for the exact solution; $C(0,s)$ is the particular solution given by Equation (27.2); $C(0,s)$ is the term for the initial pollution source concentration which is 0; and $\int_0^t G(t-\tau,s)d\tau$ is the integral of the equation.

The exaction solution can be thus written as:

$$C(t,s) = C_1(t,s) + \int_0^t C(0,s).G(t-\tau,s)d\tau$$

$$C(t,s) = \frac{c_1 \exp(-\gamma t)}{2}\left\{ \exp\left(x\frac{V_r - u_r}{2D_r}\right)erfc\left(\frac{x-u_r t}{2\sqrt{D_r t}}\right) + \exp\left(x\frac{V_r + u_r}{2D_r}\right)erfc\left(\frac{x-u_r t}{2\sqrt{D_r t}}\right)\right\}$$

$$+ C(0,s)\int_0^t G(t-\tau,s)d\tau \tag{27.53}$$

To obtain $\int_0^t G(t-\tau,s)d\tau$, we can express as given below and simplify further to get:

$$G(x,s) = e^{x\alpha_+}\left[\frac{1}{\alpha_- - \alpha_+}.\left(e^{x(\alpha_- - \alpha_+)}\right) - 1\right] \tag{27.54}$$

$$\int_0^x G(x-\tau,s)d\tau = \frac{1}{\alpha_- - \alpha_+}\int_0^x e^{(x-\tau)\alpha_+}.e^{(x-\tau)(\alpha_- - \alpha_+)} - e^{(x-\tau)\alpha_+}.d\tau \tag{27.55}$$

After integrating we obtain the following:

$$\int_0^x G(x-\tau,s)d\tau = \frac{1}{\alpha_- - \alpha_+}\int_0^x e^{(x-\tau)\alpha_+}.e^{(x-\tau)(\alpha_- - \alpha_+)} - e^{(x-\tau)\alpha_+}.d\tau$$

$$\int_0^x G(t-\tau,s)d\tau = \frac{1}{\alpha_- - \alpha_+}\left(\frac{1}{a_-}e^{\alpha_-(x-\tau)} - \frac{1}{a_+}e^{\alpha_+(x-\tau)}\right) \tag{27.56}$$

Let $(t-\tau) = y$

$$G(x-\tau,s) = \frac{1}{\alpha_- - \alpha_+}\left(\frac{1}{a_-}e^{x\alpha_-} - \frac{1}{a_+}e^{yx}\right)\Bigg|_0^x \tag{27.57}$$

Substituting Equation (27.38):

$G\left(x-\tau,s\right)$

$$
= \left[\left(\cfrac{1}{\left(\cfrac{-V-\sqrt{V^2+4D\left(s+K_1\right)}}{2D}\right)-\left(\cfrac{-V+\sqrt{V^2+4D\left(s+K_1\right)}}{2D}\right)}\right)\right.
$$

$$
\left.\times\left(\cfrac{1}{\left(\cfrac{-V-\sqrt{V^2+4D\left(s+K_1\right)}}{2D}\right)}e^{x\left(\frac{-V-\sqrt{V^2+4D(s+K_1)}}{2D}\right)}-\cfrac{1}{\left(\cfrac{-V+\sqrt{V^2+4D\left(s+K_1\right)}}{2D}\right)}e^{x\left(\frac{-V+\sqrt{V^2+4D(s+K_1)}}{2D}\right)}\right)\right]
$$

$$
-\left[\left(\cfrac{1}{\left(\cfrac{-V-\sqrt{V^2+4D\left(s+K_1\right)}}{2D}\right)-\left(\cfrac{-V+\sqrt{V^2+4D\left(s+K_1\right)}}{2D}\right)}\right)\right] \tag{27.58}
$$

The above equation is still in Laplace space and taking the inverse Laplace transform gives:

$\mathcal{L}^{-}\left(G\left(-\tau,s\right)\right)$

$$
= \left[\left(\cfrac{1}{\left(\cfrac{-V-\sqrt{V^2+4D\left(s+K_1\right)}}{2D}\right)-\left(\cfrac{-V+\sqrt{V^2+4D\left(s+K_1\right)}}{2D}\right)}\right)\right.
$$

$$
\left.\times\left(\cfrac{1}{\left(\cfrac{-V-\sqrt{V^2+4D\left(s+K_1\right)}}{2D}\right)}e^{t\left(\frac{-V-\sqrt{V^2+4D(s+K_1)}}{2D}\right)}-\cfrac{1}{\left(\cfrac{-V+\sqrt{V^2+4D\left(s+K_1\right)}}{2D}\right)}e^{\left(\frac{-V+\sqrt{V^2+4D(s+K_1)}}{2D}\right)}\right)\right]
$$

$$
-\left[\left(\cfrac{1}{\left(\cfrac{-V-\sqrt{V^2+4D\left(s+K_1\right)}}{2D}\right)-\left(\cfrac{-V+\sqrt{V^2+4D\left(s+K_1\right)}}{2D}\right)}\right)\right] \tag{27.59}
$$

$$G(x,s) = \left(\frac{D}{\sqrt{V^2 + 4D(s+K_1)}} \right) \frac{1}{\left(\dfrac{-V - \sqrt{V^2 + 4D(s+K_1)}}{2D} \right)} \cdot \left[\frac{x \exp\left(-\dfrac{(-Vt+x)^2 - tK_1}{4D_r} \right)}{\sqrt[2]{(\pi|D_r)}\, t^{3/2}} \right]$$

$$- \frac{1}{\left(\dfrac{-V + \sqrt{V^2 + 4D(s+K_1)}}{2D} \right)} \cdot \frac{x \exp\left(-\dfrac{(-Vt+x)^2 - tK_1}{4D_r} \right)}{\sqrt[2]{(\pi|D_r)}\, t^{3/2}} \qquad (27.60)$$

The exact solution to the linearized heterogeneous advection dispersion equation can be written as:

$$C(t,x) = \frac{c_1 \exp(-\gamma t)}{2} \left\{ \exp\left(x \frac{V_r - u_r}{2D_r} \right) erfc\left(\frac{x - u_r t}{2\sqrt{D_r t}} \right) + \exp\left(x \frac{V_r + u_r}{2D_r} \right) erfc\left(\frac{x - u_r t}{2\sqrt{D_r t}} \right) \right\}$$

$$\left[\left[\frac{1}{\left(\dfrac{-V - \sqrt{V^2 + 4D(s+K_1)}}{2D} \right) - \left(\dfrac{-V + \sqrt{V^2 + 4D(s+K_1)}}{2D} \right)} \right] \right.$$

$$+ \left[\frac{1}{\left(\dfrac{-V - \sqrt{V^2 + 4D(s+K_1)}}{2D} \right)} \frac{x \exp\left(-\dfrac{(-Vt+x)^2 - tK_1}{4D_r} \right)}{\sqrt[2]{(\pi|D_r)}\, t^{3/2}} \right.$$

$$\times$$

$$\left. - \frac{1}{\left(\dfrac{-V + \sqrt{V^2 + 4D(s+K_1)}}{2D} \right)} \frac{x \exp\left(-\dfrac{(-Vt+x)^2 - tK_1}{4D_r} \right)}{\sqrt[2]{(\pi|D_r)}\, t^{3/2}} \right]$$

$$- \left[\frac{1}{\left(\dfrac{-V - \sqrt{V^2 + 4D(s+K_1)}}{2D} \right) - \left(\dfrac{-V + \sqrt{V^2 + 4D(s+K_1)}}{2D} \right)} \right] \qquad (27.61)$$

27.3 NUMERICAL ANALYSIS

27.3.1 Crank–Nicolson Scheme

The Crank–Nicolson scheme is an implicit scheme. Instead of considering mesh points for performing calculations, the discretization is made at the midpoint between the ith and $(i + 1)$th levels.

The condition necessary for the convergence of some partial differential equations (PDEs) solved by the method of finite differences is given by the Courant–Friedrichs–Lewy (CFL) number. For the Crank–Nicolson numerical scheme, a low CFL number (0.5) is required for numerical accuracy. This surfaces when the explicit time scheme is used for the numerical solution. The simulation tends to produce incorrect results if the time step is longer than that for a certain explicit time-marching computer.

Forward difference approximation in time

$$\frac{\partial C_A}{\partial t} = \left(\frac{C_{A,i}^{n+1} - C_{A,i}^{n}}{\Delta t} \right) \tag{27.62}$$

Central difference dispersion

$$D\frac{\partial^2 C_A}{\partial x^2} = D\left[\frac{\left(C_{A,i+1}^{n} - 2C_{A,i}^{n} + C_{A,i-1}^{n} \right)}{\left(\Delta x\right)^2} + \frac{\left(C_{A,i+1}^{n+1} - 2C_{A,i}^{n+1} + C_{A,i-1}^{n+1} \right)}{\left(\Delta x\right)^2} \right]$$

$$+ \frac{D}{2\left(\Delta x\right)^2}\left[\left(C_{A,i+1}^{n} - 2C_{A,i}^{n} + C_{A,i-1}^{n} \right) + \left(C_{A,i+1}^{n+1} - 2C_{A,i}^{n+1} + C_{A,i-1}^{n+1} \right) \right] \tag{27.63}$$

Central difference advection

$$V\frac{\partial C_A}{\partial x} = V\left[\frac{\left(C_{A,i+1}^{n+1} - C_{A,i-1}^{n+1} \right)}{2\left(\Delta x\right)} + \frac{\left(C_{A,i+1}^{n} - C_{A,i-1}^{n} \right)}{2\left(\Delta x\right)} \right] \tag{27.64}$$

$$= \frac{V}{2\Delta x}\left[\left(C_{A,i+1}^{n+1} - C_{A,i-1}^{n+1} \right) + \left(C_{A,i+1}^{n} - C_{A,i-1}^{n} \right) \right]. \tag{27.65}$$

27.4 CENTRAL DIFFERENCE REACTION CONSTANT

$$K_1 C_A = K\left[\left(C_{A,i}^{n+1} + C_{A,i}^{n} \right) \right] \tag{27.66}$$

Hence, the finite difference approximation can be written thus:

$$\frac{C_{A,i}^{n+1} - C_{A,i}^{n}}{\Delta t} = \frac{D}{2\left(\Delta x\right)^2}\left[\left(C_{A,i+1}^{n+1} - 2C_{A,i}^{n+1} + C_{A,i-1}^{n+1} \right) + \left(C_{A,i+1}^{n} - 2C_{A,i}^{n} + C_{A,i-1}^{n} \right) \right]$$

$$- \frac{V}{2\Delta x}\left[C_{A,i+1}^{n+1} - C_{A,i-1}^{n+1} + C_{A,i+1}^{n} - C_{A,i-1}^{n} \right] + K\left(C_{A,i}^{n+1} + C_{A,i}^{n} \right) \tag{27.67}$$

where the following terms can be represented:

$$\frac{D\Delta t}{(\Delta x)^2} = \alpha, \quad \text{and}$$

$$\frac{V\Delta t}{\Delta x} = \beta$$

We expand and simplify with common terms:

$$C_{A,i}^{n+1} = C_{A,i}^{n} + \frac{1}{2}\alpha C_{A,i+1}^{n} - \alpha C_{A,i}^{n} + \frac{1}{2}\alpha C_{A,i-1}^{n} - \frac{1}{2}\beta C_{A,i+1}^{n} + \frac{1}{2}\beta C_{A,i-1}^{n} + \frac{1}{2}K_1 C_{A,i}^{n}$$
$$+ \frac{1}{2}\alpha C_{A,i+1}^{n+1} - \alpha C_{A,i}^{n+1} + \frac{1}{2}\alpha_{A,i-1}^{n+1} - \frac{1}{2}\beta C_{A,i+1}^{n+1} + \frac{1}{2}\beta C_{A,i-1}^{n+1} + \frac{1}{2}K_1 C_{A,i}^{n+1}$$

$$C_{A,i}^{n+1} = C_{A,i}^{n} - \alpha C_{A,i}^{n} + \frac{1}{2}K_1 C_{A,i}^{n} + \frac{1}{2}\alpha C_{A,i+1}^{n} - \frac{1}{2}\beta C_{A,i+1}^{n} + \frac{1}{2}\alpha C_{A,i-1}^{n} + \frac{1}{2}\beta C_{A,i-1}^{n}$$
$$+ \frac{1}{2}\alpha C_{A,i+1}^{n+1} - \frac{1}{2}\beta C_{A,i+1}^{n+1} - \alpha C_{A,i}^{n+1} + \frac{1}{2}K_1 C_{A,i}^{n+1} + \frac{1}{2}\alpha_{A,i-1}^{n+1} + \frac{1}{2}\beta C_{A,i-1}^{n+1} \qquad (27.68)$$

Opening the brackets gives:

$$C_{A,i}^{n+1} = \left(1 - \alpha + \frac{1}{2}K_1\right)C_{A,i}^{n} + \left(\frac{\alpha}{2} - \frac{\beta}{2}\right)C_{A,i+1}^{n} + \left(\frac{\alpha}{2} + \frac{\beta}{2}\right)C_{A,i-1}^{n} + \left(\frac{\alpha}{2} - \frac{\beta}{2}\right)C_{A,i+1}^{n+1}$$
$$- \left(\alpha - \frac{1}{2}K_1\right)C_{A,i}^{n+1} + \left(\frac{\alpha}{2} + \frac{\beta}{2}\right)C_{A,i-1}^{n+1}$$

Collecting together like terms and opening brackets:

$$C_{A,i}^{n+1} + \left(\alpha - \frac{1}{2}K_1\right)C_{A,i}^{n+1} = \left(1 - \alpha + \frac{1}{2}K_1\right)C_{A,i}^{n} + \left(\frac{\alpha}{2} - \frac{\beta}{2}\right)C_{A,i+1}^{n} + \left(\frac{\alpha}{2} + \frac{\beta}{2}\right)C_{A,i-1}^{n}$$
$$+ \left(\frac{\alpha}{2} - \frac{\beta}{2}\right)C_{A,i+1}^{n+1} + \left(\frac{\alpha}{2} + \frac{\beta}{2}\right)C_{A,i-1}^{n+1} \qquad (27.69)$$

The final discretized equation can then be written as:

$$C_{A,i}^{n+1}\left(1 + \alpha - \frac{1}{2}K_1\right) = \left(1 - \alpha + \frac{1}{2}K_1\right)C_{A,i}^{n} + \left(\frac{\alpha}{2} - \frac{\beta}{2}\right)C_{A,i+1}^{n} + \left(\frac{\alpha}{2} + \frac{\beta}{2}\right)C_{A,i-1}^{n}$$
$$+ \left(\frac{\alpha}{2} - \frac{\beta}{2}\right)C_{A,i+1}^{n+1} + \left(\frac{\alpha}{2} + \frac{\beta}{2}\right)C_{A,i-1}^{n+1} \qquad (27.70)$$

If we consider the following notation, the above equation can be broken down as:

$$\left(1 + \alpha - \frac{1}{2}K_1\right) = a$$

$$\left(1 - \alpha + \frac{1}{2}K_1\right) = b$$

$$\left(\frac{\alpha}{2} - \frac{\beta}{2}\right) = c$$

$$\left(\frac{\alpha}{2} + \frac{\beta}{2}\right) = d$$

$$aC_{A,i}^{n+1} = bC_{A,i}^{n} + cC_{A,i+1}^{n} + cC_{A,i+1}^{n+1} + dC_{A,i-1}^{n} + dC_{A,i-1}^{n+1} \tag{27.71}$$

Equation (27.71) becomes the first-order Crank–Nicolson finite difference approximation of the one-dimensional advection-dispersion equation.

27.5 DISCRETIZATION SCHEME FOR THE SECOND EQUATION

The second equation is stated thus:

$$\frac{\partial C_B}{\partial t} - D\frac{\partial C^2 C_B}{\partial x^2} + V\frac{\partial C_B}{\partial x} = K_2 C_B - K_1 C_A \tag{27.72}$$

Considering Equation (27.71, 27.72) can be written as:

$$C_{B,i}^{n+1} - C_{B,i}^{n} = \frac{1}{2}\left[\frac{D\Delta t}{(\Delta x)^2}\left(C_{B,i+1}^{n} - 2C_{B,i}^{n} + C_{B,i-1}^{n}\right) - \frac{V\Delta t}{\Delta x}\left(C_{B,i+1}^{n} - C_{B,i-1}^{n}\right) + K_2 C_{B,i}^{n}\right]$$
$$+ \frac{1}{2}\left[\frac{D\Delta t}{(\Delta x)^2}\left(C_{B,i+1}^{n+1} - 2C_{B,i}^{n+1} + C_{B,i-1}^{n+1}\right) - \frac{V\Delta t}{\Delta x}\left(C_{B,i+1}^{n+1} - C_{B,i-1}^{n+1}\right) + K_2 C_{B,i}^{n+1}\right] - K_1 C_A \tag{27.73}$$

Expand and simplify with common terms:

$$C_{B,i}^{n+1} = C_{B,i}^{n} + \left[\begin{array}{l}\frac{1}{2}\alpha C_{B,i+1}^{n} - \alpha C_{B,i}^{n} + \frac{1}{2}\alpha C_{B,i-1}^{n} - \frac{1}{2}\beta C_{B,i+1}^{n} + \frac{1}{2}\beta C_{B,i-1}^{n} + \frac{1}{2}K_2 C_{B,i}^{n} \\ + \left[\left(\frac{1}{2}\alpha C_{B,i+1}^{n+1} - \alpha C_{B,i}^{n+1} + \frac{1}{2}\alpha_{B,i-1}^{n+1} - \frac{1}{2}\beta C_{B,i+1}^{n+1} + \frac{1}{2}\beta C_{B,i-1}^{n+1} + \frac{1}{2}K_2 C_{B,i}^{n+1}\right)\right]\end{array}\right]$$
$$- K_1 C_{A,i}^{n+1} - K_1 C_{A,i}^{n} \tag{27.74}$$

Further simplification gives:

$$C_{B,i}^{n+1} = C_{B,i}^{n} + \frac{1}{2}\alpha C_{B,i+1}^{n} - \alpha C_{B,i}^{n} + \frac{1}{2}\alpha C_{B,i-1}^{n} - \frac{1}{2}\beta C_{B,i+1}^{n} + \frac{1}{2}\beta C_{B,i-1}^{n} + \frac{1}{2}K_2 C_{B,i}^{n} + \frac{1}{2}\alpha C_{B,i+1}^{n+1}$$
$$- \alpha C_{B,i}^{n+1} + \frac{1}{2}\alpha_{B,i-1}^{n+1} - \frac{1}{2}\beta C_{B,i+1}^{n+1} + \frac{1}{2}\beta C_{B,i-1}^{n+1} + \frac{1}{2}K_2 C_{B,i}^{n+1} - K_1 C_{A,i}^{n+1} - K_1 C_{A,i}^{n} \tag{27.75}$$

The final discretized equation can then be written as:

$$C_{B,i}^{n+1}\left(1 + \alpha - \frac{1}{2}K_2\right) = \left(1 - \alpha + \frac{1}{2}K_2\right)C_{B,i}^{n} + \left(\frac{\alpha}{2} - \frac{\beta}{2}\right)C_{B,i+1}^{n} + \left(\frac{\alpha}{2} + \frac{\beta}{2}\right)C_{B,i-1}^{n}$$
$$+ \left(\frac{\alpha}{2} - \frac{\beta}{2}\right)C_{B,i+1}^{n+1} + \left(\frac{\alpha}{2} + \frac{\beta}{2}\right)C_{B,i-1}^{n+1} - K_1 C_{A,i}^{n+1} - K_1 C_{A,i}^{n} \tag{27.76}$$

If we consider the following constants below, the above equations can be broken down as:

$$\left(1+\alpha-\frac{1}{2}K_2\right)=a$$

$$\left(1-\alpha+\frac{1}{2}K_2\right)=b$$

$$\left(\frac{\alpha}{2}-\frac{\beta}{2}\right)=c$$

$$\left(\frac{\alpha}{2}+\frac{\beta}{2}\right)=d$$

$$aC_{B,i}^{n+1}=bC_{B,i}^{n}+cC_{B,i+1}^{n}+cC_{B,i+1}^{n+1}+dC_{B,i-1}^{n}+dC_{B,i-1}^{n+1}-\left(K_1C_{A,i}^{n+1}+K_1C_{A,i}^{n}\right). \tag{27.77}$$

27.6 STABILITY ANALYSIS

Generally, the results for the explicit boundary conditions hold for time-dependent diffusion with advections provided. Given a system with periodic boundary conditions, the variation of the error may be represented as a finite Fourier series given as:

$$\epsilon(x)=\sum_{M=1}^{M}A_m e^{ik_m x} \tag{27.78}$$

where $K_m=\dfrac{\pi m}{L}$ $m=1,2,\ldots,M$ and $M=\dfrac{L}{\Delta x}$

We assume the error amplitude varies with time. The difference equation for the error is linear, that is, each term behaves as though it were the series itself. On this premise, we can comfortably formulate the error growth of a typical term:

$$\epsilon_m(x,t)=\sum_{m=1}^{M}C_{A,i}^{n}.e^{ik_m x} \tag{27.79}$$

where α is a constant.

By substituting Equation (27.79) into Equation (27.77), error variation in time steps can be described. If we consider the following notation, the above equations can be broken down as:

$$C_{A,i}^{n+1}\left(1+\alpha-\frac{1}{2}K_1\right)=\left(1-\alpha+\frac{1}{2}K_1\right)C_{A,i}^{n}+\left(\frac{\alpha}{2}-\frac{\beta}{2}\right)C_{A,i+1}^{n}+\left(\frac{\alpha}{2}+\frac{\beta}{2}\right)C_{A,i-1}^{n}$$

$$+\left(\frac{\alpha}{2}-\frac{\beta}{2}\right)C_{A,i+1}^{n+1}+\left(\frac{\alpha}{2}+\frac{\beta}{2}\right)C_{A,i-1}^{n+1} \tag{27.80}$$

If we consider the following notation, the above equations can be broken down using;

$$\left(1+\alpha-\frac{1}{2}K_1\right)=a$$

$$\left(1 - \alpha + \frac{1}{2}K_1\right) = b$$

$$\left(\frac{\alpha}{2} - \frac{\beta}{2}\right) = c$$

$$\left(\frac{\alpha}{2} + \frac{\beta}{2}\right) = d$$

$$aC_{A,i}^{n+1} = bC_{A,i}^{n} + cC_{A,i+1}^{n} + dC_{A,i-1}^{n} + cC_{A,i+1}^{n+1} + dC_{A,i-1}^{n+1} \tag{27.81}$$

First, we note the error terms expressed as:

$$\epsilon_{A,i}^{n+1} = C_{n+1} \cdot e^{ik_m \bar{x}}$$

$$\epsilon_{A,i}^{n} = C_n \cdot e^{ik_m x}$$

$$\epsilon_{i+1,A}^{n} = C_n \cdot e^{ik_m(x+\Delta x)}$$

$$\epsilon_{i-1,A}^{n} = C_n \cdot e^{ik_m(x-\Delta x)}$$

$$C_{A,i+1}^{n+1} = C_{n+1} \cdot e^{ik_m(x+\Delta x)}$$

$$C_{A,i-1}^{n+1} = C_{n+1} \cdot e^{ik_m(x-\Delta x)} \tag{27.82}$$

We insert the error term $C_{A,i}^{n} = e^{ik_m x}$ into the finite difference scheme (27.81):

$$aC_{n+1} \cdot e^{ik_m x} = bC_n \cdot e^{ik_m x} + cC_n \cdot e^{ik_m(x+\Delta x)} + dC_n \cdot e^{ik_m(x-\Delta x)} + cC_{n+1} \cdot e^{ik_m(x+\Delta x)} + dC_{n+1} \cdot e^{ik_m(x-\Delta x)} \tag{27.83}$$

We expand and simplify exponential powers:

$$aC_{n+1} \cdot e^{ik_m x} = bC_n \cdot e^{ik_m x} + cC_n \cdot e^{ik_m x} e^{ik_m \Delta x} + dC_n \cdot e^{ik_m x} e^{-ik_m \Delta x)} + cC_{n+1} \cdot e^{ik_m x} e^{ik_m \Delta x} + dC_{n+1} \cdot e^{ik_m x} e^{-ik_m \Delta x} \tag{27.84}$$

Factor out the common terms $e^{ik_m x}$ to obtain:

$$aC_{n+1} = bC_n + \left(cC_n e^{ik_m \Delta x} + cC_{n+1} e^{ik_m \Delta x}\right) + \left(dC_n e^{-ik_m \Delta x)} + dC_{n+1} e^{-ik_m \Delta x}\right)$$

$$aC_{n+1} = bC_n + c\left(C_n e^{ik_m \Delta x} + C_{n+1} e^{ik_m \Delta x}\right) + d\left(C_n e^{-ik_m \Delta x)} + C_{n+1} e^{-ik_m \Delta x}\right)$$

$$aC_{n+1} = bC_n + c\left(C_n + C_{n+1}\right) e^{ik_m \Delta x} + d\left(C_n + C_{n+1}\right) e^{-ik_m \Delta x} \tag{27.85}$$

Using the identity:

$$\sin(k_{m\Delta x}) = \frac{1}{2}\left(e^{ik_m\Delta x} + e^{-ik_m\Delta x}\right).$$

$$aC_{n+1} = bC_n + 2c\left(C_n + C_{n+1}\right)\sin(k_{m\Delta x})$$

$$C_{n+1} = \frac{b}{a}C_n + \frac{2c}{a}\left(C_n + C_{n+1}\right)\sin(k_{m\Delta x}) \tag{27.86}$$

The system stability condition can be assessed using the method of the exaggeration factor. The exaggeration factor G is given as:

$$G = \frac{C_{n+1}}{C_n}\frac{C_{n+1}}{C_n} = \frac{b}{a} + \frac{2c}{a}\left(1 + \frac{C_{n+1}}{C_n}\right)\sin(k_{m\Delta x})G = \frac{b}{a} + \frac{2c}{a}(1+G)\sin(k_{m\Delta x}) \tag{27.87}$$

Expand and make G the subject of the formula:

$$G = \frac{b + 2c\sin(k_{m\Delta x})}{a - 2c\sin(k_{m\Delta x})}$$

$$\left|\frac{b + 2c\sin(k_{m\Delta x})}{a - 2c\sin(k_{m\Delta x})}\right| \le 1 \tag{27.88}$$

Hence, the system is stable when:

$$\left|\frac{b + 2c\sin(k_{m\Delta x})}{a - 2c\sin(k_{m\Delta x})}\right| \le 1 \tag{27.89}$$

The system is stable when:

$$\left|b + 2c\sin(k_m\Delta x)\right| \le \left|a - 2c\sin(k_m\Delta x)\right|$$

$$\left|a - 2c\sin(k_{m\Delta x})\right| = \begin{cases} a - 2c\sin(k_m\Delta x) \text{ if } a - 2c\sin(k_m\Delta x) > 0 \\ -a + 2c\sin(k_m\Delta x) \text{ if } a - 2c\sin(k_m\Delta x) < 0 \end{cases} \tag{27.90}$$

where

$$\left(1 + \alpha - \frac{1}{2}K_1\right) = a$$

$$\left(1 - \alpha + \frac{1}{2}K_1\right) = b$$

$$\left(\frac{\alpha}{2} - \frac{\beta}{2}\right) = c.$$

27.7　DISCUSSION

The retardation factor will be used as the analytic parameter for the purpose of this discussion. The model approaches reactive transport by considering the possible case of two spatially isolated source pollutants with a history of interaction, in space and time. Both sources, U1 and U2, pitch the groundwater horizon at 1000 g/L along high velocity preferential trajectories, followed by an exponential decay in concentrations due to retardation. A third figure captures the process linked to the new mass contribution from U1, which is a result of mixing caused by aquifer heterogeneity. The new contributing quantity peaks at and quickly decays from 60 concentration units to 0 units. The model was also tested for a higher retardation coefficient value and the obtained results show a positive correlation with the trends expected from kinetic modelling. Using a higher retardation factor can attenuate the transport process as shown by the last eight set of images (see Figures 27.2 to 27.18). This was a case of mixing at concentrations higher than that described above, such that the new contributing quantity now peaks at 250 concentration units as opposed to 60. After reacting, the resulting system is continuous in a state of reactive transport, whose faithfulness is entirely captured by the model. Unlike the routine approach of modelling by lumping reactive process into one parameter, this model does not only provide a tool for simulating the fate of reactive species in groundwater but also provides useful insight into the mass balance relationships associated with each participating species.

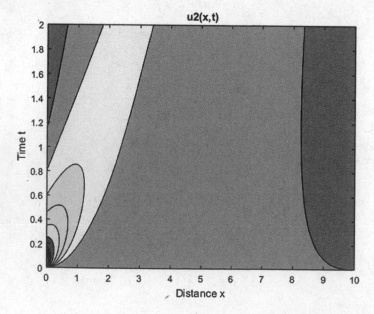

FIGURE 27.2　Section showing pollutant migration from source (U-2) along the x-axis with respect to time.

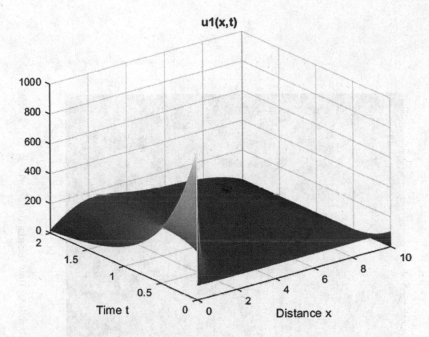

FIGURE 27.3 Surface showing U-1 concentration distribution along the X-direction in $[0, 2]$ and t in $[0, 2]$, $C(0, t) = C1\exp(-t) = 1000\exp(-t)C(x, 0) = C0$.

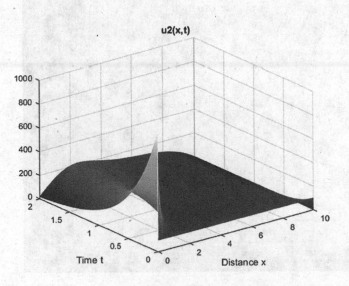

FIGURE 27.4 Surface showing U-2 concentration distribution along X-direction in $[0, 2]$ and t in $[0, 2]$, $C(0, t) = C1\exp(-t) = 1000\exp(-t)C(x, 0) = C0$.

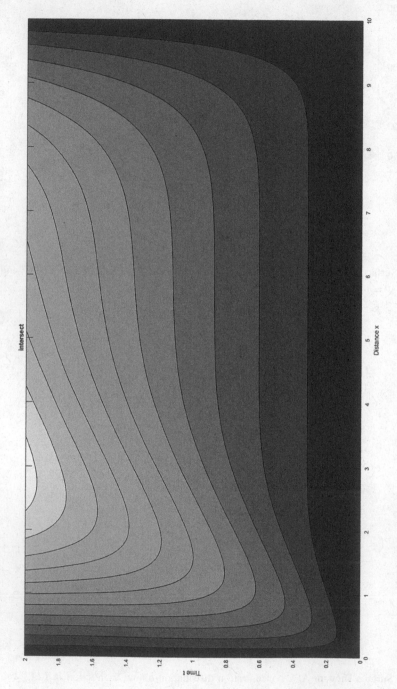

FIGURE 27.5 Contour plot showing pollution spread along reactive fronts after the mixing of the two sources (low retardation factor).

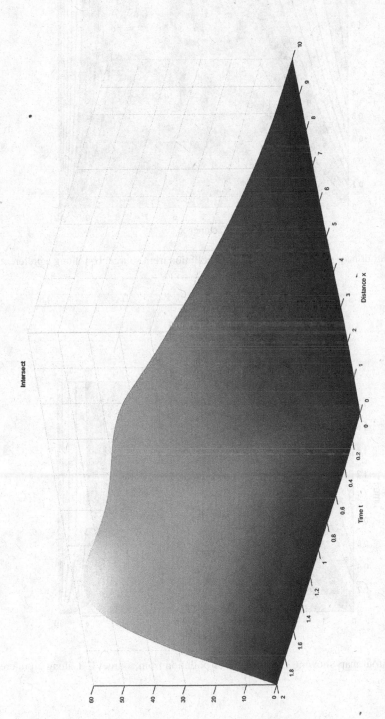

FIGURE 27.6 Surface showing concentration distribution for x in [0, 2] and t [0, 2], c (0, t) after the mixing of two sources with low retardation factor.

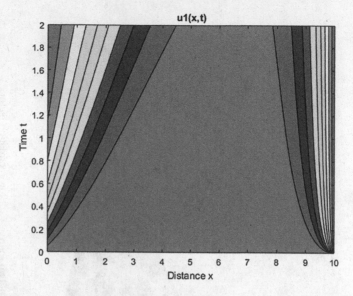

FIGURE 27.7   Contour map showing the transport of pollution from source, U-1 along a preferential.

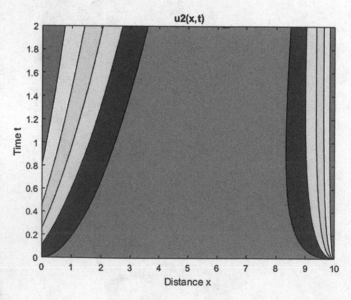

FIGURE 27.8   Contour map showing the transport of pollution from source, U-1 along a preferential path (low retardation factor).

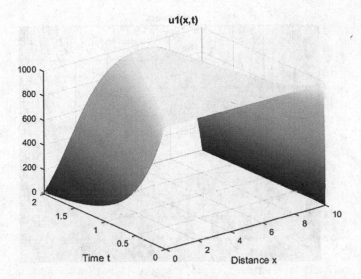

FIGURE 27.9 Surface showing U-1 concentration distribution along X-direction in $[0,2]$ *and t in* $[0,2]$, $C(0,t) = C1 exp(-t) = 1000 exp(-t) C(x,0) = C0$ (high retardation factor).

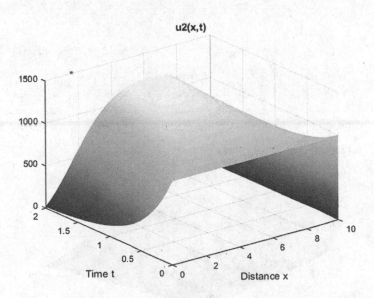

FIGURE 27.10 Surface showing U-2 concentration distribution along X-direction in $[0,2]$ *and t in* $[0,2]$, $C(0,t) = C1 exp(-t) = 1000 exp(-t) C(x,0) = C0$ (high retardation factor).

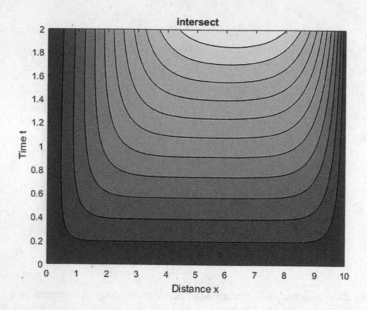

FIGURE 27.11 Contour plot showing pollution spread along reactive fronts after mixing of the two sources (the case of a high retardation coefficient).

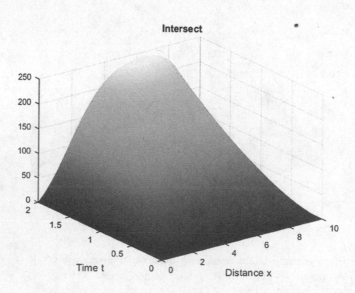

FIGURE 27.12 Surface plot showing concentration distribution after the mixing of U1 and U2 under the influence of a low retardation coefficient.

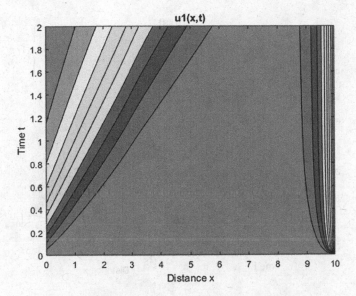

FIGURE 27.13 Contour plot showing U-1 source pollution migration along a preferential flow path influenced by a low retardation factor.

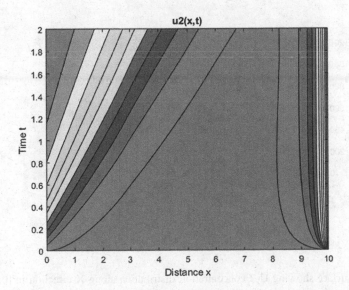

FIGURE 27.14 Contour plot showing U-2 source pollution migration along a preferential flow path influenced by a low retardation factor.

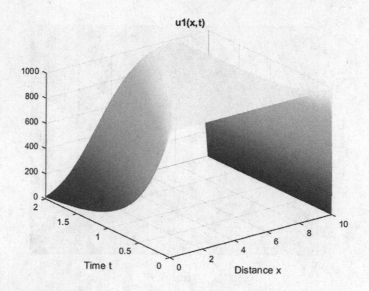

FIGURE 27.15 Surface showing U-1 concentration distribution along X-direction in $[0, 2]$ and t in $[0, 2]$, $C(0, t) = C1 exp (-t) = 1000 exp (-t) C(x, 0) = C0$ (very high retardation factor).

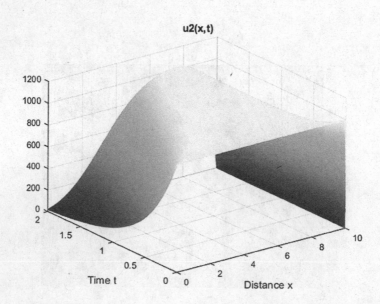

FIGURE 27.16 Surface showing U-2 concentration distribution along X-direction in $[0, 2]$ and t in $[0, 2]$, $C(0, t) = C1 exp (-t) = 1000 exp (-t) C(x, 0) = C0$ (very high retardation factor).

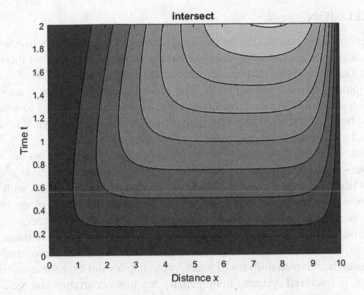

FIGURE 27.17 Contour plot showing pollution spread after intersection (high retardation factor).

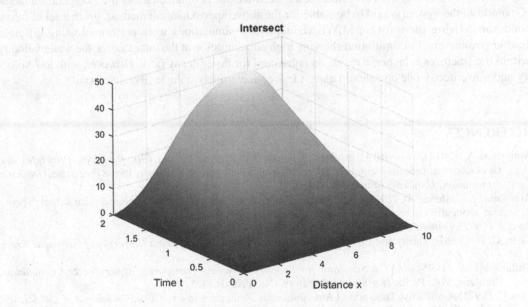

FIGURE 27.18 Surface showing the distribution of pollution after mixing (low retardation factor).

27.8 CONCLUSION

Fractured aquifers have the potential to deliver large quantities of fresh water to boreholes intersecting the fracture networks, even though they remain the most vunerable to an increasing chemical pollution thread. The basis of this chapter was to describe and to model the fate of reactive pollutants in fracture-dominated groundwater transport. The hypothetical scenario in question involved the intersection of two fractures from a regional fault system, which can form due to structural deformation of the lithosphere, caused by changing regional stress fields over geological time. Each fracture represents a travel pathway transporting pollution through the subsurface.

The movement of water in fractures is predominantly controlled by advection, and also by dispersion to a lesser extent. Routinely, the transport and transformation of reactive contaminants in groundwater has been decribed using the classical advection dispersion model with the addditon of a reaction term (sink) to capture the nature of the chemical process taking place. In reactive transport research, most studies have focused on modelling classical chemical processes involving mass transfer (adsorbtion, precipitaion, molecular diffusion, etc.) between different states of matter, from fractures to the surrounding rock matrix and vice versa. In geological systems, these processes can continue for thousands and even millions of years, providing a valid base for assuming kinetic equilibrium. However, in fractured systems, homogeneity is a rare occurrence and the closest analogue would be the case of a self-similar fractured system for which the stochastic approach is best suited.

Two point sources are monitored for a process of contaminant transport by the coupling of the linear homogeneous and the linear non-homogeneous source solutions. The exact solution of the linear system is obtained by using the Laplace transform and Green's function method.

A finite element method (FEM) numerical analysis was performed for both the homogeneous and the non-homogeneous models using the Crank–Nicolson discretized scheme which is unconditionally stable. The stability of the resultant discretized scheme is obtained using the exaggeration factor (G) method; the system is said to be stable for the above approximation method, given a set of initial conditions. Using the software MATLAB numerical simulations were performed using different aquifer parameters. For simulation showing high concentration at the intersection, the water velocity within the fractures is higher, a result also observed for faster decay rates. However, with low velocity and a low decay rate one should expect less concentration at the intersection point.

REFERENCES

Atangana, A. (2013). *General Assessment of Waste Disporsal at Douala City. Practices, Principles and Uncertainties*. Bloemfontein, Free State Province, South Africa: University of the Free State. *(Doctoral dissertation, University of the Free State)*.

Atangana, A., Baleanu, D. (2016). New fractional derivatives with nonlocal and non-singular kernel: Theory and application to heat transfer model. *Thermal Science*, 20(2), 763–769.

Bear, J. (1972). *Dynamics of Fluids in Porous Media*. Elsevier, 579–582.

Chen, C. X. (1996). Theory and model of groundwater solute transport. *China University of Geosciene Press*, 4–7.

Chilakapati, A. (1995). *RAFT a Simulator for Reactive Flow and Transport of Groundwater Contaminats*. Richland, WA: Pacific Northwest Laboratory. (No. PNL-10636).

Cho, C. (1971). Convective Transport of Amonium with nitrification in soil. *Canadian Journal of Soil Science*, 51(3), 339–350.

Clement, T. P., Sun, Y., Hooker, B. S., Petersen, J. N. (1998). Modelling multispecies reactive transport in groundwater. *Groundwater Manangement Resource*, 18(2), 79–92.

Cortis, A. A., Berkowitz, B. (2004). Analomous transportin classicl soil and sand column. *Soil Science ociety of America*, 70(4), 1539–1548.

Domenico, P. A. (1987). An analytical model for multidimensional transport of a decaying contaminant species. *Journal of Hydrology*, 91(1–2), 49–58.

Gelhar, L. (1992). A critical reviewof data of field-scale dispersion in aquifers. *Water Respurce Research*, 28(7), 1955–1974.

Gelhar, L. (1993). *Stochastic Subsurface Hydrology*. Prentice Hall.

Genuchten, M. A., Wagenet, R. J. (1989). Two-site/two-region models for pesticide transport and degradation:Theoritcal development and analytical solution. *Soil Science Society of America*, 53(5), 1303–1310.

Germon, J. (1985). Microbiology of Dentrification and Other Processes Involving the Reduction of Oxygentated Nitrogenous Compounds. *NATO Conference Series (I Ecology)*, 9. 31–46, Boston, MA: Springer.

Ginn, T. R.-V. (2017). Revisiting the analytical Solution Approach to Mising-Limited Equilibrium Multicomponent Reactive Transport Using Mixing Ratios; Identification of Basis, Fixing an erro and Dealing ith Multiple Minerals. *Advancing Earth and Space Science*. 53(11), 9941–9959.

Gouze, P. A., Melean, Y., Le Borgne, T., Dentz, M., Carrera, J. (2008). Non-Fickan dispersion in porous media explained by heterogeneous microscale matrix diffusion. *Water Rosource Research*. 44(11).

Gringarten, A.C. and Ramey, H.J. (1973). The use of source and Green's functions in solving unsteady-flow problems in reservoirs. *Society of Petroleum Engineers Journal*, 13(05), 285–296.

Helgeson, H. C. (1989). Thermodynamics and kineticconstrains on reaction rates among minerals and aqueous solutions. IV. Retrieval of rates constants and activation parameters for the hydrolysis of pyroxenes, wollastonite, olivine, andalusite, quartz and nepheline. *American Journal of Science*, 289(1), 17–101.

Holstad, A. (2000). A mathematical and numerical modelforreactive fluid flowsystem. *Computational Geoscience*, 4(2), 103–139.

Levy, M. A. (2003). Measurement and analysis of non-fickian dispersion in heterogeneous porous media. *Journal of Contaminant Hydrology*, 64(3–4), 203–226.

Lunn, M. L. (1996). Determining analytic solutions of multiple species contaminant transport with sorption and decay. *Journal of Hydrology*, 180(1–4), 195–210.

McCArtin, B. L. (2003). Accurate and efficient pricing of vanilla stock options via the Crandall Douglas Scheme. *Applied Math and Computing*, 143(1), 39–60.

Oishi, C. M. (2015). Stability analysis of Crank–Nicolson and Euler schemes. *BIT Numerical Mathematics*, 55(2), 487–513.

Pool, M., Carrera, J., Alcolea, A., Bocanegra, E. M. (2015). A comparison of deterministic and stochastic approaches for regional scale inverse modeling on the Mar del Plata aquifer. *Journal of Hydrology*, 531, 214–229.

Sharma, P. K., Mayank, M., Ojha, C.S.P., Shukla, S.K. (2018). A review on groundwater contaminant transport and remediation. *ISH Journal of Hydraulic Engineering*. doi: 10.1080/09715010.2018.1438213.

Simpson, J. M. (2015). Analytical model of reactive transport processes with spatially variable coefficients. *Royal Society of Open Science*, 2(5), 140348

Sun, Y. P., Petersen, J. N., Clement, T. P. (1999). Analytical solutions for multiple species reactive transport in multiple dimensions. *Journal of Contaminant Hydrology*, 35, 429–440.

Wilson, J. L. (1978). Two dimensional plume in uniformgroundwater flow. *Journal of the Hydraulics Division*, 104(4), 503–514.

Yunwei, S. (2003). Anlytical solutions for reactive transport of N-member radionuclide chains in a single fracyure. *Journal of Contaminant Hydrology*, 62, 695–712.

28 Stochastic Modeling in Confined and Leaky Aquifers

Sarti Amakali

University of the Free State, Bloemfontein, South Africa

Abdon Atangana

China Medical University Hospital, China Medical University, Taichung, Taiwan

University of the Free State, Bloemfontein, South Africa

CONTENTS

28.1 INTRODUCTION

Sustainable and effective groundwater management, which includes fair allocation among its users, is very necessary as its demand is ever increasing and availability is being lowered, especially in Southern Africa's driest countries, like Namibia (Kumba, 2003). Since groundwater occurs subsurface, scientists find ways to estimate groundwater flows, including aquifers' capacities. These scientific approaches are used to quantify and estimate groundwater flow, including aquifer parameters, which is done through models such as deterministic and stochastic modeling.

In this chapter we discuss the application of stochastic modeling to confined and leaky aquifers; we focus on unknown parameters which are random in nature. Deterministic modeling suits aquifers with uniform geologic stratification which are usually homogeneous; stochastic models are preferred for modeling complex heterogeneous structures. Stochastic models are based on mathematical descriptions of natural processes, which can also include statistical analysis (probabilistic). In this study we use stochastic modeling to estimate transmissivity, storativity, and leakage, which we assume can vary in confined and leaky aquifers.

Stochastic models are more highly recommended in quantitative and qualitative research for running groundwater flow models because they are unique and due to their ability to use random values to solve errors in heterogeneous environments. Stochastic models predict multiple sets of possible

DOI: 10.1201/9781003266266-28

outcomes, weighted by their likelihood or probabilities. In addition, the algorithms produced by stochastic models independently represent random functions (van Leeuwen et al., 1998).

The techniques used to solve the governing equations in modeling include those of finite difference (FD) and finite element (FE). We apply FD in our work to solve numerically partial difference equations (PDEs) to obtain an exact solution.

28.2   GROUNDWATER FLOW IN CONFINED AQUIFERS

Groundwater in confined aquifers is often found to be under pressure, rising to the potentiometric level because of a layered aquitard. The aquitard is beneficial as it protects the aquifer from contamination because polluted water cannot easily enter the confined aquifer. The aquifer is recharged at points where (i) it is exposed to the surface by means of rain or loose streams, rivers, or springs that cross the outcrops and (ii) by deep underground tributaries. Thiem's method (1906) for a steady-state flow and Theis's method (1935) for an unsteady-state flow are the solutions used to determine the parameters in confined aquifers; their consumption is also considered.

Groundwater flow equations are used to represent the properties of a groundwater system as they are assumed always to be constant. This is what is called representative elemental volume (REV). In this chapter, we challenge this concept, as the aquifer's properties may not always be constant in space, at a point in time, due to the variability in spatial distribution which can relate to the inconsistency of the geometry or pore sizes of the aquifer system. There can be changes in an aquifer's geological structures, due to hydrological cycles (including precipitation which recharges it), stream flow, and human induced activities such as over-abstraction at a particular point. Nevertheless, most authors/modelers have ignored this, so we argue that aquifer parameters cannot really be modeled as constants. Furthermore, the drawdown used may only affect the pumped well and the observation wells considered, which only represent several points (parts) of the aquifer, which may not respond to the whole aquifer system, unless it is a homogeneous environment.

Theis (1935) proposed that the groundwater flow is as shown in Equation (28.1) for confined aquifers for determining the movement of groundwater over the radial distance of a system (space) at a particular time:

$$\frac{S}{T}\frac{\partial h(r,t)}{\partial t} = \frac{\partial^2 h(r,t)}{\partial r^2} + \frac{1}{r}\frac{\partial h(r,t)}{\partial r} \tag{28.1}$$

where S represents aquifer storativity (dimensionless), T is transmissivity (m²/d) over a radial distance (r), and (t) is the pumping time.

The above equation does not consider the structure of the geological formations; it considers only the transmissivity and the storativity of the aquifer, which in a normal situation cannot be representative as the water moves slowly within the aquifer as it passes through each portion of the soil matrix. Thus, neglecting the scaling effect may probably lead to misleading results. To solve this problem, Mathobo and Atangana (2018) suggested a revision of the Theis model, starting from Darcy´s law and using the mass balance equation, without neglecting any terms in the Theis equation. They suggested a modified equation (28.2), which is apparently more complex and more informative than the existing model. Their model includes the scaling effect of the matrix soil and is given as:

$$\frac{S}{T}\frac{\partial h(r,t)}{\partial t} = \frac{\partial^2 h(r,t)}{\partial r^2}\left(1+\frac{\Delta r}{r}\right) + \frac{\partial h(r,t)}{r\partial r} \tag{28.2}$$

where S represents aquifer storativity (dimensionless) and T is transmissivity (m²/d) over a radial distance (r) at a time (t). In our work, we will consider the above equation and transform it to a stochastic model for further investigation.

28.3 A GROUNDWATER FLOW EQUATION FOR A LEAKY AQUIFER

Leaky aquifers are very complicated to investigate because additional water from an upper aquitard may enter at any time. When modeling this aquifer type, one has to be vigilant of the effects of the leakage factor. The geological materials that make up the overlying or underlying aquitard (confining layer) can be impervious, but they can also compress/bend, which may lead to a significant leakage through layers, which in some cases is ignored. The assumptions derived from Kruseman, de Rider, and Verweij (1970) for leaky aquifers in steady and unsteady states were considered. In the last decade the groundwater flow model within a leaky aquifer suggested by Hantush and Jacob (1955) has been used extensively to model such physical problems. The groundwater flow equation (28.3) in leaky aquifers derived from Hantush and Jacob's (1955) model, where the specific storage of the confining unit is assumed to be negligible and the head in the overlying aquifer is assumed to be unaffected, is:

$$\frac{\partial^2 h}{\partial r^2} + \frac{1}{r}\frac{\partial h}{\partial r} - \frac{h}{B^2} = \frac{\partial h}{\partial t} \tag{28.3}$$

Hantush (1960) modified Equation (28.3) to include the specific storage of the confining unit and to derive the following equation:

$$\frac{\partial^2 h}{\partial r^2} + \frac{1}{r}\frac{\partial h}{\partial r} - \frac{1}{B}\frac{\partial h(r,t)}{Bz} = \frac{\partial h}{\partial t} \tag{28.4}$$

where B is the dimensionless leakage parameter, h is the drawdown, r is the radial distance, t is time, and z is the vertical distance in the confining layer.

Although the model has been used with great success in recent years, one should note that some assumptions or modifications and simplifications were made to obtain a simple model. Amanda and Atangana (2018) undertook an investigation of the groundwater flowing within a leaky aquifer; they reused the Darcy law and the mass balance concept to revisit the existing model. Without neglecting any parameters and simplification of any terms, they produced a new mathematical model (Equation 28.5) for groundwater flowing within a leaky aquifer. The obtained mathematical model is:

$$\frac{S}{T}\frac{\partial h}{\partial t} = \frac{\partial}{\partial r}\left(\frac{\partial h}{\partial r}\right)\left(1 + \frac{\Delta r}{r}\right) + \frac{1}{r}\frac{\partial h}{\partial r} + \frac{h(r,t)}{\lambda^2} \tag{28.5}$$

where S represents aquifer storativity (dimensionless) and T is transmissivity (m²/d) over a radial distance at a point in time.

The above equation has a new parameter which considers the scaling of the geological formation; the model does not only depend on the transmissivity, storativity, or leakage factor but also now depends on the scale of the geological formation. In this study we use the model suggested by Amanda and Atangana (2018) and further transform it to a stochastic equation to be able to capture the leak factor and the scale factor, as well as the heterogeneity associated with the statistical setting of the aquifer parameters.

28.4 ANALYSIS OF STOCHASTIC MODELS OF GROUNDWATER FLOW: CONFINED AND LEAKY AQUIFERS

Stochastic approaches have been used in many fields of science, technology, and engineering over the last decade with great success, as they are able to capture heterogeneities that cannot be described with a normal deterministic approach. It is worth noting that several physical occurrences display statistical properties, which cannot be handled with normal deterministic equations as all the input parameters are assumed to be constant. In normal natural problems, one observes in normal situations that input parameters are stochastic and not constant. This is the same situation with aquifer parameters, including transmissivity, storativity, and hydraulic conductivity, which are geologically dependent parameters and more importantly to the discharge rate which is the force applied by the abstraction pump.

The value of each specific parameter in space is independent as geological materials vary from one point to another depending on the fracture type. Fractures are made up of void space in rocks, which affects the flow properties and physical properties of a rock. Hydraulic parameters cannot be stationary in an entire formation; like transmissivity they can only be stationary at a subunit of a geological formation in space, but not across the entire formation.

Boundaries, which might extend far away in some large-scale aquifers, cannot be modeled well because of limited knowledge about them, so they should be treated as stochastic. If these variations are neglected, it may lead to inaccurate results. It is very important to assess any uncertainty that may arise from groundwater systems.

Currently, therefore, to be more representative, we shall consider the latest version of a groundwater flow model within a confined and leaky aquifer, as proposed by Mathobo and Atangana (2018), and Amanda and Atangana (2018). The input parameters will be transformed into stochastic ones by assigning each parameter as a distribution.

28.5 ANALYSIS OF STOCHASTIC MODEL OF GROUNDWATER FLOW: CONFINED AQUIFERS

In this section we analyze the stochastic model for the groundwater flow equation (28.6) in a confined aquifer:

$$T\frac{\partial s(r,t)}{\partial t} = \frac{S}{r}\frac{\partial s(r,t)}{\partial r} + S\frac{\partial^2 s(r,t)}{\partial r^2} \tag{28.6}$$

where T represents aquifer transmissivity and S is aquifer storativity across r radial distance at a time t.

T is considered as constant in the Theis model. However, in reality T changes as a function of s (drawdown) as we argued above; in this case:

$$T \in (T_1, T_2, T_3, T_4, \cdots, T_n)$$

Drawdown data can be analyzed by using statistical methods such as an arithmetic mean or a geometric mean to obtain an accurate value, represented by distribution probability techniques assigned to a stochastic model.

With an arithmetic mean, use:

$$\overline{T_A} = \frac{1}{n}\sum_{j=1}^{n} T_j \tag{28.7}$$

Therefore, for a geometric mean, we have:

$$\overline{T_H} = \left(\prod_{j=1}^{n} T_j\right)^{\frac{1}{n}} = \sqrt[n]{T_1 T_2 T_3 \cdots T_n} \tag{28.8}$$

We consider $0 < \lambda \le 1$, and call it a stochastic parameter.
We consider X_T as a distribution associated with T.

X_T can be normal, log normal, Pareto, Poisson, a Mittag–Leffler distribution, and so on. We replace T with $\hat{T}$ to obtain:

$$\bar{T} = \bar{T} + \lambda X_T \tag{28.9}$$

Following the same approach, we obtain:

$$\bar{S} = \bar{S} + \lambda X_T \tag{28.10}$$

Now the stochastic groundwater flow model in a confined aquifer is given as:

$$\hat{T}\frac{\partial s(r,t)}{\partial t} = \hat{S}\frac{\partial s(r,t)}{r\partial r} + \hat{S}\frac{\partial^2 s(r,t)}{\partial r^2} \tag{28.11}$$

The rate of drawdown changes over time, per radial distance, due to the velocity or dispersion of water when distributed in a confined aquifer. This condition occurs during pumping at different discharge rates, the confining layers moving downwards and causing the aquifer to compress. The pumped water comes from storage within the aquifer at the beginning of pumping and drawdown increases, resulting in an unsteady state flow condition until such time that a well indicates a small change in drawdown (steady state) as storativity increases with the aquifer's thickness.

28.6 ANALYSIS OF A STOCHASTIC MODEL OF GROUNDWATER FLOW: LEAKY AQUIFERS

When analyzing hydraulic parameters in leaky aquifers, one must consider the radial distance of a pumped well and an observation well, which needs to be placed at a distance so that significant leakage in the aquitard can be measured. Theoretically, at the beginning, the well is supplied by the aquifer storage but, at a later stage, it is supplied by storage from the aquitard. The effect of leakage at this point will be minimal but needs to be considered as it also contributes to the drawdown. The value for the leakage factor should be determined from the drawdown of the pumped well and observation wells. Transmissivity and storativity values should be obtained using the drawdown data from the pumped well only because the effect of leakage is not significant and does not influence drawdown. The classical Equation (28.12) commonly used in the literature is:

$$T\frac{\partial h}{\partial t} = S\frac{\partial}{\partial r}\left(\frac{\partial h}{\partial r}\right)\left(1 + \frac{\Delta r}{r}\right) + S\frac{1}{r}\frac{\partial h}{\partial r} + S\frac{h(r,t)}{\lambda^2} \tag{28.12}$$

Following the discussion presented for the case of a confined aquifer, the above equation can be converted to the following equation but with a new parameter introduced, representing the leakage factor:

$$\hat{T}\frac{\partial h}{\partial t} = \hat{S}\frac{\partial}{\partial r}\left(\frac{\partial h}{\partial r}\right)\left(1 + \frac{\Delta r}{r}\right) + \hat{S}\frac{1}{r}\frac{\partial h}{\partial r} + \hat{S}\frac{h(r,t)}{\lambda^2} \tag{28.13}$$

Both stochastic models cannot be solved using classical analytical methods as they become nonlinear due to stochastic coefficients. Therefore, where the analytical method failed, numerical methods can be used to provide an approximated solution. This is presented in detail in the next section.

28.7   APPLICATION OF THE NEWTON METHOD ON STOCHASTIC GROUNDWATER FLOW MODELS FOR CONFINED AND LEAKY AQUIFERS

The Newton method has been regarded as very efficient in solving nonlinear equations (Burden and Faires, 2005), such as PDEs. The rule of the Newton method for an initial guess to solve an equation is $f(x) = 0$. In general, it is used to approximate and solve equations numerically by using the iteration method, discretization, a Taylor series, or the Jacobian method. Solving equations by using an iterative process, you have to keep repeating the procedure to find the root of an equation. A Taylor series is achieved by performing a series of expansions on a function of that point. In this study, we use FD to solve PDEs using discretization to reduce the data size, which corresponds with temporal and spatial variations, and applying the Newton method to approximate an exact solution of the parameters, which is then assigned to a Theis stochastic model for confined aquifers and to a Hantush stochastic model for leaky aquifers. Transmissivity is measured at a point in time, and storativity is measured per radial distance in space, which all vary.

Unlike ordinary difference equations (ODEs) which only require certain ordinary information (one variable), for PDEs we need to know the initial values which may contain multiple variables and extra information about the behavior of the solution in a temporal and spatial domain. We begin with the basic derivatives and their functions, mainly the first-order derivative of a function, the second-order derivative, and Crank–Nicholson, to prove their accuracy so that we can apply them to our model. We also include the forward, backward, and central difference techniques to approximate the derivative of the functions.

A first-order derivative with functions of one or two variables is given as:

$$f'(x) = \lim_{h \to 0} \frac{f(x+h) - f(x)}{h} \tag{28.14}$$

$$f'(x) = \frac{d}{dx} f(x) = \lim_{h \to 0} \frac{f(x+h) - f(x)}{h} \tag{28.15}$$

A second-order derivative with functions of one or two variables is given as:

$$f''(x) = f'(x)' = \frac{d^2 f(x)}{dx^2} = \lim_{h \to 0} \frac{f'(x+h) - f'(x)}{h} = \frac{\lim\limits_{h \to 0} \dfrac{f(x+h) - f(x)}{h} - \dfrac{f(x) - f(x-h)}{h}}{h} \tag{28.16}$$

$$f''(x) = \lim_{h \to 0} \frac{f(x+h) - 2f(x) + f(x-h)}{h^2} \tag{28.17}$$

The forward and backward differences for approximating the derivative of functions with one and multiple variables are:

$$f'(x_i) = \frac{f(x_{i+1}) - f(x_i)}{\Delta x}$$

$$f'(x_i) = \frac{f(x_i) - f(x_{i-1})}{\Delta x} \tag{28.18}$$

$$\frac{\partial f\left(x_i,t_n\right)}{\partial t} = \frac{f\left(x_i,t_{n+1}\right)-f\left(x_i,t_n\right)}{\Delta t} \tag{28.19}$$

$$\frac{\partial f\left(x_i,t_n\right)}{\partial t} = \frac{f\left(x_i,t_n\right)-f\left(x_i,t_{n-1}\right)}{\Delta t} \tag{28.20}$$

In cases where there is a recharge of the aquifer, the sum of the forward and backward differences is obtained to get what is called the central difference:

$$f'\left(x_i\right) = \frac{f\left(x_{i+1}\right)-f\left(x_{i-1}\right)}{\Delta x} \tag{28.21}$$

An approximation of a solution with a second-order derivative for a function with one and two variables using the forward, backward, and central difference method is given as:

$$f''\left(x_i\right) = \frac{f\left(x_{i+1}\right)-2f\left(x_i\right)+f\left(x_{i-1}\right)}{\Delta x^2} \tag{28.22}$$

$$\frac{\partial^2 f\left(x_i,t_n\right)}{\partial x^2} = \frac{f\left(x_{i+1},t_n\right)-2f\left(x_i,t_n\right)+f\left(x_{i-1},t_n\right)}{\left(\Delta x\right)^2} \tag{28.23}$$

By using the Crank–Nicholson numerical scheme, we obtain:

$$\frac{\partial^2 f\left(x_i,t_n\right)}{\partial x^2} = \frac{1}{2\left(\Delta x\right)^2}\left[f\left(x_{i+1},t_{n+1}\right)-2f\left(x_i,t_{n+1}\right)+f\left(x_{i-1},t_{n+1}\right)+f\left(x_{i+1},t_n\right)-2f\left(x_i,t_n\right)+f\left(x_i,t_n\right)\right] \tag{28.24}$$

Introducing the Newton method, we find the following:

$$\frac{dy(t)}{dt} = f\left[t,y(t)\right] \tag{28.25}$$

$$\frac{y(t_{n+1})-y\left(t_n\right)}{\Delta t} = \frac{1}{2}\left[f\left(t_n,y\left(t_n\right)\right)+f\left(t_{n+1},y^{n+1}\right)\right] \tag{28.26}$$

$$\frac{y^{n+1}-y^n}{\Delta t} = \frac{1}{2}\left[f\left(t_n,y^n\right)+f\left(t_{n+1},y^{n+1}\right)\right] \tag{28.27}$$

$$y^{n+1} = y^n + \frac{\Delta t}{2}\left[f(t_n,y^n)+f\left(t_{n+1},y^{n+1}\right)\right] \tag{28.28}$$

$$y^{n+1}-y^n - \frac{1}{2}\Delta t\left[f\left(t_n,y^n\right)+f\left(t_{n+1},y^{n+1}\right)\right] = 0 \tag{28.29}$$

From the equation above, we now replace $y^{n+1} = y$ and $y^n = y^{(1)}$:

$$y - y^{(1)} - \frac{\Delta t}{2} \Big[f\big(t_n, y^1\big) + f\big(t_{n+1}, y\big) \Big] = 0 \tag{28.30}$$

$$f(y) = y - y^1 - \frac{\Delta t}{2} \Big[f\big(t_n, y^1\big) + f\big(t_{n+1}, y\big) \Big] = 0 \tag{28.31}$$

$$f'(y) = \frac{f(y)}{dy} = 1 - \frac{\Delta t}{2} \frac{df\big(t_{n+1}, y\big)}{dy} \tag{28.32}$$

$$f'(y) = 1 - \frac{\Delta t}{2} \frac{df\big(t_{n+1}, y\big)1}{dy} \tag{28.33}$$

$$y^{n+1} = y^n - \frac{f(y_n)}{\dfrac{df(y_n)}{dy}} \tag{28.34}$$

$$y^{n+1} = y^n - \frac{y^n - \dfrac{\Delta t}{2}\Big[f\big(t_n, y^n\big) + f\big(t_{n+1}, y^{n+1}\big) \Big]}{1 - \dfrac{\Delta t}{2} \dfrac{df\big(t_{n+1}, y^{n+1}\big)}{dy}} \tag{28.35}$$

In the next section, we approximate the solution for stochastic confined and leaky aquifer models using PDEs by means of the discretization of time and space.

28.7.1 APPLICATION OF THE NEWTON METHOD TO A STOCHASTIC THEIS'S CONFINED AQUIFER

To apply the Newton method, we reformulate the groundwater flow equation in a confined aquifer to obtain a stochastic model, which covers the rate of change in transmissivity at a point in time and the storativity per radial distance:

$$\frac{\hat{S}}{\hat{T}} \frac{\partial s(r,t)}{r\partial r} + \frac{\hat{S}}{\hat{T}} \frac{\partial^2 s(r,t)}{\partial r^2} = f\big(r,t,s(r,t)\big) \tag{28.36}$$

At points r_i, t_n, we rearrange to obtain:

$$\frac{s(r_i, t_{n+1}) - s(r_i, t_n)}{\Delta t} = \frac{1}{2} \Big[f(r_i, t_n, s(r_i, t_n)) + f(r_i, t_{n+1}), s(r_i, t_{n+1}) \Big] \tag{28.37}$$

$$s_i^{n+1} - s_i^n = \frac{\Delta t}{2} \Big[f\big(r_i, t_n, s_i^n\big) + f\big(r_i, t_{n+1}, s_i^{n+1}\big) \Big] \tag{28.38}$$

From the previous equation, we then set $s_i^{n+1} = s_i$ and $s_i^n = s_i^1$ to get the root of the function, which gives us the following equations:

$$s_i = s_i^{(1)} + \frac{\Delta t}{2}\left[f\left(r_i,t_n,s_i^1\right) + f\left(r_i,t_{n+1},s_i\right)\right] \tag{28.39}$$

$$s_i - s_i^1 - \frac{\Delta t}{2}\left[f\left(r_i,t_n,s_i^1\right) + f\left(r_i,t_{n+1},s_i\right)\right] = 0 \tag{28.40}$$

According to the Newton method, these should be expressed as:

$$F\left(s_i\right) = s_i - s_i^1 - \frac{\Delta t}{2}\left[f\left(r_i,t_n,s_i^1\right) + f\left(r_i,t_{n+1},s_i\right)\right] \tag{28.41}$$

$$\frac{dF\left(s_i\right)}{ds_i} = 1 - \frac{\Delta t}{2}\left[\frac{df\left(r_i,t_{n+1},s_i\right)}{ds_i}\right]$$

$$= 1 - \frac{\Delta t}{2}\frac{df\left(r_i,t_{n+1},s_i\right)}{ds_i} \tag{28.42}$$

$$s_i^{n+1} - s_i^n = \frac{s_i^n - \frac{\Delta t}{2} f\left(r_i,t_{n+1},s_i^{n-1}\right) - \frac{\Delta t}{2} f\left(r_i,t_n,s_i^n\right)}{1 - \frac{\Delta t}{2}\frac{df\left(r_i,t_{n+1},s_i\right)}{ds_i}} \tag{28.43}$$

To further discretize the domain, we recall that:

$$f(r_i,t_n,s(r,t)) = \frac{\hat{S}}{\hat{T}}\left[\frac{\partial s(r,t)}{r\partial r} + \frac{\partial^2 s(r,t)}{\partial r^2}\right] \tag{28.44}$$

$$f\left(r_i,t_n,s\left(r_i,t_n\right)\right) = \frac{\hat{S}}{\hat{T}}\left[\frac{1}{r_i}\frac{s_{i+1}^n - s_{i-1}^n}{\Delta r} + \frac{s_{i+1}^n - 2s_i^n + s_{i-1}^n}{\left(\Delta r\right)^2}\right] \tag{28.45}$$

$$f\left(r_i,t_{n+1},s\left(r_i,t_{n+1}\right)\right) = \frac{\hat{S}}{\hat{T}}\left[\frac{1}{r_i}\frac{s_{i+1}^{n+1} - s_{i-1}^{n+1}}{\Delta r} + \frac{s_{i+1}^{n+1} - 2s_i^{n+1} + s_{i-1}^{n+1}}{\left(\Delta r\right)^2}\right] \tag{28.46}$$

$$\frac{df\left(r_i,t_{n+1},s_i\right)}{ds_i} = \frac{\hat{S}}{\hat{T}}\left[-\frac{2}{\left(\Delta r\right)^2}\right] = -\frac{2\hat{S}}{\hat{T}\left(\Delta r\right)^2} \tag{28.47}$$

And the final product after all the steps above, our groundwater flow stochastic equation for a confined aquifer with the application of the new scheme, is Equation (28.48). The solution proves that the rate of drawdown changes per radial distance in respect of the dispersion of water at a point in time in an aquifer. The discharge rate at which the well is pumped also affects the drawdown:

$$s_i^{n+1} = s_i^n - \frac{\Delta t}{2} \frac{\left\{ \left[\dfrac{1}{r_i} \dfrac{s_{i+1}^n - s_{i-1}^n}{\Delta r} + \dfrac{s_{i+1}^n - 2s_i^n + s_{i-1}^n}{(\Delta r)^2} \right] + \left[\dfrac{1}{r_i} \dfrac{s_{i+1}^{n+1} - s_{i-1}^{n+1}}{\Delta r} + \dfrac{s_{i+1}^{n+1} - 2s_i^{n+1} + s_{i-1}^{n+1}}{(\Delta r)^2} \right] \right\}}{1 - \dfrac{\Delta t}{2} \left[\dfrac{-2\hat{S}}{\hat{T}(\Delta r)^2} \right]}. \tag{28.48}$$

28.7.2 APPLICATION OF THE NEWTON METHOD TO A STOCHASTIC HANTUSH'S LEAKY AQUIFER

In order to achieve the groundwater flow stochastic model for a leaky aquifer, we also apply the Newton method using a similar procedure to the one used for a confined aquifer. We therefore reformulate the groundwater flow equation of a leaky aquifer and apply the Newton method to approximate the solution:

$$\frac{\hat{S}}{\hat{T}} \frac{\partial s}{r \partial r} + \frac{\hat{S}}{\hat{T}} \frac{\partial^2 s(r,t)}{\partial r^2} + \frac{s(r,t)}{\lambda^2} = f(r,t,s(r,t)) \tag{28.49}$$

At the points r_i, t_n, we obtain:

$$\frac{s_i^{n+1} - s_i^n}{\Delta t} = \frac{1}{2} \left[f(r_i, t_n, s_i^n) + f(r_i, t_{n+1}) \right] \tag{28.50}$$

$$s_i^{n+1} - s_i^n = \frac{\Delta t}{2} \left[f(r_i, t_n, s_i^n) + f(r_i, t_{n+1}, s_i^{n+1}) \right] \tag{28.51}$$

Since $s_i^{n+1} = s_i$ and if $s_i^n = s_i^1$, then the solution will be:

$$s_i = s_i^{(1)} + \frac{\Delta t}{2} \left[f(r_i, t_n, s_i^1) + f(r_i, t_{n+1}, s_i) \right] \tag{28.52}$$

$$s_i - s_i^1 - \frac{\Delta t}{2} \left[f(r_i, t_n, s_i^1) + f(r_i, t_{n+1}, s_i) \right] = 0 \tag{28.53}$$

By generalizing the above equation, we apply the Newton method to obtain:

$$F(s_i) = s_i - s_i^1 - \frac{\Delta t}{2} \left[f(r_i, t_n, s_i^1) + f(r_i, t_{n+1}, s_i) \right] \tag{28.54}$$

This implies that:

$$\frac{dF(s_i)}{ds_i} = 1 - \frac{\Delta t}{2} \left[\frac{df(r_i, t_{n+1}, s_i)}{ds_i} \right] = 1 - \frac{\Delta t}{2} \frac{df(r_i, t_{n+1}, s_i)}{ds_i} - \frac{s_i}{\lambda^2} \tag{28.55}$$

$$s_i^{n+1} - s_i^n = \frac{s_i^n - \frac{\Delta t}{2} f\left(r_i, t_{n+1}, s_i^{n-1}\right) - \frac{\Delta t}{2} f\left(r_i, t_n, s_i^n\right)}{1 - \frac{\Delta t}{2} \frac{df\left(r_i, t_{n+1}, s_i\right)}{ds_i}} \qquad (28.56)$$

From the previous equation, we get the solution:

$$f(r,t)\, s(r,t) = \frac{\hat{S}}{\hat{T}}\left[\frac{\partial s(r,t)}{r \partial r} + \frac{\partial^2 s(r,t)}{\partial r^2}\right] \qquad (28.57)$$

$$f\left(r_i, t_n, s\left(s_i^n\right)\right) = \frac{\hat{S}}{\hat{T}}\left[\frac{1}{r_i}\frac{s_{i+1}^n - s_{i-1}^n}{\Delta r} + \frac{s\left(r_i, t_n\right)}{\lambda^2} + \frac{s_{i+1}^n - 2s_i^n + s_{i-1}^n}{\left(\Delta r\right)^2}\right] \qquad (28.58)$$

$$f\left(r_i, t_{n+1}, s\left(s_i^{n+1}\right)\right) = \frac{\hat{S}}{\hat{T}}\left[\frac{1}{r_i}\frac{s_{i+1}^{n+1} - s_{i-1}^{n+1}}{\Delta r} + \frac{s\left(r_i, t_{n+1}\right)}{\lambda^2} + \frac{s_{i+1}^{n+1} - 2s_i^{n+1} + s_{i-1}^{n+1}}{\left(\Delta r\right)^2}\right] \qquad (28.59)$$

$$\frac{df\left(r_i, t_{n+1}, s_i\right)}{ds_i} = \frac{\hat{S}}{\hat{T}}\left[-\frac{2}{\left(\Delta r\right)^2}\right] = -\frac{2\hat{S}}{\hat{T}\left(\Delta r\right)^2} \qquad (28.60)$$

After we have discretized the stochastic groundwater flow for a leaky equation with the steps specified above, unlike that for a confined aquifer's equation, we now add a new parameter λ^2 which represents a leakage factor in an aquifer:

$$s_i^{n+1} = s_i^n - \frac{\Delta t}{2}\frac{\left\{\left[\frac{1}{r_i}\frac{s_{i+1}^n - s_{i-1}^n}{\Delta r} + \frac{s_{i+1}^n - 2s_i^n + s_{i-1}^n}{\left(\Delta r\right)^2}\right] + \left[\frac{1}{r_i}\frac{s_{i+1}^{n+1} - s_{i-1}^{n+1}}{\Delta r} + \frac{s_{i+1}^{n+1} - 2s_i^{n+1} + s_{i-1}^{n+1}}{\left(\Delta r\right)^2}\right] + \frac{s_i^{n+1}}{\lambda^2}\right\}}{1 - \frac{\Delta t}{2}\left[\frac{-2\hat{S}}{\hat{T}\left(\Delta r\right)^2} + \frac{1}{\lambda^2}\right]} \qquad (28.61)$$

The new parameter is known as the leakage factor, which shall now be considered even though its effects on drawdown are negligible. Therefore, the modified new scheme is made up of certain parameters in a leaky aquifer such as transmissivity, storativity, the leakage factor, and drawdown. The leakage factor is ignored in Hantush–Jacob's equation, thus we modified it and included it, which contributes to the storativity of the aquifer, irrespective of the thickness of the aquitard. The fact that the leakage factor should be recognized further balances the steady condition when the water is leaking through the aquitard into the aquifer, which is ideal for recharging it.

28.7.3 STABILITY OF THE STOCHASTIC CONFINED AQUIFER EQUATION

Stabilizing the model means that most of the objects used in it will be stable over time and do not need changing. The model is considered accurate when it reaches stability convergence and there is little difference between the exact and approximate solution when fitting the dataset. If the model's parameters are different during the forecast period than they were during the construction period, then the model estimated will not be very useful regardless of how well it was estimated (Sinha and Prasad, 1979). If the model's parameters were unstable over the construction period, then the model was not even a good representation of how the parameters actually occur within the groundwater flow in the aquifer at a point in time. We analyzed our model for any inconsistency by means of stability before the predictions were applied to a real system.

Stability analysis for Equations (28.75, 28.76, 28.87) is as follows:

$$
\delta_i^{n+1}\left[1+\frac{2}{(\Delta r)^2\left(1-\frac{\Delta t\hat{S}}{\hat{T}(\Delta r)^2}\right)}\right]=\delta_i^n\left[1-\frac{\Delta t}{(\Delta r)^2\left(1-\frac{\Delta t\hat{S}}{\hat{T}(\Delta r)^2}\right)}\right]+\delta_{i+1}^{n+1}\left[\left(\frac{-\Delta t}{2r_i\Delta r}+\frac{1}{(\Delta r)^2}\right)\frac{1}{1+\frac{\Delta t\hat{S}}{\hat{T}(\Delta r)^2}}\right]
$$

$$
+\delta_{i-1}^{n+1}\left[\frac{\Delta t}{2\Delta r}+\frac{1}{(\Delta r)^2}\right]\frac{1}{1+\frac{\Delta t\hat{S}}{\hat{T}(\Delta r)^2}}+\delta_{i+1}^n\left[\frac{-\Delta t}{2r_i}-\frac{\Delta t}{2(\Delta r)^2}\right]\frac{1}{1+\frac{\Delta t\hat{S}}{\hat{T}(\Delta r)^2}}
$$

$$
+\delta_{i-1}^n\left[\frac{-\Delta t}{2r_i}-\frac{\Delta t}{2(\Delta r)^2}\right]\frac{1}{1+\frac{\Delta t\hat{S}}{\hat{T}(\Delta r)^2}}
$$

If $\delta_i^n=\delta_n e^{ikmr}$, then $\delta_{i-1}^{n+1}=\delta_{n+1}e^{ikm(r-\Delta r)}$

$$
\delta_{n+1}e^{ikmr}\left(1+\frac{2}{(\Delta r)^2\left(1-\frac{\Delta t\hat{S}}{\hat{T}(\Delta r)^2}\right)}\right)=\delta_n\left(1-\frac{\Delta t}{(\Delta r)^2\left(1+\frac{\Delta t\hat{S}}{\hat{T}(\Delta r)^2}\right)}\right)
$$

$$
+\delta_{n+1}e^{ikm(r+\Delta r)}\left(\left(\frac{-\Delta t}{2r_i\Delta r}+\frac{1}{(\Delta r)^2}\right)\frac{1}{1+\frac{\Delta t\hat{S}}{\hat{T}(\Delta r)^2}}\right)+\delta_{n+1}e^{ikm(r-\Delta r)}\left(\left(\frac{\Delta t}{2\Delta r}+\frac{1}{(\Delta r)^2}\right)\frac{1}{1+\frac{\Delta t\hat{S}}{\hat{T}(\Delta r)^2}}\right)
$$

$$
+\delta_n e^{ikm(r+\Delta r)}\left(\left(\frac{-\Delta t}{2r_i}-\frac{\Delta t}{2(\Delta r)^2}\right)\frac{1}{1+\frac{\Delta t\hat{S}}{\hat{T}(\Delta r)^2}}\right)+\delta_n e^{ikm(r-\Delta r)}\left(\left(\frac{-\Delta t}{2r_i}-\frac{\Delta t}{2(\Delta r)^2}\right)\frac{1}{1+\frac{\Delta t\hat{S}}{\hat{T}(\Delta r)^2}}\right)
$$

$$
\text{(28.62)}
$$

We further convert:

$$e^{ikm(r-\Delta r)} = e^{ikmr} \cdot e^{-ikm\Delta r}$$

$$e^{ikm(r+\Delta r)} = e^{ikmr} \cdot e^{ikm\Delta r}$$

And add it to the solution to get:

$$\delta_{n+1}e^{ikmr}\left(1+\frac{2}{(\Delta r)^2\left(1-\dfrac{\Delta t\hat{S}}{\hat{T}(\Delta r)^2}\right)}\right) = \delta_n e^{ikmr}\left(1-\frac{\Delta t}{(\Delta r)^2\left(1-\dfrac{\Delta t\hat{S}}{\hat{T}(\Delta r)^2}\right)}\right)$$

$$+\delta_{n+1}e^{ikmr}\cdot e^{ikm\Delta r}\left(\left(\frac{-\Delta t}{2r_i\Delta r}+\frac{1}{(\Delta r)^2}\right)\frac{1}{1+\dfrac{\Delta t\hat{S}}{\hat{T}(\Delta r)^2}}\right) + \delta_{n+1}e^{ikmr}\cdot e^{-ikm\Delta r}\left(\left(\frac{\Delta t}{2\Delta r}+\frac{1}{(\Delta r)^2}\right)\frac{1}{1+\dfrac{\Delta t\hat{S}}{\hat{T}(\Delta r)^2}}\right)$$

$$+\delta_n e^{ikmr}\cdot e^{ikm\Delta r}\left(\left(\frac{-\Delta t}{2r_i}-\frac{\Delta t}{2(\Delta r)^2}\right)\frac{1}{1+\dfrac{\Delta t\hat{S}}{\hat{T}(\Delta r)^2}}\right) + \delta_n e^{ikmr}\cdot e^{-ikm\Delta r}\left(\left(\frac{-\Delta t}{2r_i}-\frac{\Delta t}{2(\Delta r)^2}\right)\frac{1}{1+\dfrac{\Delta t\hat{S}}{\hat{T}(\Delta r)^2}}\right)$$

$$(28.63)$$

After simplification, we obtain:

$$\delta_{n+1}\left(1+\frac{2}{(\Delta r)^2\left(1-\dfrac{\Delta t\hat{S}}{\hat{T}(\Delta r)^2}\right)}\right) = \delta_n\left(1-\frac{\Delta t}{(\Delta r)^2\left(1-\dfrac{\Delta t\hat{S}}{\hat{T}(\Delta r)^2}\right)}\right)$$

$$+\delta_{n+1}e^{ikm\Delta r}\left(\left(\frac{-\Delta t}{2r_i\Delta r}+\frac{1}{(\Delta r)^2}\right)\frac{1}{1+\dfrac{\Delta t\hat{S}}{\hat{T}(\Delta r)^2}}\right) + \delta_{n+1}e^{-ikm\Delta r}\left(\left(\frac{\Delta t}{2\Delta r}+\frac{1}{(\Delta r)^2}\right)\frac{1}{1+\dfrac{\Delta t\hat{S}}{\hat{T}(\Delta r)^2}}\right)$$

$$+\delta_n e^{ikm\Delta r}\left(\left(\frac{-\Delta t}{2r_i}-\frac{\Delta t}{2(\Delta r)^2}\right)\frac{1}{1+\dfrac{\Delta t\hat{S}}{\hat{T}(\Delta r)^2}}\right) + \delta_n e^{-ikm\Delta r}\left(\left(\frac{-\Delta t}{2r_i}-\frac{\Delta t}{2(\Delta r)^2}\right)\frac{1}{1+\dfrac{\Delta t\hat{S}}{\hat{T}(\Delta r)^2}}\right)$$

$$(28.64)$$

$$\delta_{n+1}\left[\left[\left(1+\frac{2}{(\Delta r)^2\left(1-\frac{\Delta t\hat{S}}{\hat{T}(\Delta r)^2}\right)}\right)-e^{ikm\Delta r}\left(1-\frac{\Delta t}{(\Delta r)^2\left(1+\frac{\Delta t\hat{S}}{\hat{T}(\Delta r)^2}\right)}\right)\right]\right.$$

$$\left.+e^{-ikm\Delta r}\left(\frac{-\Delta t}{2r_i\Delta r}+\frac{1}{(\Delta r)^2}\right)\frac{1}{\left(1+\frac{\Delta t\hat{S}}{\hat{T}(\Delta r)^2}\right)}\right]$$

$$=\delta_n\left[\left[\left(\frac{\Delta t}{2\Delta r}+\frac{1}{(\Delta r)^2}\right)\frac{1}{\left(1+\frac{\Delta t\hat{S}}{\hat{T}(\Delta r)^2}\right)}+e^{ikm\Delta r}\left(\left(\frac{-\Delta t}{2r_i}-\frac{\Delta t}{2(\Delta r)^2}\right)\frac{1}{\left(1+\frac{\Delta t\hat{S}}{\hat{T}(\Delta r)^2}\right)}\right)\right]\right.$$

$$\left.+e^{-ikm\Delta r}\left(\left(\frac{-\Delta t}{2r_i}-\frac{\Delta t}{2(\Delta r)^2}\right)\frac{1}{\left(1+\frac{\Delta t\hat{S}}{\hat{T}(\Delta r)^2}\right)}\right)\right] \qquad (28.65)$$

For simplicity, we put the following in Equation (28.65) to get the outputs in Equations (28.66–28.68):

$$a_1=1+\frac{2}{(\Delta r)^2\left(1-\frac{\Delta t\hat{S}}{\hat{T}(\Delta r)^2}\right)}$$

$$a_2=1-\frac{\Delta t}{(\Delta r)^2\left(1+\frac{\Delta t\hat{S}}{\hat{T}(\Delta r)^2}\right)}$$

$$a_3=\left(\frac{-\Delta t}{2r_i\Delta r}+\frac{1}{(\Delta r)^2}\right)\frac{1}{\left(1+\frac{\Delta t\hat{S}}{\hat{T}(\Delta r)^2}\right)}$$

$$a_4=\left(\frac{\Delta t}{2\Delta r}+\frac{1}{(\Delta r)^2}\right)\frac{1}{\left(1+\frac{\Delta t\hat{S}}{\hat{T}(\Delta r)^2}\right)}$$

$$a_5 = \left(\frac{-\Delta t}{2r_i} - \frac{\Delta t}{2(\Delta r)^2} \right) \frac{1}{\left(1 + \frac{\Delta t \hat{S}}{\hat{T}(\Delta r)^2} \right)} \tag{28.66}$$

$$a_1 = 1 + \frac{2}{(\Delta r)^2 \left(1 - \frac{\Delta t \hat{S}}{\hat{T}(\Delta r)^2} \right)}$$

$$a_2 = 1 - \frac{\Delta t}{(\Delta r)^2 \left(1 + \frac{\Delta t \hat{S}}{\hat{T}(\Delta r)^2} \right)}$$

$$a_3 = \left(\frac{-\Delta t}{2r_i \Delta r} + \frac{1}{(\Delta r)^2} \right) \frac{1}{\left(1 + \frac{\Delta t \hat{S}}{\hat{T}(\Delta r)^2} \right)}$$

$$a_4 = \left(\frac{\Delta t}{2\Delta r} + \frac{1}{(\Delta r)^2} \right) \frac{1}{\left(1 + \frac{\Delta t \hat{S}}{\hat{T}(\Delta r)^2} \right)}$$

$$a_5 = \left(\frac{-\Delta t}{2r_i} - \frac{\Delta t}{2(\Delta r)^2} \right) \frac{1}{\left(1 + \frac{\Delta t \hat{S}}{\hat{T}(\Delta r)^2} \right)}$$

Our equation now becomes:

$$\delta_{n+1} \left[a_1 - e^{ikm\Delta r} a_2 + a_3 e^{-ikm\Delta r} \right] = \delta_n \left[a_4 + e^{ikm\Delta r} a_5 + e^{-ikm\Delta r} a_5 \right] \tag{28.67}$$

$$\delta_{n+1} \left[a_1 + a_2 \left(\cos km\Delta r + i \sin(km\Delta r) \right) + a_3 \left(\cos km\Delta r - i \sin km\Delta r \right) \right] = \delta_n \left[a_4 + 2a_5 \cos(km\Delta r) \right] \tag{28.68}$$

$$\delta_{n+1} \left[a_1 + a_2 \cos(km\Delta r) + a_3 \cos(km\Delta r) + i \sin(km\Delta r)(a_2 - a_3) = \delta_n \left(a_4 + 2a_5 \cos(km\Delta r) \right) \right] \tag{28.69}$$

Factorizing the previous equation we get:

$$\delta_{n+1} \left[a_1 + (a_2 + a_3) \cos(km\Delta r) + i \sin(km\Delta r)(a_2 - a_3) \right] = \delta_n \left[\left(a_4 + 2a_5 \cos(km\Delta r) \right) \right] \tag{28.70}$$

Every model is regarded as stable when it converges this way:

$$\left|\frac{\delta_{n+1}}{\delta_n}\right| < 1$$

The approximate and exact solution for the real and imaginary system proves that our model is stable as it resulted in a solution which is < 1, which is accurate as theoretically predicted and as proved in Equations (28.71–28.74):

$$\frac{\delta_{n+1}}{\delta_n} = \frac{a_4 + 2a_5 \cos(km\Delta r)}{a_1 + (a_2 + a_3)\cos(km\Delta r) + i\sin(km\Delta r)(a_2 - a_3)} \tag{28.71}$$

$$\left|\frac{\delta_{n+1}}{\delta_n}\right| = \left|\frac{a_4 + 2a_5 \cos(km\Delta r)}{a_1 + (a_2 + a_3)\cos(km\Delta r) + i\sin(km\Delta r)(a_2 - a_3)}\right| \tag{28.72}$$

$$\frac{\delta_{n+1}}{\delta_n} = \frac{|a_4 + 2a_5 \cos(km\Delta r)|}{\sqrt{(a_1 + (a_2 + a_3)\cos(km\Delta r))^2 + (\sin(km\Delta r)(a_2 - a_3))^2}} < 1 \tag{28.73}$$

$$|a_4 + 2a_5 \cos(km\Delta r)| < \sqrt{(a_1 + (a_2 + a_3)\cos(km\Delta r))^2 + (\sin(km\Delta r)(a_2 - a_3))^2} \tag{28.74}$$

The numerical analysis for our stochastic equation in a confined aquifer converges with the actual analysis of a system. The parameters assigned to construct the models meet with what was theoretically predicted. When we arrive at this stage, it shows that our model will perform well and can be relied upon to make predictions in the future. Therefore, in the next part of the simulation, we should expect these parameters to be similar to the simulated ones.

28.8   STABILITY OF THE STOCHASTIC LEAKY AQUIFER EQUATION

Regarding the stability of the leaky equation, we follow a similar approach as used for the confined aquifer. We should note that for the leaky aquifer equation there is an additional parameter known as the leakage factor, which compromises the aquitard, though which is not common in a confined aquifer. This can be checked in Equations (28.75–28.77) to obtain Equations (28.78 and 28.79):

$$\delta_i^{n+1}\left[1 + \frac{2}{(\Delta r)^2\left(1 - \frac{\Delta t\hat{S}}{\hat{T}(\Delta r)^2}\right) + \frac{1}{\hat{\lambda}^2}}\right] = \delta_i^n\left[1 - \frac{\Delta t}{(\Delta r)^2\left(1 - \frac{\Delta t\hat{S}}{\hat{T}(\Delta r)^2}\right) + \frac{1}{\hat{\lambda}^2}}\right]$$

$$+ \delta_{i+1}^{n+1}\left[\left(\frac{-\Delta t}{2r_i\Delta r} + \frac{1}{(\Delta r)^2}\right)\frac{1}{1 + \frac{\Delta t\hat{S}}{\hat{T}(\Delta r)^2} + \frac{1}{\hat{\lambda}^2}}\right] + \delta_{i-1}^{n+1}\left[\frac{\Delta t}{2\Delta r} + \frac{1}{(\Delta r)^2}\right]\frac{1}{1 + \frac{\Delta t\hat{S}}{\hat{T}(\Delta r)^2} + \frac{1}{\hat{\lambda}^2}}$$

$$+ \delta_{i+1}^n\left[\frac{-\Delta t}{2r_i} - \frac{\Delta t}{2(\Delta r)^2}\right]\frac{1}{1 + \frac{\Delta t\hat{S}}{\hat{T}(\Delta r)^2} + \frac{1}{\hat{\lambda}^2}} + \delta_{i-1}^n\left[\frac{-\Delta t}{2r_i} - \frac{\Delta t}{2(\Delta r)^2}\right]\frac{1}{1 + \frac{\Delta t\hat{S}}{\hat{T}(\Delta r)^2} + \frac{1}{\hat{\lambda}^2}} \tag{28.75}$$

If $\delta_i^n = \delta_n e^{ikmr}$, then $\delta_{i-1}^{n+1} = \delta_{n+1} e^{ikm(r-\Delta r)}$

$$\delta_{n+1} e^{ikmr}\left(1 + \dfrac{2}{(\Delta r)^2\left(1 - \dfrac{\Delta t \hat{S}}{\hat{T}(\Delta r)^2} + \dfrac{1}{\lambda^2}\right)}\right) = \delta_n\left(1 - \dfrac{\Delta t}{(\Delta r)^2\left(1 + \dfrac{\Delta t \hat{S}}{\hat{T}(\Delta r)^2} + \dfrac{1}{\lambda^2}\right)}\right)$$

$$+ \delta_{n+1} e^{ikm(r+\Delta r)}\left(\left(\dfrac{-\Delta t}{2r_i \Delta r} + \dfrac{1}{(\Delta r)^2}\right)\dfrac{1}{1 + \dfrac{\Delta t \hat{S}}{\hat{T}(\Delta r)^2} + \dfrac{1}{\lambda^2}}\right) + \delta_{n+1} e^{ikm(r-\Delta r)}\left(\left(\dfrac{\Delta t}{2\Delta r} + \dfrac{1}{(\Delta r)^2}\right)\dfrac{1}{1 + \dfrac{\Delta t \hat{S}}{\hat{T}(\Delta r)^2} + \dfrac{1}{\lambda^2}}\right)$$

$$+ \delta_n e^{ikm(r+\Delta r)}\left(\left(\dfrac{-\Delta t}{2r_i} - \dfrac{\Delta t}{2(\Delta r)^2}\right)\dfrac{1}{1 + \dfrac{\Delta t \hat{S}}{\hat{T}(\Delta r)^2} + \dfrac{1}{\lambda^2}}\right) + \delta_n e^{ikm(r-\Delta r)}\left(\left(\dfrac{-\Delta t}{2r_i} - \dfrac{\Delta t}{2(\Delta r)^2}\right)\dfrac{1}{1 + \dfrac{\Delta t \hat{S}}{\hat{T}(\Delta r)^2} + \dfrac{1}{\lambda^2}}\right)$$

$$(28.76)$$

We further convert:

$$e^{ikm(r-\Delta r)} = e^{ikmr} \cdot e^{-ikm\Delta r}$$

$$e^{ikm(r+\Delta r)} = e^{ikmr} \cdot e^{ikm\Delta r}$$

Adding it to the solution to get:

$$\delta_{n+1} e^{ikmr}\left(1 + \dfrac{1}{\lambda^2}\dfrac{2}{(\Delta r)^2\left(1 - \dfrac{\Delta t \hat{S}}{\hat{T}(\Delta r)^2} + \dfrac{1}{\lambda^2}\right)}\right)$$

$$= \delta_n e^{ikmr}\left(1 - \dfrac{\Delta t}{(\Delta r)^2\left(1 - \dfrac{\Delta t \hat{S}}{\hat{T}(\Delta r)^2} + \dfrac{1}{\lambda^2}\right)}\right) + \delta_{n+1} e^{ikmr} \cdot e^{ikm\Delta r}\left(\left(\dfrac{-\Delta t}{2r_i \Delta r} + \dfrac{1}{(\Delta r)^2}\right)\dfrac{1}{1 + \dfrac{\Delta t \hat{S}}{\hat{T}(\Delta r)^2} + \dfrac{1}{\lambda^2}}\right)$$

$$+ \delta_{n+1} e^{ikmr} \cdot e^{-ikm\Delta r}\left(\left(\dfrac{\Delta t}{2\Delta r} + \dfrac{1}{(\Delta r)^2}\right)\dfrac{1}{1 + \dfrac{\Delta t \hat{S}}{\hat{T}(\Delta r)^2} + \dfrac{1}{\lambda^2}}\right) + \delta_n e^{ikmr} \cdot e^{ikm\Delta r}\left(\left(\dfrac{-\Delta t}{2r_i} - \dfrac{\Delta t}{2(\Delta r)^2}\right)\dfrac{1}{1 + \dfrac{\Delta t \hat{S}}{\hat{T}(\Delta r)^2} + \dfrac{1}{\lambda^2}}\right)$$

$$+ \delta_n e^{ikmr} \cdot e^{-ikm\Delta r}\left(\left(\dfrac{-\Delta t}{2r_i} - \dfrac{\Delta t}{2(\Delta r)^2}\right)\dfrac{1}{1 + \dfrac{\Delta t \hat{S}}{\hat{T}(\Delta r)^2} + \dfrac{1}{\lambda^2}}\right)$$

$$(28.77)$$

After simplification, we obtain:

$$
\delta_{n+1}\left(1+\frac{1}{\overset{\scriptstyle 2}{\lambda}}\frac{2}{(\Delta r)^2\left(1-\frac{\Delta t\hat{S}}{\hat{T}(\Delta r)^2}+\frac{1}{\overset{\scriptstyle 2}{\lambda}}\right)}\right)
$$

$$
=\delta_n\left(1-\frac{\Delta t}{(\Delta r)^2\left(1-\frac{\Delta t\hat{S}}{\hat{T}(\Delta r)^2}+\frac{1}{\overset{\scriptstyle 2}{\lambda}}\right)}\right)+\delta_{n+1}e^{ikm\Delta r}\left(\left(\frac{-\Delta t}{2r_i\Delta r}+\frac{1}{(\Delta r)^2}\right)\frac{1}{1+\frac{\Delta t\hat{S}}{\hat{T}(\Delta r)^2}+\frac{1}{\overset{\scriptstyle 2}{\lambda}}}\right)
$$

$$
+\delta_{n+1}e^{-ikm\Delta r}\left(\left(\frac{\Delta t}{2\Delta r}+\frac{1}{(\Delta r)^2}\right)\frac{1}{1+\frac{\Delta t\hat{S}}{\hat{T}(\Delta r)^2}+\frac{1}{\overset{\scriptstyle 2}{\lambda}}}\right)+\delta_n e^{ikm\Delta r}\left(\left(\frac{-\Delta t}{2r_i}-\frac{\Delta t}{2(\Delta r)^2}\right)\frac{1}{1+\frac{\Delta t\hat{S}}{\hat{T}(\Delta r)^2}+\frac{1}{\overset{\scriptstyle 2}{\lambda}}}\right)
$$

$$
+\delta_n e^{-ikm\Delta r}\left(\left(\frac{-\Delta t}{2r_i}-\frac{\Delta t}{2(\Delta r)^2}\right)\frac{1}{1+\frac{\Delta t\hat{S}}{\hat{T}(\Delta r)^2}+\frac{1}{\overset{\scriptstyle 2}{\lambda}}}\right) \tag{28.78}
$$

$$
\delta_{n+1}\left[\left(1+\frac{1}{\overset{\scriptstyle 2}{\lambda}}\frac{2}{(\Delta r)^2\left(1-\frac{\Delta t\hat{S}}{\hat{T}(\Delta r)^2}+\frac{1}{\overset{\scriptstyle 2}{\lambda}}\right)}\right)-e^{ikm\Delta r}\left(1-\frac{\Delta t}{(\Delta r)^2\left(1+\frac{\Delta t\hat{S}}{\hat{T}(\Delta r)^2}+\frac{1}{\overset{\scriptstyle 2}{\lambda}}\right)}\right)\right.
$$

$$
\left.+e^{-ikm\Delta r}\left(\frac{-\Delta t}{2r_i\Delta r}+\frac{1}{(\Delta r)^2}\right)\frac{1}{\left(1+\frac{\Delta t\hat{S}}{\hat{T}(\Delta r)^2}+\frac{1}{\overset{\scriptstyle 2}{\lambda}}\right)}\right]
$$

$$
=\delta_n\left[\left(\frac{\Delta t}{2\Delta r}+\frac{1}{(\Delta r)^2}\right)\frac{1}{\left(1+\frac{\Delta t\hat{S}}{\hat{T}(\Delta r)^2}+\frac{1}{\overset{\scriptstyle 2}{\lambda}}\right)}+e^{ikm\Delta r}\left(\left(\frac{-\Delta t}{2r_i}-\frac{\Delta t}{2(\Delta r)^2}\right)\frac{1}{\left(1+\frac{\Delta t\hat{S}}{\hat{T}(\Delta r)^2}+\frac{1}{\overset{\scriptstyle 2}{\lambda}}\right)}\right)\right.
$$

$$
\left.+e^{-ikm\Delta r}\left(\left(\frac{-\Delta t}{2r_i}-\frac{\Delta t}{2(\Delta r)^2}\right)\frac{1}{\left(1+\frac{\Delta t\hat{S}}{\hat{T}(\Delta r)^2}+\frac{1}{\overset{\scriptstyle 2}{\lambda}}\right)}\right)\right] \tag{28.79}
$$

For simplicity, we use the following parameters to obtain Equation (28.80):

$$a_1 = 1 + \frac{1}{\lambda^2} \frac{2}{(\Delta r)^2 \left(1 - \frac{\Delta t \hat{S}}{\hat{T}(\Delta r)^2} + \frac{1}{\lambda^2}\right)}$$

$$a_2 = 1 - \frac{\Delta t}{(\Delta r)^2 \left(1 + \frac{\Delta t \hat{S}}{\hat{T}(\Delta r)^2} + \frac{1}{\lambda^2}\right)}$$

$$a_3 = \left(\frac{-\Delta t}{2 r_i \Delta r} + \frac{1}{(\Delta r)^2}\right) \frac{1}{\left(1 + \frac{\Delta t \hat{S}}{\hat{T}(\Delta r)^2} + \frac{1}{\lambda^2}\right)}$$

$$a_4 = \left(\frac{\Delta t}{2 \Delta r} + \frac{1}{(\Delta r)^2}\right) \frac{1}{\left(1 + \frac{\Delta t \hat{S}}{\hat{T}(\Delta r)^2} + \frac{1}{\lambda^2}\right)}$$

$$a_5 = \left(\frac{-\Delta t}{2 r_i} - \frac{\Delta t}{2(\Delta r)^2}\right) \frac{1}{\left(1 + \frac{\Delta t \hat{S}}{\hat{T}(\Delta r)^2} + \frac{1}{\lambda^2}\right)} \tag{28.80}$$

Our equation now becomes:

$$\delta_{n+1}\left[a_1 - e^{ikm\Delta r} a_2 + a_3 e^{-ikm\Delta r}\right] = \delta_n\left[a_4 + e^{ikm\Delta r} u_5 + e^{-ikm\Delta r} u_5\right] \tag{28.81}$$

$$\delta_{n+1}\left[a_1 + a_2\left(\cos km\Delta r + i \sin(km\Delta r)\right) + a_3\left(\cos km\Delta r - i \sin km\Delta r\right)\right] = \delta_n\left[a_4 + 2a_5 \cos(km\Delta r)\right] \tag{28.82}$$

$$\delta_{n+1}\left[a_1 + a_2 \cos(km\Delta r) + a_3 \cos(km\Delta r) + i \sin(km\Delta r)(a_2 - a_3) = \delta_n\left(a_4 + 2a_5 \cos(km\Delta r)\right)\right] \tag{28.83}$$

Factorizing the previous equation we get:

$$\delta_{n+1}\left[a_1 + (a_2 + a_3)\cos(km\Delta r) + i \sin(km\Delta r)(a_2 - a_3)\right] = \delta_n\left[\left(a_4 + 2a_5 \cos(km\Delta r)\right)\right] \tag{28.84}$$

The model is regarded as stable when it converges this way, following the steps from Equation (28.85) to the last step at Equation (28.88):

$$\left|\frac{\delta_{n+1}}{\delta_n}\right| < 1$$

$$\frac{\delta_{n+1}}{\delta_n} = \frac{a_4 + 2a_5 \cos(km\Delta r)}{a_1 + (a_2 + a_3)\cos(km\Delta r) + i\sin(km\Delta r)(a_2 - a_3)} \tag{28.85}$$

$$\left|\frac{\delta_{n+1}}{\delta_n}\right| = \left|\frac{a_4 + 2a_5 \cos(km\Delta r)}{a_1 + (a_2 + a_3)\cos(km\Delta r) + i\sin(km\Delta r)(a_2 - a_3)}\right| \tag{28.86}$$

$$\frac{\delta_{n+1}}{\delta_n} = \frac{|a_4 + 2a_5 \cos(km\Delta r)|}{\sqrt{(a_1 + (a_2 + a_3)\cos(km\Delta r))^2 + (\sin(km\Delta r)(a_2 - a_3))^2}} < 1 \tag{28.87}$$

$$|a_4 + 2a_5 \cos(km\Delta r)| < \sqrt{(a_1 + (a_2 + a_3)\cos(km\Delta r))^2 + (\sin(km\Delta r)(a_2 - a_3))^2} \tag{28.88}$$

The numerical analysis for our stochastic equation in a leaky aquifer converges with the actual analysis of a system. This means that the model parameters assigned during construction meet exactly with the theoretically predicted ones. In the next section of the simulation, we should expect the model to perform well so that it can be used for predictions in the future.

28.9   SIMULATION

After the development of the stochastic groundwater flow models in confined and leaky aquifers, the next step was to run the model via simulation to obtain the actual representation of the behavior of the system. Simulation helps us to test the system and the conditions being investigated which may be difficult to imitate in the real world. For complex systems, simulations provide valuable solutions by providing clear insights which are necessary for predicting or forecasting the future behavior of a system and also for determining what could be done to influence that future behavior. The images obtained after the completion of simulations representing groundwater flow behavior in confined and leaky aquifers are presented below. We consider a theoretical sample of storativity and transmissivity; we compute harmonic, geometric, and arithmetic means; and we consider the normal distribution in these cases. Numerical simulations are depicted in Figures 28.1–28.9, including drawdown contour plots for confined cases.

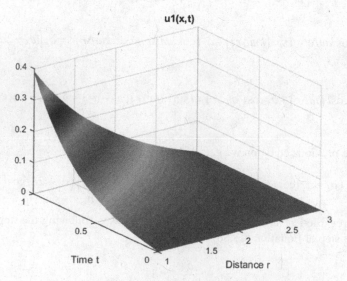

FIGURE 28.1   Numerical solutions with harmonic mean for transmissivity and storativity.

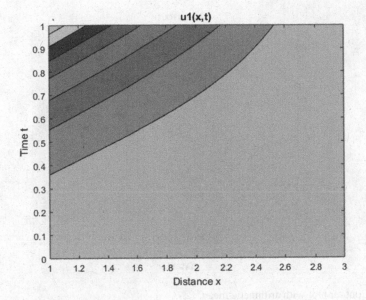

FIGURE 28.2 Contour plot for harmonic mean.

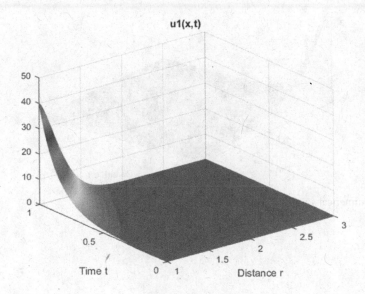

FIGURE 28.3 Numerical simulation with arithmetic mean.

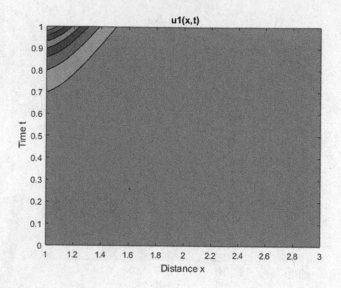

FIGURE 28.4   Contour plot with arithmetic mean.

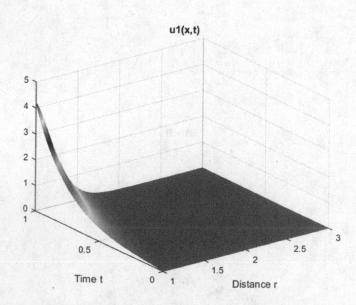

FIGURE 28.5   Numerical solution with geometric mean.

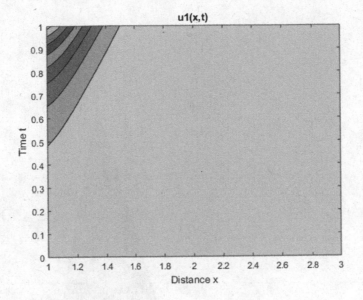

FIGURE 28.6 Contour plot of geometric mean.

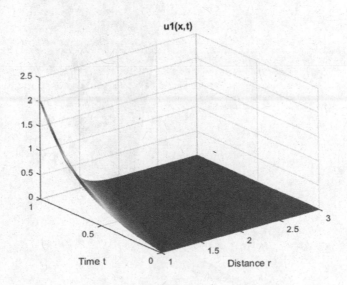

FIGURE 28.7 Numerical solution with geometric mean.

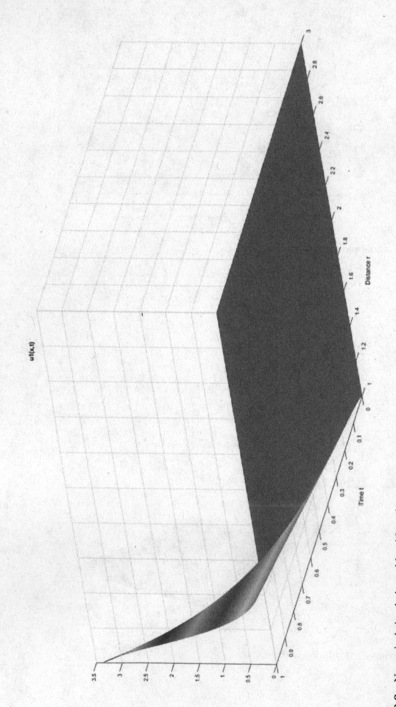

FIGURE 28.8 Numerical simulation with arithmetic mean.

FIGURE 28.9 Contour plot with arithmetic mean.

Here we present the drawdown of water level for the leaky aquifer; numerical simulations are depicted in Figures 28.10–28.17.

The numerical simulations show that modeling with a stochastic approach is more suitable as there may be many ways to capture heterogeneities. One of the most important findings in this work is the fact that the well-known arithmetic mean that is widely used in the field of geohydrology provides exaggerated results. This exaggeration can be well explained with the concept of salaries.

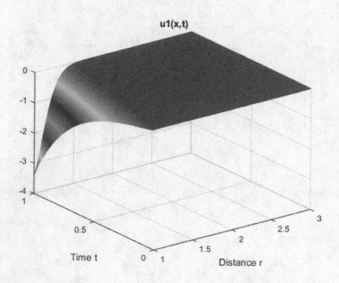

FIGURE 28.10   Numerical simulation for harmonic mean.

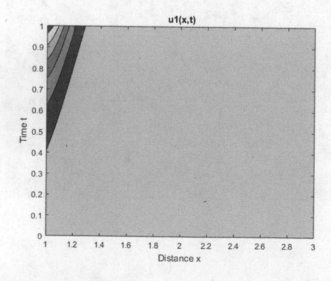

FIGURE 28.11   Contour plot with harmonic mean.

Consider a rich man living in the same room as a poor man. If one asks what the average salary in this room is obviously one will conclude that each person is rich if the average is arithmetic. Another illustrative example is that of the quantity of water within an aquifer. One could drill four boreholes at different locations and measures the quantity of water in each and then uses the arithmetic average to state that the average water within a certain portion is X. However, there could be some places within that system that have no water. On the other hand, one can see that modeling using

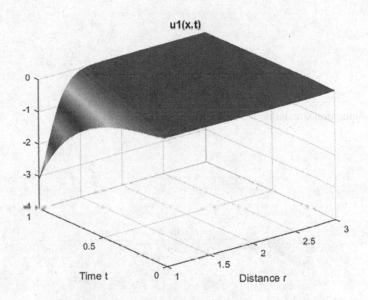

FIGURE 28.12 Numerical simulation with harmonic mean.

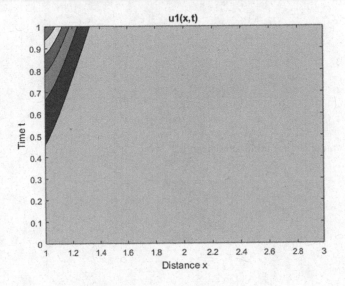

FIGURE 28.13 Contour plot with harmonic mean.

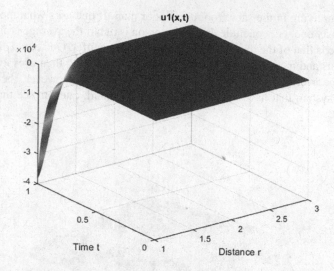

FIGURE 28.14 Numerical simulation for arithmetic mean.

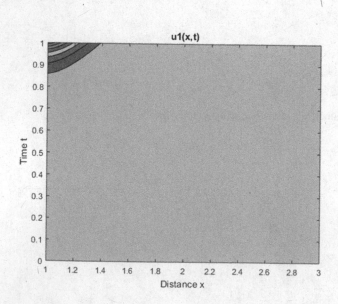

FIGURE 28.15 Contour plot for arithmetic mean.

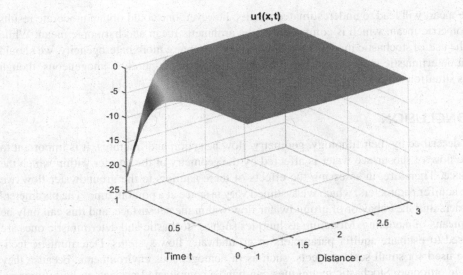

FIGURE 28.16 Numerical simulation for geometric mean.

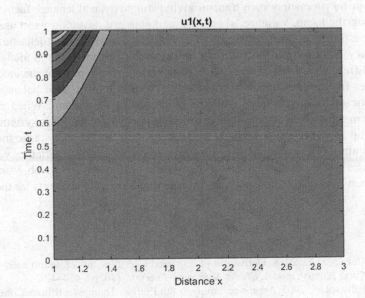

FIGURE 28.17 Contour plot with geometric mean.

the harmonic mean will lead to underestimated results; however, one could obtain moderate results using the geometric mean, which is comprised of the arithmetic mean and harmonic mean. While we suggest the use of stochastic models in geohydrology to capture more heterogeneity, we should point out that deterministic models could be useful if the situation is purely homogeneous, though of course this situation is very hard to find in real life.

28.10 CONCLUSION

Aquifers are described by their lithology, geometry, flow behavior, and fractures. It is important to recall that the flow of subsurface water is affected by the geometry of the aquifer within which the flow takes place. Therefore, in assessing the effects of these aquifers to the groundwater flow, we estimated the aquifer parameters, which we assumed vary in space at a point in time. The parameters help us to understand the behavior of groundwater flow within the subsurface, and this can only be achieved by means of modeling. Modeling techniques such as stochastic and deterministic ones are commonly used to estimate aquifer parameters in groundwater flow systems. Deterministic techniques can be used for small scale projects, such as in homogeneous environments, because they predict a single outcome. Stochastic techniques can handle complicated projects in both homogeneous and heterogeneous environments which otherwise will be impossible. The latest version by Amanda and Atangana (2018) is then modified into a new scheme, indicating that aquifer parameters change in time at a point in a system.

Based on the latest version of the equation of Amanda and Atangana (2018), and also of Atangana and Mathobo (2018), the mathematical equation was modified further to capture the statistical analysis setting of aquifer parameters such transmissivity, storativity, and leakage factor. The statistical analysis including the mean, variance, skewness, and standard deviation were used in the modified equation to estimate aquifer parameters. The numerical analysis of stochastic models for the confined and leaky equation were solved with the application of the Newton method by means of discretization. Different methods of probability distribution were applied to the models to describe appropriate better fits for each parameter. The model results show a good balance, which means that the stochastic technique captured well the prediction of the system. When comparing with other simulation methods such as deterministic models done in groundwater systems their strength relies on capturing any uncertainties which arise in any environment. Stochastic models produced in this chapter contain PDE equations with functions of one or multiple variables which have been analyzed numerically. The existing models of Theis, and of Hantush and Jacob, were reviewed and modified into a new version of Atangana and Amanda that accurately accounts for the matrix effect.

REFERENCES

Amanda, R. and Atangana, A., 2018. Derivation of a groundwater flow model within leaky and self-similar aquifers: Beyond Hantush model. *Chaos, Solitons & Fractals*, *116*, pp. 414–423.

Burden, R.L. and Faires, J.D., 2005. *Numerical Analysis* 8th Edition. Thompson Brooks/Cole.

Hantush, M.S. and Jacob, C.E., 1955. Non-steady radial flow in an infinite leaky aquifer. *Eos, Transactions American Geophysical Union*, *36*(1), pp. 95–100.

Hantush, M.S., 1960. Modification of the theory of leaky aquifers. *Journal of Geophysical Research*, *65*(11), pp. 3713–3725.

Kumba, F.F., 2003. Farmer participation in agricultural research and extension service in Namibia. *Journal of International Agricultural and Extension Education*, *10*(3), pp. 47–55.

Mathobo, M. and Atangana, A., 2018. Analysis of exact groundwater model within a confined aquifer: New proposed model beyond the Theis equation. *The European Physical Journal Plus*, *133*(10), p. 415.

Maus, G. and Nijhawan, R., 2008. Motion into and out of the blind spot: Evidence for spatial extrapolation of moving objects. *Perception*, *37*(2), pp. 310–310.

Sinha, N.K. and Prasad, T., 1979. Some stochastic modelling techniques and their applications. *Applied Mathematical Modelling*, *3*(1), pp. 2–6.

Theis, C.V., 1993. Estimating the transmissibility of aquifer from the specific capacity of wells. Methods of determining permeability, transmissibility, and drawdown. *US Geological Survey Water Supply Pater*, *1536*, pp. 332–336.

Thiem, G., 1906. *Hydrologic Methods*. Gebhardt, JM Leipzig, Germany.

van Leeuwen, M., te Stroet, C.B., Butler, A.P. and Tompkins, J.A., 1998. Stochastic determination of well capture zones. *Water Resources Research*, *34*(9), pp. 2215–2223.

Wang, Y., Ocampo-Martinez, C., and Puig, V., 2016. Stochastic model predictive control based on Gaussian processes applied to drinking water networks. *IET Control Theory & Applications*, *10*(8), pp. 947–955.

Index

Printed in the United States
by Baker & Taylor Publisher Services